International Technical Postgraduate Conference (Tech-Post 2017)

AA006667

IOP Conference Series: Materials Science and Engineering
Volume 210

Kuala Lumpur, Malaysia
5 – 6 April 2017

ISBN: 978-1-5108-4294-6
ISSN: 1757-8981

Printed from e-media with permission by:

Curran Associates, Inc.
57 Morehouse Lane
Red Hook, NY 12571

Some format issues inherent in the e-media version may also appear in this print version.

Printed by Curran Associates, Inc. (2017)

For permission requests, please contact the Institute of Physics at the address below.

Institute of Physics
Dirac House, Temple Back
Bristol BS1 6BE UK

Phone: 44 1 17 929 7481
Fax: 44 1 17 920 0979

techtracking@iop.org

Additional copies of this publication are available from:

Curran Associates, Inc.
57 Morehouse Lane
Red Hook, NY 12571 USA
Phone: 845-758-0400
Fax: 845-758-2633
Email: curran@proceedings.com
Web: www.proceedings.com

Table of contents

Volume 210

International Technical Postgraduate Conference

5–6 April 2017, University of Malaya, Kuala Lumpur, Malaysia

Accepted papers received: 6 June 2017
Published online: 26 June 2017

Preface

Biomedical Engineering

Chemical Engineering

Civil Engieering

Electrical Engineering

Mechanical Engineering

IOP Conf. Series: Materials Science and Engineering **210** (2017) 011001 doi:10.1088/1757-899X/210/1/011001

Preface

The International Technical Postgraduate Conference (Tech-Post 2017) takes place at the Faculty of Engineering, University of Malaya, Kuala Lumpur, Malaysia from 5th to 6th of April 2017. Tech-Post 2017 is organized to bring together innovative postgraduate researchers in all disciplines of Engineering to a common forum. The primary goal of the conference is to promote R&D activities and provide current global research progress in the major topics in engineering. Another goal is to promote dissemination and exchange of scientific knowledge and research know-how among postgraduate researchers working in various universities in Malaysia and abroad.

Tech-Post 2017 is fully organized by postgraduate students with technical supports from Deputy Dean of Postgraduate Studies Office, Faculty of Engineering, University of Malaya. This conference has received more than hundred participants from various institutions in Malaysia, Indonesia, India, etc. A compilation of articles that represent the significant results of the conference has been selected to publish in this volume. Five main research topics in engineering have been compiled as listed:

Biomedical Engineering
Biomechanics, Biomaterials Cell and Tissue Engineering, Medical Imaging, Systems Neuroscience, Bioinformatics and Computational, Biology Computational and Physiological Modelling, Biomedical, Microelectromechanical (BioMEMS) & Bioinstrumentation and Rehabilitation

Civil Engineering
Construction Management, Mathematics and Surveying, Hydraulics, Water and Environmental Engineering, Geotechnical Engineering, Transportation Engineering & Structural Engineering and Material

Chemical Engineering
Advanced Materials and Technology, Bioprocess Engineering, Health, Safety and Environment, Process Systems Engineering and Control, Purification and Separation Processes & Reaction Engineering

Electrical Engineering
Photonics, Power Electronics Drives, Automation and Control, Electromagnetic Radiation and Devices, Image and Signal Processing & Microsystems and Optoelectronics.

Mechanical Engineering
Applied Mechanics, Solar Energy, Heat Transfer, Combustion, Tribology, Heat Engines, Material Sciences, Engineering Design, Manufacturing Automation and Robotics, Manufacturing Management, Manufacturing Processes & Manufacturing Systems

This publication is sponsored by the Faculty of Engineering, University of Malaya, and Institute of Research Management and Services (IPPP), University of Malaya.

Editors
Professor Dr. Sulaiman Wadi Harun & Anas Abdul Latiff

IOP Conf. Series: Materials Science and Engineering **210** (2017) 011001 doi:10.1088/1757-899X/210/1/011001

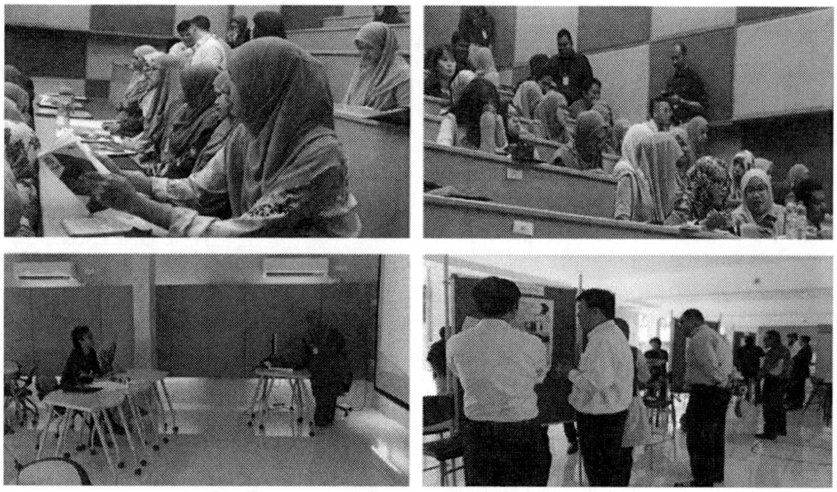

Oral presentation and poster session.

Best presenter of oral presentation award and best poster presentation award.

Organizing committee.

IOP Conf. Series: Materials Science and Engineering **210** (2017) 011001 doi:10.1088/1757-899X/210/1/011001

List of Committee

Patron
Professor Ir. Dr. Noor Azuan Abu Osman

Advisor
Professor Dr. Sulaiman Wadi Harun

Organizing Committee
Dr. Hazli Rafis Abdul Rahim (Chairman)
Nurizaz Nadia Mohamad (Secretary)
Umi Zalilah Mohamad Zaidi (Bursary)
Muhammad Basril Muhammad Asri (Technical)
Dhamayanti Nair Jaya (Registration)
Ahmad Razif Muhammad (Publicity)
Anas Abdul Latiff (Publication)
Asdani Saifullah Dolbashid (Logistic)
Nashrah Hani Jamadon (Ceremony)
Siti Zulaiha Uyub (Concession)
Mustapha Mohammed Bello (Sponsorship)

(Support Team)
Atikah Aman, Nur Hidayanti Ambrizal, Muhammad Aiman Mahadi, Nurul Fatin Nabilah Abdul Samat, Nur Dalila Salehhuddin, Muhammad Taufiq Ahmad, Ahmad Haziq Aiman Rosol, Mohd Faiz Azmi, Muhammad Afiq Dzulkifli, Nurul Salwani Mohamad Saadon, Biddatul Syirat Zainal

Publication Committee
Professor Dr. Sulaiman Wadi Harun, Anas Abdul Latiff, Muhammad Basril Muhammad Asri, Muhammad Aiman Mahadi, Asdani Saifullah Dolbashid, Mohd Faiz Azmi, Mustapha Mohammed Bello, Dr. Hazli Rafis Abdul, Siti Zulaiha Uyub, Biddatul Syirat Zainal, Ahmad Haziq Aiman Rosol, Atikah Aman, Nurizaz Nadia Mohamad, Muhammad Taufiq Ahmad, Ahmad Razif Muhammad

Keynote speaker
Professor Ir. Dr. Abdul Aziz Abdul Raman (Faculty of Engineering, University of Malaya)
"The Postgraduate Journey: What's in For You?"

IOP Conf. Series: Materials Science and Engineering **210** (2017) 011002 doi:10.1088/1757-899X/210/1/011002

Peer review statement

All papers published in this volume of *IOP Conference Series: Materials Science and Engineering* have been peer reviewed through processes administered by the proceedings Editors. Reviews were conducted by expert referees to the professional and scientific standards expected of a proceedings journal published by IOP Publishing.

3D cardiac wall thickening assessment for acute myocardial infarction

A Khalid[1], B T Chan[1], E Lim[1], and Y M Liew[1,*]
[1]Department of Biomedical Engineering, Faculty of Engineering, University of Malaya, 50603 Kuala Lumpur, Malaysia

E-mail: amirahkhalid1@um.edu.my

Abstract. Acute myocardial infarction (AMI) is the most severe form of coronary artery disease leading to localized myocardial injury and therefore irregularities in the cardiac wall contractility. Studies have found very limited differences in global indices (such as ejection fraction, myocardial mass and volume) between healthy subjects and AMI patients, and therefore suggested regional assessment. Regional index, specifically cardiac wall thickness (WT) and thickening is closely related to cardiac function and could reveal regional abnormality due to AMI. In this study, we developed a 3D wall thickening assessment method to identify regional wall contractility dysfunction due to localized myocardial injury from infarction. Wall thickness and thickening were assessed from 3D personalized cardiac models reconstructed from cine MRI images by fitting inscribed sphere between endocardial and epicardial wall. The thickening analysis was performed in 5 patients and 3 healthy subjects and the results were compared against the gold standard 2D late-gadolinium-enhanced (LGE) images for infarct localization. The notable finding of this study is the highly accurate estimation and visual representation of the infarct size and location in 3D. This study provides clinicians with an intuitive way to visually and qualitatively assess regional cardiac wall dysfunction due to infarction in AMI patients.

1. Introduction

Acute myocardial infarction (AMI) is among the most commonly diagnosed cardiovascular diseases today, affecting over 800,000 people annually [1]. Excessive build-up of plaque and blockage in the coronary arteries causes inadequate blood supply to various parts of the cardiac tissue [2], leading to localized myocardial injuries and necrosis. Patients who had AMI have higher risk of developing ventricular remodeling, a pathological process whereby the left ventricle (LV) changes in size, shape and functions. Expansion of the infarct region can adversely lead to elevated systolic and diastolic wall stress. In severe cases, acute myocardial infarction causes functional irregularities, ventricular deformation, and total heart failure that kills lives [3].

The deviation of global indices such as myocardial mass, blood volume, and ejection fraction from baseline value of healthy subjects has long been used in current clinics to identify functional abnormality in AMI patients. Researches, however, has found that these indices have very limited sensitivity to provide accurate and reliable diagnosis for such group of patients at the early stage of disease development as the changes in cardiac function is rather localized instead of globalized [4]. In addition, infarct zone of the myocardium tends to become rigid and thinner as a result of injury and necrosis [5]. Regional assessment of wall thickness and thickness changes (i.e. wall thickening) may, therefore, provide a means to improve clinical diagnosis for better patient outcome. A comprehensive analysis of these regional indices in 3D could help clinicians to assess the extent of myocardial infarction in terms of size and location, providing them a useful and convenient way to visualize and understand cardiac wall condition of individual patients [5].

IOP Conf. Series: Materials Science and Engineering **210** (2017) 012001 doi:10.1088/1757-899X/210/1/012001

To date, regional deformation of cardiac wall has been investigated using various imaging modalities, including computed tomography (CT), echocardiography, and magnetic resonance imaging (MRI) Studies utilizing CT have shown limitations with respect to image quality that leads to an overestimation of the infarct size [6-7]. Echocardiography has been used widely but primarily for 2D deformation analyses of the cardiac wall and only few studies have attempted it in 3D [8]. Due to the inherent speckle noise, wall deformation analysis from echocardiography suffers from limited accuracy, i.e. with 87% of accuracy in tracking of the infarcted region reported [9]. MRI, as a gold standard cardiac assessment modality [6], has been useful in identifying infarct location and transmural utilizing the 2D late gadolinium enhancement (LGE) technique in AMI patients [10]. The impact of infarct on wall motion dysfunction has mostly been assessed from cine and tagged MRI using eyeballing in clinical practice. Increasing number of quantitative techniques [6,11] has been proposed to date to analyze motion dysfunction from cine and tagged MRI but majority of techniques focused on 2D MR images or in-plane measurement e.g. center-surface techniques, guide-point modeling, and active appearance models. These techniques are prone to error when the imaging plane does not intersect the cardiac wall orthogonally, which is a rather common problem as the LV is known to be concave and may become deformed in shape under disease conditions. The only notable 3D mesh-based wall thickness measurement was proposed by Tobon-Gomez et al. [11] but the technique was applied to differentiate the different phenotypes of LV hypertrophy (i.e. hypertrophy cardiomyopathy and hypertensive heart disease). None of the techniques proposed to date, which is based on more detailed 3D personalized computational model, has correlated anomaly in wall thickening to infarction in AMI patients.

In the present study, we develop a novel technique to extract regional wall thickness measurements in 3D from personalized cardiac models generated in-house with prior motion correction. We investigated the regional thickening of the cardiac wall across the full cardiac cycle and compared the thickening measurements between 5 AMI patients and 3 healthy subjects. In AMI patients, the localized abnormality in wall thickening was correlated and validated with the infarct region in LGE images. We have shown that our regional wall thickening assessment technique is useful for the localization of infarct region and extent, providing a means to directly correlate cardiac wall contractility dysfunction to infarct severity.

2. Methods

2.1 MRI data acquisition

Five patients diagnosed with AMI and 3 healthy volunteers were recruited at the University of Malaya Medical Centre with written consent. For each subject, multi-breath-hold steady-state free precession short-axis cine image stack (FOV: 350 × 350 mm, image matrix: 256 × 256, pixel size: 1.37 × 1.37 mm, slice thickness: 8 mm, gap: 0 mm, TE/TR: 1.6/3.7 ms, flip angle: 55°, number of slices: 10–15, number of cardiac phases: 20, breath hold time: 15 s) covering base to apex of the LV were acquired. A set of 6 multi-breath-hold long-axis cine images radially oriented around the center of the LV chamber at uniform angular interval were also prescribed with the same acquisition parameters by using the first short-axis slice at the base for planning.

For AMI patients, collocated short-axis inverse recovery fast gradient echo LGE MRI images were obtained. For this acquisition, patients were administered 140 lg/kg/min of adenosine infusion for 4 minutes, followed by 20 ml of normal saline flushing. An intravenous injection of gadolinium-based contrast agent at 0.2 ml/kg was subsequently performed for first-pass perfusion. A further 0.2ml/ kg of the gadolinium agent was administered for rest and delayed gadolinium enhancement at 10-minute intervals. Typical delay enhancement imaging parameters were used, as follows: TE/TR: 3.0/6.0 msec, inversion time (TI): 200–300 msec (depending on null point of normal myocardium), flip angle: 20°, no gap between planes and breath-hold time: 18 seconds. The delay time was chosen to yield images in the mid- to late-systolic phase. All cine and LGE MRI images were acquired during end expiration breath-hold using a 1.5T MRI system (Sigma HDxt 1.5T, GE Healthcare, WI, U.S.A). Our study protocol has received approval by University of Malaya Medical Ethics Committee (Ref: 989.75).

2.2 Image segmentation, motion correction and reconstruction of LV model in 3D

All short-axis and long-axis images were semi-automatically contoured to extract the LV endocardial and epicardial wall [12]. The contouring was performed by using automated border recognition tool and minor manual fixes. The papillary muscles were included in the blood pool and excluded from the myocardium. The displacement between images due to motion artifacts was corrected using in-house image registration algorithms and 3D LV model specific to each subject were reconstructed based on the B- spline fitting technique [13]. For each subject, twenty 3D LV surface mesh models were generated for a full cardiac cycle. Each LV model was formed by combining 2 quadrilateral meshes, one from the endocardial wall and the other one from the epicardial wall. Each mesh contains 100 x 100 surface points and was orientated in patient coordinate system.

2.3 Development of 3D cardiac wall thickness model

We used the sphere fitting approach to accurately estimate regional wall thickness of the cardiac wall. Specifically, for each subject, all 3D LV models were aligned by using principal component analysis so that the long axis of the LV was parallel to z-axis. A small moving window was subsequently applied across the entire epicardial mesh surface. At each window, k-nearest neighbor technique was used to search for 6 nearest neighbors in the endocardial wall for each epicardial point. This forms a small localized group of points whereby a maximally inscribed sphere was fitted within the convex hull. The window size was allowed to vary by location based on a prior rough estimation of Euclidean distance between endo- and epicardial wall through direct subtraction of wall coordinates. Regional wall thickness was extracted as the diameter of the fitted sphere. Figure 1 illustrates examples of spheres fitted on the cardiac wall for wall thickness extraction. To facilitate visualization of abnormality in cardiac wall thickening, the thickness measurements were color-coded onto the epicardial surface, generating a 3D cardiac wall thickness map for each cardiac phase. The accuracy of each wall thickness model was validated against LGE image which depicts the location of infarct. This involved superimposing the 2D LGE images onto the model for visual correlation of cardiac wall contractility dysfunction and location of infarct. The sphere fitting framework was implemented in MATLAB (vR2012a, Mathworks, Natick, MA) on an Intel(R) Core(TM) i5-3570 CPU @3.40 GHz computer using parallel computing.

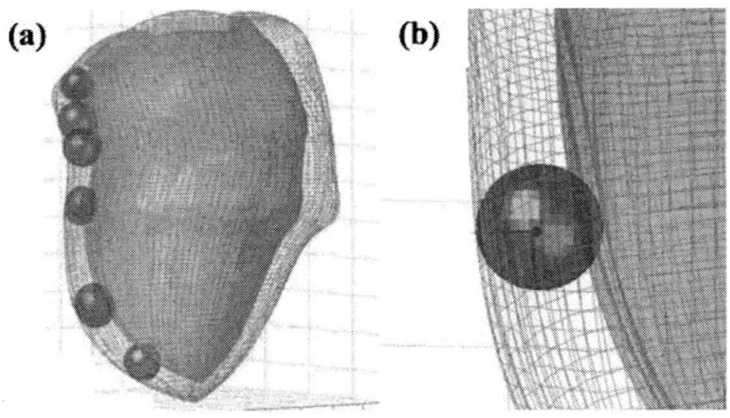

Figure 1. Three-dimensional view of the wall thickness measurement using the proposed sphere fitting approach. (a) A number of spheres were maximally inscribed between the endocardial and epicardial walls at different locations. (b) Zoomed in view of (a) where the red line within the sphere indicates the radius of the sphere.

IOP Conf. Series: Materials Science and Engineering **210** (2017) 012001 doi:10.1088/1757-899X/210/1/012001

3. Results

The present study utilizes the method of sphere fitting to extract the LV wall thickness in 3D. Figure 2 illustrates the 3D LV model of a healthy subject and an AMI patient, which has been color-coded with wall thickness measurement and overlaid with a short-axis cine MRI image. Blue color indicates thick myocardium whereas red color indicates thin myocardium.

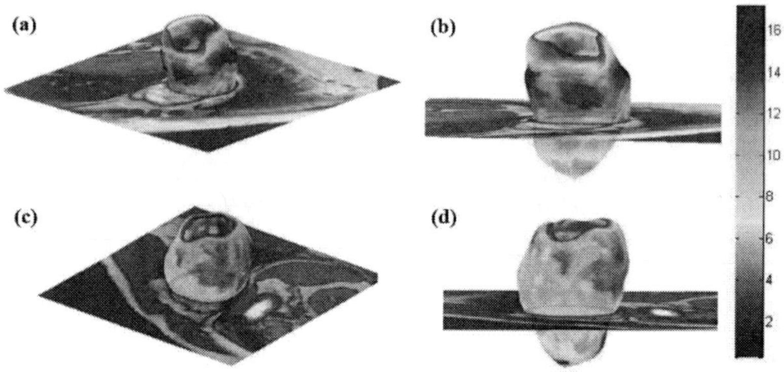

Figure 2. The 3D wall thickness models of a subject (a, b) and patient (c, d), each in isometric view (a, c) and front view (b, d). The models were superimposed on the cine MRI image in 3D space. Both the model and image were taken from the same cardiac phase i.e. systolic phase (Phase 5) in a cardiac cycle. The colour bar depicts the wall thickness in mm.

In figure 2, the LV wall thickness for a healthy subject is compared with the LV wall thickness of an AMI patient during the systolic phase. During this phase, the myocardium thickness increases as the LV contracts to pump blood. In figure 2 (a) and 2 (b), the myocardium of the healthy subject is thicker (consider the light green and dark blue shading) in comparison to the wall thickness of the AMI patient in figure 2 (c) and (d) (consider the yellow and red shading). In addition, in the figure above, the exact superimposition of the LV wall thickness models on the cine images represents the accuracy of the technique used.

Figure 3 shows the average wall thickness of a healthy subject in a full cardiac cycle. The LV was divided into three segments, i.e. base, mid, and apex. As expected, the mid-ventricular region demonstrates the highest average wall thickness followed by the base. The lowest average wall thickness was found at the apex.

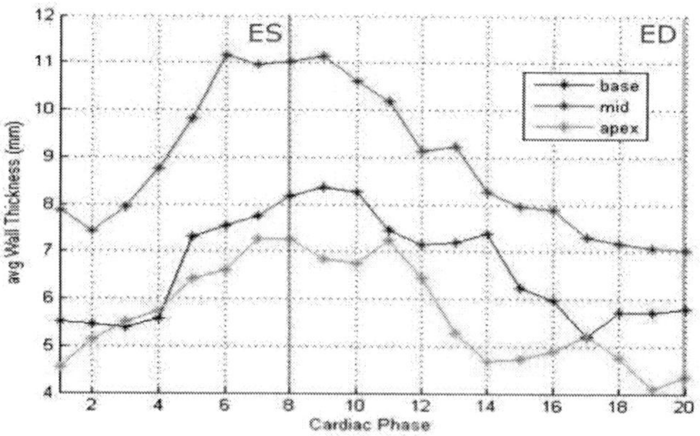

Figure 3. Average wall thickness of a healthy subject over a full cardiac cycle where ES indicates end systolic phase and ED indicates end diastolic phase. LV was divided into basal, midventricular, and apical segments.

IOP Conf. Series: Materials Science and Engineering **210** (2017) 012001 doi:10.1088/1757-899X/210/1/012001

The 3D wall thickness model was qualitatively assessed by superimposing it with the cine/LGE images and the 17-segment AHA LV division lines (figure 4). Direct correlation of anomaly in wall contractility (as conveyed by our wall thickness model) with the infarct location (as revealed by LGE image) is evidenced. The average wall thickness within the infarct region for the 5 AMI patients has been measured to be approximately 2.86±1.11 mm as compared to the healthy region 8.73±1.01 mm at end-diastole.

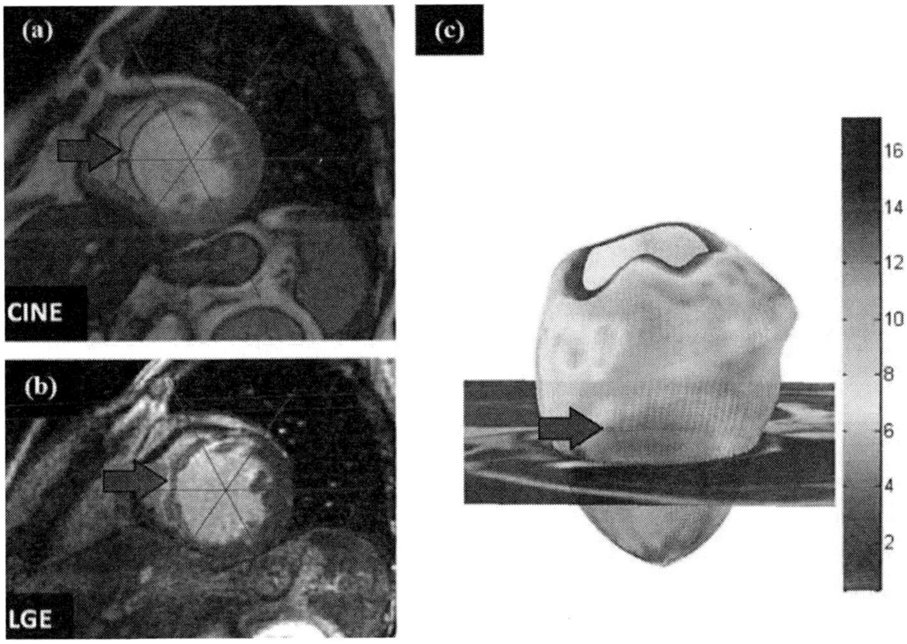

Figure 4. Correlation of infarct location on the cine (a) and LGE (b) images to thickness anomaly on 3D wall thickness model (c). The red arrows are pointing at the same infarcted region of the LV in all images. Yellow shading in A and B highlights the width of the infarct, which is located at the mid anteroseptal and mid inferoseptal region. Colour bar indicates wall thickness in mm.

4. Discussion

Wall thickness is an important clinical index to help clinician in identifying myocardial dysfunctional secondary to cardiac diseases. Regional assessment of wall thickness has recently received more attention for research especially in the case of focal myocardial disease, including acute myocardial infarction, as commonly used global indices (e.g. ejection fraction, blood volume and myocardial mass) have been shown to fail in aiding early diagnosis of this disease until later stage of disease progression [14]. While majority of the published techniques focused on in-plane wall thickness measurement in 2D, we adapted the use of sphere fitting technique to extract regional wall thickness in 3D from our in-house personalized quadrilateral LV mesh model, providing more accurate estimation of wall thickness. This technique does not suffer from error due to oblique intersection of MRI scanning plane with the cardiac wall as in 2D analysis. Our proposed method conducts over 10,000 fittings of maximally inscribed spheres over the entire LV myocardium from the base to the apex. The thickness measurements were mapped onto the 3D LV model for visualization and direct correlation of thickening abnormality to infarct region in LGE image.

For healthy subjects, the wall thickness measurement was found to fall within the normal range as reported in previous studies [15]. Furthermore, based on our assessment of the global indices, no significant differences exist among the measurements between the subjects and the patients. For instance, at the end-diastolic phase, the LV mass for a subject is 47.479 g, and the LV of the patient
50.202 g, the LV volume of another subject is 51.742 ml, and the LV volume of another patient is 68.035 ml respectively. This trend is found throughout the patient and subject database. In contrast, the average regional wall thickness between remote regions and infarcted regions varied extensively and is clearly evident in the color mapped 3D wall thickness; making the proposed technique useful for differentiating between normal and injured myocardium.

Moreover, the technique implemented in this study has shown promising results in terms of identifying localized myocardial injury via a 3D wall thickness assessment. In addition, it has potential applications in various cardiovascular diseases associated with changes in cardiac wall thickening. For example, the regional assessment of wall thickness may also be performed for patients with LV hypertrophy and dilated cardiomyopathy for identifying the extent of myocardial injury. However, this preliminary study requires a larger patient database for further validation and currently lacks a quantitative assessment of the wall thickness models. In the near future, we work to further improve our sphere fitting technique to achieve higher accuracy assessment of the localized myocardial injury.

5. Conclusions

We have described a sphere fitting approach to extract regional wall thickness from 3D motion corrected personalized LV mesh model. The proposed method was tested in 5 patients and 3 healthy volunteers. The wall thickness map was found to provide clinician with an intuitive and comprehensive means to visualize and identify wall thickening abnormality due to infarction in AMI patients in3D.

References

[1] Boateng, S., & Sanborn, T. 2013 Disease-a-Month Acute myocardial infarction **59** 83-96
[2] Thygesen, K., Alpert, J. S., & White, H. D. 2007 *Journal of the American College of Cardiology* **50** 2173-2195
[3] Rischpler, C. 2016 *Journal ofnuclear medicineandmolecularimaging* **60** 236-251
[4] Karamitsos, T. D., Francis, J. M., Myerson, S., Selvanayagam, J. B., & Neubauer, S. 2009 *Journal of the American College of Cardiology* **54** 1407-1424
[5] Jung, J.-W., Kang, H.-R., Kim, M.-H., Lee, W., Min, K.-U., Han, M.-H., & Cho, S.-H. 2012 *Radiology* **264** 414-422
[6] Ko, S. M., Kim, Y. J., Park, J. H., & Choi, N. M. 2014 *The British journal of radiology*
[7] Doltra, A., Amundsen, B. H., Gebker, R., Fleck, E., & Kelle, S. 2013 *Current Cardiology Reviews* **9** 185-190
[8] Arai, K., Hozumi, T., Matsumura, Y., Sugioka, K., Takemoto, Y., Yamagishi, H., ... & Yoshikawa, J. 2004 *The American Journal of Cardiology* **945** 552-558
[9] Bhan, A., Sirker, A., Zhang, J., Protti, A., Catibog, N., Driver, W., & Shah, A. M. 2014 *American Journal of Physiology-Heart and Circulatory Physiology* **306** H1371-H1383
[10] Doltra, A., Amundsen, B. H., Gebker, R., Fleck, E., & Kelle, S. 2013 *Current Cardiology Reviews,* **9** 185-190
[11] Tobon-Gomez, C., Butakoff, C., Yushkevich, P., Huguet, M., & Frangi, A. F. 2010 *2010 Annual International Conference of the IEEE In Engineering in Medicine and Biology Society* 2642- 2645
[12] Heiberg, E., Sjögren, J., Ugander, M., Carlsson, M., Engblom, H., & Arheden, H.

IOP Conf. Series: Materials Science and Engineering **210** (2017) 012001 doi:10.1088/1757-899X/210/1/012001

2010 *BMC medical imaging* 10 1

[13] Liew, Y. M., McLaughlin, R. A., Chan, B. T., Aziz, Y. A., Chee, K. H., Ung, N. M., ... & Lim, E. 2015 *Physics in medicine and biology* **60** 2715

[14] Christian, T. F., Gibbons, R. J., & Gersh, B. J. 1991 *Journal of the American College of Cardiology* **17** 1303-1308

[15] Kawel, N., Turkbey, E. B., Carr, J. J., Eng, J., Gomes, A. S., Hundley, W. G., ... & Lima, J. A. 2012 *Circulation: Cardiovascular Imaging* **5** 500- 508

IOP Conf. Series: Materials Science and Engineering **210** (2017) 012002 doi:10.1088/1757-899X/210/1/012002

Effect of stump flexion contracture with and without prosthetic alignment intervention towards postural stability among transtibial prosthesis users

M F Ghazali[1], N A Abd Razak[1], N A Abu Osman[1], H Gholizadeh[2]
[1] Department of Biomedical Engineering, Faculty of Engineering, University of Malaya, 50603, Kuala Lumpur, Malaysia.
[2] The Ottawa Hospital Rehabilitation Centre, Canada

E-mail: fazlighazali90@yahoo.com

Abstract. Knee flexion contracture on a stump side is a phenomenon in which the stump cannot move in normal range of motion (ROM) or cannot be fully extended. This study has been carried out by using Biodex Stability System (BSS) in order to investigate the effect of stump flexion contracture towards the postural stability among the transtibial prosthesis users with the intervention of alignment accommodation. The BSS provides the reading of anterior-posterior stability index (APSI), medial-lateral stability index (MLSI), and overall stability index (OSI). Higher reading of the index indicates lesser stability. Each of the subjects had been tested in three different sessions that were Visit 1 (before contracture improvement), Visit 2 (after contracture improvement without alignment readjustment), and Visit 3 (after contracture improvement with alignment readjustment). The APSI reading was significantly higher during Visit 2 compared to Visit 1 and Visit 3. The OSI during Visit 2 was also found significantly higher compared to Visit 3. In Visit 2, the degree of contracture was significantly improved with 44.1% less than Visit 1. The stability index in anterior-posterior aspect (APSI) was proven to be lower as the prosthetic alignment was adjusted according to the ROM of knee. This finding explained that the alignment set up based on the adaptation with the stump's ROM can contribute positively in maintaining postural stability.

1. Background

Knee contracture or flexion contracture on the stump is one of the common problems occurred among the below knee amputees. It refers to the situation in which the patient cannot fully straighten the knee or in other words, the range of motion of the joint is limited from normal [1]. It is a result of the shortening and tightness of the muscles which might be associated with their structural changes due to the long period of bending position or inactivity due to the illness and lack of exercise [2]. The significant characteristics of contracture are limitation of joint mobility and increasing resistance of the joint to passive ROM which is the movement of joint applied by external power, either by someone or by exercise machine [3]. The mobility of joint is very important for a prosthetic candidate. This is because the limitation of joint motion which refers to joint contracture will significantly affect the fitting and function of the prosthesis [4].

In overcoming the contracture, most of the efforts are focused on treating the contracture itself, decreasing the contracture angle or prevent it from becoming worse. For transtibial amputation cases, the amputees are advised to avoid placing pillow under the knee. An extension board is commonly used to be placed across the back of the knee joint to ensure the extension position [5]. Besides, stretching the soft tissues which can be done actively or passively is also very helpful to overcome the contracture.

Active stretching means the amputee actively moves the joint in full range of motion. Meanwhile, passive stretching refers to the using of a splint or orthotic device to forcing the joint in extended position [6].

For an amputee who is going to proceed with a prosthesis, the presence of stump contracture could be a barrier for him/her to achieve a successful prosthesis usage. This is because the contracture will disrupt the prosthetic alignment that could lead to various kinds of adverse effects while using the prosthesis [6]. However, the alignment of the prosthesis still can be adjusted to accommodate with the knee flexion contracture if any but if the angle exceeding 25°, the fitting might be difficult [4]. The prosthetic alignment can be defined as the positioning of the socket in relation with the other components of the prosthesis such as ankle joint and shank tube [7]. Generally, good alignment of a prosthesis will contribute to an energy efficient and smooth gait pattern. [8]. For the amputees with stump contracture, the alignment that should be concerned is the sagittal plane. It is the anteroposterior positioning of the socket with regard to the foot. The socket flexion will be increased respectively with the contracture angle which means the alignment is deviating from the standard. Otherwise, the amputee will exhibit gait deviations during walking such as rough rollover of the foot and inconsistent heel strike [9].

Prosthetic alignment is also vital in retaining postural balance either during walking or quiet standing [10]. Quiet standing can be defined as staying upright in a static condition without moving. The regulation of balance is a vital effort for the body in maintaining the stability in this position. Meanwhile, balance is the ability in regaining the Centre of Mass (CoM) within the base support in order to maintain body equilibrium. To preserve balance, human body is believed to orientate the CoM, which is a virtual point equivalent to total body mass [11][12]. This response will return the CoM within the base of support (BoS) which is the displacement region for the Center of Pressure (CoP). CoP is the point of average distribution for total pressure towards the surface of the contact area [13]. Indeed, balance of an individual can be assessed by measuring the displacement of CoP, by getting the Stability Index (SI). It is a standard deviation reading that assesses the path of sway around the zero point from the center of the platform. The unit is in degree. Therefore, it indicates the degree of foot displacement in sagittal and frontal plane motions [14]. SI which consists of OSI, APSI, and MLSI are measured by BSS. BSS has been developed with advantages and is a convenient technology in studying postural stability. The system is available for multi-range of people, easy to administer, and simple to interpret [15][16].

There were a lot of research had been done in order to study on the postural control during quiet standing among amputees. Ku et al. had done a systematic review regarding the factors that contributed to the balance instability for lower limb amputees. In their review, they found that majority of the articles revealed the increase in postural sway among the amputees compared to normal population. One of the major factors that explained the instability was asymmetry in body weight distribution between the sound and the artificial legs [17]. Katch et al. and Salsabili et al. had proven the increasing body-mass index (BMI) could bring significant impact on the postural control in terms of muscular torque, attentional cost, postural sway, energy cost, and motor reaction time [16][18]. According to Colne et al., increase in BMI could possibly reduce the ability of the muscles to generate sufficient force in order to control the CoM displacement [19]. Another potential factor that was affecting postural balance was age. The elderly population had been found to exhibit greater postural sway [17].

In rehabilitation process, stump contracture is one of the major inhibitors for an amputee to acquire a prosthesis. It has been found that 10° is the maximum amount of stump contracture for the amputee to possess a great walking ability. With the degree more than that, it could lead to the failure of the prosthesis usage for mobility. Even still can be used, there will be a lot of factors that need to be reconsidered while making the prosthesis such as alignment and components [4][6]. However, there is still no study conducted regarding the influence of prosthesis alignment intervention towards the stump contracture effect. Therefore, in this study, the postural stability of below knee prosthesis users during quiet standing was evaluated at different degrees of stump contracture, with and without the intervention of prosthesis alignment adjustment. The study was carried out at Motion Analysis Laboratory & Centre for Prosthetics and Orthotics Engineering, University of Malaya. Two of the researchers are the prosthetist with at least three years experienced in handling amputees. The tests were conducted by using

BSS which also provides the stability index readings for both medial-lateral and anterior-posterior aspect (MLSI and APSI) and overall stability index (OSI) which was the combination of these two. Every subject (n=10) underwent the same tests for three different sessions (Visit 1, Visit 2, and Visit 3). The ROM of stump or prosthetic alignment were varied accordingly for each session. In Visit 1, the tests were conducted with the existence of stump contracture and alignment readjustment according to the standard by the researcher. Visit 2 would be proceeded when the subjects had achieved at least 5^0 of contracture improvement after the intervention of therapies and proper stump positioning without alignment readjustment. Meanwhile, the tests with prosthesis alignment readjustment according to the improvised ROM of stump were then conducted in Visit 3. Those three stability index readings, APSI, MLSI, and OSI obtained for every session were collected and compared. Thus, the influence of ROM of stump with and without prosthetic alignment intervention towards the amputees' postural stability during quiet standing can be determined.

2. Methodology

2.1. Ethics Approval and Consent

The experimental protocol for this work was approved by Ministry of Health Medical Research Ethics Committee (MREC), Malaysia. Written informed consent was granted by the participants from the authors for the publication. Approval ID: NMRR-16-2106-32880 (IIR). One registered prosthetist fabricated all the prostheses to avoid alterations due to manufacturing, alignment and fitting. All the procedure of socket making and fitting involves the Certified Prosthetics and Orthotics (CPO) which had been recognized by International Society of Prosthetics and Orthotics (ISPO).

2.2. Subjects Recruitment

This study was conducted on 10 subjects who fulfilled the required criteria:
Unilateral transtibial amputee.
1) The age is from 50 to 75 years old
2) Possession of stump contracture from 10° to 25°.
3) Good compliance in effort of reducing knee contracture. The minimum amount of stump contracture reduction before proceed with the second session is 5°.
4) Able to communicate individual perception regarding the prosthetic alignment.
5) The subjects are using the same endoskeletal prosthetic system, PTB socket with pelite liner, and SACH foot.
6) Minimum change of body weight throughout the experimental period. The maximum acceptable amount of change is 3kg.
7) Able to stand using the prosthesis independently.
8) Absence of any wound or injury on the residual limb.
9) Fulfil all the requirements listed above. The failure of a subject to fulfil any of the requirement will end up with disqualification.

2.3. Instrumentation

Biodex Stability System (BSS; Biodex, Inc., Shirley, NY, USA) is a convenient and reliable system that is able to provide clinical data measurements for postural stability assessment [20]. In this study, the subjects had been tested by using BSS machine in order to measure the stability index. It consists of a screen monitor which acts as a functional display for setting and controlling purposes. It also has a circular platform for a subject to stand in order to measure the displacement of CoP.

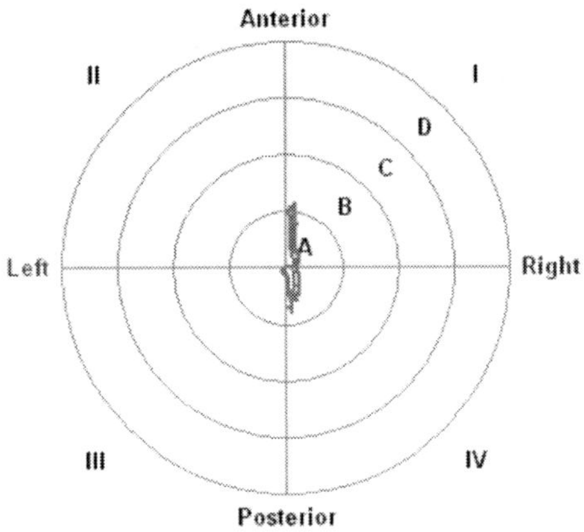

Figure 1. The displacement of CoP of a subject from the platform centre.

The software used in BSS machine to measure the stability indexes was Biodex software (version 3.1; Biodex® Medical System, Shirley, NY, USA). There were three aspects of stability indexes measured for the purpose of upright postural stability assessment which were OSI, APSI, and MLSI. The standard deviations (SD) of the CoM displacement from the zero point which was the centre of platform were presented by these three indexes [21]. The units were in degree. In other words, these stability index indicated the degree of feet displacement in sagittal and frontal plane motions. For the unstable posture, the stability index reading will be higher in correspond to the greater body movements while lower stability index indicated greater posture stability. The sampling rate was 20Hz.

2.4. Experimental Protocol
In this research, every subject underwent the experiment for three sessions (Visit 1, Visit 2, and Visit 3). For every session, the subjects were given a briefing about the experimental procedure and purpose so that they could easily cooperate throughout the experiment. Before the arrival of the subjects, the lab and the equipments were well prepared and the BSS was tested in order to make sure it was functional.

In Visit 1, demographic information especially related to amputation background were measured and recorded. In order to measure stump contracture angle, goniometer was used. The subject was asked to extend the knee as much as possible while the goniometer was placed at the mid patellar level laterally over the knee. The upper arms of the goniometer was pointed towards the greater trochanter while the lower arm was aligned parallel to the tibial bone.

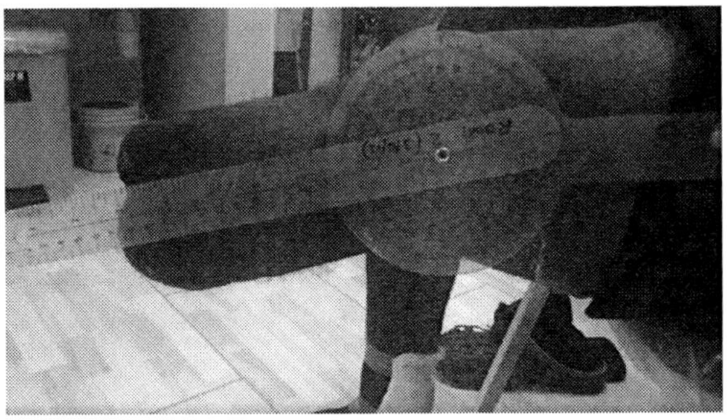

Figure 2. Positioning of goniometer during measuring knee contracture angle.

Before the experiment started, the prosthetic alignment was checked again by the researcher according to the standard alignment set up procedure even when the subject was already satisfied with the previous alignment. This was to ensure the consistency of the experiment by providing the optimum alignment set up task to a single researcher with the same procedure and references. After the completion of the prosthesis aliment set up, the subject was allowed to walk around the lab with the prosthesis in order to get used to the new alignment setup. When the subject was ready, the postural stability experiment then started. All the subjects undergone bipedic stance test (BLS) for the experiment in order to evaluate the postural balance score under static level. This study used cross over study design at which all the subjects underwent the same procedures. In the experiment, the subjects were standing on the round platform of BSS with bare feet for both prosthesis and sound legs. They were asked to stand comfortably still with their arms crossed over their chest. The opening angle and heel placement of their feet were recorded to ensure the consistency throughout the experiments.

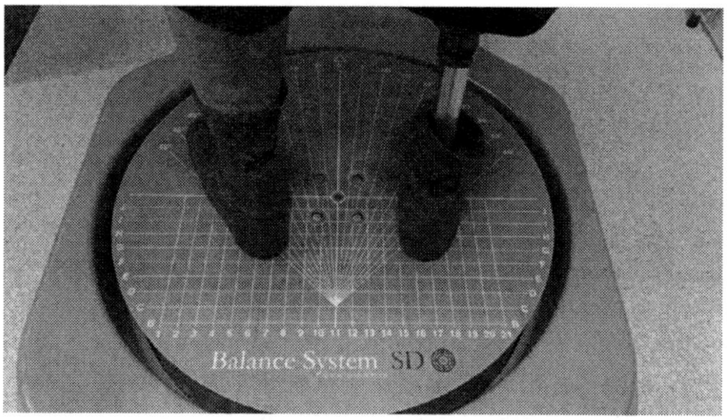

Figure 3. The comfortable feet position of a subject on the BSS platform.

Their visual were maintained by looking straight ahead at the screen monitor. During the postural stability test, the subjects were asked to place their feet in comfortable and stable position on the BSS platform while simultaneously maintained the position of a moving pointer at the center point on the screen monitor. The pointer was moving according to the displacement of CoP. To ensure the consistency and precision of this study, the subjects were allowed to practice the balance training once so that the learning effect could be minimized. Therefore, the difference between the trials would not be related to the learning effect. Three successful trials were conducted for each subject with 20 seconds duration each. The subjects were allowed to rest for 20 seconds in between the trials. The stability index readings were calculated and averaged by Biodex software. The handrails attached on the BSS were only being used to prevent falling in case if the subject lost balance. For a successful trial, handrails should not be used at all.

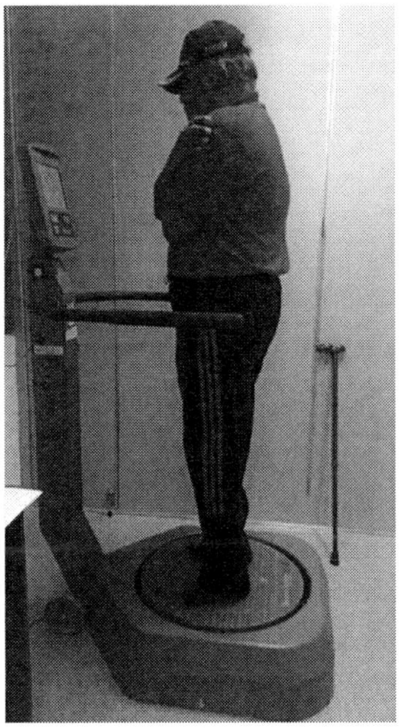

Figure 4. A subject during experiment

The results and data obtained were then recorded. After the experiment, a stump contracture splint was given to every subject together with the exercise list in order to reduce the stump contracture. The subjects were advised to wear the knee contracture splint all the time as when the prosthesis was not used while the exercise list should be done at least two sessions per day. The next session (Visit 2) was conducted at least on the next four weeks, based on the knee contracture improvement (5° minimum).

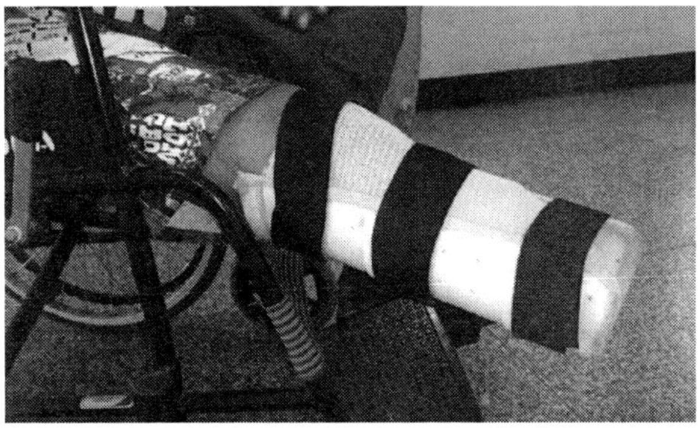

Figure 5. A subject was wearing stump splint after completing the first session experiment.

In Visit 2, the subject's body weight was measured again to ensure there was no significant difference with the reading during Visit 1 (less than 3kg difference). The angle of stump contracture was also measured again and recorded. The experiment would only be proceeded as the stump contracture improved with at least 5°. The same postural stability test as in Visit 1 was then carried out. Upon the completion of the experiment and all three stability index were obtained, the alignment of the prosthesis was adjusted again starting from static to dynamic alignment according to the improvised stump contracture angle. The next session (Visit 3) was then carried out on the following week. The purpose was to obtain the stability index readings when the subject already get used to the new alignment setup and gained full control of it.

During Visit 3, the subjects' body weight and knee contracture angle were measured again in order to ensure no significant changes. The postural stability test was then carried out as in Visit 1 and Visit 2. The OSI, MLSI, and APSI were then recorded and tabulated together with the readings obtained in the previous sessions.

2.5. Prosthetic Alignment

For transtibial prosthetic alignment setting, there were two parts that could be adjusted. Those were socket adaptor and foot adaptor. The alignment of a prosthesis was done according to two standard procedures, starting from static alignment and followed by dynamic alignment. In static alignment, the adjustment was made when the subjects wore the prosthesis and in standing position [22]. This alignment setting was based on the researcher's observation when the subjects were standing and in respond to their comments [7]. Proper alignment was determined based on the consideration of the anteroposterior and mediolateral positioning of socket and foot, prosthesis height, and foot rotation [4].

Dynamic alignment is more subjective compared to the static alignment. The researcher set up the alignment based on the observation on the subjects using prosthesis during walking. Dynamic alignment setting was also made based on the respond to the subjects' comments. There were two standard guides which had been used by the researcher to get the optimum dynamic alignment:

2.5.1. Prosthetist Judgment. The judgment was made based on the knowledge of the causes of gait deviation, experience, understanding of the loading concept on the stump, and the subjects' feedbacks [23].

2.5.2. Subjects' Evaluation. The researcher referred to the verbal feedbacks and subjects' perception regarding the prosthetic function and satisfaction in order to obtain the optimum alignment. The subjects were asked to describe perception of the prostheses in both sagittal and coronal plane based on their feelings when using the prostheses. Further verbal explanation were given by the researcher to ensure the subjects' understanding with the perception should be given by them. The evaluations regarding the perception during ambulation were made in three aspects. Those were early stance sagittal dynamic, late stance sagittal dynamic, and coronal dynamic.

2.6. Statistical Analysis

The stability indexes were presented as mean and standard deviation (mean±SD). Shapiro-Wilk's test was used to investigate the normality of the data. Based on the test, the collected data in this study was proven normally distributed. In order to determine the occurrence of significant difference of the SI readings between the three visits, the repeated measures analysis of variance (ANOVA) was performed. Meanwhile, Tukey's Honestly Significant Differences test had been used in the Post-hoc analysis to determine where the differences occurred. For the degree of stump contracture between the three visits, paired t-test was performed to determine whether the contracture readings for were significantly different with each other. This was because the population of samples were correlated with each other. The significance level was accepted at $p \leq 0.05$ for both ANOVA and paired t-test analysis. The statistical analysis was done by using statistical software Minitab Express 1.5.0 (Version 1.5.0, Minitab Inc, Pennsylvania, USA).

3. Results

The experiments had been successfully done toward 10 below knee amputee subjects in between eight weeks period. The range of activity level among the subjects was started from K1 until K3. The mean of age among them was 57.2 years old with standard deviation of 6.25. All of them were using their own prosthesis along the experiments. The prosthetic alignment setting was redo in Visit 1 by one of the researchers who is a certified prosthetist in order to maximize the consistency of the experiments. The selected subjects were fulfilled all the required criteria as mentioned in Methodology. The demographic data for all the subjects in this experiment are as shown in Table 1.

IOP Conf. Series: Materials Science and Engineering **210** (2017) 012002 doi:10.1088/1757-899X/210/1/012002

Table 1. Subjects' demographic information

Subject	1	2	3	4	5	6	7	8	9	10	Mean	SD[c]
Sex	Male	Male	Male	Female	Female	Male	Male	Male	Male	Male		
Age (years)	55	53	50	58	64	51	51	63	68	59	57.2	6.25
Body mass (kg)	64	89	88	82	67	87	58	70	64	77	74.6	11.45
Height (m)	1.71	1.79	1.68	1.62	1.54	1.63	1.69	1.75	1.66	1.60	1.67	0.074
Cause of amputation[a]	D	D	D	D	D	D	D	D	D	D		
Time (Years)[b]	2	2	2	1	2	3	2	1	4	3	2.2	0.92
Amputation side	Right	Right	Right	Left	Right	Right	Left	Left	Left	Right		
Activity level (K-Level)	K1	K3	K1	K2	K2	K2	K2	K1	K3	K3		
Length of stump (m)	0.15	0.18	0.13	0.15	0.20	0.13	0.19	0.19	0.15	0.14	0.16	0.026

[a]D=Diabetes
[b]Years since amputation
[c]Standard deviation

The mean and SD for all three stability index (MLSI, APSI, and OSI) and stump flexion contracture during Visit 1, Visit 2, and Visit 3 are as shown in Table 2.

Table 2. Mean and standard deviation of stump contracture, MLSI, APSI, and OSI.

	Visit 1	**Visit 2**	**Visit 3**
	Mean±SD	**Mean±SD**	**Mean±SD**
Stump Flexion Contracture (^{0})	14.5±3.171	8.1±2.846	7.5±3.136
MLSI (^{0})	0.99±0.67	0.85±0.29	0.73±0.23
APSI (^{0})	0.42[1]±0.27	1.26±0.54	0.57[2]±0.34
OSI (^{0})	1.20±0.71	1.77[3]±0.66	1.09±0.39

*Significant difference between APSI were indicated as [1](Visit 1 vs Visit 2) and [2](Visit 2 vs Visit 3) while between OSI was indicated as [3](Visit 2 vs Visit 3).

Based on the result of repeated measures ANOVA, during Visit 2, the stability index for anterior-posterior aspect (APSI) was significantly different from the index obtained during Visit 1 and Visit 3 in which p=0.007 and p<0.0001 respectively. Similarly for OSI between Visit 2 and Visit 3, based on the post hoc test, the index reading obtained in Visit 2 was significantly higher than Visit 3 (p<0.0001). Meanwhile, it has been found that the OSI reading was increased by 47.5% in Visit 2 compared to Visit 1. However, the increasing percentage was proven insignificant (p=0.0643). For the medial-lateral aspect of postural stability index (MLSI), the reading obtained in Visit 1 was 16.5% and 35.6% higher compared to Visit 2 and Visit 3 respectively. However, the differences were not significant with p=0.3022 between Visit 1 and Visit 2 and p=0.1573 between Visit 1 and Visit 3. The stump contracture readings for all the subjects from Visit 1 until Visit 3 are as shown in Table 3.

Table 3: The degree of stump flexion contracture of all subjects for every session (n=10).

Subject	1	2	3	4	5	6	7	8	9	10	Mean	SD
Contracture during Visit 1 (^{0})	12	16	15	11	12	21	17	16	11	14	14.5	3.171
Contracture during Visit 2 (^{0})	6	9	10	6	7	14	11	7	5	6	8.1	2.846
Contracture during Visit 3 (^{0})	6	11	10	4	5	13	11	5	5	5	7.5	3.342

*note that the contracture reading during Visit 1 was significantly lower than Visit 2 and Visit 3 ($p \leq 0.05$) while no significant difference between Visit 2 and Visit 3 ($p > 0.05$).

As mentioned in the required criteria list, the minimum requirement of flexion contracture possession was 10^0 while the maximum was 25^0 for the first session. In Visit 1, the maximum reading was exhibited by Subject 6 which was 21^0 while the minimum reading was 11^0, exhibited by Subject 4 and Subject 9. The greatest contracture improvement was achieved by Subject 8. He was able to reduce the flexion contracture by 9^0 which was 40.6% greater than the mean contracture improvement (6.4^0). Based on paired t-test statistical analysis, the mean contracture reading during Visit 2 was significantly lower than Visit 1 (p<0.0001) which indicates the good commitment of the subjects throughout this experiment. Besides that, the mean reading of contracture was insignificantly improved by 7.41 % during Visit 3 (p=0.0839). Among all the subjects, there was a subject who exhibited greater reading of flexion contracture during Visit 3 compared to Visit 2. Since the reading was still acceptable based on the required criteria (5^0 less than the reading during Visit 1), the experiment was still valid to be proceeded.

4. Discussion

In effort of studying biomechanical factors that bring significant influence towards postural stability among amputees, numerous studies has been carried out by the researchers with various conditions which act as manipulated variables while postural stability scores or stability index (SI) as responding variables [14][24][25][26]. However, there is still no study regarding the correlation between below knee stump contracture with postural stability among the amputees. Since the stump contracture is one of the most common issues among the transtibial amputees, this study had been conducted to investigate the correlation between ROM of the stump with postural stability while using the prosthesis, with and without the intervention of alignment setting. The aim of this study was to determine whether the limited ROM of stump could possibly affect the postural stability control during quiet standing among the below knee amputees. The influence of the prosthetic alignment also could be identified whether it can compensates the limited ROM of the stump by bringing back the standard postural stability as for the standard ROM. The findings of this research would be very helpful to the rehabilitation professionals and amputees themselves for the rehabilitation programme.

The study about postural control has been widely done by researchers. Biomechanical factors are the most vital components that should be taken into account in postural control study. According to Chiari et al., there are several biomechanical factors that had been proven influencing postural stability control such as body weight, height, base of support (BoS), foot width, and opening angle of the feet/out-toing angle [10]. These findings suggested that the prosthetic components and alignment setting should be considered as the vital biomechanical factors in any postural stability study among lower limb amputees. Therefore, in this study, the feet opening angle were adjusted according to the standard alignment setting with the agreement from the subjects during alignment adjustment process. This will provide maximum stability to the subjects when using the prosthesis [27].

In order to maximize the consistency of the study, the type of prosthetic foot used by the subjects in this research were fixed to Solid Ankle Cushioning Heel (SACH) foot. This was due to the potential effect caused by the prosthetic ankle activities of different type of foot towards postural stability control [24]. Besides that, the experiment had been carried out on the same day and at the same time for all three sessions for every subject to ensure the efficiency and consistency of the study. Different daily physical activities might be carried out on the different days. Therefore, the related muscles conditions, focus, and consciousness should be the same as much as possible for those three sessions [17, 28].

In this study, stump contracture improvement was necessary for every subject before being proceeded to Visit 2. In order to gain the improvement, the interventions of stump splint and stretching exercises had been implemented to the subjects. However, the positive results were only applied for those who were comply with the interventions. Ten out of 14 subjects exhibited the contracture improvement as according to the required criteria which was 5^0 minimum while the rest had failed to do so. The limitation of time was one of the possible causes. The subjects in this study were given four to eight week times to achieve the minimal contracture improvement. In fact, several previous researchers who were studying the subjects with knee contracture required more than three months before they could see the significant improvement results, depend on the severity of the cases and subjects' compliance [2, 6].

Minimum change of body weight also had become one of the vital conditions before proceed with the following sessions. The weight change between the sessions was limited until 3kg. It was due to the finding that the change in body weight would cause the change in postural stability capability. The reduction in body weight was proven to improve the postural control [29]. Besides, the falling risk would be greater as the body weight increase which indicates poor stability control [30].

Biodex Stability System (BSS) had been used to measure the displacement of foot CoP for both sound and prosthesis legs with different ROM of the stump during quiet standing. The data acquired in this study was postural balance score under static level or in other words, Stability Index (SI). SI represents the Standard Deviation (SD) that assess the path of CoP sway around zero point from the center of the platform. It consists of APSI, MLSI, and OSI. OSI is the tilt degree combination for both medial-lateral (ML) and anterior-posterior (AP) axes. This postural balance assessment with different ROM of stump could contribute a vital information in rehabilitation field since stump contracture is one of the biggest issues frequently emerge among the amputees [4, 6, 27].

Prior to experiments, the subjects were allowed to do balance training using Biodex Stability Sytem once. According to Ku et al., proprioception and vestibular equilibrium might be influenced by the learning process. Therefore, the results obtained later on might be affected. However, a practice trial would be crucial to ensure the understanding of the subjects about the experimental procedure. In order to minimize the influence of the learning factor, in one session, the test were performed for three times so that the data collected would be standardized and more accurate [14].

Previous research has found that the orientation of body plays a significant role in regaining postural balance [31]. It could be more vital as the orientation involve is at the lower segment of body which was the prosthetic alignment with respect to ROM of stump for this study. Thus, it was expected that the change in ROM of stump and the alignment set up will bring difference in postural stability index during quiet standing. This was because the extension angle of the stump and the prosthetic alignment during quiet standing will determine the position of Ground Reaction Force (GRF) vector. As the line vector falls more forward, the knee will be more stable in extended condition [4, 7]. This will significantly influence the CoP sway in AP axis.

As shown in Table 2, APSI reading was higher than MLSI during Visit 2 when there was ROM improvement of the stump while the prosthetic alignment remain unadjusted. The improvement of ROM in this study was in sagittal plane which refers to the improving degree of extension. During quiet standing, both of the knees will be in full extension position they capable of. With more degree of extension during Visit 2 session, the previous alignment setting promotes the body's CoM to sway more in sagittal plane compared to the frontal plane. This was because the alignment setup was only matched with the previous ROM which was before the contracture improvement. The socket flexion and foot dorsiflexion were more than the required angle which lead to the excessive shifting of GRF vector posteriorly, promoting knee instability in sagittal plane [8, 32]. This might explain the higher SI in anterior-posterior aspect compared to the medial-lateral aspect.

Meanwhile, as shown in Table 2, the value of MLSI exhibited by the subjects was significantly and insignificantly higher than APSI during Visit 1 ($p=0.0073$) and Visit 3 ($p=0.0582$) respectively. These findings were in agreement with Blaszczyk et al., who claimed that the postural instability is usually associated with the body sway in lateral aspect [33]. Trunk and hip are the segments responsible in controlling balance in medial-lateral direction while the medial-lateral motion towards lateral direction is generated by pelvis segment [34]. In addition, based on Mayer et al. findings, the CoP sway was more prominence on ML axes compared to AP axes among the lower limb amputees. It was due to the reaction of both sides of hip joints to stabilize the body CoM as a compensation response to the loss of an ankle [25, 26, 31]. Besides that, lower limb amputees commonly have problem with weight shifting in effort to maintain the postural stability in medial-lateral aspect [35]. Therefore, the higher reading of MLSI compared to APSI in our result was expected.

In Visit 2, the OSI readings was higher compared to Visit 1 and significantly higher compared to Visit 3 ($p<0.0001$). This finding may suggest the importance of prosthetic alignment with respect to the ROM of the knee joint. The improvement of knee ROM has been proven less effective for the amputees to gain more postural balance if the prosthetic alignment is not readjusted according to it. This assumption is also supported by the low OSI reading during Visit 3. Meanwhile, the insignificant difference of OSI between Visit 1 and Visit 3 ($p=0.344$) may also suggest that the adaptation of prosthetic alignment with the knee ROM is vital for an amputee to maximize the postural balance even there is a limitation in ROM.

Arnold and Schmitz (1998) had carried out a research towards 19 healthy subjects with no history of lower limb injury in order to find the normal pattern of postural stability using BSS. As the outcome of the research, they have found the intratester reliabilities value for APSI, MLSI, and OSI which were 0.80, 0.43, and 0.82 respectively. Figure 6 shows the line plot of the stability indexes obtained from this research and the values found by Arnold and Schmitz (1998).

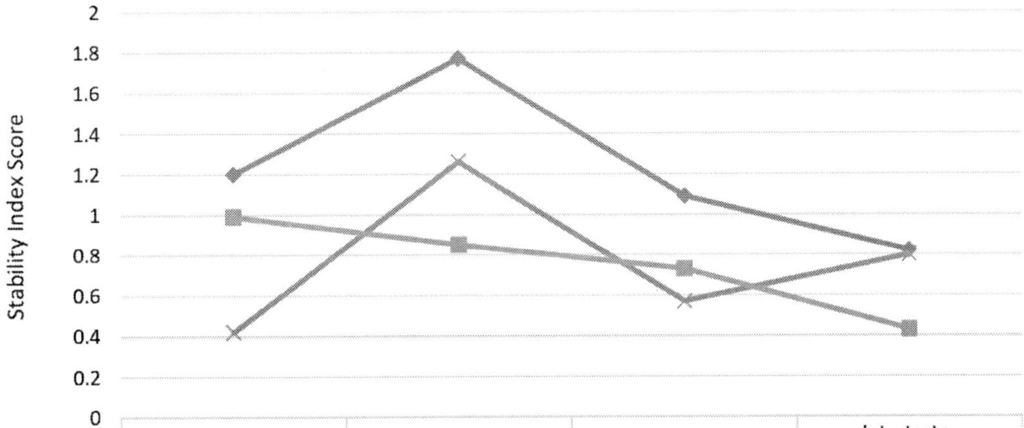

	Visit 1	Visit 2	Visit 3	Intratester Reliabilities
OSI	1.2	1.77	1.09	0.82
APSI	0.42	1.26	0.57	0.8
MLSI	0.99	0.85	0.73	0.43

Figure 6: Stability indexes (APSI, MLSI, OSI) in Visit 1, Visit 2, and Visit 3 with the intratester reliabilities value found by Arnold and Schmitz (1998).

The intratester reliabilities value show the smallest SI reading in medial-lateral aspect which indicates more stability compared to the amputee subjects in this research. Same goes to the OSI value but the difference is not significant. This might be due to the high reading of APSI since it is accounting 95% of the OSI variance and closely related with the OSI reading compared to MLSI [21].

Since the intratester reliabilities values of the postural stability score found by Arnold and Schmitz (1998) were based on the normal healthy subjects without any lower limb injuries history, it was expected that all the SI readings would be lower than those three visits. However, it should be reminded that the outcomes of the postural stability test is a subjective matter which is affected by several other factors such as ages, BMI, training effect, and body structure. Furthermore, the healthy and amputee subjects have different lower limbs structure and control which is believed to be the vital factor of the SI variation when compared.

The subjects' age in this study were limited from 50 to 68 years old and the amputation cause for all of them was diabetes mellitus. The selection based on these criteria was made on purpose in order to narrow down the optimum dynamic alignment setup so that it could be more consistent. This was because Zahedi et al. had found that wide range of alignments could satisfy the youngster subjects and those alignments can be considered as optimum alignments in clinical situation which means not only accepted by the patient, but also satisfy the prosthetist's judgment [23]. Furthermore, according to McClenaghan et al., there was a significant difference in postural stability control in medial-lateral aspect between the elderly and young adult population. In order to narrow down the variation of results obtained and at the same time avoiding high standard deviation, those were the reasons why the subjects below than 50 years old were excluded from this study [36].

The strength of this study was on the selection of the subjects and devices. These two aspects were carefully chosen and set based on fixed criteria in order to ensure the consistency of the experiments and the accuracy of the result. As mentioned, a subject who exceed the weight change limit was not being recalled for the next session. The prosthetic foot used by the subjects were also fixed to Solid Ankle Cushioning Heel (SACH) type as there were evidence that revealed the contribution of different prosthetic foot types towards SI [24]. The prosthetic alignment setup throughout this study were set by three years experienced Certified Prosthetist and Orthotist (CPO) who is also one of the authors. Besides using his experience, the adjustment were also made based on the published and acknowledged literatures [4, 7, 23].

However, the limitation of the subjects' criteria in this study cause a limitation to the collected data as well. This has become more prominent as the subjects were also from the same geographical area and most of them get the rehabilitation treatment from the same hospital. The limitation of this study towards the degree of stump contracture reduction achievement also should be improved in future research. This is because 5^0 of additional extension was not much. Furthermore, certain errors could probably happened when measuring the contracture angle such as different positioning of goniometer and variation of effort when fully extending the stump. Even the error was just 1^0 or 2^0, it is a large percentage from the degree of ROM improvement. Therefore, in future research, the higher degree of contracture improvement should be applied in the required criteria and its effect on postural stability index should be explored further.

5. Conclusion

This study has proved that the prosthetic alignment intervention was able to compensate the adverse effect of stump contracture towards postural stability during quiet standing. It also has been found that the stability could be worse if the prosthetic alignment was not properly set according to the ROM of the stump. A standard range of stability index for amputee population who had similar criteria with the subjects also has been obtained and could be referred in future study.

6. References

[1] LaRaia N 2010 Ask the physical therapist. *In Motion* **20** 33-4
[2] Finger E and Willis FB 2008 Dynamic splinting for knee flexion contracture following total knee arthroplasty: a case report *Cases J.* **1** 421.
[3] Katalinic OM, Harvey LA and Herbert RD 2011 *Phys. Ther.* **91** 11-24
[4] Bowker JH, Michael JW and Smith DG 2004 *Atlas of amputation and limb deficiencies: Surgical, prosthetic, and rehabilitation principles* 3rd ed. Rosemont, IL: American Academy of Orthopaedic Surgeons
[5] Pasquina PF and Cooper RA, editors 2009 *Care of the combat amputee.* Government Printing Office
[6] Knoope C 2011 The Flextension: Preventing flexion contractures in the knee after transtibial amputations http://resolver.tudelft.nl/uuid:efbb9305-23d2-4198-a403-be651df826e5
[7] Edelstein JE and Moroz, A 2011 *Lower-limb prosthetics and orthotics: clinical concepts* Slack
[8] Lannon, N 2003 Trans-tibial alignment: normal bench alignment *ISPO Ortholetter* **12** 12-3
[9] Saleh M and Murdoch G 1985 In defence of gait analysis. Observation and measurement in gait assessment *Bone Joint J.* **67(2)** 237-41
[10] Chiari L, Rocchi L and Cappello A 2002 *Clin. Biomech.* **17(9)** 666-77
[11] Hernández A, Slider A, Heiderscheit BC and Thelen DG 2009 *Gait Posture* **30** 217–22
[12] Alexandrov AV. Frolov AA, Horak, FB, Carlson-Kuhta P and Park, S 2005 *Biol. Cybern.* **93** 309-22

[13] Gravante G, Russo G, Pomara F and Ridola C 2003 *Clin. Biomech.* **18** 780-2
[14] Ku PX, Osman NA, Yusof A and Abas WABW 2012 *J. Biomech.* **45** 1638-42
[15] Aydoğ E, Bal A, Aydoğ ST and Çakci A 2006 *Clin. Rheumatol.* **25** 462- 7
[16] Salsabili H, Forogh B and Rajabali S 2011 *J. Rehabil. Res. Dev.* **48** 775
[17] Ku PX., Osman, NA and Abas WABW 2014 A systematic review *Gait Post.* 39(2) 672-82
[18] Katch V, Becque MD, Marks C, Moorehead C and Rocchini A 1988 *Am. J. Clin. Nutr.* **47** 26-32
[19] Colné P, Frelut ML, Peres G and Thoumie P 2008 *Gait Post.* **28** 164-9
[20] Testerman C and Griend RV 1999 *Foot Ankle Int.* **20** 317-21
[21] Arnold BL and Schmitz RJ 1998 *J. Athl. Train.* **33** 323
[22] Chen CWJ 2012 *Evaluation of Dynamic Prosthetic Alignment Techniques for Individuals with Transtibial Amputation* (Master thesis) University of Toronto, Ontario, Canada
[23] Zahedi MS, Spence WD, Solomonidis SE and Paul JP 1986 *J. Rehabil. Res. Dev.* **23** 2-19
[24] Arifin N, Osman NAA, Ali S, Gholizadeh H and Abas WABW 2015 *Proc. Inst. Mech. Eng. H. J. Eng. Med.* **229** 491-8
[25] Mayer Á, Tihanyi J, Bretz K, Csende Z, Bretz É and Horváth M 2011 *Musculoskelet. Disord.* **2** 118
[26] Buckley JG, O'Driscoll D and Bennett SJ 2002 *Am. J. Phys. Med. Rehabil.* **81** 13-20
[27] Boone DA, Kobayashi T, Chou TG, Arabian AK, Coleman KL, Orendurff MS and Zhang M 2012 *J. Rehabil. Res. Dev.* **49** 843-54
[28] Jorgensen MG, Rathleff MS, Laessoe U, Caserotti P, Nielsen OBF and Aagaard P 2012 *Gait Post.* **35** 653-7
[29] Teasdale N, Hue O, Marcotte J, Berrigan F, Simoneau M, Dore J, Marceau P, Marceau S and Tremblay A 2007 *Int. J. Obes.* **31** 153-60
[30] Corbeil P, Simoneau M, Rancourt D, Tremblay A and Teasdale N 2001 *IEEE Trans. Neural. Syst. Rehabil. Eng.* **9**126-36
[31] Horak FB and Nashner LM 1986 *J. Neurophysiol.* **55** 1369-81
[32] Kang P, Kim JH and Roh JS 2006 *Phys. Ther. Korea* **13** 71-78
[33] Błaszczyk JW, Orawiec R, Duda-Kłodowska D and Opala G 2007 *Experimental Brain Research* **183** 107-14
[34] Shumway-Cook A and Woollacott MH 2007 *Motor control: translating research into clinical practice* 3rd ed. Phyladelphia, Pennsylvania: Lippincott Williams & Wilkins
[35] Isakov E, Mizrahi J, Ring H, Susak Z and Hakim N 1992 *Arch. Phys. Med. Rehabil.* **73** 174-8
[36] McClenaghan BA, Williams HG, Dickerson J, Dowda M, Thombs L and Eleazer P 1996 *Gait Post.* **4** 112-21

Mathematical Modelling of Carbon Dioxide Exchange in Hollow Fiber Membrane Oxygenator

Hany Hazfiza Manap[1], Ahmad Khairi Abdul Wahab[1] and Fathiah Mohamed Zuki[2]

[1]Department of Biomedical Engineering,
[2]Department of Chemical Engineering, Faculty of Engineering,
University of Malaya
50603 Kuala Lumpur, Malaysia
Email: khairi@um.edu.my

Abstract. The main objective of this study is to simulate the carbon dioxide (CO_2) blood-gas exchange in membrane oxygenator, evaluate the effect of sweep gas flow rate on CO_2 partial pressure (pCO_2) and to determine the response of pCO_2 to the step change of sweep gas flow rate. In extension to the simulation of gas transfer in membrane oxygenator, open-loop response on sweep gas flow rate was applied to the developed mathematical model to prove its effect on CO_2 removal rate. Three sweep gas flow rates were set, i.e. 1 L/min (Qg/Qb= 0.5), 2 L/min (Qg/Qb= 1) and 4 L/min (Qg/Qb= 2) and the simulated pCO_2 in artery was observed. Next, one-way ANOVA test was performed to determine the significant difference in CO_2 removal rate between these gas flow rates. For the next test, sweep gas flow rate was changed at t= 60 second and t= 120 second of simulation, to evaluate how pCO_2 will response to the step change of gas flow rate. As the result, the highest CO_2 removal observed when Qg/Qb=2, as compared to gas-blood ratio of 1 and 0.5. The ANOVA test also proved the significant difference in simulated pCO2 for each flow rate. For the last test, a proportional response of pCO_2 towards the change of sweep gas flow rate was highlighted. In conclusion, the rate of CO_2 removal from blood can be manipulated by sweep gas flow rate and this rate can be adjusted to achieve the optimum performance of membrane oxygenator for CO_2 sequestration.

Keywords: Membrane oxygenator, carbon dioxide removal, mathematical modelling, open-loop control, one-way ANOVA.

1. Introduction
Blood oxygenators are commonly used as artificial lung in extracorporeal circulation circuit. This artificial lungs are used to maintain physiologic oxygen (O_2) and carbon dioxide (CO_2) level in blood [1-3] and also used as respiratory assist device to allow time for the native lung to heal by reducing mechanical ventilation setting [4, 5]. There are two types of blood oxygenators that have been invented, which are bubble oxygenators and membrane oxygenators. Unfortunately, there are various adverse effects reported for bubble oxygenators, such as significant damage of blood constituent due to direct contact between blood and air surfaces, platelet and alveolar sequestration, contribute to higher risk of air embolism (micro- and macro-) and much more [6-9]. For these reasons, more reliable membrane oxygenators dominate the current market. Hence, the focus of present study is solely on the

membrane oxygenators, in terms of mathematical modelling and variables that can be manipulated to optimize its function.

Since membrane oxygenators dominate the current market, its latest and most reliable design are always in demand. Numerous investigations are conducted to improve their design and offer the best performance. In early years, building and testing membrane oxygenator prototypes was conducted by trial and error process. This process can be time consuming and expensive. With the development of computer-assisted model, cost for trial and error in manufacturing and testing for multiple produced prototype design can be minimized [10]. Using this method, mathematical modelling and simulation of gas exchange (O_2 and CO2) in membrane oxygenator can be conducted, which enable the prediction of gas transfer in membrane oxygenator. Then, the experimental measurement is conducted for the same model to identify mass transfer coefficient and validate the model [2]. Once the model is validated, the prototype is developed based on it for the next production stage.

The theoretical idea of O_2 and CO_2 diffusion in membrane oxygenator was previously analyzed and experimentally determined by Weissman and Mockros [11, 12] by considering the reaction between oxygen and hemoglobin. Analysis of O_2 transfer rate was calculated and also the improvement in O_2 transfer rate for different configuration of oxygenator. Based on the analysis, it was concluded that tubular oxygenator is practical for this purpose. This works had been extended by Mockros and Leonard [13] for a compact cross-flow tubular oxygenators to determine partial pressure of O_2 by incorporating oxygen-haemoglobin reaction which exists in the blood.

Inspired by the hallmark work by Mockros and Leonard [13], Vaslef et. al [14] developed their own mathematical model to predict O_2 transfer, but they were using semi-empirical theoretical. The observation proved that the predicted O_2 transfer rate was closed to experiment measurement.

Apart of O_2 modelling, Svitek and Federspiel [2] developed a mathematical model that able to predict CO_2 removal exchange rate in hollow fiber membrane oxygenator using semi-empirical approach. Based on the results reported by these authors [2], it was clearly shown that the difference between predicted CO_2 exchange rates (using mathematical model) and measured experimental data did not differs much, which varies in the range of 6% to 9% accordingly to the blood flow rate. This observation thus validates the mathematical model that was developed in the reported study.

Mathematical modelling also could be used to determine the best design for membrane oxygenator. For example, Matsuda and Sakai [15] investigated the effect of number of tied hollow fiber on blood flow in terms of friction factor and mass transfer coefficient. They found an inverse proportional relationship between numbers of tied hollow fiber on O_2 transfer rate, and proved that single hollow fiber is the most effective design for optimum O_2 transfer rate. Tabesh et al. [16] then use a theoretical model to evaluate the design of hollow fiber membrane oxygenator in terms of geometric data, configuration properties and design specification for 6 commercial oxygenators. Vaslef et al. [10] also used the computer method to validate their prototype device of implantable artificial lung in order to reduce the cost for trial-and-error manufacturing using the same method employed by Hormes et al. [17].

As an advanced works in numeric analysis and designing of membrane oxygenator, Turri and Yanagihara [18] developed a two-dimensional computer-assisted numeric simulator which takes account for blood buffering capacity. In addition to their successful in predicting O_2 and CO_2 mass exchange as compared with the experimental value, the researchers also concluded there was no effect of blood base excess on CO_2 mass exchange rate. For the effect of sweep gas flow rate, it were reported that only CO_2 exchange rate was affected by the sweep gas flow rate, whilst O_2 exchange rate shows no significant increasing as the sweep gas flow rate increased. All these findings showed the great contribution of mathematical modelling and process simulation in the development of membrane oxygenators.

2. Methodology

Basically, the simulation is adopted from the mathematical model that been developed by Hexamer and Werner [19], which is the reformulated model by Hill et al. [20, 21]. This model is a mathematical description of gas diffusion process ($O_2$and CO_2) in membrane oxygenator and comprised of three main compartments: gas (*g*), plasma (*pl*) and red blood cell (*rbc*). In each compartment, volume balances are calculated for five state variables, namely O_2 partial pressure (pO_2), CO_2 partial pressure (pCO_2), bicarbonate, carbamate and hydrogenium [22, 23]. Due to the chemical binding of O_2 and CO_2 within the blood, gas transfer in this model is considered as highly non-linear. The generic model used is displayed in Figure 1.

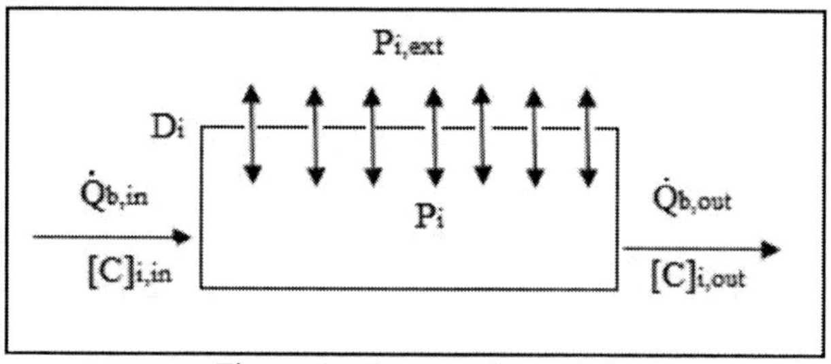

Figure 1. The generic model

From figure 1, there are few assumptions can be made:
i. Compartment's volume and its flow rate is assumed to be constant ($\dot{Q}_{b,in} = \dot{Q}_{b,out}$).
ii. There is optimal mixing inside the compartment ($[C]_i = [C]_{i,out}$).

From these assumptions, volume change rate for both gas (CO_2 and O_2) stored in a compartment can be written as:

$$V_b \frac{d\,[C]_i}{dt} = \dot{Q}_{b,in}[C]_{i,in} - \dot{Q}_{b,out}[C]_{i,out} + D_i\left(P_{i,ext} - P_i\right) + R_i \quad (1)$$

Where $[C]_i$ denotes the concentration of component 'i', V_b denotes the volume of the compartment, $\dot{Q}_{b,in}, \dot{Q}_{b,out}$ represent flow rates of blood (in and out) and $[C]_{i,in}, [C]_{i,out}$ are the concentration at inflow and outflow of the compartment. On the other hand, D_i is the bulk diffusion capacity, which its mass transfer is driven by gradient of partial pressure, Pi_{ext}-P_i. Finally, R_i then stands for gain/loss of the substrate due to chemical reactions. For the equations below, gas, whole blood, plasma and red blood cell are represents by subscript of *g, b, pl* and *rbc*, respectively. For simplicity, the charge of hydrogen ion (H^+) and bicarbonate ion (HCO_3^-) are neglected in the symbol [19].

2.1. Gas Compartment

Gas compartment is also modelled based on the equation (1), with assumption that no volume loses occur across the membrane ($\dot{Q}_g = constant$).

$$V_g \cdot \frac{Fi_{O2}}{dt} = \dot{Q}_g\left(Fi_{O2,in} - Fi_{O2}\right) - D_{O2}(pO2_g - pO2_b) \quad (2)$$

$$V_g \frac{Fi_{CO2}}{dt} = \dot{Q}_g\left(Fi_{CO2,in} - Fi_{CO2}\right) - D_{CO2}(pCO2_g - pCO2_b) \quad (3)$$

The dependency of gas fraction-partial pressure was derived from Henry's law as [24-29]:

$$p_{O2,g} = p_{bar} \cdot Fi_{O2} \quad (4)$$

$$p_{CO2,g} = p_{bar} \cdot Fi_{CO2} \quad (5)$$

Where $p_{O2,g}$and $p_{CO2,g}$ is partial pressure of O_2 and CO_2, respectively, p_{bar} is the atmospheric pressure (mmHg), F_{iO2} is O_2 fraction in mixing gases and F_{iCO2} is CO_2 fraction in mixing gases.

There are three assumptions are made in this case [29]:

1. F_{iCO2} in equation [5] is zero (since there is no CO_2 is used in sweep gas)
2. There are perfect mixing condition in gas compartment ($Fi_{O2,in} = Fi_{O2}$ and $Fi_{CO2,in} = Fi_{CO2}$)
3. No flow difference occur ($\dot{Q}_g, in = \dot{Q}_g, out = \dot{Q}_g$)

By substituting (4) and (5) into (2) and (3), the equations can be rearranged as:

$$V_g \frac{dpO2_g}{dt} = \dot{Q}_g(pO2_{g,in} - pO2_g) - p_{bar} \cdot D_{O2}(pO2_g - pO2_b) \quad (6)$$

$$V_g \frac{dpCO2_g}{dt} = \dot{Q}_g(pCO2_{g,in} - pCO2_g) - p_{bar} \cdot D_{CO2}(pCO2_g - pCO2_b) \quad (7)$$

2.2. *Carbon dioxide transfer in blood compartment*

Prior to the equation of blood-gas transfer, the volumes and flow for plasma and red blood cell are defined as:

$$V_{rbc} = V_b . hct \quad (8)$$

$$V_{pl} = V_b . (1 - hct) \quad (9)$$

$$\dot{Q}_{rbc} = \dot{Q}_b . hct \quad (10)$$

$$\dot{Q}_{pl} = \dot{Q}_b . (1 - hct) \quad (11)$$

Then, the equations for CO_2 transfer are:

$$V_{pl}\, \alpha_{co2} \frac{d\, P_{co2,pl}}{dt} = \dot{Q}_{pl}([CO2]_{pl,in} - [CO2]_{pl}) + D_{CO2,m}(P_{CO2,g} - P_{CO2,pl}) + D_{CO2,rbc}\,(P_{CO2,rbc} - P_{CO2,pl}) + V_{pl} \cdot R_{HCO3,pl} \quad (12)$$

For plasma part, and

$$V_{rbc}\, \alpha_{co2} \frac{d\, P_{co2,rbc}}{dt} = \dot{Q}_{rbc}([CO2]_{rbc,in} - [CO2]_{rbc}) + D_{CO2,rbc}(P_{CO2,pl} - P_{CO2,rbc}) + V_{pl} \cdot R_{HCO3,rbc} - V_{rbc} \frac{d\,[carb]}{dt} \quad (13)$$

For the red blood cell part;

Where;

$$[CO2]_{pl} = \alpha_{CO2} . P_{CO2,pl} \quad (14)$$

$$[CO2]_{rbc} = \alpha_{CO2} . P_{CO2,rbc} \quad (15)$$

$$R_{HCO3,pl} = -k_u \alpha_{CO2}\, P_{CO2,pl} + \frac{k_v}{k} [H]_{pl}\, [HCO3]_{pl} \quad (16)$$

$$R_{HCO3,rbc} = cat\, . (-k_u \alpha_{CO2}\, P_{CO2,rbc} + \frac{k_v}{k} [H]_{rbc} [HCO3]_{rbc} \quad (17)$$

Equations (12) and (13) are obtained from assumption that the concentration of CO_2 in both plasma and red blood cell are linked to the partial pressure via the solubility coefficient.

Next, the equations for bicarbonate transfer for plasma ($[HCO_3]_{pl}$) and red blood cell ($[HCO_3]_{rbc}$) are:

$$V_{pl}\frac{d[HCO\ ]_{pl}}{dt} = \dot{Q}_{pl}([HCO3]_{pl,in} - [HCO3]_{pl}) - D_{HCO\ ,rbc}\left([HCO3]_{pl} - \frac{[HCO3]_{rbc}}{r}\right) - V_{pl}R_{HCO3,pl}$$

(18)

$$V_{rbc}\frac{d[HCO3]_{rbc}}{dt} = \dot{Q}_{rbc}([HCO3]_{rbc,in} - [HCO3]_{rbc}) + D_{HCO\ ,rbc}\left([HCO3]_{pl} - \frac{[HCO\ ]_{rbc}}{r}\right) - V_{pl}R_{HCO3,rbc} \quad (19)$$

Alphabet '*r*' in equation (16) and (17), is the result of [*HCO3*] diffusion across red blood cell membrane and some complex biochemical effects [29]. It is defined as:

$$r = (0.058\, pH_{virt} - 0.437)S - 0.529\, pH_{virt} + 4.6 \quad (20)$$

Equation (18) then influences virtual pH-value, which is the important determinant in pO_2 measurement:

$$pH_{virt} = -\log(r[H]_{rbc}) \quad (21)$$

The diffusion capacity ($D_{CO2,rbc}$) and ($D_{HCO3,rbc}$) used in equations (10), (11), (16) and (17) are estimated from in-vitro measurements [20]:

$$D_{CO2,rbc} = \frac{0.693 \cdot \alpha_{CO2}}{\tau_{rbc}} \cdot \frac{V_{rbc}V_{pl}}{V_{rbc}+V_{pl}} \quad (22)$$

$$D_{HCO3,rbc} = \frac{0.693}{\tau_{HCO3}} \cdot \frac{V_{rbc}V_{pl}}{V_{rbc}+V_{pl}} \quad (23)$$

Carbamino reaction in equation (11) and (24) is due to direct bound between part of CO_2 within red blood cell and haemoglobin. The substance that produced from this reaction is called carbamate (*carb*):

$$V_{rbc}\frac{d[carb]}{dt} = \dot{Q}_{rbc}([carb]_{in} - [carb]) + k_a[CO2]_{rbc} \cdot V_{rbc}([Hb] - [carb]) \cdot \left(\frac{k_{zo}S}{k_{zo}+[H]_{rbc}} + \frac{k_{zr}(1-S)}{k_{zr}+[H]_{rbc}}\right) - V_{rbc}\frac{k_a[carb][H]_{rbc}}{k_c} \quad (24)$$

Lastly, exchange of hydrogen ion in plasma ($[H]_{pl}$) and red blood cell ($[H]_{rbc}$) are described by its hydrogen ion concentration:

$$V_{pl}\frac{d[H]_{pl}}{dt} = \dot{Q}_{pl}([H]_{pl,in} - [H]_{pl}) - V_{pl}\frac{2.303}{\beta_{pl}}[H]_{pl} \cdot R_{HCO3,pl} \quad (25)$$

$$V_{rbc}\frac{d[H]_{rbc}}{dt} = \dot{Q}_{rbc}([H]_{rbc,in} - [H]_{rbc}) + V_{rbc}\frac{2.303}{\beta_{rbc}}[H]_{rbc} \cdot \left(-R_{HCO3,rb} + 1.5\frac{dcarb}{dt} - 0.6 \cdot cap\frac{dS}{dt}\right)$$

(26)

Note that all the initial values, oxygenator parameters, biophysical and chemical parameters mentioned in the equations above are defined in Appendix 1. Due to the objectives and scope of this study, only CO_2 transfer will be considered, while the simulation of O_2 diffusion across hollow fiber membrane can be reviewed elsewhere [24-26, 30-32].

The simulation of gas diffusion in membrane oxygenator (equations (2) –(26)) was implemented in MATLAB/SIMULINK environment to calculate volume balance for state variables in blood-gas diffusion with respect to time.

In addition to the process simulation, open loop control study was conducted to determine the controllable factor that influence gas transfer rate in membrane oxygenator. There are many references which stated that pCO_2 is controlled by sweep gas flow rate, while pO_2 is controlled by blood flow rate and oxygen fraction in gas compartment [33-35]. In fact, this determinant factor were also had successfully proven using in both in vitro [36, 37] and in vivo study [38-40] or even through simulation via computer-assisted approach [18]. Thus, the mathematical model used for this present study was simulated using three flow rates of sweep gas, which are 1 L/min, 2 L/min and 4 L/min, while the blood flow rate was kept constant as 2 L/min (Table 1). These flow rates are chosen based on the ratio suggested by The Food and Drug Administration (FDA) in their guidance for cardiopulmonary bypass oxygenator [41].

Table 1: Ratio of flow rate.

Sweep gas flow rate (L/min)	Blood flow rate (L/min)	Qg/Qb ratio
1	2	0.5
2	2	1.0
4	2	2.0

Next, a statistical test (one-way ANOVA) was performed using two approaches, which were statistical software (Statistical Package for the Social Sciences or SPSS) and MATLAB. This is to ensure the accuracy and validity of our statistical analysis. ANOVA (analysis of variance) is an extension of t-test, which determine equality between several means by comparing their variance among groups relative to variance within groups [42, 43]. Using ANOVA, the mean between two or more samples distribution can be determined if it differ significantly from one another with significant level set as $\alpha= 0.05$ (95% confidential interval) [44].

Lastly, upon the proof of strong relationship between the gas flow rate and rate of CO_2 clearance, another test was conducted. In this test, two time periods of t=60 second and t= 120 second were isolated, where the gas flow was changed markedly. This is to see how the simulated pCO_2 respond accordingly to the change of sweep gas flow rate.

3. Results and discussions

Simulation result of CO_2 transfer across membrane oxygenator is depicted in Figure 2. From the figure, it can be seen that Qg/Qb= 2 has the highest CO_2 removal rate as compared to the other two. The increasing in pCO_2 clearance rate is proportional to the elevation of sweep gas flow rate. This finding agrees with the result reported by [18, 36-38] which advocated that the sweep gas flow rate has effect on measured pCO_2under constant rate of blood flow.

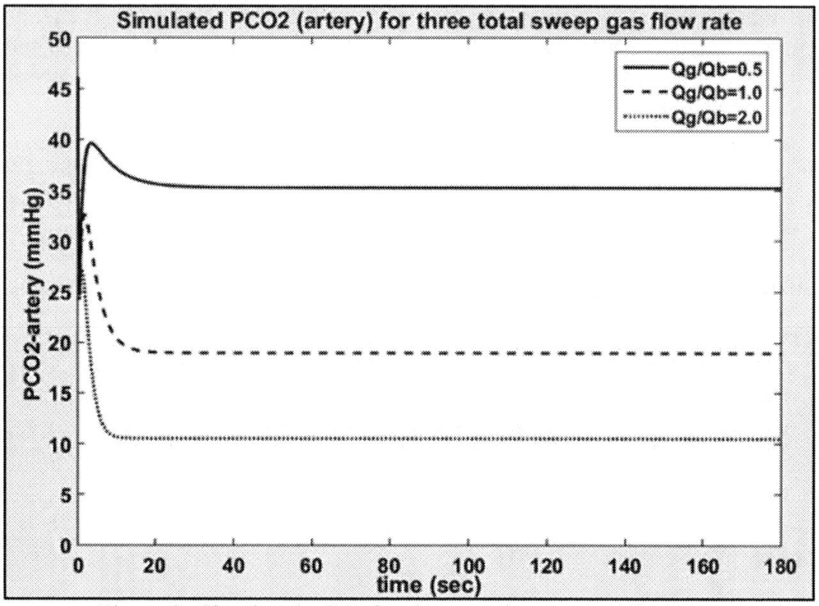

Figure 2: Simulated pCO_2 for three total sweep gas flow rate

A membrane oxygenator comprised of thousands microporous hollow fiber membranes that allows gas exchange between gas and blood compartment. In most design, sweep gas (commonly O_2) will flows through the inside lumen of the hollow fiber, while blood flows outside the hollow fiber. During extracorporeal circulation, O_2 in sweep gas area will diffuses down its concentration gradient across membrane wall into the blood, while CO_2 will diffuse down its concentration gradient from blood into the sweep gas area, which is then removed when the sweep gas exit the oxygenator [45]. According to Cove et al. [46], there are three major factors that influence the amount of gas transfer in membrane oxygenator, which are concentration gradient, contact time between membrane- blood and membrane diffusion characteristics. Adjusting the sweep gas flow rate will affect the concentration gradient of CO_2 in membrane oxygenator. The increment of sweep gas flow rate depreciates CO_2 accumulation along sweep gas pathway [36]. Consequently, the fractional concentration of CO_2 in sweep gas is reduced, while O_2 fractional concentration is elevated. Increasing in pCO_2 concentration gradient between blood and gas phase then augments CO_2 removal [47].This explains why the simulated pCO_2 in artery reduced as the sweep gas flow rate increased.

Due to the proven relationship between pCO_2 and sweep gas flow rate, statistical analysis was performed in order to provide statistical evidence between these three flow rates. As tabulated in Table 2, mean for simulated pCO_2 of three flow rates showed great deviation between each other with 36.153 mmHg, 20.070 and 12.052 for rate ratio (Qg/Qb) of 0.5, 1.0 and 2.0, respectively. Results obtained from ANOVA test then proved the significant difference, where *p*-value= 0.000.

Table 2: Statistical result for one-way ANOVA.

Qg/Qb	Mean ± standard deviation (pCO_2)	*P*-Value
0.5	35.153 ± 1.831	
1.0	20.070 ± 3.699	0.000[a]
2.0	12.052 ± 5.329	

[a] Significant difference (ANOVA, $p<0.05$)

The present work was then extended to investigate how the simulated pCO_2 in artery respond to the change of sweep gas flow rate at each step change of flow rate. The behaviour of this model is illustrated in Figure 3. During the initial sweep gas flow rate (1 L/min), initialization process occurred for simulated pCO_2 until it reaches constant at 35.27 mmHg. Then, the sweep gas flow rate was adjusted to 2 L/min for another 60 seconds. In conjunction with this change, the pCO_2 took approximately 15 seconds to become stable at 18.96 mmHg. Finally, the pCO_2 reached back to 35.27 mmHg when the flow rate was set back to 1 L/min. The simulation results agree with the pattern of pCO_2 response reported by Hexamer et al. [19], which the simulated pCO_2 change with respect to the changing of gas flow rate. It also takes some time to achieve steady state.

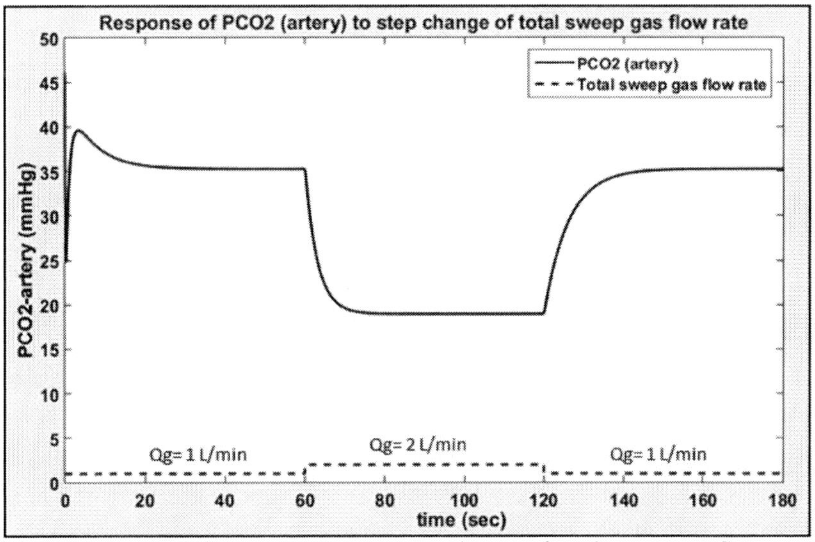

Figure 3: Simulated pCO_2 response to step change of total sweep gas flow rate

These findings are great indicators for the application of closed-loop control strategy to membrane oxygenator in controlling pCO_2in blood. According to Chung et al. [48], CO_2 is more soluble than O_2, hence it diffuse faster than O_2. Due to this advantage, CO_2 is transfers approximately 10 times more efficiently than O_2 [48]. Rapid diffusion of CO_2 that leaves blood will cause hypocapnia (state of reduced CO_2 in blood, where pCO_2< 35 mmHg at sea level [49]). To prevent this problem, a good automated controller is needed to ensure pCO_2 do not fall below 35 mmHg. This will be investigated in our future work as an extension for this present study.

4. Conclusion

There is significant difference in pCO_2reading for three sweep gas flow rates (1 L/min, 2 L/min and 4 L/min), and simulation results clearly show that the pCO_2removal rate has proportional relationship with sweep gas flow rate. Future works shall include various control and advanced strategies of this model for closed loop system in order to obtain the optimal CO_2 partial pressure for blood purification process.

Acknowledgement

Authors wishing to acknowledge financial support by the University of Malaya Research Grant (UMRG), No: RP015-2012G, and also Postgraduate Research Grant (PPP), No: PG022-2015B, Department of Biomedical Engineering, University of Malaya, Malaysia. The University of Malaya Fellowship Scheme (SBUM) award for the first author (HHM) is highly acknowledged and appreciated.

References

[1] Manap H H and Abdul Wahab A K 2016 Extracorporeal carbon dioxide removal (ECCO2R) in respiratory deficiency and current investigations on its improvement: a review *Journal of Artificial Organs* 1-10.

[2] Svitek R G and Federspiel W J 2008 A mathematical model to predict CO2 removal in hollow fiber membrane oxygenators *Ann Biomed Eng.* **36**:6 992-1003.

[3] Finney S J 2014 Extracorporeal support for patients with acute respiratory distress syndrome *Eur Respir Rev.* **23**:133 379-89.

[4] Zwischenberger J B et al. 1999 Percutaneous extracorporeal arteriovenous CO2 removal for severe respiratory failure *Ann Thorac Surg.* **68**:1 181-7.

[5] Zwischenberger J B et al. 2001 Percutaneous extracorporeal arteriovenous carbon dioxide removal improves survival in respiratory distress syndrome: A prospective randomized outcomes study in adult sheep *The Journal of Thoracic and Cardiovascular Surgery* **121**:3 542-551.

[6] Lim M W 2006 The history of extracorporeal oxygenators. *Anaesthesia* **61**:10 984-95.

[7] Akhter M 1996 Pulmonary reperfusion injury : A comparison of bubble and membrane oxygenators *Indian Journal of Thoracic and Cardiovascular Surgery* **12**:1 11-14.

[8] Pearson D T and McArdle B 1989 Haemocompatibility of membrane and bubble oxygenators. *Perfusion* **4**:1 9-24.

[9] Medtronic Incorporation 2004 Design and principle of the extracorporeal circuit *Techniques in extracorporeal circulation, 4th Edn. Ed* Kay P H and Munsch C M (USA: CRC Press).

[10] Vaslef S N et al 1994 Computer-assisted design of an implantable, intrathoracic artificial lung *Artif Organs* **18**:11 813-7.

[11] Weissman M H and Mockros L F 1967 Oxygen transfer to blood flowing in round tubes *Journal of the Engineering Mechanics Division* **93**:6 225-244.

[12] Weissman M H and Mockros L F 1969 Oxygen and carbon dioxide transfer in membrane oxygenators *Medical and Biological Engineering*,**7**:2 169-184.

[13] Mockros L F and Leonard R 1985 Compact cross-flow tubular oxygenators *Trans Am Soc Artif Intern Organs* **31** 628-33.

[14] Vaslef S N et al.1994 Use of a mathematical model to predict oxygen transfer rates in hollow fiber membrane oxygenators *ASAIO J.* **40**:4 990-6.

[15] Matsuda N. and Sakai K 2000 Blood flow and oxygen transfer rate of an outside blood flow membrane oxygenator *Journal of Membrane Science* **170**:2 153-158.

[16] Tabesh H et al. 2012 A theoretical model for evaluation of the design of a hollow-fiber membrane oxygenator *J Artif Organs* **15**:4 347-56.

[17] Hormes M et al. 2011 A validated CFD model to predict O(2) and CO(2) transfer within hollow fiber membrane oxygenators *Int J Artif Organs* **34**:3 317-25.

[18] Turri F and Yanagihara J I Computer-Assisted Numerical Analysis for Oxygen and Carbon Dioxide Mass Transfer in Blood Oxygenators *Artificial Organs* **35**:6 579-592.

[19] Hexamer M and Werner J 2003 A mathematical model for the gas transfer in an oxygenator, in *IFAC Conference on Modelling and Control in Biomedical System*, D. Feng and E. Carson, Editors. 2003: Melbourne, Australia. p. 409-414.

[20] Hill E P, Power G G and Longo L D 1973 Mathematical simulation of pulmonary O2 and CO2 exchange *American Journal of Physiology -- Legacy Content* **224**:4 904-917.

[21] Hill E P, Power G G and Longo L D 1973 A mathematical model of carbon dioxide transfer in the placenta and its interaction with oxygen. *American Journal of Physiology -- Legacy Content* **224**:2 283-299.

[22] Hexamer M, Werner J and Misgeld B J E 2009 Concepts for Simplifying Automatic Blood-Gas Control during Extracorporeal Circulation, in *World Congress on Medical Physics and Biomedical Engineering*, September 7 - 12, 2009, Munich, Germany:

[23] Hexamer M et al. 2004 Automatic control of the extra-corporal bypass: system analysis, modelling and evaluation of different control modes *Biomed Tech (Berl)* **49**:11 316-21.
[24] Walter M et al. 2012 Closed Loop Physiological ECMO Control, in *5th European Conference of the International Federation for Medical and Biological Engineering*: 14–18 September 2011, Budapest, Hungary Ed Jobbágy A (Springer Berlin Heidelberg: Berlin, Heidelberg) p. 319-322.
[25] Walter M et al. 2016 Assistive Control of Extracorporeal Oxygenation Systems, in *12th German Russian Conference on Biomedical Engineering*: Suzdal, Russia.
[26] Walter M et al. 2010 Automatisierung und Fehlerdiagnose bei der extrakorporalen Membranoxygenierung (Automation and Fault Supervision for Extracorporeal Membrane Oxygenation Systems) *Automatisierungstechnik* **58** 277-285.
[27] Walter M et al. 2009 Automation of long term extracorporeal oxygenation systems in *Control Conference (ECC) European*.
[28] Walter M et al. 2010 A physiological model for extracorporeal oxygenation controller design *Conf Proc IEEE Eng Med Biol Soc* p. 434-7.
[29] Misgeld B J E 2007 Automatic control of the heart-lung mechine. Ruhr-University Bochum: Germany.
[30] Kopp R et al. 2016 Automatic Control of Veno-Venous Extracorporeal Lung Assist *Artificial Organs* **40**:10 992-998.
[31] Walter M et al. 2009 Automation of long term extracorporeal oxygenation systems in *European Control Conference (ECC)*.
[32] Walter M et al. 2010 A physiological model for extracorporeal oxygenation controller design *2010 Annual International Conference of the IEEE Engineering in Medicine and Biology*.
[33] Gravlee G P 2008 *Cardiopulmonary Bypass: Principles and Practice* (USA: Lippincott Williams & Wilkins)
[34] Estafanous F G, Barash P G and Reves J G 2001Cardiac anesthesia: Principles and clinical practice (USA: Lippincott Williams & Wilkins)
[35] Richard C. et al. 2014 Extracorporeal life support for patients with acute respiratory distress syndrome: report of a Consensus Conference *Ann Intensive Care* **4** 15.
[36] Federspiel W J and Haulert B G 1996 Sweep gas flowrate and C02 exchange in artificial lungs *Artificial Organs* **20**:9 1050-1052.
[37] Hout M S, Hattler B G and Federspiel W J 2000 Validation of a model for flow-dependent carbon dioxide exchange in artificial lungs *Artif Organs* **24:**2 114-8.
[38] Karabulut H et al. 2002 Adjustment of sweep gas flow during cardiopulmonary bypass *Perfusion* **17**:5 353-6.
[39] Park M et al. 2016 Factors associated with blood oxygen partial pressure and carbon dioxide partial pressure regulation during respiratory extracorporeal membrane oxygenation support: data from a swine model *Revista Brasileira de Terapia Intensiva* **28** 11-18.
[40] Schmidt M et al. 2013 Blood oxygenation and decarboxylation determinants during venovenous ECMO for respiratory failure in adults *Intensive Care Medicine* **39**:5 838-846.
[41] U.S. Food and Drug Administration 2000 Guidance for Cardiopulmonary Bypass Oxygenators 510(k) Submissions; Final Guidance for Industry and FDA Staff [Retrieved on 5 January 2017]; Available from: http://www.fda.gov/RegulatoryInformation/Guidances/ucm073668.htm.
[42] Ostertagov E and Ostertag O 2013 Methodology and Application of Oneway ANOVA. *American Journal of Mechanical Engineering* **1**:7 256-261.
[43] Larson M G 2008 Analysis of Variance *Circulation* **117**:1 115-121.
[44] Manap H H, Tahir N M and Yassin A I M 2011 Statistical analysis of Parkinson disease gait classification using Artificial Neural Network. *2011 IEEE International Symposium on Signal Processing and Information Technology (ISSPIT)*.

[45] Federspiel W J and Henchir K 2008 Lung, Artificial: Basic Principles and Current Applications *Encyclopedia of Biomaterials and Biomedical Engineering, Second Edition (Online Version)* (USA: CRC Press) p1661-1672.
[46] Cove M E, MacLaren G, Federspiel W J and Kellum J A 2012 Crit Care **16** 232.
[47] Scaravilli V, Zanella A, Sangalli F and Patroniti N 2014 Basic aspect of physiology during ECMO support, eds. F. Sangalli, A. Pesenti, N. Patroniti *ECMO-Extracorporeal Life Support in Adults* (Italy: Springer-Verlag)
[48] Chung M, Shiloh A L and Carlese A 2014 Monitoring of the Adult Patient on Venoarterial Extracorporeal Membrane Oxygenation *The Scientific World Journal* **2014** 10.
[49] Solano C. M E., Ichel Castillo B., María C. Niño de Mejía 2012 Hypocapnia in Neuroanesthesia: Current Situation. *Colombian Journal of Anesthesiology* **40**:2, 137-144. .

Appendix 1

Table 3: Oxygenator venous input conditions.

Symbol	Variable	Value	Unit
$[CO2]_{pl}$	CO_2 concentration (plasma)	0.00138	M
$P_{CO2,pl}$	CO_2 partial pressure (plasma)	46	mmHg
$[CO2]_{rbc}$	CO_2 concentration (red blood cell)	0.00138	M
$P_{CO2,rbc}$	CO_2 partial pressure (red blood cell)	46	mmHg
$[HCO3]_{pl}$	Bicarbonate concentration (plasma)	0.0263	M
$[HCO3]_{rbc}$	Bicarbonate concentration (red blood cell)	0.0182	M
r	Ratio (from equation [20])	0.69	-
$[H]_{pl}$	Hydrogen concentration (plasma)	$42.3 X 10^{-9}$	M
$[H]_{rbc}$	Hydrogen concentration (red blood cell)	$61 X 10^{-9}$	M
$[carb]$	Carbamino concentration	0.00235	M
$P_{O2,g}$	O_2 partial pressure in gas compartment	-	mmHg
$P_{CO2,g}$	CO_2 partial pressure in gas compartment	-	mmHg
$\dot{Q}_b$	Blood flow rate	2	L/min
$\dot{Q}_g$	Total gas flow rate	Refer table 1	L/min
hct	hematocrit	0.3	-
F_{iO2}	O_2 fraction in mixing gas (sweep gas)	66%	-
F_{iCO2}	CO_2 fraction in mixing gas (sweep gas	0%	-

Table 4: Oxygenator parameters.

Symbol	Parameter	Unit
V_b	0.25	Liter
V_g	0.10	Liter
P_{bar}	760	mmHg
$D_{O2,m}$	11.291	μL/(mmHg sec)
$D_{CO2,m}$	414.64	μL/(mmHg sec)

IOP Conf. Series: Materials Science and Engineering **210** (2017) 012003 doi:10.1088/1757-899X/210/1/012003

Table 5: Biophysical and biochemical parameters.

Symbol	Parameter	Value	Unit
α_{O2}	O_2 solubility	1.35×10^{-6}	M/mmHg
α_{CO2}	CO_2 solubility	3×10^{-5}	M/mmHg
β_{pl}	Plasma buffer capacity	6×10^{-3}	mol/pH
β_{rbc}	Red blood cell buffer capacity	57.7×10^{-3}	mol/pH
cat	Carbonic anhydrase catalysis factor	13000	-
$[Hb]_{rbc}$	Red blood cell haemoglobin concentration	20.7×10^{-3}	M
k	Carbonic acid dissociation equilibrium constant	5.5×10^{-4}	M
k_u	CO_2 hydration reaction forward rate constant	0.12	sec^{-1}
k_v	CO_2 hydration reaction reverse rate constant	89	sec^{-1}
k_a	CO_2-hemoglobin reaction forward constant	5000	$Msec^{-1}$
k_c	Carbamate ionisation constant	2.4×10^{-5}	-
k_{zo}	Oxygenated haemoglobin amino group ionisation constant	8.4×10^{-9}	M
k_{zr}	Reduced haemoglobin amino group ionisation constant	7.2×10^{-8}	M
τ_{rbc}	Half-time of red cell membrane diffusion	0.001	sec
τ_{HCO3}	Half-time of red blood cell membrane chloride shift	0.2	sec

IOP Conf. Series: Materials Science and Engineering **210** (2017) 012004 doi:10.1088/1757-899X/210/1/012004

Toxicology Analysis of Tissue-Mimicking Phantom Made From Gelatin

A S Dolbashid, N Hamzah, W S W K Zaman and M S Mokhtar

Department of Biomedical Engineering, Faculty of Engineering, University of Malaya, 50603 Kuala Lumpur, Malaysia

E-mail: asdanisaifullah@gmail.com

Abstract. Skin phantom mimics the biological skin tissues as it have the ability to respond to changes in its environment. The development of tissue-mimicking phantom could contributes towards the reduce usage of animal in cosmetics and pharmacokinetics. In this study, the skin phantoms made from gelatin were tested with four different commonly available cosmetic products to determine the toxicity of each substance. The four substances used were; mercury-based whitening face cream, carcinogenic liquid make-up foundation, paraben-based acne cleanser, and organic lip balm. Toxicity test were performed on all of the phantoms. For toxicity testing, topographical and electrophysiological changes of the phantoms were evaluated. The ability of each respective phantom to react with mild toxic substances and its electrical resistance were analysed in to determine the toxicity of all the phantom models. Four-electrode method along with custom made electrical impedance analyser was used to differentiate electrical resistance between intoxicated phantom and non-intoxicated phantom in this study. Electrical resistance values obtained from the phantom models were significantly higher than the control group. The result obtained suggests the phantom as a promising candidate to be used as alternative for toxicology testing in the future.

1. Introduction

Tissue-mimicking (TM) phantoms or well known as tissue-mimicking materials are objects that mimic or resemble human tissue such as skin [1].The low cost phantom development were already studied extensively for spectroscopy and imaging field. A TM phantom that accurately represents human tissue, are important in safety testing and other tissue testing application [2]. Phantom has been developed recently in such a various ways in order to get the most accurate results for its similarity in mimicking the human tissue. These materials may be classified into several categories based on few ingredients [3]. Basically phantoms are made for imaging modalities related with tissue frequency and electrical properties in field such as ultrasound, magnetic resonance imaging (MRI), and computed tomography (CT).The most reliable organic material developed as phantom is gelatin because of its desirable mechanical properties [2], [4]–[7].

At certain level, all harmful chemical can cause harm to the cells and tissues. However, the chemical is considered to be non-toxic when it only causes damage with a very large amount. When one speaks about toxicity test in food or domestic product, issues that will arise is necessarily the animal testing [8]. A report has stated that at least one million of animal testing are done all over the world annually

[9]. One of the toxic substances, mercury is the most common chemicals found in lightening product. Liang and colleagues have reported that mercury is presence in most common skincare product such as eye drops, nasal drops, antibiotics and cosmetic products [10]. Women who used mercury-lightening product often have elevated mercury levels in the hair, blood and urine [11]. Parabens are also the most commonly used additives in many cosmetic products and they act as preservatives in cosmetic [12]. Methylparaben, propylparaben, and butylparaben are the compounds that can easily be found in cosmetic product especially face and body cleanser.

Electrical resistance analysis is low cost yet non-invasive commonly used in body composition approach and clinical condition assessment [13]. Electrical resistance analysis is not limited to only body composition and single sample measurement; it has also been used in clinical condition measurement. One such example is the use of electrical impedance characterization to compare normal and cancerous hepatic liver tissue [14].

In this study, phantom made from gelatin was tested for its toxicity by looking at its microscopic examination, weight reduction and electrical resistance. The topographical condition of the phantom, its percentage of weight loss, and electrical resistance are important criteria in order to distinguish between normal and intoxicated phantom.

2. Material and methods

2.1. Materials

Gelatin, ovalette, and sunflower oil were purchased from a bakery supplier store (Bake With Yen Sdn Bhd). Propylene glycol and formalin (37% formaldehyde) were obtained from Sigma-Aldrich (St.Louis,MO,USA). Whitening night cream, UV foundation make up base, acne cleanser, and organic lip balm were purchased from a cosmetic store in Kuala Lumpur, Malaysia.

2.2 Fabrication of the phantom

The phantom was fabricated according to the method described by Hahn and Noghanian with slight modification to obtain the composition intended for this study [5]. The phantom was made based on this composition; 7 g of calf-skin gelatin (200 bloom), 60 ml of distilled water, 7 ml of propylene glycol, 5 ml of sunflower oil, 0.30 g of ovalette (surfactant), and 20 drops of formalin (37% formaldehyde).

60 ml of distilled water was heated until it reached 50°C. 7ml of propylene glycol was then added as a stabilizer followed by the addition of 7 g of gelatin. The temperature was allowed to increase until it reached 70°C. In a separate container, another solution of 0.30 g ovalette and 5 ml of sunflower oil was prepared. These two solutions were then mixed together. The ovalette acts as a surfactant that will assist in the bonding between oil and water parts in the solution.

The mixture was then stirred until a compound mixture formed. Formaldehyde was then added to the mixture. The mixture was then poured in petri dishes and stored in the refrigerator for 10 hours to ensure complete solidification.

2.3 Toxicity testing

Two groups of four phantoms were prepared, with four belongs to the test group and another four belong to the control group. All of the test phantoms were test simultaneously with the control phantoms. The four cosmetic products (whitening night cream, UV foundation make up base, acne cleanser, and organic lip balm) were heated with a solvent to ensure better diffusion when the substances were introduced into the phantom. Distilled water was used as the solvent, with 1 g of each substance was added to 1 ml of distilled water. The phantom with the added cosmetic substance was then left in the drying oven at 37°C for 10 hours to allow proper reaction to take place. Equal amount of the liquefied cosmetic substance was added to the phantom every day during the 7-day course of testing.

2.3.1 Microscopic observation

The surface of each phantom was observed through light microscope. A 5X magnification was used to obtain the image.

2.3.2 Weight reduction analysis

The initial weights of all the phantoms (±73 g) were measured and the weight loss during 7 days of experiment was recorded. The readings obtained then analysed using one-way ANOVA calculations (SPSS software).

2.3.3 Electrical resistance measurement

The variation of the electrical resistance values in a continuous range of frequency is termed as galvanic profile. The general morphology obtained from the galvanic profile of the test phantoms was used and compared with the galvanic profile of the control group.

3. Results

3.1 Microscopic observation

Toxic substances in cosmetic products have particular effect to the surface of the phantom. To observe the skin phantom feasibility on toxicology analysis, the image of phantom's surface on control phantom and every test substance were obtained and compared. The images were obtained from Light Microscope with 5X magnification.

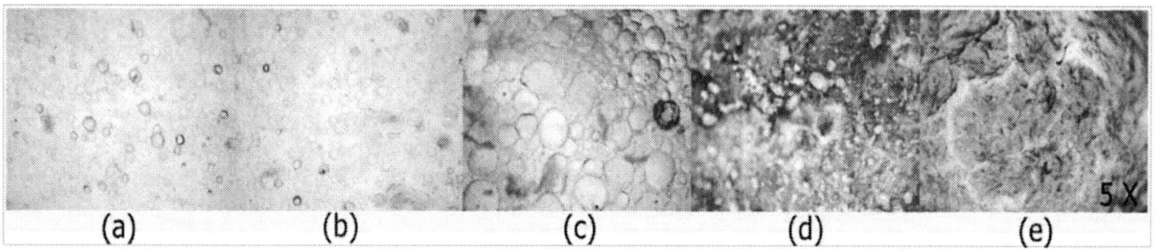

Figure 1. Surface of skin phantom models from control group (a) and after toxicity test of; organic lip balm (b), paraben-based acne cleanser (c), carcinogenic liquid foundation (d) and mercury-based face cream (e).

Figure 1(a) shows a normal microscopic image of agar/gelatin gels. The image presented was from one of the control phantoms. The light area represents the gelatin region and the dark spot indicated air bubbles and open pores. Air bubbles are trapped on the surface and in the structure itself during the mixing process of the phantom.

Figure 1(b) showed almost similar morphology as in control group, apart from hardened composition of the phantom due to the presence of beeswax in the organic lip balm. Figure 1(c-e) showed the presence bigger air bubbles and also surface lesion on the surface of the phantoms.

3.2 Weight reduction analysis

The initial weights of all phantoms were measured respectively and a mean weight of ±73g was obtained for a single phantom. Weight reduction percentage demonstrates the percentage of weight loss throughout the whole test duration for a maximum of 7 days.

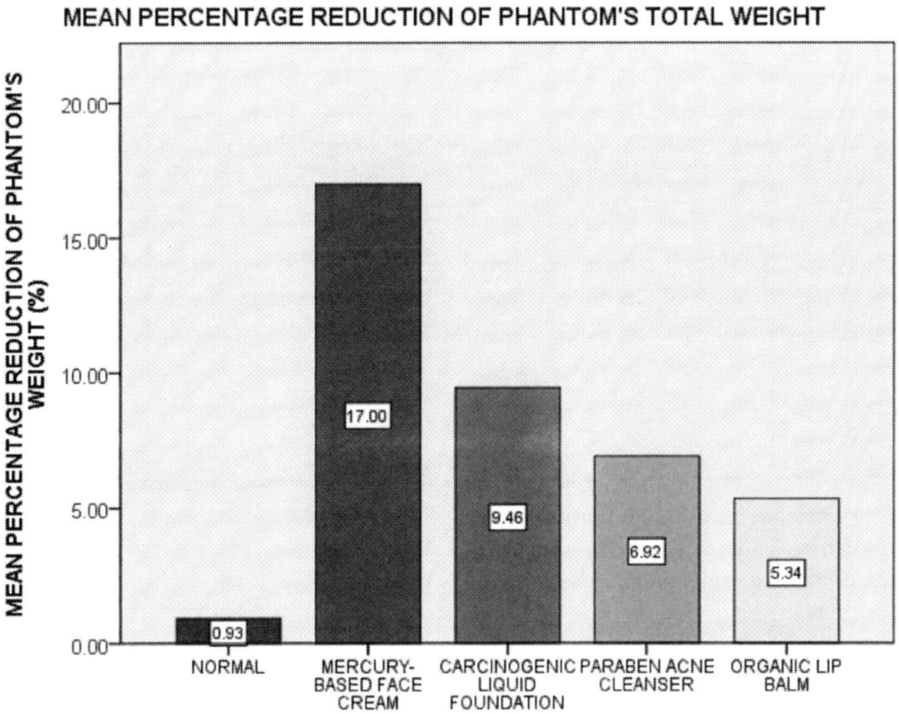

Figure 2. Phantom's weight reduction percentage (with a respective $p<0.05$ to test subjects)

A one-way measure ANOVA was ran on five set of 15 values of weight reduction percentage of skin phantom sample to determine if there was difference in phantom's total weight loss among test and control group. There were statistically significant in means among intoxicated phantom group and normal phantom group with $p< 0.05$.

3.3 Electrical resistance measurement

The general morphology of the galvanic profile, as measured in both groups, resembles an increase in the electrical resistance, corresponding with the increase of the current frequency, at a constant voltage of 7V of alternating current.

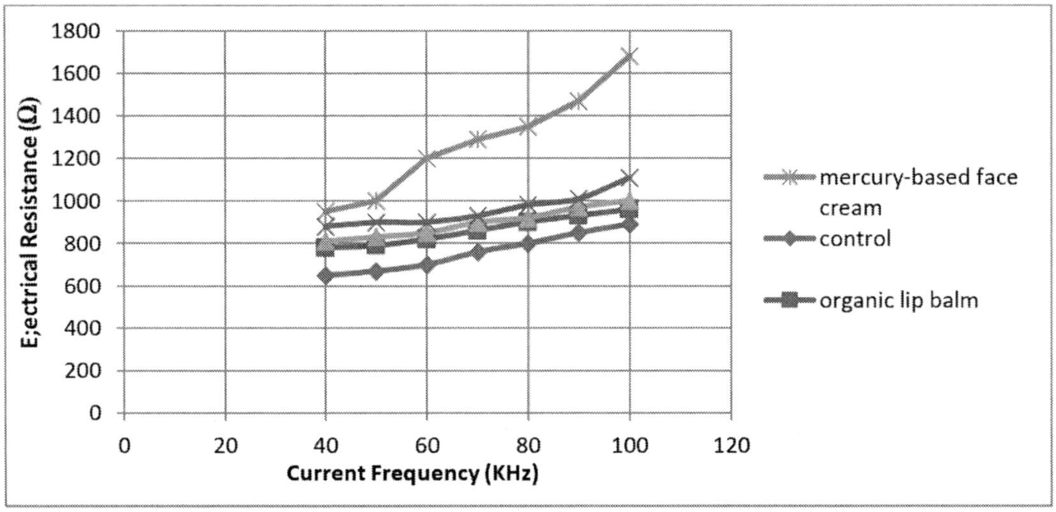

Figure 3. The electrical resistance of the skin phantom in the test group and the control group at the lower end of frequency range. Each point represents the average of measurement for one specific frequency.

When the galvanic profiles between control group and test group were compared, the electrical resistance was significantly higher in the mercury based face cream-intoxicated group, especially on high frequency measurements.

4. Discussion

After toxicity test was conducted, the topographical evaluation towards intoxicated skin phantom by acne cleanser, carcinogenic foundation and mercury-based face cream revealed a minor destruction on the surface of the phantom which can be clearly seen on Figure 1(c-e). Each test substance (Figure 1(c-e)) contained toxic substances which could leads to breast cancer and other health side effects. [15]. Paraben-based cleanser is capable of storing extra fat in human skin tissue and that can be related to the presence of bigger air bubbles beneath the surface area of the phantom (Figure 1(c)). The bigger air bubbles and surface lesions were inferred to be caused by the reaction of the highly active toxic ingredients in the test substances itself. On figure 1(e), mercury-based face cream affect the surface of the phantom by causing minor erosions and making the phantom loses more weight than other test substances. This phenomenon also happened in carcinogenic liquid foundation (Figure 1(d)), in which the whole phantom turns into yellowish-brown color due to synthetic dye used to make the substances. Organic lip balm shows a less than 5% difference when compared with the control group; no presence of potential toxic substances was made as a conclusion to this finding. Organic lip balm, however, harden the composition of the skin phantom due to the presence of beeswax in the ingredients organic lip balm.

The phantom weight loss is due to several factors. Firstly, fluid exodus happened right after the phantom is kept under 37°C. A normal phantom lost about 10% or less of its total weight through the process of evaporation with maximum of 6 days after solidification. Secondly, water exit from the phantom was due to the reaction of chemical substances contained in the test materials with the internal structure of the phantom. The test group shows increased values in weight percentage reduction especially on mercury-based face cream followed by carcinogenic liquid foundation which may be due to the presence of salicylic acid and retinol found in the test substances [16].

The purpose of measuring the resistance of skin phantom analogues for this experiment is as an attempt to conclude if the intoxicated phantom has a measurable difference of electrical properties compared to normal skin phantom. Relating to biologic tissue, another interesting attribute of the electrical properties is that the resistance of the tissue increase with tissue necrosis. It is because when the cells die, the cell membrane loses its ability to maintain its insulating properties and ionic gradients [17]. The differences in the electrical properties of normal phantom and intoxicated phantom will play a large role in determining the viability of the phantom to act as an alternative medium in toxicology analysis. Since toxicology analysis is a wide range of study, focusing in developing a skin phantom that mimics human biological tissue will be a good step towards starting a phantom-based toxicology analysis in the future. The data gathered for this thesis established the fact that intoxicated skin phantom tissue has higher electrical resistance than normal phantom. The electrical resistance was higher in the intoxicated group, especially on high frequencies range of the electrical current. Mercury-based face cream demonstrated significant effects on skin phantom particularly in the erosion of the phantom surface and high electrical resistance. This can be related to a research on mercury effect in most skin-whitening cream which demonstrated a permanent damage of rat liver, kidney and brain when treated with mercury for three times a week [18]. The main cause for the increased of electrical resistance of the intoxicated group could be the destruction of cross-linked polymeric structures. In general, fat has lower electrical conductivity than structures containing water. The measurement of high frequency is relevant for living tissue because they have the capacity to pass through membranes.

Acknowledgement

This research is supported by University of Malaya Research Grant (UMRG: RP022C-14AFR).

References

[1] N. Ortega-Quijano, F. Fanjul-Velez, I. Salas-Garcia, and J. L. Arce-Diego 2011 *Proceedings of 21st International Conference Radioelektronika 2011* 1–4
[2] A. I. Farrer, H. Odéen, J. de Bever, B. Coats, D. L. Parker, A. Payne, and D. A. Christensen 2015 *J. Ther. Ultrasound* **3** 9
[3] M. Lazebnik, E. L. Madsen, G. R. Frank, and S. C. Hagness 2005 *Phys. Med. Biol.* **50** 4245–4258
[4] P. G. Anderson, N. C. Rouze, and M. L. Palmeri 2011 *Ultrason. Imaging* **33** 134–142
[5] C. Hahn and S. Noghanian 2012 *Int. J. Biomed. Imaging* **2012** 1–12
[6] M. M. Nguyen, S. Zhou, J. Robert, V. Shamdasani, and H. Xie 2014 *Ultrasound Med. Biol.* **40** 168–176
[7] A. M. R. Pinto, P. Bertemes-Filho, and A. S. Paterno 2015 Gelatin as a Skin Phantom for Bioimpedance Spectroscopy178–182
[8] J. Tannenbaum 1999 *ILAR J.* **40** 97–110
[9] S. K. Doke and S. C. Dhawale 2015 *Saudi Pharm. J.* **23** 223–229
[10] L. Liang, J. Gilkeson, E. Swain, E. Bennett, M. Li, M. Deng, and P. Pang 2013 *J. Cosmet. Dermatological Sci. Appl* **3** 256–262
[11] P. del Giudice and P. Yves 2002 *Int. J. Dermatol.* **41** 69–72
[12] R. Winter 2009 A consumer's dictionary of cosmetic ingredients : complete information about the harmful and desirable ingredients found in cosmetics and cosmeceuticals *Three Rivers Press*
[13] S. Khalil, M. Mohktar, and F. Ibrahim 2014 *Sensors* **14** 10895–10928

[14] S. Laufer, A. Ivorra, V. E. Reuter, B. Rubinsky, and S. B. Solomon 2010 *Physiol. Meas.* **31** 995–1009
[15] P. D. Darbre, A. Aljarrah, W. R. Miller, N. G. Coldham, M. J. Sauer, and G. S. Pope 2004 *J. Appl. Toxicol.* **24** 5–13
[16] E. C. Davis and V. D. Callender 2010 *J. Clin. Aesthet. Dermatol.* **3** 20–31
[17] X. Chen, W. Li, J. Ren, D. Huang, W. He, Y. Song, C. Yang, W. Li, X. Zheng, P. Chen, and J. Han 2014 *Cell Res.* **24** 105–121
[18] I. Al-Saleh, N. Shinwari, I. El-Doush, G. Billedo, M. Al-Amodi, and F. Khogali 2004 *BioMetals***17** 167–175

IOP Conf. Series: Materials Science and Engineering **210** (2017) 012005 doi:10.1088/1757-899X/210/1/012005

Animal Robot Assisted-therapy for Rehabilitation of Patient with Post-Stroke Depression

Winal Zikril Zulkifli[1], Syamimi Shamsuddin[1] and Lim Thiam Hwee [2]
[1] Faculty of Manufacturing Engineering, Universiti Teknikal Malaysia Melaka, Hang Tuah Jaya, 76100 Durian Tunggal, Melaka, Malaysia.
[2] Psychosocial Department, SOCSO Rehabilitation Center, Bandar Hijau, Hang Tuah Jaya, Melaka, Malaysia

E-mail: syamimi@ieee.org

Abstract. Recently, the utilization of therapeutic animal robots has expanded. This research aims to explore robotics application for mental healthcare in Malaysia through human-robot interaction (HRI). PARO, the robotic seal PARO was developed to give psychological effects on humans. Major Depressive Disorder (MDD) is a common but severe mood disorder. This study focuses on the interaction protocol between PARO and patients with MDD. Initially, twelve rehabilitation patients gave subjective evaluation on their first interaction with PARO. Next, therapeutic interaction environment was set-up with PARO in it to act as an augmentation strategy with other psychological interventions for post-stroke depression. Patient was exposed to PARO for 20 minutes. The results of behavioural analysis complemented with information from HRI survey question. The analysis also observed that the individual interactors engaged with the robot in diverse ways based on their needs Results show positive reaction toward the acceptance of an animal robot. Next, therapeutic interaction is set-up for PARO to contribute as an augmentation strategy with other psychological interventions for post-stroke depression. The outcome is to reduce the stress level among patients through facilitated therapy session with PARO

1. Introduction

World Health Organization (WHO) has classified depression as common illness worldwide, with an estimated 350 million people affected. In Malaysia, the prevalence is estimated between 8-10% of the population [1]. Strong range of emotions on sadness and bereavement are often related to Clinical Depression or technically referred to as Major Depressive Disorder (MDD). MDD may disturb physiological functions due to high level of stress and low mood. Potential consequences are increased blood pressure and change in emotion [2]. Therefore, it is important to have interventions that can reduce the consequences. Animal therapy is proven to have positive effects to relieve depression and anxiety.

Early studies showed that animals are capable of decreasing stress, reduce heart and respiratory rate, and show positive changes in hormone level as well mood elevation [3]. However, the use of animals has a number of drawbacks. In many therapy institutes, an animal is not always available to the patients because of the usage within scheduled event. Animal therapy is also discouraged due to the fear of diseases, bites, or allergies. Robot therapy, specifically in mental health care through interactions with robot has been studied by many researchers. Therapy through robotic animals is expected to have physiological effects similar to animal therapy. This research aims to use a therapeutic seal robot PARO (Figure 1) which focuses on human-robot interaction (HRI) with PARO as a social companion.

Figure 1. Therapeutic animal robot PARO

2. Subjective Evaluation

To investigate the acceptance of Malaysians to an animal robot, subjective evaluation was done with animal robot PARO for twelve rehabilitation patients at SOCSO Rehabilitation Centre Melaka. The questions aim to investigate the opinions, impressions and acceptance level of PARO.

The evaluation period was done during a weekly group therapy session. All patients gave consent to participate. During the session, researcher explained about PARO's functionality followed by a demonstration for about 8 minutes. After that, the patients were given a chance to interact with PARO. Each patient got the chance to personally touch, stroke, hug and talk to PARO. A questionnaire of 10 questions was given to the patients after the interaction. The question topics are shown in Table 1. The first two questions asked about previous experiences with animals. Question 3 until Question 10 covers the patient's opinion and feelings related to PARO, the robotic animal. Survey responses were tabulated in a graph (Figure 2).

Table 1. Question for Subjective Evaluation

Evaluation Topics	
1. Ever own pets?	6. PARO looks alive
2. Do you like animals?	7. Feel calm with PARO
3. PARO is cute	8. Want to meet PARO again
4. Like to touch PARO	9. Want to own PARO
5. Happy with PARO	10. PARO is a robot

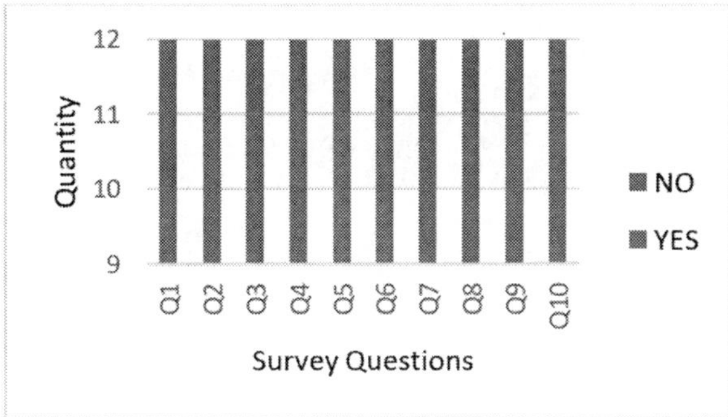

Figure 2. First interaction survey question

Overall result shows positive reaction toward the acceptance of animal robot (in this case; PARO) among rehabilitation patients. Following this, a full experiment involving therapeutic interaction between PARO and post-stroke depression patients was planned to further investigate the benefit of PARO in therapy.

3. Methodology

The experimental procedure serves as platform to investigate changes in stress level of patients with post-stroke depression before and after interacting with PARO. The experiment flow based on flow chart (Figure 3). The proposal was submitted and gain approval from UTeM Research Ethics before conducting the study. Selected of patients based on certain criteria of depression with insomia and ability to converse in Malay and English. Only two respondent can undergo therapy in one day. A short session with psychologist was conducted to explain about the experiment flow. PARO robot was socializing with patients and assisted by a psychologist. There is certain condition where the procedures are aborted, whereby patients feel restless, abnormal behavior, unable to interact, scare of animal robot or robot perform irregular action protocol and makes patients request to abort. The interaction content is to introduce the robot to patients. The patients will familiarize and encouraged to interact under psychologist supervision, in addition some topics brought up for conversation throughout the session. The interaction also involve interviewing and video recording. The session will be recorded by video recorder (Figure 4).

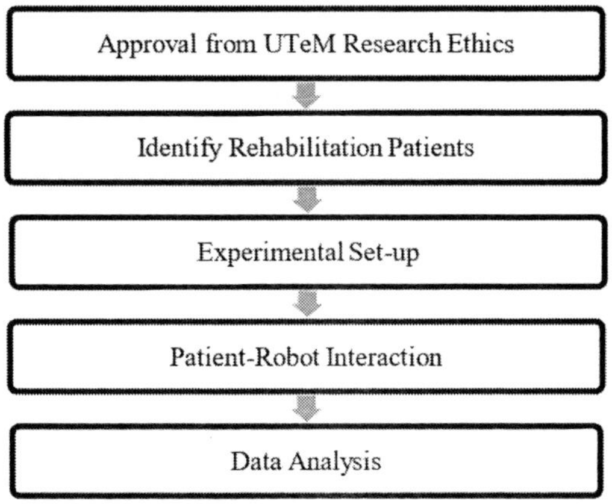

Figure 3. Flow of methodology

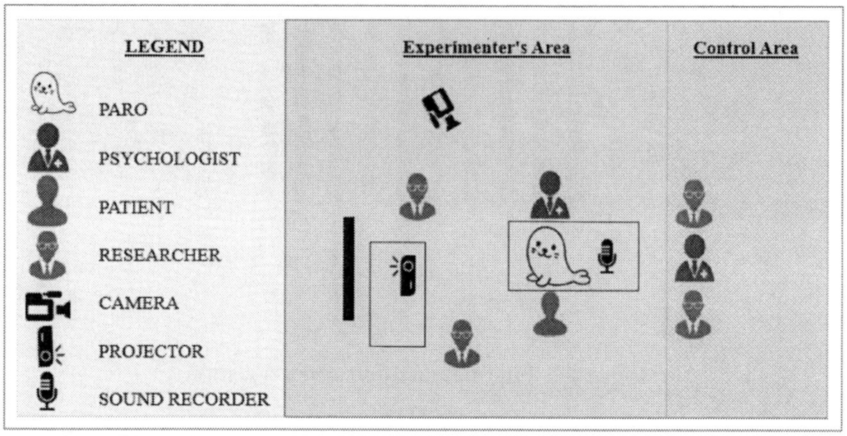

Figure 4. Proposed room layout for the experiment

The emotion gained from video camera was with analyze with OpenCV. With facial emotion detection, algorithms detect faces within video and sense micro expressions of smile by analyzing the relationship between points on the face. The number of smile was recorded and store as indication of positive emotions. There are several questionnaires that was used for the participant such as the HRI survey, Hamilton Depression Rating Scale (HAM-D), Columbia-suicide severity rating scale (C-SSRS) and Pittsburgh Sleep Quality Assessment (PSQI). HAM-D is a multiple questionnaire used to provide indication of depression, and as guide to evaluate recovery. On the other hand, C-SSRS is used to rates of suicidal ideation on a scale. It also acts as a tool for measuring behaviour in primary care. At the same time, PSQI is an instrument used to measure the quality and patterns of sleep in adults. The standardized measure designed to gather consistent information about the subjective nature of people's sleep habits.

4. HRI Survey

Robots come into contact with humans in both physical and emotional. Many potential factors motivate their social acceptance. In this study, a series questionnaire of human-robot interaction has been answered by the patient. The questionnaire divide into two section, which is before the patient meet the robot and after the patient done meeting with the robot. The questionnaires were designed to analyze the autonomy on realistic animal robot, mechanic movement of animal robot, attitude toward animal robot and human level acceptance of animal robot. During the therapy session, these animal robots were presented as therapy animal that capable to respond and comfort the patient. For autonomy on realistic animal robot, the patient was asked their opinion on the animal robot. The patient describes physical appearance of the robot. The highlighted part of the body includes the fur, limb, eyes and tail. However, the patient thought PARO was a bear instead of baby seal. The patient describes PARO fur as soft and really comfortable to touch, thus the PARO fur was created as hypoallergenic. PARO also let the patient experience it mechanic of movement, which it involves the movement of eyelashes, tail, neck and hand. During the survey, the patient mentions several times that she like the eyelashes movement. The patient was asked whether she can express the feeling toward the robot. She agrees that she can talk comfortably with PARO. The patient also expresses herself as 'mother' when she talks to PARO. PARO remind the patient's cat that she once loves. On the survey question, the patient was asked about her opinion on robot. The patient describes PARO can raise her loving experience toward animal robot.

5. Conclusion

Therapy session using animal robot PARO has huge potential to reduce depression symptom such as stress level and sleep time. PARO has an ability to calm agitated patient, feel loved, facilitate communications and lessen the need for medication. Robotic technology like PARO offers excellent opportunities to connect with individual who appears otherwise unreachable.

Acknowledgement

The authors wish to thank Ministry of Higher Education Malaysia, [FRGS/1/2016/SKK06/FKP-AMC/F00321], Universiti Teknikal Malaysia Melaka and SOCSO Rehabilitation Center Melaka for their support.

References

[1] N. C. Guan 2014 *Med J Malaysia* **69**
[2] Y. Kawaguchi, K. Wada, M. Okamoto, T. Tsujii, T. Shibata, and K. Sakatani 2011 IEEE *RO-MAN* 308-313
[3] J. Pipitpukdee, and W. Phantachat, 2011 *Proceedings of the 5th International Conference on Rehabilitation Engineering & Assistive Technology 2011* 43

IOP Conf. Series: Materials Science and Engineering **210** (2017) 012006 doi:10.1088/1757-899X/210/1/012006

Performance of Fluidized bed Fenton process in Degrading Acid Blue 113

M M Bello[1, 2] and A A Raman[1, *]
[1] Department of Chemical Engineering, Faculty of Engineering, University of Malaya, Kuala Lumpur, 50603, Malaysia
[2] Centre for Dryland Agriculture, Bayero University, P M. B. 3011, Kano, Nigeria.

E-mail: azizraman@um.edu.my and mmbello.cda@buk.edu.ng

Abstract. The performance of a fluidized bed Fenton process in degrading Acid Blue 113 (AB 113) was investigated. Fluidized bed Fenton process is a modification of conventional Fenton oxidation, aimed at reducing sludge generation and improving process performance. Response surface methodology was used to study the effects of operational parameter on the color removal from the dye. Dimensionless factors, Dye/Fe^{2+}, H_2O_2/Fe^{2+} and pH were used as the independent variables in Box-Behnken Design (BDD). Reduced quadratic model was developed to predict the color removal. The process could remove up to 99 % of the initial color. The most significant factor for color removal was found to be Dye/Fe^{2+}, followed by H_2O_2/Fe^{2+}. Unlike conventional Fenton, the initial pH of the solution does not have a significant effect on the color removal.

1. Introduction

The environmental pollution arising from the use of synthetic dyes in textile and other industries has become a source of concern globally. Textile industry produces large quantity of dyes-containing effluents which are highly polluting, posing a threat to the environment [1]. Organic dyes have complex structures and high stability, making them difficult to degrade through the conventional methods [2]. Consequently, the treatment of dyes-containing effluents has attracted researchers' interest.

Various technologies such as adsorption [3], coagulation-flocculation[4], biological process [5], membrane technology [6] have been investigated for dye removal. These methods have some limitations in textile wastewater treatment. The physicochemical methods usually concentrate the pollutants into solid/liquid side streams[1] whereas biological process is relatively slow and usually convert the dye to some intermediates[7]. Thus, the search for effective treatment technologies for dye removal has been intensified.

Advanced oxidation processes (AOPs), which are based on the generation of reactive oxidants, are considered promising technologies for the degradation of recalcitrant dyes [8]. Among AOPs, Fenton oxidation has received wide attention due to its effectiveness, simplicity of the process and availability of Fenton's reagent [9]. The conventional Fenton oxidation is based on a catalytic reaction between Fe^{2+} and H_2O_2. In a general form, the reaction can be written as shown in equation 1 [10].

$$M^{n+} + H_2O_2 \rightarrow M^{n-1} + OH^- + HO \ (M = Fe) \quad (1)$$

Despite the effectiveness of Fenton oxidation in degrading recalcitrant pollutants, it faces some limitations which have hindered its large-scale applications. Fenton oxidation is pH dependent and favors a narrow range of operational pH (3 – 4). This is because the production of HO, which are

IOP Conf. Series: Materials Science and Engineering **210** (2017) 012006 doi:10.1088/1757-899X/210/1/012006

responsible for oxidizing the pollutant, is enhanced at lower pH [11]. Another limitation is the production of ferric sludge which requires further treatment and disposal [12]. Many approaches, such as heterogeneous Fenton and Fenton-like processes, have been taken to address these limitations. Recently, the use of fluidized bed technology to reduce sludge generation in conventional Fenton has received interest [13–16]. In fluidized bed Fenton (FBF), solids are used which provide surface for the precipitation of iron oxide and hence, lower sludge generation.

In this work, the performance of FBF process in the degradation of an azo dye, AB113 was investigated. Response Surface Methodology (RSM) was used to design the experiment and investigate the effects of operational parameters. RSM is a multivariate statistical analysis tool that can be used to design experiment, develop empirical models and study the effects of operational parameter [17]. Among the design approaches in the RSM, Box-Behken Design (BBD) is considered suitable as it has a low number of required experimental runs [18]. Thus, BBD was employed in this work to design the experiments and develop the RSM model for color removal from the Acid Blue 113.

2. Methodology

2.1. Chemicals

The chemicals used were purchased from Sigma-Adrich and include H_2O_2 (33 %), $FeSO_4.7H_2O$, H_2SO_4, NaOH, SiO_2 (0.2 – 1 mm) AB113 (dye content = 50%). All the chemicals were of analytical grades and used without further purification. Distilled water was used in all samples preparation.

2.2. Experimental setup

The schematic diagram of the FBF process is shown in Figure 1. It consists of a reservoir, recirculation section, magnetic pump and a flow meter. The reactor is a glass column of 135ml capacity and 30 cm height. The magnetic pump was used to pump the dye solution from the storage tank to the reactor at a flow rate of 0.21 mL/min. Glass beads were packed at the bottom of the reactor to a height of 4 cm to act as distributors. SiO_2 was placed above the glass beads to a height of 6 cm. Internal recirculation was used to control the fluidization of the carriers. The desired amount of Fe^{2+} was added into the system in form of $FeSO_4.7H_2O$. The Fenton process was initiated by the addition of H_2O_2. Samples were collected at pre-determined intervals and the pH of the sample was raised above 10 with NaOH to stop the reaction.

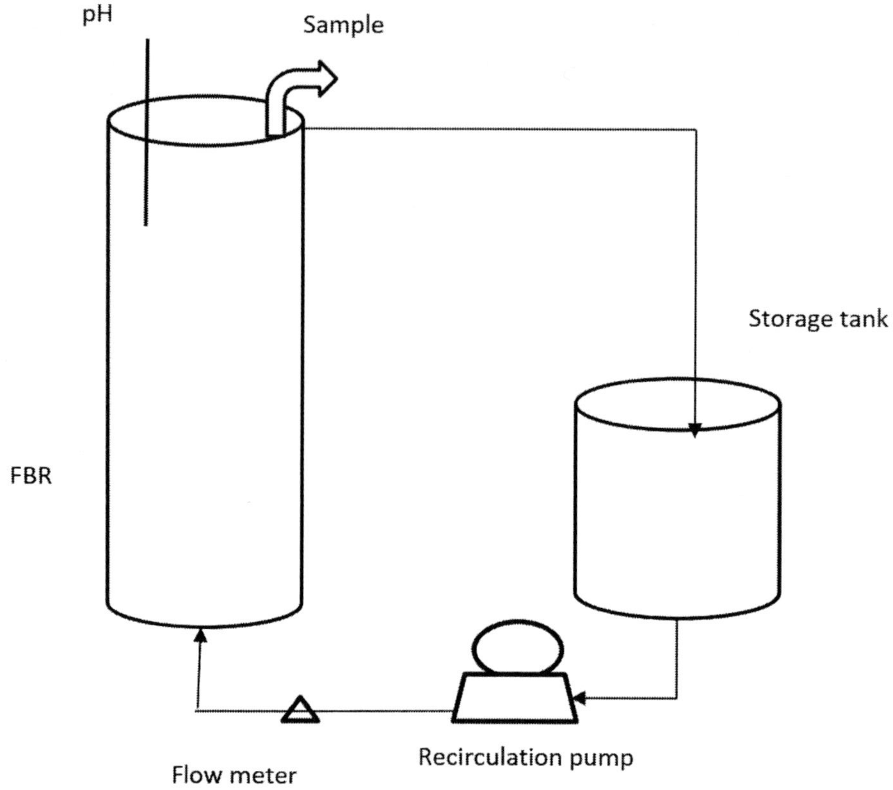

Figure 1. Experimental setup

2.3. Analysis

The analysis was done according to the Standard Methods for the Examinations of Water and Wastewater [19]. Color removal from the dye solution was measure using a UV Spectrophotometer (Spectroquant® Pharo 300, Merck, Germany). The color was measure as absorbance at 464 nm wavelength. The percent color removal is reported as the difference between the initial absorbance (A_o) and absorbance at a time, t (A_t) as shown in equation 2.

$$\%\ \text{Color removal} = \frac{A_o - A_t}{A_o} \times 100 \qquad (2)$$

2.4. Experimental Design

Design Expert software (Stat-Ease, Inc., USA; Version 8.0.7.1) was used for the RSM analysis. BBD was used to design the experiment, develop empirical models and investigate the effects of the operational parameters on the process performance. The operational parameters considered are [Dye/Fe^{2+}], [H_2O_2/Fe^{2+}] and pH. The range of these parameters and their coding are shown in table 1.

Table 1. Experimental ranges of the process variables.

Independent variables	Coded	Low Actual Value	High Actual Value
[Dye]/[Fe^{2+}] (wt/wt)	A	10	50
[Fe^{2+}]/[H_2O_2] (wt/wt)	B	5	25
pH	C	3	9

IOP Conf. Series: Materials Science and Engineering **210** (2017) 012006 doi:10.1088/1757-899X/210/1/012006

3. Results and Discussion

3.1. Model fitting and ANOVA

The BBD suggested a total of 17 runs with 5 center points and 12 non-center points. Table 2 shows the experimental and predicted response based on the BDD. The highest experimental color removal (99.51 %) is slightly lower than the highest predicted color removal (100.53 %). However, it can be observed that experimental and predicted responses are nearly the equal.

Table 2. BBD design variables with the experimental and predicted responses.

Run	Dye/Fe^{2+} (wt/wt)	Fe^{2+}/H_2O_2 (wt/wt)	pH	Color removal (%)	
				Exp	Pred
1	30.00	15.00	6.00	91.35	90.32
2	10.00	15.00	9.00	99.20	100.53
3	50.00	5.00	6.00	71.28	71.77
4	10.00	25.00	6.00	99.51	99.38
5	10.00	15.00	3.00	98.77	99.41
6	50.00	25.00	6.00	85.34	86.68
7	30.00	15.00	6.00	90.07	90.32
8	30.00	5.00	9.00	83.50	83.60
9	30.00	25.00	3.00	94.21	94.15
10	50.00	15.00	3.00	84.28	83.46
11	50.00	15.00	9.00	78.01	77.87
12	30.00	15.00	6.00	90.12	90.32
13	30.00	25.00	9.00	93.05	91.91
14	10.00	5.00	6.00	98.65	97.67
15	30.00	15.00	6.00	90.62	90.32
16	30.00	5.00	3.00	85.45	85.84
17	30.00	15.00	6.00	90.43	90.32

3.1.1. ANOVA for color removal

The color removal was found to fit a reduced quadratic model (equation 3). The ANOVA for the model is shown in table 3. The fitness of the model is determined by the P-values of the model, lack of fit, adequate precision measure and R^2. For the model to fit, the model must be significant (p value $<$ 0.0500) and the lack of fit needs to be insignificant (p value $>$ 0.1000). It can be seen that the model is highly significant (P value $<$0.0001) while the lack of fit is insignificant (P value = 0.0822). On the other hand, the adequate precision (48.917) is higher than the minimum desirable value (4) indicating that the model can be used to navigate the design space. The adjusted R^2 is also in reasonable agreement with the predicted R^2. Figure 2 shows the graph of predicted versus experimental COD removal. it can be seen that a straight line is obtained, further indicating the agreement between the predicted and the experimental response.

Furthermore, the ANOVA shows that [Dye/Fe^{2+}] and [H_2O_2/Fe^{2+}] have significant effects on the color removal. The effect of pH on the color removal is also significant. The interaction of [Dye/Fe^{2+}] and [H_2O_2/Fe^{2+}] has a highly significant effect on the color removal. The quadratic term of [H_2O_2/Fe^{2+}] also has a significant effect on the color removal.

IOP Conf. Series: Materials Science and Engineering **210** (2017) 012006 doi:10.1088/1757-899X/210/1/012006

Table 3. ANOVA for the Color removal.

Source	Sum of Squares	df	Mean square	F value	P value	Remark
Model	956.89	6	159.48	184.46	<0.0001	Highly Sig.
A-Dye/Fe	745.28	1	745.28	861.99	<0.0001	Highly Sig.
B-H_2O_2/Fe	138.00	1	138.00	156.26	<0.0001	Highly Sig
C-pH	10.01	1	10.01	11.58	0.0067	Sig
AB	43.55	1	43.55	50.37	<0.0001	Highly Sig.
AC	11.22	1	11.22	12.98	0.0048	Sig.
B^2	8.83	1	8.83	10.22	0.0096	Sig.
Residual	8.65	10	0.86			
Lack of fit	7.54	6	1.26	4.55	0.0822	Not Sig.
Pure error	1.11	4	0.28			
Cor total	965.54	16				
Adeq precision	48.917					
R^2	0.9910					
Adjusted R^2	0.9857					
Predicted R^2	0.9527					

$$\boldsymbol{Color\ removal}\ (\%) = +99.95 - 0.56[Dye/Fe] + 0.35[H_2O_2/Fe] + 0.46pH + 0.02[Dye/Fe][H_2O_2/Fe] - 0.03[Dye/Fe]pH - 0.01[H_2O_2/Fe]pH \quad (3)$$

In equation 3, positive model terms represent synergetic effect on the COD removal while negative terms indicate antagonistic effects on the color removal. Therefore, [H_2O_2/Fe], pH and the interaction of [Dye/Fe^{2+}] and [H_2O_2/Fe^{2+}] contribute positively to the color removal. On the other hand, [Dye/ Fe^{2+}], interaction of [Dye/Fe^{2+}] and pH and the interaction of [H_2O_2/Fe^{2+}] and pH have antagonistic effects on the color removal.

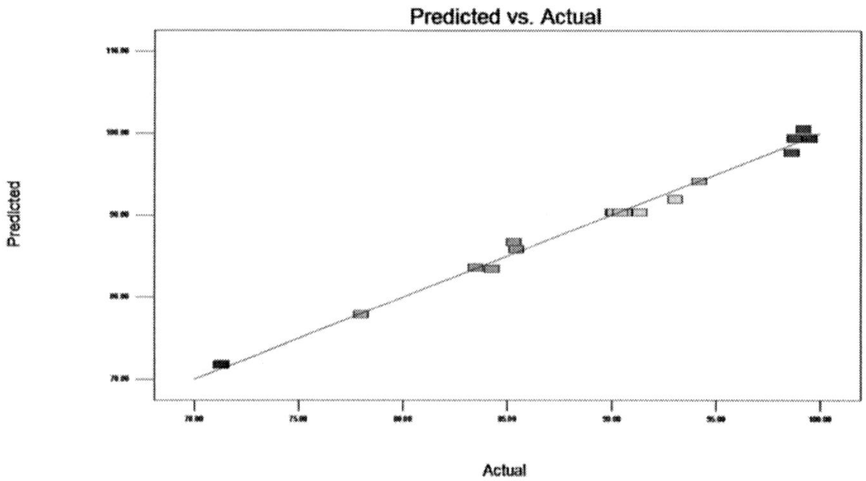

Figure 2. Predicted versus actual color removal

3.2. Effects of operational parameters

One of the most challenging aspects of Fenton process is the selection and optimization of operational parameters. Thus, understanding the effects of operational parameters on the performance of FBF is necessary for process control. In this section, the effects of operational parameters on COD and color removals by the FBF process are discussed.

The effects of Dye/ Fe^{2+}, H_2O_2/Fe and pH on the color removal were investigated. Figure 3 shows the 3D plots of the color removal as a function of (a) Dye/Fe and H_2O_2/Fe and (b) Dye/ Fe^{2+} and pH. It can be seen that the color removal decreases as Dye/ Fe^{2+} increases from 10 to 50 (Fig 3 a). This is because the amount of Fe decreases as the Dye/ Fe^{2+} increases since the dye concentration was kept constant. Thus, there is insufficient Fe^{2+} to catalyze the production of OH. Also, the color removal increases as the H_2O_2/Fe^{2+} increases from 5 to 15. However, when the H_2O_2/ Fe^{2+} was increased from 15 to 25, the color removal decreases, albeit slightly. This could be due to complexation causes by excess H_2O_2. These complexes can increase the color of the solution. On the other hand, the effect of solution pH is not very significant. From Figure 3 (b), it can be seen that the color removal did not change significantly when the pH was increased from 3 to 9. This could be due to the heterogeneous reaction induced by the iron crystallization.

Figure 4 shows the perturbation plot for the color removal. The perturbation plots compare the effects of all factors at a particular point in the design space. A steep slope or curvature in a factor shows that the response is sensitive to that factor. A relatively flat line shows insensitivity to change in that particular factor. In this case, the perturbation is plotted at the center point of the design space. From figure 4, it can be seen that the color removal is most sensitive to Dye/ Fe^{2+}, followed by H_2O_2/ Fe^{2+}. On the other hand, the perturbation of pH is nearly flat, indicating that the color removal is not sensitive to the solution pH.

(a) (b)

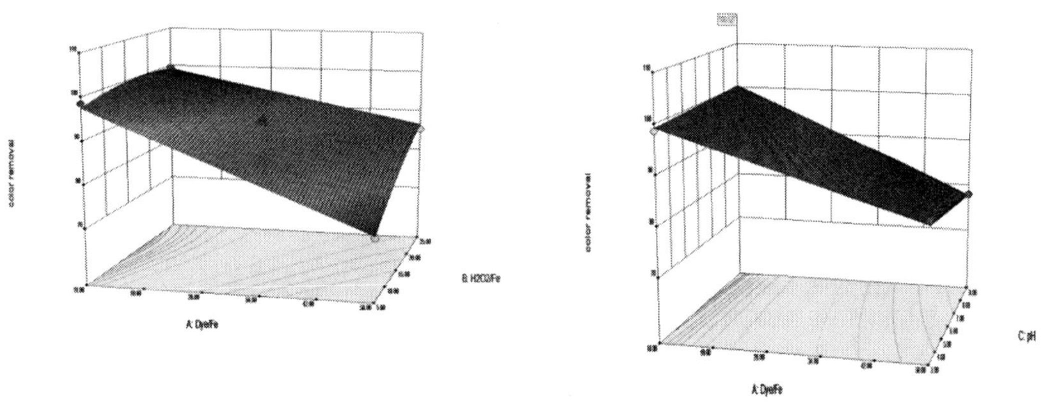

Figure 3. 3D plots for the effects of (a) Dye/Fe^{2+} vs H_2O_2 and (b) Dye/Fe^{2+} and pH

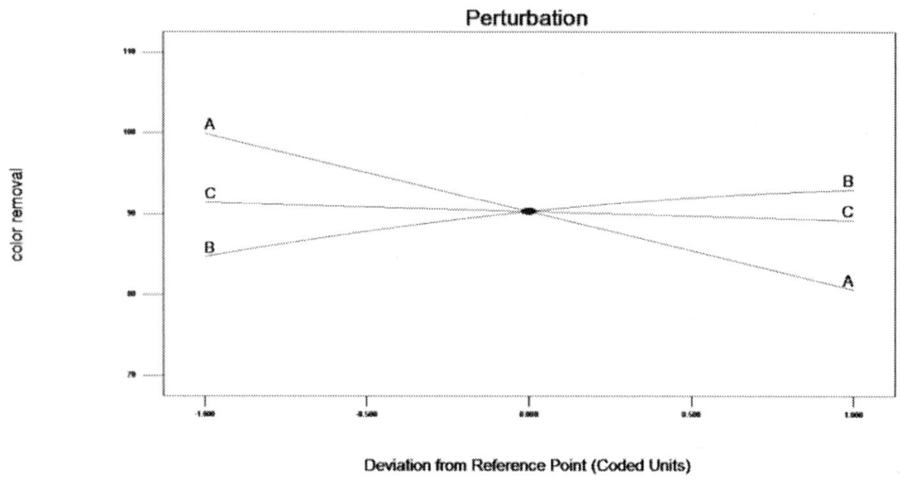

Figure 4. Perturbation plot for color removal

4. Conclusion

This study investigated the performance of FBF process in the degradation of a model dye, AB113. RSM was found to be a suitable tool for designing the FBF process and studying the effects of operational parameters. The FBF process could remove up to 99 % of the initial color of the dye. The color removal can be predicted by a reduced quadratic model developed through the BBD-RSM. The most significant parameter affecting the performance of the FBF process is Dye/Fe^{2+}. The effect of solution pH on the color removal was found to be insignificant. This study shows that FBF process can be a promising technology for the degradation of textile effluents.

References

[1] Rajasimman M, Babu S V and Rajamohan N 2017 *J. Taiwan Inst. Chem. Eng.* **0** 1–11
[2] Carvalho S S F and Carvalho N M F 2017 *J. Environ. Manage.* **187** 82–8
[3] Raghunath S, Anand K, Gengan R M, Kumar M and Maity A 2016 *J. Photochem. Photobiol. , B Biol.* **165** 189–201
[4] Lau Y Y, Wong Y S, Teng T T, Morad N, Rafatullah M and Ong S A 2014 *Chem. Eng. J.* **246** 383–90
[5] Malachova K, Rybkova Z, Sezimova H, Cerven J and Novotny C 2013 *Water Res.* **47** 7143–8
[6] Liu C, Mao H, Zheng J and Zhang S 2017 *J. Memb. Sci.* **530** 1–10
[7] Bilinska L, Gmurek M and Ledakowicz S 2016*Chem. Eng. J.* **306** 550–9
[8] Cai M, Su J, Zhu Y, Wei X, Jin M, Zhang H, Dong C and Wei Z 2015 *Ultrason. Sonochem.* **28** 302–10
[9] Chen S, Wu Y, Li G, Wu J, Meng G, Guo X and Liu Z 2016 *Appl. Clay Sci.* **136** 103–11
[10] Kouraichi S, El-Hadi Samar M, Abbessi M and Boudouh H 2015 *Catal. Sci. Technol.* **5** 1052–64
[11] Malik P K and Saha S K 2003 *Sep. Purif. Technol.* **31** 241–50
[12] Anotai J, Chen C, Bellotindos L M and Lu M 2012 *Bioresour. Technol.* **113** 272–5
[13] Bello M M, Abdul Raman A A and Purushothaman M 2017 *J. Clean. Prod.* **141** 1492–514
[14] Aghdasinia H, Bagheri R, Vahid B and Khataee A 2016 *Environ. Technol. (United Kingdom)* **37**
[15] Garcia-Segura S, Bellotindos L M, Huang Y-H, Brillas E and Lu M-C 2016 *J. Taiwan Inst. Chem. Eng.* **0** 1–15
[16] Chen T-C, Matira E M, Lu M-C and Dalida M L P 2016 *Int. J. Environ. Sci. Technol.* **300** 1–12
[17] Sukriti, Sharma J, Chadha A S, Pruthi V, Anand P, Bhatia J and Kaith B S 2017 *J. Environ.*

Manage. **190** 176–87
[18] Ain N, Talib A, Salam F, Azah N, Ainliah S, Ahmad A and Sulaiman Y 2017 *J. Electroanal. Chem.* **787** 1–10
[19] APHA 1992 Standard Methods for the Examination of Water and Wastewater *Water Environ. Fed.* **18th** 9–45

IOP Conf. Series: Materials Science and Engineering **210** (2017) 012007 doi:10.1088/1757-899X/210/1/012007

Adsorption of arsenic from aqueous solution using magnetic graphene oxide

A.I.A. Sherlala[1,2], A.A. Raman[1,*] and M.M. Bello[1]

[1]Department of Chemical Engineering, Faculty of Engineering, University of Malaya, 50603 Kuala Lumpur, Malaysia
[2]Department of Chemical Engineering, College of Engineering Technology-Janzour, Libya

E-mail: azizraman@um.edu.my

Abstract. A binary of graphene oxide (GO) and iron oxide (IO) was prepared and used for the removal of arsenic from aqueous solution. The synthesized compound was characterized using XRD analysis. The prepared composite was used for the adsorption of arsenic from aqueous solution. Central Composite Design was used to design the adsorption experiments and to investigate the effects of operational parameters (initial concentration of arsenic, adsorbent dosage, pH and time) on the adsorption capacity and efficiency. The adsorbent shows a high adsorption capacity for the arsenic. The adsorption efficiency ranges between 33.2 % and 99.95 %. The most significant factors affecting the adsorption capacity were found to be the initial concentration of arsenic and the adsorbent dosage. The initial pH of the solution slightly affects the adsorption capacity, with the maximum adsorption capacity occurring around pH 6 – 7. Thus, the developed adsorbent has a potential for effective removal of arsenic from aqueous solution.

1. Introduction

The environmental pollution caused by heavy metals has attracted concern globally due to the detrimental effects of heavy metal to human health and the environment. Heavy metals are ubiquitous and unlike organic pollutants, are stable and possess low levels of biodegradability [1]. Heavy metals have been associated with various health problems such as hematopoietic, renal, gastrointestinal, cardiovascular, and reproductive systems [2]. Discharging effluents containing heavy metals is therefore a source of serious concern. Therefore, the removal of heavy metal ions from wastewater streams elicits attention.

There are many technologies that have been investigated for heavy metal removal such as ion-exchange, coagulation, reverse osmosis, and electrochemical processes. Most of these processes have some limitations such as selectivity and sensitivity to the solution pH. Adsorption is a very effective method for the removal of heavy metals from wastewater [3]. Activated carbon is the widely used adsorbent but its high cost limits its application. Other low-cost materials such as grape bagasse [3], bentonite clay [4], cow bone [5] and palm shell [6] have been investigated as alternative adsorbents. These materials exhibit low adsorption capacities [7]. Thus, there is need to develop effective adsorbent for heavy metals removal from the environment.

Thus, this study was conducted to synthesize and magnetic graphene oxide and apply it in the adsorption of a model heavy metal from aqueous solution. One of the heavy metals of particular concern is arsenic which is known to have carcinogenic, dermatological, cardiorenal and gastrointestinal effects [8,9]. Thus, arsenic is used as the model pollutant. In order to study the effects of operational parameters, response surface methodology (RSM) was employed. RSM is a multivariate statistical tool that can be used to design experiments, develop empirical model and study the effects of the independent variables [10].

2. Methodology

2.1. Chemicals

All chemicals used were of analytical reagent grade and used without further purification. Graphite powder (particle size <20 μm) was obtained from Sigma Aldrich Malaysia. Other chemicals such as Ferric chloride (99%), ferrous chloride (99%), Potassium permanganate (99%), Ammonia solution (30%), Hydrogen Peroxide (30% w/v) and Sodium Arsenite (98.5%) was procured from R&M marketing Essex UK. Deionized water was used for samples preparation.

2.2. Synthesis of the adsorbent

Graphene oxide (GO) was synthesized according to improved Hummer's method. A graphite powder (3.0 g) was mixed with concentrated H_2SO_4 (70 ml) in an ice bath. Under vigorous agitation, $KMnO_4$ (9.0 g) was added slowly to keep the temperature of the suspension lower than 20 °C. Sequentially, the reaction system was transferred to oil bath which was maintained at 40 °C. The reaction mixture was vigorously stirred for 30 minutes. Subsequently, 150 mL of the water was added and solution was stirred for 15 min at 95 °C. Later, 500 mL of water was added followed by the slow addition of 15 mL H_2O_2. As a result, the color of the solution was turned from dark brown to yellow. The resulting mixture was filtered and wasted with 1:10 HCl aqueous solution (250 mL). The filtered solids were dried in air and as a result graphene oxide (GO) powder was obtained.

Iron oxides were prepared by chemical precipitation. 2.43 g of $FeCl_3$ and 0.99 g of $FeCl_2$ were dissolved in 80 mL of deionized water and mixed properly. Subsequently, 6 mL of ammonia solution was added and sonicated for 1 h. The resulting mixture was decanted, washed two time with deionized water and separated again. The separated solids were dried in oven at 50 °C.

The iron oxide particles were mixed with 0.2 g of graphene oxide in 50 mL of deionized water. The mixture was stirred for 18-24 h and centrifuged. The separated solid content was washed and centrifuged. The process was repeated two times and dried in oven at 50 °C to obtain GO/Iron oxide binary.

2.3. Analysis

The morphologies of the synthesized adsorbents were characterized by Field Emission Scanning Electron Microscope-E. The chemical constituents in the adsorbents were analyzed using Energy-dispersive X-ray (EDX) analysis. The crystalline phase of iron oxide was investigated using X-ray diffraction (XRD) technique. Fourier Transform Infrared spectroscopy was used to study the functional groups of the adsorbents within the frequency range of 4000-400cm^{-1}. The concentration of arsenic in the aqueous solution was determined using Inductively coupled plasma mass spectroscopy (ICP Optima 7000 DV, Perkin Elmer).

2.4. Experimental Design

Response surface methodology (RSM) was used to design the adsorption experiments with the aid of Design-Expert 8. Central composite design (CDD) was used as the design domain because of its effectiveness in the design and optimization of chemical process. Four independent variables (initial arsenic concentration, adsorbent dosage, pH and time)) were selected as the factors while the arsenic removal (%) and adsorption capacity (mg/g) were chosen as the responses. The ranges of the independent variables used are shown in Table 1.

IOP Conf. Series: Materials Science and Engineering **210** (2017) 012007 doi:10.1088/1757-899X/210/1/012007

Table 1. Range of operational parameters used in the RSM.

Independent variables	Coded	Low Actual Value	High Actual Value
Initial As Concentration (ppm)	A	10	100
Adsorbent dosage (mg)	B	0.1	0.5
pH	C	2	12
Time (min)	D	30	120

2.5. Experimental Procedure

The adsorption of heavy metals was performed in a batch experiment. 0.1-0.5 gram of synthesized adsorbents added to 100 mL of arsenic solution at desired concentrations. The arsenic solution was prepared using a deionized Millipore water. The mixture was stirred and pH of the solution was adjusted according to the desired value by adding 1.0 M NaOH or 1.0 M HNO_3. The stirring time was also varied accordingly. After a desired time, the solid phase was removed from the supernatant by decantation with magnet and centrifuge at 4000 rpm for 30 min. Arsenic content in the obtained supernatant is determined by ICP.

After separation, the treated solution was sampled and analyzed to determine the arsenic removal efficiency. The removal efficiency was calculated according to equation 1:

$$\% \, Removal = \frac{C_o - C_t}{C_o} \times 100 \quad (1)$$

Where C_o and C_t are the initial and equilibrium concentrations of arsenic in the solution (mg/L).

3. Results and discussion

3.1 Characteristics of the adsorbent

XRD patterns were obtained to investigate the phase and structure of graphite, GO, IO and hybrid of GI-IO. As shown in figure 1(a), the graphite displays a characteristic peak at $2\theta = 27°$. After oxidation, the characteristic graphite peak disappeared and was replaced by a well-defined peak at $2\theta = 11.5°$ (figure.1 (b)) indicating the introduction of oxygen-containing groups to form the GO [11]. The increased d-spacing of GO is ascribed to the presence of copious oxygen-containing functional groups. For the pattern of iron oxide (figure 1 (c)), there appears five main peaks around $2\theta = 30.38, 35.84, 43.62, 57.46, 63.02$ degrees, all of which can be indexed to the (220), (311), (400), (511), and (440) planes of ϒ-Fe2O3 (JPDS no. 39-1346). However, the hybrid of GO and ϒ-Fe_2O_2 exhibited almost the same pattern as the pristine ϒ-Fe_2O_3 (figure 1 (d)). This may be due to the presence of small amount of GO, compared to the amount of IO in the hybrid.

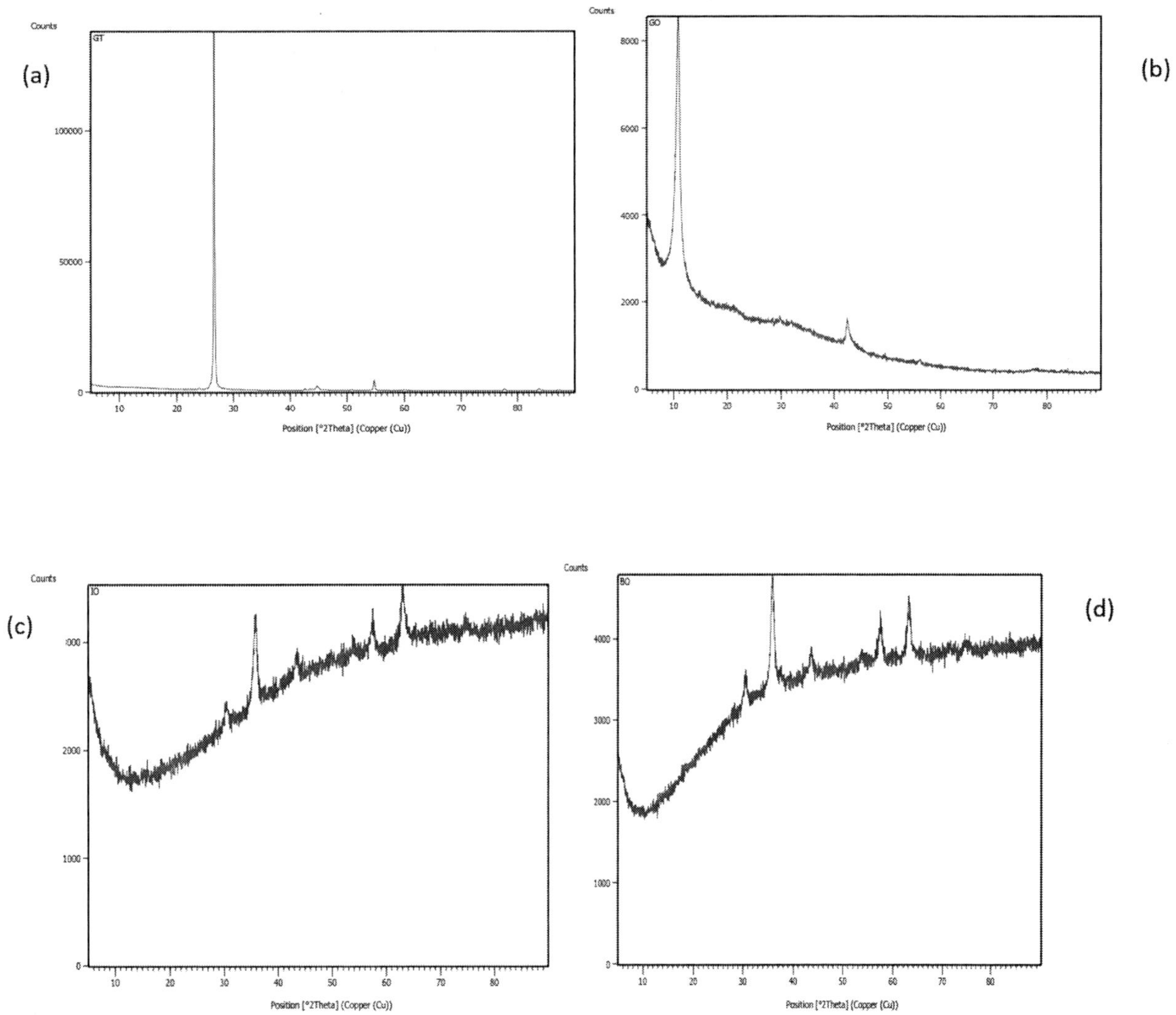

Figure 1. XRD pattern for (a) GT, (b) GO, (c) IO and (d) BO.

3.2. Adsorption studies

3.2.1. RSM-CCD Design. In order to study the performance of the synthesized adsorbent and the effects of the process variables on the adsorptive removal of arsenic using the developed adsorbents, central composite design was used to design the experiment with the aid of Design-Expert 8. The four important process parameters selected include the initial concentration of the arsenic, adsorbent dosage, pH and the adsorption time. The response selected was the adsorption efficiency. The 30 experiments suggested by the CCD were run and the observed responses are shown in table 2. The highest removal efficiency was 99.95 % obtained with run 22 while the lowest observed adsorption efficiency was 33.2 % obtained with run 13. These values were input into the Design-Expert and analysed using the designed CCD model. The responses as predicted by the model are also shown in table 2. It can be observed that, for both responses, the predicted values are in reasonable agreement with the experimental values.

Table 2. Experimental and predicted responses.

Run	Initial As conc. (ppm)	Adsorbent dosage (g))	pH	Time (min)	Adsorption Efficiency (%)	
					Exp.	Pred.
1	10	0.5	12	30	75.35	76.38
2	55	0.1	7	75	51.5272	54.29
3	10	0.3	7	75	94.64	85.26
4	55	0.3	7	75	76.3090	86.62
5	55	0.3	7	30	71.7272	78.16
6	10	0.5	2	120	94.08	87.96
7	10	0.1	2	30	51.55	46.98
8	55	0.3	12	75	47.7090	79.43
9	10	0.5	2	30	90.37	104.30
10	100	0.5	12	120	52.18	60.98
11	100	0.5	12	30	51.68	43.24
12	100	0.5	2	120	55.3	66.27
13	100	0.1	12	30	39.77	33.00
14	55	0.3	7	75	77.8545	86.62
15	55	0.5	7	75	88.7418	68.46
16	100	0.1	2	30	35.98	40.61
17	10	0.1	2	120	54.49	63.08
18	100	0.5	2	30	47.71	48.53
19	10	0.1	12	120	36.21	39.62
20	10	0.5	12	120	60.75	60.05
21	55	0.3	7	75	71.7090	86.62
22	10	0.1	12	30	34.34	23.52
23	100	0.1	2	120	91.359	90.78
24	55	0.3	7	75	99.9109	86.62
25	55	0.3	7	75	99.92	86.62
26	100	0.3	7	75	99.953	85.26
27	55	0.3	2	75	99.9109	93.80
28	100	0.1	12	120	99.949	89.95
29	55	0.3	7	75	99.870	86.62
30	55	0.3	7	120	99.8727	95.08

3.2.2. Model development and ANOVA.

The removal efficiency was found to fit a reduced quadratic model. Equation 2 shows the quadratic model proposed by the RSM in terms of the coded parameters. The positive terms indicate positive effect on the arsenic removal. In this case, adsorbent dosage, adsorption time, interaction between initial arsenic concentration and pH, and the interaction between initial arsenic concentration and time have positive effect on the arsenic removal. On the other hand, negative terms indicate negative effects on the arsenic removal. Thus, initial arsenic concentration, initial pH, interaction between initial arsenic concentration and adsorbent dosage, interaction between adsorbent dosage and pH, interaction between adsorbent dosage and time, and the quadratic effect of adsorbent dosage have negative effect on the arsenic removal.

The analysis of variance (ANOVA) was used to test the adequacy of the model. Table 3 shows the ANOVA. The first step is to check the p-value of the model. For the model to be valid, the p-value must be less than 0.0500, indicating that the model is significant. In this case the p-value is 0.0002, indicating that there is only 0.02 % chance that the model value is due to error. Thus, the model is significant. The 'lack of fit' measure is required to be non-significant (above 0.0500) for the model to be valid. For this model, the p-value for the lack of fit is 0.5306. This means that there is 53.06 % chance that the value of the lack of fit is due to error. The adequate precision which measures the signal to error ratio is required to be above 4. The adequate precision in this case is 9.674, much higher than the minimum desirable value. On the other hand, the R^2 was found to be 0.7816. Although this value is not very high, the adjusted R^2 is in reasonable agreement with the predicted R^2. This is required for the model to be valid. Also, figure 2 shows the graph for the predicted versus actual percent removal. The straight line and well distributed data shows a reasonable agreement between the experimental and predicted data.

$$Percent\ removal = +86.62 - 1.36A + 7.09B - 7.19C + 8.46D - 12.35AB + 5.66AC + 8.52AD - 1.11BC - 8.11BD - 25.24B^2 \quad (2)$$

Table 3. ANOVA for percent removal

Source	**Sum of squares**	**df**	**Mean square**	**F value**	**P value**	**Remark**
Model	12926.84	10	1296.68	6.80	0.0002	Sig.
A-Arsenic Conc.	33.26	1	33.26	0.17	0.6804	Not Sig.
B-Adsorbent dosage	903.93	1	903.93	4.75	0.0420	Sig
C-pH	*929.98*	1	*929.98*	*4.89*	*0.0394*	Sig
D-Time	*1288.35*	*1*	*1288.35*	*6.78*	*0.0175*	Sig.
AB	*2439.82*	*1*	*2439.82*	*12.83*	*0.0020*	Sig.
AC	*511.89*	*1*	*511.89*	*2.69*	*0.1173*	Not Sig.
AD	*1161.07*	*1*	*1161.07*	*6.11*	*0.0231*	Sig.
BC	*19.85*	*1*	*19.85*	*0.10*	*0.7502*	Not Sig.
BD	*1052.00*	*1*	*1052.00*	*5.53*	*0.0296*	Sig.
B^2	*4586.70*	*1*	*4586.70*	*24.12*	*< 0.0001*	Highly Sig.
Residual	3612.64	19	190.14			
Lack of Fit	*2683.75*	*14*	*191.70*	*1.03*	*0.5306*	Not Sig.
Pure Error	*928.90*	*5*	*185.78*			
Cor Total	16539.48	29				
Adeq precision	9.674					
R^2	0.7816					
Adjusted R^2	0.6666					
Predicted R^2	0.4682					

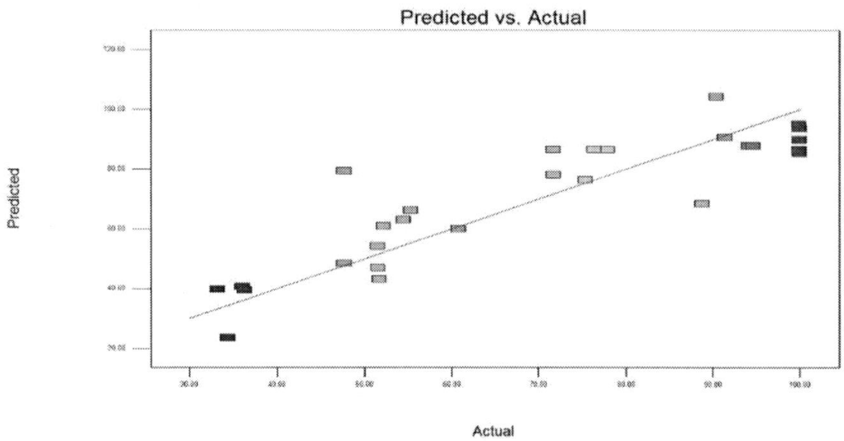

Figure 2. Predicted versus actual percent removal.

From the ANOVA in table 3, it can be seen that the initial arsenic concentration does not a have a significant effect on the arsenic removal. On the other hand, the adsorbent dosage, pH and time have significant effect on the arsenic removal efficiency. This can be clearly seen in the perturbation plot shown in figure 3. A perturbation plot compares the effects of independent variables on the response at a particular point in the design space. A steep slope shows that the response is sensitive to that variable. A relatively flat line shows that the response is not sensitive to the variables. In Figure 3, B (adsorbent dosage) has the highest curvature, indicating that the arsenic removal is most sensitive to the adsorbent dosage. This is followed by C and D. However, the curve of A (arsenic concentration) is relatively flat. This shows that the arsenic removal is not sensitive to the initial concentration of the arsenic.

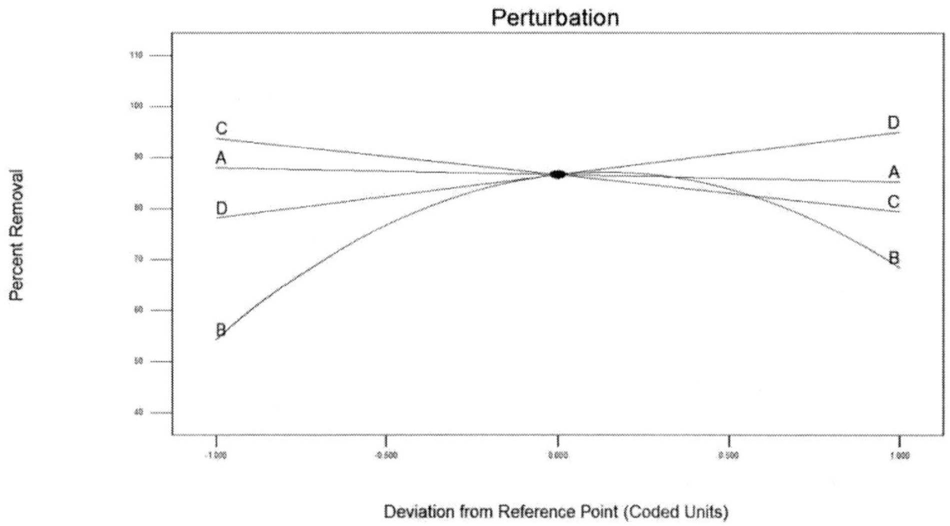

Figure 3. Perturbation plots for arsenic removal efficiency

4. Conclusion

This study was conducted to develop an effective graphene-based adsorbent for the removal of arsenic from aqueous solution. The magnetite-graphene oxide was successfully synthesized as verified by the XRD analysis. The adsorbent shows a high adsorption capacity for the arsenic. The adsorption efficiency ranges between 33.2 % and 99.95 %. The most significant factors affecting the adsorption capacity were found to be the initial concentration of arsenic and the adsorbent dosage. The arsenic removal can be predicted by a second order model developed through RSM-CCD. Although this binary has shown potential for arsenic removal, functionalization using appropriate material can enhance its adsorptive capacity.

Acknowledgment
This work was financially supported by the University of Malaya Postgraduate Research Fund (IPPP) Number PG322-2016A.

References

[1] Hegazi H A 2013 *HBRC Journal* **9** 276–282
[2] Gupta V K and Ali I 2004 *Journal of Colloid Interface Science* **271** 321–328
[3] Demiral H and Güngör C 2016 *Journal of Cleaner Production* **124**
[4] Neto F D A, Nascimento E S, Cantuaria M L and Vieira M G A 2015 *Journal of Cleaner Production* **112** 1112–1121
[5] Cechinel M A P, Ulson De Souza S M A G and Ulson De Souza A A 2014 *Journal of Cleaner Production* **65** 342–349
[6] Ismaiel A A, Aroua M K and Yusoff R 2013 *Chemical Engineering Journal* **225** 306–314
[7] Abbas A, Al-amer A M, Laoui T, Al-marri M J, Nasser M S, Khraisheh M and Ali M 2016 *Seperation and Purification Technology* **157** 141–161
[8] Chen X, Yang L, Zhang J and Huang Y 2014 *Chinese Journal of Chemical Engineering* **22** 1340–1346
[9] Cortés-arriagada D and Toro-labbé A 2016 *Applied Surface Science* **386** 84–95
[10] Kim S C 2016 *Journal of Industrial and Engineering Chemistry* **38** 93–102
[11] Wang B, Si L, Geng J, Su Y, Li Y, Yan X and Chen L 2017 *Applied Catalysi B: Environmental* **204** 316–323

Optimization of electrocoagulation process for the treatment of landfill leachate

N. Huda[1], A. A. Raman[1,*], and S. Ramesh[2]

[1]Chemical Engineering Department, Faculty of Engineering, University Malaya, 50603 Kuala Lumpur, Malaysia.

[2]Centre of Advanced Manufacturing & Material Processing (AMMP), Department of Mechanical Engineering, Faculty of Engineering, University of Malaya, 50603 Kuala Lumpur, Malaysia.

E-mail: azizraman@um.edu.my

Abstract. The main problem of landfill leachate is its diverse composition comprising of persistent organic pollutants (POPs) which must be removed before being discharge into the environment. In this study, the treatment of leachate using electrocoagulation (EC) was investigated. Iron was used as both the anode and cathode. Response surface methodology was used for experimental design and to study the effects of operational parameters. Central Composite Design was used to study the effects of initial pH, inter-electrode distance, and electrolyte concentration on color, and COD removals. The process could remove up to 84 % color and 49.5 % COD. The experimental data was fitted onto second order polynomial equations. All three factors were found to be significantly affect the color removal. On the other hand, electrolyte concentration was the most significant parameter affecting the COD removal. Numerical optimization was conducted to obtain the optimum process performance. Further work will be conducted towards integrating EC with other wastewater treatment processes such as electro-Fenton.

1. Introduction

Waste generation increases yearly due to the exponential population growth and urbanization [1]. Being the lowest ranking waste management option in the waste hierarchy, landfilling still remains as the dominant method used in many countries. At present, landfilling contributes 80% of the main waste disposal method in many countries including Malaysia and has become ubiquitous in modern society [2]. Leachates generated during the landfilling operation varies according to the nature of the landfill material, and poses hazard to the environment since it is difficult to be treated for its complex mixture [3] of biodegradable or non-biodegradable, organic or inorganic and toxic or non-toxic waste [4]. Sustainable management of leachate appears to be an enduring problem for landfill operators and regulators as landfill can continue to generate leachate for several hundred years after they have ceased to operate. Thus, it is necessary to purify the leachate before its discharge into the environment.

Usage of conventional leachate treatment methods such as aerobic, anaerobic, flotation, coagulation–flocculation, chemical precipitation, adsorption, and air stripping are declining in recent years [5,6] even though they are economical and easy to maintain, due to its incapability to eliminate the recalcitrant pollutants efficiently from the wastewater [7], while chemical oxidation produces intermediate products which remain in the solution and may entail a similar or even higher toxicity than initial compounds [8]. Advanced Oxidation Processes (AOPs) were identified to treat the wastewater, where oxidation processes using highly reactive hydroxyl radicals (•OH) are used to remove the recalcitrant organic pollutants that are difficult to treat by conventional methods [9]. However, AOPs can result in high chemical consumption, high demand of electrical energy for devices and produce secondary waste, which increases the treatment costs [9].

Electrocoagulation (EC) is an evolving alternative technology which can be used to remove various pollutants from wastewater [10] through in-situ generation of coagulants by electro-oxidation of the sacrificial anode, constructed mainly from iron (Fe) or aluminium (Al) metals [11]. EC has been applied in broad range successfully for treatment of textile dyes [12], biodiesel wastewater [13], pharmaceutical wastewater [14], oil tanning effluent [15], industrial wastewater [16], pretreated coke wastewater [17], and heavy metals in groundwater [18]. EC has gained significant attention from many researchers, owing to its advantages like reduced sludge production, no requirement for chemical use, ease of operation [19], short operating time, and low capital and operating costs [11]. Thus, the main objective of the present study is to investigate the decolorisation and degradation of a landfill leachate using iron as non-toxic and readily available electrode material. Several parameters, namely initial pH, conductivity, and inter-electrode distance were investigated for their effects on the color and COD removal efficiency. Central Composite Design (CCD) was employed to obtain optimal parameters for efficient removal of pollutants with a limited number of experiments.

2. Methodology

2.1 Chemicals
Sulfuric Acid (H_2SO_4) (96% purity, AR Grade, Fisher Scientific) and Sodium Hydroxide, NaOH (R & M Chemicals) were used for pH alteration. Hydrochloric acid, HCl (37% purity, AR Grade, Eriendemann Schmidt) was used to clean the electrodes. Sodium chloride, NaCl (R & M Chemicals) was used as electrolyte. COD Cell Test (300 – 3500 mg/l, Spectroquant-Merck) was used to measure COD of the leachate.

2.2 Effluent Preparation and Characteristic
Raw landfill leachate was collected from Jeram Municipal Solid Waste Sanitary Landfill, Selangor and used without pretreatment. Effluent dilution was done using distilled water. COD and color removal of treated effluent was measured using Spectrophotometer (Spectroquant Pharo 300). The pH of the effluent was measured using a pH meter (CyberScan pH 300, EUTECH Instruments).

2.3 Electrochemical system setup
The basic electrochemical setup was developed using iron sheet as anode and cathode (42 cm^2 effective surface area), dipped into 500 mL leachate solution in a 600-mL beaker (Figure 1). The reaction mixture was stirred using magnetic stirrer (WiseStir MS-20D) and magnetic stirring bar at 100 rpm. DC regulated power supply (EXTECH-382213) was used to supply 1.0 A of current to the electrochemical system for 60 minutes.

2.4 Electrochemical design
The experimental design and mathematical modelling were performed using Design-Expert 8.0 software. Central Composite Design (CCD) was used for optimization of operating parameters for the removal of pollutants by EC. Three important operating parameters that were optimized in this study are initial pH, inter-electrode distance and electrolyte concentration. The levels and range of factors chosen for this study is presented in Table 1. Responses included in the model were removal efficiency of color and COD.

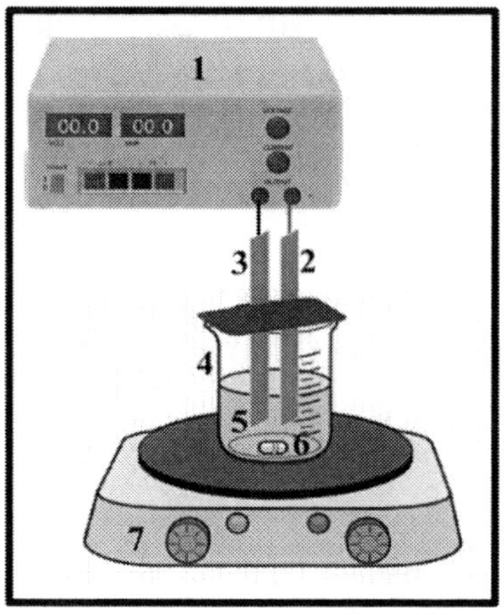

Figure 1. Schematic diagram of EC setup: (1) DC power supply, (2) anode, (3) cathode, (4) EC cell, (5) effluent, (6) magnetic stirrer and (7) magnetic stirrer.

Table 1. The level and range of factors chosen in EC study.

Variable	Coded levels		
	-1	0	+1
A, Initial pH	3.00	6.00	9.00
B, Inter-electrode distance	1.00	1.50	2.00
C, NaCl concentration	1.00	1.50	2.00

3. Results & Discussion

3.1 RSM Modal Development

Total 20 runs with 8 factorials, 6 axial, and 6 center points were suggested by Design Expert to optimize responses. All the results for color and COD removal from RSM model are presented in Table 2. Based on the experimental results, empirical second order polynomial equations showing the interaction between the proposed independent factors are presented in Table 3 as equation (1) and (2). The observed color and COD removal varied between 15.7 - 84.1 and 20.0 - 49.5, respectively.

Table 4 provides the analysis of variance for color and COD removal efficiency. For color removal, the Model F-value of 375.51 implies the model is significant. There is only a 0.01% chance that a "Model F-Value" this large could occur due to noise. Values of "Prob > F" less than 0.0500 indicate model terms are significant. In this case A, AB, AC, BC, A^2, ABC, A^2C, AB^2 are significant model terms. The "Lack of Fit F-value" of 0.87 implies the Lack of Fit is not significant relative to the pure error. There is a 47.43% chance that a "Lack of Fit F-value" this large could occur due to noise. Non-significant lack of fit is good since we want the model to fit. "Adequate precision" measures the signal to noise ratio. A ratio greater than 4 is desirable. In this case, the ratio of 69.671 indicates an adequate signal. This model can be used to navigate the design space, since the R-squared value is very close to 1.0000. Besides, for COD removal, the Model F-value of 3.40 implies the model is significant. There is only a 4.78% chance that a "Model F-Value" this large could occur due to noise. Values of "Prob > F" less than 0.0500 indicate model terms are significant. In this case C and BC are significant model terms. The "Lack of Fit F-value" of 1.01 implies the Lack of Fit is not significant relative to the pure error. There is a 46.30% chance that a "Lack of Fit F-value" this large could occur due to noise. A ratio of 7.648 indicates an adequate signal. This model can be used to navigate the design space.

Table 2. Factors and response of CCD

Run	pH	Electrode distance (cm)	NaCl (g)	Removal efficiency (%)	
				COD	Color
1	6	1.5	2.34	23.6	82.3
2	3	1	2	40.2	18.8
3	0.95	1.5	1.5	20.0	15.7
4	6	0.66	1.5	44.1	82.8
5	9	2	2	20.8	72.3
6	3	2	1	42.9	36.3
7	9	2	1	49.1	77.6
8	6	1.5	1.5	29.8	84.0
9	9	1	1	48.6	78.7
10	6	1.5	1.5	41.6	83.6
11	6	1.5	1.5	43.7	81.2
12	9	1	2	49.5	78.4
13	3	1	1	42.4	71.4
14	6	1.5	0.66	49.3	81.3
15	3	2	2	35.1	55.0
16	6	1.5	1.5	43.3	84.1
17	6	1.5	1.5	40.4	81.9
18	6	1.5	1.5	42.9	82.5
19	11.05	1.5	1.5	40.6	67.9
20	6	2.34	1.5	43.3	81.7

Table 3. Quadratic equations color and COD removal

Response	Equation	
Color removal (%)	$82.60 + 27.33*A - 0.58*B + 0.30*C - 1.04*A*B + 3.53*A*C + 8.29*B*C - 21.43*A^2 - 0.12*B^2 - 9.53*A*B*C - 5.25*A^2*C - 11.62*A*B^2$	(1)
COD removal (%)	$40.24 + 0.67*A - 2.51*B - 5.90*C - 2.94*A*B - 2.15*A*C - 4.35*B*C + 0.13*A^2 + 1.43*B^2 - 1.13*C^2 - 2.94*A*B*C$	(2)

Table 4: ANOVA output

Source	Sum of squares	df	Mean square	F value	Prob>F	Remarks
For color						
Model	5636.02	11	512.37	375.51	< 0.0001	significant
Lack of Fit	2.46	2	1.23	0.87	0.4743	not significant
R-Square	0.9983					
Adeq. Precision	69.671					
For COD						
Model	946.76	10	94.68	3.40	0.0478	significant
Lack of Fit	83.72	3	27.91	1.01	0.4630	not significant
R-Square	0.8097					
Adeq. Precision	7.648					

3.2 Effects of operating parameters

3.2.1 Color removal.

The response surface 3D plots to estimate the color removal over independent variables initial pH and inter-electrode distance are shown in Figure 2. All three factors were found to be the significant parameter affecting the color removals. Decolorization of leachate using EC process is tightly bounded by the pH of the reaction. Decolourization of effluent is very high at neutral and alkaline pH, but very low at acidic medium. This could be related to the reactions taking place at anode and cathode during EC process: Eq (3-5). When iron electrode is used as anode and cathode, the generated Fe^{2+} ions will further oxidize and react to produce corresponding hydroxides and/or polyhydroxides [17], and the formation of these complexes depends strongly on the pH of the effluent [20]. At alkaline pH, iron hydroxides formed remains in the aqueous streams as a gelatinous suspension, thus remove the pollutants from effluent by coagulation, adsorption, co-precipitation and sweep flocculation [21].

<u>Anode</u>:

$Fe\ (s) \rightarrow Fe^{2+}\ (aq) + 2e^-$ (3)

<u>Cathode</u>:

$2H^+(aq) + 2e^- \rightarrow H_2\ (g)$ (acidic medium) (4)

$2H_2O + 2e^- \rightarrow H_2\ (g) + 2OH^-\ (aq)$ (neutral or alkaline medium) (5)

Moreover, it is also observed that decolorization of effluent increases as the inter-electrode distance decreases, due to increasing conductivity. Closer inter-electrode distance enhances the electrode dissolution process due to enhancement of electrode pore size [22]. This phenomenon would indirectly reduce the power consumption and operating cost.

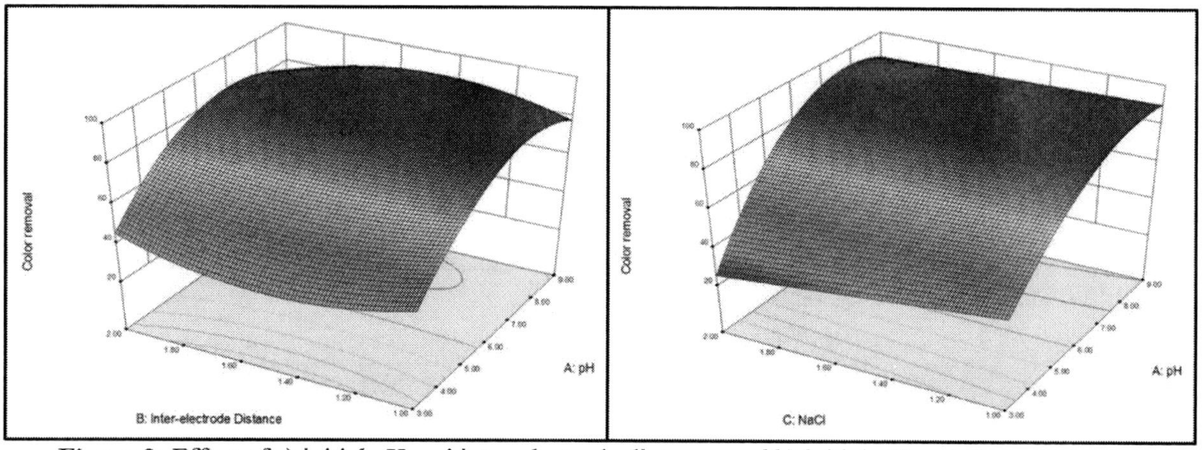

Figure 2: Effect of a) initial pH and inter-electrode distance, and b) initial pH and electrolyte concentration on color removal (%); 3D surface plots.

3.2.2 COD removal. The response surface 3D plots to estimate the COD removal over independent variables electrolyte concentration and inter-electrode distance are shown in Figure 3. Electrolyte concentration was the most significant parameter affecting the COD removal. The addition of the supporting electrolyte which allows the color and COD removal efficiency increases occurs due to the participation of active chlorine anion roles in EC reaction [23]. Sodium chloride added to the EC system increase the conductivity and generates hypochlorite ions, which act as an oxidizing agent in the pollutant degradation.

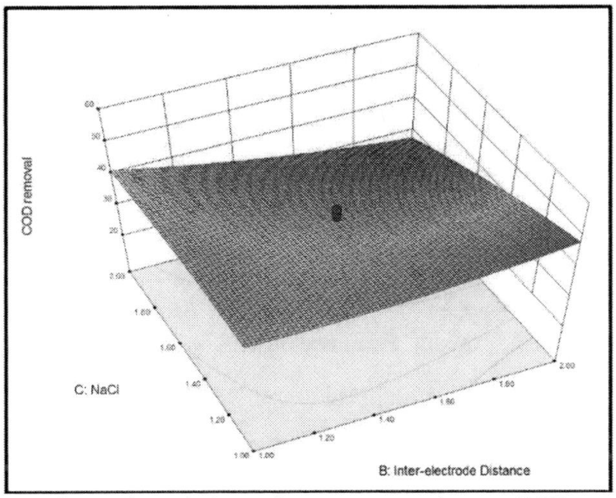

Figure 3: Effect of inter-electrode and electrolyte concentration on COD removal (%); 3D surface plots

3.3 Optimization using response surface methodology

In this study of RSM, optimization of the experimental results is keeping all factors within the range and all responses as maximum. From the analysis, the optimum conditions for initial pH, inter-electrode distance and electrolyte concentration are 7.73, 1.16 cm and 1.00 g respectively. Experiments were conducted to validate the optimized conditions and predicted results as presented in Table 5. The actual results obtained for color and COD removal are 89.01% and 46.05%, respectively which is very close to the predicted result. This indicates a good agreement of the experimental and predicted results under optimized conditions.

Table 5: Predicted and actual responses at optimized operating parameter

	pH	Electrode distance (cm)	NaCl (g)	Removal Efficiency (%)		Desirability
				Color	COD	
Predicted	7.73	1.16	1.00	90.23	46.12	0.92
Actual				89.01	46.05	

4. Conclusion

In this study, the efficiency of EC process applied to the treatment of landfill leachate effluent was investigated. It was shown that EC treatment achieves an effective removal of color and COD. The treatment efficiency was found to be function of the initial pH, inter-electrode distance and electrolyte concentration. The quadratic model developed in this study shows the presence of a high correlation between experimental and predicted values. Analysis of variance showed a high coefficient of determination value (R^2 =0.9983 for color removal and 0.946 for COD removal), thus ensuring a satisfactory adjustment of the second-order regression model with the experimental data. Under optimal values of process parameters (initial pH: 7.73, inter-electrode distance: 1.16 cm, and electrolyte concentration: 1.00g), 89.01% of color removal and 46.05% of COD removal were obtained. Since the COD removal of leachate using EC is very low, further work will be conducted towards integrating EC with other wastewater treatment processes such as electro-Fenton.

Acknowledgment: This work was financially supported by the University of Malaya Postgraduate Research Fund (IPPP) Number PG093-2016A.

IOP Conf. Series: Materials Science and Engineering **210** (2017) 012008 doi:10.1088/1757-899X/210/1/012008

References

[1] Azni, I. (2009). What is the choice: Land disposal or Biofuel? In Waste management (pp. 1–65). Universiti Putra Malaysia.
[2] Norkhadijah, S. S. I., & Latifah A. M. 2013 *Journal of Toxicology and Environmental Health Sciences* **5** 86–96.
[3] Labanowski, J., Pallier, V., & Feuillade-Cathalifaud, G. 2010 *Journal of Hazardous Materials* **179** 166–172.
[4] Brennan, R. B., Healy, M. G., Morrison, L., Hynes, S., Norton, D., & Clifford, E. 2016 *Waste Management* **55** 355–363.
[5] Thompson, G., Swain, J., Kay, M., Forster, C.F. 2001 *Bioresource Technology* **77** 275–286.
[6] Renou, S., Givaudan, J. G., Poulain, S., Dirassouyan, F., & Moulin, P. 2008 *Journal of Hazardous Materials* **150** 468–493.
[7] Jeworski M, Heinzle, E. 2000 *Biotechnology Annual Review* **6** 163-196.
[8] Sirés, I., Brillas, E., Oturan, M. A., Rodrigo, M. A., & Panizza, M. 2014 *Environmental Science and Pollution Research* **21** 8336–8367.
[9] Cesaro A, Naddeo V, Belgiorno V 2013 *Journal of Bioremediation and Biodegradation* **4**.
[10] Vepsäläinen, M., Kivisaari, H., Pulliainen, M., Oikari, A., & Sillanpää, M. 2011 *Separation and Purification Technology* **81** 141–150.
[11] Zhao, S., Huang, G., Cheng, G., Wang, Y., & Fu, H. 2014 *Desalination* **344** 454–462.
[12] Song, S., He, Z., Qiu, J., Xu, L., & Chen, J. 2007 *Separation and Purification Technology* **55** 238–245.
[13] Chavalparit, O., & Ongwandee, M. 2009 *Journal of Environment Science* **21** 1491–1496.
[14] Farhadi, S., Aminzadeh, B., Torabian, A., Khatibikamal, V., & Alizadeh Fard, M. 2012 *Journal of Hazardous Materials* **219–220** 35–42.
[15] Maha Lakshmi, P., & Sivashanmugam, P. 2013 *Separation and Purification Technology* **116** 378–384.
[16] García-garcía, A., Martínez-miranda, V., Martínez-cienfuegos, I. G., Almazán-sánchez, P. T., Castañeda-juárez, M., & Linares-hernández, I. 2015 *Fuel* **149** 46–54.
[17] Ozyonar, F., & Karagozoglu, B. 2015 *Separation and Purification Technology* **150** 268–277.
[18] You, H. J., & Han, I. S. 2016 *Journal of Environmental Chemical Engineering* **4** 1008–1016.
[19] Emamjomeh, M. M., & Sivakumar, M. 2009 *Journal of Environmental Management* **90** 1663–1679.
[20] Brillas, E., & Martínez-huitle, C. A. 2015 *Applied Catalysis B, Environmental* **166–167** 603–643.
[21] Gengec, E., Kobya, M., Demirbas, E., Akyol, A., Oktor, K. 2012 *Desalination* **286** 200–209.
[22] Yoriya, S. 2012 *International Journal of Electrochemical Science* **7** 9454–9464.
[23] Susree, M., Asaithambi, P., Saravanathamizhan, R., & Matheswaran, M. 2013 *Journal of Environmental Chemical Engineering* **1** 552–558.

Synthesis, characterization and electrochemical study of Mn-doped TiO_2 decorated polypyrrole nanotubes

M.R. Saidur[1], **A.R. Abdul Aziz**[1,*] **and W.J. Basirun**[2,3]

[1]Department of Chemical Engineering, Faculty of Engineering, University of Malaya, 50603, Kuala Lumpur, Malaysia.
[2]Department of Chemistry, Faculty of Science, University of Malaya, 50603, Kuala Lumpur, Malaysia.
[3]Institute of Nanotechnology and Catalysis (NANOCAT), University of Malaya, 50603 Kuala Lumpur, Malaysia.

E-mail: azizraman@um.edu.my

Abstract. Nanostructured conductive polymers are the growing interest in the field of electrochemistry due to their superior conductivity and environmental friendliness. The existence of transition metal oxides could improve their nanostructure as well as conductive properties. In this study, polypyrrole nanotubes are synthesized in the presence of TiO_2 and manganese (Mn)-doped TiO_2 nanoparticles (NPs) to investigate their electrochemical properties. Details characterization of the synthesized composites were done by X-Ray diffraction (XRD) and TEM. The TEM analysis shows that doping of TiO_2 with Mn decrease the grain size of the TiO_2 nanoparticles and successively its effects on the synthesis of the PPy nanotubes (PPyNTs). TEM confirmed that PPyNTs synthesized in the presence of Mn-doped TiO_2 are thinner in size compare to the PPyNTs synthesized in presence of pure TiO_2. The electrochemical effectiveness of the synthesized PPy nanocomposite was investigated using cyclic voltammetry (CV) and electrochemical impedance spectroscopy (EIS). CV and EIS both on a modified glassy carbon electrode reveal the better electron transportability for the Mn-doped TiO_2 PPyNTs due to the synergistic effect of doping and decreased the size of PPyNTs as well as increased surface area.

1. Introduction

One-dimensional (1D) nanostructured materials such as nanowires, nanorods, nanobelts, and nanotubes have produced immense attention due to their special geometries, extraordinary physical and chemical properties. Therefore, they are highly effective for the fabrication of nanodevices, nanosensors and microelectronics [1, 2]. Polypyrrole (PPy), which is a typical conductive polymer, possess unique electrical, optical and optoelectrical properties, and is environmentally friendly. It has abundant application in energy storage, chemical sensors, biosensors and corrosion protection [3, 4]. Therefore, it has attracted extra attention in 1D nanostructured materials fabrication [5]. 1D nanostructured PPy can be synthesized through both chemical and electrochemical polymerization using template or template-free approach [6, 7] and also from electrospinning method [8].

Titanium dioxide (TiO_2) is a promising nanomaterial for many emerging applications including photocatalysis, solar cell fabrication, batteries and gas sensing. Doping of transition metal ions into TiO^2 lattice conveys different strains due to the difference between ionic radius which leads a variation in the lattice parameters and modifies it's optical, electrical as well as magnetic properties [9, 10].

It has been reported that polymers synthesized in the vicinity of nanoparticles could decrease the polymer size and consequently increase the surface area [3]. It is also expected that the incorporation of TiO_2 with PPy might improve the electrochemical properties of PPy. However, only a few studies can be found in the literature. In this study, we synthesize Mn-doped TiO_2 and TiO_2 NPs decorated PPy nanotubes and studied their electrochemical properties on a glassy carbon electrode.

2. Methodology

2.1 Synthesis of Mn-doped TiO_2 nanoparticles (NPs)

Reagent grade chemicals obtained from Sigma-Aldrich and Merck, Malaysia were used to synthesize all the materials. Mn-doped TiO_2 nanoparticles were synthesized through a sol-gel method from titanium tetra-isopropoxide $Ti[OCH(CH_3)_2]_4$ where manganese acetate tetrahydrate was the doping precursor [11]. A well-mixed mixture of $Ti[OCH(CH_3)_2]_4$ and Mn precursors (3 mol %) in glacial acetic acid was poured into double distilled water and then stirred for 6 h with a constant rate. A ratio of 1: 10:200 for $Ti[OCH(CH_3)_2]_4$: glacial acetic acid: H_2O was maintained. After stirring the gel-like mixture was kept in an oven at 75 °C for 24 hours followed by grounded to a fine powder. The fine powder was finally calcinated for 5 h at 500 ° C. In a Similar way, TiO_2 NPs were prepared for comparison.

2.2 Preparation of Mn-doped Polypyrrole nanotubes (PPyNTs)

Mn-doped TiO_2/PPyNTs were synthesized according to a previous report where methylene orange (MO) was used as a soft template during the chemical oxidative polymerization [12]. Briefly, the oxidant $FeCl_3$ was dissolved in 5mM MO solution and subsequently, 0.19 g of Mn-doped TiO_2 NPs was added to the mixture. To spread the NPs on the MO template, the mixture was then stirred for 30 min. After that, 0.2 ml of pyrrole monomer was added drop by drop without stirring. The whole system was immediately placed at 0-4 ° C for complete the polymerization for 24 h. The resulting polymer composite then filtered and washed with distilled water and ethanol to fully remove the MO. Finally, the composite dried at 40 ° C in an oven. PPyNTs were also prepared in presence of TiO_2 NPs and absence of NPs.

2.3 Characterization

The existence of TiO_2 and Mn-doped TiO_2 NPs in the synthesized PPyNTs was investigated through X-Ray Diffraction (XRD) in Siemens D5000 model X-Ray diffractometer. Field emission scanning electron microscopy (FESEM) in a Hitachi SU8220 microscope. Transmission electron microscopy (TEM) were taken from MIMOS Berhad, Malaysia to confirm the morphology and calculate the size of the NPs and NTs composites.

2.4 Study of electrochemical properties

As prepared Mn-doped TiO_2/PPyNTs and TiO2/PPyNTs were dispersed in DMF (1 mg/ml) through ultrasonication. 10 μl of each dispersion were then dropped cast on a cleaned glassy carbon electrode surface and kept to dried for studied the electrochemical properties. Cyclic Voltammetry (CV) and electrochemical impedance spectroscopy (EIS) were performed in a Methrom Autolab (PGSTAT-302N) which is operated by GPES and FRA.EXE software.

3. Results and discussion

3.1 XRD and TEM analysis

The crystal phase of the synthesized TiO_2 and Mn-doped TiO_2 NPs has been identified from XRD and the corresponding lattice patterns are presented in figure 1. In pure TiO_2 NPs, all the diffraction peaks can be indexed to tetragonal anatase structure of TiO_2 with a good agreement of standard JCPDS entry code No. 00-001-0562. For Mn-doped TiO_2 NPs, no additional peak which is related to Mn or any of its oxide were found. This confirms the phase purity of the NPs. However, most of the predominance peaks intensity (triplet) (101) (103) and (200) were increased with an increase of full width at half maximum. Moreover, the peaks were shifted slightly to a lower angle when we compared to the pure TiO_2 NPs. Hence, it could be assumed that some of the Mn^{2+} ions can enter the Ti crystal site and replaced some Ti^{4+} ions. These substitutional Mn^{2+} ions perhaps doped in regular size of Ti ions and consequently decrease the grain size of the pure TiO_2 NPs [13, 14].

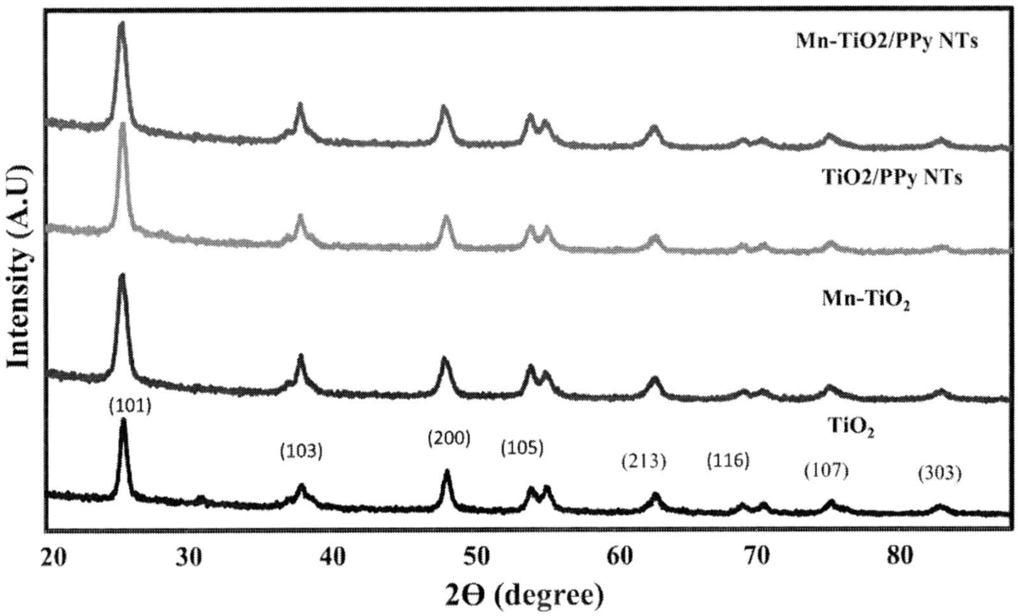

Figure 1. The powder XRD diffraction pattern of pure TiO_2, Mn-TiO_2 NPs, TiO_2/PPyNTs, and Mn-TiO_2/PPyNTs.

The mean crystalline size was calculated from Scherrer's equation [15] and was found 18.6 and 11.90 nm for pure and Mn-doped TiO_2 NPs respectively. The XRD pattern of TiO_2/PPyNTs and Mn-doped TiO_2/PPyNTs were found in a similar pattern with following the similar changes, therefore, confirm the existence of NPs on the PPyNTs.

Morphology of the nanostructured materials is an important characteristic as the performance depends on the morphology. Figure 2. (a-b) shows the TEM images of pure TiO_2 and Mn-doped TiO_2 NPs. We can see clearly a uniform spherical shaped NPs has formed and Mn dopant has no influence to change the morphology. However, a less agglomeration and reduction of crystal size were observed for Mn-doped TiO_2 NPs. The mean diameter is calculated around (17.19 ± 1.07) nm for pure TiO_2 NPs and (10.97 ± 1.70) nm for Mn-TiO_2 NPs (figure 2c, 2d), which are well matched to the size calculated from Scherrer's equation in XRD. The beneficial effect of doping of Mn in TiO_2 NPs can be seen clearly from the TEM image of synthesized PPyNTs composite (figure 2e and 2f). A much thinner nanotube-shaped PPy nanocomposites were formed in presence of Mn-doped TiO_2 and the NPs are decorated on the tube surface.

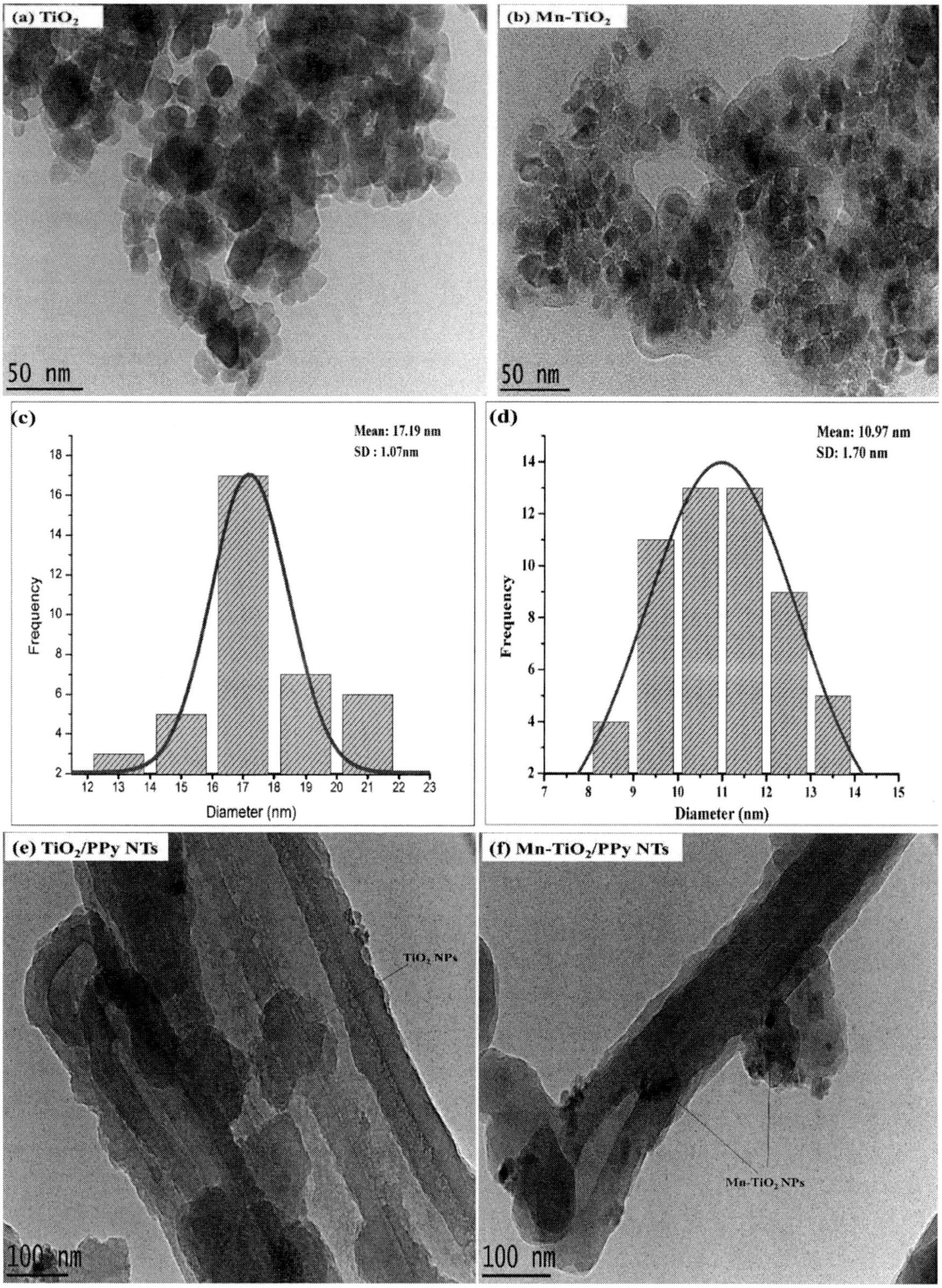

Figure 2. TEM images of TiO_2 NPs (a), Mn-doped TiO_2 (b), TiO_2/PPyNTs (e), Mn-TiO_2/PPyNTs (f) and particle size distribution of TiO_2 NPs (c), Mn-doped TiO_2 NPs (d).

3.2 Electrochemical performance of synthesized PPyNTs

CV was performed in 10 mM $Fe(CN)_6{}^{3-/4-}$ (1:1) in 0.5 M KCl as supporting electrolyte within 0 to 0.6 V at a scan rate 50 mV/s. Figure 3. shows the CV profile of bare GCE, TiO_2/PPyNTs, and Mn-TiO_2/PPyNTs modified GCE. An increase anodic (I_{pa}) and cathodic (I_{pc}) peak current were observed for TiO_2/PPyNTs modified GCE compared to bare GCE. In addition, the oxidation-reduction potential also shifted to a lower voltage, which is always beneficial. A further increase in the peak current was observed for Mn-TiO_2/PPyNTs modified GCE. This could be due to the synergistic effect of Mn-doped TiO_2 and the decrease of NTs diameter as well as the increased surface area.

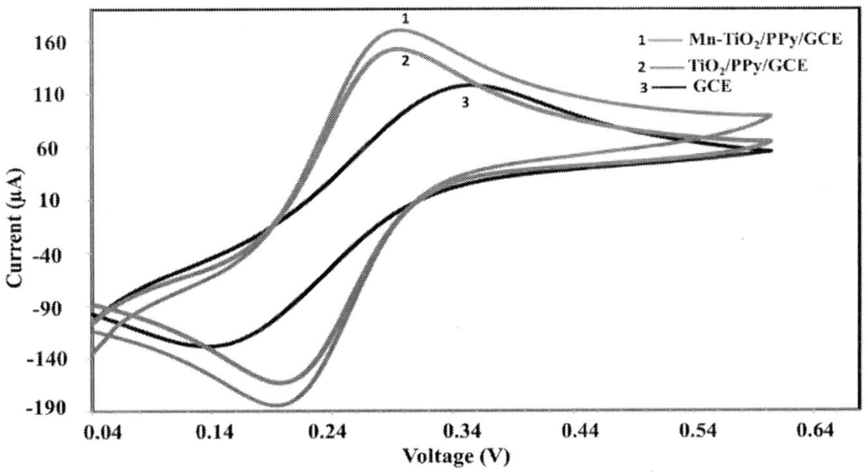

Figure 3. CV in 10 mM $Fe^{2+/3+}$ and 0.5 M KCl as supporting electrolyte.

To investigate the interfacial properties of the nanocomposite-modified GCE electrode, EIS was used as it is one of the efficient technique. Figure 4(a). shows the Nyquist plots of Mn-doped TiO_2/PPyNTs (1), TiO_2/PPyNTs (2) and bare GCE (3). In EIS, the Nyquist plot consists of a semicircle at a higher frequency which signifies the electron transfer limited process on electrode/electrolyte interface. Also, an inclined line at lower frequency correlates the mass transfer process.

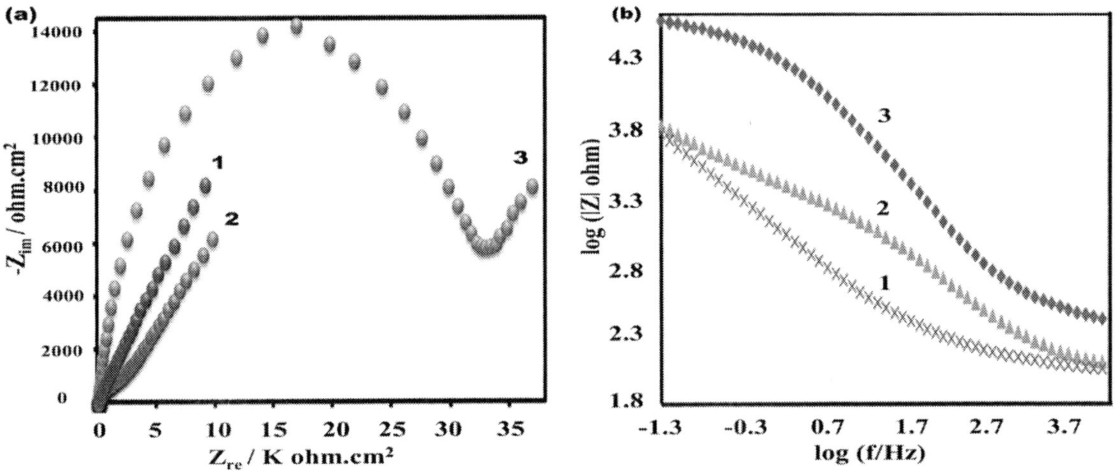

Figure 4. EIS of Mn-doped TiO2/PPy NTs (1), TiO2/PPy NTs (2) and bare GCE (3) 1 mM $Fe(CN)_6{}^{3-/4-}$ (1:1) in 0.1 M KCl as supporting electrolyte at OCP within a frequency range 100 kHz to 0.1 Hz.

The electron transfer resistant can be evaluated by comparing the diameter of the semicircle. It is well known that decrease the semicircle diameter corresponds to lower the electron transfer resistant which means higher conductivity [16]. This can exactly see in the comparison of three electrodes. The semicircle diameter decreased a lot for pure TiO_2/PPyNTs. However, a further decreased can be observed for Mn-TiO_2/PPyNTs.

The effect of Mn doping is also confirmed from Bode plots in figure (4b). Whenever we compare the log |Z| of the three electrodes it can be seen that the charge transfer resistance shows the same direction: GCE > TiO_2/PPyNTs/GCE > Mn-TiO_2/PPyNTs/GCE, that is good agreement with the result obtained from the Nyquist plot.

4. Conclusion

Nanostructured polypyrrole in nanotube morphology were synthesized in the presence of pure TiO_2 and Mn-doped TiO_2 NPs. XRD and TEM analysis showed that the average diameter of pure TiO_2 and Mn-doped TiO_2 NPs is 17.19 nm and 10.97 nm respectively. They also confirmed the presence of Mn-doped TiO_2 NPs produced significantly thinner size PPyNTs composite. CV measurement and impedance analysis on GCE electrode showed the higher electron conductivity and minimum charge transfer resistance for Mn-TiO_2/PPyNTs due to high surface area. Therefore, Mn-doped TiO_2/PPy NTs could be utilized as effective electrode materials for the development of sensitive electrochemical sensor. However, further investigation is required to establish the optimum dopant concentration for best electrochemical properties.

Acknowledgements

The authors are grateful to the University of Malaya High Impact Research Grant (UM.C/625/1/HIR/MOHE/ENG/37) from the ministry of Higher Education Malaysia, PG149-2015B and GC001C-14SBS for financially supporting this work.

References

[1] Bubnova O 2017 Printed electronics: Nanotube resolution *Nat Nano*
[2] N. Shehada, J. C. Cancilla, J.S. Torrecilla, E.S. Pariente, G. Brönstrup, S. Christiansen, D.W. Johnson, M. Leja, M.P. Davies and O. Liran *ACS Nano* **10** 7047-7057
[3] M. Ladan, W.J. Basirun, S.N. Kazi and F.A. Rahman 2017 *Materials Chemistry and Physics* **192** 361-373
[4] J.W. Park, W. Na and J. Jang 2016 *RSC Advances* **6** 14335-14343
[5] I. Sapurina, Y. Li, E. Alekseeva, P. Bober, M. Trchová, Z. Morávková and J. Stejskal 2017 *Polymer* **113** 247-258
[6] Z.D. Kojabad and S.A. Shojaosadati 2016 *Materials & Design* **96** 378-384
[7] M. Mahmoudian, Y. Alias and W. Basirun 2012 *Electrochimica Acta* **72** 53-60
[8] I.S. Chronakis, S. Grapenson and A. Jakob 2006 *Polymer* **47** 1597-1603
[9] W. Zhang, W. Zhou, J.H. Wright, Y.N. Kim, D. Liu and X. Xiao 2014 *ACS applied materials & interfaces* **6** 7292-7300
[10] M.S. Park, S. Kwon and B. Min 2002 *Phys. Rev. B* **65** 161201
[11] I.Y. Jeon and J.B. Baek 2010 *Materials* **3** 3654-3674
[12] M. Mahmoudian, Y. Alias, W. Basirun and M. Ebadi 2013 *Applied Surface Science* **268** 302-311
[13] M. Karthik, M. Parthibavarman, A. Kumaresan, S. Prabhakaran, V. Hariharan, R. Poonguzhali and S. Sathishkumar 2017 *Journal of Materials Science: Materials in Electronics* 1-8
[14] H. Zhang, T. Ji, Y. Liu and J. Cai 2008 *The Journal of Physical Chemistry C* **112** 8604-8608
[15] M. Sookhakian, Y. Amin, W. Basirun, M. Tajabadi and N. Kamarulzaman 2014 *Journal of Luminescence* **145** 244-252
[16] I.I. Suni 2008 *TrAC Trends in Analytical Chemistry* **27** 604-611

Review of water footprint components of grain

Wan Amiza Amneera Wan Ahmad[1], Nik Meriam Nik Sulaiman[1,*] and Noor Zalina Mahmood[2]

[1]Department of Chemical Engineering, Faculty of Engineering, University of Malaya, 50603 Kuala Lumpur, Malaysia
[2]Institute of Biological Sciences, Faculty of Science, University of Malaya, 50603 Kuala Lumpur, Malaysia

E-mail: meriam@um.edu.my

Abstract. Burgeoning global population, economic development, agriculture and prevailing climate pattern are among aspects contributed to water scarcity. In low and middle income countries, agriculture takes the highest share among water user sector. Demand for grain is widespread all over the globe. Hence, this study review published papers regarding quantification of water footprint of grain. Review shows there are various methods in quantifying water footprint. In ascertaining water footprint, three (green, blue, grey) or two (green, blue) components of water footprint involved. However, there was a study introduced new term in evaluating water footprint, white water footprint. The vulnerability of varying methods is difficulty in conducting comparative among water footprint. Salient source in contributing high water footprint also varies. In some studies, green water footprint play major role. Conversely, few studies found out blue water footprint most contributing component in water footprint. This fluctuate pattern influenced by various aspects, namely, regional climatic characteristics, crop yield and crop types.

1. Introduction

The significance of water on human being and environment is immense. Prevailing population growth and economic development has led to increase in freshwater demand. Water availability and agriculture is inextricably linked. As the demand for agriculture expand significantly, the menace on availability of water resource expanding as well. By 2025, it is expected that water withdrawal for agriculture will increase by 50% in developing countries and 18% in developed countries [1].

Agriculture takes the highest share among water user sector in both, low and middle income countries [2]. In order to attain the ascending food requirements of expansion global population, enlargement of irrigated agriculture is necessary. Nevertheless, waterlogging and soil salination are two consequences caused by enlargement of irrigated agriculture. These two environmental impacts however can be mitigated by implementing mitigating measures to reduce the inflow or by applying remedial measures to raise water and salt outflow [3]. In order to boost agriculture output, sufficient water resource for irrigation is pivotal. India, second largest world main paddy producers [4], has over 66 million hectares area equipped for irrigation [5].

So far, grains are the salient source of food not only for direct human consumption but also indirect use. Barley, rice, maize and wheat have become dominant crops globally [6]. For nearly 20 years, grains supply per capita has been declined for 17% resulted from shortage of freshwater [7]. Apprehensive with the future water crisis the world might face, research on water footprint (WF) exceptionally essential and gained worldwide attention. Water footprint has appeared as a pivotal technique in order to evaluate water consumption of both, goods and services. Evaluation methods proposed by two communities, namely water footprint network (WFN) and the life cycle assessment (LCA) [8].

2. Grains production

World production of maize in 2014 was 1037 million tonnes with the harvested area of 184 million hectares globally. The three top producers of maize were United States of America, China and Brazil [9]. While for rice, 741 million tonnes of rice produced in 2014 with 162 million hectares of arable land in 2014. China, India and Indonesia were the three major producers of rice [9]. Wetland and upland systems are two systems of rice production. Asia was the largest contributor for rice production with 90.8% of average production for the period of 1994-2014 [9]. In Asia, rice fields are prepared with tillage followed by puddling. During the growth period, the soil is saturated and there is standing water throughout the period [10].

World production of wheat is slightly lower than rice, 729 million tonnes and the harvested area was 220 million hectares. Similar to rice, China and India were the top two producers of wheat followed by United States of America [9]. For trading, wheat is categorized into well-defined categories of grain hardness, namely soft, medium-hard and hard and also colour based which are red, white and amber. In term of growing season, it can be sort into spring and winter wheat [11].

3. Water scarcity

Water accessibility which is below than 1000 m^3 per person per year is considered as water scarcity condition for a country or region. Nonetheless, most regions facing severe water scarcity whereby water availability is less than 500 m^3 per person per year [2]. Apart from water availability, water scarcity also related to quality of available water [12].

There are numerous factors that contribute significantly to water scarcity. Among them are burgeoning global population and economic development. Water availability per capita descended by 42% in Solomon Islands, 36% in Malaysia, Pakistan and Nepal, 29% in India and Bangladesh between 1990 and 2010. Apart from that, decreasing water availability also stem from agriculture sector as it consume 70% out of all withdrawals. Prevailing climate pattern also influence water resources. For instance few regions have been affected by extreme events such as drought, flood and cyclone. It is also accountable for rising sea level which directly increased salinization in river deltas and lake and eventually decreased the availability of freshwater [5].

4. Water footprint

Back in 1990s, in order to analyse food and water security, Prof Tony Allan coined the term 'virtual water' as amount of water needed in producing product or service by considering the whole supply chain [13]. Later, WFN defined water footprint as an indicator of water usage by considering water consumed directly and indirectly by consumers or producers [14]. WF expounds water consumption volumes by source and polluted volumes by type of pollution. All components involved in quantifying WF are specified in time and space [13].

Green, blue and grey are three components of WF. Each component indicates different source and type of water to consider. Green WF elucidates as water that evaporated from soil moisture [13] and mostly used to produce agricultural products [14]. Whilst blue WF represents water by which withdrawn from ground or surface water sources [13] and grey WF can be defined as volume of water needed in assimilating waste water [14].

5. Water footprint of grain

Agriculture water management is crucial for Iran since most regions are situated in arid and semiarid area. Pattern of water withdrawal percentage for agricultural use in Iran were fluctuate. In 1995, 91.57% of total water withdrawal was consumed by agriculture. The value then increased to 93.37% in 2001 and showed slightly decreased in 2004 (92.18%) [15]. Study conducted by [16] assessed water footprint (WF) of wheat, barley and maize in fifteen main cereal producing provinces in Iran. The quantification of WF in this study modified existing WF method by Hoekstra by introducing new component in calculation, white water footprint, which indicates irrigation water losses. For the period of 2006 to 2012, national total water footprint (NTWF) of wheat, barley and maize were 36777, 7975 and 3744 million cubic meters (MCM) respectively. For wheat, the highest contributor for high NTWF was green (17222 MCM) followed by blue (7406), grey (6809 MCM) and white (5340 MCM). Similar to wheat, greatest contributor to NTWF of barley was green WF with the value of 3363 MCM followed by blue, grey and white with the value of 1761, 1518 and 1334 MCM respectively. Conversely for maize, green WF was the smallest NTWF contributor with the value of 83 MCM. The biggest was blue (1541 MCM) followed by grey (1084) and white (1056 MCM). The blue and green WF for the three grain was 31376 MCM with wheat is the biggest consumer since it is cultivated 4.7 and 23 times larger areas compared to barley and maize. Apart from that, in term of yield, wheat was produced 4.6 and 5.9 times more than barley and maize. Hence, efficient water management and allocation in wheat production is pivotal as to mitigate insufficient freshwater. Meanwhile, grey and white WF for wheat, barley and maize was 17231 MCM which elucidate that in producing cereal, huge amount of water loss during irrigation and might never reaches water table as well. Since Iran is located in an arid area, it is vital to alleviate this matter by improving irrigation management practice as to decrease the share of white WF.

WF of grain production in North China Plain (NCP) for past 35 years was ascertained by [17]. This study assessed WF of winter wheat and summer maize for 35 years period starting from 1980 to 2014. Study was conducted at Luancheng station, northern part of NCP and received annual precipitation of 480 mm. The quantification of WF for this study involved three WF components, namely, green, blue and grey. Nexus between irrigation and agriculture is very strong as to secure crop production. In order to retained soil water content above 65%, irrigation was applied 3 to 4 times per season for winter wheat and 1 to 2 times for summer maize. Based on the outcome from this study, WF for winter wheat was 0.89 m^3/kg which found stem from yield, N fertilizer, effective precipitation and irrigation whilst 0.70 m^3/kg for summer maize which solely driven by yield. The components of WF for winter wheat were 0.21 m^3/kg (green), 0.55 m^3/kg (blue) and 0.13 m^3/kg (grey) whilst for summer maize 0.42 m^3/kg (green), 0.19 m^3/kg (blue) and 0.09 m^3/kg (grey). Major contributor for both, winter wheat and summer maize were contrast. The dominant WF component for winter wheat was blue while green for summer maize. This circumstance was driven by regional climatic characteristics in NCP.

A study by [18] quantified WF in the second largest river basin of China, Yellow River Basin (YRC). The WF for maize is the highest (8419 m^3/t) followed by rice (5803 m^3/t), barley (3095 m^3/t) and wheat (2214 m^3/t). By comparing the WF components for all crops, grey WF are the highest for rice, maize and barley, meanwhile green WF highest for wheat. This might be the consequence from rising rate of artificial fertilizer application [18]. Blue WF remains the lowest contributor for all crops. Another study regards to WF was conducted by [19] which assessed WF of Italian wheat from 2011 to 2015. The study covered Northern, Central and Southern of Italy by considering seventeen regions. WF for wheat was 5327 m^3/ha with the highest component is green (4526 m^3/ha) followed by grey and blue with the value of 740 and 61 m^3/ha respectively. All three components were recorded highest at North part of Italy while the lowest at Southern of Italy.

Apprehensive with water resources which threatened by the fluctuate patterns of precipitation periods and distribution in Taiwan, [20] conduct a WF assessment for rice and corn in three regions in central and southern Taiwan, namely Yunlin, Tainan and Chiayi. Corn was planted during autumn, October to February involving only two regions, Tainan and Chiayi whilst rice was planted from July to October involving all the three regions. Based on data collected from 2007-2011, WF of rice was 4809 m^3/t, 1521 m^3/t for Chiayi and Tainan meanwhile 3288 m^3/t for Yunlin. While for corn, WF for Tainan and Chiayi were 704 and 669 m^3/t respectively. In Taiwan, rice cultivation is predominantly depends on irrigation while corn is rain-fed, thus blue WF was high for rice (3671 m^3/t) compared to corn (538 m^3/t). In contrast, corn indicate highest grey WF (484 m^3/t) as compared to rice (220 m^3/t) stem from excessive used of nitrogen fertilizer (160-171 kg/ha) which eventually lead to abundance water usage to dilute pollutants as to comply to environmental quality standards.

Another study with regards to WF of rice conducted by [21] which contemplates two rice growing areas in Argentina, Entre Rios and Santa Fe for the period of 2009-2013. In this study, grey WF was considered inconsequential by the researchers based on several reviewed publications, claimed that the level of residue found were below established standard, hence no large anxiety for the ecosystem. WF of rice Santa Fe (987 m^3/t) was slightly higher as compared to Entre Rios (845 m^3/t). High WF in Santa Fe was due to higher blue WF (627 m^3/t) as compared to Entre Rios (477 m^3/t). Green WF was not obviously different for both regions, Entre Rios (367 m^3/t) whereas Santa Fe (360 m^3/t). This circumstance was view of the fact that rice yields were higher in Entre Rios compared to Santa Fe in all season analysed.

Be in distress of high water scarcity since located in arid country in Mediterranean, has caused Tunisia to face constraints in producing food stem from shortage of water resource [22]. A study conducted by [22] determine WF of wheat in Tunisia from 1996-2005. The findings show that the contrast in WF for North (2550 m^3/t) and Central Tunisia (2710 m^3/t) is small, however values in South Tunisia (4220 m^3/t) contrast significantly. The difference for green and grey WF was not significant for all three areas. But for blue WF, the difference is clearly noticeable in the southern Tunisia where blue WF is almost three times more than the north and five times more than central. This was mainly due to difference in production systems, whereby in South, irrigation is the dominant production system whilst rain fed is dominant in North and Central. Another study regarding WF in China conducted by [23] discovered WF of rice, wheat and corn for 8 regions of China by considering data in 2010. Among the three crops, rice (1.39 m^3/kg) had the highest WF while corn (0.91 m^3/kg) indicates the lowest. Although water consumption is highest for rice, but if viewed in terms of WF component, corn (the lowest WF) has the highest green WF while wheat has the greatest impact on blue WF. This circumstance was result from various crop pattern implemented in each region involved.

Beijing, located in North China Plain is an exceptionally water-scarce region in China [24]. A study on WF was carried out by [24] in Beijing municipality. Based on the study, for 2009, WF of wheat was 711.8 m^3/t while 868.2 m^3/t for maize planted in Beijing. A very significant gap can be seen for blue WF as the highest was from wheat (377.5 m^3/t) compared to maize (4.8 m^3/t). Wheat had the largest blue WF stem from excessive amount of irrigation water consumed as to cater large planted area. Contras for maize, the blue WF was small but had the largest green WF (422.3 m^3/t) due to large-scale cultivation of rain fed maize and its growth period is during rainy season. A global WF assessment for wheat was conducted by [11] for the period of 19966-2005. The average global WF was 1830 m^3/t with the major contributor was green WF (1279 m^3/t) followed by blue (343 m^3/t) and grey (208 m^3/t). The first three salient contributors were Morocco (3710 m^3/t), Iran (3690 m^3/t) and Kazakhstan (3629 m^3/t) even though the contribution towards global wheat production less than 2.0%.

6. Conclusion

Based on reviews, the most contributor component in contributing high water footprint fluctuating. Type of crop is among major sources that influence blue or green component of WF. Crop that predominantly depends on irrigation will indicates high blue WF as compared to green whilst rain-fed crops will result in high green WF. Climate pattern for particular region also play an important role in determining which component of WF will contribute highest. For instance, for arid regions, irrigation is crucial for crop growth which leads to high blue WF. From economic view, blue WF should be scrutinized as blue water costly as compare to green water. Implementation of efficient water management is pivotal in controlling blue WF. For grey WF, in most cases, grey component is lower than blue and green. In spite of that, it's still required attention too. Crop producers should not only engrossed in what increase crop yield but also comprehend and implement sustainable agriculture without effect the quality of crops.

Acknowledgment

We acknowledge the support from University of Malaya (UM) through postgraduate research fund (PPP) grant (PG088-2016A) under the project titled 'Quantification and analysis of water and carbon footprint of rice production in Malaysia.

References

[1] S.H. Gheewala, T. Silalertruksa, P. Nilsalab, R. Mungkung, S.P. Perret and N. Chaiyawannakarn 2014 *Water* **6** 1698-1718.
[2] L.S. Pereira, I. Cordery, and I. Iacovides 2002 *Coping with water scarcity*. Technical Documents in Hydrology, no. 58: Unesco, Paris
[3] A. Singh 2016 *Agricultural Water Management* **174** 2-10.
[4] DOA 2014 *Paddy statistics of Malaysia 2013*. Malaysia: Department of Agriculture.
[5] FAO 2014 FAO Statistical Yearbook 2014 Asia and the Pacific Food and Agriculture. Food and Agriculture Organization of the United Nations Regional Office for Asia and the Pacific. Retrieved from www.fao.org/3/a-i3590e.pdf
[6] V. Fantin, S. Righi, I. Rondini and P. Masoni 2017 *Journal of cleaner production* **140** 631-643.
[7] D. Pimentel, B. Berger, D. Filiberto, M. Newton, B. Wolfe, E. Karabinakis, E. Poon, E. Abbett and S. Nandagopal 2004 *BioScience* **10** 909-918.
[8] S. Pfister, A.M. Boulay, M. Berger, M. Hadjikakou, M. Motoshita, T. Hess, B. Ridoutt, J. Weinzettel, L. Scherer and P. Döll 2017 *Ecological Indicators* **72** 352-359.
[9] FAO 2017 FAOSTAT database, production, crop. Retrieved from http://www.fao.org/faostat/en/#data/QC/visualize
[10] A. Chapagain, and A. Hoekstra 2011 *Ecological Economics* **4** 749-758.
[11] M. Mekonnen and A. Hoekstra 2010 *A global and high-resolution assessment of the green, blue and grey water footprint of wheat.*
[12] E. DeNicola, O.S. Aburizaiza, A. Siddique, H. Khwaja and D.O. Carpenter 2015 *Annals of global health* **3** 342-353.
[13] A. Chapagain and K. James 2011 *The water and carbon footprint of household food and drink waste in the UK.* Waste & Resources Action Programme (WRAP), Banbury, Oxon, UK & WWF, Godalming, Surrey, UK.
[14] Y.J. Lee 2015 *Environmental Impact Assessment Review* **54** 1-8.
[15] FAO 2013 FAOSTAT database, Agri-environmental indicators, water. Retrieved from http://www.fao.org/faostat/en/#data/EW
[16] B. Ababaei and H.R. Etedali 2017 *Agricultural Water Management* **179** 401-411.
[17] Y. Lu, X. Zhang, S. Chen, L. Shao and H. Sun 2016 *Journal of cleaner production* **116** 71-79.
[18] L. Zhuo, M.M. Mekonnen, A.Y. Hoekstra and Y. Wada 2016 *Advances in Water Resources* **87** 29-41.
[19] N. Casolani, C. Pattara, and L. Liberatore 2016 *Land Use Policy* **58** 394-402.
[20] M.H. Su, C.H. Huang, W.Y. Li, C.T. Tso and H.S. Lur 2015 *Journal of cleaner production* **88** 132-138.
[21] R.P. Marano and R.A. Filippi 2015 *Ecological Indicators* **56** 229-236.
[22] H. Chouchane, A.Y. Hoekstra, M.S. Krol and M.M. Mekonnen 2015 *Ecological Indicators* **52** 311-319.
[23] Y. Wang, P. Wu, B. Engel and S. Sun 2014 *Science of The Total Environment* **497** 1-9.
[24] J. Huang, H.L. Zhang, W.J. Tong and F. Chen 2012 *Journal of cleaner production* **21** 45-50.

Effect of varying the amount of binder on the electrochemical characteristics of palm shell activated carbon

Hawaiah Imam Maarof[1,2], Wan Mohd Ashri Wan Daud[1], and Mohamed Kheireddine Aroua[1,*]

[1] Department of Chemical Engineering, Faculty of Engineering, University of Malaya, 50603 Kuala Lumpur, Malaysia.
[2] Faculty of Chemical Engineering, Universiti Teknologi MARA, 13500 Permatang Pauh, Pulau Pinang, Malaysia.

E-mail: mk_aroua@um.edu.my

Abstract. Polytetrafluoroethylene (PTFE) is among the most common binders used in the fabrication of an electrode, which is used for various electrochemical applications such as desalination, water purification, and wastewater treatment. In this study, the amount of the binder was varied at 10, 20, 30, and 40 wt% of the total mass of palm shell activated carbon (PSAC). The PSAC was used as the active material and carbon black was used as the conductive agent. The effect of different amounts of binder was observed by evaluating the electrochemical characteristics of the electrode through cyclic voltammetry (CV) and potentio electrochemical spectroscopy (PEIS). The CV analysis was employed to determine the geometric area normalised electrode double layer capacitance, C_E, and the electrode reaction of the prepared electrode. Meanwhile, the common redox probe, ferro/ferricyanide in 0.5 M NaCl, was employed to estimate the electron transfer resistance through PEIS. The electrochemical characterisation proved that the optimum amount of PTFE was 20 wt% for the 4:1 ratio of active material to conductive agent. On increasing the amount of the binder to 30 wt% and 40 wt%, the estimated value of C_E decreased and remained almost equivalent. Adding more than 30 wt% of binder resulted in pore blockage and reduced the available active site on the PSAC electrode. In addition, the electron transfer resistance of the prepared electrode was found to be in the range of 4-5 $\Omega\cdot cm^2$.

1. Introduction

There has been growing interest among researchers to explore the performance of porous electrodes in water and wastewater treatment by means of capacitive deionisation (electrosorption), electro-oxidation and electrodeposition [1-8]. A porous electrode provides a high surface area per unit mass and helps increase the rate of electron transfer [9] as well as the electrochemical active site. The use of activated carbon-based electrodes derived from waste and industrial by-products is gaining attention because it is essential to have such a precursor to ensure the economic feasibility of the electrode fabrication process However, there are only limited works that explore the potential of waste-derived activated carbon electrodes for water and wastewater treatment. Among them are whitewood biochar [10] and palm kernel shell [8, 11, 12]. The waste-derived activated carbon electrodes are in either granular or powdered form as they are placed in a pack bed reactor or preferably fabricated in a form of sheet.

A binder is an essential component in electrode preparation as it helps in the binding of discrete porous particles to form an electrode that facilitates the electrochemical activities in a system. Fluorine-containing resin materials such as polytetrafluoroethylene (PTFE) are widely used as a binder, as they have excellent chemical and thermal resistance and are best known for their binding properties. PTFE is widely used as a binder for the fabrication of cathode for energy storage, particularly in microbial fuel cell [13-16] and capacitor/supercapacitor [17, 18]. On one hand, Zarei, et al. [19] developed carbon nanotube-PTFE cathode for dye removal under several different operating conditions, while on the other hand, Park, et al. [20] prepared activated carbon-PTFE electrode sheet for desalination. PTFE as a binder was reported to provide high reproducible performance of electrode at low cost as compared to Nafion [14]. Nafion is commonly used as binder in preparing the electrode by brushing method. However, a uniform mixture of active material and PTFE electrode is generally difficult to achieve because of excessive fibrillation due to the PTFE properties [21]. To a certain extent, PTFE is likely to impair the characteristics of the electrode if added in an amount more than adequate. Additionally, PTFE is used in the form of an aqueous dispersion rather than a powder in order to obtain a more uniform mixture.

In this work, a local source of industrial waste, namely palm shell activated carbon (PSAC), was used as the precursor of a porous carbon-based electrode. An optimum amount of PTFE from its aqueous dispersion was determined to bind porous powdered PSAC. Although Misnon, et al. [22] derived an electrode from PSAC and evaluated its electrochemical properties, they merely focused on preparing the electrode as a supercapacitor in which PVDF was used as the binder. In another study by Abbas, et al. [23], PTFE was reported as a better binder compared with PVDF. They reported that PVDF resulted in pronounced reduction of pore volume of activated carbon (Norit® DLC Supra) and lower capacitance was obtained than with PTFE. Hence, the present work aims to explore the feasibility of the local industrial waste, namely PSAC, as the active material for an electrode in which PTFE was used as the binder. There is no available open literature on the electrochemical properties of PSAC-PTFE composite electrode, particularly on the electrochemical active surface area. Therefore, the main objective is to prepare an adequate strength of electrode with an optimum amount of PTFE as the binder. The electrochemical properties of the prepared electrode were further examined to ensure its potential for using as free-standing electrode sheet in water and wastewater treatment.

2. Methodology

2.1. Electrode preparation

The granular PSAC was procured from a local supplier, Bravo Green Sdn. Bhd., Sarawak. It was ground to fine particles and subsequently sieved to obtain the desired particle size (<100 μm). The electrode was prepared by mixing a specific amount of powdered PSAC, carbon black (Super P, Alfa Aesar), and 60 wt% PTFE suspension (Sigma-Aldrich). The mass ratio of PSAC and carbon black was fixed at 4:1. An equal total mass of active material and conductive agent (ca. 0.5 g) was prepared for each type of electrode while varying the percentage of PTFE accordingly. Four different electrodes with pre-determined mass percentage of PTFE were prepared. The mass of PTFE was varied at 10, 20, 30, and 40 wt% and these electrodes were identified as 10PTFE, 20PTFE, 30PTFE, and 40PTFE, respectively. Each mixture was thoroughly hand mixed to produce an electrode paste by the addition of 1,3-propanediol as the solvent. The paste was then kneaded and pressed in a stainless-steel mould prior to drying at 80°C for 2 h, at 125°C for 1 h, and at 250°C for 1 h. When the electrodes are taken out from the oven, they were hand-pressed to increase the mechanical strength. The dried electrodes were then packed into a 5-mm diameter empty electrode tip. The geometrical surface area of the electrode was fixed at 0.196 cm^2 (5 mm diameter). Prior to evaluating the electrochemical characteristics, the electrode surface was polished using the micropad polishing cloth. Each electrode was held vertically and 8-motions were performed for 20 times on the wet polishing pad, which was pre-wetted by distilled water. The electrodes were then dried using hot air blower, cooled at room temperature, and made ready for electrochemical characterisation studies. The morphology of the electrode surface was examined using field emission scanning electron microscopy (FESEM) (AURIGA, Carl Zeiss), equipped with energy dispersive X-ray (EDX) analyser.

2.2. Chronoamperometry (CA)
The experiments were conducted in a single compartment of 100 mL glass cells at room temperature. A three-electrode cell configuration was employed in which a platinum wire and Ag/AgCl (3M KCl) reference electrodes were used for the measurement. The solution was 5 mM potassium ferrocyanide electrolyte in 0.5 M NaCl that acted as the supporting electrolyte. The CA analysis employed a potentiostat (Bio-Logic, SP-300), and the data were analysed using the EC-Lab® software. The electrochemical active surface area was obtained by Cottrell equation (Eq. 1), where I is the current (A), n is the number of electrons transferred, F is Faraday's constant (96485 C/mol), A is the electrochemical active area (cm^2), C is the concentration of active species (mol/cm^3) and D is the diffusion coefficient (cm^2/s).

$$I = \frac{nFACD^{\frac{1}{2}}}{\pi^{\frac{1}{2}}t^{\frac{1}{2}}} \qquad \text{Eq. 1}$$

2.3. Cyclic voltammetry (CV)
The CV analysis that was carried out in this study had an identical setup as the CA analysis. The electrolyte was 0.5 M NaCl. The CV experiments and analyses were performed by a potentiostat instrument (Metrohm, Autolab PGSTAT101) controlled by the NOVA 1.10 software. In addition, the solution was stirred in between the experiments to restore initial conditions; however, it was not stirred during the experiment to maintain the flux to the electrode under diffusion-controlled behaviour. Triplicate runs were carried out for each electrode, and the polishing step was performed between each run. The third cycle of each CV was reported unless otherwise stated.

2.4. Potentio electrochemical impedance spectroscopy (PEIS)
The analysis was conducted using a setup and instruments that were similar to that used in the CA analysis. The solution was 5 mM potassium ferrocyanide in 0.5 M NaCl that acted as the supporting electrolyte. PEIS was performed at a potential bias of 0.26 V (vs. Ag/AgCl) over the frequency range of 1 MHz to 1 Hz with an AC perturbation of 10 mV.

3. Result and discussion

3.1. Electrochemical active surface area
Porous properties of the PSAC led to a higher active electrochemical surface area of an electrode when compared with its geometrical area. CA analysis produced a current-time curve as shown in Figure 1. The relationship between the current and the time can be described by the Cottrell equation with the specific known value of the diffusion coefficient, D, of ferrocyanide, i.e. $D = 7.6 \times 10^{-5}$ cm^2/s [24]. A linear plot of i vs. $t^{-0.5}$ was obtained, and the electrochemical active surface area of the electrode was estimated from the slope. The electrochemical active surface area was compared to the geometrical surface area, which was 0.196 cm^2 and listed in Table 1. The 10PTFE was found to have the highest amount of electrochemical active surface area among others with 1.572 cm^2, which was 8 times higher than its geometrical surface area. The electrochemical active surface area decreased by 42% when the amount of PTFE was increased to 20%. Such an outcome suggests that increasing the amount of PTFE decreases the available high surface area of porous PSAC for electrochemical activities. In addition, almost a similar value of the electrochemical active surface area was observed for 20PTFE, 30PTFE, and 40PTFE, which was between 0.92 and 0.96 cm^2. It was about more than 5 times greater than its geometrical surface area.

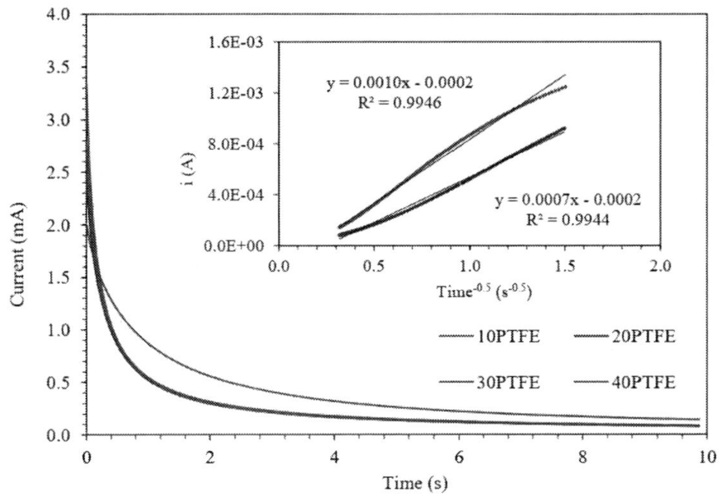

Figure 1. A single-step CA plot of the PSAC electrode using electrodes prepared using different amounts of PTFE at a voltage step from 0 to 0.4 V. Inset shows the plot of i vs. $t^{-0.5}$, which has been employed to deduce the electrochemical active surface area of electrodes.

Table 1. Electrochemical active surface area of prepared electrodes deduced from CA analysis.

Electrode	**10PTFE**	**20PTFE**	**30PTFE**	**40PTFE**
Electrochemical active surface area, cm^2	1.572	0.917	0.959	0.940
Electrochemical / Geometrical active surface area, cm^2/cm^2	8.0	4.7	4.9	4.8

3.2. Surface morphology

Figure 2 illustrates the FESEM-EDX elemental mapping for the electrodes with a different composition of PTFE. Accordingly, EDX detects the emission of four major elements on the surface of the PSAC electrode, namely carbon, oxygen, fluorine, and a small percentage of silica, as listed in Table 2. These compositions are with reference to the same mapping area as shown in Figure 2. The fluorine element represents the distribution of PTFE on the electrode surface. Generally, all the elements are heterogeneously scattered on the surface of the electrode. It is clearly seen that fluorine dots are sparse (blue colour) in the 10PTFE and are the least concentrated compared to the other electrodes. As the amount of PTFE increases to 40%, the intensity of fluorine dots also increases as the blue spots become denser. This is in agreement with the atomic percentage of fluorine, which increases proportionally as the amount of PTFE increases from 10% to 40%, as listed in Table 2. On the contrary, the percentage of carbon present on the surface of the electrode is inversely proportional to the atomic percentage of fluorine. For instance, in the 10PTFE electrode, the least amount of PTFE introduced in the electrode results in a larger amount of PSAC available on the same area of imaging. This trend is reversed for 40PTFE. The high intensity of fluorine dots in 40PTFE suggests an excess amount of PTFE on the surface which hinders the electrochemically active surface area of the PSAC. Additionally, Figure 3 illustrates a higher magnification of emission scanning on the surface of the PSAC-PTFE composite electrode. It is clearly seen that this composite electrode consists of a wide range of pore sizes, from micro (< 2 nm), meso (2-50 nm), to macro (> 50 nm) pores.

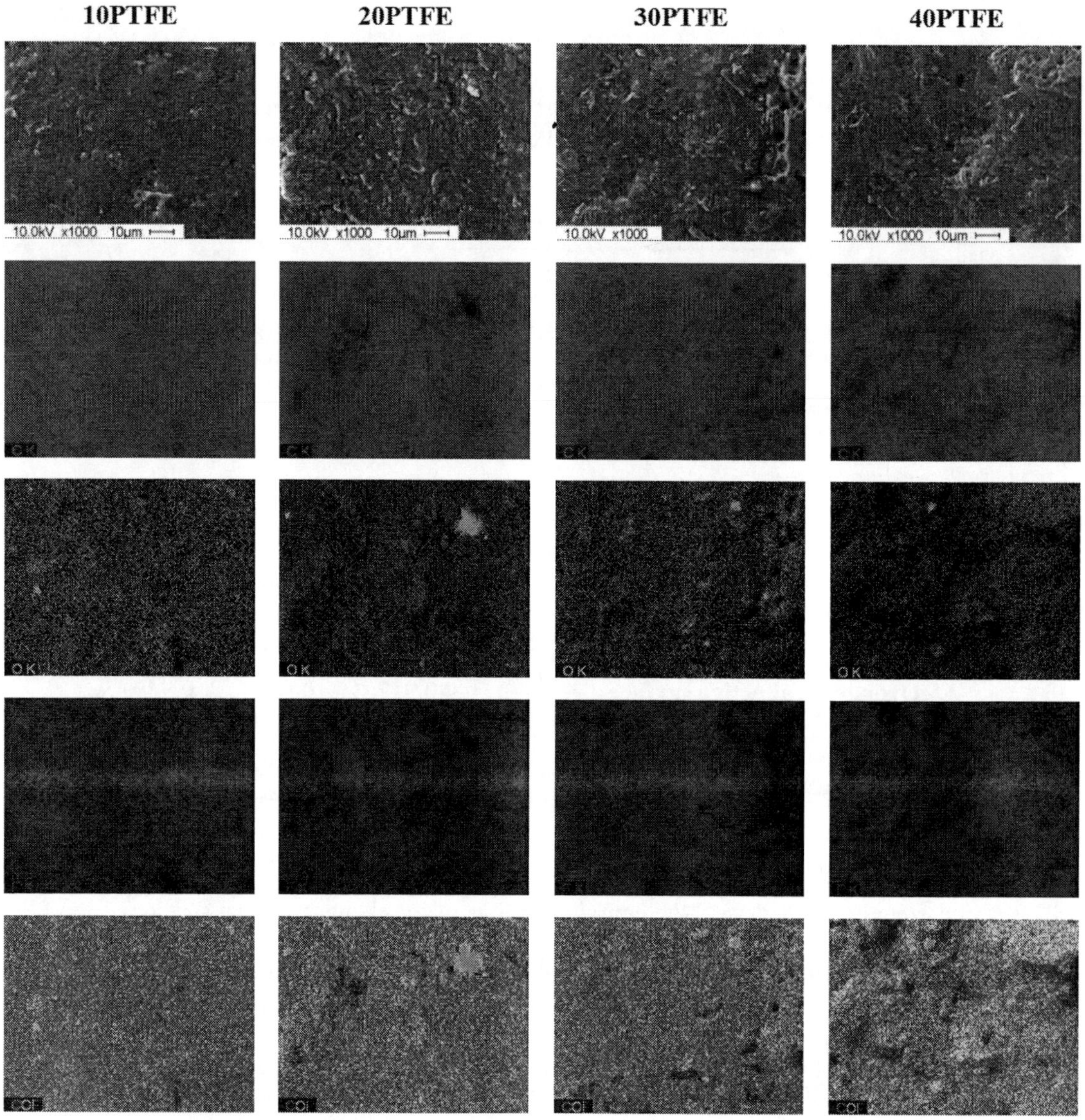

Figure 2. FESEM-EDX elemental mapping of prepared electrodes using different amount of PTFE.

Figure 3. FESEM imaging showing a wide range of pore size available in PSAC-PTFE composite electrode.

IOP Conf. Series: Materials Science and Engineering **210** (2017) 012011 doi:10.1088/1757-899X/210/1/012011

Table 2. Elemental composition of electrodes with different composition of PTFE.

	Atomic percentage, %			
Elemental	**10PTFE**	**20PTFE**	**30PTFE**	**40PTFE**
C	91.47	86.24	84.40	79.81
O	3.79	4.58	3.00	2.82
F	4.58	9.04	12.41	17.11
Si	0.15	0.14	0.18	0.26

3.3. Double layer capacitance, C_E

Figure 4 shows the cyclic voltammogram of the prepared electrodes. The error bar is deduced from the standard deviation of the triplicates run. In between each run, the electrode surface was subjected to a polishing step, as describe in Section 2.3. The average standard deviation for each electrode is listed in Table 3. 10PTFE shows the highest value of average standard deviation (± 91 μA). After every polishing step in between the CV test, a new surface area on 10PTFE has been introduced, but it leaves a noticeably large amount of carbon particles on the polishing pad. This suggests that 10% of PTFE is not sufficient to bind the PSAC and carbon black. The PSAC and carbon black easily come out from the electrode during typical polishing steps. A lower value of the average standard deviation is found for the electrode with a higher percentage of PTFE, suggesting that sufficient binder is applied in preparing the PSAC-PTFE composite electrode. An about similar finding was reported by Dong, et al. [14], where, when the ratio of PTFE to activated carbon was less than a certain amount, the paste cannot be firmly bound. For this reason, in their study, the ratio of activated carbon/PTFE was varied between 3 and 11. This ratio is equivalent to 9 to 33% of PTFE in each electrode. However, carbon black is not one of the electrode components.

Double layer capacitance for each electrode is estimated from the average value of current produced by the CV analysis. The effect of varying the amount of PTFE on the geometrical area normalised electrode double layer capacitance, C_E is presented by

Figure 5. C_E is calculated using Eq. 2, where E_1 and E_2 are the lower and upper limits of potential, $i(E)$ is the instantaneous current, $\int_{E_1}^{E_2} i(E)dE$ is the total voltammetric charge obtained by integration of forward and backward sweep in CV, (E_1-E_2) is the selected potential width (e.g. 0.2-0.3 V), A is the geometrical surface area of the electrode (0.196 cm^2), and v is the scan rate.

$$C_E = \frac{1}{2(E_2 - E\,)Av}\int_{E_1}^{E_2}(E)\,dE \qquad \text{Eq. 2}$$

The highest C_E is obtained by 10PTFE and it has the lowest amount of PTFE among the others. As for 20PTFE, the amount of C_E is found to be decreased by 42%. Further increase of PTFE from 30% to 40% does not show a steep reduction in capacitance where it remains in the range of 33 to 34 mF/cm^2. Thus, this proves that the reduction of C_E is mainly due to the reduction of the electrochemical active surface area. As the amount of PTFE is increased, it impairs the available surface area of PSAC, including its pores. However, all these C_E values are apparently overestimated as the electrode has a large surface area due to its porous characteristics. To take into account the effect of the electrochemical active surface area on the double layer capacitance, the electrochemical active area normalised electrode double layer capacitance, C_E* is deduced and presented in Figure 5. The relationship between the capacitance and the electroactive surface area is clearly seen as the CE* lies between a small range of CE* (6.7 to 8.0 mF/cm2) for all the electrodes which have different compositions of PTFE.

Although the 10PTFE electrode gives the highest active surface area and C_E, the requirement for sufficient strength of binding between the powdered PSAC and carbon black is not achieved. The new surface introduced by 10PTFE after the polishing step, with the apparently high average standard deviation is claimed due to an insufficient amount of binder. On the contrary, Park, et al. [20] reported that the best carbon sheet was prepared using 4 wt% of binder for the desalination purpose, according to the optimum obtained capacitance calculated from the CV analysis. If the optimum capacitance is to become the main criterion by ruling out the requirement of minimum mechanical strength, then 10% of PTFE seems to be the best amount for preparation of the PSAC-PTFE composite electrode. However, due to the physical characteristics of the PSAC and the amount of carbon black used, 20PTFE is better than 10PTFE in providing sufficient binding strength between the active material and conductive agent.

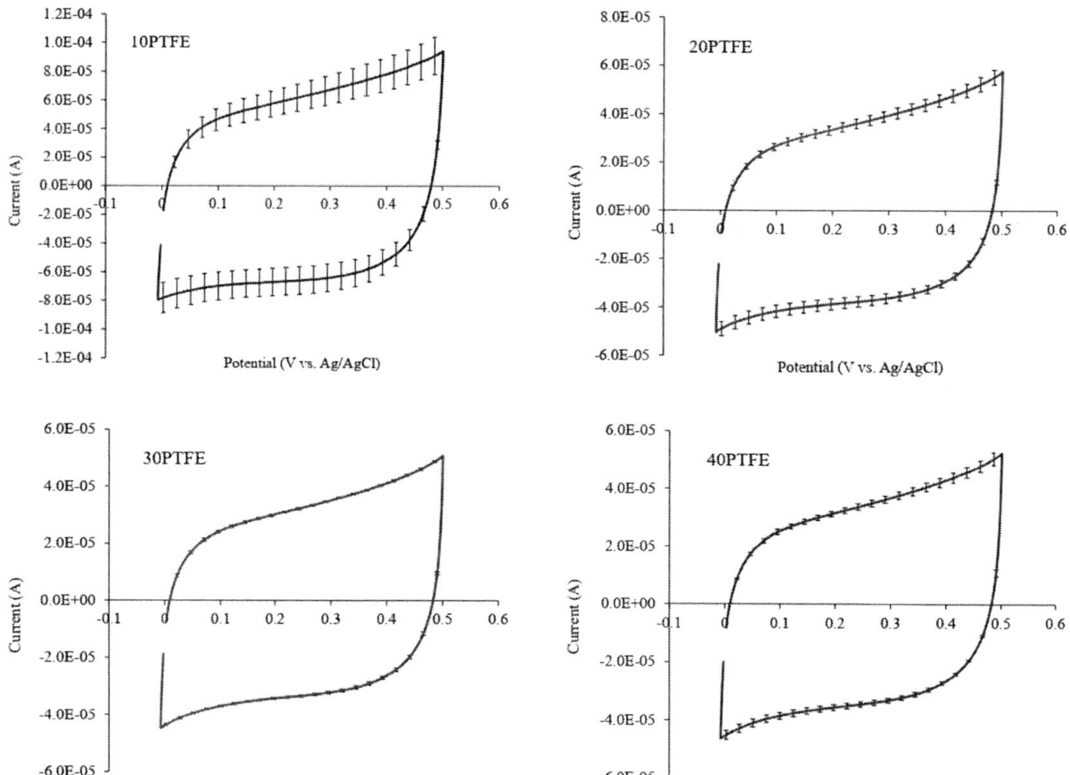

Figure 4. Cyclic voltammograms of 10PTFE, 20PTFE, 30PTFE and 40PTFE.

IOP Conf. Series: Materials Science and Engineering **210** (2017) 012011 doi:10.1088/1757-899X/210/1/012011

Table 3. Average standard deviation deduced from the triplicates run of CV by different composition of PTFE.

	10PTFE	20PTFE	30PTFE	40PTFE
Average standard deviation, μA	± 91	± 21	± 0.4	± 10

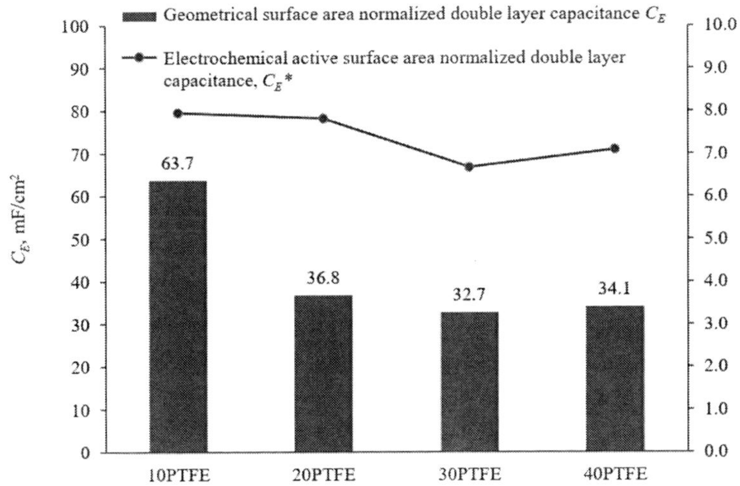

Figure 5: Estimated C_E and C_E^* by using PSAC electrodes with different amount of PTFE.

3.4. Electron transfer resistance

The impedance spectra for 20PTFE, 30PTFE and 40PTFE are compared in Figure 6. The PEIS data are analysed by fitting them into the Randles equivalent electrical circuit, as illustrated in the inset of Figure 6. All the three Nyquist plots similarly display a depressed semicircle over a high frequency range with the centre lying some distance below the x-axis. A depressed semicircle can be explained by the Constant Phase Element (CPE). CPE is the deviation from true capacitor behaviour and it can be attributed to the presence of double layer capacitance [25] and electrode surface roughness [26]. This is in agreement with the rough and non-uniform surface area of PSAC-PTFE electrode, as illustrated by Figure 3. Additionally, the diameter of the depressed loops represents the electron or charge transfer resistance, R_{ct} due to the Faradaic process. R_{ct} is increasing in the following trend, 40PTFE (3.7 $\Omega\cdot cm^2$) < 30PTFE (4.6 $\Omega\cdot cm^2$) < 20PTFE (5.2 $\Omega\cdot cm^2$). The highest electron transfer by 40PTFE can be described by the high conductivity of the prepared electrode due to the durable binding of PSAC and carbon black using a higher amount of PTFE. However, the value of R_{ct} is considered as comparable between the other two. In addition, a straight long line with an angle of 45° to the real axis for all the three electrodes represents semi-finite diffusion impedance, also called the Warburg impedance.

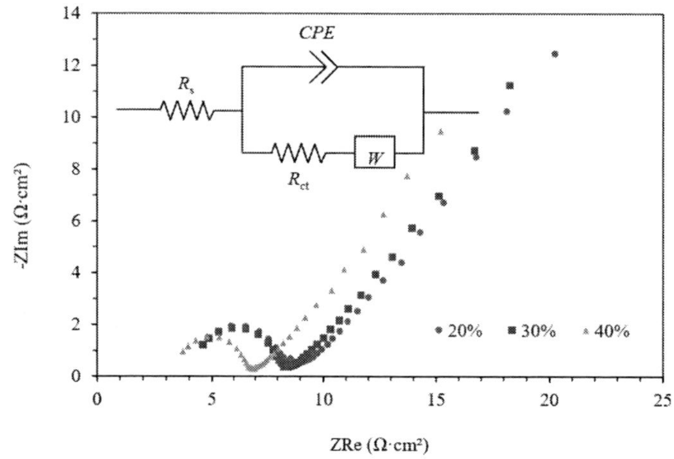

Figure 6: Nyquist plot of PEIS for 20PTFE, 30PTFE and 40PTFE. Inset is the equivalent Randles circuit.

4. Conclusion

This work demonstrates the effect of PTFE composition as a binder in the preparation of the PSAC-PTFE composite electrode. The 10PTFE electrode gives the highest electrochemical active surface area and C_E. However, it was observed that 10% PTFE was not sufficient to firmly bind the PSAC and carbon black for the purpose of preparing a durable free-standing electrode sheet for water or waste-water treatment. On the contrary, 20PTFE has sufficient binding strength between PSAC and carbon black. It produces an acceptable reproducibility in the voltammogram to deduce the double layer capacitance properties of a PSAC-based electrode. It is proven that increasing the amount of PTFE from 10% to 20% in the electrode preparation does impair the electrochemical properties in terms of the electrochemical active surface area and C_E. Further increase of PTFE up to 40% does not significantly help in improving the electrochemical properties of the PSAC-PTFE hybrid electrode. Meanwhile, a comparable electrochemical active surface area, C_E and R_{ct}, were observed between 20PTFE, 30PTFE, and 40PTFE.

Acknowledgement

The authors would like to acknowledge the financial support from the Ministry of Education (MOE), Malaysia for High Impact Research Grant (UM.C/HIR/MOHE/ENG/43), the University of Malaya for PPP grant (PG149-2016A) and the facilities provided at the Centre for Separation Science and Technology (CSST). H.I. Maarof gratefully acknowledges the Ministry of Higher Education (MOHE), Malaysia and Universiti Teknologi MARA, Malaysia for the postgraduate scholarship.

References

[1] L. Yan, Y.F. Wang, J. Li, H.D. Shen, C. Zhang and T.T. Qu 2016 *Bulletin of the Chemical Societyof Japan* **89** 50-7
[2] R.G. Saratale, K.J. Hwang, J.Y. Song, G.D. Saratale and D.S. Kim 2016 *Journal ofEnvironmental Engineering* **142**
[3] L. Xu, Y.K. Sun, L.C. Zhang, J.J. Zhang and F. Wang 2016 *Desalination and Water Treatment* **57** 8815-8825
[4] E.J. Bain, J.M. Calo, R. Spitz-Steinberg, J. Kirchner and J. Axen 2010 *Energy Fuel* **24** 3415-3421
[5] F. Duan, Y. Li, H. Cao, Y. Wang, J.C. Crittenden and Y. Zhang 2015 *Chemosphere* **125** 205-211
[6] C.H. Hou and C.Y. Huang 2013 *Desalination* **314** 124-129
[7] S. Nadakatti, M. Tendulkar and M. Kadam 2011 *Desalination* **268** 182-188

[8] G. Issabayeva, M.K. Aroua and N.M. Sulaiman 2006 *Desalination* **194** 192-201
[9] C. Punckt, M.A. Pope and I.A. Aksay 2013 *The Journal of Physical Chemistry C* **117** 16076-16086
[10] A.M. Dehkhoda, E. Gyenge and N. Ellis 2016 *Biomass and Bioenergy* **87** 107-121
[11] M.H. Chakrabarti, M. Saleem, M.F. Irfan, S. Raza, D.B. Hasan and W. Daud 2011 *International Journal of Electrochemical Science* **6** 4470-4480
[12] F. Offei, A. Thygesen, M. Mensah, K. Tabbicca, D. Fernando, I. Petrushina and G. Daniel 2016 *Energies* **9** 1-14
[13] S. Cheng and J. Wu 2013 *Bioelectrochemistry* **92** 22-26
[14] H. Dong, H.B. Yu, X. Wang, Q.X. Zhou and J.L. Feng 2012 *Water Research* **46** 5777-5787
[15] H. Dong, H.B. Yu and X. Wang 2012 *Environmental Science and Technology* **46** 13009-13015
[16] X.Y. Zhang, X. Xia, I. Ivanov, X. Huang and B.E. Logan 2014 *Environmental Science and Technology* **48** 2075-2081
[17] A. Gambou-Bosca and D. Belanger 2015 *Journal of The Electrochemical Society* **162** A5115-5123
[18] Z.T. Zhu, S.H. Tang, J.W. Yuan, X.L. Qin, Y.X. Deng, R.J. Qu and G.M. Haarberg 2016 *International Journal of Electrochemical Science* **11** 8270-8279
[19] M. Zarei, A. Niaei, D. Salari and A.R. Khataee 2010 *Journal of Electroanalytical Chemistry* **639** 167-174
[20] K.K. Park, J.B. Lee, P.Y. Park, S.W. Yoon, J.S. Moon, H.M. Eum and C.W. Lee 2007 *Desalination* **206** 86-91
[21] T. Shimizu, Y. Higashihata and T. Nakamura (1998). United States Patent No. 5707763.
[22] I.I. Misnon, N.K.M. Zain, R.A. Aziz, B. Vidyadharan and R. Jose 2015 *Electrochimica Acta* **174** 78-86
[23] Q. Abbas, D. Pajak, E. Frąckowiak and F. Béguin 2014 *Electrochimica Acta* **140** 132-8
[24] J.A. Bard and L.R. Faulkner 2001 *Electrochemical Methods: Fundamentals and Applications* 2nd Ed. (New York: John Wiley & Sons, Inc.) pp 813
[25] V.F. Lvovich 2012 Phenomenon *Impedance Spectroscopy* ed V.F. Lvovich (New Jersey: John Wiley & Sons, Inc.) pp 37-7
[26] R. De Levie 1965 *Electrochimica Acta* **10** 113-30

IOP Conf. Series: Materials Science and Engineering **210** (2017) 012012 doi:10.1088/1757-899X/210/1/012012

Production of Fatty Acid Methyl Ester from Low Grade Palm Oil Using Eutectic Solvent Based on Benzyltrimethylammonium Chloride

Shahidah N. Rashid[1,2] , Adeeb Hayyan[1,2*], Mohd Ali Hashim[1]

[1]Department of Chemical Engineering, University of Malaya Centre for Ionic Liquids (UMCiL), University of Malaya, Kuala Lumpur 50603, Malaysia
[2]Institute of Halal Research University of Malaya (IHRUM), Academy of Islamic Studies, University of Malaya, 50603, Kuala Lumpur, Malaysia

E-mail: adeeb.hayyan@yahoo.com

Abstract. In this study, high content of free fatty acid (FFA) from low grade palm oil (LGPO) was treated using deep eutectic solvent (DES). This DES was prepared via mixing of *p*-toluenesulfonic acid monohydrate (PTSA) with benzyltrimethylammonium chloride (BAC). Various effect of reaction conditions were studied such as catalyst dosage (0.5% to 4.5%), molar ratio (2:1 to 12:1), reaction time (10 to 120 min) and reaction temperature (30 °C to 80 °C). This study exhibited that 2% catalyst dosage, 10:1 molar ratio of methanol to oil, 30 min of reaction time and reaction temperature of 60 °C were the optimum condition to reduce the FFA content to less than 2%.

1. Introduction

Numerous studies were motivated by a demand for renewable source of energy and fluctuating prices of petrol diesel in the current market [1,2]. A mixture of acidic crude palm oil (ACPO) and sludge palm oil (SPO) showed excellent result in producing biodiesel [3]. The low grade oil mixture was the waste oil rejected from palm oil refineries due to its high free fatty acid value of over 5% [3]. It is a perfect candidate for biodiesel production due to its low cost and it is available domestically.

Deep eutectic solvent (DES) can be used for the treatment of free fatty acid (FFA) and for biodiesel production [3]. DES is a eutectic mixture which can easily synthesize from combination of either salt, salt, hydrogen bond donor (HBD): HBD, and salt: HBD [4,5]. An ammonium based deep eutectic solvent; N,N-diethylenethanol ammonium chloride with *p*-toluenesulfonic acid monohydrate (PTSA) was applied as a novel recyclable catalyst [3]. PTSA was mixed with allyltriphenylphosphonium bromide as a phosphonium-based deep eutectic solvent in previous study and also shown a great impact on recyclability study where it can be used up to four runs [6].

Another type of salt benzyltrimethylammonium chloride (BAC) was proposed in this study to prepapre novel recyclable DES. This research work presents the catalytic activities of BAC-DES catalyst; BAC mixed with PTSA in esterification of low grade palm oil (LGPO). The optimum condition of the esterification reaction is proposed. The parameters are to be the parameters studied are dosage of catalyst, reaction temperature and time and molar ratio. A multi-unit reactor with reflux condenser using feedback controlled system was used to conduct the experiment.

2. Experimental

2.1. Materials and Method
Low grade palm oil (LGPO) with FFA content of 9.2% was collected from a local palm oil mill, Malaysia. Methanol, potassium hydroxide, benzyltrimethylammonium chloride, *p*-toluenesulfonic acid monohydrate were purchased from Sigma Aldrich.

2.2. Synthesized of DES based catalyst
The BAC-DES catalyst was prepared using method of Hayyan et al. [3]. The molar ratio of BAC to PTSA was 1:3. The required amount of BAC powder was mixed with PTSA gradually until it becomes a clear homogenous solution. The mixing condition is at 60 °C and the magnetic stirrer was set at 350 rpm using hot plate and magnetic stirrer. The mixture was sealed and placed in desiccator with silica gel to prevent absorption of moisture.

2.3. Synthesized of FAME from LGPO
LGPO was pre-heated at 80 °C before esterification. The required amount of oil sample was placed in multi-unit reactor with reflux condenser and speed controller using feedback controlled system. The specific dosage of BAC-DES catalyst was mixed with methanol until homogeneous prior addition to multi-unit reactor. Esterification was carried out with different catalyst dosage (0.5% to 4.5%, based on oil weight), molar ratio of methanol to oil (2:1 to 12:1), reaction time (10 to120 min) and various reaction temperature (30 °C to 80 °C). The standard batch was run with 30 g of oil, 12 g of methanol and 0.3 g of catalyst placed in the jacketed vessel. The reaction conditions were at 60 °C, 30 min, and under stirring at 350 rpm. After esterification, the solvent and catalyst was separated using centrifugation. After separation, treated oil was further heated at 75°C in oven to remove excess methanol. The reusability of BAC based DES catalyst was investigated using selected optimum condition and was run without addition of fresh catalyst. The FFA content was evaluated using titration according to AOCS official method Ca 5a-4 [14]. The transesterification of treated oil was fixed as follows: 1 % of KOH solubilize in methanol (10:1 molar ratio of methanol to oil) at 60 °C, 30 min, and 350 rpm stirring speed [10].

3. Results and discussion

3.1. Effect of BAC-DES catalyst and molar ratio
This is the first time for BAC based DES to be used for esterification of LGPO to produce biodiesel. BAC is an off-white quaternary ammonium compound with a structural resemblance to acetylcholine [7]. As mentioned above, the targeted FFA after esterification in this study was to be at $\leq$ 2%. From Figure 1, the reduction of FFA was proportional with increasing dosage of catalyst. The result shows that BAC based DES possessed high catalytic activity. Dosage of catalyst at value 1.5% is sufficient to reduce the FFA content from 9.2% to 1.8% and gives 88.5% FAME conversion. Though it used higher amount compared to other studies, 2% is still considered acceptable dosage to be applied in esterification at industrial scale. Thus, 2% of BAC based DES was selected as the optimum dosage in esterification of LGPO. From figure 1, the conversion of FFA to FAME was found to increase with increasing of dosage used.

Molar ratio was varied from 2:1 to 12: of methanol to oil. From Figure 2, 6:1 molar ratio was sufficient to reduce the FFA content to $\leq$ 2%. There was no enhancement in the reduction of FFA content after molar ratio 10:1. Thus, 10:1 was selected as the optimum molar ratio to esterify LGCPO with the reduction of FFA content from 9.2% to 1.06%. This study slightly advantage compared to other study done by Corro et al where 12:1 molar ratio of methanol was used to esterify Jatropha curcas crude oil with the aid of heterogeneous catalyst [8].

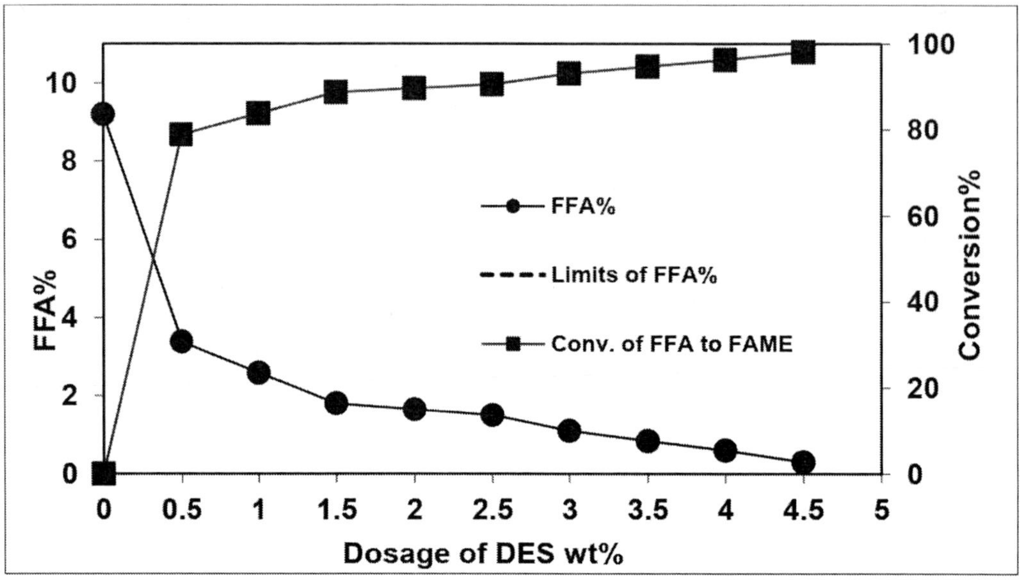

Figure 1. Effect of BAC based DES dosage on the yield of treated LGCPO and the correspondence reduction of FFA content at 30 min reaction time, 10:1 molar ratio, 60 °C, and 300 rpm.

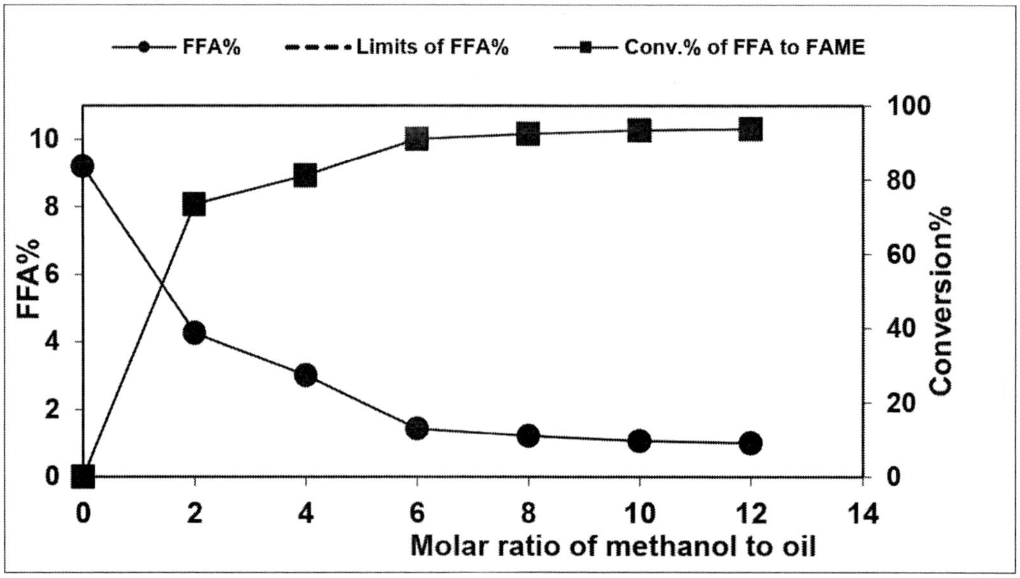

Figure 2. Effect of molar ratio on the yield of treated LGPO and the correspondence reduction of FFA content at 30 min reaction time, 60 °C, 1.5% catalyst dosage and 300 rpm

3.2. Effect of reaction time and reaction temperature

As shown in Figure 3, the reduction of FFA to targeted limit was achieved even only with 10 min reaction time. It revealed that BAC based DES possessed strong catalytic activity and was able to react with reactant in short period during esterification. Although prolonged the reaction time could lower the FFA content from 9.2% to 0.39%. A period of 30 min was deemed as the optimum reaction time since it reduced the FFA content to 1.61%. A short period of time is preferred to save operating cost of biodiesel production [9,10].

From Figure 4, reaction temperature of 70 °C showed the highest reduction of the FFA content from 9.2% to 1%. However, 60 °C was chosen as the optimum reaction temperature to reduce energy consumption and for economic feasibility. At that temperature, it reduced FFA up to 1.3% which lower than industrial limit [6]. The reaction temperature selected for this study is below the boiling point of the methanol, 64.7°C. As can be seen from Figure 4, the FFA reduction was increased at 80°C and reduced the conversion of FAME to 86.25%. Too high reaction temperature may cause the reaction yield to decrease. The amount of loss is believed to be caused by the vaporization of methanol during reaction [11].

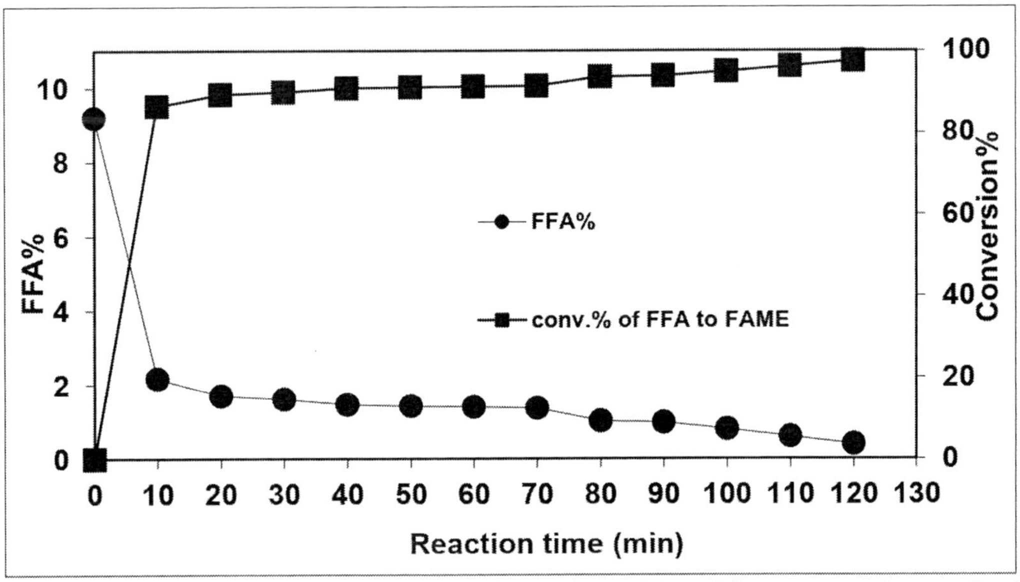

Figure 3. Effect of reaction time on the yield of treated LGCPO and the correspondence reduction of FFA content at10:1 molar ratio, 60 °C, 1.5% catalyst dosage and 300 rpm

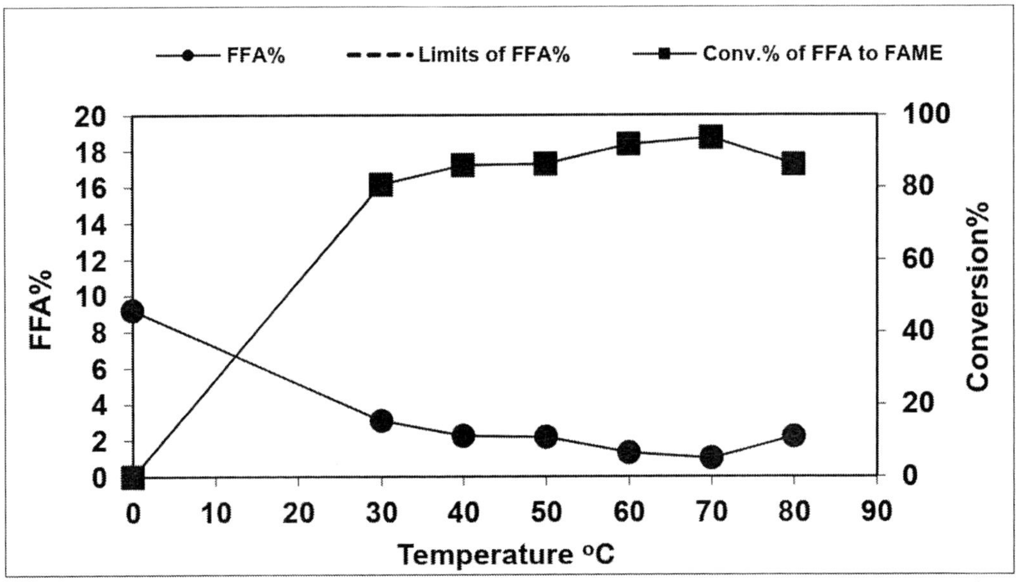

Figure 4. Effect of reaction temperature on the yield of treated LGCPO and the correspondence reduction of FFA content at10:1 molar ratio, 30 min reaction time, 1.5% catalyst dosage and 300 rpm

3.3. *Validity of optimized condition and recyclability*

In the present study, 2% of catalyst dosage, 10:1 molar ratio of methanol to oil, 30 min of reaction time and reaction temperature of 60 °C was selected as the optimum operating condition for esterification of LGPO, using BAC-based DES. The optimum condition was able to reduce the FFA content from 9.2% to 1.23%. The 6 consecutive runs were conducted without adding fresh catalyst. According to Figure 5, the percent of reduction of FFA decreased to 50% in the 2nd run, and constantly showed 50% reduction of FFA.

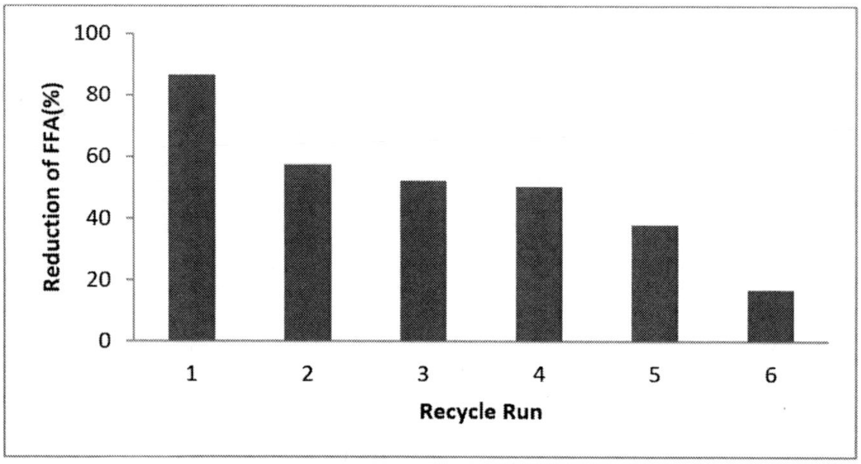

Figure 5. Recyclability study using BAC-DES catalyst in optimum reaction condition

4. Conclusions

In this study, BAC-based DES was used for the treatment of FFA content in LGPO. It proved to have high catalytic activity as of homogeneous acid catalyst. It has lower cost and easy to prepare in comparison to ionic liquids. This study exhibited that 2% catalyst dosage, 10:1 molar ratio of methanol to oil, 30 min of reaction time and reaction temperature of 60°C were the optimum condition as it gives high FAME conversion.

References

[1] S. Jayakumar, M M. Yusoff, M.H.A. Rahim, G.P. Maniam, and N. Govindan 2017 *Renewable and Sustainable Energy Reviews* **72** 33–47
[2] P.S. Bisen, B.S. Sanodiya, G.S. Thakur, R.K. Baghel, and G.B.K.S. Prasad 2010 *Biotechnology Letters* **32** 1019–1030
[3] A. Hayyan, M.A. Hashim, M. Hayyan, F.S. Mjalli, and I.M. AlNashef 2013 *Industrial Crops and Products* **46** 392–398
[4] M.K. AlOmar, M. Hayyan, M.A. Al-Saadi, S. Akib, A. Hayyan, and M.A. Hashim 2016 *Journal of Molecular Liqiuds* **215** 98–103.
[5] B.S. Singh, H.R. Lobo, D.V. Pinjari, K.J. Jarag, A.B. Pandit, and G.S. Shankarling 2013 *Ultrasonics Sonochemistry* **20** 287–293
[6] A. Hayyan, M.A. Hashim, F.S. Mjalli, M. Hayyan, and I.M. AlNashef 2013 *Chemical Engineering Science* **92** 81–88
[7] B.I. Ghanayem 2000 *NTP technical report on the toxicity studies of methacrylonitrile (CAS No. 126-98-7). Administered by gavage to F344/N rats and B6C3F1 mice* NH Publication Research Triangle Park, NC.
[8] G. Corro, F. Bañuelos, E. Vidal, and S. Cebada 2014 *Fuel* **115** 625–628
[9] N. Lamba, J.M. Modak, and G. Madras 2017 *Energy Conversion and Management* **138** 77–83
[10] A. Hayyan, M.Z. Alam, M.E.S. Mirghani, N.A. Kabbashi, N.I.N.M. Hakimi, Y.M. Siran, and S. Tahiruddin 2010 *Journal of Energy and Power Engineering* **4** 11–17

[11] H.J. Cho, S.H. Kim, S.W. Hong, and Y.K. Yeo 2012 *Fuel* **93** 373–380

IOP Conf. Series: Materials Science and Engineering **210** (2017) 012013 doi:10.1088/1757-899X/210/1/012013

A low sludge generated anode by hybrid solar electrocoagulation for the removal of lead

F. Hussin and M.K. Aroua*

Department of Chemical Engineering, Faculty of Engineering, University of Malaya, 50603 Kuala Lumpur, Malaysia

E-mail: *mk_aroua@um.edu.my

Abstract In this work, perforated zinc is proposed as a new anode for lead removal by hybrid solar electrocoagulation. The characteristics of the sludge were investigated to understand the behaviour of lead removal during electrocoagulation. Sludge products formed were characterised using X-ray diffraction (XRD), X-ray fluorescence (XRF) and Field Emission Scanning Electron Microscopy (FESEM). In addition, the pH variation during electrocoagulation and effects on the sludge products were examined. At optimum conditions showed that the perforated zinc electrode produced better performance with high removal efficiency, low sludge volume index and less energy consumption.

1. Introduction

Water is vital in human life. It is used in various applications including in agricultural, industrial, shipping, household, recreational, and environmental activities. Hence inadequate infrastructure and management systems for industrial and domestic wastewater have contributed to escalating wastewater crisis, which adversely affects human health [1]. Globally, the total of sewage, agricultural waste, and industrial effluent discharged into the water are about two million tonnes every day [2]. Water polluted by heavy metal ions is extremely harmful to human beings, and it has been a major concern to scientists and engineers around the world.

Lead (Pb(II)) has been extensively documented as one of the most toxic heavy metals. In 2015, it is ranked second in the list of prioritized hazardous substances by The Agency for Toxic Substances and Disease Registry [3]. High concentration of lead in wastewater would harm living organisms and ecosystems. It can damage brain, liver, kidney, basic cellular processes, and reproductive system [4]. Despite these adversities, lead is still a significant contaminant produced in numerous industries such as in the production of batteries, paintings, pigments, fertilizers, cables, alloys, steel, plastics and glass [5, 6]. Based on statistics published by the Statistic Portal, the world's lead consumption has increased significantly since 2004 [7]. Therefore, it is imperative to reduce lead concentration to an acceptable value before being discharged to the environment. The maximum acceptable limit of lead in drinking water has been outlined by several organisations. The European countries for instance, outlined the limit at 0.01mgL^{-1}. The United States Environmental Protection Agency limited it to 0.015 mgL^{-1}, and the World Health Organization limited it to 0.01 mgL^{-1} [8, 9]. The Malaysian Department of Environment has set the allowed level of lead for sewage and industrial effluents for upstream to downstream discharge at 0.1mgL^{-1} and 0.5mgL^{-1} respectively.

Numerous methods have been introduced to remove lead from water and wastewater such as chemical precipitation, adsorption, ion exchange, reverse osmosis, electrochemical, membrane separations, and solvent extraction. Each method has its own advantages and disadvantages [10, 11]. Compared to other alternatives, the electrochemical processes offer several promising approaches for preventing and treating of water polluted by heavy metals. The technique is commended for its versatility, energy efficiency, amenability to automation, environmental compatibility, and cost effectiveness [12, 13]. In an electrochemical process, anode electrodes are dissolved through an electrolysis process to form a range of coagulant species and metal hydroxides [14]. These coagulant species and metal hydroxides are capable of destabilising and adsorbing the dissolved contaminants present in the water or wastewater to be treated [15]. During the process, anodic dissolution of the metal electrodes involves the evolution of hydrogen gas at the cathodes [16]. The hydrogen bubbles then captures and brings afloat any suspended solids formed, before removing the contaminants [17]. There were a few studies conducted to treat the wastewater and synthetic solution containing lead ions involving electrochemical treatment [4, 10, 11, 18-21]. Based on their studies, the most commonly used electrode materials in the electrochemical process were aluminum (Al) and iron (Fe) plate.

Therefore, the main objectives of this study are: (1) to investigate the effect of operating conditions such as interaction between initial-final pH, distance between electrode, sludge volume and energy consumption and (3) to examine the sludge surfaces morphological features by employing the XRD, FESEM surface image and EDAX analysis methods at different pH solution.

2. Materials and methods

2.1. Chemicals

In testing the ion removal for heavy metal, lead nitrate ($Pb(NO_3)_2$) was purchased from Sigma-Aldrich Chemical Company. Sulfuric acid (H_2SO_4), hydrochloric acid (HCl), sodium hydroxide (NaOH), and sodium chloride (NaCl) were bought from Merck. The pH levels of the solutions were adjusted by adding the suitable amounts of 0.01 M H_2SO_4 or 0.01 M NaOH.

2.2. Electrochemical system set-up

The experiments were conducted by using a 1000 mL borosilicate glass beaker placed on a magnetic stirrer. The electrochemical system was set up with a monopolar configuration. Zinc perforated electrode (dimensions: 7 cm × 5 cm) with hole diameter of 0.5 cm was used as the anode and cathode pairs electrode. The inter-electrode distance was kept constant at 1.0 cm in every experiment conducted. The solar photovoltaic (PV) module (KINVE, KV250-60P) was purchased from BOS Asia Company (Malaysia). The PV panels were fixed to the roof of the Chemical Engineering Department, University of Malaya, Malaysia, located at (3° 7' 1") north latitude and (101° 39' 12") west longitude. The current and voltage of the DC-DC charge controller were within 0–10A and 0–15V ranges, respectively. The pH of the solution was measured by using a pH meter (HANNA instruments). The solution was agitated with a magnetic stirrer at 250 rpm during the process. In addition, 0.01 M of sodium chloride (NaCl) was used as the supporting electrolytes. To determine the system performance, 5-mL samples were taken from the reactor at time intervals of 5 minutes. The samples were later filtered at 0.45 μm to eliminate the sludge formed during the analysis.

2.3. Analytical methods

The residuals of heavy metal concentrations were analysed by Inductively Couple Plasma-Optical Emission Spectrometry (ICP-OES) (Perkin Elmer, Optima 7000 DV). Stock standard solutions (1000 mg/L) for Pb (II) ions were prepared by dissolving lead nitrate $Pb(NO_3)_2$ in distilled water. All electrolysis experiments were performed at room temperature and the analyses were carried out in triplicate Removal efficiency, energy consumption and sludge volume was determined using the following equations [22, 23, 24]:

$$\text{Pb(II) removal (\%)} = \frac{Co - Ct}{Co} \times 100 \quad (1)$$

Where Co is the initial Pb(II) concentration (mg/L) and Ct is the Pb(II) concentration after treatment (mg/L).

$$E\,(kWh/m^3) = \frac{U \times I \times t}{V} \quad (2)$$

Where E is the energy consumption (kWh/m^3), U is the voltage (V), I is the current (A), t is the treatment time (h) and V is the volume of aqueous solution (m^3).

The settleability of the precipitates is indicated by the sludge volume index (SVI), which refers to the volume in millilitres occupied by 1 gram of suspension after 30 min.

$$SVI\,(mL/g) = \frac{H_{30}}{H_0 SS} 1000 \quad (3)$$

Where SVI is the sludge volume index (mL/g), SS is the initial sludge concentration after treatment (g/L), H_0 is the initial height of the sludge in the settling column (cm) and H_{30} is the height of the sludge once it has settled after 30 min (cm).

3. Results and discussion

3.1. Effect of pH and treatment time

The initial pH is one of the most sensitive operating parameters influencing the performance of the electrocoagulation process. In order to determine the optimum pH value, a series of batch experiments in the pH range 4.0–9.0 were carried out at a constant initial Pb(II) concentration of 10mg/L, current density of $1.13 mA/cm^2$ and stirring speed of 250 rpm. The effect of different initial pH level on the Pb(II) removal efficiency is shown in Figure 1. Our findings clearly demonstrate that an increase in the pH from 4.0 to 8.0 leads to an increase in the efficiency of Pb(II) removal from 50% to 99% after 20 minutes of treatment. However, the efficiency of Pb(II) removal substantially decreased at pH > 8.0 after 10 minutes of the process. This decrease in removal efficiency at strongly acidic and strongly alkaline pH was due to the amphoteric (at acidic pH) behaviour that leads to soluble cations and monomeric anions (at alkaline pH), in addition, at high pH more sludge was produced this will affect on the performance of Pb(II) removal [19]. The importance of the pH value in the removal of heavy metals from aqueous solutions has been confirmed by Escobar et al. [10], who asserted that the removal efficiency of Pb(II) increased considerably when pH increased. They claimed that, for low pH solutions (i.e. strongly acidic), the oxidation of metal ions diminishes, and therefore the metal removal decreases. At high pH values (i.e.

weakly alkaline), metal removal increased due to the precipitation that occurred in the solution. Figure 2 shows the change of pH after electrocoagulation process. It can be seen that, the final pH for all experiments were higher than initial pH, but for pH 9 the final pH was lower than initial (Figure 2). The results also showed that, sludge volume and energy consumption increased with increasing the pH values. This is because of the enhancement in pH value will cause more sludge was produced and high energy consumption. Selection of optimum pH is important to reduce high operating cost. In our study, the highest Pb(II) removal efficiency was obtained at pH 8, but pH 7 was selected for subsequent experiments due to low energy consumption and sludge volume index (Table 1).

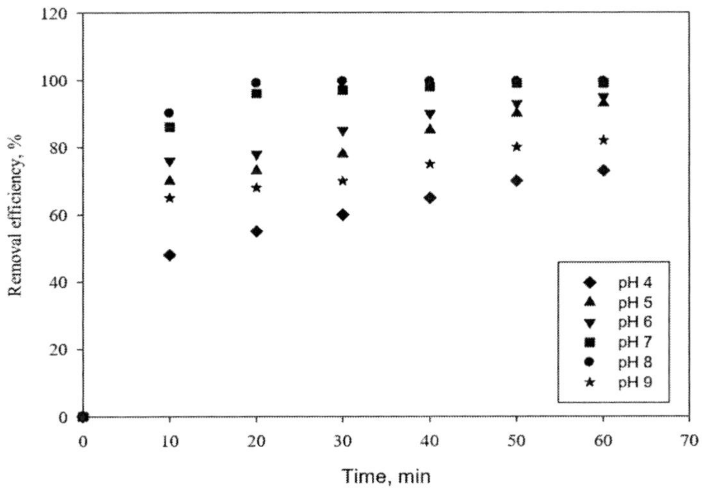

Figure 1. Effect of pH on the removal of lead by electrocoagulation (initial lead concentration = 10 mg/L, current density = 1.13 mA/cm^2, stirring speed = 250 rpm).

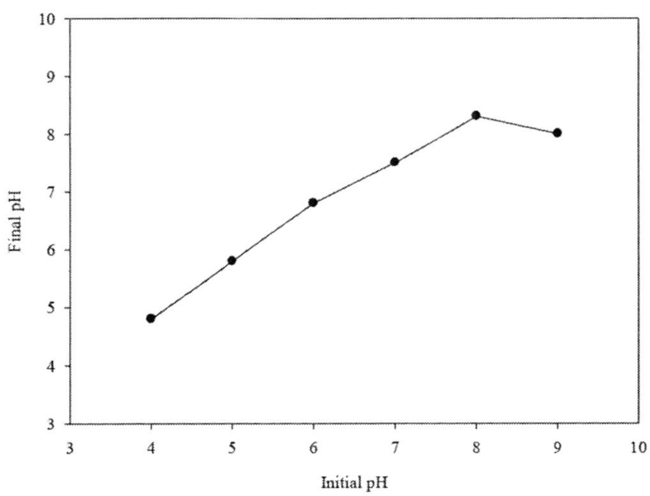

Figure 2. Effect of pH variation after electrocogulation process (initial lead concentration = 10 mg/L, current density = 1.13 mA/cm^2, stirring speed = 250 rpm).

Table 1. A Comparison of the final pH, sludge production and energy consumption.

Initial pH	Removal efficiency, %	Sludge volume, mL/g	Energy consumption, (kWh/m^3)
3	60	28.5	0.192
4	73	30.8	0.202
5	93	35.2	0.216
6	95	58.5	0.230
7	99	62.5	0.249
8	99.5	68.5	0.268
9	82	71.5	0.285

3.2. Characterisation of sludge

Figure 3 presents the XRD analysis of sludge after electrocoagulation treatment. It can be seen that a high peak at 25° to 28° is indicates the presence of zinc oxide. Other compound that which is appears in the image is zinc hydroxide carbonate which show at peak 25° to 28°. Figures 4 and 5 show the EDX and FESEM image, which confirms that, the presence of ultrafine zinc, lead, oxygen and carbon on the electrode surface. This indicates that lead ion has been adsorbed onto zinc oxide.

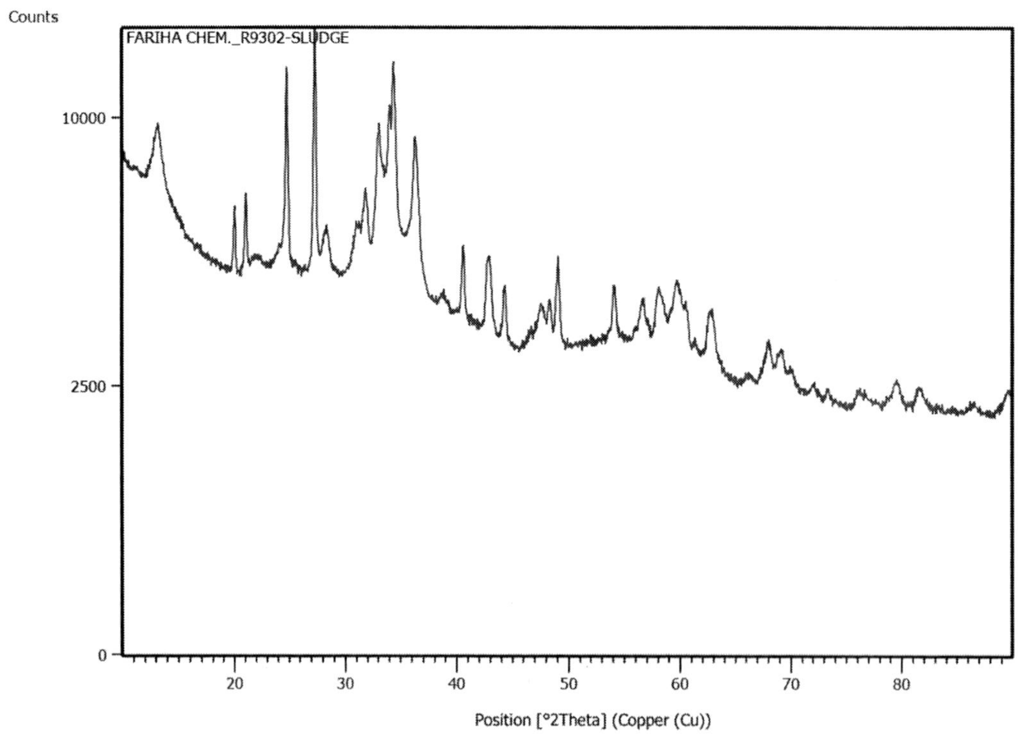

Figure 3. XRD image of sludge produced after treatment.

IOP Conf. Series: Materials Science and Engineering **210** (2017) 012013 doi:10.1088/1757-899X/210/1/012013

Figure 4. EDX image of components in the sludge after electrocoagulation process

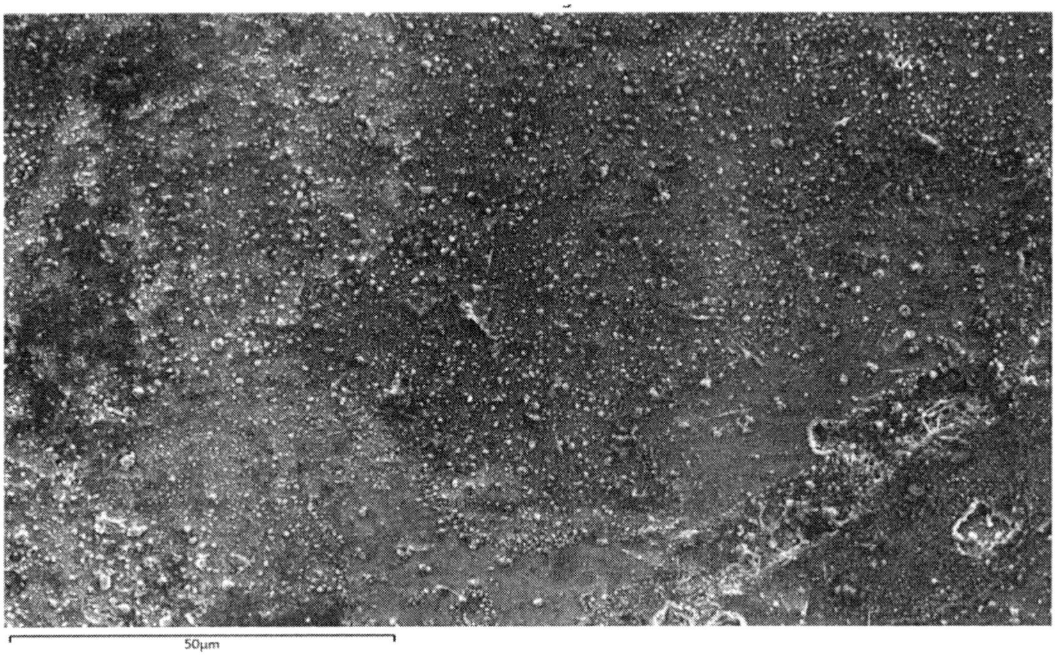

Figure 5. FESEM image of components in the sludge after electrocoagulation process

IOP Conf. Series: Materials Science and Engineering **210** (2017) 012013 doi:10.1088/1757-899X/210/1/012013

3.3. Effect of electrode distance

To study the effect of the electrode distance on the Pb(II) removal efficiency, several experiments were carried out for varying electrode distances between 0.5, 1.0, 1.5 and 2 cm. When the electrode distance increased the ohmic loss in relation to the anode and cathode over voltages and the resistance to mass transfer became larger, besides the kinetics of both charge transfer was slowed down [22]. The results also showed that, increase in the electrode distance will decrease removal efficiency. This is due to the rate of particle aggregation and adsorption of Pb(II) became lower. Figure 6 clearly show that the removal efficiency of Pb (II) increased at 0.5-1.5 cm electrode distance and then decreased with the increase of inter-electrode distance. The optimum Pb(II) removal percentage as well as treatment cost was achieved for the electrode distance value of 1.0 cm. Therefore, it was beneficial to choose an optimum small electrode distance of 1.0 cm to minimise energy consumption and increase the Pb(II) removal efficiency.

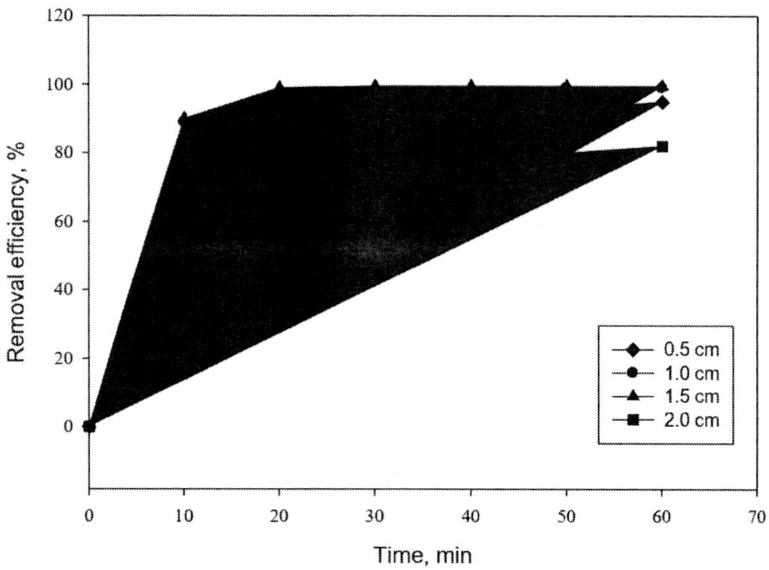

Figure 6. Effect of distance between electrode on the removal of lead by electrocoagulation (initial lead concentration = 10 mg/L, current density = 1.13 mA/cm^2, stirring speed = 250 rpm).

4. Conclusion

Presence of lead in produced water is one of the reasons that hinder its reusability for other purposes such as irrigation and drinking. The electrocoagulation technique employed in this study is able to reduce the Pb (II) concentration to below the WHO permissible level. Low SVI value was observed by using perforated zinc electrode. Large sludge volume caused longer landfill area usage. Therefore, this electrode was effective in removing Pb (II) with producing low sludge volume and efficient energy consumption. Optimal operating conditions of pH 7.0, current density of 1.13mA/cm^2, electrode distance of 1.0 cm, initial concentration of 10mg/L and 20 minutes treatment time resulted in 99.1% of Pb (II) removal. The sludge produced containing only zinc oxide, zinc hydroxide and lead oxide. This is indicates that the sludge was not toxic and safe to the environment.

Acknowledgments
We are grateful for financial support provided by the University of Malaya Scholarship Scheme which made this research possible. The authors also appreciate the financial support provided by High Impact Research Grant UM.C/HIR/MOHE/ENG/43

References

[1] Etorki A, El-Rais M, Mahabbis Mand Mohamed M 2014 *American Journal of Analytical Chemistry* **5** 225-234
[2] UNDESA 2015 Retrived from http://www.un.org/waterforlifedecade/quality.shtml
[3] ATSDR 2014 Retrived from http://www.atsdr.cdc.gov/
[4] Vasudevan S, Lakshmi J and Sozhan G 2012 *Environmental of Science and Pollution Research* **19** 2734-2744
[5] Patterson J 1985 *Industrial wastewater treatment technology* (Butterworth Publishers: United States)
[6] Günay A, Arslankaya E and Tosun İ 2007 *Journal of Hazardous Materials* **146** 362-371
[7] Statista 2013 Retrived from http://www.statista.com/statistics/264877/world-consumption-of-lead-metal/
[8]Balaria A and Schiewer S 2008 *Separation and Purification Technology* **63** 577-581
[9] Bhattacharjee S, Chakrabarty S, Maity S, Kar S, Thakur P, Bhattacharyya G 2003 *Water Research* **37** 3954-3966
[10] Escobar C, Soto-Salazar C and Inés T 2006 *Journal of Environmental Management* **81** 384-391
[11] Liu Y, Yan J, Yuan D, Li Q and Wu X 2013 *Chemical Engineering Journal* **218** 81-88
[12] Rajeshwar K, Ibanez J and Swain G 1994 *Journal of Appled Electrochemistry* **24** 1077-1091
[13] Qin X, Yang B, Gao F and Chen G, 2013 *Journal of Environmental Engineering* **139** 1004-1016
[14] Kobya M, Hiz H, Senturk E, Aydiner C and Demirbas E 2006 *Desalination* **190**
[15] Ersoz M and Barrott L 2012 *Best Practice Guide on Metals Removal From Drinking Water By Treatment* (IWA Publishing)
[16] Vasudevan S, Florence E, Jothinathan L, Subbiah R, Swaminathan M and Ganapathy S 2010 *CLEAN – Soil, Air, Water* **38** 225-229
[17] Jiang J-Q, Graham N, André C, Kelsall G and Brandon N 2002 *Water Research* **36** 4064-4078
[18] Khosa M, Jamal M, Hussain A, Muneer M, Zia K and Hafeez S 2013 *Journal of The Korean Chemical Society* **57**
[19] Mansoorian H, Mahvi A and Jafari A 2014 *Separation and Purification Technology* **135** 165-175
[20] Wided B, Béchir H, Afef B and Limam A 2014 *International Journal of Engineering Research & Technology* **3**
[21] Mohammed A and Al-Mureeb M 2010 *Journal of Engineering* **16** 5811-5821
[22] Ghosh D, Medhi C and Purkait M 2008 *Chemosphere* **73** 1393-1400
[23] Beyazit N 2014 *Int. J. Electrochem. Sc* **9** 4315-4330
[24] Hussin F, Abnisa F, Issabayeva G and Aroua MK 2017 *Journal of Cleaner Production* (in press)

IOP Conf. Series: Materials Science and Engineering **210** (2017) 012014 doi:10.1088/1757-899X/210/1/012014

An efficient near infrared spectroscopy based on aquaphotomics technique for rapid determining the level of Cadmium in aqueous solution

Alfian Putra [1*], MariaVassileva[2], Ryoko Santo[2] and RouminaTsenkova [2]
[1]Department of Chemical Engineering, Lhokseumawe Polytecnic State, Indonesia
[2] Departement of Enviroment and Bioproduction Engineering, Graduate School of Agriculture Science, Kobe University,1-1,Rokkodai,Nada,Kobe 657-8501, Japan

E-mail: *putraalf@yahoo.com

Abstract. Cadmium (Cd) is a common industrial pollutant with long biological half-life, which makes it as a cumulative toxicant. Near-infrared spectroscopy has been successfully used for quick and accurate assessment of Cd content in agricultural materials, but the development of a quick detection method for ground and drinking water samples is equal importance for pollution monitoring. Metals have no absorbance in the NIR spectral range, thus the methods developed so far have focused on detection of metal-organic complexes (move to intro). This study focuses on the use of Aquaphotomics technique to measure Cd in aqueous solutions by analyzing the changes in water spectra that occur due to water-metal interaction. Measurements were performed with Cd (II) in 0.1 M HNO_3, in the 680-1090 nm (water second and third overtones) and 1110-1800 nm (water first overtone) spectral regions, and were subjected to partial least-square regression analysis. It was found/determined that A concentration of Cd from 1 mg L^{-1} to 10 mg L^{-1} could be predicted by this model with average prediction correlation coefficient of 0.897. The model was tested by perturbations with temperature and other metal presence in the solution. The regression coefficient showed consistent peaks at *728, 752, 770, 780, 1362, 1430,1444, 1472/1474 and 1484* nm *under various* perturbations, indicating that metal to influence the water spectra. The residual predictive deviation values (RPD) were greater than 2, indicating that the model is appropriate for practical use. The result suggested that this newly proposed approach is capable of detecting metal ion in a much simpler, rapid and reliable way.

1. Introduction

Cadmium (Cd) is a non-degradable heavy metal with long biological half-life (30 years) and untill today there is no proven treatment for chronic cadmium intoxication (Malec.et.al.,2009) The main sources of pollution are improper waste dumping, agricultural chemicals and industrial effluents (Chen et al., 1998). After been released into the environment, it ultimately gets transported into water and enters the food chain (Malec et al. 2009). Divalent Cd (II) ion is highly toxic when compared to its hexavalent counterparts. Cd readily dissolves in dilute nitric acid, HNO_3, making it a good solvent to study Cd in aqueous solution. Furthermore, 0.1 M HNO_3 protects the dissolved Cd from air oxidation (Mihit et al., 2010).

Traditional methods for trace metal ion detection include inductively coupled plasma atomic emission spectroscopy (ICP-AES) (Campilo.et.al., 1999), electrochemical method (Tsalev et.al., 1984), stripping potentiometry (Forrer et.al., 2005). Although these methods are excellent for Cd(II) detection, the instrumentation is expensive and not ideal for real time monitoring. Simple, sensitive sensors that are easy to work with would be a great significance for wide scale monitoring of metal presence. Near infrared spectroscopy (NIRS) could be an alternative for monitoring Cd presence. The most attractive features of **NIRS** analysis are its speed, minimal sample preparation and its being a non-destructive method, making it possible to conduct large numbers of analysis in short time.

Cadmium as such have no absorbtion in NIR range, but dissolved in organic matter, they actively form complexes with organic molecules. Alteration in the vibrational mode of organic complexes has been exploited for accurate detection of metals in agriculture materials and food, such as wines (Sauvage et.al.,2002), legumes (Cozzolino and Moron, 2004), forage (Clark et.al,1989), grasses and hays (Saiga,et.al.,1989), heavy metal pollution (include Cd) (Pieponen,1989), Cadmium and lead in the mussel *Mytilus galloprovincialis* (Font,et.al,2006). NIRS has been also used to predict trace metal content in sediments or soils (Malley and Williams, 1997; Kemper and Sommer, 2002).

Water associates strongly with ions, organic monomers and polymers through hydrogen bonds. Therefore, the change in water absorption bands in the NIR region permits measurement of small quantities or of structural changes in other molecules in the system. This concept has been explored by Aquaphotomics, which has become a powerful strategy to understand water spectral changes related to small concentration of solutes (Tsenkova,2007). Sakudo et al. recently reported successful detection of Cu(II), Mn(II), Zn(II) and Fe(III) in aqueous solution by NIRS (Sakudo et.al..,2006). Despite using only direct NIR measurements of metal solutions, their findings are a good illustration of NIRS potential for metal detection without organic matrices.

The present study examines Cd(II) effect on the vibrational mode of water in the absence of organic solutes and apply Aquaphotomics concept as a tool to predict low metal concentration (1-10 mgL^{-1}). For this purpose two particularly informative spectral regions have been selected, 680-1090 nm (water second and third overtones) and 1110-1800 nm (water first overtone). Furthermore, Cd detectection under various perturbations has been performed, thus mimiking real life pollution monitoring conditions.

2. Procedure

2.1 Sample Preparation

To minimize metal contamination of the sample, all glass or vessels to be used were immersed during one day in HNO_3 and then washed using deionized water from milliQ water purification system (Millipore, Molsheim, Germany). Standard solution of Cd(II) (1000 mgL^{-1}) was purchased from Wako Pure Chemical Industries Japan (Tokyo, Japan). Working stock solutions of Cd(II) at 10 mgL^{-1} were prepared by direct dilution of the standard solution with 0.1 M HNO_3.

2.2 NIR Spectra collection

The transmittance spectra were recorded by NIR System 6500 spectrophotometer (FossNIR-System, Laurel, USA) fitted with a quartz cuvette with 2 mm optical path length. The cuvette was positioned in a cell holder in conjunction with a temperature bath to maintain temperature at 25, 30 and 37^0C. Three consecutive spectra for each metal concentration over the wavelength region of 400 - 2500 nm, in 2 nm steps, were registered. The spectral data were collected as absorbance value [log (1/T)], where T= transmittance.

2.3 Data processing

Three consecutive spectra for each solution were used to develop partial least squares (PLS) regression model (Pirrouette 3.11, Infometrix Inc., Woodinville, WA, USA). A matrix data set was constructed with rows representing metal samples and the columns corresponding to the absorbance in 400-2500 nm range (data not shown). Prior to calibration, spectral data were mean centered and transformed using none and smooth transformation with 5-25 data-point windows. In the development of all calibration models twenty PLS factors were set up as maximum. The optimum number of PLS factors used in the models was determined by step-validation. To make the analysis more focus , absorbance range was divided into two ranges: 680-1090 nm (second and third water overtone) [27] and 1110-1800 nm (first water overtone) [27]. The optimum calibration models were determined by the lowest standard error of calibration (SEC) and standard error of prediction (SEP) and the highest correlation coefficient (R^2). The ratio of standard error of Performance to Standard Deviation (RPD) was used to evaluate the accuracy of prediction. Perturbations. Cd solutions were perturbed with temperature at 25, 30 and 37^{0}C, and other metals (Mg (II), Mn (II) and Zn (II)). Metals were added as 5 mg L^{-1} concentration into increasing concentrations of Cd solutions (0-10 mgL$^{\cdot}$1 in 0.1M NHO with 1 mgL^{-1} step at 25^{0}C). Respective spectral data of all perturbed solutions were taken.

3. Results

3.1 Calibration model

The average correlation coefficient (R^2pred.) is reported to be an indicator of a model successful practical usage when it is higher than 0.70.(Moron,*et.al.*,2003). The average R^2pred. of PLS model developed for Cd(II) in 0.1M NHO avg 0.945 (Table 1) and avg 0.88 for Cadmium and other metal as perturbations (Table 2). Chang et al. define the residual predictive deviation (RPD) >2.0 as indicator of good prediction models[21]. The average RPD value of the developed model is 2.725 (Table 1), thus the model presented can be considered as an acceptable accuracy for analytical purposes.

Table 1. Calibration and prediction statistics for Partial Least-squares Regression (PLS) Model for Cadmium (II) in 0.1 M NHO_3 aqueous solution, indicating the coefficient of determination in Prediction (R^2_{Pred}) =Avg 0.945 and RPD (SD/SEV) >2 that the models are appropriate for practical use.

Metal Ion	Wavelength Range	Factor	Calibration		Validation		Prediction		RPD
			R^2	SEC	R^2	SEC	R^2	SEC	
Cd(II)	680-1090	9	0.97	0.56	0.89	1.05	0.95	2.08	3.09
	1110-1800	16	0.98	0.48	0.82	1.38	0.94	4.29	2.36
Average			0.975	0.52	0.855	1.215	0.945	3.185	2.725

IOP Conf. Series: Materials Science and Engineering **210** (2017) 012014 doi:10.1088/1757-899X/210/1/012014

Table 2. Calibration and prediction statistics for Partial Least-squares Regression (PLS) Model for Cadmium (II) and (Mg,Zn and Mn) perturbation in 0.1 M NHO_3 aqueous solution, these results indicate that the coefficient of determination in Prediction (R^2_{Pred}) = Avg 0.88 and RPD (SD/SEV) >2 that the models are appropriate for practical use.

Metal Ion	Wavelength Range	Factor	Calibration		Validation		Prediction		RPD
			R^2	SEC	R^2	SEC	R^2	SEC	
CdMg	680-1090	9	0.93	0.91	0.83	1.31	0.9	4.3	2.44
	1110-1800	8	0.92	0.93	0.71	1.52	0.84	1.73	2.1
CdZn	680-1090	15	0.99	0.18	0.95	0.71	0.81	2.71	4.5
	1110-1800	11	0.97	0.61	0.84	1.30	0.82	3.33	2.46
CdMn	680-1090	13	0.93	0.94	0.73	1.58	0.97	1.47	2.02
	1110-1800	15	0.98	0.41	0.76	1.57	0.96	1.41	2.03
Average			0.95	0.66	0.80	1.33	0.88	2.49	2.59

Concentration of Cadmium: 0 - 10 mg.L^{-1}(1 mg.L^{-1},step)
CdNx, Cd(II) + 5 mg/L concentration of metal Nx
Factor, Maximum factor used
R^2, correlation coefficient
SEC, standard error of calibration, RPD, standard error of performance to standard deviation

3.2 Metals as perturbations

Perturbation can be defined as evaluation of a physical system by applying small changes to selected parameters and re-estimating the resulted system (Tsenkova,2007). In the short NIR range (680-1090 nm), the regression coefficient for the different metal perturbations exhibit a consistent pattern of positive and negative peaks at 728, 752, 768, and 780 nm (Figure 1a), making this area become most important for investigation of metal influence on water spectra. Similar characteristic absorbance bands were obtained by A.Sakudo et.al. in 710-750 nm short wavelength region (Sakudo et a. 2006). Important wavelengths in the middle NIR (1110-1800 nm) range were found at 1362, 1404, 1408, 1434, 1444, 1472, and 1484 nm, negative and positive peaks (Figure 1b). These results showed the consistency which Cd affects water spectra at certain wavelengths, so its presence has been predicted by using these wavelengths as important variables in the regression vector for Cd.

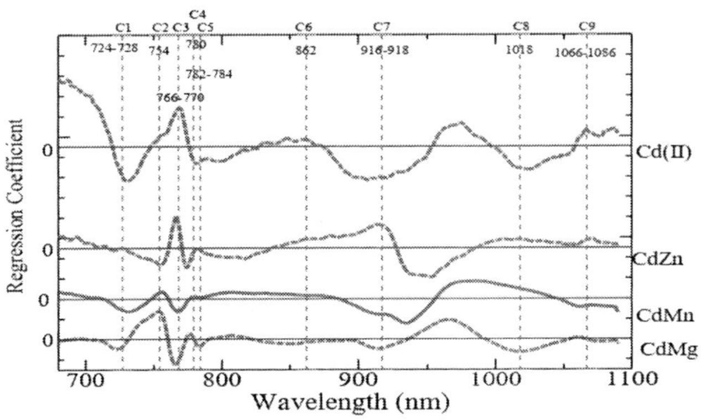

(a)

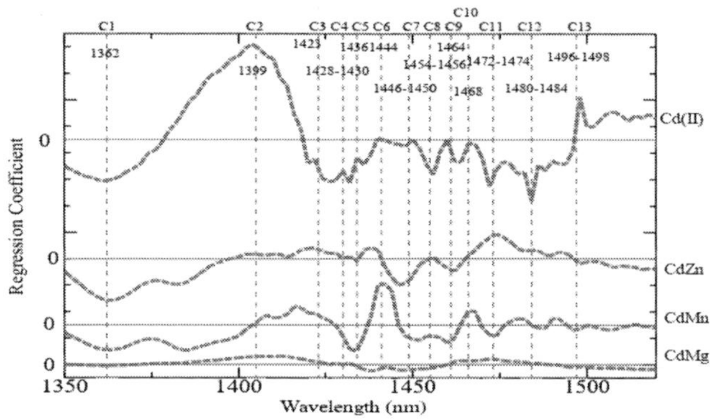

(b)

Figure 1. Spectral Cadmium in water: regression vector of the NIR Spectroscopy model for Cadmium in 0.1 M HNO_3 with other metals presence as perturbations, the Water Matrix Coordinates (WAMACS) describes that consistency coordinates of Cadmium under various perturbations. (a) 680-1090 nm range, and (b) 1110-1800 nm range.

3.2 Data processing
Three consecutive spectra for each solution were used to develop partial least squares (PLS) regression model (Pirrouette 3.11, Infometrix Inc., Woodinville, WA, USA). A matrix data set was constructed with rows representing metal samples and the columns corresponding to the absorbance in 400-2500 nm range (data not shown). Prior to calibration, spectral data were mean centered and transformed using none and smooth transformation with 5-25 data-point windows. In the development of all calibration models twenty PLS factors were set up as maximum. The optimum number of PLS factors used in the models was determined by step-validation. To make the analysis more focus , absorbance range was divided into two ranges: 680-1090 nm (second and third water overtone) [27] and 1110-1800 nm (first water overtone) [27]. The optimum calibration models were determined by the lowest standard error of calibration (SEC) and standard error of prediction (SEP) and the highest correlation coefficient (R^2). The ratio of standard error of Performance to Standard Deviation (RPD) was used to evaluate the accuracy of prediction.
Perturbations. Cd solutions were perturbed with temperature at 25, 30 and 37^{0}C, and other metals (Mg (II), Mn (II) and Zn (II)). Metals were added as 5 mg L^{-1} concentration into increasing concentrations of Cd solutions (0-10 mgL-1 in 0.1M NHO with 1 mgL^{-1} step at 25^{0}C). Respective spectral data of all perturbed solutions were taken.

4. Discussion

4.1 Calibration model
The average correlation coefficient (R^2pred.) is reported to be an indicator of a model successful practical usage when it is higher than 0.70.(Moron,*et.al.*,2003). The average R^2pred. of PLS model developed for Cd(II) in 0.1M NHO avg 0.945 (Table 1) and avg 0.88 for Cadmium and other metal as perturbations (Table 2). Chang et al. define the residual predictive deviation (RPD) >2.0 as indicator of good prediction models[21]. The average RPD value of the developed model is 2.725 (Table 1), thus the model presented can be considered as an acceptable accuracy for analytical purposes.

4.2 Temperature as perturbations

The temperature change affects the vibration intensity of molecular bonds, therefore it changes the spectrum. (Osborn,*et.al.,*1993). Temperature influences considerably spectral measurements. Baxter *et al.* has reported that by increasing temperature the broad band, it can be seen as an overlay of many bands belonging to different cluster sizes of molecules formed by hydrogen bonding, it is shifted towards lower energies as the degree of hydrogen bonding decreases (Baxter,et.al.,1996). These findings agree with our experimental data; NIR spectra of Cd (II) aqueous solution at different temperatures shifted towards lower absorbance as temperature increased (Figure 3).

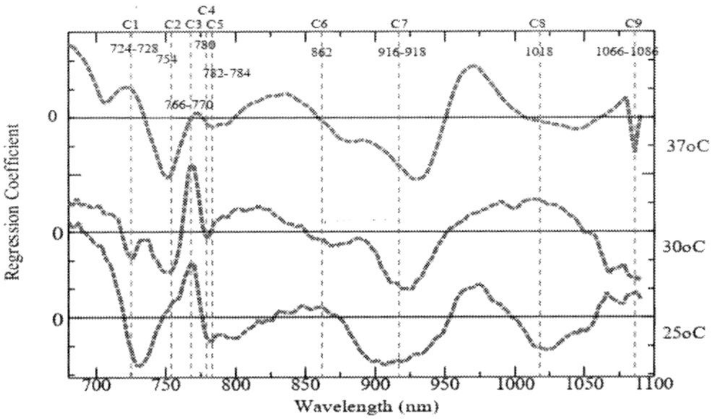

(a)

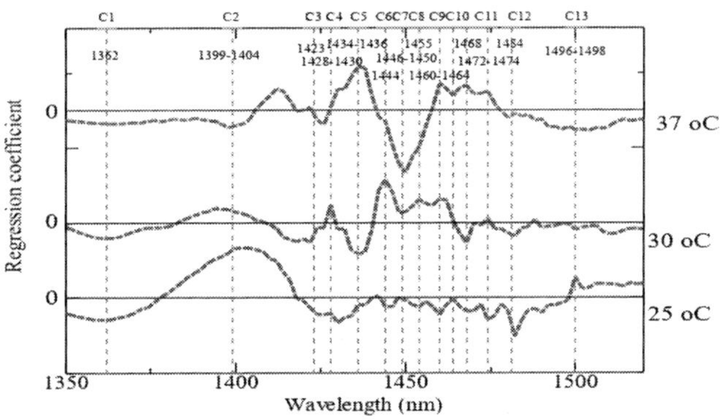

(b)

Figure 2. Spectral Cadmium in water: regression vector of the NIR Spectroscopy model for Cadmium in 0.1 M HNO_3 with temperatures as perturbations, the Water Matrix Coordinates (WAMACS) describes that consistency coordinates of Cadmium under various temperatures. (a) 680-1090 nm range, and (b) 1110-1800 nm range.

In 680-1090 nm range, the spectral regression coefficients for different temperatures of Cd solution display consistent positive and negative peaks at 728,752, 770 and 778 nm (Figure 2a). Consistency is also visible in 1110 – 1800 nm range, with negative and positive peaks at 1362, 1399, 1408, 1430, 1444, 1448, 1462, 1468, 1472, and 1498 respectively (Figure 2b). Figure 2 shows that, with a different temperature, in particular wavelengths its consistency have negative and positive peaks, which indicates the existence cadmium in the sample. It displays Cd perturbates water matrix in characteristic way at different temperatures. These results are consistent with the data on presence of other metals as perturbation in the regression coefficient (Figure 1). Therefore, it can be speculated that the commonly observed peaks may be related to the interaction between Cd (II) and water.

IOP Conf. Series: Materials Science and Engineering **210** (2017) 012014 doi:10.1088/1757-899X/210/1/012014

Water matrix absorbance coordinates, WAMACS. The term Water Matrix Coordinates (WAMACS) represents the influence of different perturbations on the water absorbance in NIR range.[4] Finding common absorbance bands with WAMACS in the regression vector for Cd (II) is important for understanding water-metal interaction.

Table 3. Common wavelengths with WAMACS for Cd under various perturbations indicates the presence of cadmium with the occurrence of water spectra.

680-1090 nm Range							
Wavelength	WAMACS	CdMg	CdMn	CdZn	Cd 25°C	Cd 30°C	Cd 37°C
724-728 nm	C1*	724(-)	724 (-)	728 nm(+)	728 nm(-)	724 nm (-)	724 nm(+)
754 nm	C2*	754 nm (+)	754 nm (+)	754 nm (-)	752 nm (+)	752 nm (-)	752 nm (-)
766-770 nm	C3*	766 nm (-)	766 nm (-)	766 nm (+)	768 nm (+)	768 nm (+)	770 nm (+)
778-780 nm	C4*	780 nm(-)	778 nm(+)	780 nm(+)	778 nm(-)	778 nm(-)	778 nm(-)
782-784 nm	C5*	782 nm (-)	782 nm (+)	782 nm (+)	784 nm (-)	784 (+)	784 (-)
862 nm	C6*	(-)	(+)	(+)	(+)	(-)	(-)
916-918 nm	C7*	916 nm (-)	916 nm (-)	916 nm (+)	916 nm (-)	916 nm (-)	916 nm (-)
1018 nm	C8*	(-)	(-)	(+)	(-)	(+)	(-)
1066-1086 nm	C9*	1066 nm(-)	1066 nm(-)	1066 nm (+)	1066,1086 nm(+)	1086 nm (-)	1086 nm (-)
* Tsenkova, 2007							

1110-1800 nm Range							
Wavelength	WAMACS	CdMg	CdMn	CdZn	Cd 25°C	Cd 30°C	Cd 37°C
1362 nm	C1*	(-)	(-)	(-)	(-)	(-)	(-)
1399-1405 nm	C2*	1404 nm(+)	1404 nm(+)	1404 nm(+)	1404 nm(+)	1399 nm (+)	1399 nm (+)
1424 nm	C3*	(+)	(+)	(+)	(-)	(-)	(-)
1428-1430 nm	C4*	1424 nm (+)	no band	1424 nm (+)	1424 nm (-)	1430 nm (-)	1430 nm (+)
1434-1436 nm	C5*	1432 nm(+)	1436nm (-)	1434 nm(-)	1430 nm(-)	1430 nm(-)	1430 nm(-)
1444 nm	C6*	(-)	(+)	(+)	(+)	(+)	(-)
1446-1452 nm	C7*	1452 nm(+)	1452 nm(-)	1452 nm (+)	1450 nm (-)	1448 nm (+)	1448 nm (+)
1454-1456 nm	C8*	1454 nm (-)	1454 nm (-)	1454 nm(+)	1456 nm (-)	1456 nm (+)	no band
1460-1464 nm	C9*	1460 nm (-)	1460 nm (-)	1460 nm (-)	1460 nm (-)	1462 nm (+)	1462 nm (+)
1468 nm	C10*	(+)	(-)	(+)	(-)	(-)	(+)
1472-1474 nm	C11*	1474 nm (+)	1474 nm (-)	1474 nm (+)	1472 nm (-)	1472 nm (+)	1472 nm (+)
1480-1484 nm	C12*	no band	no band	1484 nm (+)	1484 nm (-)	1484 nm (-)	1482 nm (-)
1496-1498 nm	C13*	1498 nm (-)	1498 nm (-)	1498 nm (-)	1498 nm (+)	1496 nm (-)	1496 nm (-)
* Tsenkova, 2007							

(+), Positive Peaks No band, No Positive nor Negative Peaks.
(-), Negative peaks C, Water matrix coordinate

Cd (II) affects the water spectra on a consistent basis (Table 3). Regression coefficients reveal several consistent wavelengths activated under perturbations with temperature and other metals (728, 752/754, 780, 1362, 1407/1408, 1432, 1452/1455, 1472/1474 and 1486/1488 nm). The regression vector (Figure 1 and 2) shows the influence of different perturbations on water absorbance. The presence of these perturbations also affects the spectra of water, however, some specific wavelengths showed consistency in the water matrix coordinates. These wavelengths showed high correlation value with Cd(II) (Figure 1 and 2), and in general they matched with positive and negative peaks in the correlation plots corresponding to coordinates C_1,C_2,C_3,C_4,C_5,C_6,C_7,C_8,C_9 and C_{10} in Water matrix coordinates (WAMACS) (Table 3). The finding of common absorbance bands with water in the regression vector for Cd (II) indicates that water-metal interaction plays very important role to understand metal presence in water. In this study, spectral analysis revealed the existence of specific WAPs for each perturbation at the same bands (WAMACs). Thus, using the consistency in the absorbance patterns, the presence of metal in the sample can be detected.

5. Conclusion

Considering Cd toxicity, quick and reliable environmental monitoring is primordial. The potential of NIR spectroscopy was investigated for identification and quantification of Cd(II) in aqueous HNO_3. The results showed that NIRS could measure low concentrations of Cd by PLS model and using different perturbations that provided a useful tool for investigating the interaction of metal with water and its quantitative detection, with average coefficient of prediction (R^2_{Pred}) = 0.945. This can be explained with aquaphotomic concept where Cd could be detected at particular wavelengths corresponding to the identified WAMACS. These results indicate that the interaction of NIR light and water is a useful tool for detectingng metal in water and analyzing water samples for cadmium contamination. The explanation is an excellent prove to the Aquaphotomics and this concept could be used for other purposes too such as finding new water bands. However, further investigations are needed to determine the effect of metal valence, as interaction between cations and water may be influenced by valence.

Acknowledgement
This work is financially supported by Lhokseumawe Polytecnic state Indonesia and Kobe University Japan

References

[1] Afkhami,A.,Madrakian,T.,Siampour,H., 2006 *Hazardous Materials* B **138** 269-272
[2] B.G. Osborne, T. Fearn, P.H.Hindle, 1993 *Logman Scientific and Technical, Essex* UK, **716** 227
[3] Bart J C J ,Plastic Additives: Advance Industrial Analysis, IOS Press, Amsterdam 2006
[4] Berentsen,S. ,Stolz,T.,Molt,K., 1997 Journal of Molecular Structure 410-411
[5] Chang,C.W.,Laird,D.A., 2002 N. Soil.Sci. 167
[6] Clark,D.H., Cary,E.E., Mayland, H.F. 1989 *Agron.J.,* **81** 91-95
[7] Cozzolino,D. ,Moron,A. 2004 *Anim.Feed Sci.Technol.* **111** 161-173
[8] Cozzolino,D.,Kwiatkowski,M.J.,Damberg,R.G. 2008 *Talanta* **74** 711-716
[9] Cozzolino,D.,Kwiatkowski,M,J.,Dambergs,R,G.,Cynkar,W.U.,Janik,L,J.,Skououmounis,G.,Gishen,M., Analysis of elements in wine using near infrared spectroscopy and partial least squares regression, Talanta, **74** 711-716
[10] Font, R.,Velez, D.,Celestino,M., Bailon A.H., Montoro,A, 2007 *Food Chemistry* **105** 667-674
[11] Frost, R.L.,Wain,D.L.,Martens,W.N.,Reddy,B.J., 2007 *Spectrochimia Acta.* **66** 1068-1074
[12] M.J.Baxter, H.M,Crews, M.J. Dennis, I Goodall, D. Anderson, *Food Chem.* 60 443 (1996).
[13] Malley, D.F., and William, P.C., Use of Near-infrared reflectance spectroscopy in prediction of heavy metals in freshwater sediment by their association with organic matter. *Environ. Sci. Technol.,* 31, 3461-3467 (1997).
[14] Moron, A.,and Cozzolino, D., 2003 *J.Near Infrared Spectrosc.* **11** 145-154
[15] Malec, P.,Maleva,M.G.,Prasad,M.N.,Strazalka,K., 2009 Aquatic Toxiology. **95** 213-221
[16] Mihit,M,.Laarej,K,.El Makarim,H,A,.Bazzi,L,.Salghi,R,.Hammouti, B 2010 Arabian J.Chemical **3** 57-66
[17] Osborne,B.G., Fearb,T. and Hindle,P.T., 1993 "Practical NIR Spectroscopy with Aplication in Food and Beverage analysis (Logman Food Technology)," Longman Group United Kingdom, Essex,1-35
[18] Ozaki Y; McClure W F; Christy A 2007 A Near-Infrared Spectroscopy in Food Science and Technology, Wiley-Intercience
[19] Pieponen ,S., 1989 *Chemometrics and Intelligent Laboratory System,* **7** 163-170
[20] R.Tsenkova, 2000 Method for information Extraction from Analyzed Samples. *Patent Aplication* 2000-183427
[21] R.Tsenkova, 2004 Method for Analysis with visible and Near-infrared Spectroscopy, *Patent Aplication* 2000-065975
[22] R.Tsenkova, Aquaphotomics : extended water mirror approach reveals peculiarities of prion protein alloform, *NIR news* **18** , 14 (2007) R.Tsenkova, Aquaphotomics:the extended water mirror effect explaints why small concentrations of protein in solution can be maesured with near infrared ligth. *NIR news* 19 , 4 (2008)
[23] R.Tsenkova, 2009 *Journal of Near Infrared Spectroscopy* **17** 303-314

[24] Sakudo,A.,Yoshimura,E.,Tsenkova,R.,Ikuta,K.,Onodera,T., 2007 *Toxicological Science* **132** 135-141

[25] Sakudo,A.,Tsenkova,R.,Tei,K.,OnozukaT.,Ikuta,T.,Yoshimura,E.,Onodera,T., 2006 *Bioschi. Biotechnol. Biochem* **70** 1578-1583

IOP Conf. Series: Materials Science and Engineering **210** (2017) 012015 doi:10.1088/1757-899X/210/1/012015

Mechanical properties of bioplastics cassava starch film with Zinc Oxide nanofiller as reinforcement

Harunsyah*, M Yunus and Reza Fauzan

Department of Chemical Engineering, Lhokseumawe State Polytechnic, Buketrata, Lhokseumawe 24300, Indonesia.

Email: *harunsyah@pnl.ac.id, mysri_son@yahoo.com, reza.fauzan@gmail.com

Abstract: This study focuses on investigating the influence of zinc oxide nanofiller on the mechanical properties of bioplastic cassava starch films. Bioplastic cassava starch film-based zinc oxide reinforced composite biopolymeric films were prepared by casting technique. The content of zinc oxide in the bioplastic films was varied from 0.2%, 0.4%, 0.6%, 0.8% and 1.0% (w/w) by weight of starch. Surface morphologies of the composites bioplastic films were examined by scanning electron microscope (SEM).The result showed that the Tensile strength (TS) was improved significantly with the additional of zinc oxide but the elongation at break (EB %) of the composites was decreased. The maximum tensile strength obtained was 22.30 kgf / mm on the additional of zinc oxide by 0.6% and plastilizer by 25%. Based on data of FTIR, the produced film plastic did not change the group function and it can be concluded that theinteraction in film plastic produced was only a physical interaction. Biodegradable plastic film based on cassava starch-zinc oxide and plasticizer glycerol showed that interesting mechanical properties being transparent, clear, homogeneous, flexible, and easily handled.

1. Introduction

Plastics have been used widely for packaging material since long time ago[1]. However, plastic wastes can pollute the environment because of its persistency to be biodegraded by microorganism. Efforts have been conducted to develop environmental friendly plastic from renewable resources. In recent years, a large progress has been achieved in the development of biodegradable products based on agricultural materials, due to a growing concern on the environmental impact and also the uncertainty of petroleum supply [1]. One of the most studied and promising raw materials for the production of biodegradable plastics is starch, which is a natural renewable carbohydrate polymer and an available low cost material. [2]. However, films formed from starch are brittle and difficult to handle; plasticizers are normally added to the film-forming solution before casting and drying procedures, as a way to overcome films brittleness [3]. To improve starch-based plastic characteristics and the mechanical resistance, many researchers have demonstrated the interest of using filler as reinforcement in thermoplastic matrixes and have shown that fibers incorporation can increase films tensile strength and elasticity modulus and decreases their elongation capacity [4]. Among the promising Nano fillers that have stirred much interest among researchers are including organo clay, nano silica, carbon Nano tube and Nano calcium carbonate. Studies have shown that the large surface area possessed by these Nano fillers promote better interfacial interactions with the polymer matrix compared to conventional micrometer sized particles, leading to better property enhancement [5]. There are number of inorganic mineral fillers used in polypropylene. The most common of these fillers are talc, calcium carbonate and barium

sulphate; other mineral fillers used are wollastonite and mica. Mineral fillers are generally much cheaper than polypropylene resin itself. Mineral fillers reduce the costs of the compound formed with polypropylene and also increase the stiffness. Mineral fillers also provide reinforcement to the polymer matrix as well. Some mineral filler are surface treated to improve their handling and performance characteristics [6]. Silanes, glycols, and stearates are used commercially to improve dispersion and processing, as well as to react with impurities.
In the present work, we have investigated the fabrication of bioplastic film base on cassava starch (biopolymer) by using glycerol as plasticizer obtained by casting metho. The bioplastic film is reinforced by zinc oxide nanofiller to enhance mechanical properties. The mechanical properties of bioplastic film were determined by tensile strength tests, surface morphology was examined by using scanning electron microscopy and thermo mechanical was examined by using thermogravimetric analysis.

2. Procedure

A. Starch Extraction
Starch was extracted from cassava and tree cassava tubers, grown in Indonesia and purchased from the local market. The cassava roots are ripe after 12 months. For each species, the tubers were washed, peeled and grated. The resulting paste was mixed with water and the solution was filtered on a clean cloth. The collected filtrate was then allowed to stand for 6 hours followed by the removal of the supernatant. The white precipitate (starch) was then recovered, sun-dried and stored in polyethylene bags at room temperature.

B. Preparation of ZnO Nanoparticle Bioplastic Film
ZnO nanoparticles was purchased from US Research Nanomaterials, Inc. The preparation of bioplastic packaging films and the casting was based on Alebooyeh et al [7] with some modifications. First, zinc oxide nanoparticles were dispersed in distilled water solution at 0.2 to 1.0% (w/w) of the total starch and stirred for 1 hour and ultrasonicated for 30 minutes. Then the solution was heated to 85 ± 5°, held for 15 minutes for gelatinization. Next, glycerol plasticized addition is carried out with the concentration variation of 25, 30 and 35%, and it is agitated until homogeneous. Then, the homogeneous solution is casted above a plate with 2.0 mm thickness. Furthermore, drying is carried out in oven at temperature of 60 ^{o}C for 5 hours. The dry bioplastic films were removed from the oven and stored at controlled conditions (25 ^{o}C and 75% of relative humidity) for at least 48 hours before measurements. Control films were also prepared but without the additional of nanoparticles. The dried films were peeled and cut to have an average dimension of 7cm x 5cm. The thickness was also measured and samples were further used.

C. Tensile Strength (TS) and Elongation at Break (%E) Analysis
In general, the physical and electrical properties of plastics are influenced by temperature and relative humidity in a manner that materially affects test results. In order to make reliable comparisons between different materials and between different laboratories, it is necessary to standardize the humidity conditions, as well as the temperature, to which specimens of these materials are subjected prior to and during testing. Therefore, all films plastic were conditioned prior to subject them to permeability and mechanical tests according to Standard method, ASTM-D618-61. Films used for testing Water Vapor Permeability (WVP), Tensile Strength (TS) and Elongation (E) were conditioned at 75% relative humidity and 25 ^{o}C by placing them in a desiccators over a saturated solution of $Mg (NO3)_2.6H2O$ for 24 hours or more. For other tests, film samples were transferred to plastic bags after peeling and placed in desiccators. The tensile strength and elongation at break of the films were measured using a computer type universal testing

machine (HUNG TA, TH-8503). The tensile samples were cast in a collapsible aluminum mold in based on ASTM standard D638 for tensile tests and the creep specimen was prepared as dumbbells by compression molding based on ASTM D2990 for tensile creep tests.

D. Scanning Electron Microscopy Analysis (SEM)
Scanning electron microscopy film plastic surface morphology was examined by using scanning electron microscopy. The samples were mounted on stub with double-sided adhesive tape (5x5 mm) and coated with a thin layer (150-180 A) of gold (JEOL JFC-1600 auto fine coater). Images were taken using a JEOL JSM-6510-LA Japan with an accelerating voltage of 0.5 to 30 kV. TEM images were recorded with a JEOL model transmission electron microscope, operating at 200 kV, with a point-to point resolution of 0.3 nm.

E. Thermal Properties Analysis
The thermo mechanical (TM) test of the films plastic was taken by using computer controlled Thermogravimetric analysis (TGA) (Model: DSC-60 Supplier: Shimadzu Corp.). The temperature range was maintained at 30°C to 600°C and the temperature was increased at a rate of 10°C/min. The flow rate of nitrogen gas was 20 ml/min. Sample weights were 5.0 mg.

3. Results
In order to know the quality of resulted bioplastic film, several parameters which usually used to analyse resulted bioplastic are measured such as tensile strength, elongation at break. Mostly, bioplastic materials are used because they have desirable mechanical properties such as tensile strength and elongation at break. For this reason, the mechanical properties may be considered as the most important of all the physical properties of bioplastic for most applications.

Table 1. Tensile strength of the bioplastic film with variation zinc oxide filler.

Plastilizer (Glycein)	Tensile Strength (kgf/mm^2) ZnO nanofiller (%)					
	0.0	**0.2**	**0.4**	**0.6**	**0.8**	**1.0**
25 %	20.90	17.30	20.50	22.30	18.60	12.90
30 %	17.00	19.20	20.00	20.60	18.80	14.30
35 %	12.60	20.10	19.30	20.20	20.50	12.40

Table 2. Elongation at break of the bioplastic film with variation zinc filler .

Plastilizer (Glycerin)	Elongation at Break (%) ZnO nanofiller (%)					
	0.0	**0.2**	**0.4**	**0.6**	**0.8**	**1.0**
25 %	266,20	332,20	252,00	198,40	153,20	236,80

IOP Conf. Series: Materials Science and Engineering **210** (2017) 012015 doi:10.1088/1757-899X/210/1/012015

30 %	250.80	234.80	123.40	122.80	139.60	123.20
35 %	474,00	196,80	182,60	206,00	285,20	269,20

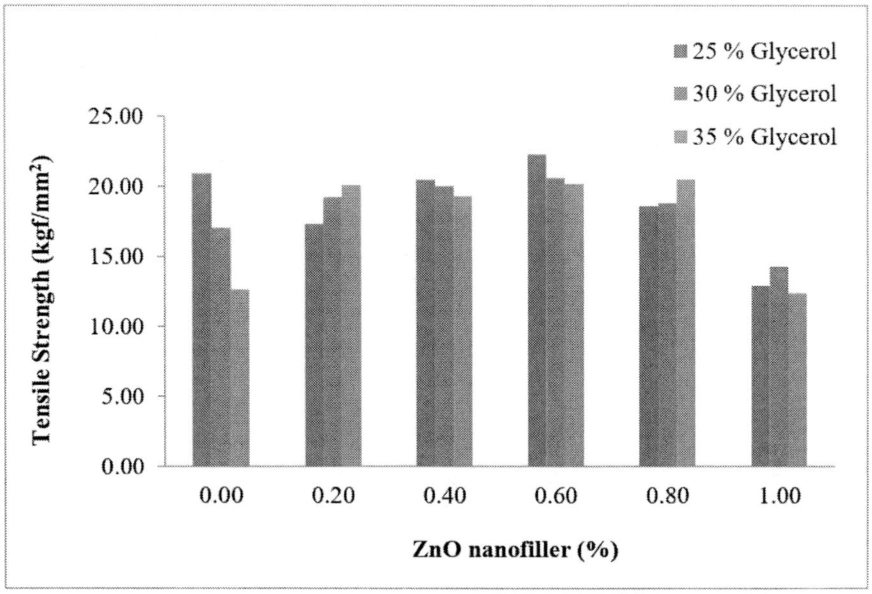

Figure 1. Efect of ZnO filler concentration on tensile strength

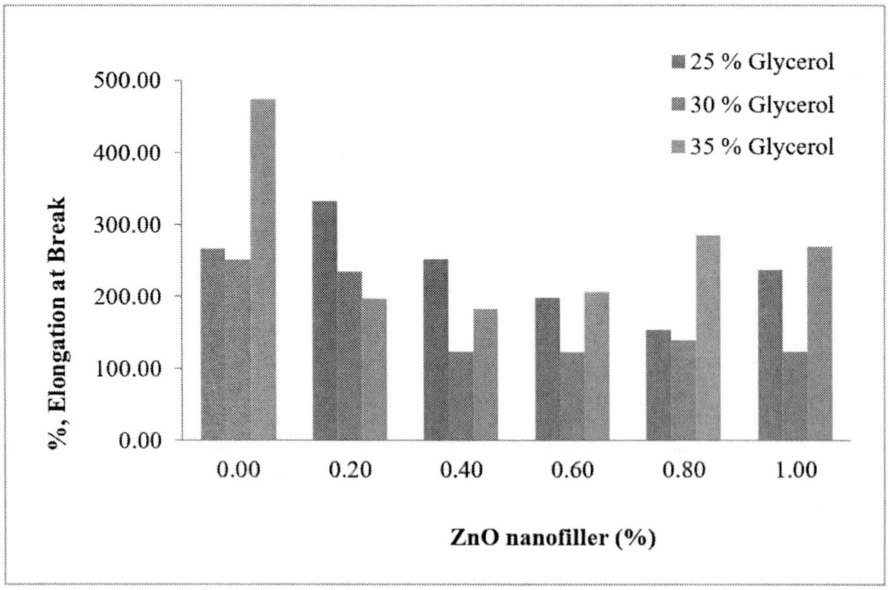

Figure 2. Efect of ZnO filler concentration on elongation at break

4. Discussion

4.1 Effect of Zinc Oxide Filler Concentration on Tensile Strength and Elongation

Tensile strength is the maximum load large unity initial cross-sectional area of the sample. Tensile strength indicates the ability to accept load or tension without damaging the composite or broken which is stated with a maximum tension before breaking namely ultimate tensile strength. Tensile strength of composite material can be affected by several factors, including the relative comparison between the matrix and the reinforcement materials in composite materials, namely how many zinc oxide is added to the polymer matrix compared with composite materials. Elongation at Break indicates the quantity of the change of maximum film length while obtaining tensile strength until the film breaks, compared to the initial length. The result of tensile strength and elongation at break is given in Table 1and table 2.

Figure 1 and figure 2 presented the influence of zinc oxide reinforcement on tensile strength and elongation at break, where the increase of zinc oxide composition quantity indicates the tendency to increase tensile strength value and the further will be declined, in other words, the higher the concentration of nano zinc oxide, it will generate a tensile strength diminishing. This is because the particles have more space in the matrix (starch), thus affects the tensile strength of the bioplastic film, as obtained in the study of Wang et al [8] and Waryat et al [9]. While for the percentage elongation, the value will decrease with increasing concentration of zinc oxide. Table 1 and table 2 above illustrate the tensile strength properties that indicates the concentration addition tendency of glycerol as plasticizer could increase elongation percentage and reduce tensile strength. The maximum tensile strength obtained was 22.30 kgf/mm2 on addition of zinc oxide by 0.6 percent and plastilizer by 25%.

4.2 Scanning Electron Microscopy Analysis (SEM)

Morphology, hygroscopicity and mechanical characteristics of the bioplastic film produced is closely related to the concentration of glycerol and ZnO that were added in synthesizing of the film plastic produced. Effect of ZnO nanofiller was added to the on morphologytic bioplafilms can be seen in Figure 3.

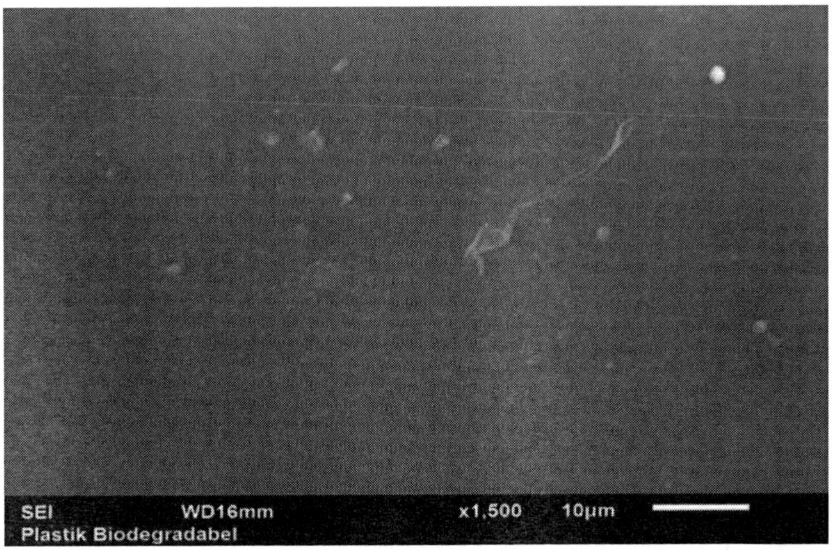

Figure 3. SEM analysis bioplastik film with ZnO nanofiller at optimum condition

Visually one of the results of scanning electron microscopy film plastic produced for optimum condition (plastilizer glycerol 25% and ZnO 0.6%) is not porous and it looks smoother, no cracks or air bubbles and in general, the film appearance is transparent. By using SEM, the morphology of the resulting film packaging there are a number of starches that this clumping occurs because starch insoluble depleted due to the effects of complaining and temperature conditions are not uniform when the plastic film is made.

4.3 Thermogravimetric analysis (TGA)

Thermogravimetric analysis (TGA) was used to investigate experimentally the thermal stability of cassava starch-ZnO blends. The thermal stability of any polymeric material is largely determined by the strength of the covalent bonds between the atoms forming the polymer molecules. Figures 4 show the initial TGA thermograms (without ZnO) and the corresponding rate of reaction curves of with concentration of ZnO blends. Based on the TGA study in figure 4 and figure 5, few points may be concluded that the major loss in weight (20-70%) for without ZnO or with ZnO blends occurs within the range of 200-350° C, in which with ZnO blends possess higher thermal stability than without ZnO blends.

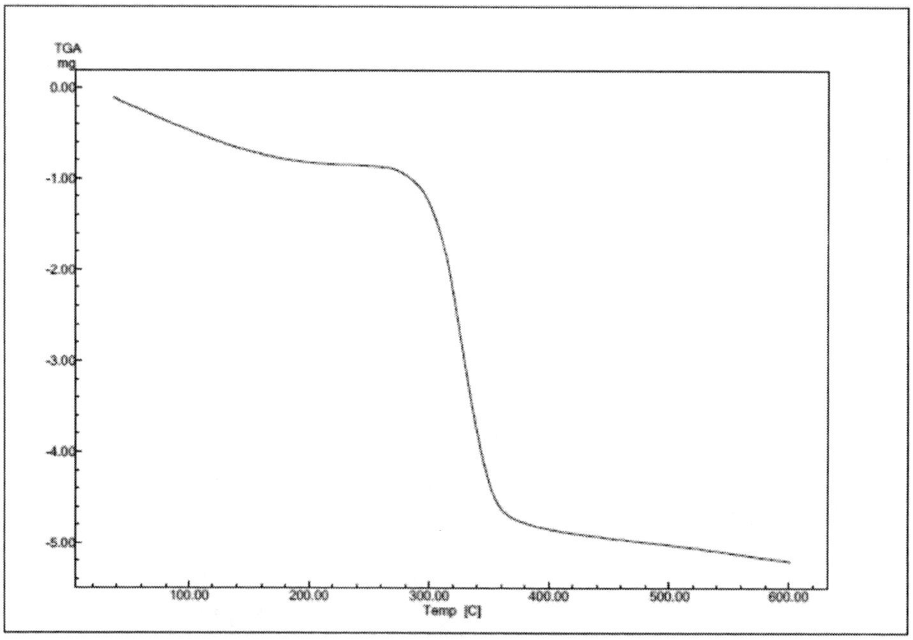

Figure 4. TGA analysis of bioplastik film with ZnO nanofiller on optimum condition

4.4 Fourier Transform Infra-Red Analysis of Bioplastic Films (FT-IR)

The infrared analysis was carried out using Fourier-Transform Infrared Spectroscopy (FT-IR) model Mattson-Genssis, made by Unicam, England, over the range 400- 4000 cm-1. FT-IR spectroscopic analysis can give useful information about chemical changes occurring in polymer systems due to blending process, casting process. However, the detection of such changes may be restricted because often different polymer compounds in the polymer blends are chemically similar, so their absorption peaks are masked. To facilitate the detection of changes, a comparison between the characteristic bands of the compounds is investigated.

FT-IR spectroscopic analysis can give useful information about chemical changes occurring in polymer systems due to blending process. However, the detection of such changes may be restricted because often different polymer compounds in the polymer blends are chemically similar, so their absorption peaks are masked. To facilitate the detection of changes, a comparison between the characteristic bands of the compounds is investigated.The interaction of polymer starch with ZnO nanoparticle could be identified with FT-IR spectra shown in Figure 5, the FTIR spectra of blend substances should show no appreciable changes in the spectral band positions.

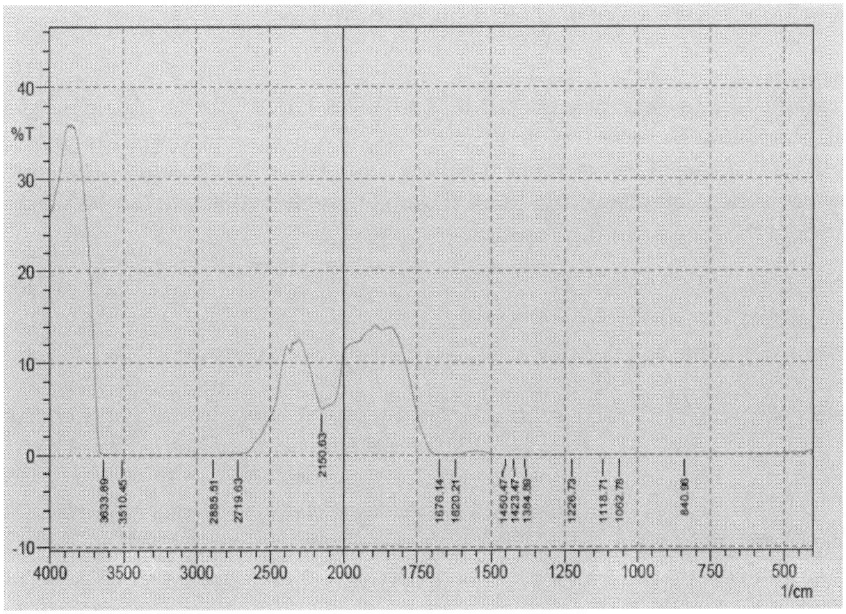

Figure 5. FTIR spectra analysis of bioplastik film with ZnO nanofiller on optimum condition

Results of this study suggested that addition of ZnO nanofiller on starch-based bioplastics interact strongly on fabricated films. Data revealed that the increasing concentration of ZnO nanparticles in bioplastic films involved stronger alkane bond, C-O bond, C=C bond, and C-C bond formations. This suggests a stronger attachment and stronger material strength as shown in films mechanical properties.

5. Conclusion

This study described the effect of ZnO nanofiller on optimum mechanical properties of bioplastic film based on cassava strach. The goal of the research was to obtain an early identification of formulation composition to further design on a complete experimental study. Results of the study showed that the addition of ZnO nanofiller resulted in some changes in the mechanical behaviour of bioplastic film as outlined below:

1) Cassava starch based films plasticized with glycerol showed interesting mechanical properties such as being transparent, clear, homogeneous, flexible, and easily handled.
2) The results establish that films plastic based on cassava starch and plasticized with glycerol can be considered as an interesting biodegradable alternative packaging material.
3) Bioplastic film made from cassava starch by using glycerol and the addition of Zinc Oxide filler has mechanical properties such as white transparent, clean, homogeneous, easily bent and handled.

4) The optimum formulation composition has tensile strength value of 22.30 kgf/mm2 at ZnO concentration of 0.6% and plasticizer concentration of 25%.
5) The best value of Elongation at Break is as large of 122.80% ZnO concentration of 0.6% and plasticizer concentration of 30%.
6) Bioplastic film hydrophilic or not resistant to water and the resulting functional groups together with the constituent components of starch. In the synthesis of plastic film that happening is the process of blending physics and the group C = O and CO Ester carbonyl makes environmentally friendly plastic. The morphology structure of bioplastic, both top and cross section, has not given the homogeneous morphology structure.

Acknowledgement
The authors would like to acknowledge SIMLITABMAS-DIKTI for being financed and given the opportunity to conduct this research under The Competitive Research Grant Program 2017-2018, to the Department of Chemical Engineering, Lhokseumawe State Polytechnic – Buketrata Campus, for lending them the needed equipment and to all those not mentioned above that in their own little way have contributed to the success of this research.

References
[1] Nattakan, S., Nittaya, L., Atitaya, N., Natthawut,Y. and Tawee,T. 2012 *Composites Science and Technology* 72 845–852
[2] Xiaofei ,M., Jiugao ,Y and John , Kennedy 2005 *Carbohydrate Polymers* **62** 19–24
[3] Carmen, M.O., Joao, B.L., Fabio,Y. 2009 *Food Hydrocolloids* **23** 1328–1333
[4] Avérous, L., Fringant,C., Moro,L. 2001 *Polymer* **42** 6565-6572
[5] Guo, F. 2004 Clay/Polymer Composites: The story Materials Today November 50-55
[6] Sudhin Datta and David, J.L. 1998 Polymeric Compatibilizers:Usesand Benefits in Polymer Blends.Munich;Vienna and New York: Hanser/Gardner Publications, Inc,Cincinnati
[7] Alebooyeh R., Nafchi A.M, Jokar, M. 2012 *Journal of Chemical Health Risks* **2** 13-16
[8] Wang ZL. 2012 Toward self-powered nano system; from nanogenerators to nanopiezotronics. Advanced Functional Material.
[9] Waryat, Romli M, Suryani A, Yuliasih I, Johan S. 2012 *Jurnal Teknologi Pertanian Agritech* **33** 197 – 207

IOP Conf. Series: Materials Science and Engineering **210** (2017) 012016 doi:10.1088/1757-899X/210/1/012016

The effect of the natural bentonite to reduce COD in palm oil mill effluent by using a hybrid adsorption-flotation method

Ratni Dewi, Ratna Sari and Syafruddin
Department of Chemical Engineering, Lhokseumawe State of Polytechnic, Indonesia.

E-mail: *raihan_annisa@yahoo.co.id, akmalsalwa@yahoo.co.id, irsyafruddin@yahoo.com

Abstract—Palm oil mill effluent is waste produced from palm oil processing activities. This waste are comingfrom condensate water, process water and hydrocyclone water. The high levels of contaminants in the palm oil mill effluent causes the waste becomes inappropriate to be discharged to water body before processing, one of the most major contaminants in wastewater is fats, oils and COD.This study investigated the effectiveness of chemically activated bentonite that serves as an alternative to reduce the COD in adsorption and floatation based palm oil effluent waste processing. Natural bentonite was activated by using nitrit acid and benzene. In the existing adsorption material to improve COD reduction capability whereas the flotation method was used to further remove residual effluent which is still remain after the adsorption process. An adsorption columns which operated in batch was used in the present study. By varying the circulation time and adsorbent treatment (activated and non-activated), it was shown that percentage of COD reduction reached 75% at the circulation time of 180 minutes for non activated adsorbent. On the other hand the percentof COD reduction in adsorption and flotation process using activated bentonite reached as high as 88% and 93% at the circulation time of 180 minutes.

1. Introduction

Indonesia is currently acountry with the largest oil fields in the world. The area of oil palm in Indonesia in 2009 amounted to 8127 million hectares [2] and continues to increase every year. Increasingoil palm plantations will result in increasing the amount of palm oil mill to process the palm fruit. Palm oil mill effluent will produce a number ranging between 600-700 liters/ton of fresh fruit bunches (FFB). This waste comes from drab water, condensate water, process water and water hydrocylon. The materials contained in the waste, among others, oil-grease, NH3-N, TOC, COD, BOD, and TSS with high concentrations which need to be processed before discharged to the environment.

Palm oil wastewater treatment is conventionally done biologically by using pond system, where the liquid waste is processed in a pond by utilizing anaerobic and aerobic microbes. This process takes 120-140 days to break down the organic material with efficiency of 60-70%. Required land area reachs up to 7 ha to process 30 tons/hour of fresh fruit bunches (FFB) and it limitate the availability of land, more over, in the rainy season the pond will be overflow. Therefore it is necessary to have a process that can cultivate the oil palm waste water efficiently. One of the methods that can be used is the adsorption which using natural adsorbent (in this case the bentonite) combined with flotation process.

Research adsorption using bentonite as the adsorbent has been done by several researchers [4,6-8,11-13,17-21]. Weishan (2013) used bentonite to remove polyvinyl alcohol in water, so did Ishq, M (2015), he also used bentonite to remove oil in water. Similarly, the preliminary study of hydrocarbon compounds in the wastewater by flotation process has also been carried out [1,3,5,11,14,16]. The study of adsorption studies using other adsorbent has been done to remove oil in water using activated carbon and zeolite as adsorbents [14]. While the use of organic clay to eliminate petroleum hydrocarbons in industrial waste have also been studied [10]. From the above explanation, it can be seen that the adsorption method is suitable for setting aside hydrocarbon compounds in the water, especially forlevels of COD.

In this study, a modified natural bentonite is used as an adsorbent hydrocarbons (such as fats and oils) in the palm oil mill effluent (POME). Based on the results of previous studies, it is known that the adsorption method was only able to adsorb free hydrocarbons in the water, but for bounded hydrocarbons in water it is necessary to give an additional alternative process, namely the flotation method.

2. Procedure

The study was conducted in a laboratory scale, using adsorption column with a circulatory system and proceed with the flotation process. The sample used is palm oil mill effluent from Cot Girek.

2.1 Materials and Equipment

Materials used in the study is the effluent waste palm oil, Bentonite (activated and non-activated), Crystal K_2CrO_7 0.1 N H_2SO_4 (Merck), HNO_3 0.5N (Merck), benzene (Merck), Ag_2SO_4 (Merck), $HgSO_4$ (Merck), indicator Ferroin (Merck), and $Fe(NH_4)_2(SO_4)_2.6H_2O$ (Merck). While the tool used is the adsorption column and flotation (adsorption column made of acrylic), pumps, flow meters, sewage tank, aerator, diffuser as well as the supporting tools of analysis.

2.2 Bentonite Adsorbent Preparation

Natural bentonite originated from Nisam area of North Aceh is crushed and sieved to obtain the appropriate size (± 1 cm). Bentonite is dried in a furnace at a temperature of 600^{0}C for 2 hours, cooled to room temperature and stored in a desicator. Bentonite then was activated with 0.5 N nitric acid and benzene, each for two hours, washed with distilled water until the pH is neutral. Bentonite is dried in an oven at temperature of 105^{0}C for two hours thus ready to use as an adsorbent.

2.3 Experimental Procedure

In the adsorption process the waste tank, liquid waste was included as much as 10 liters of POME. Adsorption column was filled with bentonite with a 30 cm bed height. Waste was inserted through the top of the column by using a pump with a flow rate (1 L/min). COD changes were observed in the variation of circulation time for 15-180 minutes during the adsorption process was takingplace. While in the flotation process palm oil effluent was included from the adsorption process results in a flotation tank as much as 5 liters, aerator was include which has been connected with the diffuser using a hose. The changes were observed for each waste COD adsorption process results. The series of adsorption and flotation unit is shown in Figure 1.

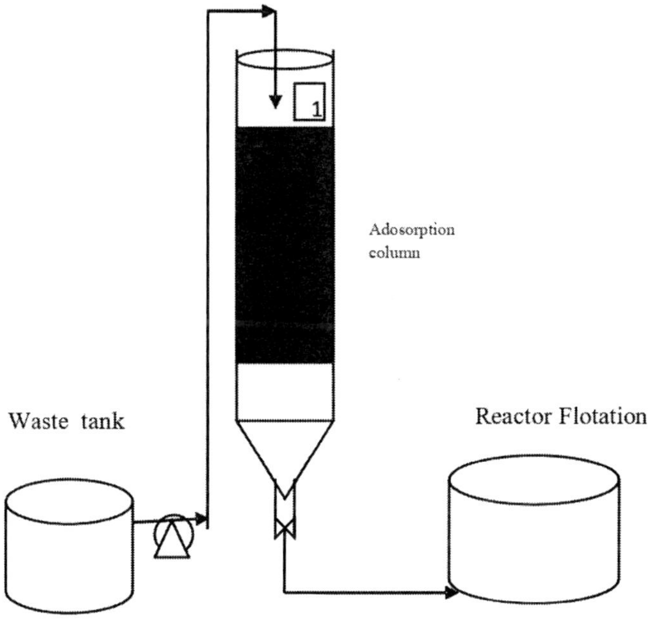

Figure 1. Schematic Adsorption Process Equipment and Floatation

2.4 Analysis

In this research, analysis of the waste after adsorption and flotation processes, namely COD levels measured at the time variation of circulation. The methods used are closed reflux method (titration) with the analysis tool COD reactor.

3. Results

A. The characteristic of palm oil mill waste

Based on the study conducted in laboratory of unit operatiom and water procesing, the result showed the analysis of COD level and fat in the water of palm oil mill waste in the waste reservoir tank in Cot Girek before and after the process of adsorption and flotation using activated bentonite. The analyisis is as shown in table 1.

Table 1. Characteristic of palm oil liquid waste (before process)

Parameter	Unit	value
TOC	mg/	5065,13
COD	mg/L	18.739,776
pH		5
TSS	mg/L	15.000

B. Characteristic of Bentonite

Characteristic of bentonite was conducted by measuring the value of Cation Exchange Capaticity (CEC) and the particle size of bentonite. CEC was done by using methilen blue method, while the size of bentonite pore was measued by using SEM. The result of CEC measurement is displayed in table 2.

Table 2. Result of Bentonite CEC measurement

No.	Type of Adsorbent	Cation Exchange Capacity /CEC (meq/100 gr)
1.	Bentonite after heating process in 600^0C	30
2.	Bentonite after being activated by using nitric acid and heating process in 600^0C	42
3.	Bentonite after being activated by using benzene and heating process in 600^0C	48

The following characteristic bentonite beside by using the measurement of CEC value is by determining the pore size of bentonite by using SEM in 4000x enlargement, and the result gained is as in figure 2.

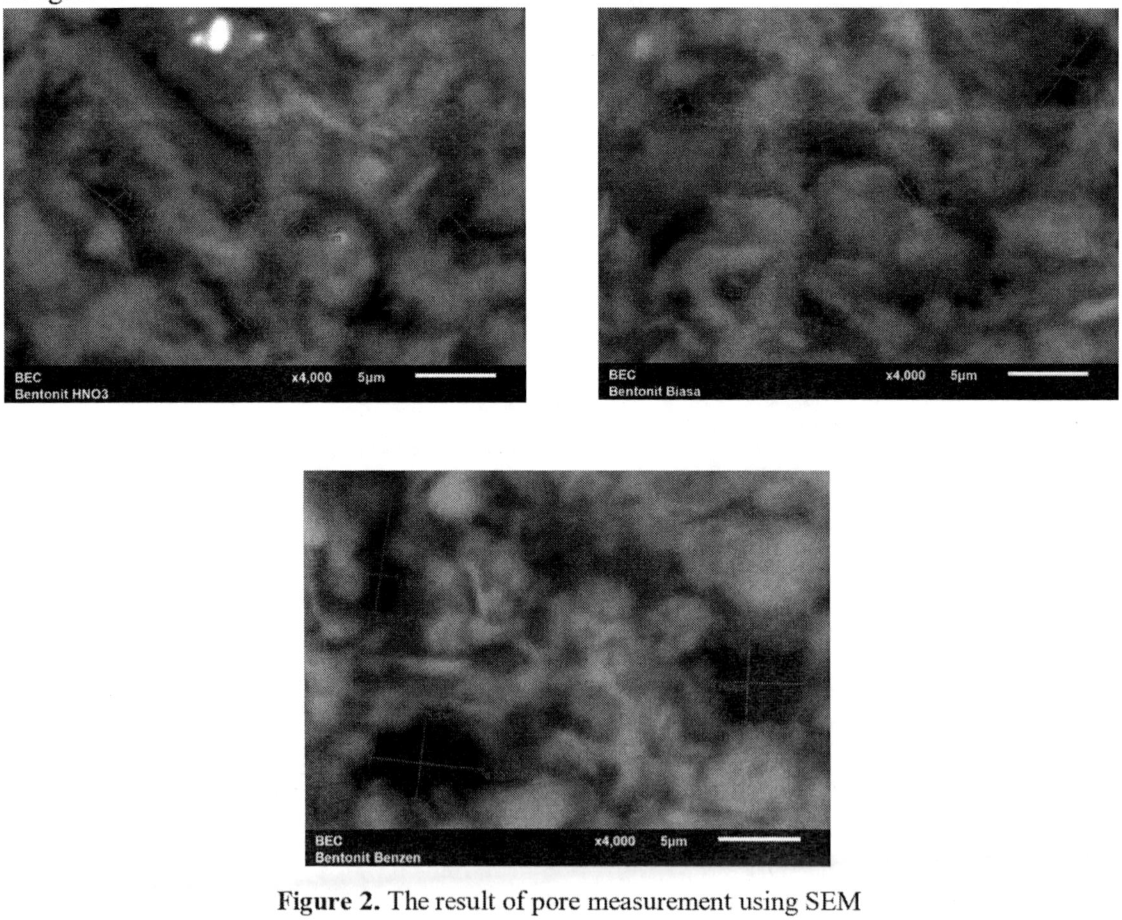

Figure 2. The result of pore measurement using SEM

Based on the measurement of pore size by using SEM, the result gained is as displayed in table 3.

Table 3. Size of Bentonite Pore.

No.	Type of Bentonite	Range of Pore (μm)
1.	Bentonite after heating process in 600^{0}C	0,610 – 2,030
2.	Bentonite after being activated by using nitric acid after heating process in 600^{0}C	0,728 – 3,780
3.	Bentonite after being activated by using benzene after heating process in 600^{0}C.	1,861 – 6,990

From the adsorption and flotation process obtained the results of COD removal as shown in Figure 3. From the Figure it is seen that the longer the contact time given, the removal efficiency COD liquid waste palm mill obtained greater.

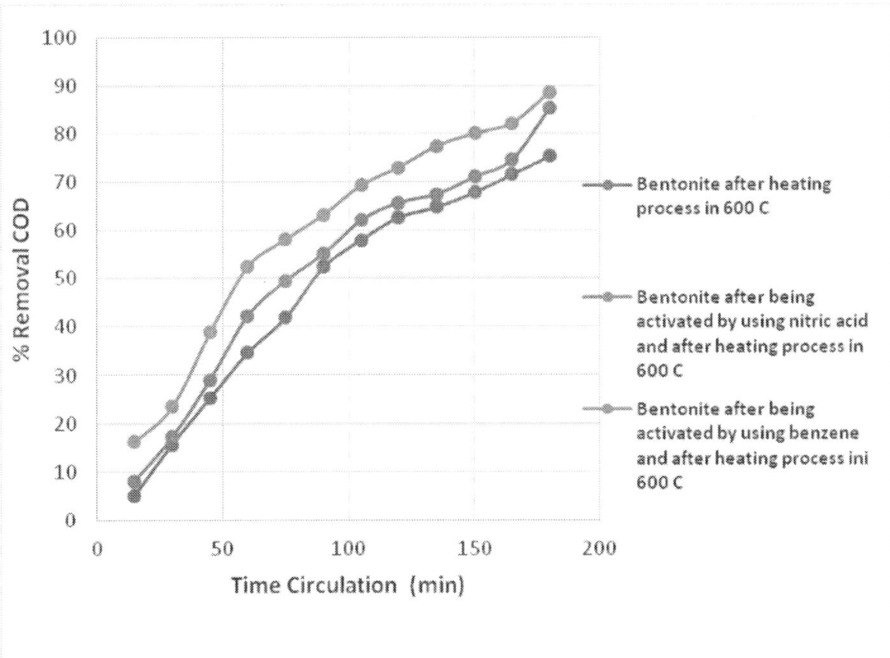

Figure 3. Graphic of COD removal based on circulation time variatiom in adsorpstion process

In flotation process, flotation tank with six liters capaticity is used and oxygen is in through difuser.The waste that will be processed is originated from the output of effluent in adsorption coloumn with contact time for 30 minutes. The result of COD removal in flotation process is explained in figure 3.

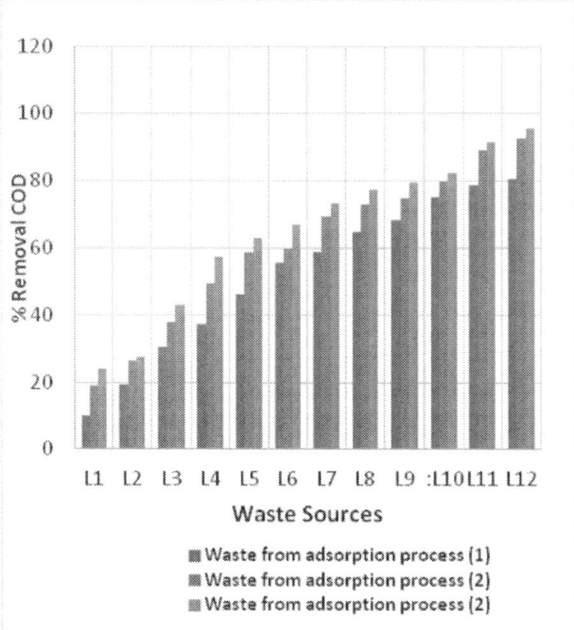

Figure 4. Graphic of COD removal in flotation process by varying the waste sources.

L1	= Waste from adsorption process in 15 minutes circulation time
L2	= Waste from adsorption process in 30 minutes circulation time
L3	= Waste from adsorption process in 45 minute circulation time etc
L12	= Waste from adsorption process in 180 minutes circulation time

4. Discussion

From the analysis of table 1, it can be seen that the value of TOC and COD from the liquid waste of palm oil mill is still beyond the maximum limit of the standard quality for liquid waste of palm oil mill that has been set by Bapedal, KEPMEN LH No.51/MenLH/10/1995 that 500 mg/l for COD.While the TOC value if referred to the standard quality of oil and gas waste, it has the limit for 100 mg/L. Based on those TOC and COD values, it is necessary to conduct the process for waste water of Palm Oil Mill and in his study, the removal of COD is observed through the adsorption and flotation process by using bentonite as the adsorbent.

In table 2, it showed the varies value of CEC of bentonite. Based on physical and chemical activation, the bentonite has CEC as much as 30, 42 and 48 meq/100 gr for the bentonite that has been heated in 600^{0}C, by using nitric acid and benzene respectively.Chemical activation to the heated bentonite raised up the value of CEC where benzene is better than nitrit acid. The value of CEC is relatively small due to the heating process in 600^{0}C in the furnace that caused the the broken of misel in the bentonite. However the heating in the furnace is neccesary because the used bentonite is in adsorption coloumn, without the heating process, the bentonite will be smashed and cause the blockage in the coloumn. In pre-research, it has been done vary of heating temperature treatment to get the stable bentonite to the adsorption process especially to the liquid waste of palm oil mill. The result of pre-research showed that the heating temperature in 600^{0}C produced the stable bentonite and it is not smashed during the adsorption process.

From figure 3. It showed the percentage of COD removal after adsorption process by varying contact time from 15 minutes to 180 minutes. From the graphic, it can be seen clearly that in 180 minute. It also can be seen that in 180 menit the biggest removal of COD was gained for the three types of adsorbents, however the optimum time is still not be reached considering the given trend is continously raising. For the activated bentonite physically through the heating process in 600^0C, it is gained the COD removal as much as ± 75%, while for the both physical and chemical activated bentonite, the removal reached up to ± 86% and ± 89%, each for nitric acid and benzene as the activator. The duration of contact time between adsorbat and adsorbent would effect the capaticy of adsorption.The longer contact time, the more organic compound will be adsorpted untill it become balance. On the other hand, when it has been balanced, the contact time will no longer give effect to the capacity of adsorption.

This study used three types of treatment for the bentonite, there are ; bentonited activated physically through the heating process in 600^0C (adsorben 1), adsorbent activated phisically and chemically by using nitric acid (adsorben 2) and benzene (adsorben 3) as activators. The heating process in furnace is conducted to keep the form of bentonite stable in adsorber coloumn during the adsorption process. From Figure 3, it showed that the adsorbent is only activated physically which is by heating in furnace in 600^0C for one hour, it gave the smallest percentage of removal.While the bentonite that has been activated both physically and chemically gave the better result, especially the activation using benzene. This is also supported by the each CEC values of bentonite as displayed in table 2. In thermal activation, high temperature can omit water molecule and other impurities. The change of structure and composition during the heating process can vary based on the chemical composition of natural bentonite and time of heating. The over heating can cause the short space in structure and interlayer of bentonite, the small interlayer space can cause the particle diffusion closer to another atom that can affect the space surface and CEC value of the bentonite. This is supported from CEC data in table 2.

The purpose of chemical activation by using nitric acid is to exchange Ca^{2+} that exist in Ca-bentonite become ion $H^+$and release ion Al, Fe, dan Mg and other impurities in the bentonite structure, so that the bentonite become active. For that purpose, beside nitric acid, sulfate acid and choride acid are chemical compound that also commonly used. During the activation process, Al, Fe, and Mg are dissolved in solution, then the acid adsorption occured to the bentonite structure, so the structure series have wider space.

Beside activated with acid, in this study the bentonite is also activated by using organic compound, benzene. Generally, bentonite is modified with organic compound to raise its afinity to organic compunds that trigger COD and TOC in liquid waste. The organic cation that is used shoul have the positive contain. The organic cation will replace the anorganic cation in bentonite interlayer area so it can increase its activity to the organic compound.

The modified bentonite with organic compound is namely organo bentonite. The addition of organic compound can also create organo bentonitewhich has spesific character, one of its important characters is thermal stability. In this study, the activated bentonite using benzene has higher thermal stability compared to the activated bentonite using nitric acid and heated in furnace.The CEC value of the organo bentoniteis as much as 48 meq/100gr, bigger that the bentonite activated with HNO_3 and heated in furnace which is as much as 42 meq/100gr and 30 meq/100gr (table 2.).

From Figure 4, It was resulted the COD removal after flotation process.The coloumn effluent of adsorption in 180 minutes with activated bentonite after being processed in flotation tank and resulted the average COD removal for ± 93%. Studies about flotation method have been done by several reseachers, Hong Zhong et al, 2015 studied the removal of mine ore by using flotation method and managed to remove up to 100%. So did Jinzhou Qu et al (2015), he observed the characterization of floatation in distributing coal particle, while I Wayan Budiarsa (2012) analyzed the COD removal of industrial waster which resulted the efficiency more than 90%. From the several previous researchers and the result from this study, it showed that the flotation method is the approprate method to remove COD in liquid waste of palm oil mil.

5. Conclusion

The combination between adsorption and flotation process showed the significant effect to the drecreading of COD level in liquid waste of palm oil mill. The percentage of COD removal of Palm oil mill liquid waste after adsorption process gained up to 88,76 %, that is in contact time 180 minutes with bentonite activated with benzene. The using of nitric acid and benzene as the bentonite activator gave the very good removal COD, where benzene as the activator was able to adsorb better. The floatation process was able to remove COD in liquid waste of palm oil mill mainly 93% with contact time 30 minutes.

Acknowledgement

The Authors gratefully acknowledges RISTEKDIKTI for funding this research under Hibah Bersaing Grant.

References

[1] A. Vidyadhar 2013 *Separation and Purifications Technology* **118** 305-312
[2] Department of Plantations 2010 Annual Data Local Government.
[3] E. Jorjani et al 2013 Fuel **114** 10-15
[4] Guerra, D.J.L et al 2013 Water Resources and Industry **4** 32-50
[5] Jinzhou Qu et al 2015 Procedia Engineering **102** 159-166.
[6] Jian Zhong Guo et al 2012 Journal of colloid and interface science 61-66
[7] Kovo G. Akpomie et al 2015 Journal of Basic and Applied Sciences **4** 1-13
[8] Khaled Okiel et al 2011 Egyptian Journal of Petroleum **20** 9-15
[9] The decision of the State Minister of Environment No. Kep-51 / MenLH / 10/1995 BAPEDAL (1999). About the Liquid Waste Quality Palm Oil Factory.
[10] Masamah Sharaft et al 2010 *Journal of Applied Chemical Researches*
[11] Mohammad Ishaq et al 2015 *Journal of Saudy Chemical Society*
[12] Mohd. Remy Rozainy et al 2014 *APCBEE Procedia* **10** 229-234
[13] Manjot Toor et al 2012 *Chemical Engineering Journal* **187** 79-88
[14] S. Zaidi et al 2015 *Arabian Journal of Chemistry*
[15] Sawsan A.M. Mohammed, et al 2011 Journal of Chemical and Petroleum Engineering
[16] Shobana Dey et al 2013 *Powder Technology* **246** 689-694
[17] Weishang Tan et al 2014 Bioresources echnology **160** 115-118
[18] Weishan Wang et al 2013 *Chemical Engineering Journal* **214** 343-354
[19] Yaxin Zhang et al 2012 *Journal of Environmental Sciences* **24** 1525-1532
[20] Zhiguan Yang et al 2013 *Chemical Engineering Journal* **223** 31-39
[21] Yuksel Sarikaya 2013 *Applied Surface Science* **283** 678

Effect of Elevated Temperature on Mechanical Assets of Metakaolin Base Steel Fiber Reinforced Concrete

Vijay Anand M[1], Azmi Ibrahim[2], Anand A. Patil[3] and K U Muthu[3]
[1]Dept. of Architecture, MS Ramaiah Institute of Technology, Bangalore, India
[2]Faculty of Civil Engineering, Universiti Teknologi Mara, 40450 Shah Alam, Selangor, Malaysia.
[3]Brindavan College of Engineering, Dwarakanagar, Banglore, India.

E-mail: kumuthu64@gmail.com

Abstract. The fact of vast usage of concrete leads to important problems regarding its design and preparation of eco-friendly to obtain an economic cost of the product on varieties of time periods. Conventional ordinary Portland concrete may not able to meet its functional requisites as it found inconsistency in high temperature. The exposing of concrete structure to elevated temperature may be in case of rocket launching space ships, nuclear power plants. In this experiment, to enhance the high temperature resistance, pozzolanic materials and steel fibres are added to preserve the strength characteristics of concrete structure. In this analysis, the pozzolanic admixture MK is used as partial replacement of cementatious materials. The volume fraction of steel fibre is varied 0.25%, 0.5%, 0.75% and 1% by preserving MK as stationary for 10% replacement of cement. The strength parameters of concrete such as compressive strength, split tensile strength and flexural strength are studied.

1. Introduction

Concrete is characterized by quasi-brittle seizure, the nearly complete loss of loading capacity, once failure started. With spurt in modern technology and immeasurable usage of concrete and mortars, the strength workability, durability and other major characteristics of the normal concrete require modifications to make it most suitable for all circumstances. Conventional ordinary Portland cement concrete which is designed on the basis of its strength characteristics may not able to meet its functional prerequisites or demands as it found discrepancy in aggressive environment, time of constructions, energy absorption Capacity, repair and retrofitting jobs etc. As a result of which usage of admixtures became complementary solution. In recent decade, the civil concrete structures have been subjected to fire accidents or blasts which lead to significant loss of assets. As a result of unpredictable misfortunes, structures are exposed to high temperatures that will produce considerable adaptations in the physical and mechanical properties with irreversible loss of strength and toughness. The material integrity and porosity is important aspects in fire resistance of concrete, where heat and mass transfer process is concerned. Especially, the pore pressure rise and temperature gradient related to spalling or separation of surface layers, which could significantly decrease the fire resistance of concrete structures. The main illustration for exposing concrete assemblies to high temperature are concrete foundations in case of rocket launching platform, structures in nuclear power station or sometimes those exposed to fire accidents or else blasts. As a result of rendering to the high temperature, explosive spalling has been noticed by many researchers often causing in serious deterioration of concrete possessions. High heat causes various physical (e.g., evaporation, condensation, water and vapour advection, phase expansion), chemical (e.g., dehydration, thermo-chemical damage) and mechanical (e.g., thermo-mechanical damage, cracking, spalling) activities takes place resulting in worsening of concrete properties.

In order to overcome the heat impairments on concrete properties, addition of admixtures is one of finest way which increases pozzolanic activity due to the presence of finer particles upgrades concrete characteristics. The enhancement of strength and durability of concrete carried out by adding cementitious materials such as Fly Ash (FA), GGBS, Silica Fume (SF), Metakaolin (MK) during construction results in better performance even in aggressive environment due to is filler effect and pozzolanic reactions. To enrich fire resistance, addition of fibres recommended since at elevated temperature, melting of fibres will create additional networks to allow water vapour pressure generated in structure. The inclusive of fibres will dominates over plastic and drying shrinkage cracking, in addition to enlarging energy absorption capacity of the concrete sample.

2. Literature Review

A lot of work has been done and going on the use of MK as a cementatious material in enhancing different properties of FRC. Since MK reduces porosity of harden concrete in addition to densifies and reduces thickness of interfacial zone, thus optimising the adhesion between the cement and aggregates. The steel fibres (ST) boost the energy absorption capacity of unheated specimen. In the research, various variations studied; such as effect of different upraised temperature exposure, effect on duration of heating period to record the physical observations immediately after heating and during testing of specimens like colour, propagation of cracks and mode of failure [1]. The reduction in strength at 200$^{\circ C}$ (not in case of specimen tested immediately after heating for one hour). There is no definite behaviour pattern could be observed with reference to four heating periods and three modes of testing conditions considered [1].

The thermal conductivity is maximum in room temperature and is gradually decreases at higher temperature of 600$^{\circ C}$ and 800$^{\circ C}$ respectively. The reduction of thermal conductivity was accompanied by the declining of density so that magnitude of the falling of thermal diffusivity was lower, approximately twice to trice, even though it seems to be a very good result [2]. It is pronounced that rise in strength of the plain cement mixes cured adiabatically at the elevated temperatures when compared to those cured isothermally at the same temperatures. Addition of FA and slag causes better confrontation to chloride permeability at all temperatures compared to OPC sample [3].

The residual strength of concrete losses as the exposure temperature rises, and prolonging the heating time is inversely proportional to residential strength [4]. The spalling frequency upsurges with higher MK content, mainly because of its dense pore structure seems to be the reason for such spalling [5]. Results providing conclusion that cement replaced by 10% of MK as admixture achieves maximum strength at pre and post mature stages for composite fibres (ST and PPF) reinforced HPC. For generation of optimal values, 10% MK replaced by cement and 1.25% of composite fibres for accomplishing maximum benefits [6].

When specimens exposed to elevated temperature increases, the strength of concrete is retarded due to spalling effect and at same time it began to deteriorate. However, the addition of fibres into the concrete mixture minimised the cracking effects that occur in the concrete [7]. The density of concrete increases marginally with upsurges in percentage of fibre content. Density of SFRC with MK has increased a bit over normal concrete and this is due to partial replacement of cement by MK, which densifies the concrete because of its micro- filler effect since having relatively finer particle size [8]. The desirable content of MK is 10% for development of High Strength Concrete as it gives reasonable mean target strength and workability than any other proportions. The mechanical strength characteristics values will increase up to 100$^{\circ C}$ and thereafter strength values fall with higher temperature [9].

The literature survey reveals that only rare number of experimental studies are carried on combined effect of high temperature and heating time on residual strength of concrete, which needs to be more investigation that will be beneficial in engineering practices.

3. Experimental Programme

3.1 Materials

In this work the locally available JK cement of 43 grade confirming to IS: 8112-1989 has been used. The specific gravity of cement was 3.15. The coarse aggregates obtained from local crushing unit confirming the IS 2383. The aggregate of perishable through IS 12.5 mm sieve and retained on IS 6.5 mm sieve was used in experimental practice. The specific gravity of coarse aggregate was 2.7 and bulk

IOP Conf. Series: Materials Science and Engineering **210** (2017) 012017 doi:10.1088/1757-899X/210/1/012017

density was 1600 kg/m^3. Conventionally locally obtained river sand was used as fine aggregates in this tentative study with fineness modulus and specific gravity of 3.1 and 2.7 respectively. The MK obtained from JITMULL MANUFACTURES located at Chennai, Tamilnadu having bulk density 0.4-0.5 kg/l. The crimped ST used in trials is obtained from STEWOLS INDIA (P) LTD. located at Nagpur, Maharashtra. The aspect ratio, specific gravity and density of 50, 7.9 and 7840 kg/m^3 respectively.

3.2 Mix Proportion

Table 1. Tabulation for mix design for M40 grade concrete.

Serial number	Components	Mass (kg/m^3)
1	Cement	337.5
2	Metakaoline	37.5
3	Water	150
4	Fine sand particles	758.93
5	Coarse aggregates	1237.7
6	Water – binder ratio	0.4

4. Experimental Procedure

The various dimensions of moulds to be prepared are arranged as listed below.

Table 2. Different mould dimensions arranged.

Sample	Tests	Dimensions (mm)
Cubes	Compression test	100×100×100
Cylinder	Split tensile test	L=300; d=100
Beams	Flexural test	100×100×100

While fraternising MK based FRC, primarily the dehydrated mixing of basic ingredients such as cementatious materials such as cement and MK, sand, dry coarse aggregates is carried out. After uniform colour of mix obtained, calculated amount of ST is spread uniformly over mixture. This process was followed by addition of water which was a mixing media. For different combinations, same casting procedure was pursued with different volume fractions namely 0.25%, 0.50%, 0.75% and 1% of Steel fibres. This wet mix is poured in predetermined dimensional moulds. After 28 days of water healing of specimens is followed by subjecting those to heavy raised temperature in muffle chamber. In current work, temperature variation is controlled to ranges of 200°C, 400°C, 600°C and 800°C for constant duration of one hour for each sample as shown below.

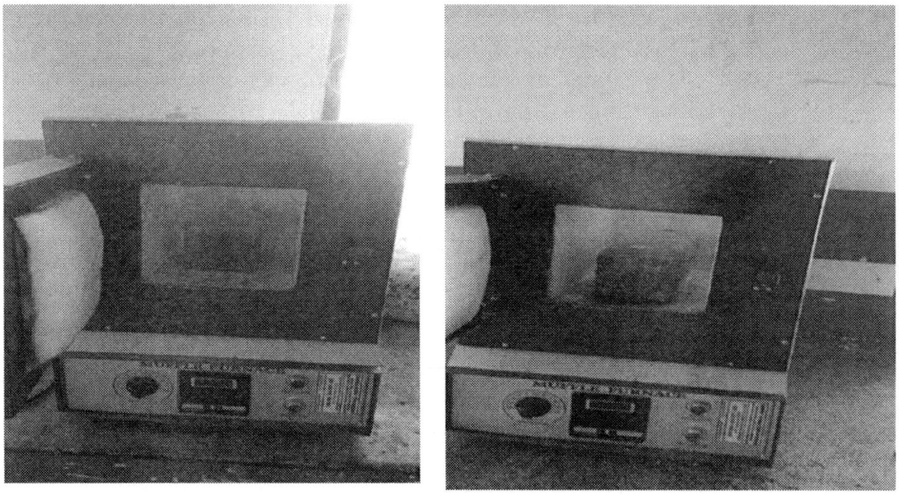

Figure 1. Specimens exposed to elevated temperature in muffle furnace.

4.1 Remarks on exposure to elevated temperature

The alteration in color of specimen due to heating after fire exposure may give a sense of temperature of concrete element has faced. The change of color of elements mainly depends upon the constituents with which it is made up of. In addition to this, surface cracking is remarked when elements is faced a temperature above 600°C. The cracking reveals the rate of heating and duration heat faced by concrete elements and it is well known fact that rapid heating to extreme temperature cause additional internal cracks.

Table 3. Tabulation of color changes of specimen

Temperature variation (°C)	Color changes
27°C	White
200°C	Off white / buff grey
400°C	Dusky grey
600°C	Dark grey
800°C	White

Figure 2. Color change and surface cracks due to exposure to elevated temperature.

The mechanical attributes of heat affected specimens are analysed by conducting compression, split tensile and flexural test.

5. Test Results and Discussion

In this experimental analysis, the pozzolanic admixture MK is used as partial replacement of cementatious materials. The volume fraction of steel fibre is varied 0.25%, 0.5%, 0.75% and 1% by preserving MK as stationary for 10% replacement of cement. The strength parameters in concrete such as compressive strength, split tensile strength and flexural strength are studied. Main intension is to determine optimal range of volume fraction of ST providing extreme strength behavior after heated condition. The cast concrete elements for compressive strength test (100×100×100 mm cubes), split tensile test (length =300mm, diameter =100mm cylinders), flexural strength test (100×100×300 mm beams) for 28 days of water curing and are subjected to heavy heat in muffle furnace. From test results each combination are compared with each other to determine efficient volume fraction of fibres and to study the behavior of elements for variety of temperature ranges.

5.1 Compressive Strength Results

In muffle furnace these elements were exposed to pre-determined temperature levels, primary to subjection to compression test. In current analysis, cubes of dimensions 100×100×100 mm^3 were tested in Compression Testing Machine (CTM) of maximum capacity of 3000KN to determine its strength and investigated accordance to IS: 516-1959. In CTM, load is applied sluggishly awaiting failure of cubes occurs.

Table 4. Mean compressive strength values of concrete specimens.

Temperature exposure	Combinations based of proportions of fibres			
	C1 (0.25%)	C2 (0.50%)	C3 (0.75%)	C4 (1.0%)
27^oC	50.16	53.03	53.16	53.16
200^oC	36.53	37.8	43.23	38.5
400^oC	34.33	36.80	37.66	36.00
600^oC	26.96	29.26	33.20	32.03
800^oC	20.40	19.96	19.83	25.03

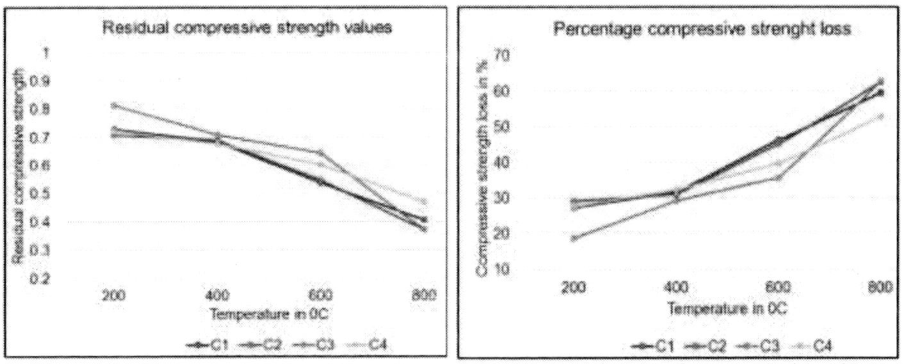

Figure 3. Compressive strength performance of the specimens after exposed to elevated

5.2 Split-tensile test results

The evaluation of tensile strength is compulsory to determine the load at which the concrete members may crack, where cracking is a form of tension failure. The experiment is conceded as per IS: 5816-1999. Cylinders of diameter 100mm and length 300mm were cast for different proportions of ST. Each element is tested in CTM of adequate capacity and applied with elevated loads at specified rate. Procedure is carried out until element develops tension cracks on surface for failure load.

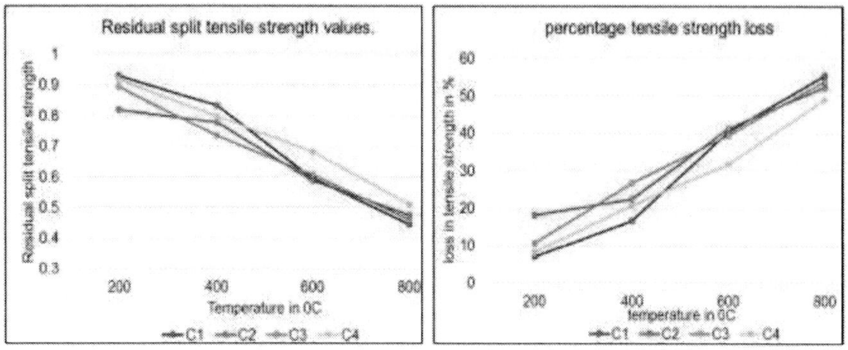

Figure 4. Tensile strength performance of the specimens after exposed to elevated temperature.

5.3 Flexural Test Results

For the flexural strength test, pre-defined dimensional prismatic specimens were cast. The size of beams in current assessment as defined in previous chapters, 100×100×300 mm^3 tested in Universal Testing Machine (UTM). Since its longitudinal dimension is reduced as per convenience to work with heating in muffle furnace. This experiment helps in studying modulus of rupture, bending capacity, in addition to providing flexural strength of prisms. Usually flexural strength is directly proportional to failure load and depends on aspect ratio. The test is approved as per IS: 516-1959.

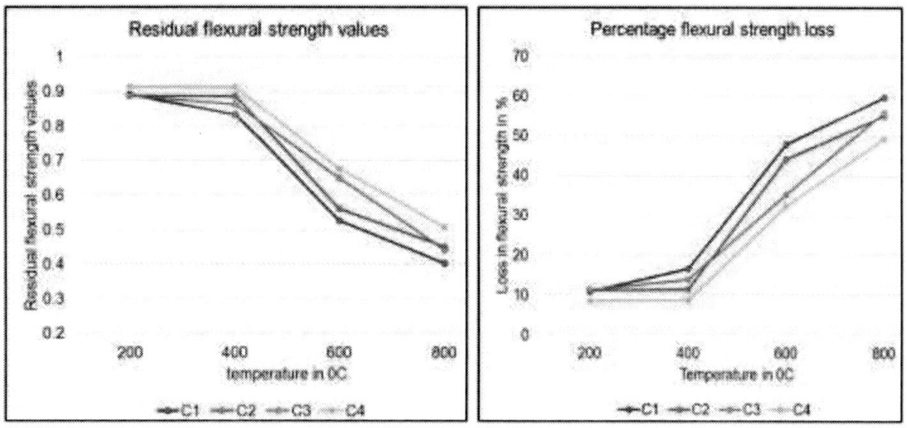

Figure 5. Flexural strength performance of the specimens after exposed to elevated temperature.

6. Conclusion

The MK composite concrete will show increasing strength by 6% with adding volume fraction of steel fibres. For volume fraction of 0.75% Steel fibres is providing optimal strength for primary stage of heating up to 600$^{\circ C}$. At extreme heat, addition of proportion of fibres will upgrades its strength characteristics even at 800$^{\circ C}$. Thus, it is finalized that 1% of volume fraction of fibres will be optimal at extreme temperature. It is well known fact that strength attributes will decline with improve in temperature. Overall remarks conclude that combination C4 with 1 % of steel fibres gives maximum residual strength and minimum strength loss at temperature 800$^{\circ C}$.

References

[1] G.K. Suneetha, H. Narendra and K.U. Muthu 2004 *The National Conference on Materials and Structures* 118-122
[2] R. Cerny, J. Podebradska, M. Totova, J. Toman, J. Drchalova 2004 *Materials and Structures* **37** 597-607.
[3] R. Chini and L. Acquaye 2005 *Materials and Structures* **38** 673-679.
[4] Toumi, M. Resheidat, Z. Guemmadi and H. Chabil 2009 *Jordan Journal of Civil Engineering* **3** 322-330.
[5] K.S. Kulkarni, S.C. Yaragal and K.S. Babu Narayan 2011 *International Journal of Applied Engineering and Technology* **1** 48-60 .
[6] M. Adams Joe and A. Maria Rajesh 2014 *International Journal of computational engineering research* **4** 56-59.
[7] Narayana Suresh, Vivek Bindiganavile and Mahesh Prabhu 2014 *Internatioanl journal of engineering technology and advanced engineering* **4** 220-224
[8] Aminuddin Jameran, Izni S. Ibrahim, Siti Hamizah S. Yazan, Siti Nor A. A. Rahim 2015 *The 5th international conference of euro Asia civil engineering forum* 818-824
[9] Amol Patil and P.D. Kumbar 2015 *International journal of research in advent technology* **3** 291-296.

IOP Conf. Series: Materials Science and Engineering **210** (2017) 012018 doi:10.1088/1757-899X/210/1/012018

Finite element analysis on the structural behaviour of square CFST beams

M.F. Javed*[1], N.H. Ramli[1], S. Kashif-ur-Rehman[1], N.B. Khan[1]
Department of Civil Engineering, University Malaya, Malaysia

E-mail: arbabfl@gmail.com

Abstract. This paper presents the finite element (FE) analysis and modelling of square concrete-filled steel tube (CFST) members subjected to a flexural load. A parametric study is conducted using the verified FE model to study the effect of the depth-to-thickness (D/t) ratio (18.75, 25, and 30), the compressive strength of infilled concrete (60, 80, and 100 MPa), and the yield strength of the steel tube (410, 500, and 600 MPa) on the flexural behaviour of the square CFST members. Decreasing the D/t ratio (from 30 to 18.75) can significantly increase the ultimate capacity of the square CFST members (up to 25%) while having a marginal effect on the initial stiffness of the CFST members. The ultimate bending capacity of the CFST members increases by up to 55% when the yield strength of the outer steel tube increases from 410 MPa to 600 MPa. However, the flexural capacity increases by only 12% when the compressive strength of the infilled concrete increases from 60 MPa to 100 MPa, hence showing a marginal effect. Results of the parametric studies are used to assess the current design models, and Han's model predicts the most accurate flexural capacity.

1. Introduction

Several researchers presented different procedures in order to study the effect of both static and dynamic loading on different composite systems [1], [2], [3]. Numerous problems arise while dealing with modelling of composite systems that combine two different materials, ductile steel and brittle concrete. The modelling of such composite sections should capture the relative stiffness of each material properly. Concrete filled steel tube (CFST) is a composite material which is composed of steel tube filled with concrete. The use of CFST columns and beams, in construction of buildings, has been increased exponentially in recent decades [4], [5], [6]. As suggested by structural engineers, the CFST members are the most interesting composite members for several modern building projects [7], [8]. Compared to conventional hollow steel/concrete members, this type of composite member has more advantages; such as high speed of construction work due to the omission of reinforcing bars and framework, low structure cost, conservation of environment, high ductility and strength capacity [9], and also provide a good damping merits and are excellent seismically resistant [10],[11]. Furthermore, different types of concrete including recycled aggregates can be used as an infill in CFST, thus helps in cleaning the environment [12], [13], [14], [15].

Several researchers have investigated the structural behaviour of circular CFST. However, only few studies are available in the literature regarding the numerical analysis on the structural performance of circular CFST. Hu et al., [16] proposed material constitutive model for circular CFST columns subjected to pure bending. They performed finite element analysis (FEA) and validated the theoretical results with the experimental data, and concluded that the concrete acts as ideal material to resist compressive loading in the typical applications, only when the depth-to-thickness (D/t) ratio is greater than 74. In addition, the infilled concrete has no significant effect on the strength of CFST columns when the D/t ratio is less than 20.

Lakshmi and Shanmugam [17] proposed a semi-analytical method by using an iterative process, the relationship between moment-curvature-thrust is generated to investigate the behaviour of CFST columns. They considered different cross-sections including square, rectangle and circle of compact section in the FE analysis. The pin-ended columns subjected to bi-axial or uniaxial loads were studied. They verified the theoretical and experimental results and concluded that the moment capacity of columns decreases with an increase in axial load. Whereas, Liang et al. [18] studied the FEM behaviour of simply supported composite beams subjected to combined shear and flexure loading. A 3D FE model was developed to consider the material and geometric non-linear behaviour of composite beams and is verified by experimental results. The verified FE model was then used to study the effects of different factors effecting the combined moment and shear capacities of concrete slab and composite beams. In addition, the effect of the degree of shear connection on the vertical shear strength of deep composite beams loaded in shear was studied. Design models were proposed for vertical shear strength including the contributions from composite action and concrete slab. For the design of simply supported composite beams under combined bending and shear, ultimate moment-shear interaction was proposed. The proposed design models provided economical and consistent design procedure for simply supported composite beams.

Lu and Kennedy [19] examined the effects of different D/t ratio and different shear span to depth ratio on twelve square and rectangular steel beams. They reported that the flexural strength of the CFST is increased by 10-30% over that of hollow steel sections, depending on relative proportions of concrete and steel. The flexural stiffness is also increased due to concrete infill. The shear-span to depth ratio has no significant effect on ultimate strength of CFST. Formulae for the strength of square and rectangular CFST under flexure load were suggested. However, Gho and Liu [20] studied the flexural behaviour of rectangular CFSTs by using high strength steel and concrete. It was concluded that AISC, ACI and EC4 substantially underestimated the flexural strengths of high -strength CFST. Whereas, Elchalakani et al., [21] performed experiments on circular CFSTs subjected to pure bending. They reported that the concrete filling in the steel tubes increases the ductility, energy absorption and strength of thinner sections. Han [22] proposed a model that can predict the structural behaviour of CFST after conducting a series of experiments on CFST beams. Both square and rectangular CFST were tested. The author concluded that the moment capacities of CFST beams predicted by BS5400 (1979), LRFD-ASIC (1999), EC4 (1994) and AIJ (1997) are lower than the experimental values. However, the model is only valid for D=100-2000 mm; f_{scy}=200-500 MPa and f_{ck}=20-80 MPa and cannot be used for high strength and ultra-high strength concrete. Recently, Zhou, Fan [23] studied the behaviour of CFST under tensile loading. It was observed that the stiffness of hollow steel tube is increased 31.8% by adding concrete infill in hollow steel tube.

As the available literature lacks the numerical model for the flexural behaviour of square and rectangular CFST, therefore, the main aim of this study is to investigate the flexural behaviour of CFST beam numerically by using the commercial FEA package ANSYS [24]. This investigation contains the FEA modelling technique to analyse the flexural behaviour, interaction of concrete and steel and load-deflection curves of different types of steel and concrete for square and rectangular CFST's under pure bending. FEA model is verified with the experimental results available in the literature. After FEA model verification, the numerical analysis is extended to perform the parametric study like compressive strength of concrete, D/t ratio and yield strength of steel on the performance of CFST beams under flexure load.

IOP Conf. Series: Materials Science and Engineering **210** (2017) 012018 doi:10.1088/1757-899X/210/1/012018

2. Details of finite element model

General

The commercial finite element tool ANSYS is used to study the flexural behaviour of the CFST beams. Two basic materials are considered to model the flexural behaviour of CFST beams. Steel is used to model the outer tube while the inner core is defined with concrete material.

Finite element type and boundary conditions

A simply supported hollow-steel beam filled with concrete under two-point loading is used to investigate the flexural behaviour and strength. Mesh independency study has been performed to get optimized mesh with less computational cost. Element type for concrete core and steel tube is selected from element library in ANSYS. Based on the geometric features of steel and concrete, a suitable element type for the analysis is selected. The steel tube part was divided into 1344 elements and 9312 nodes while the concrete model included 13080 elements and 78952 nodes. Accurate boundary conditions have to be applied on the nodes lying on the plane of symmetry, in order to reflect the accurate flexure behaviour. Hinge supports are defined by constraining the nodes in y-direction and allow free to move in X and Z-direction. Both supports are located at the distance of 100 mm from free ends. No imperfection of loading or boundary conditions are taken into consideration.

Materials Modelling

A steel model for structural steel as suggested by Han et al., [25] is used for uni-axial stress-strain relation of steel. In this model, hardening of the structural steel is considered. The deformation of steel includes elastic, elastic-plastic, plastic, hardening and fracture are shown in figure 1. Where f_p, f_y, and f_u represents the proportional limit, yield, and ultimate strength of steel, at their respective strains, and $\varepsilon_e = 0.8\varepsilon_{e1}$, f_y/E_s, $\varepsilon_{e1} = 1.5\varepsilon_e$, $\varepsilon_{e2} = 10\varepsilon_{e1}$, $\varepsilon_{e3} = 100\varepsilon_{e1}$.

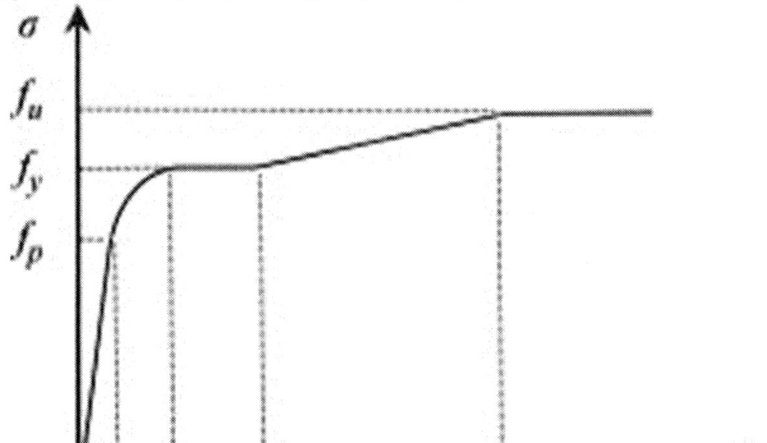

Figure 1. Schematic sketch of uniaxial stress–strain relation for steel [25]

The Von-Mises yield function with associated plastic flow is used in multi axial stress states. The structural steel is assumed to have isotropic hardening behaviour, so that yield stresses increase or decrease in all stress direction when plastic straining occurs. When the Mises stress reaches the yield stress of the steel, stress can still increase when subjected to further plastic straining, because there is a hardening stage in the stress-strain relationship. The modulus of elasticity and the Poisson's ratio for the steel are taken as 2x105 N/mm2 and 0.3, respectively. Poisson's ratio of concrete under flexural stresses is in the range of 0.15-0.22, with a representative value of 0.20. Table 1 shows the material properties of steel and concrete used in the analysis.

Table 1. Material Properties of steel and concrete

Materials	Dimension of section (mm)	Yield stress (f_y), (MPa)	Ultimate stress (f_u), (MPa)	Modulus of elasticity (E), (MPa)	Poisson's Ratio
Steel	72 x 72 x 3.2	345	510	2.1×10^5	0.3
Concrete	65.6 x 65.6	32.3	-	2.842×10^4	0.2

3. Model verification

The accuracy and efficiency of the developed FE model are demonstrated through the comparisons between the FE and experimental results with different parameters as performed by Soundararajan and Shanmugasundaram [26]. The ultimate flexure load, moment-curvature curves and load-deflection curves of CFST beams are considered in the verification of the FE model developed. The details of the experimental setup and the boundary conditions in the FE model are shown in figure 2 and 3, respectively.

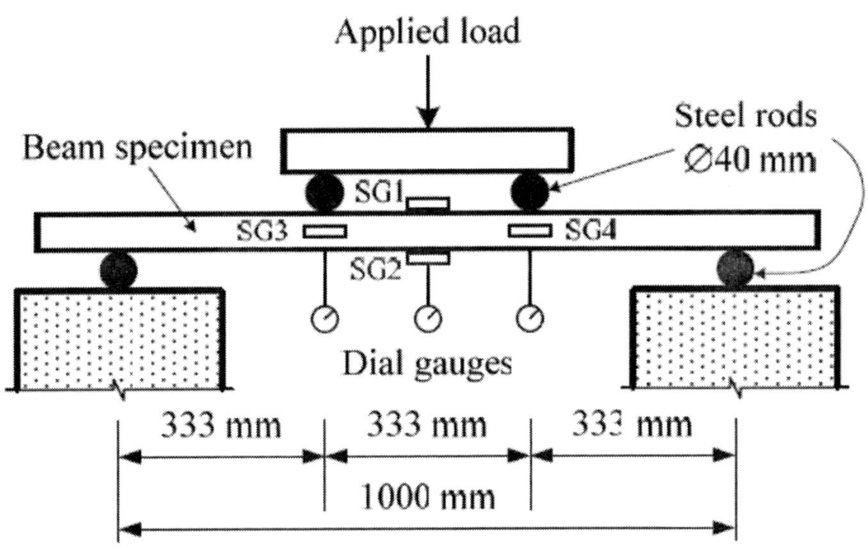

Figure 2. The details of the experimental test setup [26]

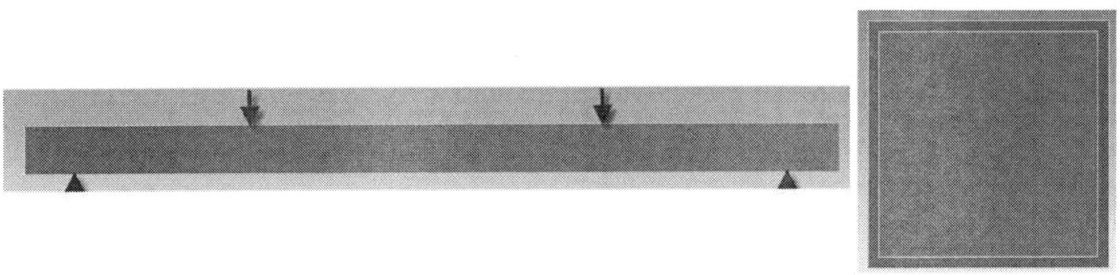

Figure 3. Boundary conditions and Cross-section of the modelled beam

4. Ultimate Strengths of CFST beams

The geometry, material properties and experimental results of CFST beams tested by Soundararajan and Shanmugasundaram [26] are mentioned in table 2. The dimensions of steel tube are kept constant in all experiments. The specimens were made of normal mix concrete (NMC), Fly-ash concrete (FAC), quarry waste concrete (QWC) and low-strength concrete (LSC). The depth-to-thickness ratio of the steel tube was kept as 20.5. All Specimens were tested under the two-point flexure loads. CFST beams were constructed using concrete of strengths in the range of 21 MPa to 32.6 MPa. The yield strength of steel tubes was 345 MPa while the ultimate stress was 510 MPa as shown in table 1.

Table 2. Geometry, Material properties and Ultimate moment of tested specimen

Specimen designation	Dimensions (mm)	Area of steel, A_s (mm^2)	Area of concrete, A_c (mm^2)	Types of filled concretes	Compressive strength of concrete (MPa)	Average ultimate moment
NMC	72 x 72 x 3.2	881	4303	Normal Mix concrete	32.6	10.01
FAC	72 x 72 x 3.2	881	4303	Fly Ash Concrete	32.5	10.12
QWC	72 x 72 x 3.2	881	4303	Quarry Waste Concrete	21.63	9.99

Table 3. Ultimate Flexure Moment

Specimen Designation	$P_{e,Exp}$ (kN)	Average, P_e (kN)	P_{eFE} (kN)	Mu_{Exp} (kN-m)	Average ultimate moment exp. (kN-m)	Ultimate Moment FE (kN-m)	Mu_{Num}/Mu_{exp}
NMC-1	61			10.06			
NMC-2	58	60.67	61	9.57	10.01	10.07	1.005
NMC-3	63			10.40			
FAC-1	60			9.90			
FAC-2	61	61.33	61	10.07	10.12	10.07	0.994
FAC-3	63			10.40			
QWC-1	59			9.74			
QWC-2	60	60.33	59	10.00	9.99	9.73	0.974
QWC-3	62			10.23			
Mean							0.991
Standard Deviation (SD)							0.016
Co-efficient of Variation (COV)							0.016

Table 3 shows the experimental and computational ultimate flexure strengths of CFST beams. Where Pu,exp denotes the experimental ultimate flexure strength and Pu,fe represents the ultimate flexure strength predicted by the FE model. It can be seen from table 3 that the predicted ultimate flexure strengths of tested specimen are in good agreement with experimental results. The mean ultimate flexure strength predicted by the numerical model is 1.01 times the experimental value with a standard deviation of 0.016 and a coefficient of variation of 0.016.

IOP Conf. Series: Materials Science and Engineering **210** (2017) 012018 doi:10.1088/1757-899X/210/1/012018

5. Load-deflection curves

The load-deflection curves for CFST beams predicted by FE model were compared with the experimental results provided by [26]. The load-deflection curves for NMC predicted by FE model and obtained from the experiments are shown in figure 4. It can be observed that the load-deflection curve predicted by the numerical model is in good agreement with the experimental results. The difference between the experimental and numerical ultimate flexure load is only 3%. Whereas, figure 5 gives the predicted and experimental axial load-deflection curves for the specimens of QWC concrete as tested by the same author. It can be seen that the numerical model predicts very good relationship of the load-deflection curve for the tested specimen up to the ultimate flexure load. However, the computed ultimate load is slightly higher than the experimental value due to the uncertainty of the initial imperfections and material properties.

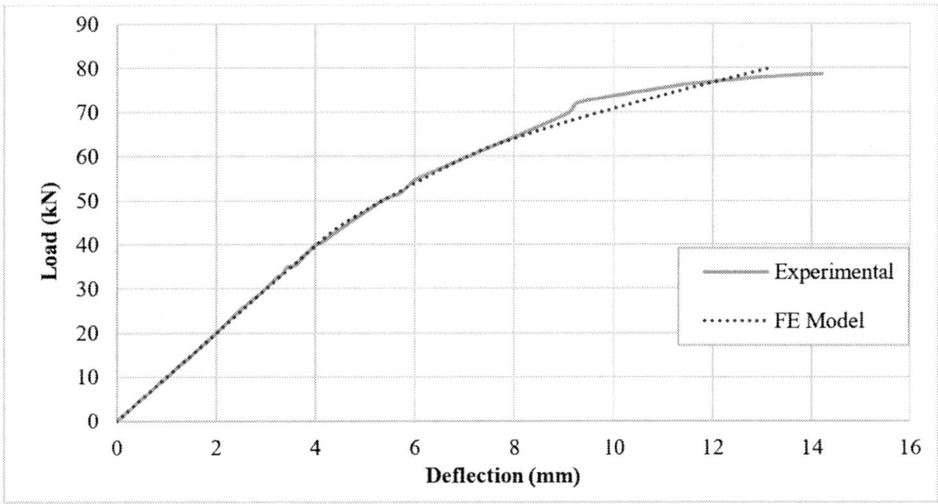

Figure 4. Load-Deflection Curve for NMC

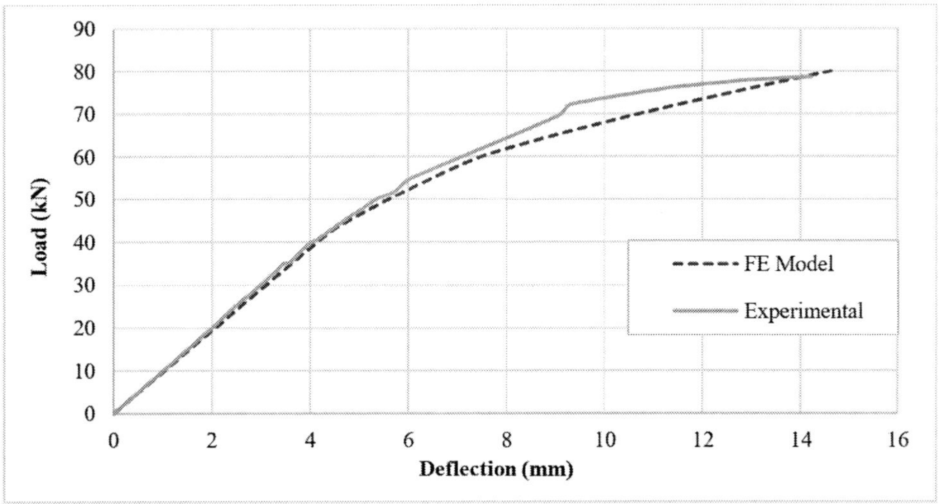

Figure 5. Load-Deflection curve for Quarry Waste Concrete (QWC)

6. Parametric Study

An extensive parametric study was performed to investigate the influences of depth-to-thickness ratio, concrete compressive strengths and steel yield strengths on the fundamental behavior of CFSR beams under flexure load only. Only one variable was considered at a time to assess its individual effect. Three different concrete strengths with a range of steel yield strength and D/t ratio were selected that represent CFST members currently used in Malaysia.

6.1 Effect of Depth-to-thickness ratio

The strength of CFST beams depends on depth-to-thickness ratio (D/t). The FE was used to examine the effects of D/t ratio on the ultimate moment/load capacity and load deflection curves of CFST beams. The dimensions of the beams for the analysis were selected as a cross-section of 80 x 80 mm and length of 1200 mm. The D/t ratios of the beam sections were calculated as 18.75, 25 and 30 by changing the thickness of steel tubes. The yield and tensile strengths of steel tubes were 345 MPa and 510 MPa, respectively, and the modulus of elasticity was about 200 GPa. Whereas, the compressive strength and modulus of elasticity of the in-filled concrete was about 30 MPa and 29000 MPa, respectively.

The influence of D/t ratio on the flexure load-deflection curves for CFST beams are illustrated in figure 6. As can be seen in the figure 6 that by increasing the D/t ratio of the CFST beams have slightly reduces their initial stiffness. However, increasing the D/t ratio significantly reduces the ultimate flexure load carrying capacity of CFST beams. This is attributed to the fact that the beam having larger D/t ratio has a lesser area of steel and it may have undergone local buckling which reduces the ultimate load carrying capacity of the beam. Similar conclusions were drawn by different author Chen and Wang [27], after performing experiments on thin-walled walled dodecagonal section double skin CFST [27] and circular CFST [28] under bending.

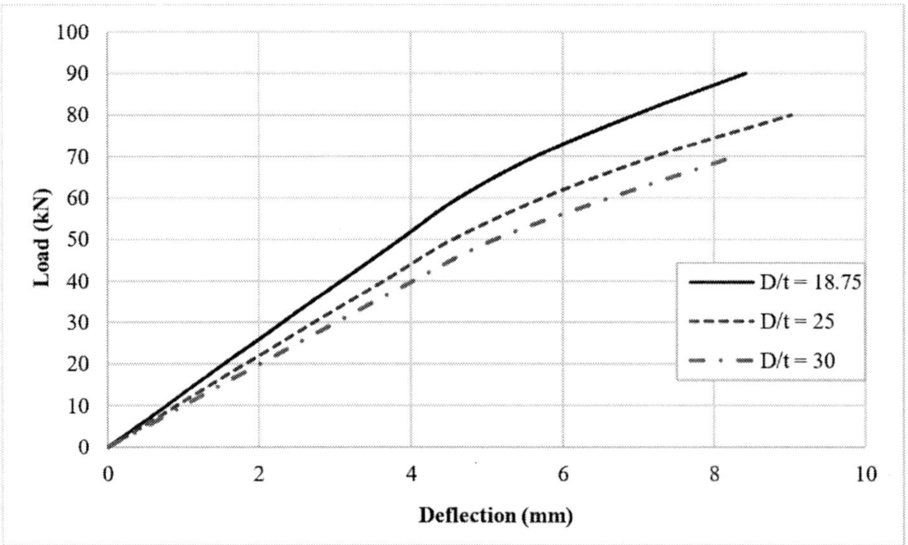

Figure 6. Load-deflection curve of different D/t ratio

6.2 Effect of grade of concrete

The effect of compressive strengths on the ultimate capacity and behaviour of CFST beams were studied by the FE Model. In the parametric study, the compressive strength of concrete was varied from 60 to 100 MPa. Both depth and width of the steel tube was placed constant as 80 mm. The steel tube wall was 2.5 mm thick, so that its D/t ratio was 30. The steel yield and ultimate strengths were 345 and 510 MPa, respectively, and the modulus of elasticity was 200 GPa.

The flexure load-deflection curves for CFST beams with different concrete strengths are shows in figure 7. As can be seen in figure 7 that the ultimate flexure strengths of square CFST beams does not increase significantly with an increase in the compressive strength of concrete. However, by increasing the compressive strength of concrete from 60 to 80 MPa and 100 MPa, the ultimate flexural strength was increases by about 4% and 12%, respectively. Figure 7 also describes that the increase in the compressive strength of concrete resulted very low improvement in the initial stiffness of the square CFST beams. Similar observations were made by number of researchers after conducting experimental study on the behaviour of ultra-high strength square CFST [29] and for circular CFST [28] under flexural load.

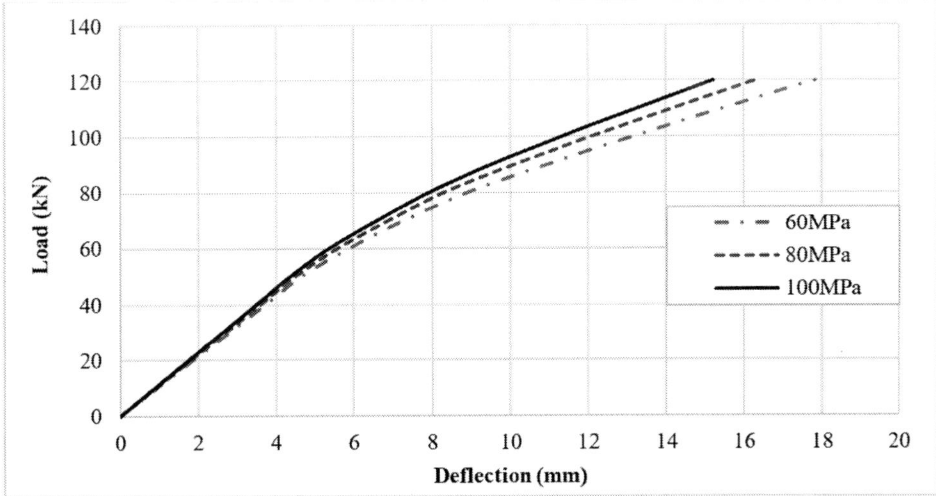

Figure 7. Load-deflection curve for different compressive strengths of concrete

6.3 Influences of steel yield strengths

The square CFSR beams with different steel yield strengths and a cross-section of 80 x 80 mm with same D/t ratio of about 25 were analysed using the FE model. The Yield strengths of the steel tubes were 410 MPa, 500 MPa, and 600 MPa the corresponding ultimate strengths were 520 MPa, 590 MPa and 690 MPa, respectively. The Young's modulus of steel was 200 GPa. The steel tubes were infilled with 60 MPa concrete have modulus of elasticity of 32600 MPa.

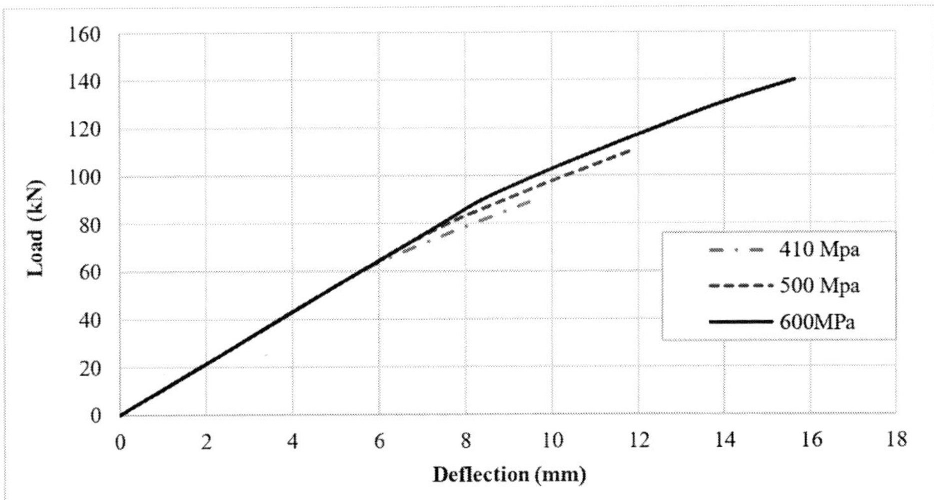

Figure 8. Load-deflection curve for different types of steel

Figure 8 illustrates the influences of steel yield strengths on the flexure load deflection curves of square CFST beams. It can be observed from the figure 8 that the yield strength of steel does not have an effect of the initial stiffness of beams. However, the ultimate flexure strength of square CFST beams is found to increase significantly with an increase in the yield strength of steel. By increasing the yield strength of steel from 410 MPa to 500 MPa and 600 MPa, the ultimate flexure load of the square CFST beam is found to increase by 22% and 55% respectively. Similar conclusions were drawn by Duarte et al., [30] for rubberized CFST members.

7. Comparison of moment capacities with design codes

7.1 Eurocode

As mentioned in Eurocode [31], the ultimate capacity of CFST columns subjected to combined bending and compression is determined from an interaction curve. An interaction curve between moment and axial compression can be obtained for column by assuming several possible positions of neutral axis with the cross-section, and determining the moments and internal forces from the resulting plastic stress blocks. In the absence of axial load and steel reinforcement in CFST, the equation for moment becomes.

$$M_{pl,Rd} = f_{yd}\left(W_{pa} - W_{pan}\right) + 0.5f_{cd}\left(W_{pc} - W_{pcn}\right) \quad (1)$$

$$W_{pc} = \frac{(b-2t)(h-2t)^2}{4} \quad (2)$$

$$h_n = \frac{A_c f_{cd} - A_{sn}(2f_{sd} - f_{cd})}{2bf_{cd} + 4t(2f_{yd} - f_{cd})} \quad (3)$$

$$W_{pan} = 2th_n^2 \quad (4)$$

$$W_{pcn} = (b-2t)h_n^2 - W_{psn} \quad (5)$$

Where, h_n is the distance from compression region to the center line of the CFST cross-section, W_{pc} and W_{pa} are the plastic section modulus for concrete and steel respectively W_{pan} and W_{pcn} are the plastic section modulus of the corresponding components within the region of 2hn from the centre-line of the composite cross-section.

As seen in table 4, Eurocode design rules give mean, standard deviation and Co-efficient of Variation (COV) values of 0.73, 0.08 and 0.12 and are conservative. However, it predicts the moment capacity more accurately than AISC-LRFD and CIDECT design rules.

7.2 AISC-LRFD

The flexural strength of CFST beams and columns according to LRFD [32] were calculated based only on the steel hollow section. The Plastic moment capacity of a CFST beam may be evaluated as;

$$M = Zf_y \quad (6)$$

where, Z is he plastic section modulus and fy is the yield strength of steel tube.

As shown in table 4, AISC-LRFD design rules give mean, standard deviation and COV values of 0.65, 0.08 and 0.13 and are most conservative.

7.3 CIDECT

The Ultimate moment capacity for CFST beams according to the CIDECT [33] can be defined as;

$$M_{u,CIDECT} = M_{ratio}\frac{D^2B-(D-2t)^2(B-2t)}{4}f_y \quad (7)$$

where, M_{ratio} is a ratio of the baring capacity of composite hollow section to that of the hollow section, D, t and B is the depth, thickness and width of composite section respectively, f_y is the yield stress.

CIDECT design rules give mean, standard deviation and COV values of 0.72, 0.08 and 0.11, respectively, and the values are conservative, as can be seen in table 4.

7.4 Han (2004) model

According to Han [22], the ultimate moment capacity of the CFST beam is given by;

$$M_u = \gamma_m f_{scy} W_{scm} \quad (8)$$

$$f_{scy} = (1.18 + 0.85\xi) f_{ck} \quad (9)$$

$$W_{scm} = \frac{B^3}{6} \quad (10)$$

$$\xi = \frac{A_s f_{yk}}{A_c f_{ck}} \quad (11)$$

$$\gamma_m = 1.04 + 0.48\ln(\xi + 0.1) \quad (12)$$

where, M_u is the moment capacity of the CFST beam, f_{scy} is the nominal yielding strength of the steel tube, W_{scm} is the section modulus of CFST cross-section, ξ is the constraining factor, and γ_m is the flexural strength index. However, Han model is only valid for D=100-2000 mm; f_{scy}=200-500 MPa and f_{ck}=20-80 MPa and cannot be used for ultra-high strength concrete.

The mean, standard deviation, COV and capacity reduction factors of the ratio of moment capacitates from FE model and HAN model are shown in table 4. Han Model values are the best in comparison to all codes with its overall mean, standard deviation and COV values of 0.80, 0.08 and 0.10, respectively.

Table 4. Comparisons of Results from different codes and FEA

Cross Section	Thickness	Yield Strength	Compressive Strength	Euro Code	Factor	AISC	Factor	CIDECT	Factor	Han	Factor	FEA Results
80 x 80	3	410	60	11.86	0.80	10.57	0.71	11.83	0.80	13.11	0.88	14.85
80 x 80	3	500	60	14.27	0.79	12.89	0.71	14.23	0.78	15.55	0.86	18.15
80 x 80	3	600	60	16.93	0.73	15.47	0.67	16.83	0.73	18.27	0.79	23.10
80 x 80	2.5	345	60	8.68	0.66	7.56	0.57	8.55	0.65	9.59	0.73	13.20
80 x 80	2.5	345	80	8.62	0.65	7.56	0.57	8.70	0.66	10.20	0.77	13.20
80 x 80	2.5	345	100	8.52	0.57	7.56	0.51	8.78	0.59	10.73	0.72	14.85
80 x 80	2.5	345	30	8.64	0.75	7.56	0.65	8.27	0.72	8.58	0.74	11.55
80 x 80	3	345	30	10.00	0.76	8.89	0.67	9.61	0.73	10.37	0.79	13.20
80 x 80	4	345	30	12.54	0.84	11.41	0.77	12.19	0.82	14.33	0.96	14.85
		Mean			0.73		0.65		0.72		0.80	
		SD		-	0.08	-	0.08	-	0.08	-	0.08	-
		COV			0.12		0.13		0.11		0.10	

8. Conclusions

The verification and applications of a three-dimensional finite element model developed for the non-linear analysis of CFST square beams at ambient temperature have been presented in this paper. The finite element model was verified by comparisons of computational solutions with experimental results present in [26] for three different types of concrete. It has been shown that the load-deflection curves and ultimate flexure load capacity of square CFST beams predicted by the numerical model are generally in good agreement with the experiential results. Therefore, the verified numerical model was used for parametric studies to investigate the effects of D/t ratio, yield strength of steel and compressive strength of concrete on the load-deflection curves and ultimate flexure load carrying capacity of CFST beams under flexure load.

This paper provided new FE results on the basic behaviour of square CFST beams filled with concrete with different parameters including depth-to-thickness ratio, concrete compressive strengths and steel yield strengths. The study concluded that depth-to-thickness of steel tube has significant impact on the performance of the CFST beams. It is also observed that compressive strength of concrete and yield strength of steel used in CFST beams, have very marginal effect on the structural behaviour of CFST. The FE results presented can be used for the verification of other nonlinear analysis techniques and to modify composite design codes for square CFST beams. Furthermore, the FE model developed can be used in the design and analysis of high strength concrete subjected to flexure load in practice. In addition, it can assist in generating the required data needed to develop design recommendations for CFST members with fewer limitations imposed.

References

[1] Abbas H S, Bakar S A, Ahmadi M and Haron Z 2015 *Građevinar* **67** pp 225-33
[2] Kabir M H, Fawzia S and Chan T H Effects of layer orientation of CFRP strengthened hollow steel members
[3] Milanović M, Cvetkovska M and Knežević P 2016 *Građevinar* **67** pp 1187-97
[4] Kiymaz G and Seckin E 2014 *Steel and Composite Structures* **17** pp 253-69
[5] Hua W, Wang H-J and Hasegawa A 2014 *Steel and Composite Structures* **17** pp 517-33
[6] Han L-H, Li W and Bjorhovde R 2014 *Journal of Constructional Steel Research* **100** pp 211-28
[7] Liew J R, Xiong M and Xiong D 2014 *J High-rise Build* **3** pp 1-7
[8] Liew J R and Xiong D 2012 *Advances in Structural Engineering* **15** pp 1487-503
[9] Qu X, Chen Z and Sun G 2015 *Steel and Composite Structures* **18** pp 71-90
[10] Yu T, Hu Y M and Teng J G 2016 *Journal of Constructional Steel Research* **124** pp 12-22
[11] Yang Y-F 2015 *Steel and Composite Structures* **18** pp 213-33
[12] Silva A, Jiang Y, Castro J, Silvestre N and Monteiro R 2016 *Journal of Constructional Steel Research* **122** pp 557-70
[13] Hamidian M R, Shafigh P, Jumaat M Z, Alengaram U J and Sulong N R 2016 *Journal of Cleaner Production* **129** pp 282-91
[14] Chen Z, Xu J, Liang Y and Su Y 2014 *Steel and Composite Structures* **17** pp 929-49
[15] Yang Y-F and Ma G-L 2013 *Thin-Walled Structures* **66** pp 62-75
[16] Hu H, Su F and Elchalakani M 2010 *Steel and Composite Structures* **10** pp 415-28
[17] Lakshmi B and Shanmugam N E 2002 *Journal of Structural Engineering* **128** pp 922-33
[18] Liang Q Q, Uy B, Bradford M A and Ronagh H R 2005 *Journal of Structural Engineering* **131** pp 1593-600
[19] Lu Y Q and Kennedy D L 1994 *Canadian Journal of Civil Engineering* **21** pp 111-30
[20] Gho W-M and Liu D 2004 *Journal of Constructional Steel Research* **60** pp 1681-96
[21] Elchalakani M, Zhao X-L and Grzebieta R 2001 *Journal of Constructional Steel Research* **57** pp 1141-68
[22] Han L-H 2004 *Journal of Constructional Steel Research* **60** pp 313-37
[23] Zhou M, Fan J-S, Tao M-X and Nie J-G 2016 *Journal of Constructional Steel Research* **121** pp 202-15
[24] Guide A C-S M 2010 Ansys Inc *Canonsburg, PA*
[25] Han L-H, Zhao X-L and Tao Z 2001 *Steel and Composite Structures* **1** pp 51-74
[26] Soundararajan A and Shanmugasundaram K 2008 *Journal of Civil Engineering and Management* **14** pp 107-14
[27] Chen J, Wang J, Xie F and Jin W-l 2016 *Thin-Walled Structures* **98** pp 293-300
[28] Chitawadagi M V and Narasimhan M C 2009 *Journal of Constructional Steel Research* **65** pp 1836-45
[29] Guler S, Copur A and Aydogan M 2012 *Structural Engineering and Mechanics* **43** pp 225-37

[30] Duarte A, Silva B, Silvestre N, de Brito J, Júlio E and Castro J 2016 *Composite Structures* **150** pp 28-40
[31] EN C 2004 1–1. Eurocode 4: design of composite steel and concrete structures—Part 1.1: General rules and rules for buildings *Brussels: European Committee for Standardization*
[32] LRFD A 1994 *Chicago: American Institute of Steel Construction*
[33] Wardenier J, Dutta D and Yeomans N 1995 *Design guide for structural hollow sections in mechanical applications*: Verlag TÜV Rheinland)

IOP Conf. Series: Materials Science and Engineering **210** (2017) 012019 doi:10.1088/1757-899X/210/1/012019

Assessment of water quality of Sembilang River receiving effluent from controlled municipal solid waste (MSW) landfill in Selangor

T.N.B. Tengku Ibrahim[1*], F. Othman[2] and N.Z. Mahmood[1]

[1]Institute of Biological Science, Faculty of Science, University of Malaya, 50603 Kuala Lumpur, Malaysia.
[2]Department of Civil Engineering, Faculty of Engineering, University of Malaya, 50603 Kuala Lumpur, Malaysia.

E-mail: kunielam@gmail.com

Abstract. Most of the landfills in Malaysia are situated near to the main river basin that supplies almost 90% of water requirement. This includes landfills in Selangor where a total of 20 landfill sites are situated in 5 main river basins and the highest number of operating landfills (three) are at the Selangor River Basin (Jeram, Bukit Tagar and Kuang Inert landfills). This situation has caused wide concern over the water safety, even the leachate has been treated. The leachate itself still contains contaminants that are difficult to treat. The main objective of this study is to investigate the effect on water quality of Sembilang River that receives effluent from the nearby landfill. In this study, we analyzed samples of water from ten sampling stations starting from the upstream to downstream of Sembilang River. The water quality was evaluated by the Water Quality Index (WQI) depending on in-situ and laboratory analysis. 11 water quality variables are selected for the quality assessment; temperature, pH, turbidity, salinity, dissolved oxygen, biochemical oxygen demand, chemical oxygen demand, total suspended solid, ammoniacal nitrogen, phosphate and nitrate. The result indicated that, when the effluent mixed with the river water, the water quality decreased gradually and was found to be lower at a few stations. The water quality of Sembilang River falls under Class III of Water Quality Index with ranges between 68.03 to 43.46 mg/L. It is revealed that the present scenario of water quality of Sembilang River is due to the effect of effluent from the landfill.

1. Introduction

Since the last century, landfilling has been the common method for the disposal of solid waste in many cities in developing Asian countries. Despite all the issues related to existing landfill sites and the increasing volume of waste generation, landfills remain the integral part of solid waste management plans [1-4]. Due to its economic advantages, landfills can potentially lead to a wide range of compounds with environmental, wildlife and human health concern. Gases and liquid leachate are the main pollution issues associated with landfill sites where leachate from MSW landfills is a threat to the quality of groundwater and surface water [5-7]. Liquid waste or leachate is produced during the stabilization in a landfill with a mixture of liquid from layers of waste and undergoes several hydrological and biogeochemical reactions. The characteristic of the landfill leachate can be varied depending on landfills characteristics, such as; the type of waste and its decomposition degree, the climatic variation during the waste disposal, the waste compaction, landfill cover, height of landfill layers, as well as the landfill environment: waste degradation phase, humidity, precipitation, temperature, etc. All these factors represent the characteristics of the landfill leachate which normally contain high Chemical Oxygen Demand (COD), total organic carbon (TOC), biochemical oxygen demand (BOD), pH, ammonical nitrogen (NH_3-N) and heavy metals content [8,9]. Landfill can be classified as young (less than 5 years) which is in acidogenic phase and has large amounts of degradable organic substances and heavy metals. While old leachate (more than 10 years) is highly contaminated with NH3-N and has a lower BOD/COD ratio lower than 0.1. Study shows that unionized ammonia and high concentration of nitrate and nitrite can cause fatalities in several species of fish and serious methemoglobinemia in young children [9, 12, 13]. This has become worldwide concern since most of the landfill sites are located near to the river with no facilities for collection and/or treatment of leachate and landfill gas. There are about 251 landfills with different sizes and ages in Malaysia and it is estimated that landfills in Malaysia can generated 3.0 million liters leachate per day [14]. Leachate then is released into the river after full or partial treatment. Selangor itself have 20 landfills that are situated at 5 main river basins and Selangor River Basin has the highest number of operating landfills, which are a major source of agriculture and productivity, industrial and domestic water supply [10, 11]. The list of landfill sites in Selangor is shown in Table 1. However, to ensure no watercourses is affected by the disposal of pollutants especially to the aquatic ecosystem, economic resource potential, public health and welfare, Malaysia have its very own wastewater discharge standard and the receiving water quality standard. Objective of this study is to assess the Sembilang River water quality for the disposal of effluent generated from the Jeram Sanitary Landfill.

Table 1. List of landfills sites a long Selangor River Basin

Landfill	River Basin	Nearest River	Year	Type of Landfill	Operator
Kuang Inert	Selangor	Dungon River	June 2007-Present	Non-sanitary Active	Worldwide Landfill Sdn. Bhd
Jeram	Selangor	Sembilang River	Jan 2007-Present	Sanitary Active	Worldwide Landfill Sdn. Bhd.
Bukit Tagar	Selangor	Bangkar River	2005-Present	Sanitary Active	KUB-Berjaya Enviro Sdn. Bhd.
Kundang	Selangor	Kuang River	2002-April 2006	Non-sanitary Closed	Alam Flora Sdn. Bhd
Kubang Badak	Selangor	Kubang Badak River	1985-March 2007	Sanitary Closed	Worldwide Landfill Sdn. Bhd.
Bukit Beruntung	Selangor	Sabai River	NA	Non-sanitary	Majlis Daerah Hulu Selangor
Sungai Sabai	Selangor	Beletak River	2000-present		
Ulu Yam Bharu	Selangor	Liam River	1997-April 2007	Non-sanitary Closed	Majlis Daerah Hulu Selangor

2. Materials and Method

2.1. *Description of the study area*

The study was carried out at Sembilang River near Jeram Town, Kuala Selangor, Malaysia. The main usage of land along the Sembilang River is for palm oil plantation and heading downstream there are industrial areas such as plactic factory and timber factory. 20 km to the north there is Pantai Remis which is a popular tourist attraction especially seafood. Sembilang River has also received attention when the prime minister has allocated RM3.8 million to rebranded fishing activity centre to serve the local fisherman. Another land usage near to Sembilang River is for Jeram Sanitary Landfill which is located at Lot No. 1595, 2598, 2959 with approximate area of 160 acres. Jeram Sanitary Landfill is operating under Worldwide Landfills Sdn. Bhd. since 1st of January 2007. This landfill has received 2,100 tonnes of waste per day with dominating waste type are domestic (95%), bulky waste and garden waste.

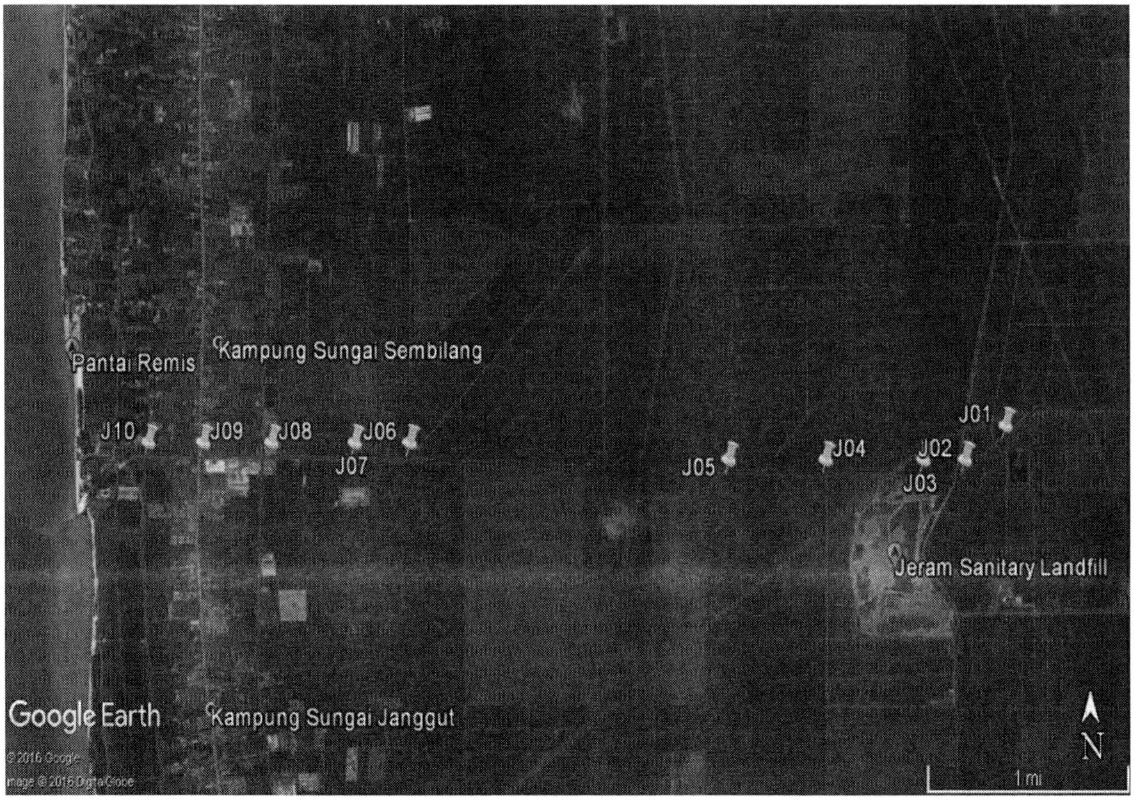

Figure 1. Study area: Sembilang River

2.2. *Data collection*

Figure 1 shows the sampling station of Sembilang River water quality. 2 stations were located upstream of Sembilang River and before the Jeram Sanitary Landfill. 8 other sampling stations were located downstream of Sembilang River and located at landfill effluent discharge, palm oil plantation, industrial zone and near to Pantai Remis on a monthly basis from September 2015 to September 2016 (Table 2). Water samples were then frozen to keep the temperature at 4°C for transportation for further analysis in laboratory. 11 water quality variables were selected for the quality assessment; temperature, pH, turbidity, salinity, dissolved oxygen (DO), biochemical oxygen demand (BOD), chemical oxygen demand (COD), total suspended solid (TSS), ammonia nitrogen, phosphate and nitrate. Standard Methods of Water and Wastewater (AHPA 2005) were followed for water samples collection and water samples analysis.

2.3. *Data analysis*

To obtain the results, Water Quality Index (WQI) were used by entering the average of six water quality parameters including DO, BOD, COD, pH, ammonia (NH_3-N) and total suspended solid (TSS) into the WQI formula (DOE):

WQI= [0.22*SIDO] + [0.19*SIBOD] + [0.16*SICOD] + [0.15*SIAN]+ [0.16*SISS] + [0.12*SIpH]

where;
SIDO = Subindex DO
SIBOD = SubIndex BOD
SICOD = SubIndex COD
SIAN = SubIndex NH3-N
SISS = SubIndex SS
SIpH = Subindex pH

Table 2. Sampling points along the Sembilang River

Location Point	Status	Coordinate		Remarks
		Latitude (Deg.)	Longitude (Deg.)	
J01	Upstream	3.196	101.373	Upstream of Jeram Landfill
J02		3.194	101.370	Upstream of Jeram Landfill
J03	Downstream	3.194	101.367	Where landfill effluent discharge
J04		3.194	101.360	Just after the landfill
J05		3.194	101.353	Palm oil plantation site
J06		3.195	101.330	School
J07		3.195	101.326	Factories
J08		3.195	101.320	Industrial zone
J09		3.195	101.315	Highway culvert
J10		3.195	101.311	Nearer to the sea

3. Results and discussion

The acidity and alkalinity of water can be measure by pH parameters which is simple but important where most chemical process in aquatic environment are controlled by any change in its value [11]. All the water samples at Sembilang River showed slightly acidic pH with the pH values varied from 3.98 to 6.45. pH value is less than 5 at upstream stations (J01 and J02) for all samples indicating the

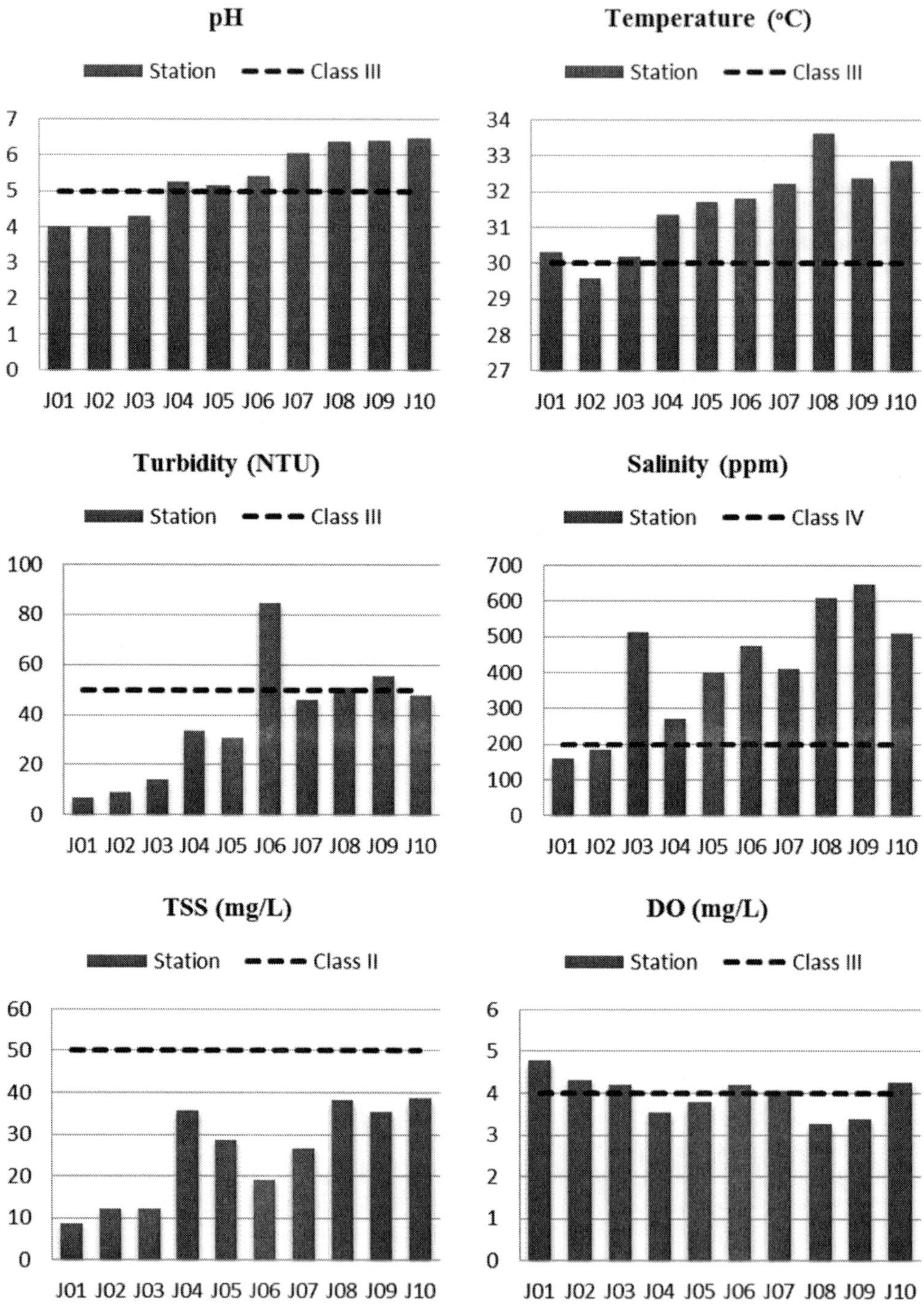
pH
Station Class III
0 1 2 3 4 5 6 7
J01 J02 J03 J04 J05 J06 J07 J08 J09 J10
Temperature (°C)
Station Class III
27 28 29 30 31 32 33 34
J01 J02 J03 J04 J05 J06 J07 J08 J09 J10
Turbidity (NTU)
Station Class III
0 20 40 60 80 100
J01 J02 J03 J04 J05 J06 J07 J08 J09 J10
Salinity (ppm)
Station Class IV
0 100 200 300 400 500 600 700
J01 J02 J03 J04 J05 J06 J07 J08 J09 J10
TSS (mg/L)
Station Class II
0 10 20 30 40 50 60
J01 J02 J03 J04 J05 J06 J07 J08 J09 J10
DO (mg/L)
Station Class III
0 1 2 3 4 5 6
J01 J02 J03 J04 J05 J06 J07 J08 J09 J10

IOP Conf. Series: Materials Science and Engineering **210** (2017) 012019 doi:10.1088/1757-899X/210/1/012019

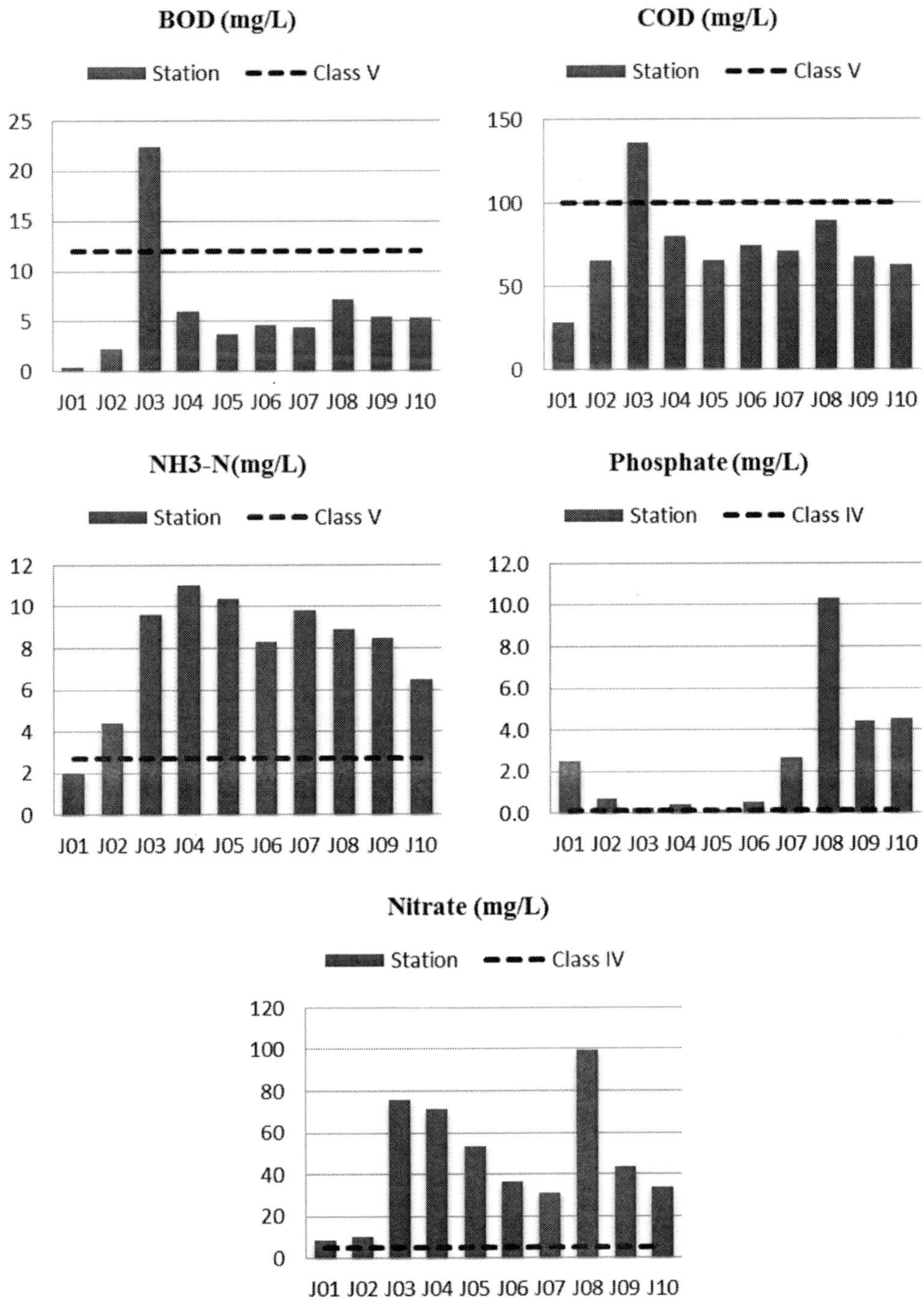

Figure 2. Average value of water quality measurement along the Sembilang River

acidic conditions is under Class III of DOE Water Quality Index Classification. This shows that the pH level was influenced by the leachate, where after point J01 and J02 the pH values increased to 6.45 at station J10. The mean pH value for Sembilang River was found to be 5.34 (Figure 2). Past studies had equally revealed the acidic nature of Sembilang River due to oxidation of pyrite (FeS_2) from the river soil which is common characteristic reported for acidic sulfate soil and could increase the mobility of heavy metals from young leachate. The higher reading of pH at the downstream of Sembilang River was caused by the palm oil plantation near the landfill where NPK fertilizer was used [12].

The temperature has a major role in dissolution and precipitation reactions and is major factor in determining water chemistry as well as biological conditions. These can be caused from various environmental processes: natural and anthropogenic [13, 14]. Global climate change, regional land-use, heated effluents from power generation plants, river flow, depth of water, cloud cover and solar radiation are those that can affect the river temperature. Temperature is important parameter for water quality, for example, most chemical and biological activity is a function of temperature because it may affect the river in so many ways such as the geographical distribution, growth rate and reproduction of aquatic life and this is dependent on both natural and anthropogenic energy exchange process [15]. Surface water is usually within the temperature range of 0 °C to 30 °C. For Sembilang River, the surface water temperature varied between 29.56 to 33.62℃. The mean value of temperature measured along the river of Sembilang River was 31.69℃ (Figure 2). Moreover, most of the time during the ten sampling dates the weather was sunny and dry. These conditions affect the temperature the most.

Natural processes such as weathering rocks and rain takes place where there is ample opportunity for the mineral crystals that constitute rock to oxidize. However, all natural waters contain salts. High level of salts not only affects the ecological health of rivers and estuaries, but also the drinking water and irrigation with serious economic, social and environmental consequences for both rural and urban communities [16]. For Sembilang River, the salinity level is between 160 ppm from the upstream and 646 ppm at the downstream near the Pantai Remis. The mean value for salinity for Sembilang River was found to be 418 ppm (Figure 2). Turbidity plays a relevant role in controlling the penetration of light into the water column where it is one of the water quality parameters that can shape the physical landscape and regulating ecological systems. Turbidity can be caused from suspended particles, either organic or inorganic and dissolved organic matter in coastal and estuarine areas. High level of turbidity can influence underwater light transmission and therefore altering the productivity and the living conditions of aquatic, animal and vegetation. From ten number of sampling at Sembilang River we found that the turbidity level is between 6.92 NTU and 84.70 NTU. The highest turbidity level was found at J07DS where the river is of grass water. The mean value for turbidity of Sembilang River is 37.99 NTU (Figure 2).

TSS is one of the most commonly used indicators of surface water quality problems. This is because TSS has a direct (physical, biological and ecological) and indirect (toxicology) impact on aquatic ecosystems. Thus, TSS is considered a good proxy for current water conditions and is useful to assess the risk of water quality hazard. The TSS level of Sembilang River for station J01US and J02US exhibits <25mg/l indicating the category of Class-I. After the landfill, when effluent is mixed with the river water, suspended solid increases gradually and are found to be higher at station J03DS, J05DS and J11DS. The value is fluctuating at downstream of Sembilang River and most of the value are of Class I-II. Past studies also show a strong correlation between TSS and turbidity. This coloration is also found in water sample at station J09 where both results for turbidity and TSS were high. Location of sampling station can be the factor for high result for TSS at J09 where it is located at the highway culvert, at which the river is mostly filled with garbage and there is no water stagnation. The mean value for TSS of Sembilang River is 25.61 mg/l (Figure 2).

Dissolved oxygen (DO) is frequently used to evaluate the water quality as the effect of industrial and municipal effluents as well as indicators of the environmental quality of watersheds. By the combination of physical, chemical and biological characteristics of the streams, the oxygen demand substances including algal biomass dissolved organic matter, ammonia, volatile suspended solids and sediment oxygen demand [17, 18]. Sembilang River recorded the level of DO in range of 3.27 mg/l to 4.77 mg/l (Figure 2). The average levels of DO were 3.98 mg/l where at sampling station J09DS the value is slightly low with 3.27 mg/l, which may be affected by the effluent from the landfill and the land usage around that area which is near to the industrial zone. Excessive algae growth exists at station J01US for the third and fourth sampling (Figure 2).

5-day BOD (BOD_5) is the most widely used parameter of organic pollution applied to both wastewater and surface water. BOD involves the measurement of the dissolved oxygen (DO) used by microorganism in the biological oxidation of organic matter. BOD test is used to determine the quantity of oxygen that is required to biologically stabilize the organic matter present. The level of BOD at station J01US (0.43 mg/l) and J02US (2.27 mg/l) show the lower concentration of BOD compared to other stations (Figure 2). As the landfill effluent is disposed at station J03DS, the BOD value exhibits higher value, which is 22.46 mg/l. From station J04DS to J10DS, it shows that the range of BOD value is between 3.72 mg/l and 6.00 mg/l (Figure 2). The mean value of BOD for Sembilang River is 6.20 and it is classified in Class III of DOE Water Quality Index Classification. COD represents the organic pollution level in water, and it is therefore one of the most commonly utilized parameters for monitoring water quality. It is defined as a number of oxygen equivalence consumed in the oxidation of organic compounds using strong oxidizing agents. The landfill effluent is disposed at station J03DS and the COD value is higher among other sampling points with 135.64 mg/l. The COD value ranges from 28.29 mg/l to 135.64 mg/l. For station J05DS to J11DS, most of the value are greater than 60 mg/l due to higher rate of oxygen consumption from water which is represented the Class IV, indicating unsuitability for any purposes. These phenomena may be the consequences resulted from the organic and inorganic suspended materials runoff from agricultural lands, in this case, is palm oil plantation where, in past studies there is a direct correlation of COD and agricultural discharges [19]. The mean value of COD for Sembilang River is 73.98 mg/l (Figure 2).

The microbial degradation of nitrogenous organic material is one of the most important sources for producing NH_3-N in a river. These compounds are transferred into the environment through different sources including byproduct of sewage decomposition. The aqueous ammonia concentration above 0.2 mg/l may be hazardous to many aquatic organisms [20]. Before the landfill, the level of NH_3-N at two stations J01US and J02US show low concentration and the values range from 2.01 to 4.42 mg/l. Whereas after the landfill, all stations show higher value (>2.7 mg/l) which is in ranges of 6.48 to 11.01 mg/l. This is indicate that the river pollute by ammonia from the fertilizers which is from the palm oil plantation at stations located downstream of the river. The concentration greater than 2.7 mg/l, indicates Class-V categorization. The mean value of NH_3-N for Sembilang River is 7.94 mg/l (Figure 2). Phosphorus (P) and nitrogen (N) are the primary nutrients that in excessive amounts pollute the lakes, streams and wetlands. Phosphate and Nitrate are compound containing nitrogen and phosphorus at elevated levels, can have harmful effects on humans and animals. Nitrates in water can cause severe illness in infants and domestic animals. In inorganic form, phosphate is preferred for plant growth but other forms can be used when phosphate is unavailable. Excess nitrate in water can cause the low concentration of DO [21]. Station J08 recorded the highest level of phosphate that is 10.33 mg/l. The range of phosphate for all 10 stations is between 0.15 to 10.33 mg/l. While for nitrate, the values fluctuated from station to station. The range is between 8.66 to 99.36 mg/l. The increase of NO_3 and PO_4 will also increase BOD and COD where the result for all the parameters shows the same results for all sampling stations. The mean value for both phosphate and nitrate are 2.65 mg/l and 46.45 mg/l (Figure 2).

3.1. Water Quality Index

Water quality index (WQI) combines the measures of several water quality variables in such way as to produce a single score that is representative of quality impairments or suitability of use. Overall WQI for Sembilang River was calculated by averaging WQI from all sampling dates in each station. On the basis of six parameters such as DO, pH, COD, BOD, NH_3-N and suspended solid the WQI value ranges from 68.03 to 43.46 mg/L. Most of the value of the data set is classified under Class III. From pollution index, it reveals that most of the stations show that water quality parameters level in station J03 is more polluted as compared to other stations (Figure 3). Discharged effluent from the landfill was the point source of water pollution and have higher pollution rate.

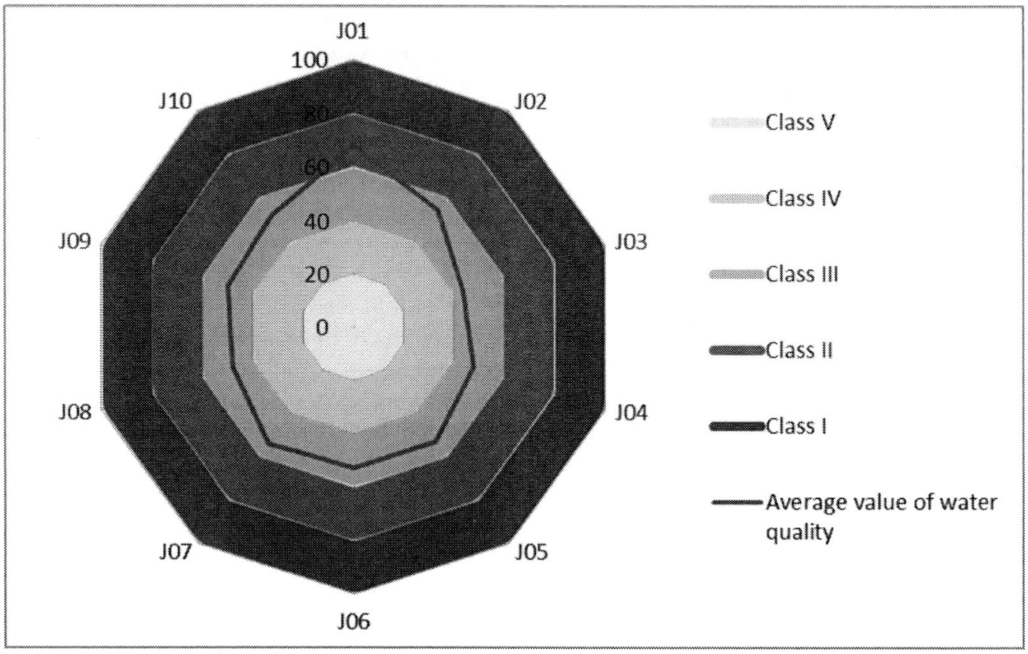

Figure 3. Average value of water quality of Sembilang River based on DOE Water Quality Index classification

4. Conclusion

Systematic study conducted to determine the impact of municipal solid waste disposal at Jeram Sanitary Landfill has revealed that Sembilang River is being contaminated by the effluent from the landfill. The data indicates the landfill as the point source for all the contaminants because the concentration of pollutants increases radically from upstream to downstream. As can be seen from the result, the WQI for Sembilang River was categorized under Class III of Malaysian Water Quality Standards, indicating the water is not suitable to be used as water supply and in need of extensive treatment. The results show that the effluent discharge from the landfill was lower as compared to raw leachate but still exceeded the standard, as can be seen in Sembilang River water quality. Thus, in general, the overall observation made in this study is that Sembilang River quality is affected by landfill operation.

5. Acknowledgement

We would like to thank the Sustainability Science (SuSci) Research Cluster, University of Malaya for the UMRG-Programme & Flagship fund FL001-13SUS and MOHE, Malaysia for their financial support in My Brain PhD on this project.

6. Reference

[1] I. Azni, I. Bulent and H. Mohd Nassir 2004 *Journal of Material Cycles and Waste Management* **6** 104-110.
[2] Z. Dongqing, S.K. Tan and M.G. Richard 2010 *Waste Management* **30** 921-933
[3] A. Allahyar and M.R. Sabour M R 2014 *Waste Management* **34** 2528- 2536
[4] O. Al- Jarrah and H. Abu- Qdais 2006 *Waste Management* **26** 299- 306
[5] N. Yang, A. Damgaard, P. Kjeldsen, L.M. Shao and P.J. He 2015 *Waste Management* **46** 362-372
[6] D.L. Jones, K.L. Williamson and A.G. Owen 2006 *Waste Mangement* **26** 825- 837
[7] T. Eggen, M. Moeder and A. Arukwe 2010 *Science of the Total Environment* **408** 5147- 5157
[8] K. Kaur, S. Mor and K. Ravindra 2016 *Journal of Colloid and Interface Science* **469** 338-343

[9] A. Fernandes, M.J. Pacheco, L. Ciriaco and A. Lopes 2015 *Applied Catalysis B: Environmental* **176-7** 183-200
[10] T.N. Tengku Ibrahim, N.Z. Mahmood and O. Faridah 2017 *Asian Journal of Microbiology Biotechnology and Environmental Science* **19** 43-48
[11] R.S. Lokhande, P.U. Singare and D.D. Pimple 2011 *International Journal of Ecosystem* **1** 1- 9
[12] Muhammad Umar, H. Abdul Aziz and M.S. Yusoff 2010 *International Journal of Chemical Engineering* 1-6
[13] N. Yusof, A. Haraguchi, M.A. Hassan, M.R. Othman, M. Wakisaka and Y. Shiraj 2009 *Waste Management* 2666-2680
[14] I. Yusoff, Y. Alias, Mohamad Yusof and M.A. Ashraf 2013 *Science Asia* **39** 392-409
[15] T. Kinouchi, H. Yagi and M. Miyamoto 2007 *Journal of Hydrology* **335** 78-88
[16] D. Caissie, N. El-Jabi and M.G. Satish 2001 *Journal of Hydrology* **251** 14-28
[17] T. Kinouchi and Z. Xin 2013 *Journal of Hydrology* **489** 16-25
[18] A.F. Pillsbury 1981 *Scientific American* **245** 32-43
[19] P.R. Kannel, S. Lee, Y.S. Lee, S.R. Kanel and S.P. Khan 2007 *Environmental Monitoring Assessment* **132** 93-110
[20] E. Sanchez, M.F. Colmenarejo, J. Vicente, A. Rubio, M.G. Garcia, L. Travieso and R. Borja *Ecological Indicators* **7** 315-328
[21] S.A. Al-Shami, C.S. Md Rawi, A.H. Ahmad, S. Abdul Hamid and S.A. Mohd Nor 2011 *Ecotoxicology and Environmental Safety* **74** 1195-1202

Preliminary study of soil liquefaction hazard at Terengganu shoreline, Peninsular Malaysia

H. Hashim*, **M. Suhatril and R. Hashim**

Department of Civil Engineering, 50603 Kuala Lumpur, Malaysia

E-mail: huzaifahashim@siswa.um.edu.my

Abstract. Terengganu is a shoreline state located in Peninsular Malaysia which is a growing hub for port industries and tourism centre. The northern part offers pristine settings of a relax beach areas whereas the southern part are observed to be a growing centre for development. The serious erosion on soil deposit along the beach line presents vulnerable soil condition to soil liquefaction consists of sandy with low plasticity and shallow ground water. Moreover, local earthquake from nearby fault have present significant tremors over the past few years which need to be considered in the land usage or future development in catering the seismic loading. Liquefaction analysis based on field standard penetration of soil is applied on 546 boreholes scattered along the shoreline areas ranging 244 km of shoreline stretch. Based on simplified approach, it is found that more than 70% of the studied areas pose high liquefaction potential since there are saturated loose sand and silt deposits layer ranges at depth 3 m and up to 20 m. The presence of clay deposits and hard stratum at the remaining 30% of the studied areas shows good resistance to soil liquefaction hence making the area less significant to liquefaction hazard. Result indicates that liquefaction improving technique is advisable in future development of shoreline areas of Terengganu state.

1. Introduction

Soil liquefaction effect in built environment often relates to lack of information and knowledge of the regional setting especially in the sense of geological aspect and the seismic characteristic. The deadly events reported in moderate to high seismic region around the world present fatalities and catastrophic failure of the buildings and civil engineering structures [1-4].

Furthermore, some human activities such as land reclamation and alteration of natural landscape contribute to the probability occurrence of soil liquefaction events and their sudden negative impact [5]. Anthropogenic impacts can cause limited aspect of resistance and capabilities of the soil environment towards such hazard [1]. Careless action and incentive on land development due to missing information and limited knowledge of regional setting often contribute to the multiplication of economic collapse. Therefore, we are exposed, unnoticeably or under the demand of land usage and development, to the risk of soil liquefaction.

We should be aware and take note that there are no absolute solutions against soil liquefaction. It is recommended to make a preliminary investigation as simple, representable and easily access overview Hence integrated measures can be taken to prepare proper planning and design in any land usage and development in adapting to the unforeseen events in the near future [6]. Towards promoting safe built environment and future development, implementation of existing workflow in assessing soil liquefaction potential is very important but most significantly an increasing role from the authority should be addressed in assessing soil liquefaction information follow by the prevention and mitigation program.

The common approach to define soil liquefaction potential is that it is the product of resistance of the soil and loading of ground motion [7]. Based on this definition or namely simplified method, the evaluation of soil resistance and earthquake loading can be conducted separately in the first place, but have to be combined for the final result which is termed as safety factor against soil liquefaction.

There are very few scientific literatures from local scale however reliable information can be observed from neighboring countries such as Indonesia and Thailand [8]. The analysis conducted involves different complexity levels and variety of methods. Examples of similar studies dealing with soil liquefaction and soil liquefaction assessment at the near about regional geological setting and seismicity characteristics promotes the use of simplified method in quantifying the hazard.

In this paper, the aim is to present a study on soil liquefaction assessment at Terengganu shoreline, Peninsular Malaysia. A total of 6 illustrations summarizing the findings related to soil liquefaction are presented: soil composition, SPT-N distribution, zone of saturation, grain size distribution plot in liquefaction margin, liquefaction susceptibility margin and liquefaction potential layer. In general, vulnerable soil towards soil liquefaction consists of saturated sandy soil type and when combines with intense ground motion, the soil will lose its shear strength and behaves as a viscous liquid in split second. The first five illustrations including charts are preliminary screening which notes all fundamental factor that influence liquefaction susceptibility. The last illustration presents results that have been analyzed using the simplified method. By this study, we try to introduce illustrations in which could be interesting to be applied to other prone areas of the region.

2. Study area

Terengganu offers a wide coverage of pristine beaches stretching approximately 244 km distance. The beach is quiet and is a home to scattering peaceful fishing village. A number of resorts located in the shoreline areas are constructed a very simple way as to accommodate tourist and local travellers. The shoreline areas are well preserved in the northern coast district: Besut and Setiu as there are very few development and changes in the natural environment. As the stretch line reaches the capital state of Terengganu, the beaches are no longer picture-perfect due to serious level of erosion. The erosion is caused by strong waves during monsoon season, coastal development projects and various made-made structures. Table 1 presents the abbreviation and general information of shoreline district in Terengganu state. A collection of 546 borehole reports at 95 locations along the shoreline of Terengganu state made it possible in addressing the depth of the studied areas. The type of beach is sandy type for all districts. The population are observed to be high in developing areas and the main city (Kuala Terengganu). Figure 1 presents the Terengganu state map and studied location. The dotted line indicates the studied shoreline areas.

Table 1. Abbreviation and general information of shoreline district of Terengganu state

No	Shoreline District	Code Name	Type of Beach	District Population	Main Economic Sector	Future Development
1	Besut	T1	Sandy	136,563	Tourism/Fishery	Tourism Hub
2	Setiu	T2	Sandy	54,563	Tourism/Fishery	Tourism Hub
3	Kuala Terengganu	T3	Sandy	337,553	Tourism/Fishery	Conurbation
4	Marang	T4	Sandy	95,283	Tourism/Fishery	Tourism Hub
5	Dungun	T5	Sandy	149,851	Main Port	Port Expansion
6	Kemaman	T6	Sandy	166,750	Main Port	Port Expansion

IOP Conf. Series: Materials Science and Engineering **210** (2017) 012020 doi:10.1088/1757-899X/210/1/012020

(a) Terengganu regional map

(b) Northern Terengganu shoreline photo

(c) Southern Terengganu shoreline photo

Figure 1. Terengganu state general information

3. Method

Figure 2 presents the flowchart of the study development. Site visit on the studied location have been conducted to record general information of the areas. Prior to site visit, standard penetration test (SPT) report is collected from various local municipalities and geotechnical engineering consultancies for infrastructures purposes. Later, related soil parameters is extracted and by using linear stratigraphy correlation method, the illustration of soil profile, SPT-N distribution, zone of saturation and liquefaction potential layer is produced. The grain size properties from laboratory reports is also extracted to further examine soil using gradation curves of liquefiable margin and liquefaction susceptibility margins.

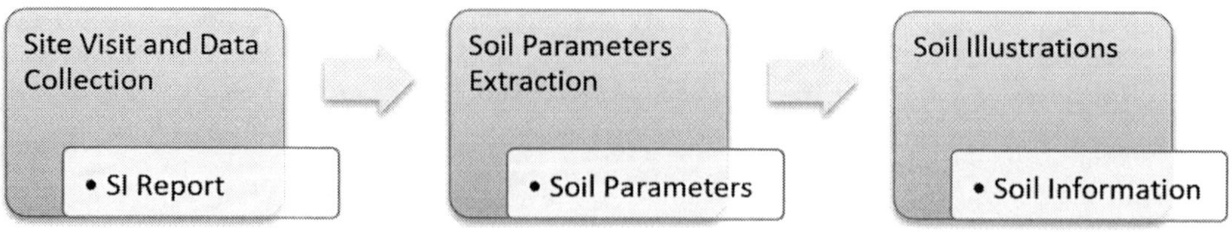

Figure 2. Flowchart of study development

3.1. Soil liquefaction screening

The screening was conducted by extracting basic information of the soil from site investigation (SI) report by means of SPT. 5 illustrations is developed for this purpose to investigate into soil liquefaction susceptibility of the region. Soil composition illustration presents general layer of soil type containing gravel, sand, clay and silty layers. The soil hardness is observed by developing the SPT-N distribution by defining 6 level of hardness ranging from soft soil to hard stratum which correlates to the number of blows. Another governing factor is the ground water table. The zone of saturation illustration defines two different areas of saturation and non-saturation zone.

The grain size distribution plot in liquefaction margin presents limit curve for soil liquefaction possibility [9]. Another chart adapted for this study is the liquefaction susceptibility margins[10]. According to researchers [11], by using simple terms relating moisture content (w_c), liquid limit (LL) and plasticity index (PI) the margins define whether or not finer particles are prone to liquefy. Soils with (i) wc/LL > 0.85 and PI < 12 are vulnerable to liquefaction, soils having (i) wc/LL > 0.80 and (ii) 12 < PI < 18 are moderately susceptible to liquefaction and further laboratory testing for fine-grained soils located in this range is recommended whereas, soils having PI > 18 are considered to be non-liquefiable.

3.2. SI report, soil sampling, SPT-N correction

The SI report collected presents information of the ground according to B.S 1377: Part 9: 1990, "Determination of the penetration resistance using split-barrel sampler", using a self-tripping hammer of 63.5 $\pm$ 0.5 kg weight of designated design. Soil samples were taken in the form of undisturbed or disturbed but representative when drilling. The disturbed samples are used for laboratory classification tests. The samples were sealed in polythene bags before sending to laboratory for further investigation whereas the undisturbed samples were collected by applying hydraulic thrust on thin wall sampling tubes of 60 mm diameter for very soft cohesive soils. The sampling tubes are later secured with wax to maintain water content. All the samples were placed in cushioned boxes and transported to laboratory to ensure minimum disturbance.

The SPT-N value is subjected to a large number of variables that affect the results. SPT-N values are standardized to $N_{(1)60}$ values in reducing the significant variability [12]. Therefore correction factors are adapted study regardless of the equipment used at site. The approach is to ensure SPT-N data used is close representation of the actual subsurface condition. The equation for $N_{(1)60}$ is as follows:

$$N_{(1)60} = N_{SPT}\ x\ C_n\ x\ C_e\ x\ C_b\ x\ C_r\ x\ C_s \quad (1)$$

where $N_{(1)60}$ is the final corrected SPT value; N_{SPT} is the raw SPT data measured at site; C_n is the overburden correction factor; C_e is the energy correction factor; C_b is the borehole diameter correction factor; C_r is the rod length correction factor; C_s is the sampler correction factor and C_b is the borehole diameter correction factor.

3.3. Evaluation of liquefaction potential

The simplified method was adapted in this study to calculate the factor of safety against liquefaction. In general terms, the stress generated by ground motion which results in soil liquefaction is defined as the cyclic stress ratio (CSR), and the strength of soil to resist the ground motion is defined as cyclic resistance ratio (CRR). There are various methods in evaluating the liquefaction potential, commonly used Boulanger Idriss (2014) [13], Cetin et al. (2004) [14] and Vancouver Task Force (2007) [15] . In this present study, Boulanger Idriss (2014) is used in tabulating the CSR and CRR. The equation for CSR and CRR as follows:

$$CSR = \frac{\tau_c}{\sigma'_{vo}} = \frac{0.65\tau_{max}}{\sigma'_{vo}} = \frac{0.65a_{max}\,\sigma_{vo}r_d}{(g)\,\sigma'_{vo}} \quad (2)$$

where a_{max} is the maximum horizontal acceleration at the ground surface; τ_{max} is the maximum horizontal shear stress in the liquefiable layer; σ_{vo} is the total vertical normal stress before the earthquake; σ'_{vo} is the effective vertical normal stress before the earthquake and r_d is the stress reduction coefficient

$$CRR_{\sigma=1,\alpha=0} = exp\left\{\frac{N_{(1)60}}{14.1}+\left(\frac{N_{(1)60}}{126}\right)^2+\left(\frac{N_{(1)60}}{23.6}\right)^3+\left(\frac{N_{(1)60}}{25.4}\right)^4-2.8\right\} \quad (3)$$

where $N_{(1)60}$ is the final corrected SPT value

The 8.0 magnitude earthquake was assumed for all the analysis and through non-linear site response analysis 0.20 g was assumed for peak surface acceleration (PSA). The equation mentioned above is used to calculate the factor of safety against liquefaction (F_s) at each layer in the boreholes. The equation for F_s as follows:

$$F_s = \frac{CRR}{CSR} \quad (4)$$

where CRR is the cyclic resistance ratio and CSR is the cyclic stress ratio. $Fs \geq 1$ indicate the soil has no potential of liquefaction however when $Fs < 1$ the soil indicate possibility of liquefaction potential.

4. Results and Discussions

The findings are presented using simple illustrations and charts in highlighting the governing factor of soil liquefaction susceptibility. The importance of each findings related to soil liquefaction is discussed in this section.

4.1. Soil Composition

4 types of soil namely clay, silt, sand and gravel are highlighted in Figure 3. The soil composition shows deep layer of sand running top from the ground surface to the bottom of 30 meter. Hard stratum is found at depth of 11 meter onwards. The unique layers also show existence of silt type soil concentration in the middle of the shoreline distance which is located in coastal city of Kuala Terengganu. Clay type soil is found to be scattered along the shoreline which highlight resistance against soil liquefaction. It can be found at few places in Setiu, Kuala Terengganu, Dungun and Kemaman district.

The site visit conducted along Terengganu shoreline reveals most areas are consisted of sandy type beaches which run from the northern district to the southern district. The existence of such natural formation are due to the high tide from South China Sea bringing in uniformly granular type deposits which in many literatures governs the soil liquefaction susceptibility [16]. Hence land expansion by land reclamation should be avoided at this particular shoreline as demonstrated by similar geo-environment setting [17]. The design of piling for future structures along Terengganu shoreline is advisable for this type of area.

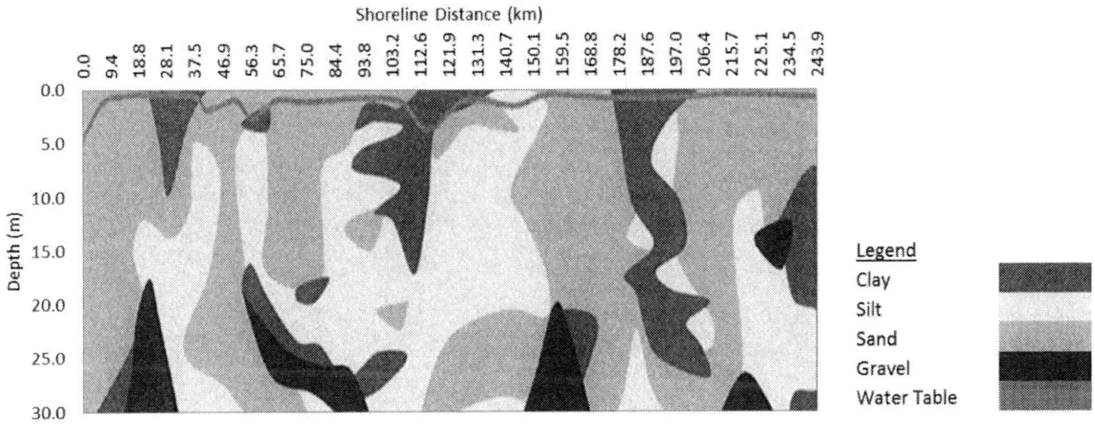

Figure 3. Soil composition of Terengganu shoreline

4.2. SPT-N Distribution

Figure 4 presents the SPT-N distribution of Terengganu shoreline. Loose and soft layer are found in great abundance in the first layer of the soil strata and this is presented with the low SPT-N values. For $N_{(1)60} > 30$, granular soils are unlikely to liquefy. The hard stratum is found at the starting depth of 7 meter from the ground surface.

In the context of earthquake, soft soil tends to amplify seismic wave resulting in increased energy and ground acceleration leading to destructive environment on the ground surface [18]. Recent literatures documented intense shaking due to the amplification of soft soil during earthquake. Therefore such basic information of the soil condition is a necessity in any land development.

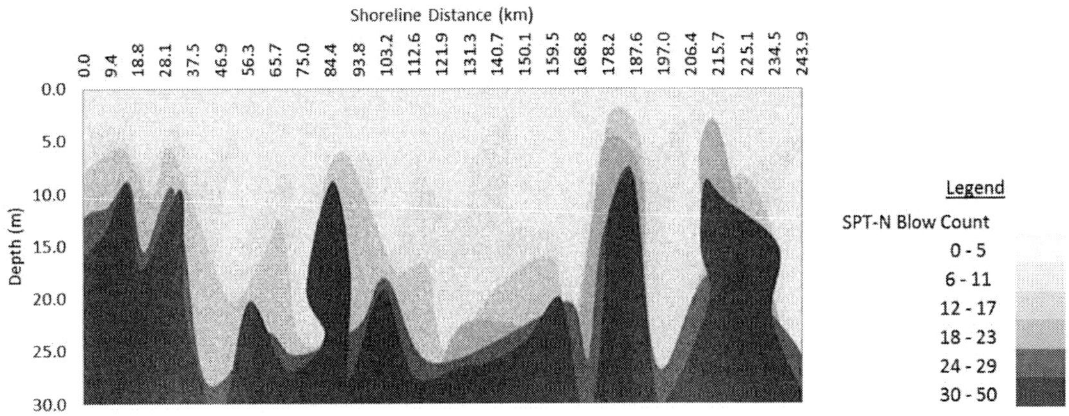

Figure 4. SPT-N Distribution of Terengganu shoreline

4.3. Zone of Saturation

Figure 5 presents the saturation zone of Terengganu shoreline. The yellow color indicate unsaturated zone whereas the blue indicate saturated zone. It was found from the compilation data that over 90% of the areas are saturated.

Liquefaction phenomenon is most likely to occur in condition where the water table is very close to the surface as reported by previous study [19]. Monitoring extensive aquifers along the shoreline is important as to observe how the water flow underground especially when is come to the construction of substructure.

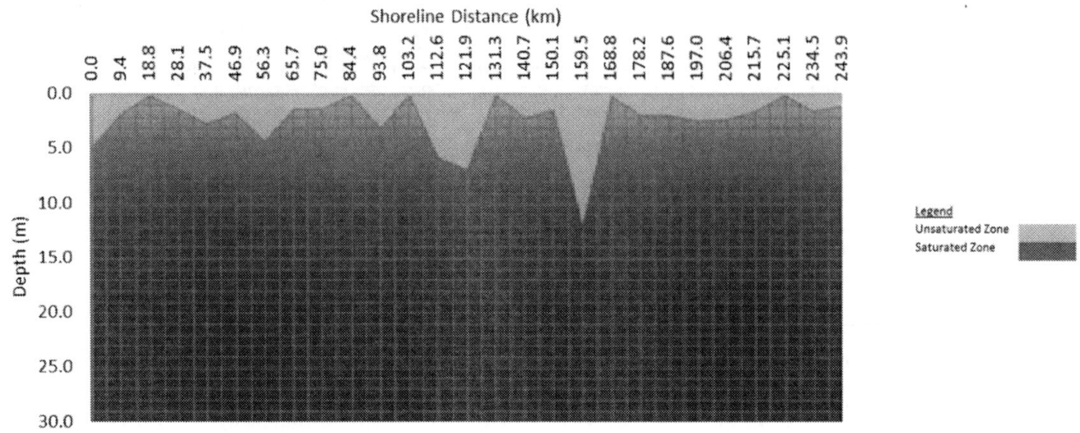

Figure 5. Saturation zone of Terengganu shoreline

4.4. Gradation curves of liquefiable margin and liquefaction susceptibility margins

A more detail soil liquefaction screening is presented in Figure 6 and Figure 7. The soil have high tendency to liquefy when fit in the curve indicated by the red line in the gradation curves of liquefiable margin. Another similar approach is to plot the soil parameters in liquefaction susceptibility margins. The value which falls in the yellow and red border indicate possibility of liquefy soil.

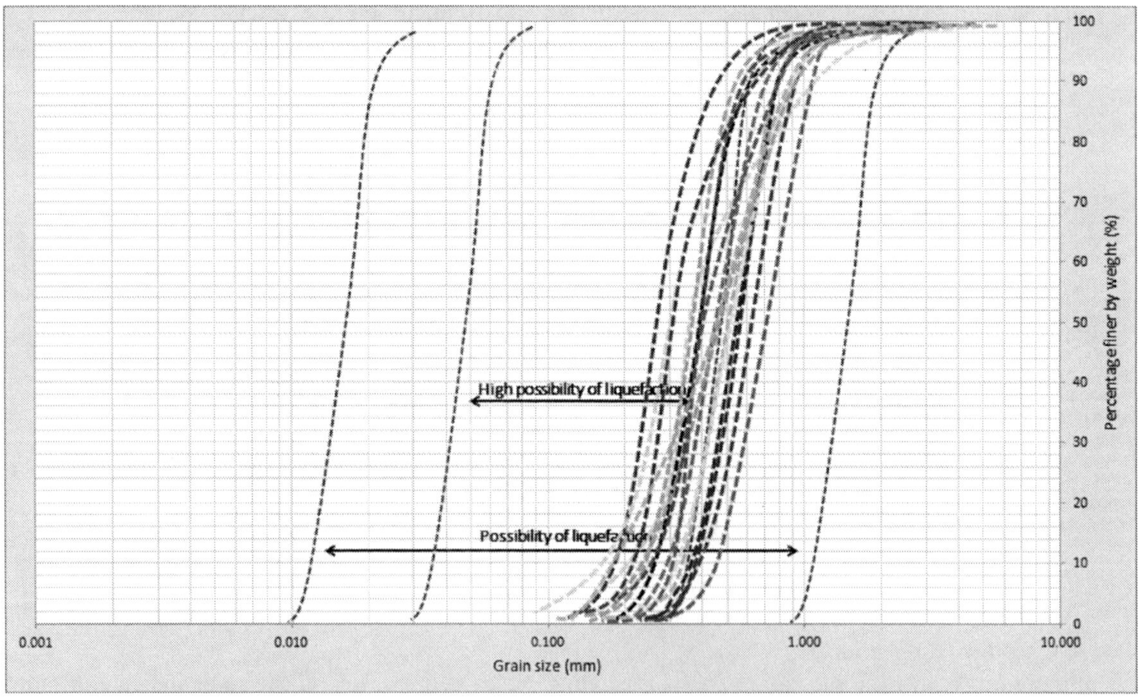

Figure 6. Gradation curves of liquefiable margin of Terengganu shoreline

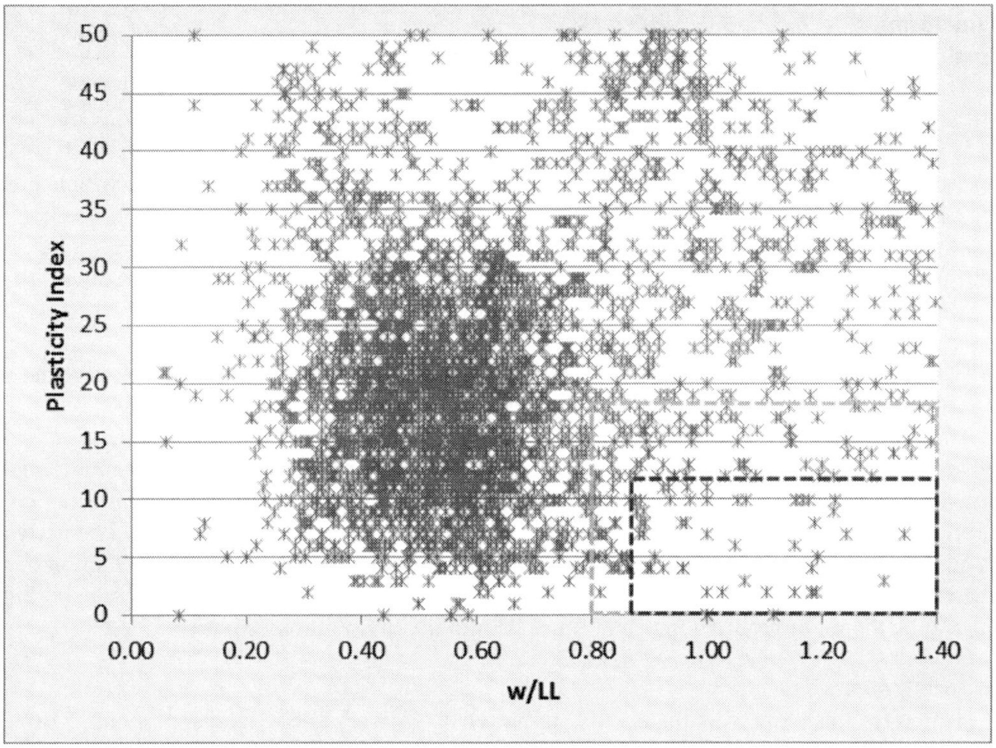

Figure 7. Liquefaction susceptibility margins

4.5. Liquefaction Potential Layer

By using simplified approach the liquefy layer is presented in Figure 8. The green zone indicate safe layer to hazard whereas the red zone indicate dangerous area with possibility of soil liquefaction occurrence. It is noticed that the soil layers prone to liquefaction have safety factor less than 1 and this occurs for soil layers lying between 0 to 30 m in scattered pattern. Some areas in the green zone denotes non liquefy soil mainly due to the existence of clay layer and the unsaturated zone.

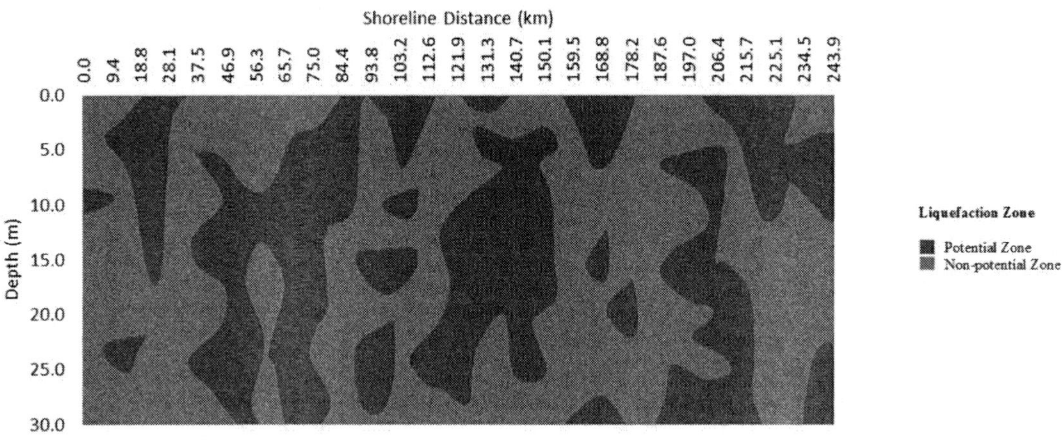

Figure 8. Liquefaction potential layer

5. Conclusion

The main conclusions drawn are:

1. Soil liquefaction screening was investigated and presented using illustrations and charts. The soil composition, SPT-N distribution and zone of saturation reveals deep layer of sand up to 30 m, soft layer of sandy deposits and saturation zone near ground surface which governs the liquefaction potential possibility.

2. The charts presented shows vulnerable deposits existence in the shoreline areas of Terengganu state in which the soil are of uniformly graded deposits indicating existence of clean sand.

3. The applied analysis based on SPT presents potential of liquefaction in Terengganu shoreline due to the presence of loose sandy silt layers in the first 20 m of soil profile.

Based on the findings of soil liquefaction screening and liquefaction potential analysis for Terengganu shoreline it clearly indicates the significant role of geotechnical information through illustrations and charts to be adopted in land usage, management and development. The early access to such information will lead proper preparation and mitigation in the sense of soil liquefaction or any other geo-hazards. Lacking information on regional setting which allows potential areas to be highlighted appears to be the main problem in promoting safe and quality built environment.

Acknowledgments

The research for this paper was financially supported by University of Malaya, grant no. PV022/2012A. In developing the ideas presented here, we have received helpful input from Kumpulan IKRAM Sdn. Bhd. (Selatan). This paper benefited from valuable comments and suggestions from the Editor and anonymous reviewers.

References

[1] Pradel D, Wartman J, Tiwari B. 2014 *Natural Hazards Review* **15** 13-26
[2] Kishida H. 1966 *Soils and Foundations* **6** 71-88
[3] Yamaguchi A, Mori T, Kazama M, Yoshida N. 2011 *Soils and Foundations* **52** 811-829
[4] Reyners M. 2011 *Seismological Research Letters* **82** 371-372
[5] Bhattacharya S, Hyodo M, Goda K, Tazoh T, Taylor CA. 2011 *Soil Dynamics and Earthquake Engineering* **31** 1618-1628
[6] Maurer BW, Green RA, Taylor O-DS. 2015 *Soils and Foundations* **55** 778-787
[7] Youd, Idriss I, Andrus RD, Arango I, Castro G, Christian JT. 2001 *Journal of geotechnical and geoenvironmental engineering* **127** 817-833
[8] Hakam A, Suhelmidawati E. 2013 *Procedia Engineering* **54** 140-146
[9] Iwasaki T. 1986 *Soil Dynamics and Earthquake Engineering* **5** 2-68
[10] Ibrahim KMHI. 2014 *Ain Shams Engineering Journal* **5** 647-655
[11] Bray JD, Sancio RB. 2006 *Journal of geotechnical and geoenvironmental engineering* **132** 1165-1177
[12] Bolton Seed H, Tokimatsu K, Harder L, Chung RM. 1985 *Journal of Geotechnical Engineering* **111** 1425-1445
[13] Idriss IM, Boulanger RW. 2015 *Soil Dynamics and Earthquake Engineering* **68** 57-68
[14] Cetin KO, Seed RB, Der Kiureghian A, Tokimatsu K, Harder Jr LF, Kayen RE. 2004 *Journal of Geotechnical and Geoenvironmental Engineering* **130** 1314-1340
[15] Anderson D, Byrne P, DeVall R, Naesgaard E, Wijewickreme D, Adebar P. 2007 *Vancouver Liquefaction Task Force Report*
[16] Lee OA. 2010 *Ocean & Coastal Management* **53** 439-446
[17] Yasuda S, Harada K, Ishikawa K, Kanemaru Y. 2012 *Soils and Foundations* **52** 793-810

[18] Papathanassiou G, Mantovani A, Tarabusi G, Rapti D, Caputo R. 2015 *Engineering Geology* **189** 1-16
[19] Van Ballegooy S, Malan P, Lacrosse V, Jacka M, Cubrinovski M, Bray J. 2014 *Earthquake Spectra* **30** 31-55

IOP Conf. Series: Materials Science and Engineering **210** (2017) 012021 doi:10.1088/1757-899X/210/1/012021

Effects of Lateral Bracing and Stiffeners on the CFRP Failure of Strengthened Steel Beams

M. Kamruzzaman[1], M.Z. Jumaat[1*], N.H.R. Sulong[1], I.M.I. Qeshta[1] and K. Narmashiri[2].

[1]Department of Civil Engineering, Faculty of Engineering, University of Malaya, 50603 Kuala Lumpur, Malaysia.
[2]Department of Civil Engineering, College of Engineering, Zahedan Branch, Islamic Azad University, Zahedan, Iran.

E-mail: zamin@um.edu.my

Abstract: In this paper, the effects of lateral bracing and web stiffeners on the Carbon Fibre Reinforced Polymer (CFRP) failure modes and buckling strength of the CFRP strengthened wide-flange steel I-beams are investigated experimentally. The study consisted of eight beams tested under static gradual load until failure. The main test variables were steel plate stiffeners, lateral bracings, and bonding of CFRP plates to beam soffits. The results showed that the use of steel plate stiffeners did not only prevent stress concentration below the point load, but it could also help to delay debonding of the externally bonded CFRP plate. The use of lateral bracing indicated a significant effect in preventing the CFRP splitting failure mode. In addition, the use of stiffeners with lateral bracing simultaneously, showed improvement in the in-plane flexural strength, stiffness and ductility of the CFRP strengthened I-beams.

1. Introduction

In recent decades, the strengthening and retrofitting of the structural elements of steel has become essential for enhancing their performance. Structures may need to be strengthened to avoid deterioration as a result of design or construction error; or to address any changes in function. Various methods have been used to strengthen steel structures. The most common strengthening technique of steel beams includes bolting or welding steel plates to the beam tension face. However, this approach has some drawbacks, such as corrosion in the steel plates. Another drawback is the weakness of the rehabilitated system of damaged beams to fatigue due to stress concentration in the damaged area as a result of welding or bolting joints. The use of carbon fibre reinforced polymer (CFRP) has recently gained worldwide attention in the field of strengthening and retrofitting steel structures because of its high strength-to-weight ratio, ease of placement and handling on site, and excellent resistance to corrosion [1]. Recently, CFRP strips used in the United Kingdom, United States of America, Switzerland and Japan showed that there was great potential for upgrading steel bridges and buildings with CFRP [2-6].

Steel beams strengthened with CFRP strips bonded to their tension flange have shown an increase in their in-plane bending strength [7-10]. Due to the strengthening arrangement of the steel beam, changes are usually observed in the failure mode. Failure modes observed are typical in-plane bending or sometimes buckling failure. While most studies have focused on local and global buckling [11-14], few investigations have been carried out on lateral buckling in strengthened steel beams [15-18].

In-plane bending, stress concentration below the loading point and lateral torsional buckling (LTB) are key limitations in strengthened steel beams, which are liable for premature failure of the CFRP plates. LTB occurs in laterally unrestrained strengthened steel beams that are bending about their strong axis, especially for the open sections that are under greater warping moments during torsion such as I or W sections.

The primary factors that influence the lateral torsional buckling (LTB) of steel W-beams include effective unsupported length, boundary conditions, type and position of loads, initial imperfections of geometry, and eccentricity of loading [19, 20].The eccentricity of loading initiates second-order bending moment along the longitudinal axis of the steel beam, consequently promoting buckling failure [21]. The beams subjected to flexure have strength and stiffness in the plane associated with bending about their major principal axis, much greater than that associated with bending about their minor principal axis [22]. Unless these members are properly braced against lateral displacement and twisting, they would be subjected to failure by LTB before achieving full in-plane bending capacity [23].

Ghafoori and Motavalli [24]investigated the influence of LTB on pre-stress levels of CFRP strips, and on bonded and un-bonded conditions of strengthened steel beams. They observed that strengthening steel beams using CFRP strips enhanced elastic stiffness for both un-bonded and bonded techniques. In addition, pre-stressing CFRP strips did not have significant results on the elastic stiffness but substantially influenced buckling strength. The effect of LTB in the plastic region of the beams was not studied.

The capacity of steel beams could be governed by buckling failure before developing their full plastic resistance, when these elements are subjected to compressive stresses. Narmashiri, Ramli Sulong [25] conducted an experimental study on unrestrained steel I-beams reinforced with CFRP laminates to investigate CFRP failure modes. The compressive flange of the beams was adhesively bonded with CFRP composite laminates and the beams were tested under four-point bending, simply supported using hinge and roller supports. The unbraced length that had the potential to cause LTB failure in all specimens was 2,000 mm. Failure modes observed from test results were CFRP splitting and debonding below the loading point, end delamination, and end debonding. Researchers also reported that the behaviour of unrestrained retrofitted beams was affected by LTB.

Despite the number of studies on the use of stiffeners for preventing web crippling with local buckling in slender beams, less attention has been given to the effect of stiffeners on the CFRP failure mode due to stress concentration in the flange below the loading point and on LTB [26-28]. The collapse of structural elements due to stress concentration could be extremely disastrous, in terms of human life and damage to infrastructure. Stress concentration could be defined as damage at the point load section due to concentrated loads, which eventually leads to LTB. Failure of strengthening CFRP plate was found to be the most significant problem affecting the service life of existing strengthened steel structures. Therefore, the objective of this research is to investigate the influence of LTB on the flexural behaviour of strengthened wide-flange steel I-beams. The effect of stiffeners below loading points, and lateral bracing on strengthening CFRP plate and buckling failures with load carrying capacity, deformations, and strain variation of strengthened steel beams are also examined.

2. Materials and methods

2.1. Materials

2.1.1. Steel beam

The steel beams used in this research were A6-ASTM mild steel W 150×100×24 sections. Figures 1 and 2 show the dimensions of the specimens. Table 1 shows the material properties of beams. The mechanical properties of the steel beams were obtained in accordance with ASTM A370.

2.1.2. CFRP strips

In this study, Sika®CarboDur S1014 [29] pultruded carbon fibre reinforced polymer strips with cross sections of 100 × 1.4 mm were used. Table 2 shows the material properties of the CFRP strips.

2.1.3. Adhesive

Sikadur®-30 epoxy adhesive [30] was used to bond the CFRP strips onto steel beams. The properties of the adhesive are shown in table 3.

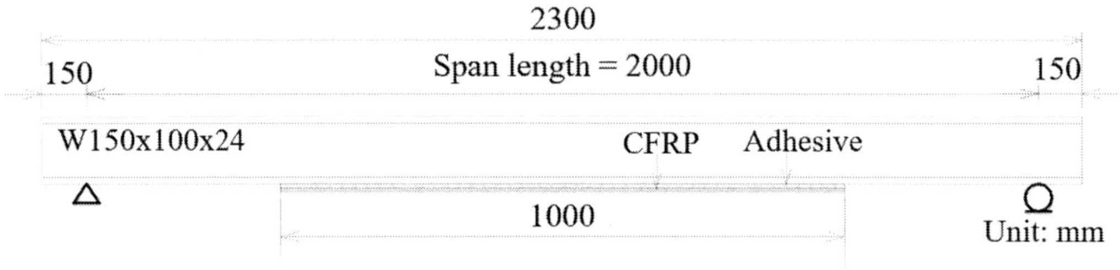

Figure 1. Elevation projection of the specimens.

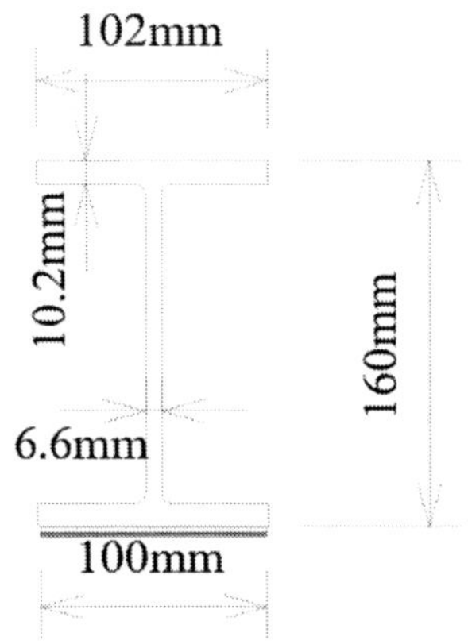

Figure 2. Cross section of the specimen

Table 1. Properties of wide-flange steel I-beams

E-Modulus (MPa)	Poison ratio	Stress (MPa)		Strain (%)	
Mean value	ϑ_{Steel}	Yielding (F_y)	Ultimate (F_u)	Yielding (ε_y)	Ultimate (ε_u)
199,600	0.3	306	457	0.12	13.5

Table 2. Properties of CFRP plates [29]

E- Modulus* (MPa)				Tensile Strength* (MPa)				Strain (%)
Mean Value	Min. value	5% fracture value	95% fracture value	Mean Value	Min. value	5% fracture value	95% fracture value	Strain at break
165,000	>160,000	162,000	180,000	3,100	>2,800	3,000	3,600	1.7 [±0.01]

Table 3. Properties of Adhesive [30]

Compressive (MPa)		Tensile (MPa)		Shear (MPa)	Bond Strength on Steel (MPa)	
E-Modulus	Strength (7 days)	E-Modulus	Strength (7 days)	Strength (7 days)	Mean Value	Min. Value
9600	70-95	11200	24-31	14-19	>30	>21

2.2. Experimental test

2.2.1. Preparation of the specimens

The experimental programme involved strengthening and testing a total of eight beam specimens. Four specimens were considered as reference beams (UB1, UB2, UB3 and UB4), while the other four specimens were strengthened with CFRP strips (SB1, SB2, SB3 and SB4). The strengthened specimens were bonded with 1,000 mm long CFRP strips. Table 4 presents the specifications of the beams used in this research. Figures 1 and 2 show the details of the strengthened beams. Steel plate stiffeners with a dimension of 47.7 × 10 mm were welded to the flanges and web on both sides of the beam below the loading point in specimens UB3, UB4, SB3, and SB4.

Table 4. Specifications of the specimens

No.	Specimen ID	Specification			Method
		Stiffeners	Lateral bracing	CFRP	
1	UB1	N/A	N/A	N/A	Exp.
2	UB2	N/A	√	N/A	Exp.
3	UB3	√	N/A	N/A	Exp.
4	UB4	√	√	N/A	Exp.
5	SB1	N/A	N/A	√	Exp.
6	SB2	N/A	√	√	Exp.
7	SB3	√	N/A	√	Exp.
8	SB4	√	√	√	Exp.

The surface of the bottom flange of each specimen was treated by a grinding wheel to remove the rust. They were then washed with acetone to remove any material that might affect bonding. The study by Teng, Fernando [31] found that sand blasting was the most effective surface treatment for steel beams prior to bonding with external strengthening materials. Hence, the surface of all beams was sandblasted to SA 2.5. The CFRP strips were then cleaned and glued by adhesive to the tension flange of the specimens.

2.2.2. Test set-up

The beams had a total length of 2,300 mm. The specimens were tested under four-point bending until failure at a span of 2,000 mm. The distance between two point loads applied by a spreader beam was 600 mm. Figure 3 shows the test set-up and the specimen details. Mid-span vertical and horizontal deflections were measured by linear variable displacement transducers (LVDT). Strain readings were monitored by strain gauges attached to different locations of the mid-span, at the loading point and at the tip of the CFRP.

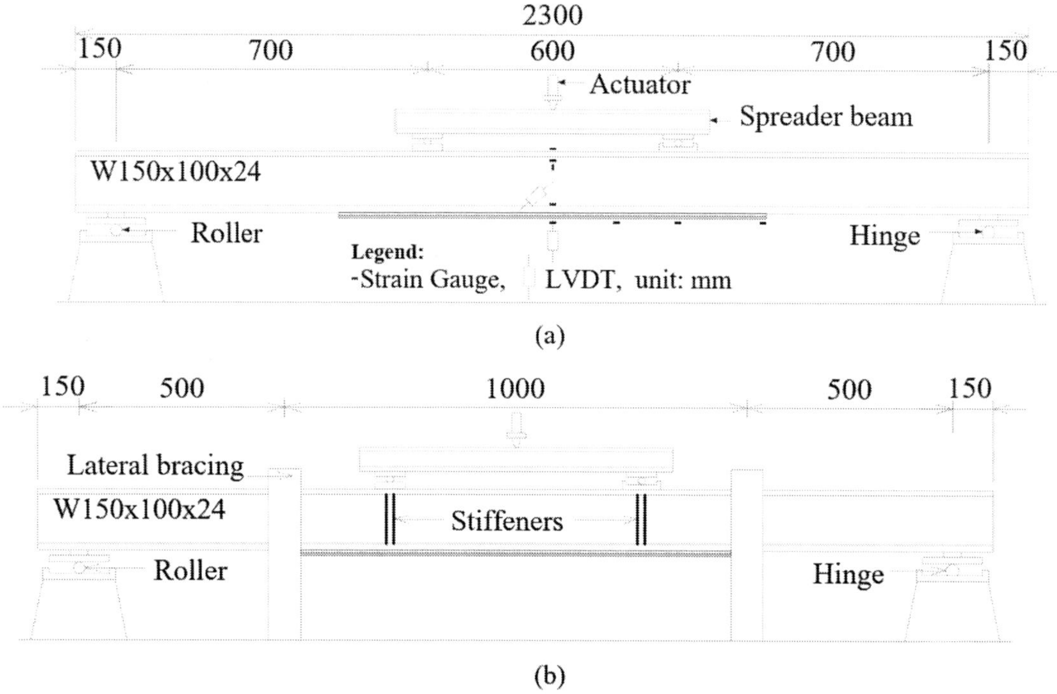

Figure 3. Schematic four-point bending setup (a) without and (b) with lateral bracing.

3. Results and Discussion

3.1. Load-deflection relationship and failure mode

The load-deflection relationships of specimens UB1, UB2, UB3 and UB4 are shown in figure 4. The variation of the trend of load-deflection curves provide a useful measure for the behaviour of beams. The figure shows that specimen UB1 had the lowest load carrying capacity, when compared with UB2, UB3 and UB4. The difference in load carrying capacity among the specimens was mainly related to the type of failure mode. Specimen UB1 failed as a result of stress concentration and due to the compression flange buckling below the loading plate, which in turn led to LTB. Table 5 presents a summary of test results. The load carrying capacity of specimen UB1 was mainly controlled by stress concentration and local buckling failure below the loading plate (figure 5). The presence of the stiffeners in specimen UB3 prevented stress concentration failure and hence increased the load carrying capacity up to 13.8% over specimen UB1. The increase in load carrying capacity was most apparent when the stiffener was welded such that it was touching the compression flange of the beam. LTB contributed to premature failure in specimens UB1 and UB3, which was due to the absence of lateral support for the effective length of the beam. Buckling failure involved twisting about a point below or near the tension flange [32]. In addition, the use of stiffeners and lateral bracing in specimen UB4 exhibited an increase in load carrying capacity up to 22.2% over specimen UB1.

IOP Conf. Series: Materials Science and Engineering **210** (2017) 012021 doi:10.1088/1757-899X/210/1/012021

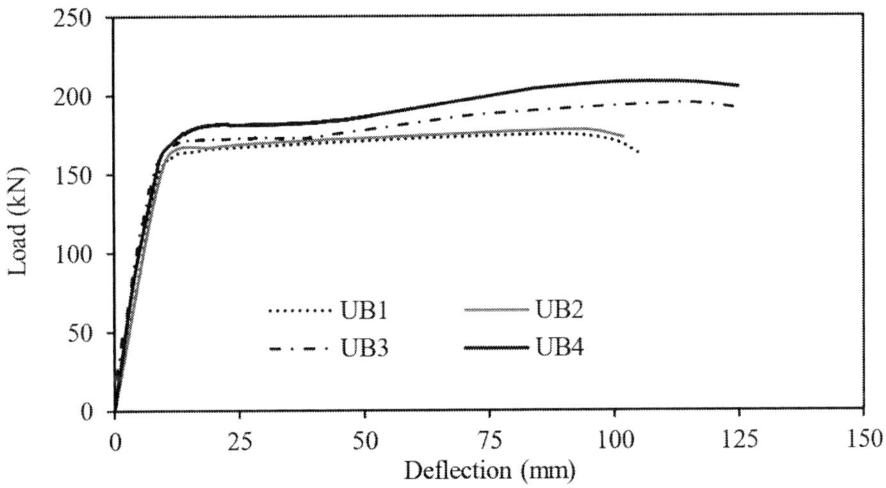

Figure 4. Load-deflection diagram of un-strengthened specimens.

Figure 5. Stress concentration and local buckling failure mode at below the loading plate.

IOP Conf. Series: Materials Science and Engineering **210** (2017) 012021 doi:10.1088/1757-899X/210/1/012021

Table 5. Summary of test results

Test unit	Yield			Peak			Failure			Ductility index								CFRP Failure mode
										Peak				Failure				
	L_y (kN)	Δ_y (mm)	E_y (kN-mm)	L_p (kN)	Δ_p (mm)	E_p (kN-mm)	L_f (kN)	Δ_f (mm)	E_f (kN-mm)	μ_{dp}	Ratio to UB	μ_{Ep}	Ratio to UB	μ_{df}	Ratio to UB	μ_{Ef}	Ratio to UB	
UB1	148.3	9.1	674.80	174	92	14390.0	167	105	16604.0	10.10	1	21.30	1	11.5	1	24.7	1	
SB1	163.0	9.0	733.50	198	32.1	5017.0	163	105	16886.0	3.56	0.35	6.80	0.32	11.6	1.01	22.4	0.94	S-ED
UB2	149	9.15	681.70	178	93	14730.0	169	105	16944.0	10.15	1	21.60	1	11.5	1		1	
SB2	165	8.95	739.00	202.5	39.5	6624.0	164	105	17152.0	4.41	0.43	8.97	0.41	11.7	1.02		0.94	
UB3	160.4	9.2	736.80	193.5	109	18923.0	192	120	21035.0	11.90	1	25.70	1	13.0	1	26.6	1	
SB3	168.5	9.08	765.00	206	40.5	6901.50	185	120	21371.0	4.46	0.37	9.02	0.35	13.2	1.01	26.5	0.98	S-ED
UB4	163.5	9.18	750.50	201.5	115	20796.5	200	120	21705.5	12.60	1	27.71	1	13.1	1	30.1	1	
SB4	171.5	9.05	777.00	215	48.9	8590.8	196	120	22164.1	5.30	0.45	11.07	0.43	13.3	1.07	30.5	1.05	ED

* L_y: Yield load; Δ_y: Yield deflection; L_p: Peak load; Δ_p: Peak deflection; L_f: Failure load; Δ_f: Failure deflection; S: Splitting; ED: End debonding.

The sequence of CFRP failure modes was generally dominated by LTB and stress concentration below loading points. Figure 6 shows a comparison of load-deflection diagrams of specimens SB1, SB2, SB3 and SB4. The strength of specimen SB4 was about 9% and 7% larger than strengthened specimens SB1 and SB3 respectively, which showed that bracing and stiffeners could increase the strength but only slightly in this case due to debonding problems. The load-deflection diagram also indicated that the load dropped when the CFRP strips split and debonded from the steel beams.

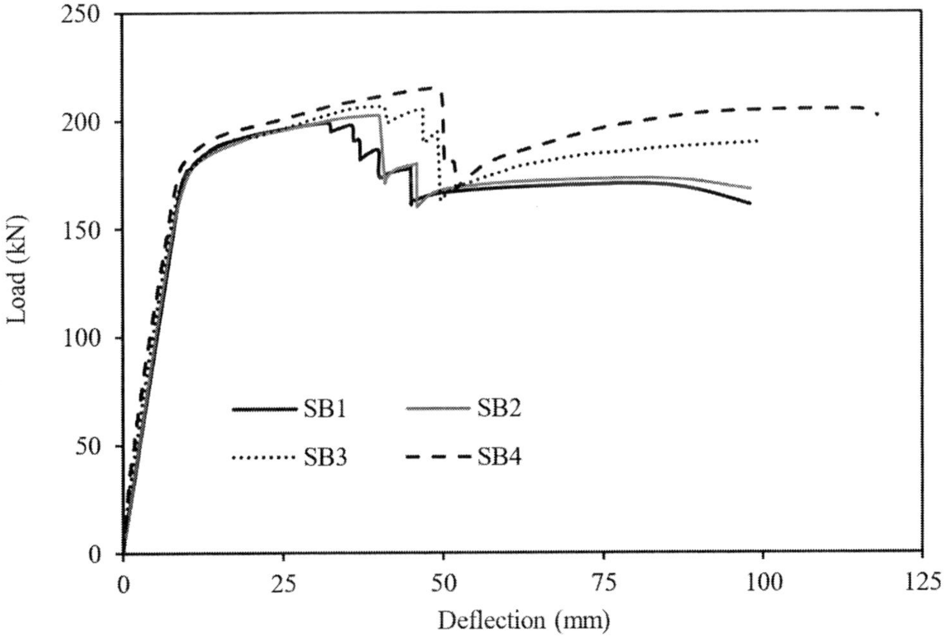

Figure 6. Load-deflection diagram of strengthened beams.

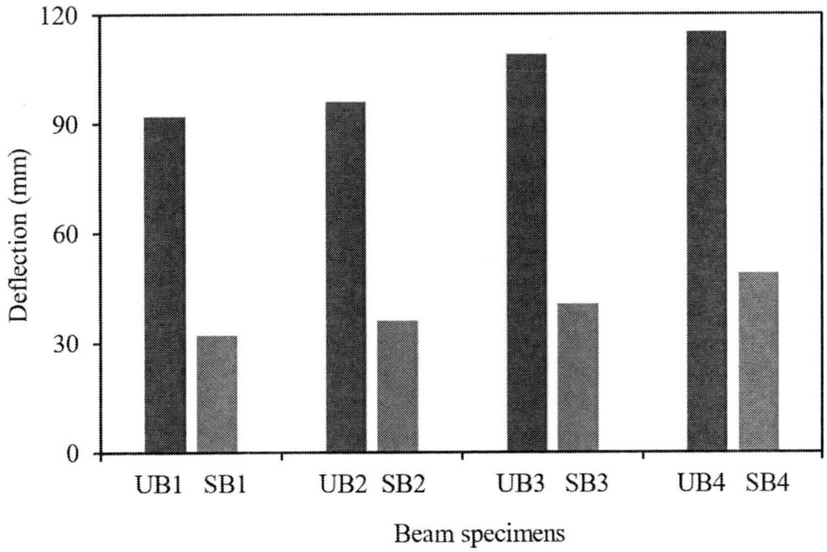

Figure 7. Variation of deflection values of beams at peak load

IOP Conf. Series: Materials Science and Engineering **210** (2017) 012021 doi:10.1088/1757-899X/210/1/012021

Figure 7 shows the deflection at peak load in the strengthened beams before the failure of CFRP by debonding. Specimens SB1, SB2, SB3 and SB4 showed that the strengthened CFRP plates failed at deflection 34%, 37.5%, 37% and 43% compared to their corresponding un-strengthened beams. Specimens SB1 and SB3 failed as a result of CFRP splitting due to the influence of LTB, which led to debonding; this was in line with findings by Narmashiri, Ramli Sulong [25]. Specimen SB1 was also affected by stress concentration below the point load that caused premature failure through the earlier splitting of the CFRP plate. A long steel beam under static load, simply-supported and without bracing, can twist or buckle out-of-plane due to the lack of unbraced length [32]. The buckling strength in specimen SB2 and SB4 increased due to the use of lateral bracing, but suddenly dropped when the CFRP plate failed as a result of debonding.

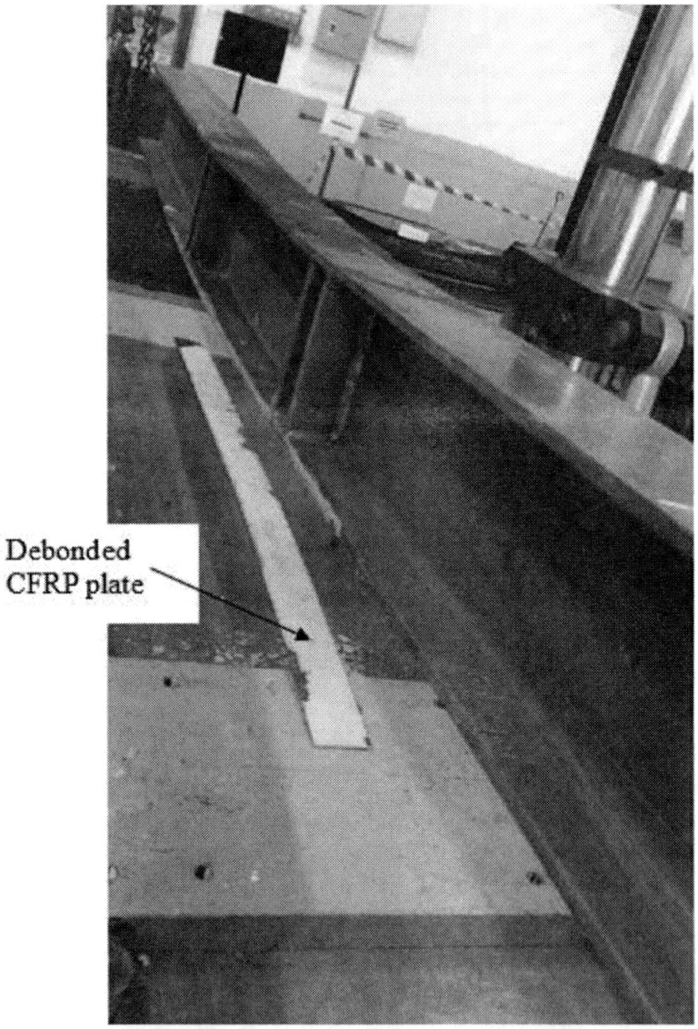

Figure 8. CFRP debonding failure in specimen SB3.

The presence of high interfacial stress near the tips of the FRP plate, which normally initiates failure in the bonding system, is considered one of the main problems in strengthened steel beams [33]. Figure 8 shows debonding of the CFRP plate in specimen SB4. Failure here was due to loss of the bond between the adhesive and CFRP, which resulted in the separation of the CFRP plate from the bottom flange. Previous studies have explained that failure of this type occurred as a result of stress from peeling and interracial shear at the FRP tips in concrete and steel structures that have been strengthened using CFRP plates [27, 33, 34].

The failure resulting from the splitting of CFRP plate started at the loading point and spread towards the entire length of the plate. This failure mode occurred as a result of LTB failure and weakness of the plate in the transverse direction, as well as splitting caused by asymmetrical stress concentration below the loading point, as observed by Narmashiri, Ramli Sulong [25]. Researchers have reported that laterally unrestrained beams were subjected to LTB, which caused the failure of CFRP plate by splitting. Figure 9 shows the splitting of the CFRP plate in specimen SB1. The CFRP plates used in this research are unidirectional, they thus have high longitudinal (axial) strength and are weak in the transverse direction. Failure of strengthened beams usually occurred as a result of sudden lateral deformation, but it was less common in concrete beams as they have greater lateral rigidity [35, 36].

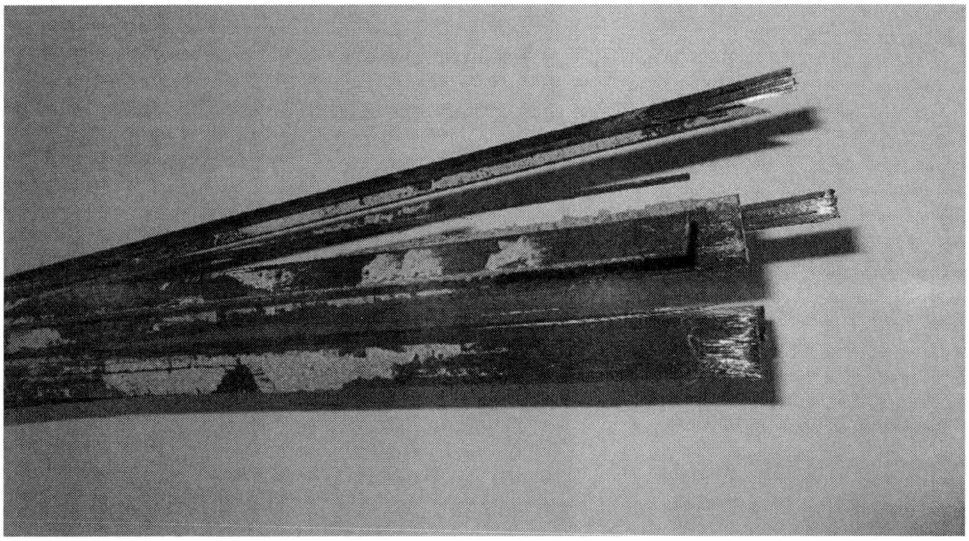

Figure 9. CFRP splitting (specimen SB1).

One of the most important parameters in strengthened/non-strengthened steel beams was boundary condition, which included support systems and types of loading. Figure 10 shows the peak load variation of the strengthened beams SB1, SB2, SB3 and SB4 and the un-strengthened beams UB1, UB2, UB3 and UB4. Specimens UB1, UB3, SB1 and SB3 were simply supported and failed by LTB, resulting in the compression flange experiencing sudden lateral deflection. The specimen experienced torsion and was twisted out of the plain of loading. As the load reached the buckling strength of the beam specimens, substantial lateral deformation was observed. Just then the external load was decreased. The lateral bracing in specimen SB4 increased the load bearing capacity to 16.42% more than UB4, at the level of 48.9 mm deflection when end-debonding (ED) initiate. It also prevents the splitting of the CFRP plate at below the point loads. A CFRP plate with ED is illustrated in figure 11. The majority of the cross-section stiffness came from the stiffener preventing stress concentration below the point load. Lateral bracing also significantly increased stiffness by increasing the buckling strength and preventing splitting of the plate at below the point load. Therefore, the use of a proper

IOP Conf. Series: Materials Science and Engineering **210** (2017) 012021 doi:10.1088/1757-899X/210/1/012021

system of stiffeners and lateral bracing could lead to remarkable enhancement in section stiffness and load bearing capacity.

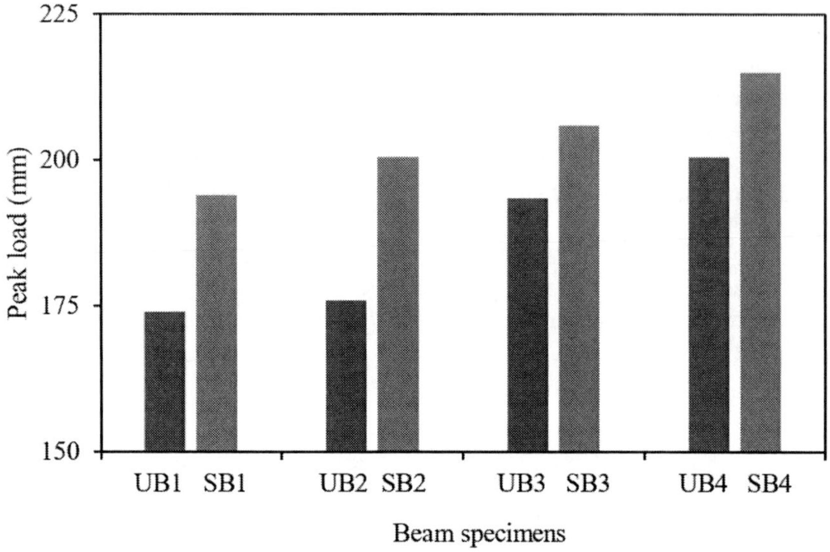

Figure 10. Peak loads of unstrengthend and strengthened specimens.

Figure 11. CFRP End-debonding (specimen SB3).

3.2. Lateral displacement

Figure 12 shows load-lateral displacement relationships at the mid-span of strengthened beams. It illustrates that after initiation of splitting of the CFRP strip on specimen SB1, the movement changed to the opposite direction. This explains why the splitting of one CFRP strip was followed by the lateral movement returning to the original direction until failure. The horizontal movement of the specimen changed due to stress concentration below the point load and LTB. The horizontal deflection of the specimen SB3 was movement in one direction and was followed by the occurrence of LTB. Failure of the strengthened specimens SB1 and SB3 began with the splitting of the CFRP plate at a deflection of 32.3 mm and 41.1 mm, respectively (figure 7). The increase in load carrying capacity of specimens SB1 and SB3 just before failure of the CFRP plate was about 20.1% and 17.2%, compared to the un-strengthened beams UB1 and UB3 at the same deflection values, respectively. Figure 13 shows the effect of lateral bracing on failure modes. Lateral bracing increases buckling strength by restraining the lateral displacement of the beam.

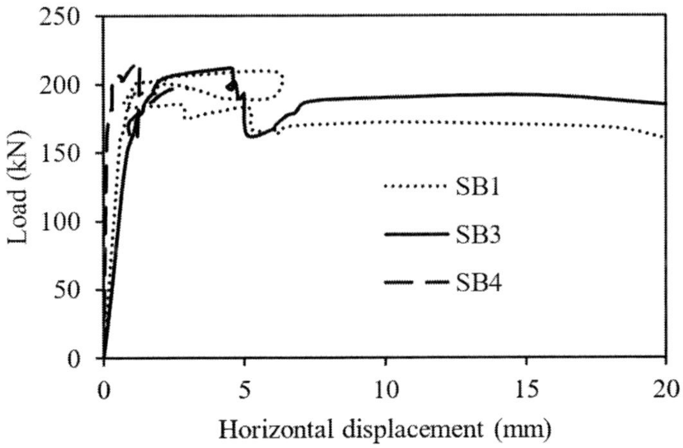

Figure 12. Load-lateral displacement relationships of strengthened beams.

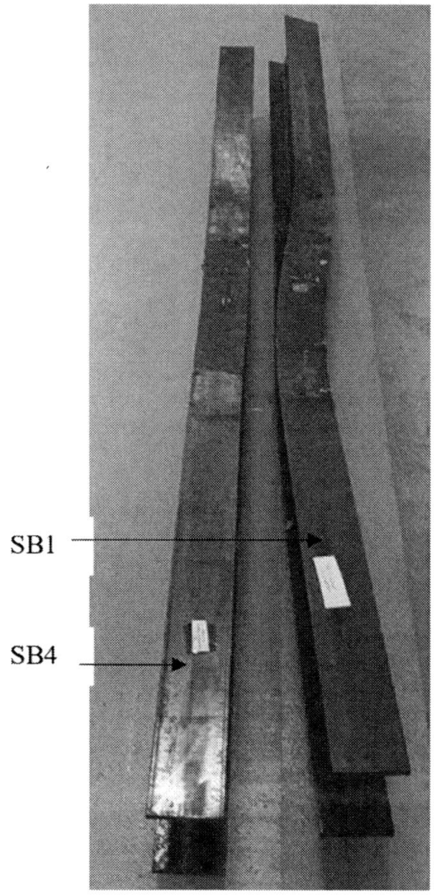

Figure 13. The effect of lateral bracing on failure mode.

Therefore, it can be concluded that applying lateral bracing to strengthened steel I-beams can decrease the instability of the beams as a result of LTB failure. In addition, the use of steel plate stiffeners can be effective in preventing stress concentration below the point load. Lateral bracing increases buckling strength by significantly reducing the influence of LTB.

3.3. Measured strains

The measured strain distribution of specimens SB1, SB3 and SB4 is shown in figure 14. The largest values of strain were found on the beams after yielding at loading points, which was in agreement with Al-Emrani, Linghoff [37] who found that the bonding performance of CFRP reinforced steel beams were different before and after yielding. For higher load levels, strain readings increased until maximum values were recorded at the mid-span. Strain readings then decreased gradually until the minimum values were recorded at the CFRP tip. Strain also increased abruptly due to the discontinuity of the CFRP strip. Results of this study were also similar to that by Deng, Lee [38]who showed that the large amount of strain concentration in epoxy adhesive at the tips of the strengthening strip, caused the sudden termination by debonding of the strip.

The effect of the application of stiffeners below the loading point and lateral bracing on the CFRP strip is also shown in figure 14. The strain readings show that at the load level of 150 kN and 175 kN, no significant difference was observed in strain for SB1 and SB3. At the load level of 198 kN, the strain below the loading point for specimen SB1 was higher than for specimen SB3. This showed that strain in the CFRP was significantly reduced after enhancement of the bending stiffness through the existence of stiffeners. Figure 14 also shows that the use of lateral bracing in specimen SB4 increased the strain at the mid-span of the CFRP strip but was significantly reduced below the loading point. At 198 kN the use of stiffeners and lateral bracing in specimen SB4 caused a reduction of strain at the CFRP strip end and below the point load.

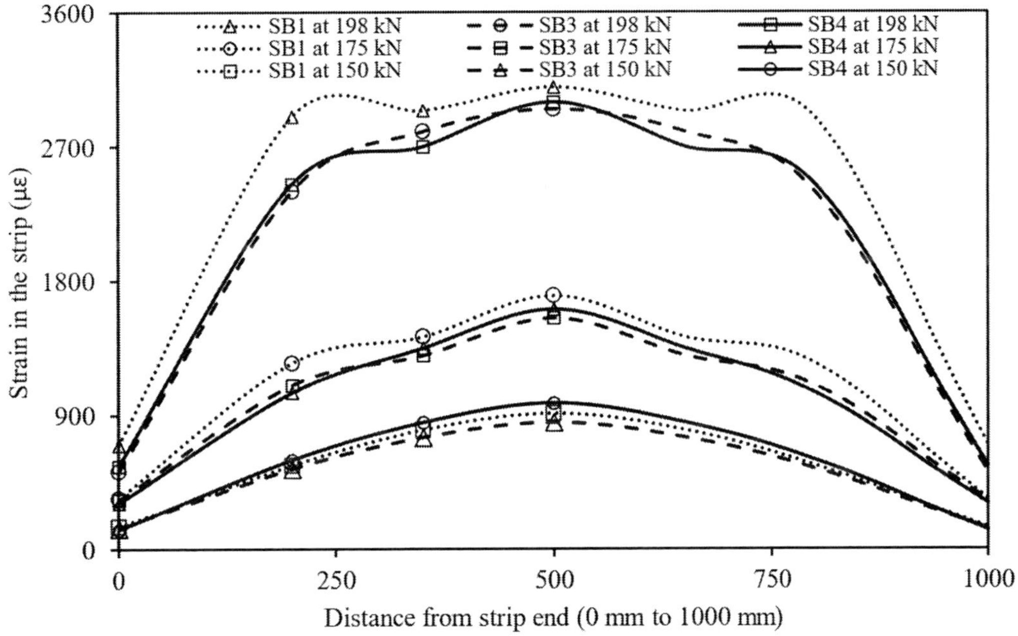

Figure 14. The measured strain distribution along the CFRP strip for specimens SB1, SB3 and SB4 at different load levels.

3.4. Ductility

The ductility of a strengthened steel beam ensures the capability of a structure to endure considerable plastic deformation without substantial loss of strength. In this research, the ductility indices of strengthened beams were obtained based on displacement (μ_{d}) and energy (μ_{E}) ratios [39]. The displacement ductility index at peak load (μ_{dp}) and at failure load (μ_{df}) are expressed as follows:

$$\mu_{\mathrm{dp}} = \frac{\Delta_u}{\Delta_y} \quad (1)$$

$$\mu_{\mathrm{d}f} = \frac{\Delta_{\mathrm{f}}}{\Delta_y} \quad (2)$$

Where, Δ_y,$\Delta_p$and Δ_f are displacements at yield, peak and failure loads, respectively.

The energy ductility index (μ_{E}) at peak load (μ_{Ep}) and at failure load (μ_{Ef}) are expressed as follows:

$$\mu_{\mathrm{Ep}} = \frac{E_p}{E_y} \quad (3)$$

$$\mu_{E\mathrm{f}} = \frac{E_{\mathrm{f}}}{E_y} \quad (4)$$

Where, E_y,$E_p$and E_f are energy at yield, peak and failure loads, respectively.

The calculated ductility indices of all beams are presented in table 5. The ratio of the ductility indices for displacement of specimens SB1, SB2, SB3 and SB4 were 0.35, 0.43, 0.37 and 0.45; while that for energy were 0.32, 0.41, 0.35 and 0.43, respectively, which corresponded to control specimens at peak loads. This indicated that strengthened beams with the CFRP plate suffered from the loss in ductility.

Several researchers have reported that the ductility of strengthened beams could be reduced as a result of reinforcements by externally bonded FRP plates [40, 41], which could be attributed to the brittle behaviour of the CFRP plate. FRP materials also have a linear stress–strain relationship up to the point of failure. It was observed that the ductility ratio for strengthened specimens SB1, SB2, SB3 and SB4 at failure load increased for the displacement ductility by 1.01, 1.02, 1.01 and 1.07; while that for energy ductility increased by 0.94, 0.94, 0.98 and 1.05, mainly as a result of the ductile properties of steel beams. The ductility of specimen SB1 was less than specimens SB3 and SB4 due to stress concentration and LTB failure. The effect of stiffeners and lateral bracing on beam behaviour indicated an increase in ductility, caused by a delay in the debonding of the reinforcing plate from the surface of steel beams. Therefore, stiffeners prevented stress concentration below the loading points. Lateral bracing delayed the failure of the CFRP plate by preventing below load-splitting (BS) CFRP failure, consequently increasing displacement and energy ductility of the strengthened beams.

4. Conclusions

The buckling strength of CFRP strengthened wide-flange steel I-beams was experimentally examined. The effect of different parameters, such as stiffeners and lateral bracing, on the CFRP failure mode and ultimate load bearing strength of steel beams was also studied. Results indicate that the stress concentration of the compression flange in beams without steel plate stiffeners increased greatly during loading and caused local buckling failure below the loading points and initiates early end-debonding. This failure was also affected by the lateral torsional buckling (LTB) of strengthened and non-strengthened beams. The use of steel plate stiffeners prevented stress concentration below the point load and improved delay in the debonding of the externally bonded CFRP plate. The use of lateral bracing showed a significant effect in avoiding the initiation of the below load-splitting CFRP failure and increased the ductility of the strengthened beams. Additionally, the prevention of LTB failure was found to delay the end debonding of CFRP strips. The stiffeners with lateral bracing increased in-plane flexural strength and stiffness. In addition, the stiffeners and lateral bracing also contributed to retard the failure of the CFRP plate. Beams that were externally bonded with CFRP plates showed a reduction in ductility compared to their corresponding non-strengthened beams. This caused the brittle properties and premature debonding failure of the FRP strips. The flexural strength and ductility of beams were found to be generally limited by end-debonding failure.

Acknowledgements

The authors gratefully acknowledge the financial support by the University of Malaya High Impact Research Grant (HIRG) No. UM.C/625/1/HIR/MOHE/ENG/36 (16001-00-D000036) –"Strengthening Structural Elements for Load and Fatigue".

References

[1] CNR D 2007 Guidelines for the design and construction of externally bonded FRP systems for strengthening existing structures – Metallic structures – Preliminary study. *National Research Council, Rome.*
[2] Zhao X-L and L Zhang 2007 *Eng. Struct.* **29** 1808-1823
[3] Suzuki H 2005 *Advanced materials for construction of bridges, buildings and other structures.*
[4] Kamruzzaman M, M Z Jumaat, N Ramli Sulong and A Islam 2014 *Sci. World J.* **2014**.
[5] Ghafoori E, M Motavalli, A Nussbaumer, A Herwig, G Prinz and M Fontana 2015 *Compos. Pt. B-Eng.* **68** 1-13
[6] Ghafoori E, G S Prinz, E Mayor, A Nussbaumer, M Motavalli, A Herwig and M Fontana 2014. *Polymers.* **6** 1096-1118
[7] Zhao X-L 2013 FRP-strengthened metallic structures. CRC Press.
[8] Bennati S, D Colonna and P S Valvo 2016 *P. Struct. Int.* **2** 2682-2689

[9] Deng J, Y Jia and H Zheng 2016 *Compos. Struct.* **136** 450-459
[10] Lin W, T Yoda, N Taniguchi and M Hansaka 2013 *J. Constr. Steel. Res.* **85** 130-139
[11] Naderian H, H Ronagh and M Azhari 2014 *Thin-Walled Struct.* **84** 289-301
[12] Dahmani L and A Boudjemia 2014 *Strength Mater-Engl. Tr.* **46** 429-432
[13] Wang Y, L Yang, B Gao, Y Shi and H Yuan 2014. *Int. J. Steel Struct.* **14** 411-420
[14] Ragheb W F 2015 *J. Constr. Steel. Res.***107** 81-93
[15] Kala Z 2013 *P. Eng.* **57** 504-514
[16] Kabir M and A Seif 2010 *J. Scientia Iranica Trans. A-Civil Eng.* **17** 262-272
[17] Nguyen T, T Chan and J T Mottram 2013 *Compos. Struct.* **100** 233-242
[18] Shaat A and A Z Fam 2009 *J. Compos. Constr.* **13** 2-12
[19] Benyamina A B, S A Meftah, F Mohri and E M Daya 2013 *Eng. Struct.* **56** 1207-1219
[20] Ozbasaran H 2014 A Parametric Study on Lateral Torsional Buckling of European IPN and IPE Cantilevers.
[21] Humbel. T 2014 Ultra-high modulus CFRP for fatigue strengthening of steel beams - experimental and theoretical investigation : Empa/ETHZ, Dübendorf, Switzerland.
[22] Galambos T V 1998 Guide to stability design criteria for metal structures. John Wiley & Sons.
[23] Ziemian R D 2010 Guide to stability design criteria for metal structures. John Wiley & Sons.
[24] Ghafoori E and M Motavalli 2015 *Constr. Build. Mater.* **76** 194-206
[25] Narmashiri K, N Ramli Sulong and M Z Jumaat 2012 *Constr. Build. Mater.* **30** 1-9
[26] Moy S and F Nikoukar 2002 *Advanced Polymer Composites for Structural Applications in Construction: Proceedings of the First International Conference, Held at Southampton University, Uk, on 15-17 April 2002*. Thomas Telford.
[27] Deng J and M M Lee 2007 *Compos. Struct.* **78** 232-242
[28] Teng J, T Yu and D Fernando 2012 *J. Constr. Steel. Res.* **78** 131-143
[29] Sika®CarboDurPlates 2013 Product Data Sheet- Pultruded carbon fiber plates for structural strengthening, [Edition 2013-03_1].
[30] Sikadur®-30 2014 Product data sheet, 2- part epoxy impregnation resign for bonding reinforcement [Edition 2014-01_1].
[31] Teng J, D Fernando, T Yu and X Zhao 2011 *Advances in FRP Composites in Civil Engineering*. 2011, Springer. p. 865-868.
[32] Yura J and B A Phillips 1992 Bracing requirements for elastic steel beams. Center for Transportation Research, University of Texas.
[33] Haghani R, M Al-Emrani and R Kliger 2009 *Constr. Build. Mater.* **23** 1413-1422
[34] Narmashiri K and M Z Jumaat 2011 *Simul. Model. Pract. Theory.* **19** 564-585
[35] Buyukozturk O, O Gunes and E Karaca 2004 *Constr. Build. Mater.* **18** 9-19
[36] Narmashiri K 2011 *Carbon Fibre Reinforced Polymer Strips for Flaxural and Shear Strengthening of Steel I-beams*, University of Malaya.
[37] Al-Emrani M, D Linghoff and R Kliger 2005 Bonding strength and fracture mechanisms in composite steel-CFRP elements. *International Symposium on Bond Behaviour of FRP in Structures (BBFS 2005),* International Institute for FRP in Construction.
[38] Deng J, M M Lee and S S Moy 2004 *Compos. Struct.* **65** 205-215
[39] Hawileh R A, H A Rasheed, J A Abdalla and A K Al-Tamimi 2014 *Mater. Des.* **53** 972-982
[40] Rasheed H A, R R Harrison, R J Peterman and T Alkhrdaji 2010 *Compos. Struct.* **92** 2379-2390
[41] Morsy A M, K M Helmy, N H El-Ashkar and M Nada 2014 *Conc. Sol.* 313

IOP Conf. Series: Materials Science and Engineering **210** (2017) 012022 doi:10.1088/1757-899X/210/1/012022

Quality control of lightweight aggregate concrete based on initial and final water absorption tests

M. Maghfouri[1,*], P. Shafigh[2, 3], Z. Binti Ibrahim[1], V. Alimohammadi[4]

[1]Department of civil Engineering, Faculty of Engineering, University of Malaya, 50603 Kuala Lumpur, Malaysia.
[2]Department of Building Surveying, Faculty of Built Environment, University of Malaya, 50603 Kuala Lumpur, Malaysia.
[3]Center for Building, Construction & Tropical Architecture (BuCTA), Faculty of Built Environment, University of Malaya, 50603 Kuala Lumpur, Malaysia
[4]Department of civil Engineering, Faculty of Engineering, University of Qom, Qom, Iran.

E-mail: m.maghfouri@gamil.com

Abstract. Water absorption test is used to evaluate overall performance of concrete in terms of durability. The water absorption of lightweight concrete might be considerably higher than the conventional concrete due to higher rate of pores in concrete and lightweight aggregate. Oil palm shell is a bio-solid waste in palm oil industry, which could be used as lightweight aggregate in the concrete mixture. The present study aims to measure the initial and final water absorption and compressive strength of oil palm shell lightweight concrete in order to evaluation of quality control and durability performance. Total normal coarse aggregates were substituted with coarse oil palm shell in a high strength concrete mixture. The quality of concrete was then evaluated based on the compressive strength and water absorption rates. The results showed that fully substitution of normal coarse aggregates with oil palm shell significantly reduced the compressive strength. However, this concrete with the 28-day compressive strength of 40 MPa still can be used as structural concrete. The initial and final water absorption test results also showed that this concrete is not considered as a good concrete in terms of durability. Therefore, it is recommended that both compressive strength and waster absorption tests must be performed for quality control of oil palm shell concretes.

1. Introduction

In construction industry, concrete is the most demandable material, which is obtained from combination of binding material, fine and coarse aggregate, chemical admixtures and water. It is considered a cheapest and most readily material with high resistance to water which can be formed easily when it is fresh [1]. Concrete is considered as a durable construction material. Durability in concrete is defined as the ability to last a long time against various environmental conditions without any significant deterioration[2]. Generally, the durability of concrete is controlled by specifying some certain requirements such as properties of concrete constituents, procedure of casting and concrete compaction, curing and compressive strength. The standard compressive strength of concrete is used as such a significant criterion in most of the specifications and technical codes, whilst only some potential of durability can be determined by the compressive strength; and it cannot be only valid criterion for evaluation of concrete quality control [3].

It has been well established that the evaluation of the quality and performance of a concrete mixture is not limited to its mechanical properties since it is important to characterize the material in terms of the factors that rate its durability [4]. There are some tests that commonly specified for determination of the concrete durability such as water permeability, rapid chloride permeability, surface absorption and waster absorption. Water absorption test according to BS 1881-122: 2011 is a simple method of determining the water tightness by submersion of the concrete specimen. Low absorption of the water can be defined as a positive result for the test.

Oil palm shell (OPS) as a bio-solid waste obtained in the processing of palm oil in the tropical regimes. The use of OPS as a natural lightweight aggregate has been investigated by several researchers. The results of recent studies showed that high strength up to 53 MPa could be achieved for lightweight oil palm shell concrete with dry density of 1830-1920 kg/m^3 [5]. Although OPS lightweight concrete has sufficient compressive strength as a principal criterion of acceptance of concrete, it does not necessarily have a good water absorption and durability. It has been reported that 24 hr water absorption of the OPS is in the range of 21- 33%. However, the normal aggregate water absorption value is less than 2% [6]. In regard to higher absorption capacity of OPS, durability of lightweight concrete using OPS must be investigated. In this regard, initial and final water absorption for high strength concrete as control mix (C1) and lightweight concrete by fully substitution of normal coarse aggregates with OPS (C2) were investigated.

2. Materials

Ordinary Portland cement complying MS EN 197-1:2007 with specific gravity 3.14 and blain specific surface area 3410 cm^2/gr was used as binder. The fine aggregate with maximum size 4.75 mm from local mine with a fineness modulus 2.72 and specific gravity 2.66 was applied. Crushed oil palm shell (OPS) and granite obtained from local mine with maximum nominal size 12 mm were used. The physical properties of crushed OPS and crushed granite are represented in table 1. In addition, Sika ViscoCrete as a chloride free superplasticizer with maximum dosage 1% of total cement mass was used.

Table 1. Physical properties of OPS and Granite stone

Physical property	OPS	Crushed granite
Specific gravity	1.17	2.62
Fineness modulus	5.78	6.60
Bulk density (compacted) (kg/m^3)	618	1421
Water absorption (10 and 30 min) (%)	7.4 and 8.9	< 1
Water absorption (1 and 24 h) (%)	10.1 and 19.7	< 1

2.1. Mix proportions

Mix proportion of normal weight high strength concrete (C1) as control mix and lightweight concrete with using of OPS (C2) are shown in table 2. In OPS lightweight concrete, total normal coarse aggregates were replaced 100% with coarse oil palm shell by volume. The cement content, water and superplasticizer for both mix designs were the same.

Table 2. Mix proportions of concretes in one batch

Mix code	Cement (kg)	Water (kg)	Superplasticizer (kg)	Sand (kg)	Coarse aggregate (kg)	
					Crushed granite	OPS
C1	55	18.2	0.55	94	103	0
C2	55	18.2	0.55	94	0	46.6

3. Results and discussion

3.1. Workability

Slump test is one of the common test for evaluation of the workability. According to ACI 116R 2000, slump test is the way to measure the consistency of concrete mixture [7]. In this regard, the workability of mix C1 and C2 was measured. The slump value for both mixtures is shown in table3. As can be seen in table 3, the incorporation of OPS in high strength concrete decreased significantly the slump value from 205 mm to 40 mm. The reason for this remarkable lower slump value of mix C2 is due to the higher water absorption of OPS comparing to the normal coarse aggregate. Another reason for higher water absorption of mix C2 could be due to high content of pores [6]. Okpala [8] reported that the porosity of the shell is 37% although the quality of OPS can be improved using pre-treatment methods such as apply 20% poly vinyl alcohol (as a PVA solution) which can decrease the water absorption of the shell from 23.3 to 4.2% [9].

Table 3. Slump value and densities of mixes

Mix code	Slump (mm)	Density (kg/m^3)	
		Demoulded	Oven dried
C1	205	2400	2340
C2	40	1940	1900

3.2. Density

Density of the hardened lightweight concrete (LWC) for structural application is often more important than the compressive strength [10] and considered as an important variable for design of concrete structures. The range of density for LWC normally is defined between 1400 to 2000 kg/m^3 compared to the NWC with 2400 kg/m^3 [11]. Table 3 shows the density of 2 mixes. It could be concluded that fully substitution of normal coarse aggregates with oil palm shell in mix C2 reduced 17% of concrete density. This is due to low density of the oil palm shell aggregate in comparison to normal weight aggregate. The practise concrete with dry density less than 2160 kg/m^3 is considered as LWC [12]. Therefore, mix C2 with an oven density of 1900 categorised as LWC.

3.3. Compressive strength

The compressive strength is a significant property of the structural concrete, which affects other mechanical properties such as splitting tensile strength, flexural strength and modules of elasticity. According to ASTM C330-89 the 28-day cylindrical compressive strength should not be less than 17 MPa. The results of compressive strength test on 100 mm cube are presented in figure 1. As shown in figure 1, it can be observed that the compressive strength achieved for HSC (C1) at the age of 28-day was 69 MPa, whereas for OPS light weight concrete (C2) at same W/C ratio the average of the compressive strength was 40 MPa which is approximately 41% lower than the control mix.

With increasing of the OPS content in the mixture along replacement of normal weight coarse aggregate, the specific area of aggregates was increased due to lower density of the OPS, thus requiring more cement paste for better and effective bonding with the shells. Since the binder (cement) remain the same, the bonding to achieve proper mechanical properties is inadequate. The failure mechanism of OPS concrete was investigated. It was reported that failure is depending on breakdown of interfacial transition zone (ITZ). Due to the smooth surface of the shell on both concave and convex there is a weakness in the bond between cement matrix and OPS aggregate to sustain against high loads. Hence, the failure of OPS concrete is mainly caused by easy debonding between the shells and cement paste that less than contribution of the aggregates strength [7, 8, 13, 14].

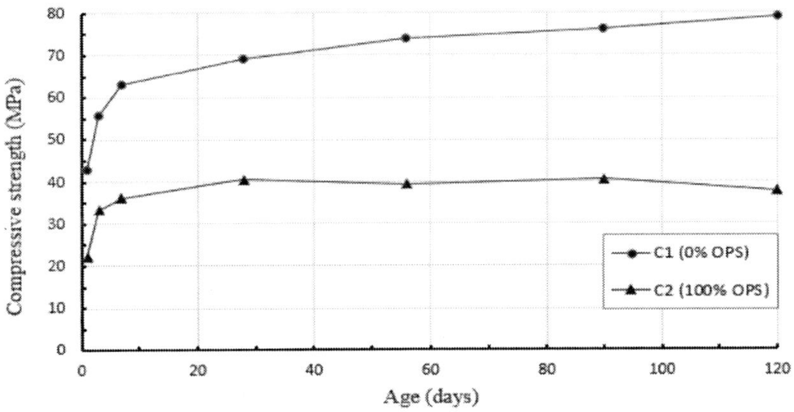

Figure 1. Compressive strength versus time

3.4. Water absorption

Water absorption is water flow into the unsaturated pores of concrete due to the differences of pressure cussed by gravitational and capillary forces [13]. Water absorption of both mix C1 and C2 was measured at the age of 28-day for 30 min and 72 h. as shown in Figure 2, it could be seen that both initial and final water absorption of OPS lightweight concrete is significantly higher than the control HSC. Due to high rate of water absorption of OPS, fully substitution of normal coarse aggregates with this pores aggregate increased its water absorption capacity. Therefore, with reducing volume of OPS in concrete mixture, it can be expected that the initial and final water absorption reduce.

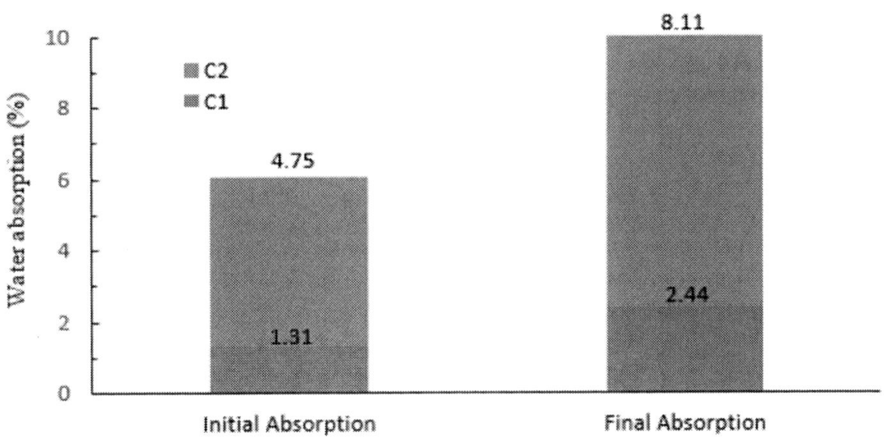

Figure 2. Water absorption of concrete mixes

Teo et al. [14] reported that water absorption rates for structural lightweight concrete using OPS in air and water curing conditions were 10.64 and 11.23 % respectively. Water absorption ranges for other types of lightweight concrete such as pumice aggregate concrete and expanded polystyrene aggregate were 14-22% and 3-6%, respectively [15, 16]. Based on 30 min water absorption value of 0-3%, 3-5% and 5% above, the quality of concrete was categorised as good, average and poor respectively by CEB-FIP [17].

Accordingly, mix C2 had initial absorption of 4.75% that could be classified as "average" quality. Furthermore, final absorption was 8.1%, which is relatively higher than the control mix by 2.4%. It has been reported that water absorption of high quality concrete usually is lower than 5% [17]. Therefore lightweight concrete using OPS as fully replacement of aggregate cannot be considered as a high durable concrete due to the high rate of water absorption although compressive strength was 40 MPa at the age of 28-day.

4. Conclusion

In this study, normal coarse aggregate was fully replaced (by volume) with oil palm shell (OPS) lightweight aggregate in a high strength concrete. Water absorption test as indicator of the quality of concrete was used to assess durability properties of the new concrete containing OPS.

Experimental test results showed that fully substitution of normal coarse aggregate with OPS in a high-strength normal-weight concrete significantly reduced the slump value of the concrete due to high water absorption of the OPS aggregates. Although, grade 40 could be achieved for OPS lightweight concrete, however, it cannot be used for constructing structural elements. This is due to the OPS concrete with high initial and final water absorptions is not classified as good and durable concrete. It can be concluded that just compressive strength test is not a good indicator to assess the quality of OPS concrete, and, therefore, durability test should be considered in the assessment. Water absorption test has a simple procedure and can be used as a first durability test for OPS concrete.

References

[1] Aslam, M., Shafigh, P., Jumaat, M. Z., & Lachemi, M. 2016 *Journal of Cleaner Production* **119** 108-117.

[2] Zulkarnain, Fahrizal, and Mahyuddin Ramli. 2008 *2nd International Conference on Built Environment in Developing Countries* 541-551.

[3] Hilsdorf, H.K. 1989 *(iabse report volume 57/1)*..

[4] Evangelista, L. and J. De Brito, 2010 *Cement and Concrete Composites* **32** 9-14.

[5] Shafigh, P., Alengaram, U. J., Mahmud, H. B., & Jumaat, M. Z. 2013 *Materials & Design* **49** 613-621

[6] Shafigh, P., M.Z. Jumaat, and H. Mahmud, 2010 I*nternational journal of the physical sciences* **5** 2127-2134.

[7] Aslam, M., P. Shafigh, and M.Z. Jumaat, 2016 *Journal of Cleaner Productio*, **126** 56-73.

[8] Okpala, D., 1990 *Building and environment* **25** 291-296.

[9] Teo, D.C., M.A. Mannan, and J.V. Kurian, 2006 *Journal of Advanced Concrete Technology* **4** 459-468.

[10] Rossignolo, J.A., M.V. Agnesini, and J.A. Morais, 2003 *Cement and Concrete Composites* **25** 77-82.

[11] Chen, B. and J. Liu, 2005 *Cement and Concrete Research* **35** 913-917.

[12] Lee, J. C., Shafigh, P., Mahmud, H. B., & Lee, S. K. (2016). *Mechanical properties of high strength semi-lightweight aggregate concrete containing high volume waste materials*. In M. H. R. O. Abdullah, Y. H. Min, N. A. M. Bashar, & S. M. F. S. A. Nasir (Eds.), *AIP* Conference Proceedings (Vol. 1774, No. 1, p. 030006). AIP Publishing..

[13] Shafigh, P., M. Jumaat, and H. Mahmud, 2012 *Arabian Journal for Science & Engineering*, **37**

[14] Teo, D. C. L., Mannan, M. A., Kurian, V. J., & Ganapathy, C. 2007 *Building and Environment* **42** 2614-2621.

[15] Babu, K.G. and D.S. Babu, 2003 *Cement and Concrete Research* **33** 755-762.

[16] Gündüz, L. and İ. Uğur, 2005 *Cement and Concrete Research* **35** 1859-1864.

[17] Ranjbar, M. M., Madandoust, R., Mousavi, S. Y., & Yosefi, S. 2013 *Construction and Building Materials* **47** 806-813.

IOP Conf. Series: Materials Science and Engineering **210** (2017) 012023 doi:10.1088/1757-899X/210/1/012023

Heat Pre-Treatment of Beverages Wastewater on Hydrogen Production

S Z Uyub*, N S Mohd[1] and S Ibrahim
Department of Civil Engineering, Faculty of Engineering, University of Malaya, Kuala Lumpur, Malaysia.

E-mail: sitizulaihauyub@gamil.com

Abstract. At present, a large variety of alternative fuels have been investigated and hydrogen gas is considered as the possible solution for the future due to its unique characteristics. Through dark fermentation process, several factors were found to have significant impact on the hydrogen production either through process enhancement or inhibition and degradation rates or influencing parameters. This work was initiated to investigate the optimum conditions for heat pre-treatment and initial pH for the dark fermentative process under mesophilic condition using a central composite design and response surface methodology (RSM). Different heat treatment conditions and pH were performed on the seed sludge collected from the anaerobic digester of beverage wastewater treatment plant. Heat treatment of inoculum was optimized at different exposure times (30, 90, 120 min), temperatures (80, 90 and 100°C) and pH (4.5, 5.5, 6.5) in order to maximize the biohydrogen production and methanogens activity inhibition. It was found that the optimum heat pre-treatment condition and pH occurred at 100°C for 50 min and the pH of 6.00. At this optimum condition the hydrogen yield was 63.0476 ml H_2/ mol glucose (H_2 Yield) and the COD removal efficiency was 90.87%. In conclusion, it can be hypothesized that different heat treatment conditions led to differences in the initial microbial communities (hydrogen producing bacteria) which resulted in the different hydrogen yields.

1. Introduction

Due to the climate change issues and concerns regarding the limited non-renewable energy systems (fossil-based), many researchers have been working on new sustainable energy sources that could substitute fossil fuels. At present, a large variety of alternative fuels have been investigated and explored and hydrogen gas is considered as one of possible solution for future due to its unique characteristics.

Hydrogen can be produced through various methods such as reforming of hydrocarbons, coal gasification, electrolysis, photochemical process and biological hydrogen production. Among the different approaches to generate hydrogen, the dark fermentation is the simplest process in obtaining biohydrogen, which can be done through anaerobic digestion of organic matter [1].

In the recent biohydrogen studies, many of the studies used seed sludge from wastewater treatment plants (WWTPs) or pure culture as the inoculum [2-8]. However, very limited studies have used seed sludge from the industrial wastewater. Other than the different type of wastewater used as inoculum, environmental and operation factors play important roles in hydrogen production efficiency during dark fermentation process. The factors are initial pH, substrate and inoculum concentration, temperature and nutrient were discussed in several review articles [9-13]. Although, many studies have demonstrated the role of variables on fermentative hydrogen production, discrepancies are still exists with regard to optimum of initial pH, pre-treatment of inoculum, temperature and inoculum size.

With regard to pre-treatment of the inoculum, among the commonly studied technologies are heat pre-treatment, repeated aeration, chloroform, acid as well as alkali pre-treatment. Among all, heat pre-treatment is the most commonly used pre-treatment of inoculum as it is proven to be able to suppress non-sporulating bacteria such as methanogens. Moreover, under appropriate conditions, H_2 producing endospore becoming active and multiplying which then resulted in increase the biohydrogen production.

Therefore, due to the disagreement arises from this issues this study was initiated in order to find the optimum conditions for heat pre-treatment that will maximize biohydrogen production, hence making anaerobic digestion feasible alternative in generating biohydrogen.

2. Materials and Methods

2.1 Substrate and seed sludge

The sludge used as the inoculum was obtained from an anaerobic digester of beverage factory located at Bangi, Selangor. The collected sludge is stored in a cold room at 4°C prior to use for experiment. This is to inhibit the degradation of the effluent from microbial activities.

The physical and chemical characteristics of the wastewater, which were pH, total suspended solid (TSS), volatile suspended solid (VSS), chemical oxygen demand (COD) and Total Kjeldahl Nitrogen (TKN) were analyzed according to APHA Standard Method [14]. Table 1 showed the physical and chemical characteristics of the inoculum.

In order to find the optimum heat pre-treatment condition of the sludge and initial pH, temperature with the exposure time and pH were chosen as variables. The temperature was varied from 80°C to 100°C and exposure time was chosen at 30, 90, and 120 min. The pH was varied as 4.5, 5.5 and 6.5.

Table 1. Chemical Characterization of Beverage Sludge.

Parameter	Beverage Sludge
Total suspended solids	14996.67 ± 128.97 mg/L
Volatile suspended solids	10497.67 ± 90.28 mg/L
Chemical Oxygen Demand	3625.67 ± 60.68 mg/L
TKN	296.17± 2.80 mg/L
pH	6.58± 0.01

2.2 Experimental design

The 158ml serum bottles with working volume of 100ml were used as batch reactor. For each experiment, the sludge was prepared according to experimental design. The inoculum was added, and followed by the addition 1g of glucose and pH adjustment solution (1 N HCl or 1 N NaOH). The bottles were sealed with rubber stopper and flushed with N_2 gas (99.99%) for 5 minutes at 10 ml / min in order to provide anaerobic condition. All experiments were conducted in an incubator shaker at 150 rpm for 24 hours in anaerobic mesophilic conditions (35°C).

Full factorial design allows for simultaneous study of several factors and the interaction between factors on response. A full factorial experimental design was employed in running experiments to optimize selected parameters for efficient H_2 production. Design Expert version 7.0 was designated for this study to

IOP Conf. Series: Materials Science and Engineering **210** (2017) 012023 doi:10.1088/1757-899X/210/1/012023

generate statistical analysis, experimental designs, and optimization of the variables. The variables are coded according to equation (1).

$$x_i = \frac{X_i - X_i^*}{\Delta X_i} \quad (1)$$

Where x_i is the coded value of ith test variable, X_i is the real value of ith test variable, X_i^* is the value of X_i at the center point and ΔX_i is the step size.

Table 2. Levels and variables for full factorial design.

Symbols	Variables	Levels		
A	Temperature	80°C	90°C	100°C
B	Time	30 min	90 min	120 min
C	pH	4.5	5.5	6.5

2.3 Analytical Method

Biogas volume was determined using gas tight syringe at room temperature and pressure (760 mm Hg). The biogas composition was measured using gas chromatograph (Perkin Elmer, AutoSystem GC) equipped with a thermal conductivity detector and data acquisition system. The temperature of the oven, injector and detector were 80°C, 200°C, and 200°C respectively with Argon as a carrier gas. The analytical procedure of APHA Standard Methods was used to determine COD, TSS, VSS and pH of the sample [14].

3. Result and Discussion

3.1 The optimization for maximum hydrogen yield.

The optimal values of the studied factors, i.e., heat treatment conditions (exposure time and temperature); pH and their interaction on hydrogen production were further analyzed using RSM. 20 experiments were carried out according to the schedule in Table 3 designated using Design Expert 7.0. to find the optimum heat treatment conditions and pH.

Table 3. Central composite design.

Std	Run	Temp (°C)	Exposure Time (min)	pH
13	1	90	75	4.5
14	2	90	75	6.5
7	3	80	120	6.5
8	4	100	120	6.5
6	5	100	30	6.5
18	6	90	75	5.5
9	7	80	75	5.5
12	8	90	120	5.5
16	9	90	75	5.5
1	10	80	30	4.5
2	11	100	30	4.5
15	12	90	75	5.5
5	13	80	30	6.5
20	14	90	75	5.5
11	15	90	30	5.5
17	16	90	75	5.5
19	17	90	75	5.5
10	18	100	75	5.5
4	19	100	120	4.5
3	20	80	120	4.5

For the beverage sludge, multiple regression analysis using Design Expert 7.0 software showed the importance of main effects and interaction of the three variables based on the response which was H_2 yield.

Equation (2) described the quadratic regression model in terms of actual factors based on analyses of variance (ANOVA) of the model.

$$H2\ Yield = +56.48 + 9.53A + 10.63B + 5.46C + 1.04AB + 2.00AC - 8.35B^2 - 10.58C^2 - 14.00A^2B - 9.49AB^2 \quad (2)$$

A, B and C are the coded values of temperature of heat treatment, exposure time and pH respectively. The ANOVA of quadratic regression model demonstrated that the model was statistically significant. The coefficient of multiple determinations, R^2 of the model was 0.834 which means that the model could explain 82.34% of the total variation in the system.

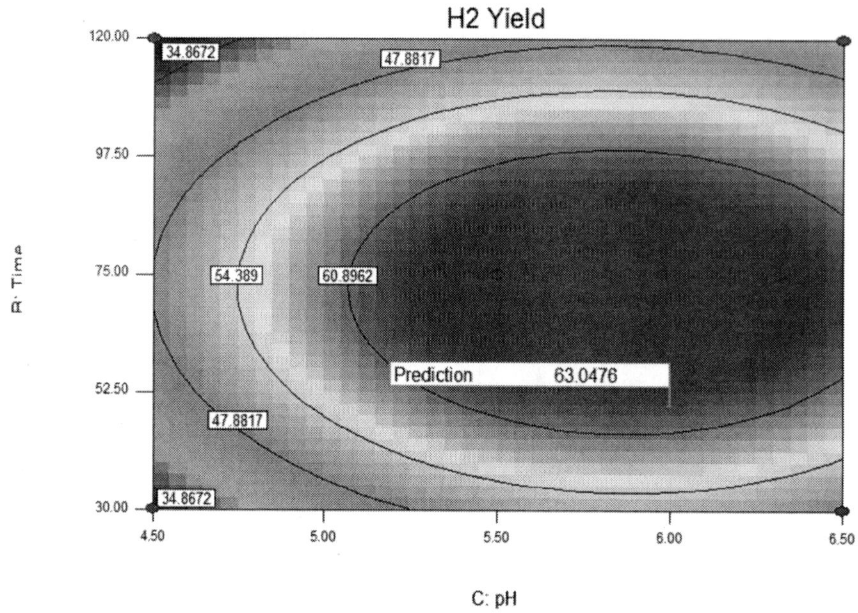

Figure 1. Contour lines.

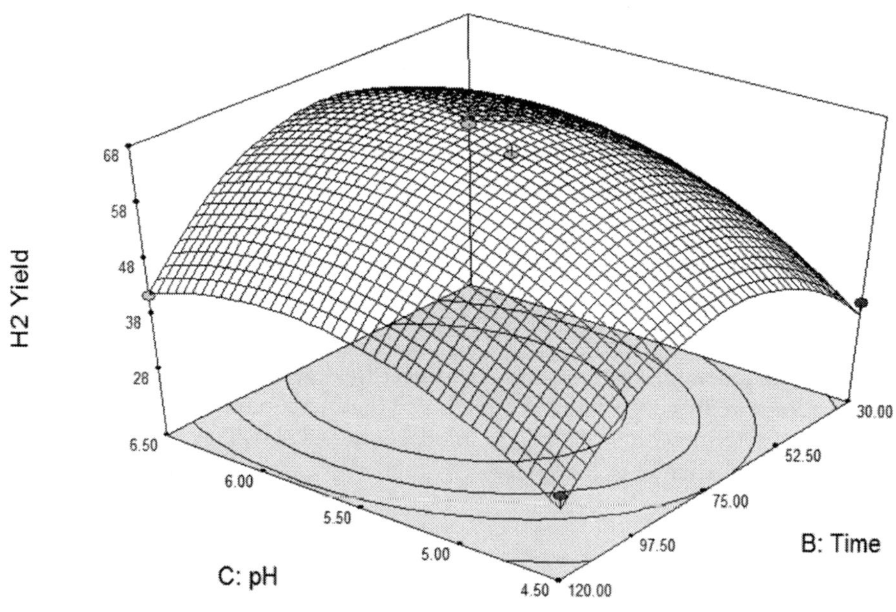

Figure 2. Response surface plot.

To generate three-dimensional response surface plots, one variable was kept as the central level and the other two variables were varied within the experimental range. Hydrogen yield over pH and exposure time were estimated using two-dimensional contour lines and three-dimensional response surface. The heat treatment conditions chosen for this study that ranged from 80°C to 100°C, did not show any significance effect, probably due to its small range.

Data analyses using RSM showed that maximum hydrogen yield and COD removal efficiency was achieved when heat pre-treatment was done at 100°C for 50 min at pH 6.00. At optimum condition, the predicted hydrogen yield was 63.0476 ml H_2/ mol glucose and the COD removal efficiency was 90.87%. The composite desirability was 0.904 which means that the probability to get predicted response was 90 times out of 100. However, in comparison to the other studies they were differences in the optimum heat treatment conditions and optimal pH for biohydrogen production and the possible reason behind the disagreement was the differences in term of source, temperature and other environmental factors [15].

Figure 1 showed that at 100°C, hydrogen yield increase with the increases of pH and exposure time. However, as the exposure time get longer, the hydrogen yield started to decrease. This was probably because high temperature and longer exposure time may have suppressed hydrogen producing bacteria.

Apparently, pre-treatment had a great influence on the biohydrogen production. Among all pre-treatment technologies, heat pre-treatment is the most commonly used pre-treatment of inoculum and it was proven to be able to suppress hydrogen consuming bacteria and at the same time enriching the hydrogen producing bacteria, thus maximize the hydrogen yield [16-17]. In agreement with other studies, the heat treatment applied in this study had successfully suppressed the HCB based on the absence of methane gas detected in the GC analysis [18].

In this study, it was found that at pH 6.00 is the maximum hydrogen production and COD removal efficiency was at the maximum. This was in agreement with a study done by Ren et al, indicated that pH 6 and 6.5 is the optimum operating pH that can improve acidogenesis and to avoid excessive propionate production [19]. Additionally, a study by Manyi-Loh et al. reported that lower pH slowed down the growth rate of methanogens and methanogenic biomass activities [20].

3.2 Verification of optimum conditions
The triplicate batch tests were conducted to verify the optimum condition analyzed by RSM. At given optimum heat treatment conditions and pH; 100°C for 50 min at pH 6.00, the hydrogen yield and COD removal efficiency slightly lower than the expected value (88.76% and 60.1766 ml H_2/ mol glucose).

4. Conclusion
This study proved that heat treatment of sludge can significantly increase biohydrogen production as well as suppressing the methanogenic bacteria efficiently. It can also be concluded that the different heat treatment conditions led to differences in the microbial communities i.e. hydrogen producing bacteria which resulted in the different hydrogen yields. Data analysis from RSM showed that the optimum heat pre-treatment condition and pH was 100°C for 50 min at pH 6.00. In future, further tests will be conducted to identify the different combinations of different environmental and operational parameters in order to determine the most ideal conditions of biohydrogen production.

References

[1] Cardoso V, Romão B B, Silva F T M, Batista F R X, Ferreira J S and Santos J G 2014 *Chemical Engineering Transactions* **38** 481-486
[2] Ghimire A, Frunzo L, Pontoni L, d'Antonio G, Lens P N, Esposito G and Pirozzi F 2015 *Journal of Environmental Management* **152** 43-48
[3] Chun-Min Liu, Jin-Long Zheng, Shu-Yii Wu and Chen-Yeon Chu 2016 *International Journal of Hydrogen Energy* **41** 4466–4473
[4] Gioannis G D, Friargiu M, Massi E, Muntoni A, Polettini A, Pomi R and Spiga D 2014 *International Journal of Hydrogen Energy* **39** 20930-20941
[5] Wang J and Wan W 2008 *International Journal of Hydrogen Energy* **33** 2934-2941
[6] Ren N Q, Guo W Q, Wang X J, Xiang W S, Liu B F, Wang X Z, Chen Z B 2008 *International Journal of Hydrogen Energy* **33** 4318–4324
[7] Agrawal P, Hema R and Kumar S M 2006 *Proceedings of the 1st International Conference on Natural Resources Engineering & Technology 2006 (Putrajaya)*
[8] Lai Z, Zhu M, Yang X, Wang J and Li S 2014 *Biotechnology for biofuels* 7:114
[9] Urbaniec K and Bakker R R 2015 *International Journal of Hydrogen Energy* **40** 3648-3658.
[10] Pandu K and Joseph S 2012 *International Journal of Advances in Engineering & Technology* 342-356
[11] Kapdan I K and Kargi F 2006 *Enzyme and Microbial Technology* **38** 569–582
[12] Bundhoo M Z, Mohee R and Hassan M A 2015 *Journal of Environmental Management* 20–48
[13] Sangyoka S, Reungsang A and Lin C Y 2016 *Sustainable Environment Research* **26** 235-242
[14] APHA 1992 *Standard Methods for the Examination of Water and Wastewater* 18th ed. (Washington, DC: American Public Health Association)
[15] Jianlong Wang and Wei Wan 2009 *International Journal of Hydrogen Energy* **34** 799-811
[16] Alemahdi N, Man H C, Rahman N A, Nasirian N and Yang Y 2015 *International Journal of Hydrogen Energy* **40** 16033-16044
[17] Mohammadi P, S Ibrahim 2011 *Journal of Cleaner Production* **19** 1654-1658
[18] Mansor M F 2016 *Journal of Engineering Science and Technology* pp 1174 - 1182
[19] Nanqi Ren, Baozheng W, Ju-Chang H. 1997 *Biotechnology and Bioengineering* 428-33
[20] Manyi-Loh C E, Mamphweli S N, Meyer E L, Okoh A I, Makaka G and Simon M 2013 *International Journal of Environmental Research and Public Health* **10** 4390–4417

IOP Conf. Series: Materials Science and Engineering **210** (2017) 012024 doi:10.1088/1757-899X/210/1/012024

Application of Water Evaluation and Planning Model for Integrated Water Resources Management: Case Study of Langat River Basin, Malaysia

W K Leong and S H Lai*

Department of Civil Engineering, Faculty of Engineering Building, University of Malaya, 50603, Kuala Lumpur, Malaysia.

E-mail: laish@um.edu.my

Abstract. Due to the effects of climate change and the increasing demand on water, sustainable development in term of water resources management has become a major challenge. In this context, the application of simulation models is useful to duel with the uncertainty and complexity of water system by providing stakeholders with the best solution. This paper outlines an integrated management planning network is developed based on Water Evaluation and Planning (WEAP) to evaluate current and future water management system of Langat River Basin, Malaysia under various scenarios. The WEAP model is known as an integrated decision support system investigate major stresses on demand and supply in terms of water availability in catchment scale. In fact, WEAP is applicable to simulate complex systems including various sectors within a single catchment or transboundary river system. To construct the model, by taking account of the Langat catchment and the corresponding demand points, we defined the hydrological model into 10 sub-hydrological catchments and 17 demand points included the export of treated water to the major cities outside the catchment. The model is calibrated and verified by several quantitative statistics (coefficient of determination, R^2; Nash-Sutcliffe efficiency, NSE and Percent bias, PBIAS). The trend of supply and demand in the catchment is evaluated under three scenarios to 2050, 1: Population growth rate, 2: Demand side management (DSM) and 3: Combination of DSM and reduce non-revenue water (NRW). Results show that by reducing NRW and proper DSM, unmet demand able to reduce significantly.

1. Introduction

In recent years, water resources management has become more complicated and controversial due to the impacts of various different parameters affecting water systems. For example, growing demands due to arising from human activity, changing of socio-economic situations, climate change, environmental considerations, and hydrologic and hydraulic conditions. All the issues together with the fact that water has a vital role in virtually every aspect of human life make water resources an important source of conflict. Thus, decision makers crucially require reliable models to allocate water resources between different stakeholders effectively and efficiently [1-3].

As a solution to this problem, researchers and scholars have emphasized in applying various simulation or optimization modeling techniques to develop decision support systems for improved management of the water resources. Rees et al. 2006 presented practical software to help in making a water balance between natural water resources and water demands [4]. Giupponi 2007 developed a DSS for integrated water resource management (IWRM) to help the decision makers in water allocation between different sectors [5]. Letcher et al. 2007 proposed a generalized conceptual framework which considered water allocation, agricultural production and water use decisions and their interaction with the river system [16]. Van Cauwenbergh et al. 2008 developed a DSS to grade different sustainable planning and management options according to their socio-economic and environmental performance in fulfilling water demand, water price, technical and economic efficiency, and social and environmental impacts [17]. Liu et al. 2010 addressed the IWRM using an optimization-based approach [18]. Gaivoronski et al. 2011 addressed the problem of water management and obtained a "robust" decision policy [19]. Coelho et al. (2012) proposed a DSS tool for regionalization in support of IWRM through Geographic Information System (GIS) processing, fuzzy set theory and a modified dynamic programming clustering algorithm. It provides the option for decision makers to include socioeconomic, political, and environmental aspects into the analysis [20]. Mianabadi et al. 2014 develop a bankruptcy rule for water resources problems that considers agents' contribution to the total resources as well as their claims for reallocation [21]. Roozbahani et al. 2015 introduced a multi-objective model for sustainable water allocation of transboundary watersheds taking into account economic, social and environmental aspects [12]. Jorda-Capdevila et al. 2016 modelled ecosystem services (ES) for recurrent socio-environmental conflicts in water management decisions [13]. Simulation model can be used to resolve conflicts indirectly as well as providing stakeholders and decision-makers with a systematic decision-making process by improving their understanding of the problem by investigating the consequences of their decisions and actions on the results to achieve a sustainable solution for a conflict [14], [15].

In the present study, we aimed to demonstrate the advantages of an integrated management planning network system as a tool for water conflict resolution. We employed a simple integrated water resources management model developed based on Water Evaluation and Planning (WEAP) to demonstrate how efficient the complexity of a water system can be managed by this methodology. The models illustrate the application of the suggested method to a specific catchment area (Langat River basin, Malaysia). The model seeks to simulate the behaviors of demand and supply components, and at the same time optimize the water allocation to the demand sites at every time step along the horizon under three different possible scenarios. Scenarios analyzes are, 1: Population growth rate, 2: Demand side management (DSM) and 3: Combination of DSM and reduce non-revenue water (NRW). The model was then run for thirty years to calculate the impact on the supply-demand gap by the year 2050. Finally, the model performances were evaluated by model evaluation statistics.

2. Study Area

The Langat River basin is located in the southern and south-eastern parts of the Selangor state of Malaysia. Figure 1 shows the boundary of Langat River Basin basin and the major tributaries of the basin: Langat River, Semenyih River and Labu River. The entire basin has an area of 2350 km^2, with main tributary, Langat River length of 182 km. This basin has a tropical climate, with average annual rainfall of 2145 mm and annual evapotranspiration of 1500 mm. The present study focused on the upper Langat River basin, which is up to the Dengkil river monitoring station (Station No.2816441). The catchment has been chosen as the study area due to its importance to the country. The catchment has a key role in providing water for domestic and commercial usage to about 2.5 million people living inside and the surrounding area outside of the basin, including Kuala Lumpur, the capital city of Malaysia with highest population density in the country and Federal Territory of Putrajaya and Cyberjaya. There two (2) reservoirs and eight (8) water treatment plants within the basin, which provide treated water to the user. Although it is a very important catchment in the Malaysia, it suffers from flooding situation, water pollution, and water shortage that makes the basin a good case study for an integrated water resources management planning network between different stakeholders and decision makers.

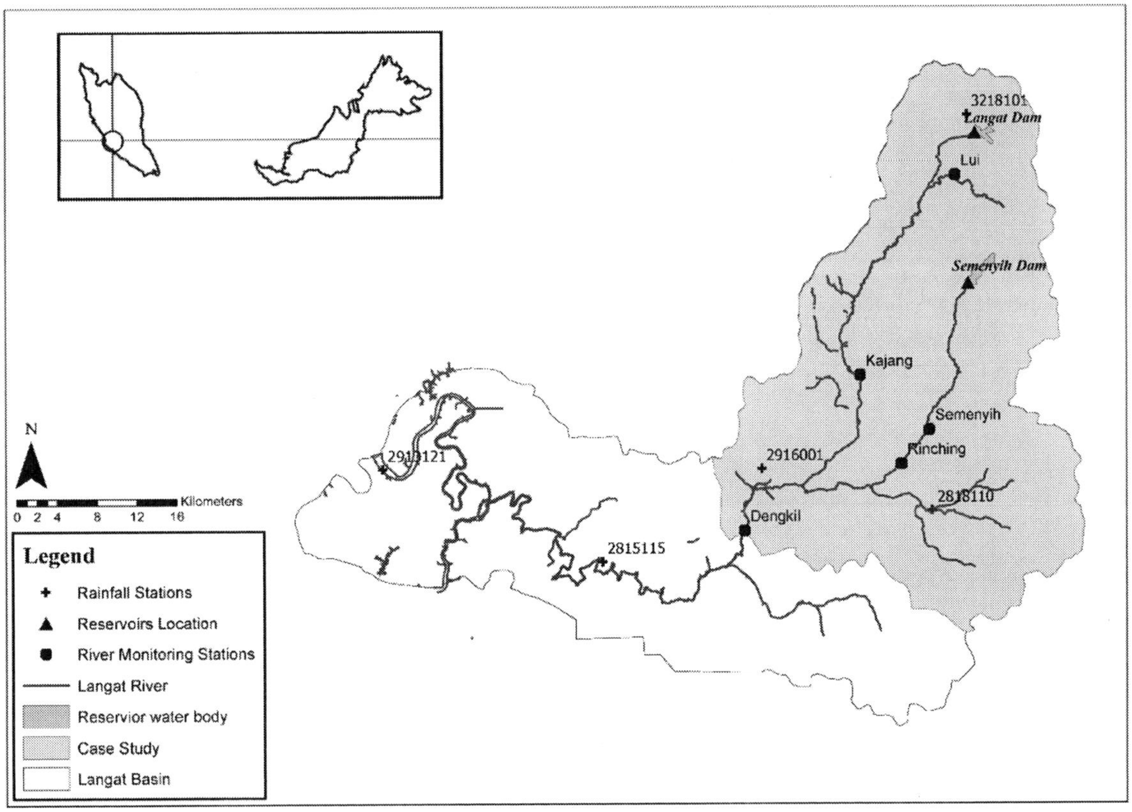

Figure 1. Langat River Basin

3. Methodology

3.1. Model Development

WFAP has a module to model hydrologic processes. The hydrological model is semi-theoretical, continuous time, semi-distributed, and deterministic. As the model is semi-theoretical, it needs calibration and verification. The WEAP does not automatically calibrate the hydrological model; therefore, the calibration must be implemented manually. The Standard method computes the water demand as the product of activity level and water use rate for different sectors, such as, industrial, municipal and agricultural. Demand for all sectors, except agricultural, can be computed only with this method. WEAP uses a one period linear programming routine to allocate resources between demands [16, 17].

Figure 2 shows the overall hydrologic modeling framework. To develop the model structure, the whole study area is divided into ten (10) hydrological catchments (according to available hydrological data) and seven (7) sub-districts which may have considered domestic and non-domestic demands. Besides, we considered two (2) demand sites located outside the study area as exports which supplied by water treatment plants in Langat river basin.

Next, the model structure developed considering involved elements including River (stream), Demand site (service area), Reservoir, Transmission link, Diversion (water treatment plant), Catchment, Streamflow gauge, for monthly time steps between 2005 to 2050 (Figure 3). From the period which flow data are available, data sets for year 2006-2013 are used for model calibration and validation. Figure 4 is the schematic of the model showing WEAP node-network topology overlaid over a few GIS layers.

Last but not least, several scenarios were developed and defined in the model to investigate the water resources problems and probable water issues which may be occur in the near future. Those scenarios

IOP Conf. Series: Materials Science and Engineering **210** (2017) 012024 doi:10.1088/1757-899X/210/1/012024

were analyzed and the scenario that showed the best model performance for the Langat River Basin will be figured out.

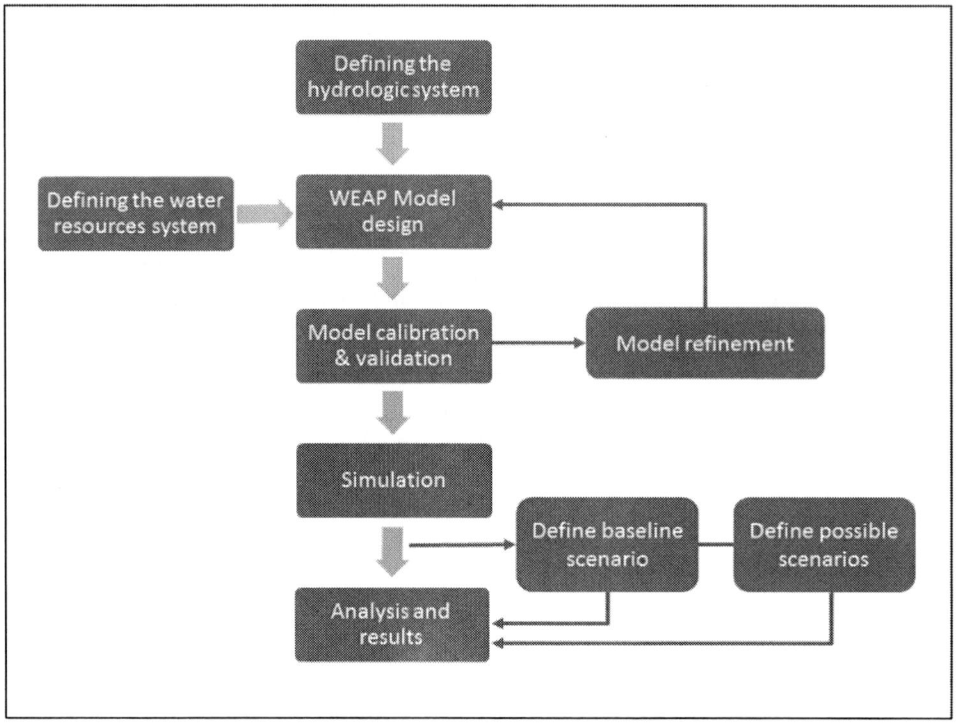

Figure 2. WEAP model framework

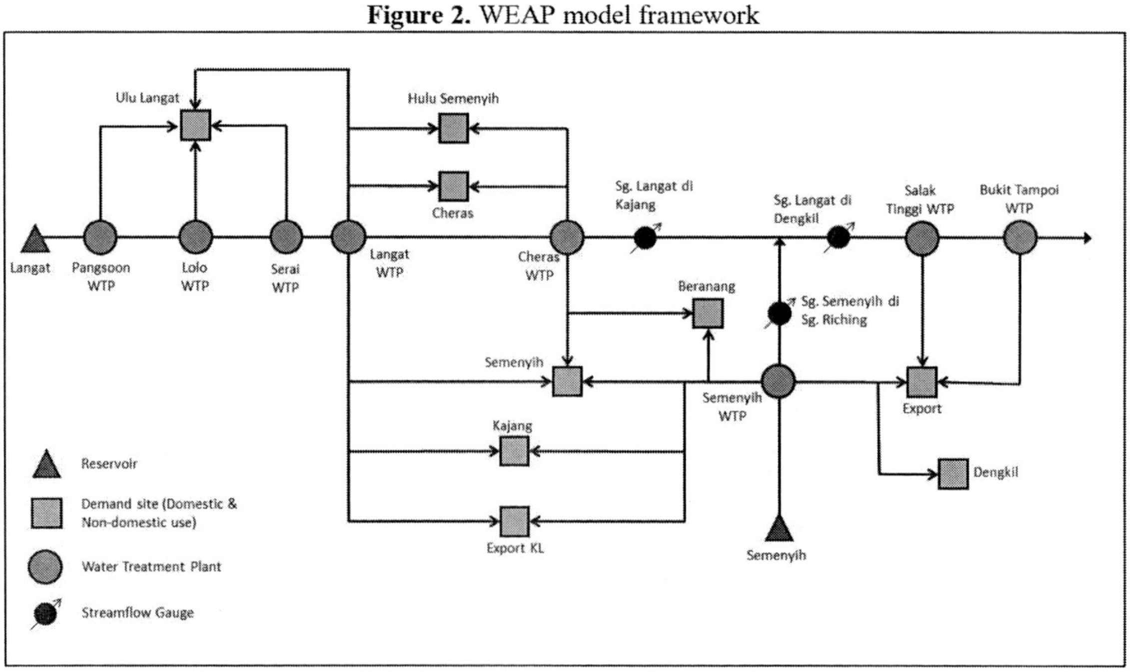

Figure 3. Schematic of the main components of the model

IOP Conf. Series: Materials Science and Engineering **210** (2017) 012024 doi:10.1088/1757-899X/210/1/012024

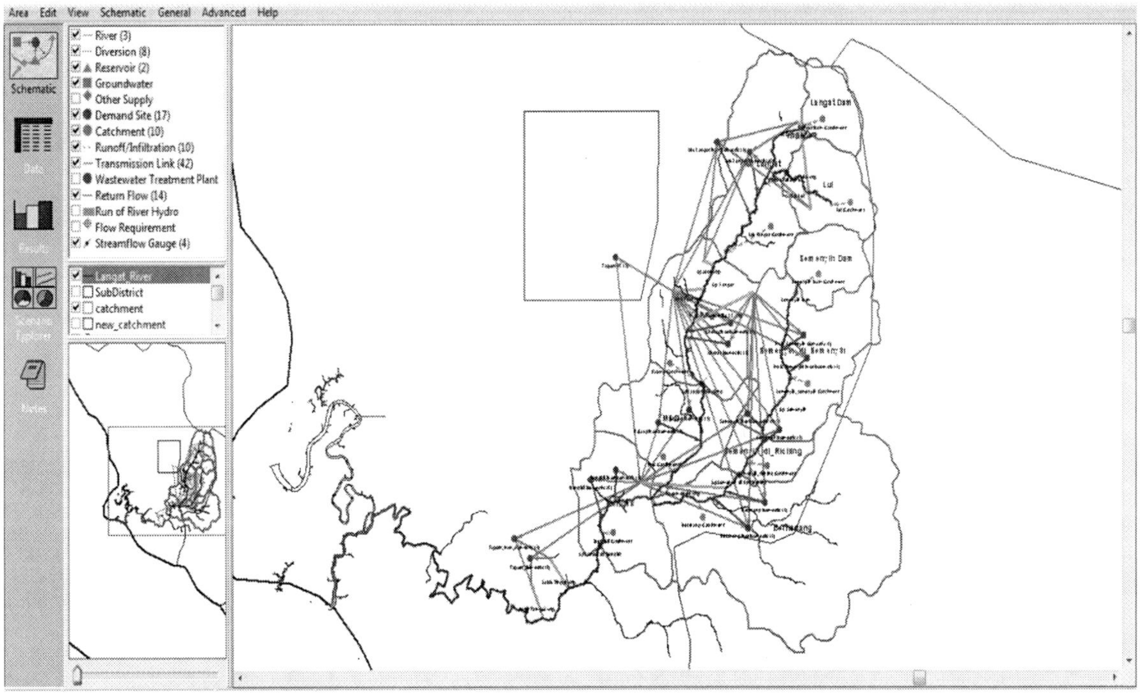

Figure 4. Schematic illustration of the Langat catchment in WEAP application

3.2. Model Calibration and Validation

In the present study, the parameters controlling the generation of runoff from climate inputs were calibrated and validated using the historical measurement of streamflow obtained from 3 gauging stations located on the Lui River, Semenyih River and main stem of the Langat River. Figure 5 is the locations of the streamflow stations on GIS map. The longest available continuous streamflow data from the Sg. Lui, Sg. Semenyih and Sg. Langat gauging stations have been recorded since August 1965, May 1975 and August 1960, respectively. Thus, the streamflow records at these stations are sufficient and appropriate for calibration and validation purposes. However due to availability of other related input data especially climate and population inputs, we chose to use data sets from the period 2006-2013 for calibration and validation of the model. The first 6 years are dedicated for calibration of the model from 2006-2011 while the following 2 years are dedicated for validation of the model from 2012-2013. The adjustment parameters of the WEAP model were calibrated by trial and error. The quantitative statistics (coefficient of determination, R^2; Nash-Sutcliffe efficiency, NSE and Percent bias, PBIAS) were computed for each set of simulated and historical streamflow over the period 2006-2013.

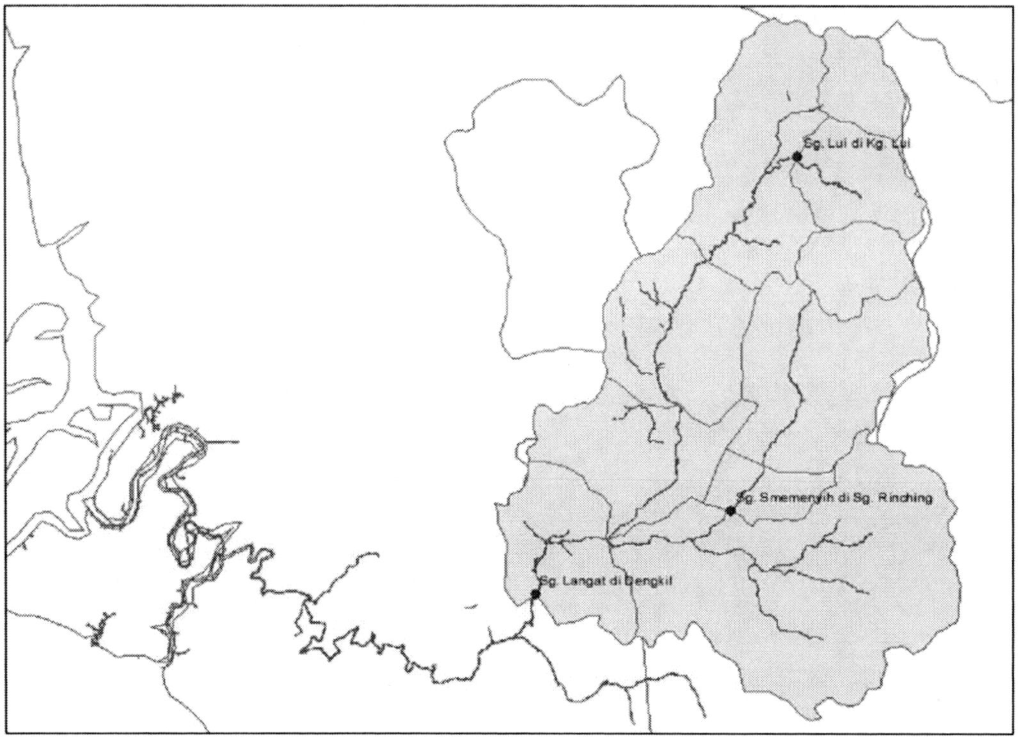

Figure 5. Locations of the streamflow stations

3.3. Model Evaluation Statistics

The quantitative statistics used for the evaluation of model performance are the coefficient of determination (R^2), the Nash-Sutcliffe model efficiency (NSE) and the Percent bias (PBIAS).

3.3.1. Coefficient of determination (R^2). The coefficient of determination (R^2) outlines the degree of collinearity between simulated and measured data. R^2 describes the proportion of the variance in measured data explained by the model. R^2 ranges from 0 to 1, given higher values indicating less error variance. Values which is greater than 0.5 are considered acceptable **[18, 19]**. The computation of R^2 is shown as below:

$$R^2 = \left[\frac{\sum_{i=1}^{n}\left(Y_i^{sim}-\bar{Y}^{sim}\right)\left(Y_i^{obs}-\bar{Y}^{obs}\right)}{\sqrt{\sum_{i=1}^{n}\left(Y_i^{sim}-\bar{Y}^{sim}\right)^2\sum_{i=1}^{n}\left(Y_i^{obs}-\bar{Y}^{obs}\right)^2}}\right] \quad (1)$$

Where,

Y_i^{sim}: the ith simulated streamflow

Y_i^{obs}: the ith observed streamflow

$\bar{Y}^{sim}$: the mean of simulated streamflow

$\bar{Y}^{obs}$: the mean of observed streamflow

IOP Conf. Series: Materials Science and Engineering **210** (2017) 012024 doi:10.1088/1757-899X/210/1/012024

3.3.2. Nash-Sutcliffe efficiency. The Nash-Sutcliffe efficiency (NSE) is a normalized statistic that determines the relative magnitude of the residual variance ("noise") compared to the measured data variance ("information") [20]. NSE indicates how well the plot of observed versus simulated data fits the 1:1 line. NSE ranges between $-\infty$ and 1.0 (1 inclusive), with NSE = 1 being the optimal value. Values between 0.0 and 1.0 are generally viewed as acceptable levels of performance, whereas values <0.0 indicates that the mean observed value is a better predictor than the simulated value, which indicates unacceptable performance. NSE is computed as shown below:

$$\mathrm{NSE} = 1 - \left[\frac{\sum_{i=1}^{n}\left(Y_i^{obs} - Y_i^{sim}\right)^2}{\sum_{i=1}^{n}\left(Y_i^{obs} - \bar{Y}^{obs}\right)^2}\right] \tag{2}$$

Where,
Y_i^{obs}: the ith observed streamflow
Y_i^{sim}: the ith simulated streamflow
$\bar{Y}^{obs}$: the mean of observed streamflow

3.3.3. Percent bias (PBIAS). Percent bias (PBIAS) measures the average tendency of the simulated data to be larger or smaller than their observed counterparts. The optimal value of PBIAS is 0.0, with low-magnitude values indicating accurate model simulation. Positive values indicate model underestimation bias, and negative values indicate model overestimation bias [21]. PBIAS is calculated as shown below:

$$PBIAS = \left[\frac{\sum_{i=1}^{n}\left(Y_i^{obs} - Y_i^{sim}\right)*(100)}{\sum_{i=1}^{n} Y_i^{obs}}\right] \tag{3}$$

Where,
Y_i^{obs}: the *i*th observed streamflow
Y_i^{sim}: the *i*th simulated streamflow

3.4. Future Possible Scenarios
In order to assess the capability of water supply in meeting the demand for treated water in Selangor, three types of future projections were investigated: population growth, demand side management and supply side management. The key results investigated were water supplied and demand coverage.

3.4.1. Reference scenario. Under this scenario, the projection assumes the supply and use of water same as current situation and associated population growth. The production of water treatment plant is at its current design capacity with population growth rate of 2.57 % (Intercensal Mid-Year Population Estimates of Malaysia and States for the year 2001-2009). The main question asked of the model at this stage is: "How much of the current demand could be met, if the water treatment plant production is at current design capacity and the population is increasing in the future?"

3.4.2. Scenario 1: Population Growth Rate. This scenario assumes a higher population growth rate of 4 % with treated water production capacity at current design capacity to evaluate the impact of population growth towards the water supply and demand of Langat in the future. "How much of the demand for treated water could be met, if the production capacity is at current design capacity and population growth is higher in the future?"

3.4.3. Scenario 2: Demand Side Management (DSM). This scenario evaluates a demand side management strategy and was applied to both the "reference" and "high population growth" scenarios above. Water savings initiatives were taken into the calculation of the model in reducing 10% of the amount of water consumption per capital from 235 ld to 211.5 ld. "If water usage per capital is reduced from 235 ld to 211.5 ld, how much of the water demand could be met by the supply with current design capacity?"

3.4.4. Scenario 3: Combination of DSM and reduce Non-Revenue Water (NRW). Under this scenario, a demand side management strategy incorporates with a supply-side management strategy – reducing water consumption per capital, as well as reducing the non-revenue water (NRW) losses from 33.35 % to 16.68 %, to be assessed against the "reference" and "high population growth" scenarios. "What would be the impact on the water supplied and demand coverage in the future conditions, given plans to reduce NRW for the supply-side and plans to reduce water consumption per capital for the demand side?"

4. Results and Discussion

4.1. Model Performance

Figure 6 reveals the hydrologic calibration and validation of the reference model. The monthly streamflow recorded at each of the gauging stations was compared against the simulated values of the model from 2006-2011 for calibration process and 2012-2013 for validation process. The simulated streamflow in Sg. Lui, Sg. Semenyih and Sg. Langat show that the model replicates the observed flows reasonably well.

At Sg. Lui, the simulated monthly flows match the observed values very closely with a Nash-Sutcliffe efficiency of 0.96 and $R^2 = 0.96$ during calibration, 0.97 and and $R^2 = 0.98$ during validation. The Pbias was estimated as 0.81 % during calibration and 3.14 % during validation (Figure 6a).

At Sg. Semenyih, monthly flows were simulated with a Nash-Sutcliffe efficiency of 0.93 and $R^2 = 0.93$ during calibration, 0.87 and $R^2 = 0.98$ during validation. The Pbias was estimated as 0.41 % during calibration and 7.96 % during validation (Figure 6b).

At Sg. Langat, monthly flows were simulated with a Nash-Sutcliffe efficiency of 0.93 and $R^2 = 0.94$ during calibration, 0.92 and $R^2 = 0.97$ during validation. The Pbias was estimated to be 4.14 % during calibration and 4.39 % during validation (Figure 6c).

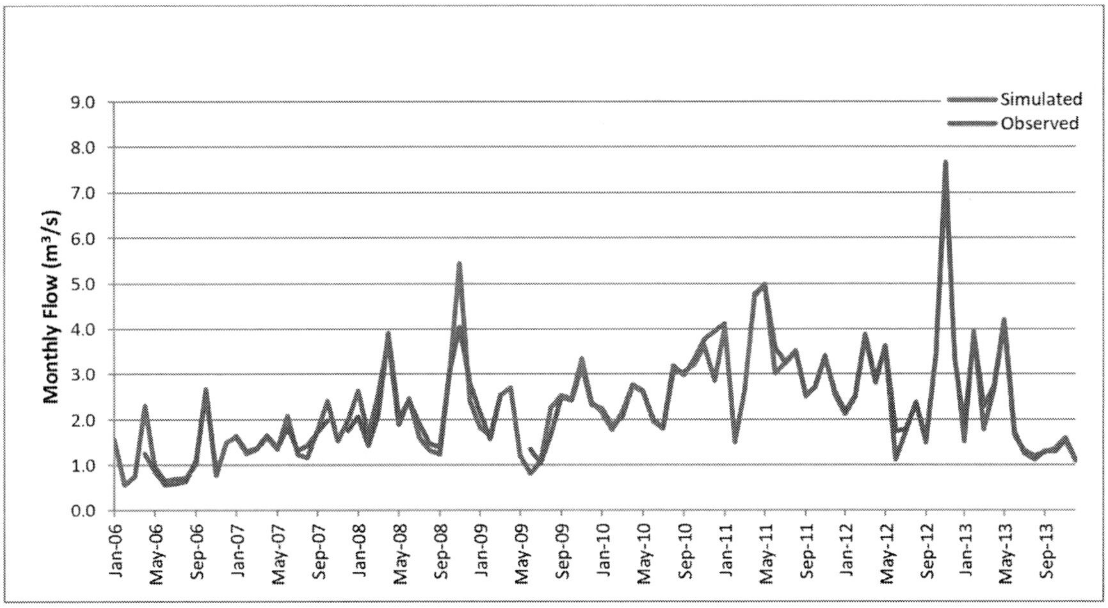

(a) Sg. Lui di Kg. Lui station

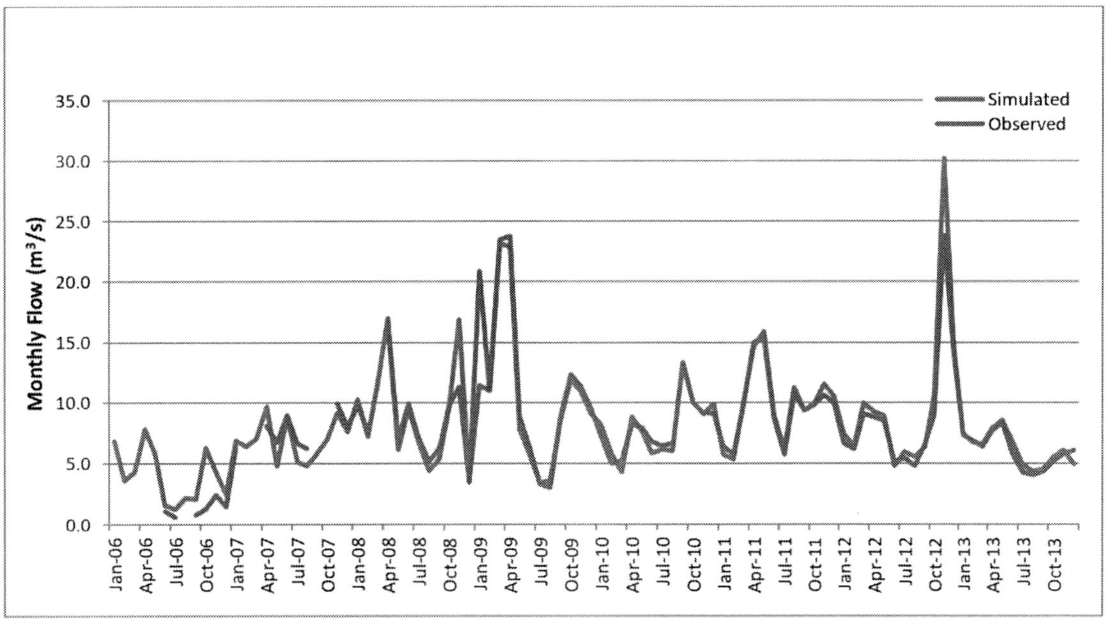

(b) Sg. Semenyih di Kg. Rinching station

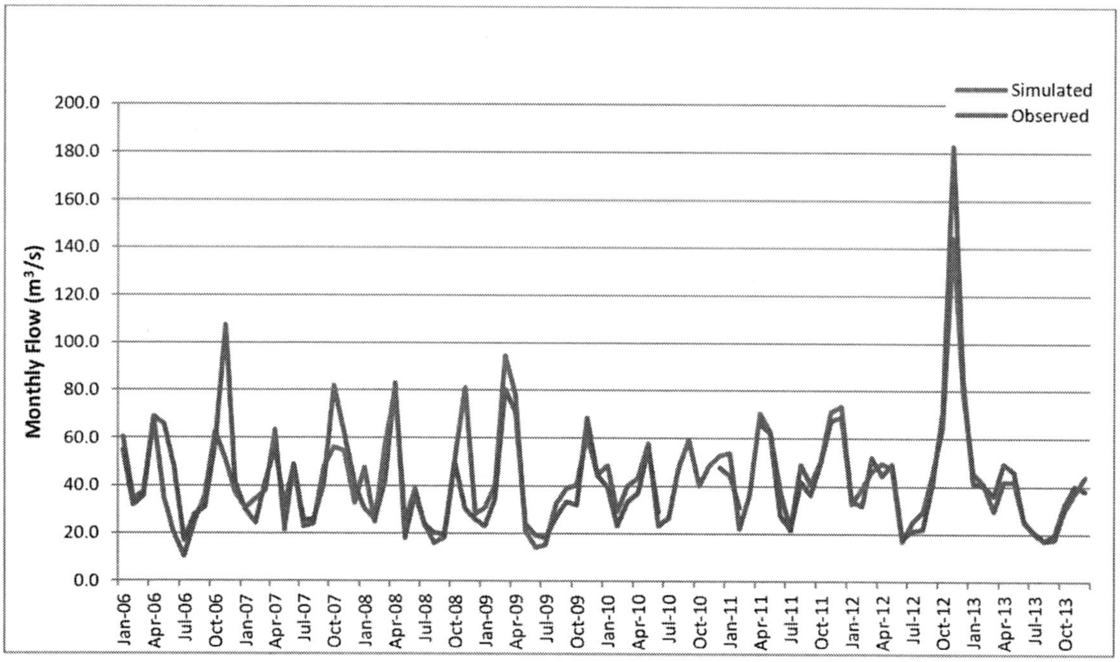

(c) Sg. Langat di Dengkil station

Figure 6. Historical monthly streamflow along with the simulated values

4.2. Reference Scenario

Water demand would steadily rise as a result of population increasing at 2.57 % annually from 2005 population of 1,085,670 to reach the capacity of 3,606,835 by 2050. The accompanying projected water demand rises to a maximum of 599.43 MCM compared to the existing demand of less than 300.00 MCM (see Figure 7), which is double of the demand of current situation.

Under the current system of water use and supply, the projection showed that Langat basin area will face water deficits for both domestic and non-domestic sectors by the year of 2019. This is probably due to the supply of treated water from the basin to areas outside the basin especially Kuala Lumpur as more than 38 % of the treated water from Sg. Langat water treatment plant and Sg. Semenyih water treatment plant were supplied for areas outside of Langat basin.

4.3. Scenario 1: Population Growth Rate.

This scenario assumes population growing at a higher rate, 4 % to forecast the impact of population growth towards the supply and demand of water for the period of 2013 to 2050 in Langat basin area with the current system of water usage and supply maintain. The accompanying projected water demand raises to a maximum of 910.16 MCM instead of 599.43 MCM under the reference scenario (see Figure 7).

Figure 7 also shows the projection of annual unmet water demand based on both scenarios: high population growth rate and the reference. The simulation results demonstrate that with high population growth rate, Langat basin area will face water deficits in 2017, which is only 2 years earlier than the reference scenario. However, the annual unmet demand for high population growth rate indicated a maximum of 621.00 MCM in 2050, which is twice the amount of the annual unmet demand of reference scenario (303.01 MCM). This implies that population growth rate shown significant impact on the demand of water in a long-term perspective and reflects the need to develop new technologies, new cooperation, or better water management plans to offset this anticipated shortfall.

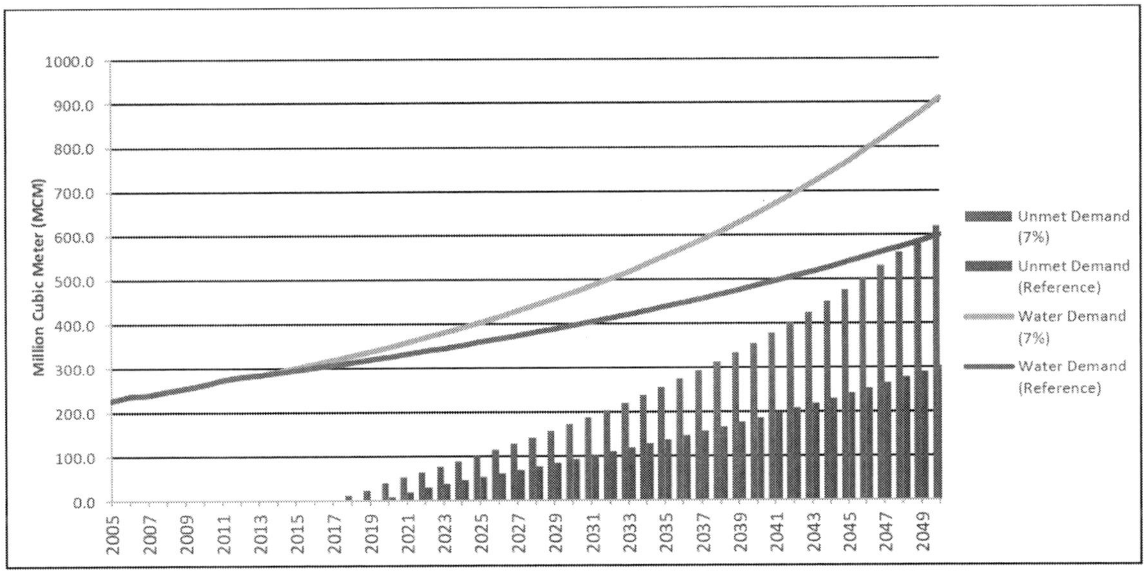

Figure 7. Annual total water demand and unmet demand of reference and population growth scenarios

4.4. Scenario 2: Demand Side Management (DSM)

This scenario was evaluated with the model, by employing both the "reference" and "high population growth" scenarios. Through this scenario, the impact of the application of water conservation and demand side strategy was assessed. Saving of 10% of the water consumption per capital from 235 liter/day to 211.5 liter/day was considered. Figure 8 and Figure 9 show the amount of water that can be saved through the implementation of DSM via the "reference" and "high population growth" scenarios.

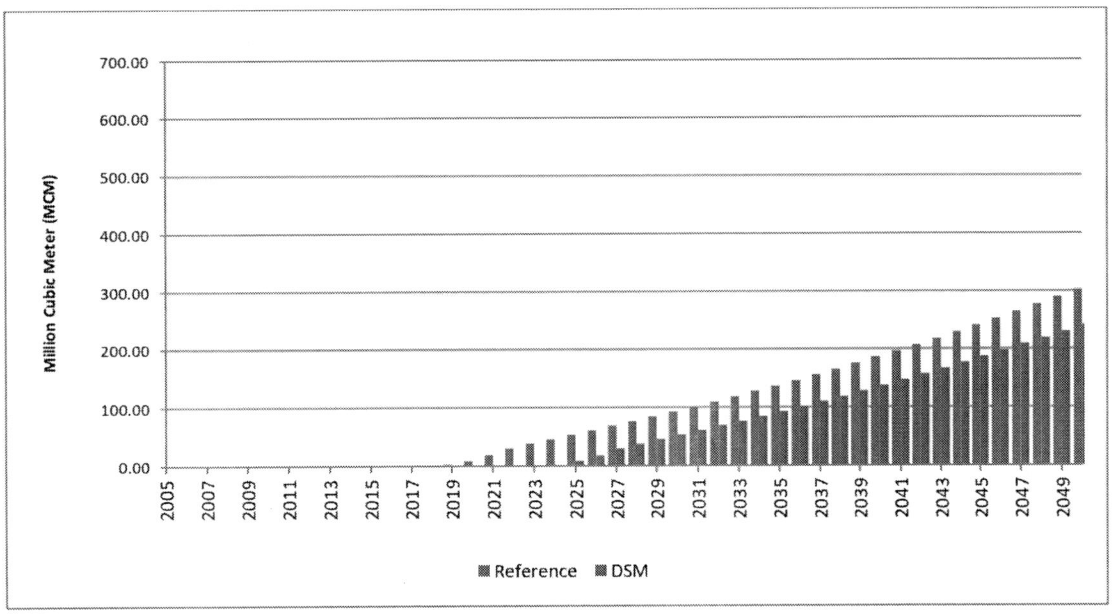

Figure 8. Annual total unmet demand based on implementation of DSM on the reference scenario

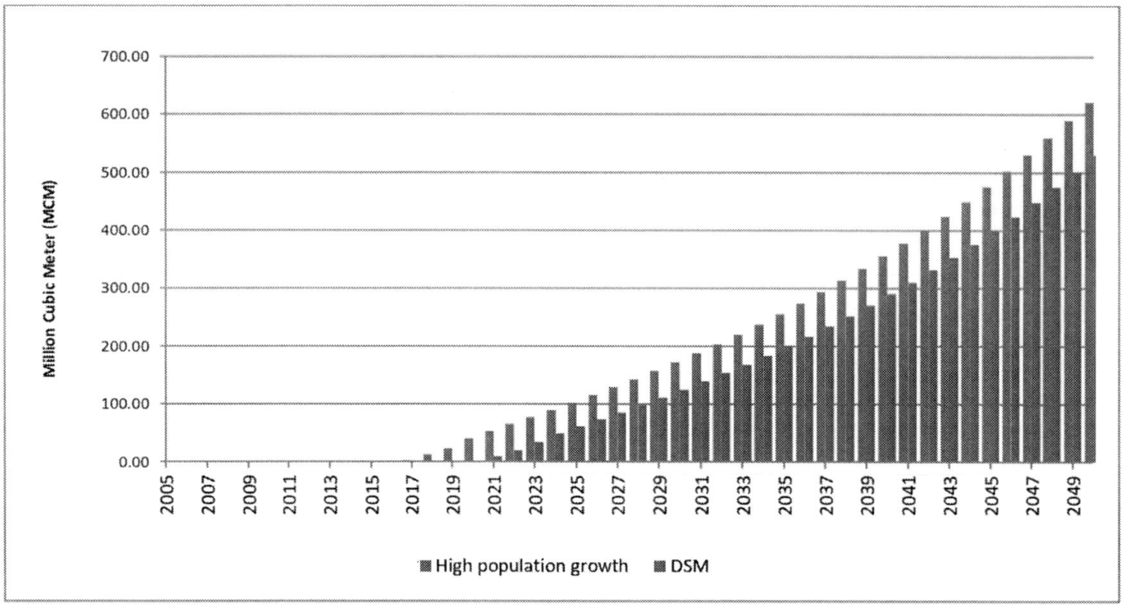

Figure 9. Annual total unmet demand based on implementation of DSM on the population growth rate scenario

Under the reference scenario, the results shown in Figure 8 reveals that with implementation of DSM, the Langat basin water supply is able to keep up with the demand until 2023 by which time demand growth will outstrip gains from the reduced water consumption by DSM. The maximum annual total unmet demand in 2050 is also able to reduce to 243.07 MCM compared to the reference condition (303.01 MCM). Overall, the total amount of unmet demand over the period 2005-2050 was decreased from 4,491.52 MCM to 3,113.88 MCM. On the other hand, the WEAP simulations showed that with demand side management effort of just reducing 10 % of the water consumption per capital, a total of 1,568.28 MCM of water is able to be saved for projection of water demand until 2050.

Under high population growth scenario, the supply is able to keep pace with the demand until 2020 in which after that the demand is greater than the supply (see Figure 9). The model also projected a reduction in the total unmet demand from 8,788.16 MCM to 6,928.18 MCM. Besides, the amount of water saving able to achieve a total number of 1,982.78 MCM for projection of water demand until 2050, indicated that more water is able to be saved as compared to the reference scenario (1,568.28 MCM). This result shows that the saving was proportional to the demand; therefore it had a greater impact in the higher population growth scenario, the one with the higher water demand.

4.5. Scenario 3: Combination of DSM and reduce Non-Revenue Water (NRW)

This scenario incorporates a demand side management and a supply side management strategies-- reducing in NRW from current condition of 33.35 % to 16.68 % as well as reduced water consumption per capital, assessed against the "reference" and "higher population growth" scenarios. Figure 10 and Figure 11 compare the water unmet demand that can be reduced through the implementation of DSM together with reduced NRW via both the "reference" and "higher population growth" scenarios.

Under the reference scenario, as a result of reduced losses and water consumption per capital, water supply is able to keep up with demand until 2032, marked a delay of 13 years compared to the reference scenario without DSM and reduced NRW (see Figure 10). In overall, through the implementation of DSM water saving and reduced NRW policies, the shortage is able to reduce by about 62.85 % from 4,491.53 MCM to 1,688.70 MCM considering the population and industrial growth.

Under the high population growth scenario, the results in Figure 11 showed that the supply able to meet 100% of the demand until 2025 which is 9 years later than high population growth scenario without

DSM and reduced NRW. The model also projected 42.11 % reduction in the total unmet water demand, from 8,878.16 MCM to 5,139.85 MCM due to implementation of DSM water saving and reduced NRW policies.

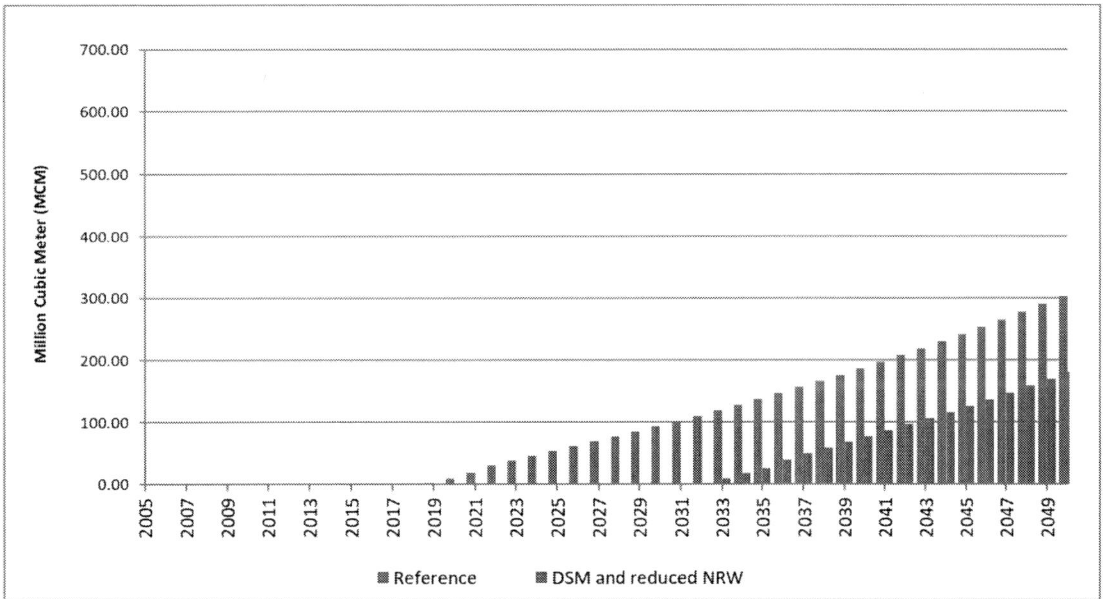

Figure 10. Annual total unmet demand based on implementation of DSM and reduced NRW on the reference scenario

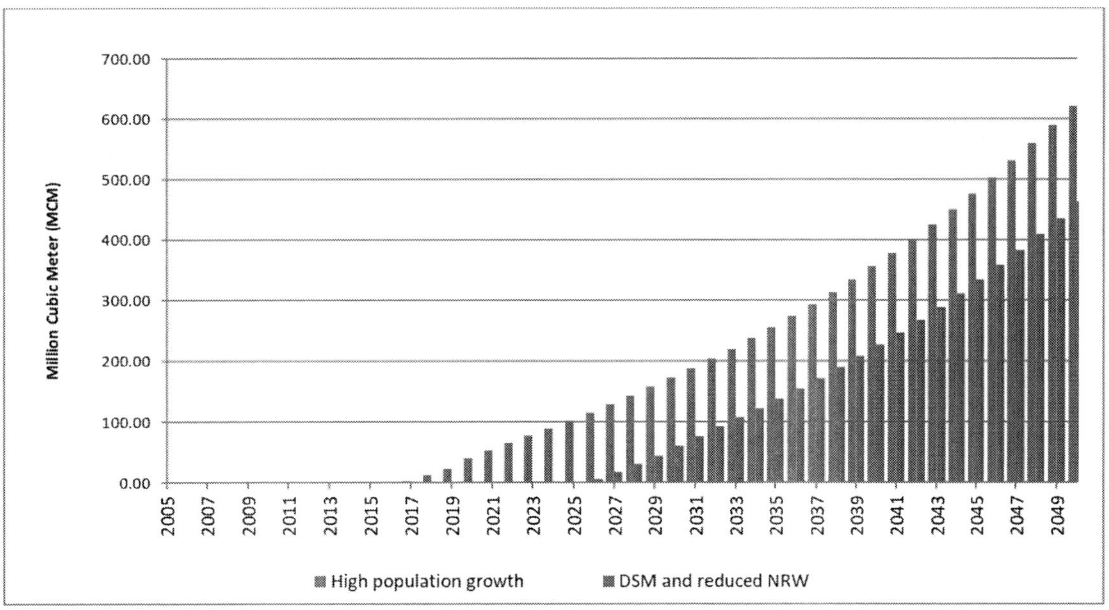

Figure 11. Annual total unmet demand based on implementation of DSM and reduced NRW on the population growth rate scenario

5. Conclusion

In the present study, an integrated hydrologic model was introduced by using WEAP for assessing the water conflict. The model developed through WEAP is highly capable and showed great performance to manage available water resources with water demand. The calibration and validation of model evaluated by the quantitative statistics (coefficient of determination, R^2; Nash-Sutcliffe efficiency, NSE and Percent bias, PBIAS) assured the model outputs scientifically sound, robust and defensible. The model able to simulate various possible future scenarios to evaluate current and future water management system of Langat River Basin and acts as an integrated management planning network for decision makers to resolve conflicts over water allocation and policies implied.

References

[1] Zarghami, M. and Akbariyeh, S., 2012 *Resources, Conservation and Recycling* **60** 99-106.
[2] Zarghami, M., Safari, N., Szidarovszky, F. and Islam, S., 2015 *Water Resources Management* **29** 4285-4303.
[3] Weng, S.Q., Huang, G.H. and Li, Y.P., 2010 *Expert Systems with Applications* 37 8242-8254.
[4] Rees, H.G., Holmes, M.G., Fry, M.J., Young, A.R., Pitson, D.G. and Kansakar, S.R., 2006 *Environmental Modelling & Software* **21** 1001-1012.
[5] Giupponi, C., 2007 *Environmental Modelling & Software* **22** 248-258.
[6] Letcher, R.A., Croke, B.F. and Jakeman, A.J., 2007 *Environmental Modelling & Software* **22** 733-742.
[7] Van Cauwenbergh, N., Pinte, D., Tilmant, A., Frances, I., Pulido-Bosch, A. and Vanclooster, M., 2008 *Environmental Geology* **54** 479-489.
[8] Liu, S., Gikas, P. and Papageorgiou, L.G., 2010 *Computer Aided Chemical Engineering* **28** 1075-1080.
[9] Gaivoronski, A.A., Sechi, G.M. and Zuddas, P., 2012 *Physics and Chemistry of the Earth Parts A/B/C* **42** 98-107.
[10] Coelho, A.C., Labadie, J.W. and Fontane, D.G., 2012 *Water resources management* **26** 1325-1346.
[11] Mianabadi, H., Mostert, E., Zarghami, M. and van de Giesen, N., 2014 *Journal of environmental management* **144** 152-159.
[12] Roozbahani, R., Schreider, S. and Abbasi, B., 2015 *Environmental Modelling & Software* **64** 18-30.
[13] Jorda-Capdevila, D., Rodríguez-Labajos, B. and Bardina, M., 2016 *Environmental Modelling & Software* **79** 22-34.
[14] Nandalal, K.D.W. and Simonovic, S.P., 2003 *Water Resources Research* **39**.
[15] Simonovic, S.P., 2012 *Managing water resources: methods and tools for a systems approach*.
[16] Yates, D., Sieber, J., Purkey, D. and Huber-Lee, A., 2005 *Water International* **30** 487-500.
[17] Sieber, J., Swartz, C. and Huber-Lee, A.H., 2005 *Water evaluation and planning system (WEAP): User guide. Stockholm Environment Institute.*
[18] Santhi, C., Arnold, J.G., Williams, J.R., Dugas, W.A., Srinivasan, R. and Hauck, L.M., 2001 *Journal of the American Water Resources Association* **37** 1169-1188.
[19] Van Liew, M.W., Arnold, J.G. and Garbrecht, J.D., 2003 *Transactions of the ASAE* **46** 1539-1551.
[20] Nash, J.E. and Sutcliffe, J.V., 1970 *Journal of hydrology* **10** 282-290.
[21] Gupta, H.V., Sorooshian, S. and Yapo, P.O., 1999 *Journal of Hydrologic Engineering* **4** 135-143.

Review: Application of coal bottom ash as aggregate replacement in highway embankment, acoustic absorbing wall and asphalt mixtures

Syakirah Afiza Mohammed and Mohamed Rehan Karim

Department of Civil Engineering, Faculty of Engineering, University of Malaya, 50603 Kuala Lumpur, Malaysia

E-mail: rehan@um.edu.my

Abstract. Worldwide annual production of coal bottom ash waste was increased in the last decade and is being dumped on landfill over the years. Its improper disposal has become an environmental concern and resulted in a waste of recoverable resources. There is a pressing and on-going need to develop new recycling methods for coal bottom ash. The utilization of coal bottom ash in highway engineering is one of the options to reduce the environmental problems related to the disposal of bottom ash. The present review describe the physical and chemical properties of coal bottom ash waste and its current application as highway embankment material, as acoustic absorbing material and as aggregate replacement in asphalt mixtures. The purpose of this review is to stimulate and promote the effective recycling of coal bottom ash in highway engineering industry.

1. Introduction

Coal ash is a co-combustion products obtained from coal fired power plant station. Type, source, fineness of the parent fuel and the operating conditions of the power plant influenced the physical and chemical properties of bottom ash produced [1]. Fly ash is extracted from the boiler flue gases and bottom ash is the coarser part collected at bottom of the furnace [2].

Compared to fly ash, commercial usage of coal bottom ash is still left behind. However, it was reported that in developed country, about 50% of coal bottom ash has been utilized in civil engineering industry such as structural fill, road constructions, aggregate replacement for embankments and cement based composites [3].

Realizing the potential of coal bottom ash to be used in civil engineering especially in highway engineering, many researcher have studied ways to convert these waste into wealth. However there are limitations that need to be considered before consuming bottom ash waste in any construction. Its irregular structural characteristics, unburned carbon particles [4, 5, 6] and the possibility to leach heavy metals can bring harm to the environment and human being if not well treated [7, 8, 9].

2. Engineering properties

Generally, coal bottom ash has similarity with conventional aggregate used in construction industry. Physical and chemical properties influence coal bottom ash subsequent use and disposal. In environmental aspect, the treatment of coal bottom ash is subjected to regulations as a hazardous waste since it is not classified as general solid waste.

2.1 Physical properties

Physically, bottom ash particles is angular, porous, and sizes ranges from gravel to fine sand with very low percentages of silt-clay sized particles [10]. Usually bottom ash sizes ranged from sands to small sizes gravels [15]. By using soil classification system (USCS), coal bottom ash had been classified as well graded sand and falls into group A-1-a using AASHTO classification system [11].

2.2 Chemical properties

Generally, bottom ash is mainly composed of silica, alumina, iron, etc. The relative elemental abundance in coal was found to be in the order of Si>Al>Fe>Ca>Mg >K>Na>Sr>V>Zn>Mn>Cr>Cu>Pb>Ni>Co>As>Cd while the order of relative elemental concentration in bottom ash was found to be Si>Al>Fe>Ca>Mg>K>Na>Mn>Sr>V>Zn>Ni>Cr>Cu>Pb>Co>Cd>As [12]. Overall, bottom ash waste was safe to be used in industrial products especially in civil engineering industry. However, the presence of toxic elements such as Cr and Pb should be considered before recycling coal bottom ash waste.

3. Comprehensive application

3.1 Highway embankment

Research on the utilization of coal bottom ash in highway embankment has been presented in previous studies [13, 14, 15]. The suitability of bottom ash in highway embankment has been investigated. Material used in highway embankment should have good drainage properties, lower plasticity, and high in strength, stability and compacted density.

From previous study, findings that bottom ash has higher specific gravity compare to fly ash [13] due to its higher iron oxide content. Its helps bottom ash to transmit less dead load to the soil that supports the embankment.

Normally higher compaction applied, will leads to higher maximum density and lower the optimum water content [15]. The increasing of the dry density was contributed by the crushed particle of bottom ash during compaction [13]. However, the addition of fly ash to the bottom ash mixture leads to increase the well graded distribution. As a result, maximum dry density increased as fly and bottom ash particles were pack more closely [15]. However, the finding showed that compacted ash mixtures are potentially corrosive. Therefore, as precaution, it is recommended to use corrosion resistant materials if any pipes or structural members are embedded in ash fill [16].

Bottom ash particles have bigger voids compare to fly ash. The values of hydraulic conductivity decreasing gradually by increasing of coal ash content as the fines included in coal ash especially bottom ash has a predominant effect on the permeability [15]. In addition, high permeability of coal bottom ash and natural soils mixture is suitable for the drainage in backfill materials in embankments where a high amount of annual rainfall occurs [14].

In compressibility aspect, regardless of the ash sources, a general observed trend is the ash samples become more compressible than conventional sands with the increasing of bottom ash. Both surface texture and particle crushing in coal bottom ash had influenced the increasing of compressibility.

Besides, the ability of soils or any materials to resist loading without failing, is also important characteristic that should be considered in highway embankment construction [15]. Therefore shear strength of the soils or materials should be determined before proceed to construction works. Due to the angularity of bottom ash particle, the friction angle increased with the increasing of bottom ash content. This provides higher resistance to particle rearrangement for sustained shearing [13].

It has been reported that, shear strength of coal bottom from Tanjung Bin Power Plant, Malaysia has been decreased without the addition of fly ash as the bottom ash exhibits small cohesion characteristics [13]. The addition of bottom ash increases critical state friction angles gradually while peak friction angles do not change significantly with increasing bottom ash content. The critical state shear strength of ash mixtures is in a very similar range to that of typical sand.

3.2 Acoustic Absorbing Materials
Other potential use of coal bottom ash waste is as acoustic absorbing materials in highway pavements. Findings show that, finest particles of bottom ash produced material with good mechanical properties, while, larger particle produced excellent sound absorption coefficients [17].

Reverberation test and transmission loss index curve from previous study prove that bottom ash as absorbent materials can be classified in the same category with conventional road noise barriers concrete [18].

Larger particles size of bottom ash decreased the surface area which then requires less water to lubricate the particles. On the other hands, the void ratio and the density of the mortar tends to increase with the increasing of bottom ash particles size. Increasing of the open void ratio had cause the decreasing of compressive strength of mortars [17]. It has been reported in latest study, that density and compressive strength of acoustic absorbing material made from 80% of bottom ash are similar to those found in barriers made of conventional or recycle materials [18].

In addition, the replacement of fine and coarse aggregates with co-combustion products (fly ash and bottom ash) have a remarkable influence on fire resistance without affecting the mechanical properties of the acoustic absorbing material produced [19].

3.3 Asphalt Mixture

3.3.1 Filler in bituminous mixture Optimum binder content increased with the present of bottom ash and/or hydrated lime due to high specific surface of both and the high porosity of the ashes. High porosity of the ashes leads to higher absorption values [20, 21]. Densities of mixes with bottom ash were higher than control mixes due to the lower specific density of lime and ashes. The highest densities are reached for a combination of around 70% bottom ash and 30% hydrated lime. Higher density values indicate better performance and longer lifespan of the pavement. This mixture is recommended for road surfaces and intermediate layers with light traffic and small infrastructure [21]. However, previous research has been reported that the replacement of bottom ash in asphalt mixtures produce a lower dynamic modulus and higher rutting potential compared to control mixture [20].

3.3.2 Aggregate replacement Bottom ash has been used as base course in cold mix asphalt with low traffic volume. Utilizing coal waste and its ash was proven in improving the mechanical properties in cold recycled asphalt mixture which Marshall stability, tensile strength and resilient modulus were increased due to its higher pozzolanic content [22]. Previous studies also show that, bottom ash more suitable to be used in the base course and shoulder area due to the lower durability and lower abrasion resistance [23]. Besides, it also might be used in secondary roads with low traffic volumes and in parking lots and driveways which has low effect on short term tensile strength [24].

Recent studies indicate that, when bottom ash was used as partial replacement of fine aggregate, the bituminous performance was equal to conventional mixes [25]. However, the accepted percentage of bottom ash to be used as aggregate replacement must be less than 30% as mixes with 50% or more were found to have low stabilities [26]. The asphalt content increased with increasing of percentage of bottom ash [27]. However, even though bottom ash were coated with asphalt, the present of pyrites tend to weather during life service and causing deep red stains in pavement surface [28].

4. Conclusion
The production of coal bottom ash is expected to be increased year by year. The potential to utilized coal bottom ash in engineering field seem to be bright. Based on the review of previous studies, it has been reported that the utilization of bottom ash are still in the early stage of commercialization. Therefore further studies are needed to turn bottom ash waste into commercial reality. If proper design and construction procedure were followed, it appears that coal bottom ash is a suitable waste to be recycled into wealth especially in highway engineering field. However, it must first meet the environmental requirements set by state regulatory agencies.

Acknowledgement
We acknowledge the support provided by University Malaya through Postgraduate Research Grant (PG272-2016A).

References
[1] Özkan, Ö., Yüksel, I., & Muratoğlu, Ö. 2007 *Waste management* **27** 161-167
[2] Singh, M., & Siddique, R. 2015 *Journal of Cleaner Production* **91** 269-278
[3] Kim, H.-K. 2015 *Construction and Building Materials* **91** 57-64
[4] Kurama, H., & Kaya, M. 2008 *Construction and Building Materials* **22** 1922-1928
[5] Kurama, H., Topcu, I., & Karakurt, C. 2009 *Journal of materials processing technology* **209** 767-773
[6] Trifunovic, P. D., Marinkovic, S. R., Tokalic, R. D., & Matijasevic, S. D. 2010 *Thermochimica Acta* **498** 1-6
[7] Bin-Shafique, S., Benson, C. H., Edil, T. B., & Hwang, K. 2006 *Environmental Engineering Science* **23** 53-67
[8] Shah, P., Strezov, V., Prince, K., & Nelson, P. F. 2008 *Fuel* **87** 1859-1869
[9] Wang, Y., Ren, D., & Zhao, F. 1999 *International Journal of Coal Geology* **40** 103-108
[10] Lokeshappa, B., & Dikshit, A. K. 2012 *APCBEE Procedia* **1** 34-39
[11] Marto, A., Hassan, M. A., Makhtar, A. M., & Othman, B. A. 2013 *APCBEE Procedia* **5** 116-122.
[12] Tiwari, M., Sahu, S., Bhangare, R., Ajmal, P., & Pandit, G. 2014 *Applied Radiation and Isotopes* **90** 53-5
[13] Awang, A. R., Marto, A., & Makhtar, A. M. 2011 *EJGE* **16** 1515-1531
[14] Jorat, M. E., Marto, A., Namazi, E., & Amin, M. 2011 *Electronic Journal of Geotechnical Engineering* **16** 841-849
[15] Kim, B. 2003 *Properties of coal ash mixtures and their use in highway embankments.*
[16] Kim, B., Prezzi, M., & Salgado, R. 2005 *Journal of Geotechnical and Geoenvironmental Engineering* **131** 914-924
[17] Arenas, C., Leiva, C., Vilches, L. F., & Cifuentes, H. 2013 *Waste management* **33** 2316-2321
[18] Arenas, C., Leiva, C., Vilches, L. F., Cifuentes, H., & Rodríguez-Galán, M. 2015 *Construction and Building Materials* **95** 585-591
[19] Arenas, C. G., Marrero, M., Leiva, C., Solís-Guzmán, J., & Arenas, L. F. V. 2011 *Waste management* **31** 1783-1789
[20] Goh, S. W., & You, Z. 2008 *Canadian Journal of Civil Engineering* **35** 1114-1119
[21] López-López, E., Vega-Zamanillo, Á., Calzada-Pérez, M., & Taborga-Sedano, M. 2015 *Materiales de Construcción* **65**
[22] Modarres, A., & Ayar, P. 2014 *Journal of Cleaner Production* **83** 263-272
[23] Churchill, E. V., & Amirkhanian, S. N. 1999 *Journal of materials in civil engineering* **11** 295-301
[24] Ksaibati, K., & Stephen, J. 1999 Utilization of bottom ash in asphalt mixes: Department of Civil and Architectural Engineering, University of Wyoming.
[25] Ksaibati, K., & Sayiri, S. R. K. 2006 Utilization of Wyoming bottom ash in asphalt mixes: Mountain Plains Consortium.
[26] Moulton, L. K., Seals, R. K., & Anderson, D. A. 1973 *Transportation Research Record* **430** 26-39
[27] Ksaibati, K., & Conner, G. L. 2004 Laboratory evaluation of bottom ash asphalt mixes: Citeseer.
[28] Usmen, M., & Anderson, D. A. 1976 *Use of power plant aggregate in asphaltic concrete.* Paper presented at the Proceedings of 4th International Ash Utilization Symposium, St. Louis, Missouri, March 24-25, 1976.

IOP Conf. Series: Materials Science and Engineering **210** (2017) 012026 doi:10.1088/1757-899X/210/1/012026

Flocculation of Turbid Water Using Polyferric-Based Composite Coagulant

K. H. Tan and S. H. Lai

Department of Civil Engineering, Faculty of Engineering, University of Malaya, 50603 Kuala Lumpur, Malaysia.

E-mail: laish@um.edu.my

Abstract. The flocculation of turbid water using polyferric chloride-polydimethyldiallylammonium chloride (PFC-PDMDAAC) has been studied. Effect of preparation parameters basicity ratio (B ratio) of PFC and PDMDAAC/PFC ratio and operating parameters pH and dosage were investigated. PFC-PDMDAAC displayed maximum turbidity removal of 94.8% at 4.0mg/L when B=0.5 and PDMDAAC/PFC ratio= 7%.The best turbidity removal efficiencies by PFC-PDMDAAC were 84.7% at pH 7.5. These results reveal that PFC-PDMDAAC is efficient for flocculation of turbid water.

1. Introduction

Over recent decades, Malaysia faces many challenges in controlling the quality and quantity of stormwater runoff [1]. One way of controlling increased runoff due to urbanization and social-economic development is to build stormwater management facilities, such as retention/detention ponds. it is found that many of the ponds in Malaysia which are high in turbidity are suffered from a number of water pollution problems such as siltation and sedimentation and causes a number of social, health and environmental problems, such as disease, odor smell, poor aesthetic value in the surrounding area. Ponds can often dampen the impact of increased turbidity events by acting as points in a stream or river where particles can settle before being drawn into the intake of a treatment plant. Hence removal of turbidity is one of the major challenge in retrieving 'dead pond' in Malaysia.

Coagulation-flocculation is an important process in water treatment. Coagulation-flocculation is one of the conventional physico-chemical methods used to destabilize disperse particles in water [2]. Coagulants are added to aggregate destabilized colloidal particles into larger-sized flocs that can be effectively removed in subsequent separation processes [3]. Coagulants may be classified into inorganic and organic coagulants. With superior coagulation efficiency and high resistance to breakage in comparison to traditional metal salts, inorganic polymeric coagulants are most commonly used in water treatment. Conventional aluminum and ferric-based salts, such as polyaluminum chloride (PAC) and polyferric chloride (PFC), are widely used nowadays. Nonetheless high dosages are required and large volume of sludge is usually produced and high cost incurred [4]. some researchers have focused on the use of organic polymer in water treatment, such as polydiallyldimethylammonium (PDADMAC). PDADMAC has shown superior coagulation performance in their wider optimum pH range and the lower dosage required for water treatment [5]. Organic polymeric flocculants have been used in water treatment for decades as coagulant aids after the addition of inorganic coagulants.

Conventionally, inorganic polymer and organic polymer are added to wastewater separately, requiring two reagent addition systems [6]. In order to utilize the advantages of both inorganic and organic coagulants to prepare more cheap, effective and environmentally friendly flocculants, research is recently focused on the development of composite inorganic-organic coagulants. Numerous researchers were using composite flocculant, produced by combining inorganic coagulant with organic coagulant. [7-13].

In this study, turbdity removal effiency of composite flocculant PFC-PD which is composed of polyferric chloride (PFC) and PDADMAC was measured in treating synthetic turbid water under different initial turbidity and pH. Besides the effects of basicity ratio (B, oH/Fe ratio) of PFC and weight percentage (w(P)) of PDADMAC was also investigated.

2. Materials and methods

2.1. Materials
Iron (lll) Chloride, Polydiallyldimethylammonium chloride, sodium carbonate and disodium phosphate were obtained directly from Sigma-Aldich (Malaysia) Sdn. Bhd.

2.2. Preparation of coagulant
Firstly, a pre-determined amount of $FeCl_3 \cdot 6H_2O$ was dissolved in deionized water. After that, Na_2CO_3 powder was added to the ferric chloride solution to achieve desired basicity (B, OH/(Fe) ratio). Then, stabiliser Na_2HPO_4 (AR) was added to the solution. Finally, a measured amount of polydimethyl diallylammonium chloride (PDMDAAC) was injected to the above solution under strong stirring until all foam disappeared to prepare the composite coagulant PFC-PD. Three percentages (3.5%, 7.0% and 14%) of PDMDAAC by weight (Wp) in the flocculants were prepared

2.3. Preparation of turbid water sample
Kaolin clay purchased from Kaolin (Malaysia) Sdn. Bhd. used to prepare turbid water samples. 1.5 g of kaolin was prepared in 5 L of distilled water by mixing for 24 hours to prepare the stock kaolin suspension. The turbidity of this sample was determined to be around 220-240 NTU. Lower turbidity samples were prepared by diluting this stock turbidity suspension of 150, 100 and 50 NTU turbidity values.

2.4. Coagulation Experiment
The jar test experiments were carried out using VELP Scientific JLT6 Leaching Jar Test with 6 1L beakers. 0.5L of turbid water sample was added into beakers and predetermined dose coagulant solution was introduced. The suspension solutions were mixed rapidly at 120 rpm for 3 minute and stirred slowly at 40 rpm for 15 minutes. The samples allowed to settle for 30 minute. The supernatant samples were taken and measured for final turbidity.

3. Results and discussions

3.1. Turbidity Removal Evaluation of PFC-PDADMAC under different initial turbidity
Clay suspension was prepared at 50NTU, 100NTU and 150 NTU turbidity values to represent low, medium and high turbidity and a series of experiments were conducted on them. To get better understanding of the potential usability of ferric chloride as coagulant and to determine the effect of the ferric chloride on different turbidity, ferric chloride concentration from 0.5mg/L to 3.0mg/L were dosed.

Figure 1 summarizes turbidity removal efficiency increased as the dose of PFC-PD increases until optimum turbidity removal was achieved. For low initial turbidity of 50NTU, when the PFC-PD dosage less than 2.0mg/L, the initial turbidity decreased sharply but beyond this optimum value, the final turbidity seems increased. The most effective removal of turbidity could be obtained there is just sufficient coagulant absorption on the fine particles. The presence of overdose would make the floc less settle able. Therefore turbidity removal efficiency decreased to 84.21% when the PFC-PD dosage increased to 3.0mg/L.

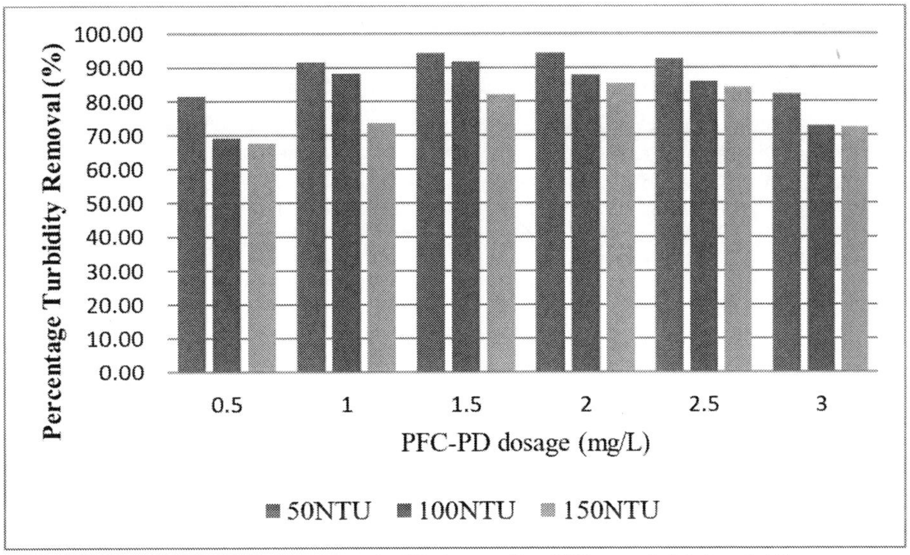

Figure 1. Turbidity removal efficiency for different initial turbidity using PFC-PD.

Even all three initial turbidity had the same optimum PFC-PD dosage range of 1.5mg/L to 2.0mg/L, it is investigated that the optimum range are boarder with the increasing of initial turbidity. There was a significant increasing of 23.9% in removal efficiency when the PFC-PD dosage increased to 2.0mg/L for 150NTU initial turbidity whereas there were only 16.17% and 12.97% removal efficiency for 100NTU and 50NTU. 50NTU initial turbidity had a narrower optimum range but showed a better effect on turbidity removal, that was 94.33% removal efficiency which is 2.6% and 9.0% higher than 100NTU and 150NTU initial turbidity.

3.2. Effect of basicity of PFC

In order to identify the optimum removal efficiency for PFC-PDADMAC, different [OH−]/[Fe], B value of polyferric chloride were prepared and investigated under constant pH 7 and initial turbidity of 100NTU as shown in Figure 2.

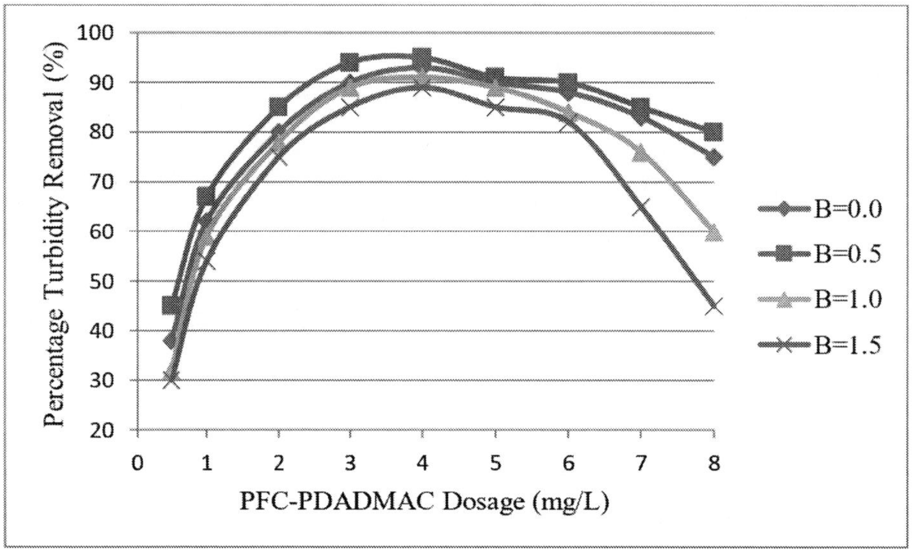

Figure 2. Turbidity removal efficiency of PFC-PDMDAAC for different basicity ratio .

It is noted that the turbidity removal of PFC–PD was slightly influenced by basicity of PFC-PD. The maximal turbidity removal (94.8 %) was achieved by $FeCl_3$–PDMDAAC at a dosage of 4.0 mg/L. Figure 2 showed that the coagulation and charge neutralization ability decreased with basicity ratio is increasing until it reached optimum turbidity removal efficiency at dosage 4.0mg/L. Most of the Fe(III) hydrolysate shows in monomer and lower polymerization at lower basicity value. Fe(III) hydrolysate which is positively charged would be easier to be adsorbed on negatively charged particles by charge neutralization. When the B value increases, the PFC hydrolysate transformed to hydroxide which has less positive charge and hence the turbidity decreased due to weak charge neutralization.

PFC-PD with higher B value achieved gradual decrease in coagulation efficiency when PFC-PD was overdosed. Organic polymer PDMDAAC is a dominant positive charge in PFC-PD and the charge neutralization ability of PFC drops when B value increases. Difference in charge neutralization ability between PFC and PDMDAAC extended and the proportion of particle surface sites which reacted with PDMDAAC increase. Therefore charge reversal was obtained by PFC-PD with higher B value.

3.3. Effect of w(P) of PDMDAAC

Figure 3 illustrates the turbidity removal efficiency of PFC (w(P) = 0) and PFC–PD (B = 0.5) with three different w(P) (3.5%, 7.0%, 14.0%). It showed that turbidity removal increased dramatically with the increase of w(P). This is because the higher weight percentage of PDMDAAC in PFC-PD could neutralize more negative charge on the particle surface and gave stronger charge neutralization compared to lower PDADMAC content. Besides, the PFC–PD with higher w(P) showed stronger turbidity removal in a much narrower dosage range, which increased the difficulty in determining the optimum dosage in water treatment. Thus the PFC–PD with w(P) = 14% and above is not applicable.

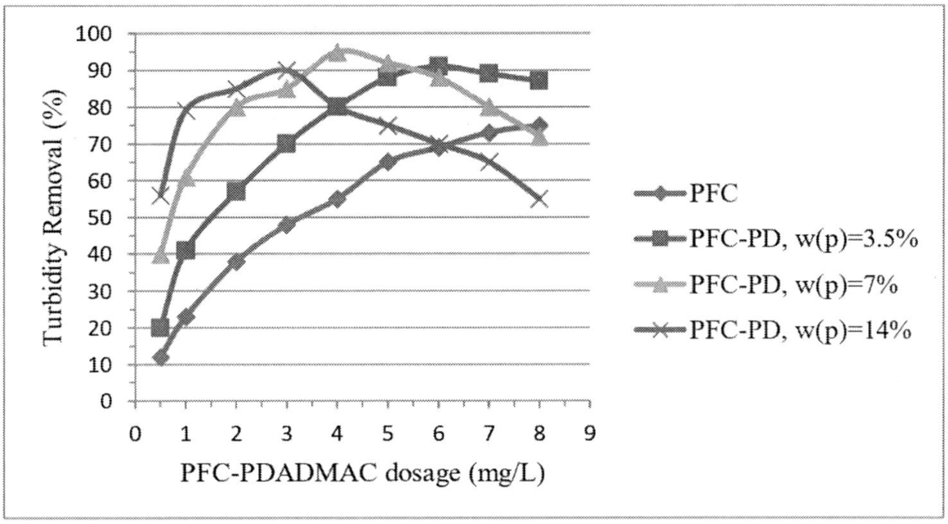

Figure 3. Turbidity removal efficiency of PFC-PDADAAC for different PDADMAC weight content.

3.4. *Effect of pH on Turbidity Removal Efficiency*

Turbid water sample of 150NTU initial turbidity was examined at various pH conditions (5.0, 6.0, 7.0, 7.5, 8.0 and 9.0). $FeCl_3$, PDMDAAC with high M_w, PFC with B=0.5 were selected for testing turbidity removal efficiency. The constant dosage used for all coagulants are 2.0 mg/L.

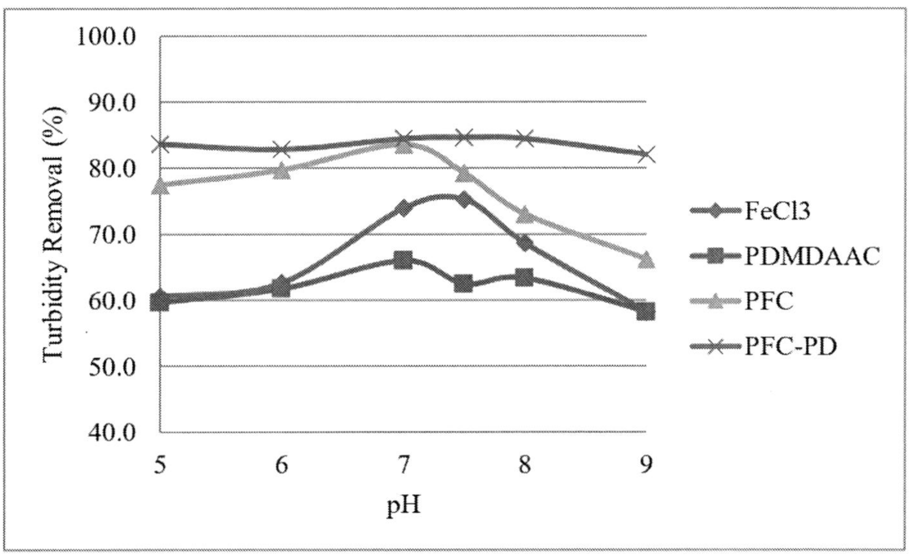

Figure 4. Turbidity removal efficiency of PFC-PDMDAAC for different PDADMAC weight content.

The highest turbidity removal efficiencies of PFC were 84.7% at pH 7 and it showed that pH would affect the turbidity removal significantly. This instance was in good agreement with the previous finding. The floc patterns of Fe (III) coagulants were different under different pH conditions. The ability of Fe (III) coagulants to neutralize the negative charge and to sweep the contamination is optimal in the middle pH range.

For PDADMAC, the turbidity removal efficiencies were mostly invariable in all pH range. All of the turbidity removal efficiencies were between 58.2% and 66.0% within the pH range of 5.0–9.0. The above phenomenon was in good agreement with previous study that PDMDAAC was not sensitive to pH within a fairly wide range of application.

It also could be observed that there were small changes in the turbidity removal rate of composite coagulant in all pH range. And the best turbidity removal efficiencies for treating high turbid water by PFC-PDMDAAC were 84.7% at pH 7.5. The results above indicate that using PFC–PDADMAC composite coagulant for treating high original pH would be more advantageous than using PFC coagulant.

3.5. Comparison of coagulation efficiency of PFC-PD with PFC,PDADMAC and $FeCl_3$

The coagulation efficiency of PFC-PD was compared with PFC, PDMDAAC and $FeCl_3$ were compared in terms of turbidity removal efficiency as shown in Figure 5. Constant initial turbidity 150NTU was used throughout this session. The turbidity removal of $FeCl_3$ and PFC increased gradually when the coagulant dosages increased. The coagulation efficiency of $FeCl_3$ and PFC was much lower than PFC-PD composite coagulant for lower coagulant dosage. Only about 60% of turbidity removal could be achieved when the dosage increased to 3.0 mg/L. on the other hand, PDMDAAC showed better turbidity removal efficiency compared to $FeCl_3$ and PFC within the dosage range 0.5–1.5 mg/L.

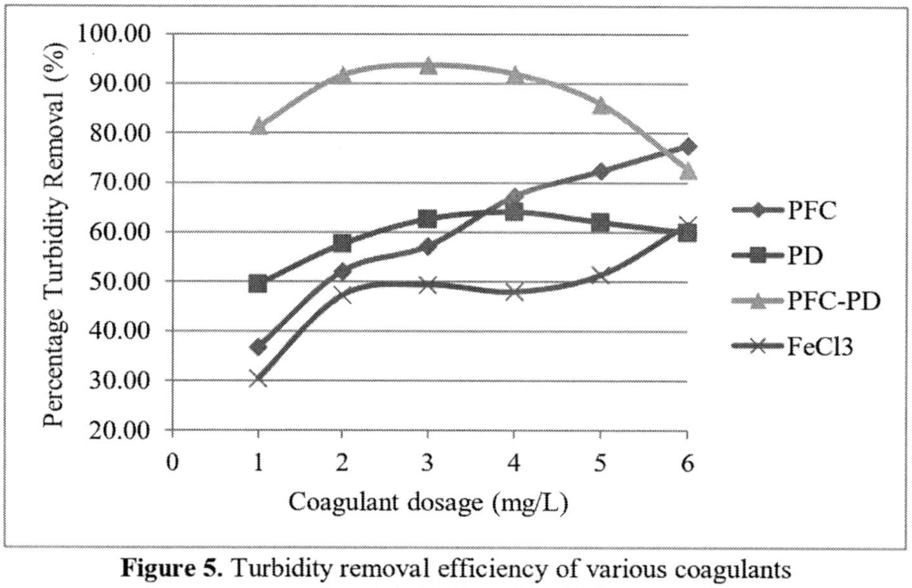

Figure 5. Turbidity removal efficiency of various coagulants under 150NTU initial turbidity

4 Conclusions

The composite coagulant PFC-PD was more efficient compared to $FeCl_3$, PFC and PDADMAC in term of turbidity removal efficiency. Therefore, PFC-PD was more suitable for surface water treatment than conventional metal coagulants and organic polymers. Besides, the composite coagulant PFC-PD performed better coagulation performance at lower B value and higher w(P).

Acknowledgement

The authors would like to acknowledge UM FRGS grant RP008B-13SUS by Universiti Malaya that has resulted in this study.

References

[1] Hern, T. K., Hin, L. S., Ibrahim, S., Sulaiman, N. M. N., Sharifi, M., & Abe, S. 2014 *Journal of Geoscience and Environment Protection* **2** 1
[2] Lee, K. E., Teng, T. T., Morad, N., Poh, B. T., & Hong, Y. F. 2010 *Separation and Purification Technology* **75** 346-351
[3] Hu, C., Liu, H., Qu, J., Wang, D., & Ru, J. 2006 *Environmental Science & Technology* **40** 325-331
[4] Sun, C., Yue, Q., Gao, B., Cao, B., Mu, R., & Zhang, Z. 2012 *Chemical Engineering Journal* **185** 29-34
[5] Kang, Q. 2007 *Separation and Purification Technology* **57** 356-365
[6] Ovenden, C., & Xiao, H. 2002 *Colloids and Surfaces A: Physicochemical and Engineering Aspects* **197** 225-234
[7] Gao, B. Y., Wang, Y., Yue, Q. Y., Wei, J. C., & Li, Q. 2008 *Separation and Purification Technology* **62** 544-550
[8] Gao, B., Yue, Q., & Wang, B. 2002 *Chemosphere* **46** 809-813
[9] Moussas, P. A., & Zouboulis, A. I. 2008 *Separation and Purification Technology* **63** 475-483
[10] Wang, Y., Gao, B., Yue, Q., Wei, J., & Li, Q. 2008 *Chemical Engineering Journal* **142** 175-181
[11] Lee, K. E., Teng, T. T., Morad, N., Poh, B. T., & Hong, Y. F. 2010 *Separation and Purification Technology* **75** 346-351
[12] Lee, K. E., Teng, T. T., Morad, N., Poh, B. T., & Mahalingam, M. 2011 *Desalination* **266** 108-113
[13] Moussas, P. A., & Zouboulis, A. I. 2009 *Water Research*, **43** 3511-3524

Performance of solvent-borne intumescent fire protective coating with Palm oil clinker as novel bio-filler on steel

S. A. S. Mustapa and N. H. Ramli Sulong*

Department of Civil Engineering, Faculty of Engineering, University of Malaya, Lembah Pantai, 50603 Kuala Lumpur, Malaysia

E-mail: *hafizah_ramli@um.edu.my

Abstract. This research deals with contribution of hybrid fillers with palm oil clinker (POC) as a novel bio-filler in solvent-borne intumescent fire protective coating for steel. The hybrid fillers with POC were mixed in appropriate amount of additives and acrylic binder to produce the intumescent coatings. The intumescent coatings were characterized by using Bunsen burner test, surface spread of flame, thermogravimetric analysis, field emission scanning electron microscopy, static immersion and Instron micro tester equipment. Specimen with POC as a single filler has significantly enhanced the fire protection performances of the intumescent coating due to the high thermal stability of POC, where less than 10% of temperature different when compared to specimens with hybrid fillers. From the flame spread classification, class 1 is the best classification while Class 4 is the worst and considered high risk. All specimens was classified as class 1 since the final spread of flame was less than 165 mm. For hybrid fillers composition, specimen consist of POC/$Al(OH)_3$/TiO_2 has significantly improved the water resistance of the coating due to the low solubility of $Al(OH)_3$ in water, while specimen contain of $Mg(OH)_2$ had higher mechanical strength due to the strong bonding between the metal surface and acrylic binder/$Mg(OH)_2$ filler. It was found that coating with the incorporation of all hybrid fillers gives excellent fire protection performance with good thermal stability, water resistance and mechanical properties. It can be concluded that, the selection of appropriate composition of fillers and binder in intumescent coating was highly influence the intumescent coating performance.

1. Introduction

Steel starts to lose its strength and stiffness when temperature reached 550°C in fire incident, where collapse of steel structures may occur [1-2]. Intumescent fire retardant coatings was used as passive fire protection materials to protect the steel structures from collapse when exposed to high temperature [3]. It was applied to the steel structure for fire protection, which can save a thousand of precious life and assets if the building can last for longer period and extending time for evacuation. When exposed to a sufficiently high temperature, the intumescent coating forms a porous char that thermally insulates the steel substrate from heat and prevent it from reaching critical temperature [4-5]. Three flame-retardant additives are employed in the formulation of intumescent coatings, i.e. an acid source (ammonium polyphosphate, APP), a carbon source (pentaerythritol, PER) and a blowing agent (melamine, MEL) mixed with flame-retardant fillers and binder. The most commonly used flame retardant fillers are metal hydroxides. As the temperature rises, these inorganic fillers decompose endothermically and therefore absorb energy and release non-flammable molecules such as H_2O and/or CO_2, which dilute combustible gases, and also promote the char formation of a protective ceramic or vitreous layer [6]. Unfortunately, this inorganic filler is expensive and will increase the production cost of intumescent coatings.

Therefore, there are several attempts in using by-product waste such as chicken eggshells (CES), rice husk and palm oil clinker (POC) as bio-fillers in intumescent coating which was believe can lower the production cost of intumescent coating.

From the previous research Yew et al. 2013, studied the influence of CES as bio-filler in intumescent coating [7]. Before CES can be used as a filler, it has to be cleaned thoroughly and dried at 90°C for 12 h in oven after the membrane was removed. The dried CES were mechanically triturated to a powder form and then milled at a milling speed of 280 rpm for 20 h to 40 h. It takes a lot of effort in processing the CES to obtain appropriate particle size before it can be used as filler compare to by-product waste such as palm oil clinker (POC) which can be used directly from the sources without going through any process and it is available in bulk quantities. Malaysia has to deal with a problem of by-product waste that is generated from palm oil processing as being the second largest country in the world that produce palm oil [8]. POC was produced in large quantities and treated as disposal waste. POC was produced after the burning of oil palm shell and palm oil fibre with 30:70 ratios at high temperature of 850°C in energy generating to run the plants [9]. Using POC as a bio-filler in intumescent coating is another way to lower the production cost as well as protecting the environment from the by-product waste. Many researchers have studied the effectiveness of intumescent coating based on their performance in fire retardancy, mechanical strength and water resistance [10-14]. This research focus on developing the best composition of intumescent fire protective coating using hybrid fillers with POC as a bio-filler on steel structures in the event of fire. The intumescent coatings was characterized based on fire protection performances, mechanical strength and water resistance.

2. Materials and methods

The intumescent coating was prepared using three flame retardant additives, i.e. ammonium polyphosphate (APP), pentaerythritol (PER) and melamine (MEL) in ratio of (2:1:1). The flame retardant additives was blended in acrylic binder of 45 wt. %. Then the mixture was added to four different hybrid fillers, i.e. palm oil clinker (POC), magnesium hydroxide [$Mg(OH)_2$], aluminium hydroxide [$Al(OH)_3$] and titanium dioxide (TiO_2). POC was taken from the Sri Ulu Langat Palm Oil Mill factory. The POC which looks like a gray porous stone with flaky and irregular shaped was crushed into required size. There are four specimens were prepared; one specimen are having POC as single filler and three specimens with hybrid fillers. The compositions of the intumescent coatings are presented in table 1. The composition of intumescent coatings were blended with a high-speed disperse mixer until completely homogenous. The mixture was coated on one-side of steel plate using brush and this step was repeated until dry film thickness of 1.5±0.2 mm was achieved. The thickness of coating layer was measured using a gauge Elcometer model A456. The coatings were considered ready for characterization when it is fully dried.

Table 1. The composition of intumescent coatings.

Specimen		SBP	SBPM	SBPA	SBPMA
Flame retardant additives: Acrylic binder (wt. %)			37 : 45		
Fillers	POC	/	/	/	/
	TiO_2		/	/	/
	$Mg(OH)_2$		/		/
	$Al(OH)_3$			/	/

2.1. Thermogravimetric analysis (TGA)

TGA was carried out in the temperature range of 30-1000°C at the heating rate of 20°C/min under air flow using a TGA/SDTA851e model. This test was used to measure the weight change of the coating as a function of temperature while being heated at a constant rate.

IOP Conf. Series: Materials Science and Engineering **210** (2017) 012027 doi:10.1088/1757-899X/210/1/012027

2.2. Bunsen burner test

Bunsen burner test was carried out to characterize the char formation and determine the temperature development on the single-side coated steel plate of each coating. A coated plate was mounted vertically and exposed to high temperature of Bunsen burner about 1000°C for 1 hour. A thermocouple wire that is connected to a digital thermometer for temperature measurement was attached to the back side of the steel plate during the fire test. The temperature readings were taken every 1 min until the steel reach its critical temperature. Temperature of 400°C has been chosen as the failure temperature for steel structure to ensure higher level of safety and protection [6].

2.3. Surface spread of flame test

The surface spread of flame test was performed according to the procedure specified in the 'BS 476: Part 7' standard. The test involves measuring the lateral spread of flame over the surface of a specimen oriented in a vertical position and classifying the system based on the rate and extent of flame spread. The steel plate (885 mm × 270 mm × 2.3 mm) coated with intumescent coating (thickness: 1.5 ± 0.2 mm) was exposed to specific heating conditions during the test. The test specimen was mounted vertically and placed at a 90° angle from the radiation panel, and was exposed to a radiant panel over a duration of 10 min. Pilot flame was applied to the bottom corner of the specimen during the first minute of the test. The time required for the flame front to reach the reference marks on the specimen and the extent of flame spread were recorded over a period of 1 min 30 s as well as at the end of the test. The burning behaviour of the coating was also observed. The materials were classified according to the test performance, as shown in table 2. Class 1 is the best classification while Class 4 is the worst classification and is considered high risk.

Table 2. Flame spread classification.

Classification	Spread of flame at 1.5 min		Final spread of flame (10 min)	
	Limit (mm)	Tolerance for one specimen in sample (mm)	Limit (mm)	Tolerance for one specimen in sample (mm)
Class 1	165	25	165	25
Class 2	215	25	455	45
Class 3	265	25	710	75
Class 4	Exceeding the limits for class 3			

2.4. Field emission scanning electron microscopy (FESEM)

After the Bunsen burner test, a piece of char was cut from the centre of char layer for FESEM test in order to examine the surface morphology of the char layers. Low beam energy of 1 kV was used to observe the morphology of the char layer using the GEMINI® Based Field Emission SEM launched by ZEISS.

2.5. Static immersion test

Water resistance of thin films was evaluated using static immersion test. The intumescent coating films were immersed in distilled water at 25°C and were dried with a piece of paper towel to remove excess water then weight change was recorded. This step was repeated at a specific time intervals. Weight change was calculated based on equation (1).

$$\Delta W = \frac{W_e - W_o}{W_o} \times 100\% \qquad (1)$$

where ΔW is water intake ratio of film (%), W_e is the weight of the film after water immersion (g) and W_o is the weight of the film before water immersion (g).

2.6. Adhesion strength test

The adhesion strength between the intumescent coating layer interfaces with steel substrate was determined by using the pull-off adhesion tester (PosiTest-AT-A Automatic, DeFelsko). The coating was applied on one side of a 50 mm × 50 mm × 2.6 mm steel plate to obtain a film thickness of 0.5 ± 0.05 mm. The flat face of a pull stub (20 mm dolly) is adhered to the coating using epoxy glue (thickness of 0.5 ± 0.05 mm). The force of peeled-off area of the samples were calculated and classified according to ASTM D4541 standard classification. Force was calculated by equation (2).

$$f_b = F/A \quad (2)$$

where f_b is the adhesion strength (MPa), F is the crack charge (N) and A is the sticking area (mm^2).

3. Results and discussion

3.1. Thermal analysis of intumescent coatings

The TGA curves of intumescent coatings was presented in figure 1. The curves of the coatings were similar between 100°C to 250°C. After temperature higher than 250°C, the curves became slightly different from each other. The residual weight for specimen SBP, SBPM, SBPA and SBPMA at 750°C was 42.17 wt. %, 46.84 wt. %, 41.98 wt. % and 47.42 wt. %, respectively. The results show that the percentage different between specimen contain POC as single filler and specimens with hybrid fillers is less than 6%, indicate that POC alone significantly enhanced thermal stability of the intumescent coating. It can be explained based on the thermal stability of POC filler. TGA curve of the POC is shown in figure 2. After the temperature reached 500°C, the weight loss for POC is only about 2%. The high thermal stability of POC was attributed to the process of burning of oil palm shell and palm oil fibre at high temperature of 850°C during energy generating. For hybrid fillers composition, specimen SBPM containing POC/$Mg(OH)_2$ had 5% increase in residual weight which contribute to a better thermal stability compared to specimen SBPA containing POC/$Al(OH)_3$, which specifies that incorporation of $Mg(OH)_2$ significantly increased the thermal stability of the intumescent coating than $Al(OH)_3$. It was found that the incorporation of all hybrid fillers in specimen SBPMA gives the highest thermal stability to the intumescent coating which indicates that the best combination of all hybrid fillers with flame retardant additives and acrylic binder had improved anti-oxidation, thermal stability and fire protection performance of the intumescent coating [15].

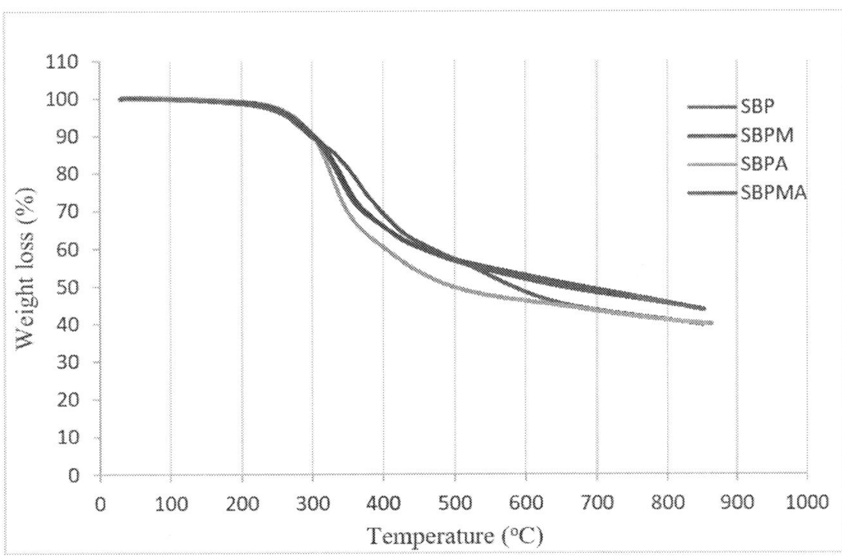

Figure 1. TGA curves of intumescent coatings.

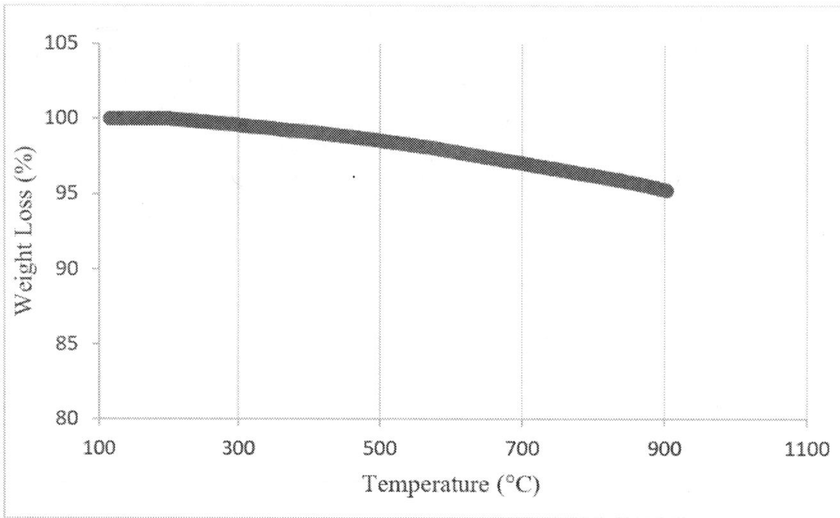

Figure 2. TGA curve of the POC.

3.2. Bunsen burner test

The temperature curves for specimens SBP, SBPM, SBPA and SBPMA are presented in figure 3. The curves for each coating during first 5 to 10 minutes had increased rapidly after exposed to fire. After about 30 minutes, the temperature reached an equilibrium value and almost unchanged. The equilibrium temperature and thickness of char layer was shown in table 3. Specimen SBP having POC as a single filler shows very effective fire protection performance of intumescent coating, where less than 10% of temperature different when compare to other samples with hybrid filler. It proved that, POC as a single filler significantly enhanced the fire protection performance of intumescent coating because of its high thermal stability. The equilibrium temperature of specimen SBPM was 4% lower than specimen SBPA due to the addition of $Mg(OH)_2$ which indicates that addition of $Mg(OH)_2$ slightly increase the fire protection performance of intumescent coating. Meanwhile, for specimen SBPMA contain all hybrid fillers was found to be optimum combination as it gives the lowest equilibrium temperature due to its effectives char protection that reduced the heat transfer to the substrate. The thicker the char layer leads to the lower equilibrium temperature which resulted in the better fire protection performance. Figure 4 show the images of intumescent coating specimen before and after Bunsen burner test.

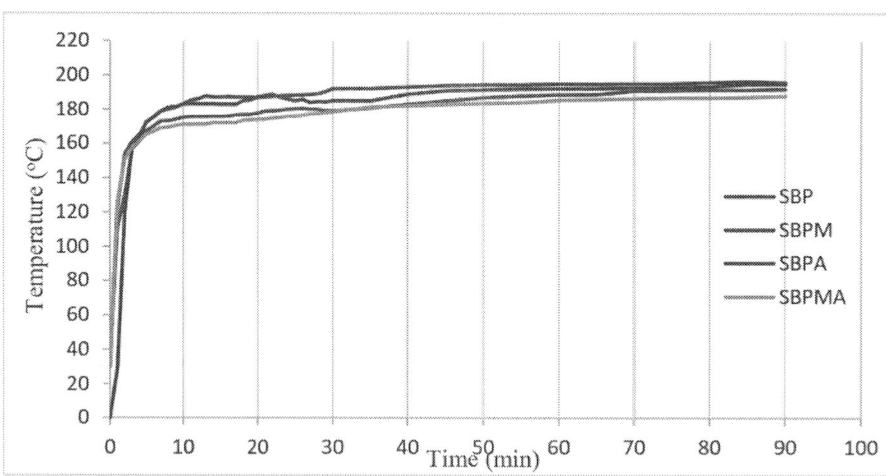

Figure 3. The time-temperature curves of intumescent coatings.

IOP Conf. Series: Materials Science and Engineering **210** (2017) 012027 doi:10.1088/1757-899X/210/1/012027

Table 3. Char thickness and equilibrium temperature of intumescent coatings.

Specimen	Equilibrium temperature (°C)	Thickness of char layer (mm)
SBP	187.7	17.21
SBPM	175.3	19.17
SBPA	183.1	17.53
SBPMA	172.0	20.02

Figure 4. The image of intumescent coatings before and after Bunsen burner test.

3.3. Surface spread of flame test

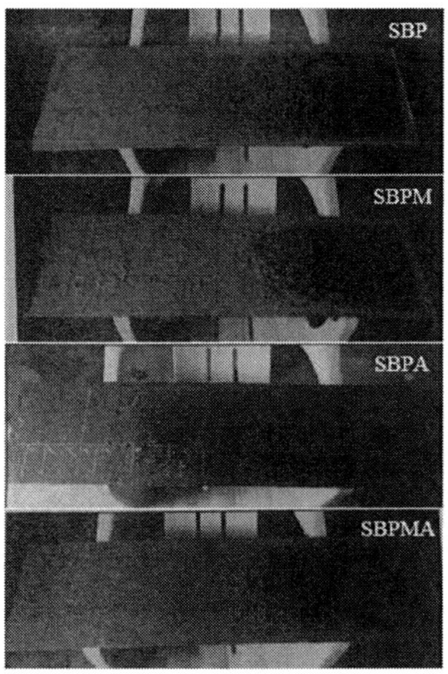

Figure 5. Specimen SBP, SBPM, SBPA and SBPMA after surface spread of flame test.

The surface spread of flame test was performed according to the procedure specified in the 'BS 476: Part 7' standard. The test involves measuring the lateral spread of flame over the surface of a specimen oriented in a vertical position and classifying the system based on the rate and extent of flame spread. The results for the test specimens are shown in table 4. From the flame spread classification, class 1 is

IOP Conf. Series: Materials Science and Engineering **210** (2017) 012027 doi:10.1088/1757-899X/210/1/012027

the best classification while Class 4 is the worst classification and is considered high risk. From the results, all specimens was classified as class 1 since the final spread of flame was less than 165 + 25 mm. Specimen SBP and SBPM had 165 mm surface spread of flame at 59 s and 55 s, respectively. Meanwhile, specimen SBPA with addition of $Al(OH)_3$ had 190 mm surface spread of flame at 1 min. Specimen SBPMA with the best combination of hybrid fillers with POC had 0 mm surface spread of flame compare to other specimen, proved that the best combination of all hybrid fillers with POC resulting in a significant improvement in lowering the rate and extend of flame spread. The char formation after the test are shown in figure 5.

Table 4. Results of surface spread of flame test.

Specimen No.	SBP	SBPM	SBPA	SBPMA
Spread of flame at 1 1/2 minutes (mm)	165	165	190	0
Distance (mm)	Time of spread of flame to indicated distance (min.s)			
75	0.56	0.34	0.47	
165	0.59	0.55	0.58	
190			1.00	
215				
240				
265				
290				
375				
455				
500				
525				
600				
675				
710				
750				
785				
825				
865				
Time of maximum spread of flame (min.s)	0.59	0.55	1.00	-
Distance of maximum spread of flame (mm)	165	165	190	0
Classification	Class-1	Class-1	Class-1	Class-1

3.4. Morphology of intumescent char layer.

The surface morphologies of char layer for intumescent coatings are shown in figure 6. It was believed that the physical structure of the char layer has significant effect to the fire protection performance of coating [16]. The high porosity surface structure was clearly observed for specimen SBP and SBPA which led to the reduction of fire protection because of the heat may penetrate to steel substrate through the char structure that is high porosity and nonuniform. Meanwhile, the char layer for specimen SBPMA had dense and uniform foam structure which could prevent the heat transfer to the steel substrate and improved fire protection efficiency. Hence, from this result it proved that specimen SBPMA with the best combination of hybrid fillers with POC resulting in homogeneous and dense foam structure had increased the fire protection performance compared to other samples.

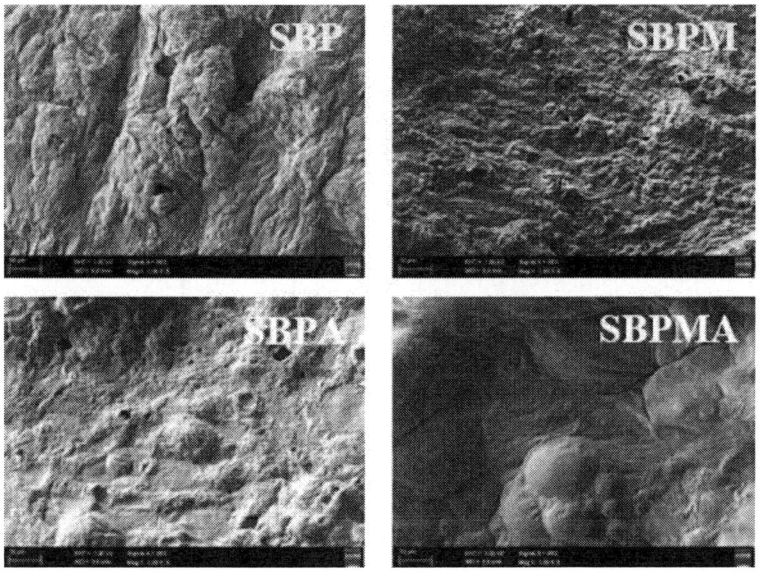

Figure 6. The surface morphologies of char layer of intumescent coatings.

3.5. Static immersion test

The weight change rate curves for specimens SBP, SBPM, SBPA and SBPMA are shown in figure 7. The experimental results show that the weight of specimen SBP was gradually increased due to the permeation process which means water could infiltrate into high porosity structure of the coating which led to the weight gain of the sample. Moreover, this situation has also occurred due to high solubility of POC in water. Moreover, the weight of coatings SBPM and SBPMA was decreased due to the migration process. As reported from previous study, the migration process may occur when some hydrophilic fire retardant additives dissolved in water during the migration process, hence reduce the weight of the coating [17]. For coating SBPA with contribution of $Al(OH)_3$ as filler, it seems that both migration and permeation process was occurred. The permeation of water and dissolution of fire retardant ingredients could be slow down with the incorporation of $Al(OH)_3$ in the intumescent coating due to its low solubility in water, which resulting in the improvement of water resistance of the coatings [18].

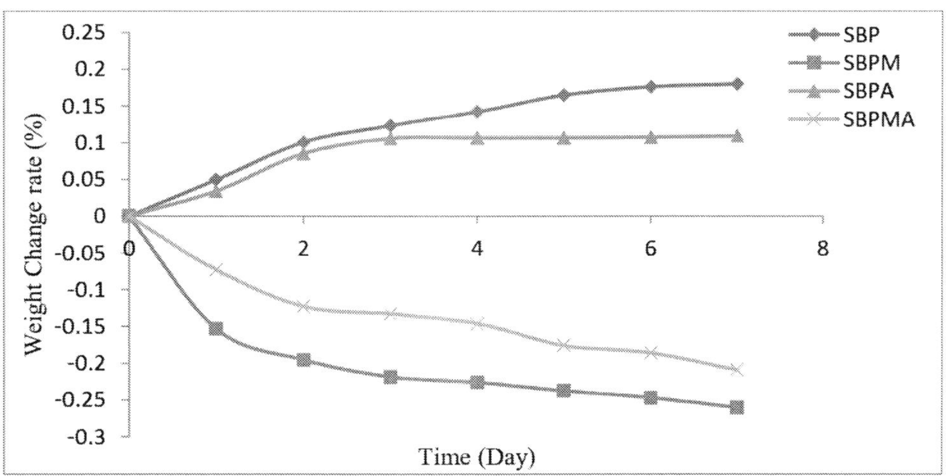

Figure 7. The weight change rate curves of intumescent coatings.

3.6. Adhesion strength test

The bonding strength of intumescent coating was investigated using the Instron microtester. The adhesion strength of specimens SBP, SBPM, SBPA and SBPMA was 4.39, 4.68, 4.03 and 5.21 MPa, respectively as in table 5. Specimen SBP with POC alone, shows better adhesion strength when compared to specimen SBPA which consist of $Al(OH)_3$ filler. Meanwhile, the addition of $Mg(OH)_2$ to the specimen SBPM showed improvement compared to other specimens due to the strong bonding strength between the metal surface and acrylic binder/$Mg(OH)_2$ filler interface adhesion for effective stress transfer [19]. Meanwhile, the significant improvement in the adhesion strength for specimen SBPMA, proved that this coating contained optimal combination of POC/$Mg(OH)_2$/$Al(OH)_3$ that led to the highest bonding strength. It can be concluded that the coating with addition of POC and $Mg(OH)_2$ had improved the adhesion strength of the coating and appropriate combination of hybrid fillers strongly influence the adhesion strength of intumescent coating.

Table 5. Adhesion strength of intumescent coatings.

Specimen	Crack charge, F (N)	Adhesion strength, fb (MPa)
SBP	1378.46	4.39
SBPM	1469.52	4.68
SBPA	1265.42	4.03
SBPMA	1635.94	5.21

4. Conclusion

In this research, different formulation of intumescent coatings was formulated using hybrid fillers POC/TiO_2/$Mg(OH)_2$/$Al(OH)_3$. Specimen SBP having POC as a single filler improved the fire protection performance and thermal stability of the intumescent coatings. From the result of Bunsen burner test, the temperature different between specimens contain single POC and hybrid fillers is less than 10%, indicated that the use of POC as a single filler is very effective in fire protection performance of the intumescent coating due to its high thermal stability. From the flame spread classification, all specimens was classified as class 1 as prescribed by the BS 476 part 7, since the final spread of flame was less than 165 mm. The combination of hybrid fillers with $Al(OH)_3$ to flame-retardant additives and acrylic binder could slow down the permeation of water and dissolution of fire retardant ingredients due to its low solubility in water, which led to an improvement in the water resistance of the coating SBPA. Furthermore, specimen SBPM shows the highest bonding strength due the addition of $Mg(OH)_2$ in the coating led to its effective interface adhesion between the coating and metal surface. Specimen SBPMA consist of the best combination of POC with all hybrid fillers shows an excellent fire protection, which produced the highest thickness of char layer, good bonding strength and water resistance of intumescent coating. From this research, it can be concluded that the appropriate combination of hybrid fillers with POC has highly influenced the fire protection performance, water resistance and adhesion strength of the intumescent coatings. Moreover, the usage of POC in intumescent coating can contribute in preserving the environment from industrial waste as well as reduced the production cost by minimizing the usage of synthetic fillers in intumescent coatings.

Acknowledgements

This work was supported by the Ministry of Education, Malaysia under FRGS Grant FP047-2013B and University of Malaya under PPP grant PG139-2016A.

References

[1] Dai X H, Wang Y C and Bailey C G 2009 *Fire Safety Journal* **44** 376-386
[2] Norgaard K P, Dam-Johansen K, Catala P and Kiil S 2013 *Progress in Organic Coatings* **76** 1851-1857
[3] Liang S Y, Neisius N M and Gaan S 2013 *Progress in Organic Coatings* **76** 1642-1665
[4] Duquesne S, Magnet S, Jama C and Delobel R 2004 *Surface & Coatings Technology* **180** 302-307
[5] Jimenez M, Duquesne S and Bourbigot S 2006 *Surface & Coatings Technology* **201** 979-987
[6] Yew M C 2011 Investigation on water-borne intumescent fire protective coatings for steel (Unpublished master's thesis), University of Malaya, Kuala Lumpur, Malaysia.
[7] Yew M C, Sulong N H R, Yew M K, Amalina M A and Johan M R 2013 *Progress in Organic Coatings* **76** 1549-1555
[8] Ahmmad R, Jumaat M Z, Alengaram U J, Bahri S, Rehman M A and Hashim H 2016 *Journal of Cleaner Production* **112** 566-574
[9] Jumaat M Z, Alengaram U J, Ahmmad R, Bahri S and Islam A B M S *Construction and Building Materials* **101** 942-951
[10] Yew M C and Sulong N H R 2012 *Mater Design* **34** 719-724
[11] Bodzay B, Bocz K, Barka Z and Marosi G 2011 *Polymer Degradation and Stability* **96** 355-362
[12] Enescu D, Frache A, Lavaselli M., Monticelli O and Marino F 2013 *Polymer Degradation and Stability* **98** 297-305
[13] Pan L L, Li G Y, Su Y C and Lian J S 2012 *Polymer Degradation and Stability* **97** 1801-1806
[14] Wang G J and Yang J Y 2010 *Surface & Coatings Technology* **204** 1186-1192
[15] Yew M C, Sulong N H R, Yew M K, Amalin M A and Johan M R 2014 *Materials Research Innovations* **18** 384-388
[16] Wang Z Y, Han E H and Ke W 2006 *Surface & Coatings Technology* **200** 5706-5716
[17] Yeh J M, Huang H Y, Chen C L, Su W F and Yu Y H 2006 *Surface & Coatings Technology* **200** 2753-2763
[18] Harlan Laboratories Ltd. UK 2010 Aluminium Hydroxide. Determination of water solubility. Project no.: 2962/0003 UK A Study Conducted at the Request of the Aluminium REACH Consortium
[19] Devendra K and Rangaswamy T 2012 *International Journal of Engineering Research and Appl*ications **2** 2028-2033

IOP Conf. Series: Materials Science and Engineering **210** (2017) 012028 doi:10.1088/1757-899X/210/1/012028

Time Series Analysis and Forecasting of Wastewater Inflow into Bandar Tun Razak Sewage Treatment Plant in Selangor, Malaysia

Taher Abunama and Faridah Othman*

Department of Civil Engineering, Faculty of Engineering, University of Malaya, Kuala Lumpur, Malaysia.

E-mail: *faridahothman@um.edu.my

Abstract. Analysing the fluctuations of wastewater inflow rates in sewage treatment plants (STPs) is essential to guarantee a sufficient treatment of wastewater before discharging it to the environment. The main objectives of this study are to statistically analyze and forecast the wastewater inflow rates into the Bandar Tun Razak STP in Kuala Lumpur, Malaysia. A time series analysis of three years' weekly influent data (156weeks) has been conducted using the Auto-Regressive Integrated Moving Average (ARIMA) model. Various combinations of ARIMA orders (p, d, q) have been tried to select the most fitted model, which was utilized to forecast the wastewater inflow rates. The linear regression analysis was applied to testify the correlation between the observed and predicted influents. ARIMA (3, 1, 3) model was selected with the highest significance R-square and lowest normalized Bayesian Information Criterion (BIC) value, and accordingly the wastewater inflow rates were forecasted to additional 52weeks. The linear regression analysis between the observed and predicted values of the wastewater inflow rates showed a positive linear correlation with a coefficient of 0.831.

1. Introduction

Sewage treatment plants (STPs) are among the most valuable infrastructures in countries' development. The main objective of STP's is to treat the collected wastewater sufficiently to prevent negative impacts to human health, aquatic life, and the surrounding environment. STP's capacity and treatment processes must be designed and operated carefully in order to provide reliable treatment despite fluctuating characteristics, such as inflow and organic loading of the influent waste stream, in order to maintain compliance environmental permit limits and effluent standards.

For that, it is significant to evaluate and predict the design loads periodically (e.g. 3-5 years), because influent hydraulic and loadings parameters can vary considerably depending on the population that is being served, vacations and even tourist inflow can affect the inflow rate of wastewater.

Forecasting and simulating STP's inflow rates are valuable in order to define the average as well as peak flow rates, which assess in future planning of collection and treatment facilities. This can be conducted based on the previous observed and recorded inflow rate values at regular time intervals, through time series analysis of sewage inflow rates into treatment plants.

Box-Jenkins [1] or Autoregressive Integrated Moving Average (ARIMA) models are able to fulfil this task, and give an accurate prediction. ARIMA model consists of an integrated component (d), which performs differencing of the time series to make it stationary [2].

Another two components are autoregressive AR (p) and moving average MA (q); AR component correlates the relation between the current value and the past value of time series, while, MA captures the duration of random shock in the series.

These techniques have been well established and used for predicting hydro meteorological parameters in various studies [3]–[8].

In this study, the for ARIMA model has been applied for a time series of Bandar Tun Razak STP's sewage inflow data. And best fitting ARIMA model were selected and assessed using linear regression analysis between the observed and predicted values.

2. Methodology

2.1. Bandar Tun Razak (BTR) STP

Bandar Tun Razak STP is located on Jalan 11/118b, Desa Tun Razak, 56000 Kuala Lumpur, Wilayah Persekutuan Kuala Lumpur, Malaysia. The sewage plant is operated and managed by the Indah Water Konsortium Sdn Bhd (IWK). Total area of the plant is about ten acres, while the reserved area is six acres (Figure 1).

BTR STP built to serve part of Kuala Lumpur, with daily design capacity of 25,000 m3 and equivalent to 100,000 populations. Currently, the plant receives and treats about 11,700 m3 per day which equivalent to 52,000 populations. Sequential Batch Reactor (SBR) treatment system is equipped in the plant.

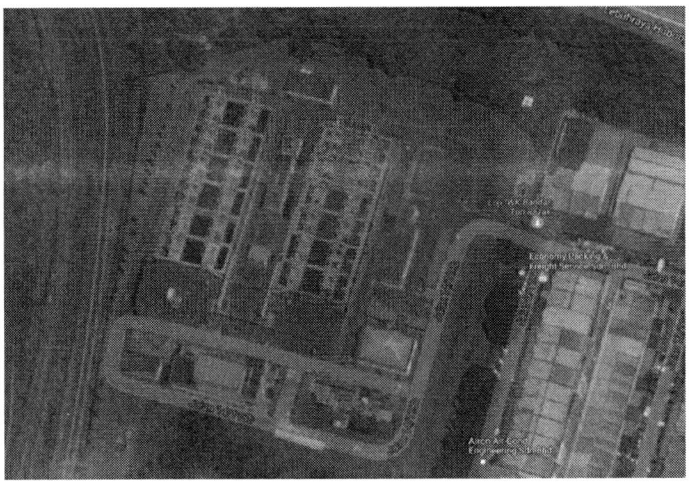

Figure 1. Plan view of BTR STP (Google earth).

2.2. Data collection

The wastewater inflow rates (Q) data was obtained from Bandar Tun Razak STP management, and it covered three continuous years on a weekly basis between 2011 and 2013, and the average inflow rate was about 16,711 m3. The weekly laboratory measurements of sewage inflow rate and loading parameters such as biological oxygen demand (BOD), chemical oxygen demand (COD), suspended solids (SS) and ammoniacal nitrogen (NH3-N), are authorized by the Department of Environment (DOE) to guarantee meeting the DOE standards of STP's.

2.3. Autoregressive Integrated Moving Average (ARIMA)

ARIMA or Box-Jenkins [12] model is considered the most popular and effective statistical models for time series forecasting. It based on generating a liner function extracted from the past observations of a time series in order to forecast the future values [9].

The linear function is consisting of three parametric components, Auto-Regression (AR), Integration (d) and Moving Average (MA) [1]. This can be illustrated in the form ARIMA (p, d, q).

In auto-regression (AR) or ARIMA (p, 0, 0) model, of order "p", the value of current output Zt (Observed value) depends upon the prior outputs "p" and the current inputs "et" (independent random shock). Therefore, the AR (p) equation can be written as:

$$Zt = \phi_1 Z_{t-1} + \phi_2 Z_{t-2} + \dots + \phi_p Z_{t-p} + e_t \quad (1)$$

While, in moving average (MA) or ARIMA (0, 0, q) model of order "q", the current output Zt (Observed Value) depends on the current input and prior inputs "q". MA (q) is represented as:

$$Zt = e_1 - \theta_1 e_{t-1} - \theta_2 e_{t-2} - \dots - \theta_q e_{t-q} \quad (2)$$

However, Autoregressive Moving Average Model (ARMA) of order (p, q) combines both AR and MA elements. An ARIMA (p, 0, q) or ARMA (p, q) is a model for a time series that depends on p past values of itself and on q past random terms e_t. This method has the form of:

$$Zt = \phi_1 Z_{t-1} + \phi_2 Z_{t-2} + \dots + \phi_p Z_{t-p} + e_t - \theta_1 e_{t-1} - \theta_2 e_{t-2} - \dots - \theta_q e_{t-q} \quad (3)$$

The Box-Jenkins models require a stationary time series data; therefore, a non-stationary data is always transformed to induce mean stationarity. A difference of order one leads to the subtraction of each observed value with the neighbouring value, which gives the new time series. Hence term "d" is referred TO the degree of ordinary differencing, applied to achieve series stationarity:

$$Yt = Z_t - Z_{t-1} \quad (4)$$

After applying the ARMA model to the differenced time series, the differencing transformation is reversed to reclaim the original values obtained by the modelled values and "integration" ("d" times) is done. A process in which the dth order differencing is involved is called an Integrated process of order d, it is denoted by the notion I (d). A combination of AR, MA and I models is called an ARIMA (p, d, q) model of order (p, d, q).

2.4. Model Development

Time series analysis and forecasting of waste water inflow in Bandar Tun Razak sewage treatment plant (STP) were performed using a historical record of 3 years (2011-2013). Sewage inflow data was provided on weekly interval basis during the study period (156 weeks).

ARIMA model was applied in this study through the following steps: model identification and estimation, diagnostic checking and forecasting. The identification test is done to obtain the value of order of differencing 'd' in ARIMA (p, d, q) and also the values of AR and MA operators. The appropriate orders of the ARIMA (p, d, q) model are usually determined through the Box-Jenkins model building methodology [8]. IBM SPSS statistics 22 software was used in this study. In addition, linear regression analysis was used to compare between the observed and predicted values.

2.5. Model performance tests

In order to judge the modelling accuracy and select the most fitted ARIMA model configurations, different performance criterions such as R-square, stationary R-square, root mean square error (RMSE), mean absolute percentage error (MAPE), and normalized Bayesian Information Criterion (BIC) were used to select the best fitting.

3. Results and discussion

3.1. Model Identification and estimation

A sequence graph of sewage inflow data (156 weeks) in Bandar Tun Razak was plotted to check the stationarity of analysed data as shown in Figures 1. By computing the autocorrelation and partial autocorrelation coefficients (ACF and PACF), the data was found to be non-stationary as shown in Figures 2.

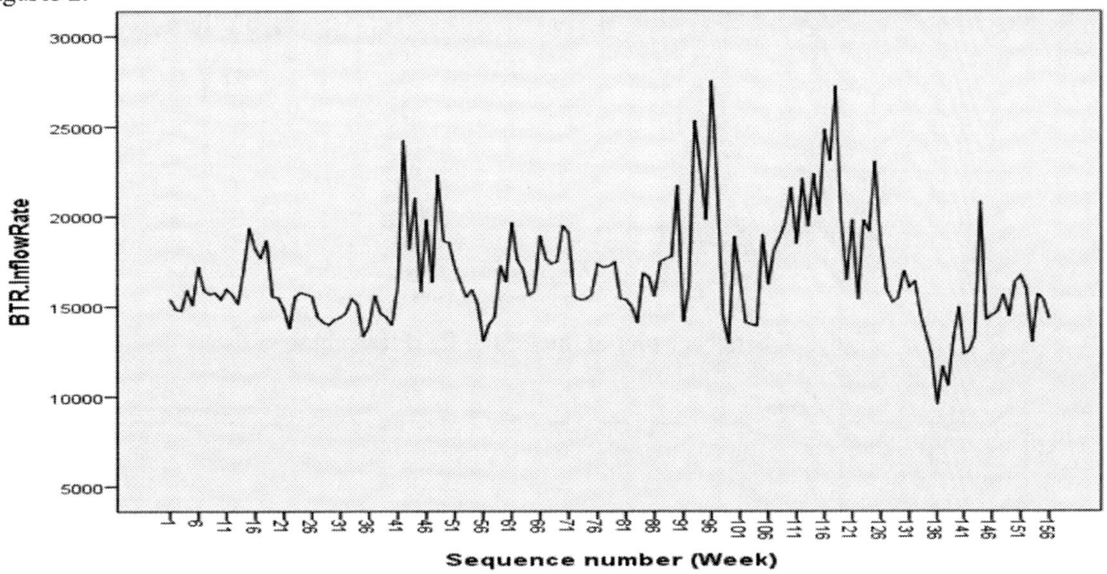

Figure 2. Sequence plot of the weekly inflow at BTR STP (152weeks).

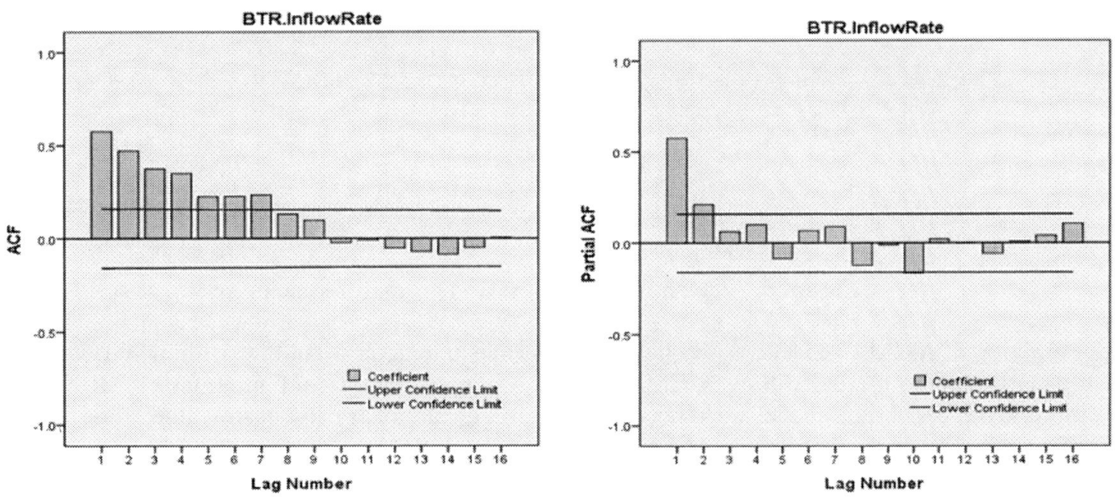

Figure 3. ACF and PACF of weekly inflow data.

Therefore, the first order differencing of the data series was applied (Figures 3). The obtained differenced data was tested for stationarity by ACF and PACF as shown in Figure 4. After examining (ACF and PACF) plots and its associated tables (Tables 7 in Appendix), it was concluded that the data become stationary in order to start applying ARIMA models.

IOP Conf. Series: Materials Science and Engineering **210** (2017) 012028 doi:10.1088/1757-899X/210/1/012028

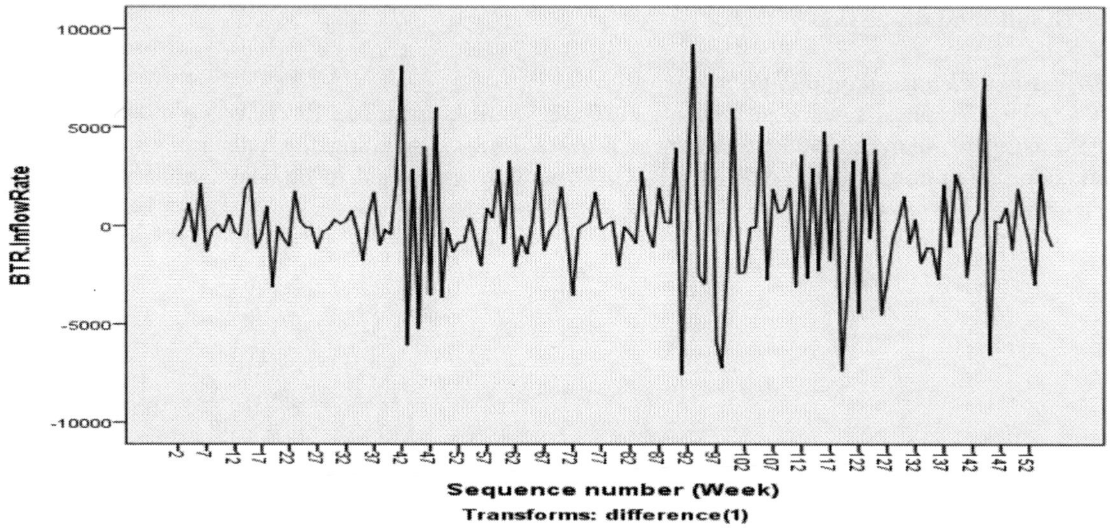

Figure 4. Sequence plot of the first order differencing of data.

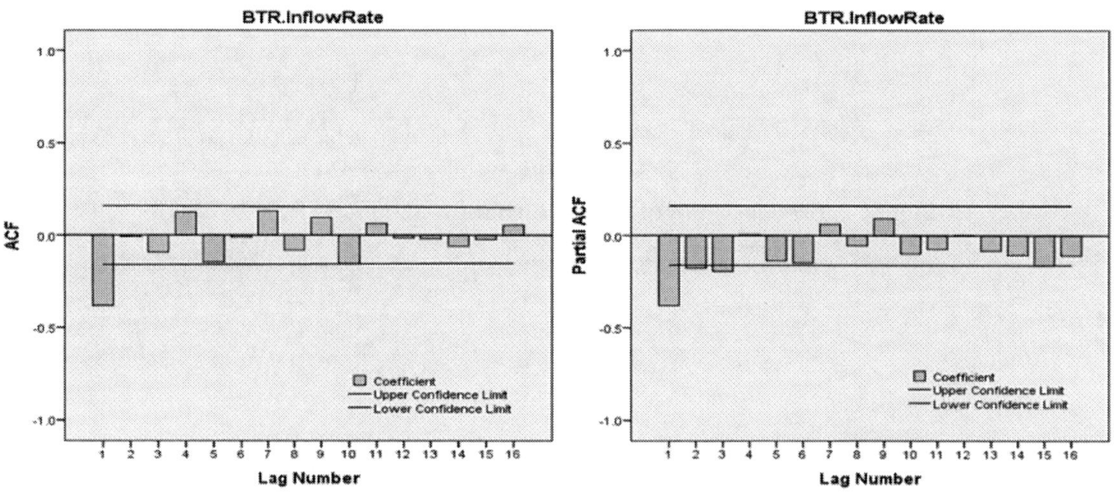

Figure 5. ACF and PACF of the first order differencing series.

Applying of ARIMA model was through trying various orders of 'p' and 'q' with the difference of one (d=1) to choose the best fitting ARIMA model. Among different configurations of ARIMA models orders, the best fitting model was chosen based on high stationary R-Square value, good value of R-Square and low values of RMSE, MAPE and Normalized BIC as illustrated in Table 1.

Table 1. Suggested ARIMA models.

Model (p, d, q)	Stationary R-Squared	R-Squared	RMSE	MAPE	Normalized BIC	Outliers
(0, 1, 1)	0.502	0.578	1990.0	8.79	15.42	5
(0, 1, 2)	0.525	0.598	1955.7	8.63	15.45	6
(0, 1, 3)	0.608	0.669	1788.8	8.08	15.34	7
(1, 1, 0)	0.573	0.639	1853.5	8.24	15.34	7
(1, 1, 1)	0.525	0.598	1955.7	8.64	15.45	6
(1, 1, 2)	0.505	0.581	1996.8	8.58	15.49	5
(1, 1, 3)	0.486	0.566	2041.0	8.93	15.57	5
(2, 1, 0)	0.507	0.583	1985.8	8.73	15.45	5
(2, 1, 1)	0.51	0.586	1986.4	8.50	15.48	5
(2, 1, 2)	0.486	0.566	2040.9	8.92	15.57	5
(2, 1, 3)	0.583	0.648	1850.8	8.18	15.44	6
(3, 1, 0)	0.593	0.656	1823.1	8.10	15.37	7
(3, 1, 1)	0.601	0.662	1812.4	8.01	15.40	7
(3, 1, 2)	0.511	0.586	1998.6	8.52	15.56	5
(3, 1, 3)	0.615	0.675	1791.3	8.01	15.44	7

The best suitable model for inflow rate of Bandar Tun Razak STP was found to be ARIMA (3, 1, 3). Main parameters of the selected model are given in the following tables:

Table 2. Statistics of the selected ARIMA (3, 1, 3) model.

Best Fit Model Statistics						Ljung-Box			Outliers
Stationary R-Squared	R-Squared	RMSE	MAPE	Max APE	Normalized BIC	Statistics	df	Sig.	7
0.615	0.675	1791.2	8.012	39.50	15.437	27.1934	12	0.007	

Table 3. Parameters of the selected ARIMA model

Parameters		Estimate	SE	t	Sig.
Constant		10.01	90.10	0.111	0.912
AR	Lag 1	-0.704	0.166	-4.228	0
	Lag 2	-0.694	0.183	-3.782	0
	Lag 3	-0.359	0.154	-2.328	0.021
Difference		1			
MA	Lag 1	-0.272	0.173	-1.574	0.118
	Lag 2	-0.650	0.154	-4.212	0
	Lag 3	0.234	0.158	1.483	0.140

IOP Conf. Series: Materials Science and Engineering **210** (2017) 012028 doi:10.1088/1757-899X/210/1/012028

Table 4. Outliers statistics

Outliers (Week No.)	Type	Estimate	SE	t	Sig.
42	Additive	5732.9	1257.7	4.558	0
90	Additive	6448.2	1405.3	4.589	0
93	Additive	12253.6	1727.7	7.093	0
94	Level Shift	7027.3	1949.0	3.606	0
98	Level Shift	-9826.9	1589.3	-6.183	0
118	Additive	6282.8	1217.3	5.161	0
144	Additive	7163.8	1233.3	5.809	0

3.2. Diagnostic checking

The selected model was tested and verified by examining the residuals ACF and PACF of various orders, which indicated a "good fit" of the model as shown in Figure 5. Autocorrelations up to 24 lags were evaluated and their significance was verified by Box-Ljung statistic as illustrated in Table 8 in Appendix.

Clearly, we can notice that almost all lags were within the reasonable level in residual ACF and residual PACF. Therefore, this refers that the selected ARIMA (3, 1, 3) model can be used for inflow rate analysis in Bandar Tun Razak STP.

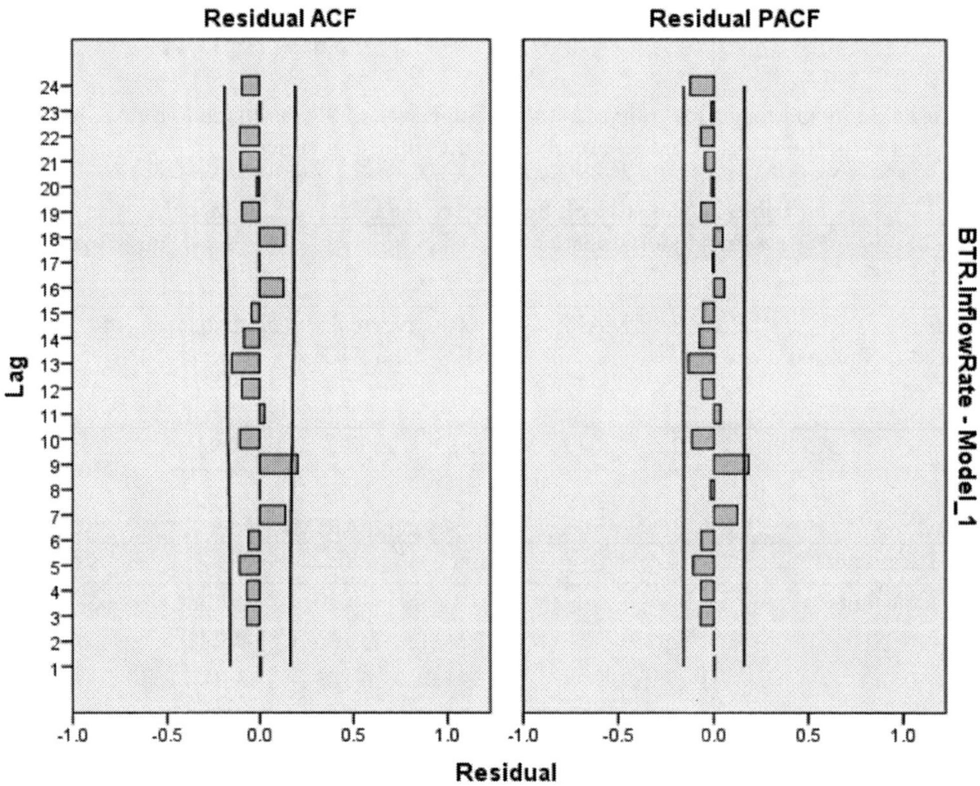

Figure 6. Residuals of ACF and PACF of the selected ARIMA model

The linear regression model was carried out between the observed inflow and predicted inflow values of ARIMA model. As shown in Table 5, the correlation coefficient of predicted inflow was 0.83, which suggests a good positive linear correlation.

Table 5. Linear Regression Model statistics.

Model	R	R Square	Adjusted R Square	Std. Error of the Estimate	Change Statistics					Durbin-Watson
					R Square Change	F Change	df 1	df2	Sig. F Change	
1	0.83	0.691	0.689	1607.519	0.691	342.11	1	153	0	1.857

The Normal P-P Plot of Regression Standardized Residual showed a random scatter of the points with a constant variance without any outliers. Since the points are close to the diagonal line (Figure 6), it is understood that the residuals are approximately normally distributed.

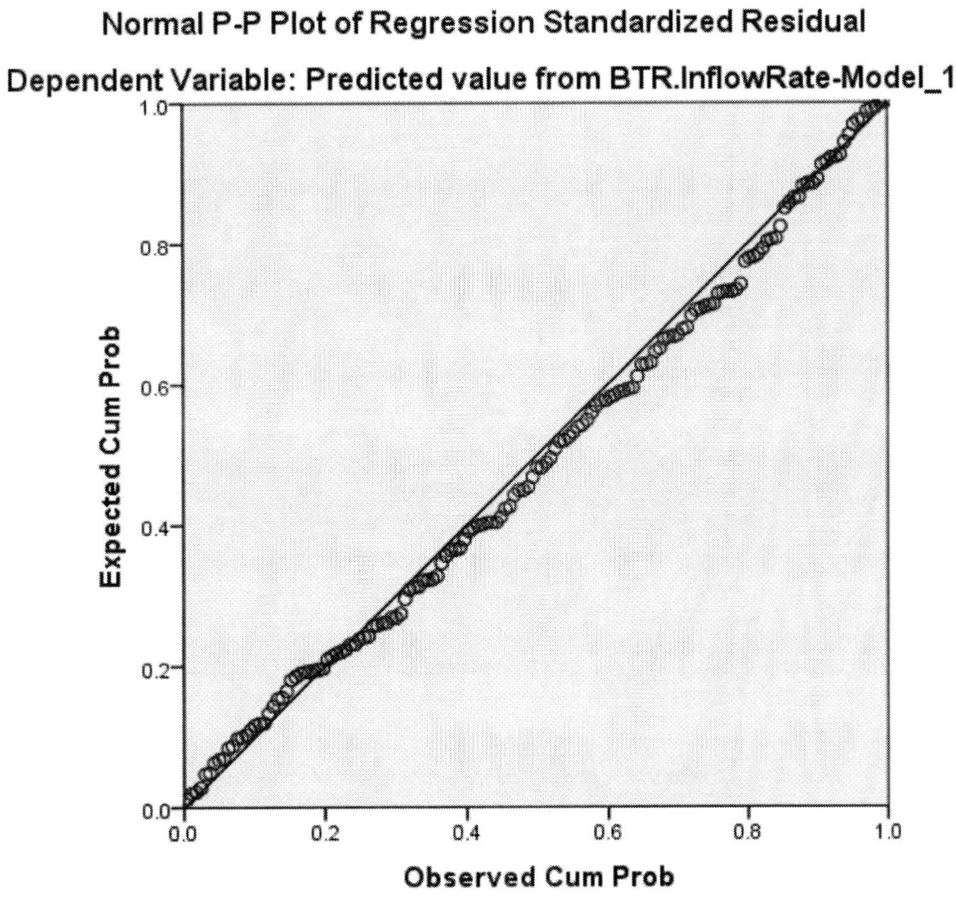

Figure 7. Normal P-P plot of regression standardized residual.

3.3. Forecasting

The best fitted ARIMA (3,1,3) was used to forecast the inflow rate till 208 weeks (4 years). The forecasted values are tabulated in Table 6, while the observed and predicted values with the confidential limits are shown in the Figure 7.

IOP Conf. Series: Materials Science and Engineering **210** (2017) 012028 doi:10.1088/1757-899X/210/1/012028

Table 6. Summary of Bandar Tun Razak STP inflow rate forecasting.

Period	Observed Inflow		Predicted Inflow		Lower Confidential Limit		Upper Confidential Limit	
Week 1-26	15,416	15,577	15,426	16,152	10,950	12,613	19,902	19,692
Week 27-52	14,413	16,352	14,809	18,604	11,270	15,066	18,348	22,141
Week 53-78	15,511	17,252	17,476	18,053	13,939	14,516	21,014	21,591
Week 79-104	17,479	13,926	16,935	15,174	13,398	11,636	20,473	18,711
Week 105-130	18,987	16,980	15,467	16,520	11,929	12,982	19,004	20,057
Week 131-156	16,087	14,314	15,724	16,578	12,186	13,041	19,261	20,116
Week 157-182	-	-	13,382	14,325	9,844	2,862	16,919	25,788
Week 183-208	-	-	14,332	14,582	2,667	(1,334)	25,998	30,498

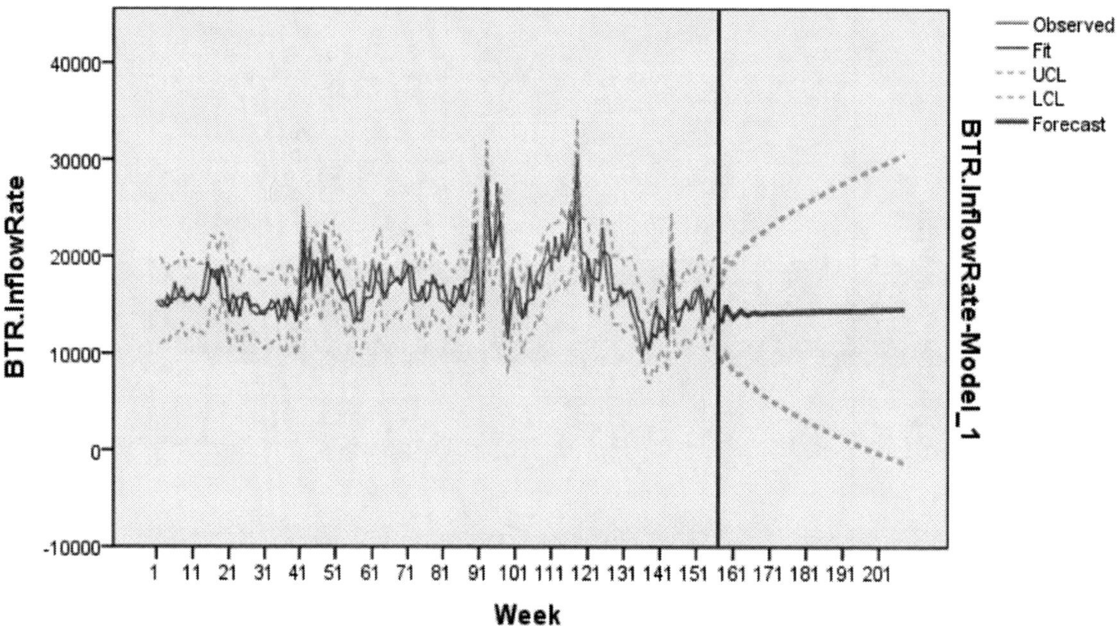

Figure 8. Forecasted ARIMA (3, 1, 3) model

4. Conclusions

In this study, time series ARIMA modelling of a weekly sewage inflow into one of the main STP's in Kuala Lumpur, Malaysia was successfully conducted. The three continues years (152weeks) data collected by the Bandar Tun Razak (BTR) STP's management was found to be non-stationary, thus it was transformed to the first order differencing (d=1) to make it stationary. Fifteen ARIMA models of various orders of 'p' and 'q' were applied on to the transformed data to select the best fitted model. Based on the diagnostics like high R^2 value and low normalized Bayesian Information Criterion (BIC), the ARIMA (3, 1, 3) was found to be the best fitted model. The linear regression model was applied between the observed and predicted values, and it showed a positive linear correlation with a correlation coefficient of 0.83. By this linear regression analysis, it was understood that there was no much variation between the observed and predicted data. The best fitted ARIMA (3, 1, 3) model forecasted the inflow

IOP Conf. Series: Materials Science and Engineering **210** (2017) 012028 doi:10.1088/1757-899X/210/1/012028

rates to additional one year (52weeks); therefore, this study can be considered for future design planning of the BTR STP.

Acknowledgment
The authors would like to acknowledge the University of Malaya Research Grant (FL001-13SUS) and Fundamental research Grant by Ministry of Higher Education (FP016-2014A) for the financial support. We are most grateful and would like to thank the reviewers for their valuable suggestions, which have led to substantial improvement of the article.

References

[1] G. E. P. Box and G. M. Jenkins 1976 *Time series analysis : forecasting and control.* Holden-Day
[2] A. Pankratz, 2009 *Forecasting with univariate Box-Jenkins models: Concepts and cases.*
[3] H. Yan and Z. Zou, 2013 *Journal of Convergence Information Technology* **8** 59–70.
[4] M. Valipour, M. E. Banihabib, and S. M. R. Behbahani 2013 *Journal of Hydrology* **476** 433–441.
[5] A. O. Pektaş and H. Kerem Cigizoglu 2013 *Journal of Hydrology* **500** 21–36.
[6] P. Narayanan, A. Basistha, S. Sarkar, and S. Kamna 2013 *Comptes Rendus Geoscience* **345** 22–27.
[7] J. R. Kim, J. H. Ko, J. H. Im, S. H. Lee, S. H. Kim, C. W. Kim, an T. J. Park 2006 *Water Science and Technology* **53** 185–19.
[8] G. P. Zhang 2003 *Neurocomputing* **50** 159–175.
[9] K. Yürekli, H. Simsek, B. Cemek, and S. Karaman 2007 *Building Environment.*

IOP Conf. Series: Materials Science and Engineering **210** (2017) 012028 doi:10.1088/1757-899X/210/1/012028

Appendix

- ACF and PACF statistics of the first order differencing series:

Autocorrelation			Box-Ljung Statistic			Partial Autocorrelation		
Lag	Value	Std. Error	Value	df	Sig.	Lag	Value	Std. Error
1	-0.380	0.080	22.77	1	0	1	-0.380	0.08
2	-0.007	0.079	22.77	2	0	2	-0.176	0.08
3	-0.091	0.079	24.09	3	0	3	-0.194	0.08
4	0.124	0.079	26.57	4	0	4	0.006	0.08
5	-0.145	0.079	29.96	5	0	5	-0.135	0.08
6	-0.010	0.078	29.98	6	0	6	-0.147	0.08
7	0.129	0.078	32.72	7	0	7	0.062	0.08
8	-0.081	0.078	33.81	8	0	8	-0.054	0.08
9	0.096	0.077	35.35	9	0	9	0.094	0.08
10	-0.154	0.077	39.33	10	0	10	-0.097	0.08
11	0.063	0.077	40.01	11	0	11	-0.072	0.08
12	-0.014	0.077	40.04	12	0	12	0.003	0.08
13	-0.018	0.076	40.10	13	0	13	-0.081	0.08
14	-0.059	0.076	40.70	14	0	14	-0.105	0.08
15	-0.022	0.076	40.78	15	0	15	-0.161	0.08
16	0.056	0.076	41.33	16	0	16	-0.108	0.08

- Residual of ACF and PACF of the selected ARIMA model:

Lag	ACF		PACF	
Lag 1	.005	.080	.005	.080
Lag 2	.002	.080	.002	.080
Lag 3	-.069	.080	-.069	.080
Lag 4	-.067	.081	-.067	.080
Lag 5	-.107	.081	-.107	.080
Lag 6	-.059	.082	-.065	.080
Lag 7	.130	.082	.122	.080
Lag 8	.000	.084	-.018	.080
Lag 9	.197	.084	.182	.080
Lag 10	-.105	.087	-.115	.080
Lag 11	.025	.087	.034	.080
Lag 12	-.095	.087	-.061	.080
Lag 13	-.147	.088	-.133	.080
Lag 14	-.083	.090	-.075	.080
Lag 15	-.042	.090	-.057	.080
Lag 16	.122	.090	.053	.080
Lag 17	-.001	.091	-.006	.080
Lag 18	.124	.091	.046	.080
Lag 19	-.092	.092	-.067	.080
Lag 20	-.013	.093	-.011	.080
Lag 21	-.101	.093	-.047	.080
Lag 22	-.104	.094	-.068	.080
Lag 23	.004	.094	-.013	.080
Lag 24	-.092	.094	-.123	.080

- Histogram chart of liner regression analysis:

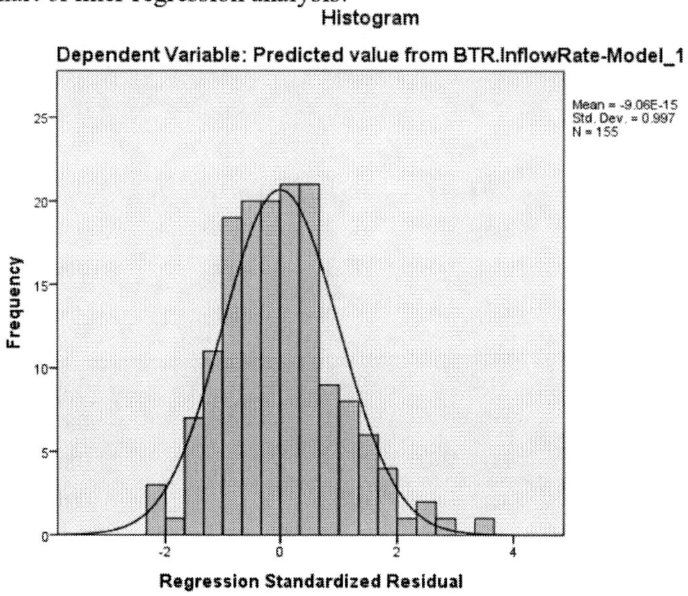

IOP Conf. Series: Materials Science and Engineering **210** (2017) 012029 doi:10.1088/1757-899X/210/1/012029

Polymer microlens fabricated on fiber tip for optical interconnect

Y Yaacob[1*], O Mikami[1], C Fujikawa[2], S Nakajima[2] and S Ambran[1]

[1] Dept. of Electronic Systems Engineering, Malaysia-Japan International Institute of Technology, Universiti Teknologi Malaysia, Jalan Sultan Yahya Petra, 54100, Kuala Lumpur, Malaysia

[2] Course of electro Photo Optics, Graduate School of Engineering, Tokai University, 4-1-1, Kitakaname, Hiratsuka, Kanagawa, 259-1292, Japan

Email: *yuzayaacob@gmail.com

Abstract. An optical interconnect has attracted much interest to many researchers as it seemed as a solution to meet the requirement of future demand due to the limitations of the electrical interconnect. Low power consumption, small signal variation, and wide bandwidth are the example of the advantages of the optical interconnect. Nevertheless, the optical interconnect requires a high optical coupling which is a hurdle for the optical interconnect. The implementation of the microlens in optical interconnect seems as one of the possible ways which can solve the problem. It is expected that the microlens can focus the spreading light beam and improved the optical coupling. Thus, the polymer microlens has been fabricated on two different platforms; (i) on a ferrule insert with fiber inserted and (ii) directly on the fiber end. The Far Field Pattern measurement system has been used to evaluate the light collecting for both microlens and the divergence angle was measured. The results show that the divergence angle of the microlens on fiber end is better compared to the microlens on the ferrule.

1. Introduction

The development of the information technology has lead to the demand for the high network capacity and data transfer. However, the conventional copper based electrical interconnects incapable of fulfilling the requirement due to their high propagation loss, high power consumption, and inter-channel crosstalk during high-speed data transfer [1, 2]. Thus, to satisfy the demand, the introduction of optical interconnects to the printed wiring board is seemed as a promising candidate to resolve the issues. The optical interconnects are recommended to replace electrical interconnects due to its low power consumption, wide bandwidth and high data rate density, less crosstalk, light weight and small variation in signal delay [3,4]. With these advantages, the possibility to apply the multi-layer optical interconnect for optical printed wiring board can be realized [4]. Nevertheless, to maintain high optical coupling efficiency between planar surface devices and the multichannel waveguide is quite difficult. This is because the light beam easily spreads from the devices or waveguide itself [5]. Thus, by implementing the microlens in optical interconnect, it is expected the light beam can be focused and improved the optical coupling.

In this paper, the preliminary study has been done to evaluate the possibility to apply the microlens into the optical interconnect. The microlens is fabricated in two different platforms; microlens on

IOP Conf. Series: Materials Science and Engineering **210** (2017) 012029 doi:10.1088/1757-899X/210/1/012029

ferrule and microlens on the fiber end. The process of the fabrication is briefly explained. Some results from Far field pattern (FFP) measurement system and divergence angle are presented.

2. Fabrication and Analysis

Several fabrication technologies to fabricate the microlens for prototype or small quantities has been created and used by many researchers such as lithography, laser- and e-beam writing, deep-proton lithography, micro-jet printing, LIGA, and laser ablation [6-9]. In this paper, the syringe method is proposed to fabricated the microlens on the ferrule due to its easiness, simplicity and low fabrication cost.

2.1. Microlens on ferrule

A ferrule is a ring or cap attached to an object to protect against damage, splitting or wear. In the optical fiber, a ferrule is one of the important component to align and protect the strip end of the fiber. Ferrule can be made of plastic, metal, glass and ceramic materials. In this experiment, a ceramic ferrule is chosen as a platform to fabricate the microlens. The dimension of the ferrule is 2500µm of diameter with 1250µm of length.

At first, the bare single mode fiber (SMF) is inserted into a ferrule. Then, the Epo-Tek adhesive is applied at the end of the ferrule to ensure the bare fiber and the ferrule are attached together. After that, the UV curable resin (NP-206) with the refractive index 1.5437 (@1551nm), 1.5456 (@1305nm), and 1.5523 (@850nm) from Nissan Chemical is dropped on top of the ferrule with SMF using a syringe [10]. Subsequently, the droplet forms into partially sphere when they touch the ferrule end due to the surface tension. Later, the droplet transform into a solid microlens after being exposed to the UV light radiation with the wavelength, $\lambda = 365$nm.

Figure 1 illustrates the fabrication process of the microlens using syringe method. The dimension of the fabricated microlens is as follow; the diameter and height of the microlens are 2500µm and 992µm respectively as shown in Figure 2. Three samples of the microlens had successfully fabricated.

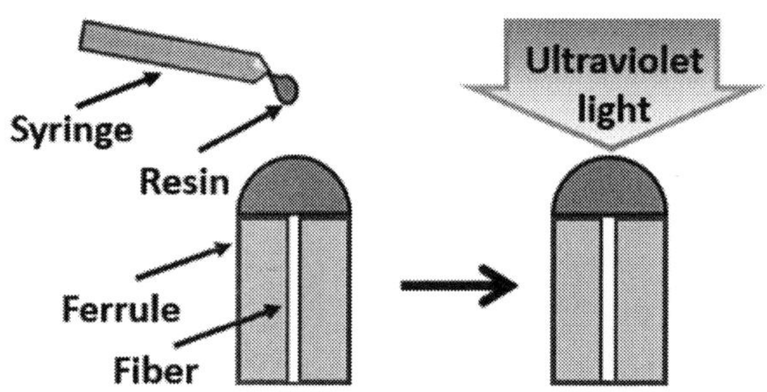

Figure 1. The fabrication process of microlens on a ferrule using a syringe.

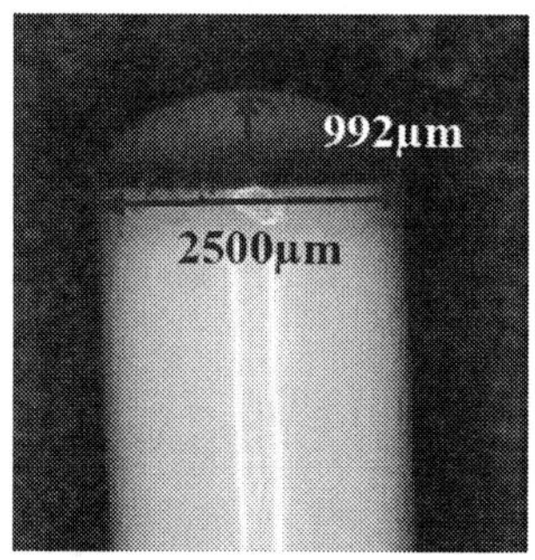

Figure 2. Example of fabricated microlens on a ferrule.

2.2. Microlens on fiber tip

The end face of the fiber tip was cut using a cleaver to obtained a perpendicular and smooth surface to avoid the diffused reflection of light emitted from the fiber before the fabrication of the microlens as shown in Figure 3 (a). The fabrication process was started with dropping a small amount of UV-curable resin onto a slide glass. Then, the resin was attached to the tip of the fiber with a sense that the tip of the fiber slightly contacted to the resin from above. After that, the resin was irradiated and cure to the ultraviolet light with the wavelength, $\lambda = 365$nm. Figure 3 (b) illustrates an example of the microlens fabricated on fiber tip. Seven samples of the microlens have been successfully fabricated.

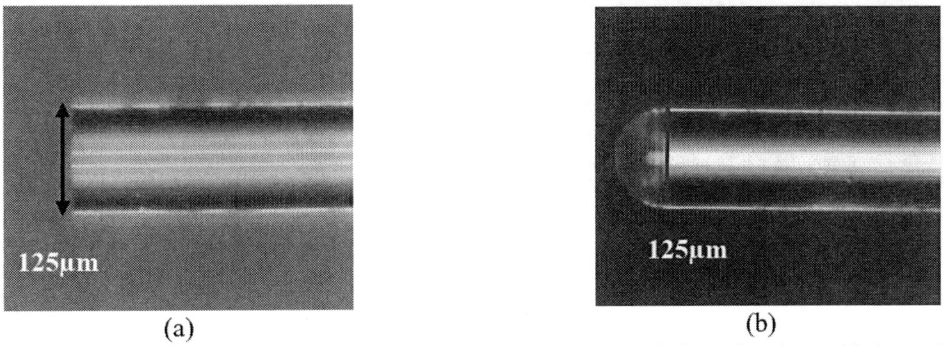

Figure 3. Example of (a) perpendicular and smooth fiber end and (b) microlens fabricated on fiber end.

2.3. *Analysis*

After the fabrication, both microlens were measured using Far Field Pattern (FFP) measurement method to measure the light collecting as one of the characteristics of the fabricated microlens. The 1550nm of wavelength is used in this measurement. The measurement setup is shown in Figure 4. Before start, the end of the fiber which has been fastened with SC connector is connect the laser diode (LD). Whilst, the fiber end with microlens is connect to the FFP measurement instrument. Based on the measurement result, the divergence angle of the fabricated microlens can be obtained.

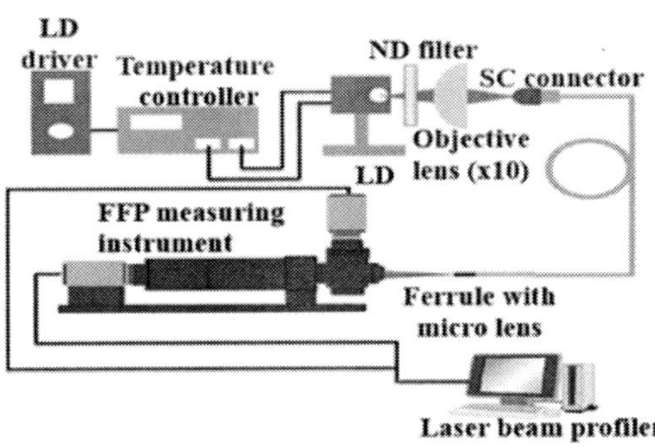

Figure 4. FFP measurement system is used to evaluate the fabricated microlens.

3. Results and discussion

Figure 5 (a) and (b) display the FFP image and profile takes from the FFP measurement for both conditions. The profile image from the x- axis shows the intensity of the light along the x-axis while the image profile from the y-axis indicates the intensity the light from the source along the y-axis. The crossing point in the figure refers to the highest intensity of light that can be detected from the light. From the profile image, the divergence angle is obtained. To obtain the divergence angle, the $1/e^2$ method is used.

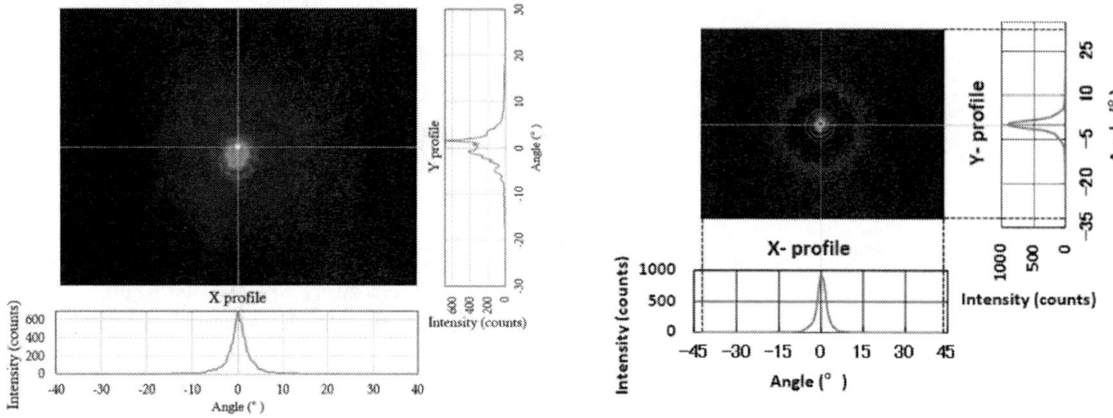

(a) (b)

Figure 5. An example of the FFP image and profile (a) microlens on the ferrule and (b) microlens on the fiber tip.

Table 1 and 2 below show the divergence angle of the output light without using the microlens, using the microlens with ferrule and using the microlens on fiber tip. Based on the tables, the divergence angle of the light without using the microlens is 13.9° at x-axis and 12.3° at y-axis while when using the microlens on the ferrule the divergence angle is smaller especially Sample 3 with 7.5° at x-axis and 8.4° at the y-axis. From the divergence angle, the light can focus more by using the microlens, thus it is seemed possible to use the microlens on optical interconnect to improve the coupling efficiency.

Thus, the divergence angles of the microlens on ferrule and on the fiber tip are compared to distinguish which microlens can perform better to focus the light. Based on the comparison, the divergence angle of the microlens fabricated on the fiber tip is much smaller compare to the microlens fabricated on the ferrule. The reason behind this result is due to the size of the microlens and the radius of the curvature produced. The radius of the curvature of the microlens on ferrule is approximately 1200μm while for the microlens on the fiber is about 65μm.

Table 1. Size and divergence angle of the microlens fabricated on the ferrule and the divergence angle without using the microlens.

Sample No.	Diameter (μm)	Height (μm)	Radius of curvature (μm)	Divergence angle (deg.)	
				X	Y
Without micro lens	—	—	—	13.9	12.3
Sample 1	2100	630	1190	10.8	10.9
Sample 2	2500	1000	1281.3	10.0	8.1
Sample 3	2500	992	1283.6	7.5	8.4

Table 2. Size and divergence angle of the microlens fabricated on the fiber tip.

Sample No.	Diameter (μm)	Height (μm)	Radius of curvature (μm)	Divergence angle (deg.)	
				X	Y
Sample 1	125	58.3	62.7	7.5	8.0
Sample 2	125	50.0	64.1	8.5	10.7
Sample 3	125	41.7	67.7	14.9	10.7
Sample 4	125	44.2	66.3	3.2	8.8
Sample 5	125	19.4	67.0	5.3	6.5
Sample 6	125	41.8	67.7	5.2	3.8
Sample 7	125	40.2	64.2	5.8	2.7

4. Conclusion

As a conclusion, the microlens has been successfully fabricated on two different platforms; (i) on the ferrule and (ii) directly on the fiber tip. The syringe method has been used to fabricate the microlens due to the easiness, simplicity and low cost procedures. Based on the result, by using the microlens the spreading of the light can be more focus. Thus, the microlens can help to improve the coupling efficiency on the optical interconnect. In addition, the size of the microlens and the radius of curvature produce can affect the divergence angle of the output light.

Acknowledgments

This work was supported in part by the JKA and its promotion funds from KEIRIN RACE (27-133) and in part by the Ministry of Higher Education (MOHE) and Universiti Teknologi Malaysia (UTM) under Fundamental Research Grant Scheme 4F881.

References

[1] Matsuoka Y. et al. 2011 *IEEE Photonics Technology Letters* **23** pp.1352-1354
[2] Ciftcioglu B. et al 2011 *IEEE Photonics Technology Letters* **23** pp. 164-166
[3] Keishi Ohashi et al. 2009, 'On-Chip Optical Interconnect' *IEEE* 2-3
[4] Hsu, S., Tsou, C., Wang, C. and Tseng, S. 2013. *Optics and Photonics Journal* **3** 252-255
[5] Matsuoka Y. et al 2010 *National Fiber Optic Engineers Conference*, OSA Technical Digest (CD) (Optical Society of America, 2010) paper JThA58
[6] Kris Naessens, Peter Van Daele and Roel G. Baets 2002 *Second International Symposium on Laser Precision Microfabrication* Singapore 1-4
[7] Kopitkovas G., Urech L., and Lippert T 2006 *Recent Advances in Laser Processing of Materials, Elsevier, Kidlington*
[8] Wang, Weisong and Ji Fang 2006 *Journal of Micromechanics and Microengineering* **16** 1221-1226
[9] Biehl S, Danzebrink R, Oliveira P and Aegerter M.A. 1998 *Journal of Sol-Gel Science and Technology* **13** 177-182
[10] Nawata H 2013 *2013 3rd IEEE CPMT Symposium Japan* 1-4

IOP Conf. Series: Materials Science and Engineering **210** (2017) 012030 doi:10.1088/1757-899X/210/1/012030

Demonstration of passive saturable absorber by utilizing MWCNT-ABS filament as starting material

S N F Zuikafly[1], F Ahmad[1*], M H Ibrahim[2], A A Latif[3] and S W Harun[4]

[1] Malaysia-Japan International Institute of Technology (MJIIT), Universiti Teknologi Malaysia, 54100 Kuala Lumpur, Malaysia
[2] Faculty of Electrical Engineering, Universiti Teknologi Malaysia, 81310 Skudai, Malaysia
[3] Department of Physics, Faculty of Science, Universiti Putra Malaysia, 43400 Serdang, Malaysia
[4] Photonics Research Centre, Univeristy of Malaya, 50603 Kuala Lumpur, Malaysia

Email: *fauzan.kl@utm.my

Abstract. This work demonstrated a stable passively Q-switched laser with the employment MWCNTs dispersed in acrylonitrile butadiene styrene (ABS) resin (MWCNTs-ABS) based filament as passive saturable absorber. The simple fabrication process of the SA is further explained, started from the process of extruding the filament through a 3D printer nozzle at 210 °C to reduce the diameter from 1.75 mm to 200 μm. It is then weighed to about 25 mg and mixed with 1 ml acetone before sonicated for 5 minutes to dissolve the ABS. The resultant MWCNTs-acetone suspension is dropped on a glass slide to be characterized using Field-Emission Scanning Electron Microscope (FESEM) and Raman spectroscopy. It is also drop-casted on the end of a fiber ferrule to be integrated in the laser cavity. The proposed work revealed that the laser oscillated at about 1558 nm with threshold input pump power of 22.54 mW and maximum input pump power of 108.8 mW. The increase in pump power resulted in the increase in repetition rate where the pulse train increases from 8.96 kHz to 39.34 kHz while the pulse width decreases from 33.58 μs to 5.14 μs. The generated pulsed laser yields a maximum of 1.01 mW and 5.53 nJ of peak power and pulse energy respectively. The signal-to-noise ratio of 40 dB indicates that the generated pulse is stable.

1. Introduction

The Multiwalled carbon nanotubes (MWCNTs) are basically rolled up graphene, made up of carbon atoms that formed hexagonal networks that eventually build up a long cylindrical structure [1]. It possesses high thermal conductivity [2] and high mechanical strength [3] offering excellent features of ultrafast laser development. It is seen and proved to be a good candidate for saturable absorber (SA). It is an excellent choice for SA integration as it can easily be employed into various fibre configurations while maintaining, an alignment free and all fiber format [4].

On another note, three-dimensional (3D) printing technology since its first invention in the 1980s has been commonly used by companies for the production of conceptual prototypes upon producing the products in mass quantities [5-6]. The printer works by depositing a material such as plastic or metal layer by layer to produce 3 dimensional objects, following the instruction set from the digital designs from computer aided design files (CAD). The science behind the material used for 3D printing is as crucial to that of the 3D printer technology itself. Companies often used pure plastic such as acrylonitrile butadiene styrene (ABS) and polylactic acid (PLA) as 3D printer filament but the incorporation of such material with another can give better ink performance [7].

Due to the interesting properties of carbon nanotubes (CNTs) which are mechanically strong, conductive and heat resistance [8], MWCNTs is incorporated with a plastic component, ABS to produce a 3D printer filament. Since MWCNTs have structural defects, the interaction with the polymers is rather strong. The 3D printer filament used in the study is easily obtainable from 3Dxtech. Such printer filament is more focused on industrial usage spanning from automobile to semiconductor industries which often requires strong, conductive and light-weight material in their works. The founder of 3Dxtech [9] also believe that MWCNTs-ABS printer filament can be used in hard drives due to its conductive properties, to replace traditional material, carbon black, with the advantage of a cleaner and more efficient conductivity.

Normally, MWCNTs powder or sheets are used as starting material for SA fabrication. Mechanical exfoliation, metal supported growth, chemical vapour deposition (CVD), and liquid phase exfoliation (LPE) are among the many methods used to realize the SA. Chemically modified graphene from mechanical exfoliation and metal supported growth provide better electrical and structural robustness but the process is rather costly, time consuming, and yield low end material [10-11]. For larger scale and lower cost fabrication, the focus moved towards CVD techniques which involve the process of depositing gaseous reactants to the substrate. However, the even dispersion of the MWCNTs is typically hard to be controlled and maintained. The LPE of MWCNTs is another example of fabrication method there is. The process usually involves three steps, the first being the dispersion of MWCNTs in a liquid medium, second being exfoliation of the dispersed material by using ultrasonication and the last step is the purification process. LPE route requires the nanomaterial to undergo several processes including ultrasonic to disperse the nanomaterials, centrifugation to remove any undispersed suspension and then mix with a host polymer. The use of the commercially available MWCNT-ABS 3D printer filament as the starting material in fabricating the saturable absorber (SA) in this work will demonstrate a simple and timesaving fabrication procedures involved, along with the fabricated SA's performance in pulse laser generation.

2. Material Preparation

The fabrication of the MWCNTs slurry used in the study was started by using MWCNTs in acrylonitrile-butadiene-styrene (ABS) 3D printer filament that is commercially available offline or online. The one used in the study is obtained online from 3Dxtech. When received, the filament diameter and weight is 1.75 mm and 750 g, respectively. The first step to the fabrication of the SA is by extruding the MWCNTs-ABS printer filament by using a 3D printer through its nozzle diameter of diameter 0.4 mm at 210°C. This is to reduce the filament's diameter to 200 μm. We then weighed the filament to about 25 mg to be mixed with 1 ml of acetone before it is sonicated for 5 minutes to dissolve the ABS. The MWCNTs-acetone suspension is finally produced. The suspension is dropped casted on the end of a fiber ferrule to later be integrated in the laser cavity. Another portion is dropped on a glass slide for characterization purposes. When dropped, the acetone on the suspension will evaporate leaving only the MWCNTs slurry. The fabricated MWCNTs slurry is investigated by using Field-Emission Scanning Electron Microscope (FESEM). Figure 1 shows the FESEM image of the MWCNTs slurry with discerning bundle of MWCNTs. The peak shift of the slurry is investigated using the Raman spectroscopy performed using LabRAM HR Evolution. The Raman shift of the MWCNTs slurry is shown in Figure 2 with D-band, G-band and G'-

band at 1346 cm^{-1}, 1574cm^{-1} and 2694cm^{-1}, respectively. The measured insertion loss is approximately 4.5 dB.

3. Material Characterization

The morphology of the fabricated MWCNTs slurry is investigated by using Field-Emission Scanning Electron Microscope (FESEM). Figure 1 shows the FESEM image of the MWCNTs slurry with discerning bundle of MWCNTs. The peak shift of the slurry is investigated using the Raman spectroscopy performed using LabRAM HR Evolution. The Raman shift of the MWCNTs slurry is shown in Figure 2 with D-band, G-band and G'-band at 1346 cm^{-1}, 1574cm^{-1} and 2694cm^{-1}, respectively. The measured insertion loss is approximately 4.5 dB.

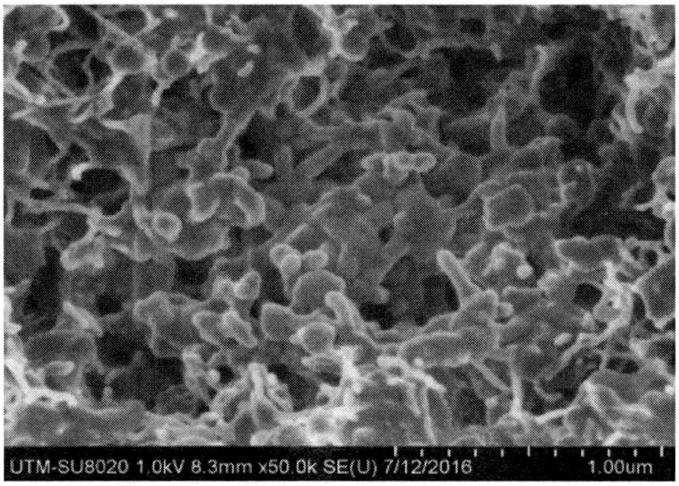

Figure 1. FESEM image of MWCNTs slurry.

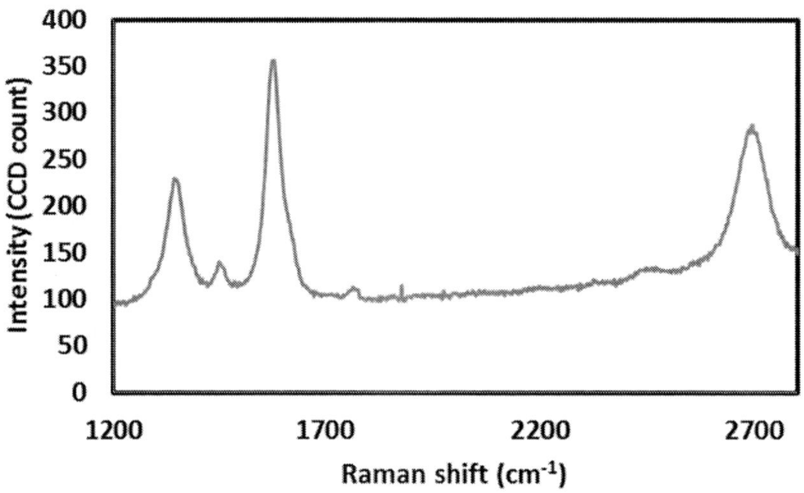

Figure 2. Raman spectroscopy of MWCNTs slurry.

IOP Conf. Series: Materials Science and Engineering **210** (2017) 012030 doi:10.1088/1757-899X/210/1/012030

4. Experimental Setup

The Figure 3 below shows the experimental setup of the passively Q-switched EDFL. The ring cavity of the overall setup consisted of a 1 m long Erbium-doped fiber (EDF) which is used as the gain medium, a 980/1550 nm wavelength division multiplexer (WDM), an isolator, the newly fabricated MWCNTs slurry as SA, and an 80/20 output coupler. The core and cladding diameter of the EDF is 8 μm and 125 μm respectively. The numerical aperture of the EDF is 0.16 and has Erbium ion absorptions of 45 dB/m at 1480 nm and 80 dB/m at 1530nm. The EDF was pumped by a 980 nm laser diode via the WDM. An isolator is incorporated in the laser cavity to ensure no back propagation of the oscillating laser. The output of the laser was tapped from the cavity through a 80/20 coupler while keeping 80% of the light to oscillate in the ring cavity. An optical spectrum analyser (OSA) with a spectral resolution of 0.05 nm is used to inspect the spectrum of the EDFL, whereas the oscilloscope was used to observe the output pulse train via a 460 kHz bandwidth photo-detector.

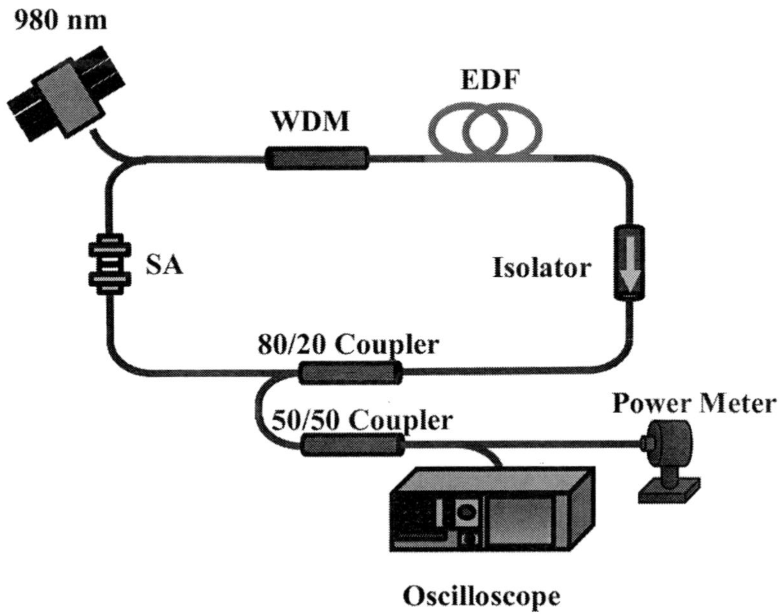

Figure 3. Experimental setup of the proposed passively Q-switched EDFL with MWCNTs slurries based SA.

5. Results and Discussions

The proposed EDFL started to lase with passively Q-switched mode at 22.54 mW of input pump power. The threshold pump power is notably lower than that reported in [1, 12-13] using similar two-dimensional (2D) materials. The low pump power threshold may be caused by the low saturation intensity properties of the material itself. Figure 4 shows the optical spectrum analyser trace at maximum input pump power of 108.8 mW with two peaks occurring at 1558.0 nm and 1558.4 nm. The observed peaks are due to mode competition of the gain medium used in this work. Figure 5 shows the oscilloscope trace of the Q-switched pulse at maximum input pump power of 108.8 mW with corresponding repetition rate of 39.34 kHz with the separation of adjacent pulse is around 24.5 μs. No distinct amplitude modulation is observed, indicating the stability of each Q-switched envelope. The single pulse envelope of the shortest pulse width at maximum input power of 5.14 μs is as shown in Figure 6.

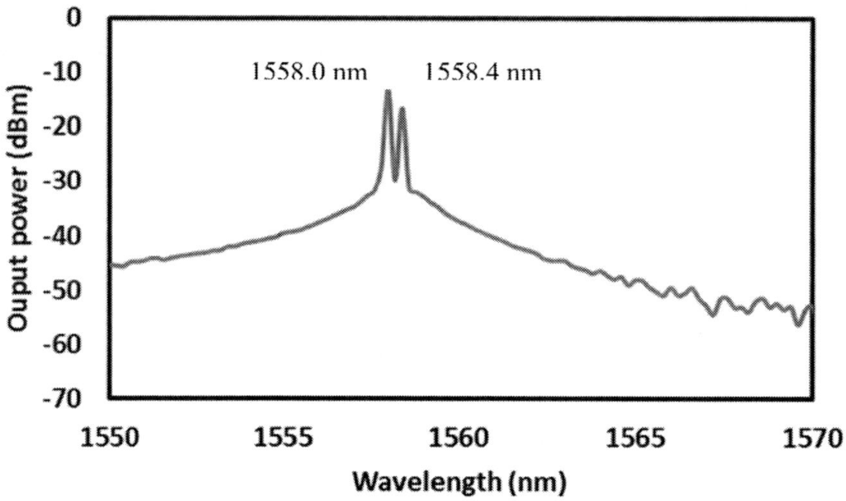

Figure 4. OSA trace at at maximum input power of 108.8 mW.

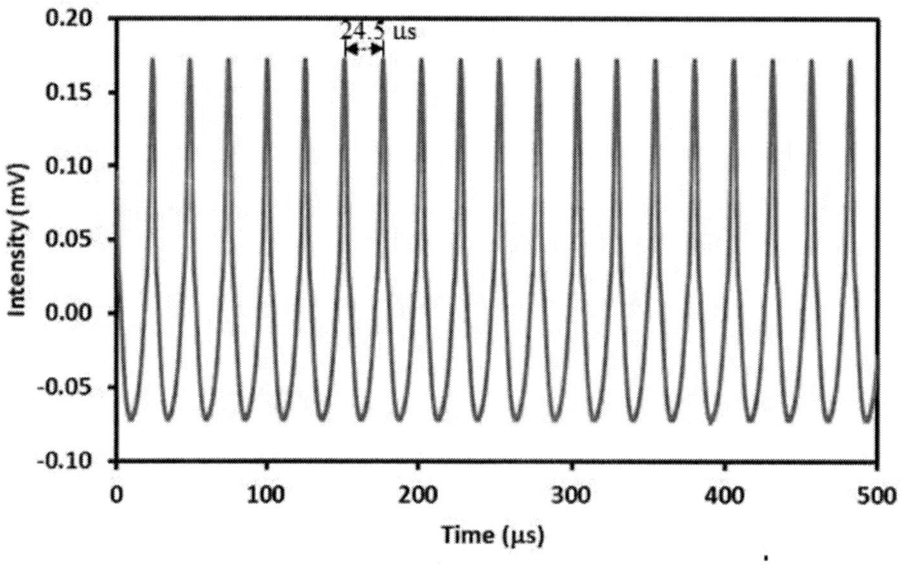

Figure 5. Pulse train of 39.34 kHz at maximum input power.

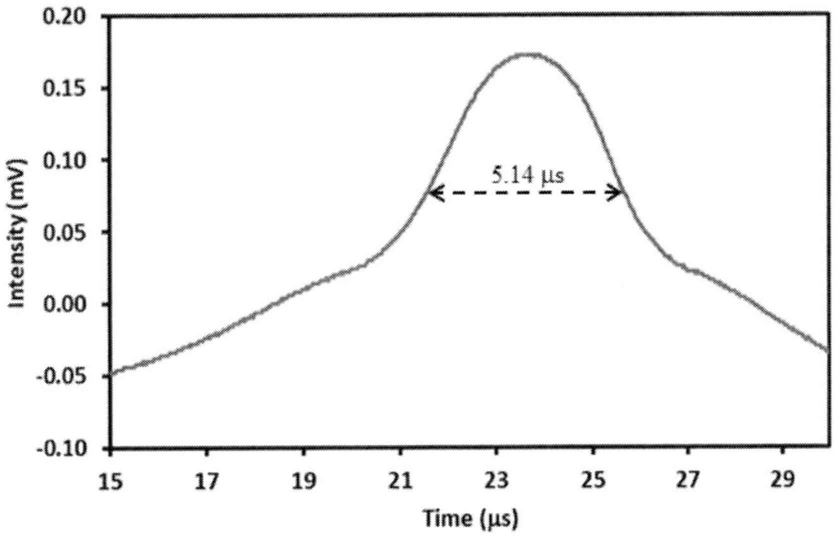

Figure 6. Single pulse envelope of the shortest pulse width of 5.14 μs.

The dynamics of the pulsed laser is investigated by tuning the pump power. As seen in Figure 7, increasing the pump power from 22.54 mW to 108.8 mW makes the repetition rate higher and the pulse width narrower. As function of the pump power, the repetition rate increases from 8.96 kHz to 39.34 kHz while the pulse width reduces from 33.58 μs to 5.14 μs, a trend typical of a passively Q-switched pulsed laser [14-16]. This is caused by the shorter time for the inversion number of the gain medium to reach the threshold due to the increasing pump power.

Figure 8 shows the peak power and pulse energy as a function of the input pump power, both increases with the increasing pump power. By increasing from threshold to maximum input pump power, the peak power and pulse energy monotonically increases yielding a maximum of 1.01 mW and 5.53 nJ, respectively with an average output power of around 0.2 mW.

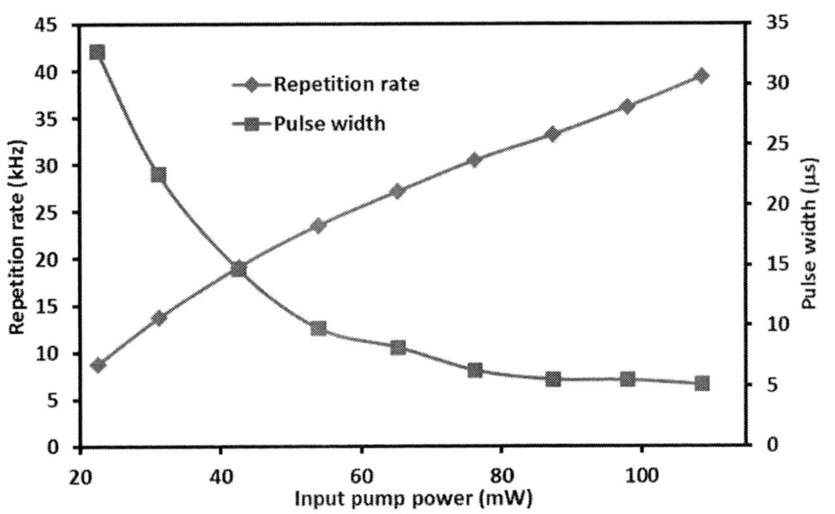

Figure 7. Repetition rate and pulse width as a function to input pump powers.

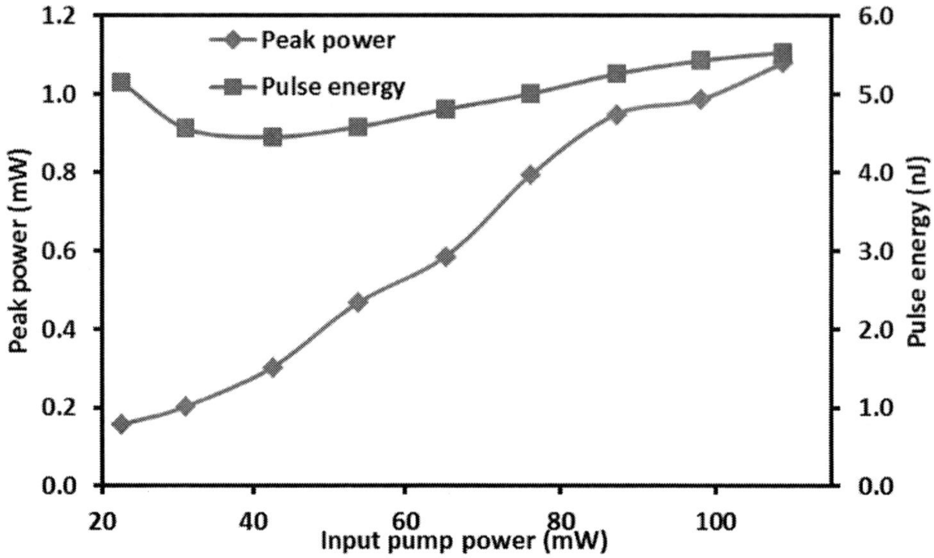

Figure 8. Instantaneous peak power and pulse energy as a function to input pump powers.

Radio frequency spectrum analyser (RFSA) is used to measure the signal-to-noise ratio (SNR) of the generated pulse. The value of the SNR is often an indicative of the pulse stability level. Figure 9 shows the recorded RFSA measurement at maximum input pump power of 108.8 mW. The first beat node at 39.34 kHz is around 40 dB and no peak occurs other than the harmonics of fundamental frequency, showing the stability of the pulse.

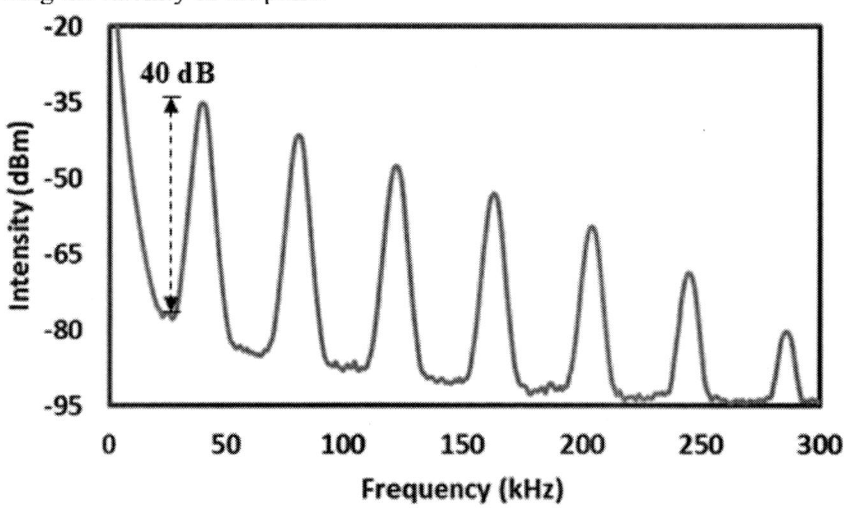

Figure 9. RFSA measurement of SNR at 108.8 mW.

6. Conclusion

To conclude, the proposed passively Q-switched EDFL employing the saturable absorber based on MWCNT-ABS printer filament was successfully demonstrated. The proposed work revealed lower threshold pump power as compared to previously reported works using SA based on MWCNT as starting material. The fabrication process is relatively simpler, involving three simple and timesaving steps; extruding the filament, dilution of the material using acetone and drop casting of the MWCNTs slurry.

IOP Conf. Series: Materials Science and Engineering **210** (2017) 012030 doi:10.1088/1757-899X/210/1/012030

References

[1] Haris H, Anyi C L, Muhammad A R, Ahmad F, Nor R M, Zulkepely N R, Ali N M, Harun S W and Arof H 2015 *Microwave and Optical Technology Letters* **57**1897
[2] Sobon G 2016 Application of 2D Materials to Ultrashort Laser Pulse Generation. *Two-dimensional Materials - Synthesis, Characterization and Potential Applications* **1**
[3] Han S A, Bhatia R and Kim S W 2015 *Nano Convergence* **2** 1
[4] Martinez A and Zhipei S 2013 *Nature Photonics* **7** 842
[5] Leigh S J, Bradley R J, Purshell C P, Billson D R and Hutchings D A 2012 *PLoS ONE* **7** e49365
[6] Schubert C, Mark C V L and Larry A D 2013 *British Journal of Ophthalmology* **98** 159
[7] Acquah S F A, Leonhardt B E, Nowotarski M S, Magi J M, Chambliss K A, Venzel T E S, Delekar S D and Al-Hariri L A 2016 *Carbon Nanotubes - Current Progress of their Polymer Composites*
[8] Kukovecz A, Gábor K and Zoltán K 2013 *Springer Handbook of Nanomaterials* 147
[9] Krassenstein B 2016 3Dxtech's Carbon Nanotube 3D Printer Filament is Here: Exclusive Images and Details *3DPrint.com*
[10] Quintana, Mildred, Jesús I T and Maurizio P 2014 *Beilstein Journal of Nanotechnology* **5** 2328
[11] Ciesielski, Artur and Paolo S 2014 *Chem. Soc. Rev.* **43** 381
[12] Tiu Z C, Ahmad F, Tan S J, Ahmad H and Harun S W 2014 *Journal of Modern Optics* **61** 1133
[13] Azooz S M, Ahmed M H M, Ahmad F, Hamida B A, Khan S, Ahmad H and Harun S W 2015 *Optik.* **126** 2950
[14] Liu L, Zheng Z, Zhao X, Sun S, Bian Y and Su Y 2013 *Optics Communications* **294** 267
[15] Lou F, Zhao R, He J, Jia Z, Su X and Wang Z 2015 *Photonics Research* **3** A25
[16] Liu J, Tian J, Guoyu H, Xu R, Li K and Song Y 2016 *Chinese Physics B* **25** 034207

IOP Conf. Series: Materials Science and Engineering **210** (2017) 012031 doi:10.1088/1757-899X/210/1/012031

Pupil Segmentation of Abnormal Eye using Image Enhancement in Spatial Domain

R. A. Ramlee[1, 2*], A. R. Ramli[1] and Z. M. Noh[2]
[1]Faculty of Engineering, Communications and Network Engineering
Universiti Putra Malaysia, 43400 UPM Serdang, Selangor Darul Ehsan, Malaysia.
[2]Faculty of Electronics and Computer Engineering (FKEKK),
Universiti Teknikal Malaysia Melaka (UTeM), Hang Tuah Jaya, 76100 Durian Tunggal, Melaka, Malaysia

Email: *ridza@utem.edu.my

Abstract. Segmentation is one of the important elements in image processing. There are various types of algorithms that have been developed by researchers, for segmenting the interesting area for classification and identification purposes. This paper, presents the pupil segmentation using logarithmic transformation (LT) and power law transformation (PLT). It is obviously seen in most of the work on pupil segmentation where assumption is made that the pupil has a homogeneous circular. However, there are cases where the shape of pupil is inhomogeneous, for instance, in synechia case. Therefore, the use of circumference equation such as circular Hough transforms (CHT) and Daugman's Integra-differential operator (DIDO) for section the non-uniform pupil will produce inaccurate segmentation. We propose a new method for pupil segmentation using the combination of LT and PLT algorithm in order to enhance the pupil segmentation. The morphological operators and black white (BW) removal are used the segmentation of pupil. The proposed system uses CASIA V1, CASIAV3 Interval and MMU1 iris database. The results show his method gives accurate segmentation compared to CHT or DIDO technique.

1. Introduction

In the iris recognition system, the segmentation process is an important step to get correct selection of pupil and iris region. Through accurate segmentation, the valuable information of iris patterns can be achieved and will increase the accuracy of the iris recognition system. However, there are cases where the pupil is seemed to be inhomogeneous circular, thus causing the segmentation to become difficult. The condition of non-uniform shape of the pupil is known as synechia. This situation also can occur due to some reasons, such as diseases and infection at pupil and iris regions. This paperwork on the abnormal pupil region segmentation which caused by eye diseases. In additional this proposed segmentation method can yield accurate pupil segmentation compare to segmentation using DIDO [1] or CHT [2].

Figure 1 shows the condition of pupils in various irregular shape. The pupil such in Figure 1, (a) and (b) are not in circular shape, where it is absolutely seen in (a) the pupil is slightly irregular oval while in (b) the pupil shape look like a star shape. These non-circular pupil condition will cause challenging to make correct segmentation, especially when using circular method such as circular CHT or DIDO.

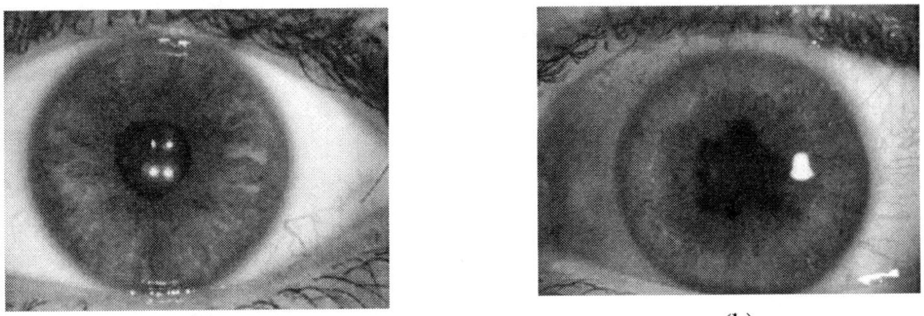

(a) (b)

Figure 1. Image of non-circular pupil, (a) pupil in oval shape, (b) pupil like star shape [3]

2. Literature Review

In this section various studies of pupil segmentation are studied. Amongst these studies, Abdullah *et al.* [4] proposed a segmentation using morphological operation and Chan-Vese active contour model to determine the pupil and iris boundary. The same authors [5] in another work propose the segmentation method using combination morphological operations and snake active contour. Their proposed system capable to detect the pupil boundary successfully.

The authors [6] proposed the angular integral projection function (AIPF) for determining the pupil boundary. For limbus boundary, they used AIPF applied in between two rectangles on both iris side. Romaguera *et al.* [7] proposed a method using histogram equalization, threshold and morphological operations. Next to detect pupil they used canny edge detection and Hough transform.

Santis *et al.* [8] indicated the segmentation of biomedical images is vital and can be used as non-invasive diagnosis. They studied on the pupil geometrical features such as diameter, area, centroid coordinates, and implemented these values in their algorithm.

In [9] the authors have similar opinion with [4]–[6] about non-circular shape of pupil. According to [9] the error in detecting the pupil boundary cause the quality of the identification system become low. Thus they proposed accurate pupil boundary using graph cuts. According to them the gray level values is used to compute the weights of the link in graph. The graph is divided into two areas pupil and the background. They have tested 756 images using their proposed system. All of these images were segmented successfully. However they did not stated exactly the accuracy percentage yield by their system. Another work related to pupil segmentation as reported by Bastos et al. [8]. They indicate most of the pupil segmentation algorithms assumed the pupil has a uniform circular. These assumption will cause inaccurate segmentation for the non-circular pupil. For this problem they propose a new method based on pull and push method and batch-SOM neural network to improve the pupil segmentation.

3. Methodology

The aim of this system is to design the segmentation of the pupillary. This system consists of the following stages such as: (a) Original image, (b) filtering, (c) Illumination removal, (d) Logarithm operator, (e) Power law transformation (exponential), (f) Morphological, (g) Binary removal and (h) segmentation pupil. The overall of our proposed system is illustrated in Figure 2 beginning with the original image and the whole process until segmentation of the pupil is achieved. For this system we had used the images from CASIA [10] and MMU [11] database.

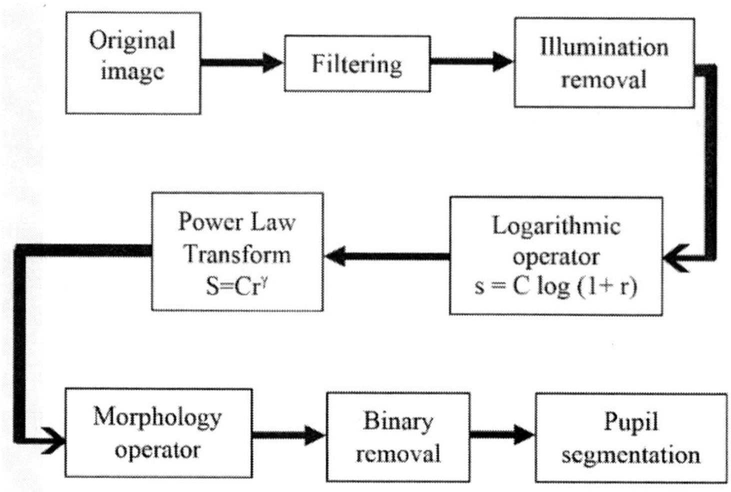

Figure 2. Flow chart of the overall steps for this pupil detection system.

3.1 Filtering

The purpose of filtering is to remove noise from the image. Therefore in this proposed system the image is filtered using the image processing filter function in the Matlab [12]. The function is 'filter2' will perform 2D finite impulse response (FIR) filter. Figure 3 shows the different between the original image in (a) and the output of the 2D filter FIR in (b). More detail of the images in Figure 3 are zoomed out by Figure 4. In Figure 3 (a), is the original image taken from (MMU 2005 database. This database contains 46 folders, where in each folder include two sub folder for left and right eye. The sub folder contains five samples of eyes with size of 320 x 240 pixels; the type of image is BMP file. In Figure 3 (b), is the image that has been filtered, in which there are significant differences in terms of texture images that are smoother than ever before. While the image of Figure 3 (c) is the result of removing illumination. Refraction of light contained in the original image was successfully removed. The aim is to avoid localization and segmentation of mistakes by detecting pixel dots of light as the pupil edges. Fig. 4 is to give an overview of refraction of light on the pupil area. Figure 4 (a) is an original pupil before pre-processing. While Figure 4 (a) is the pupil after the filtering process and lastly is refracted light on the circumstances of the pupil has been removed (Figure 4 (c)).

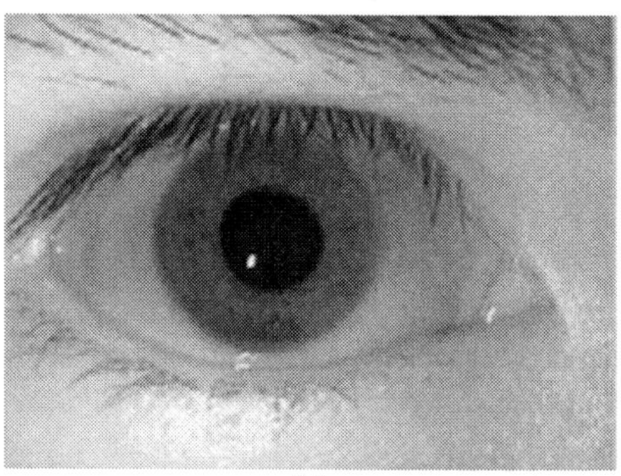

(a)

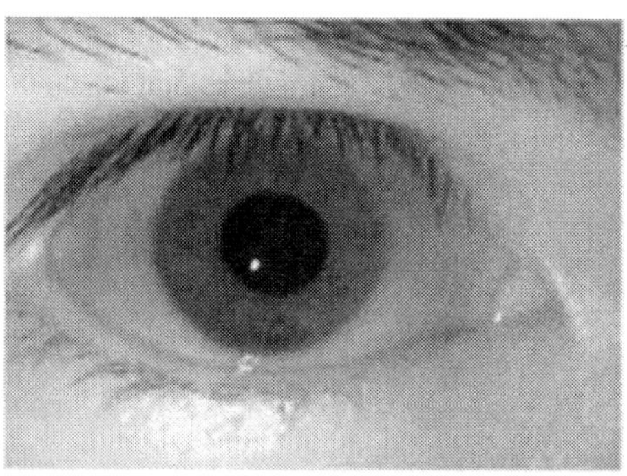

(b)

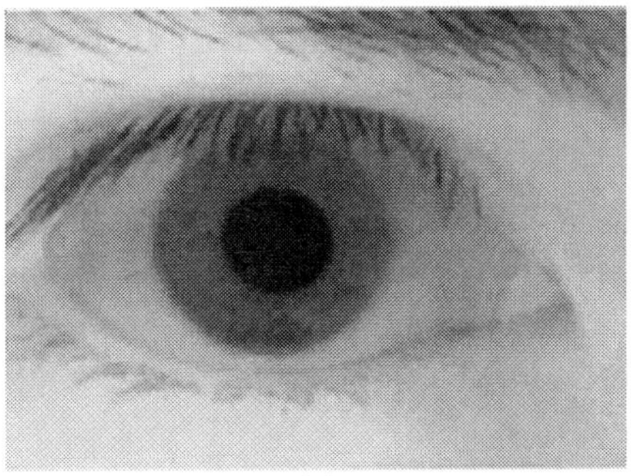

(c)

Figure 3. Image from MMU1[11] (a) original image (b) image filtering (c) image after illumination removing

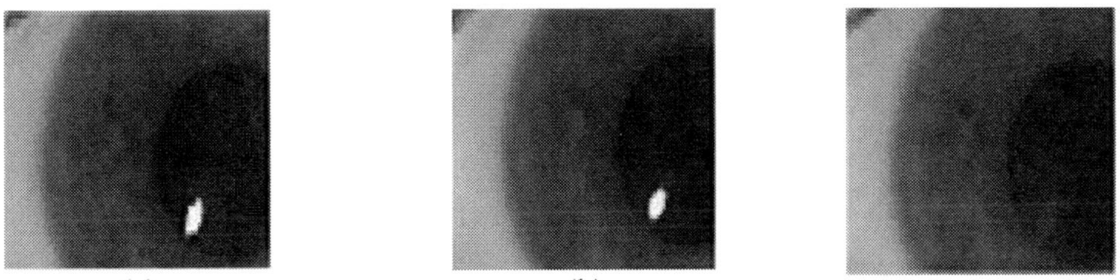

(a) (b) (c)

Figure 4. Image zooming from Figure 2 (a) original image (b) image filtering (c) image after illumination removing

3.2 Illumination removal

Illumination effect can reduce the quality of segmentation if not be eliminated in the early process of segmentation. Therefore, in this proposed system, the illumination effect is removed using the Matlab function, namely 'incomplement' in the binary image. Where, this function makes all the zero's value converted to ones, and all one's values will become zeros; white and black are reversed. As a matter of fact, the illumination consists of bright pixel, which assuming around 240 to 255.

3.3 Logarithmic transformation

Logarithmic transformation (LT) is used to reduce the contrast of the brighter regions. Elements such as the eyebrow, sclera and part of eyelid texture are removed. However the eyelash is normally still existed. The LT basically operates using logarithmic function as stated in (1), where r representing the image processed, s representing as the output and c is the constant value. The equation of LT is given as:

$$s = c\log(1 + r) \quad (1)$$

3.4 Power law Transformation

Power-law transformations (PLT) have the basic form as in equation (2), below:

$$s = cr^{\gamma} \quad (2)$$

Where, s is the output image given by manipulation of c and γ where both are positive constants. Basically PLT enhancing contrast of brighter regions of the image. By increasing the value of gamma the image

IOP Conf. Series: Materials Science and Engineering **210** (2017) 012031 doi:10.1088/1757-899X/210/1/012031

become darker and oppositely the lesser value of gamma the image become brighter, (as shown in Figure 5).

3.5 Morphology

Morphology was originally used for binary images and later prolonged to grayscale images. For this proposed system we use basic morphological operations such as dilation and erosion. Basically dilation is the process adding pixels to the images boundaries while the erosion act opposite to dilation by removing the pixels from the images boundaries. Thus we use erosion operator to remove the eyebrow from the black and white (BW) images.

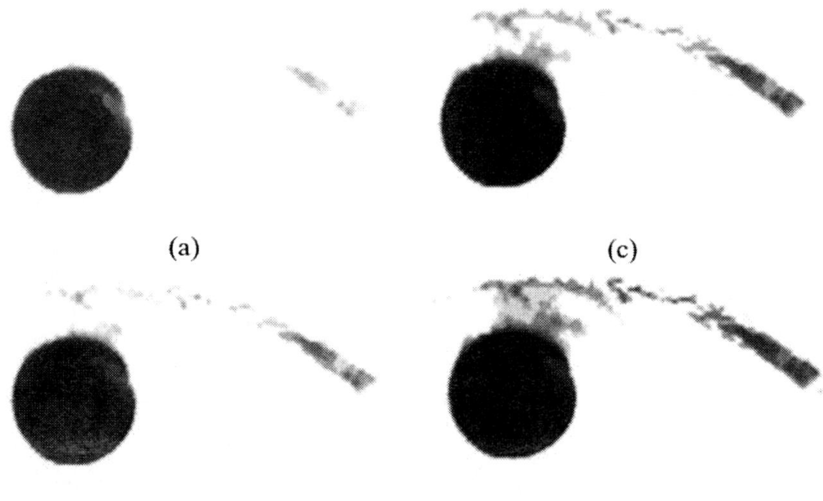

(a) (c)

(b) (d)

Figure 5. Image of pupil using PLT enhancement, starting from the top left clock-wise (a) γ=**2**, (b) γ=**3**, (c) γ=**4** and (d) γ=**5**

3.6 Removal object

The image black and white (BW) also known as binary image, where each pixel is represented by the value either zeros or ones. The value zero will represent black pixels whereas value one represent as white pixels. In this work we use the operation of BW to eliminate the small object or pixel according to number of pixel fixed in algorithm. For removing the unwanted pixel the 'bwareaopen' operation by Matlab is used. This operation will cause all connected components (objects) which values are less than setting pixels will be removed.

3.7 Segmentation.

The remaining BW pixel will be segmented as a mask. This mask later will be attached to original image to remove the non-interested area from the original image.

(a) (b)

Figure 6. Image output (a) Logarithm operator image, (b) power law transform (PLT) output

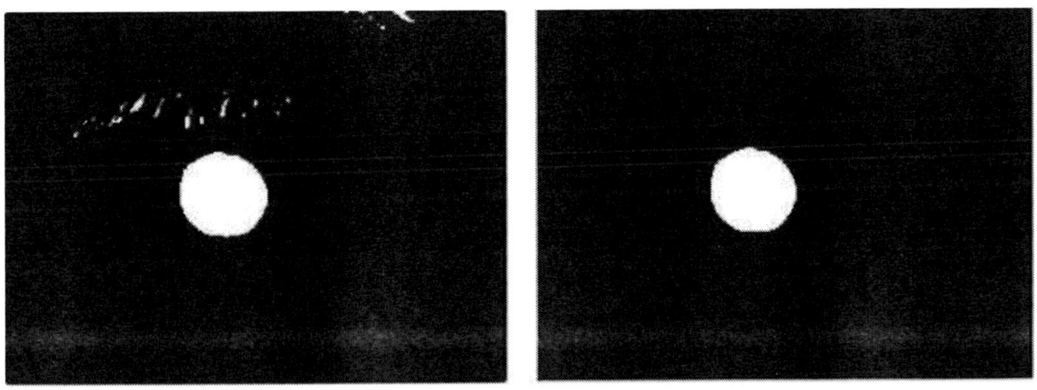

(a) (b)

Figure 7. Image output (a) black and white image (b) morphology output image

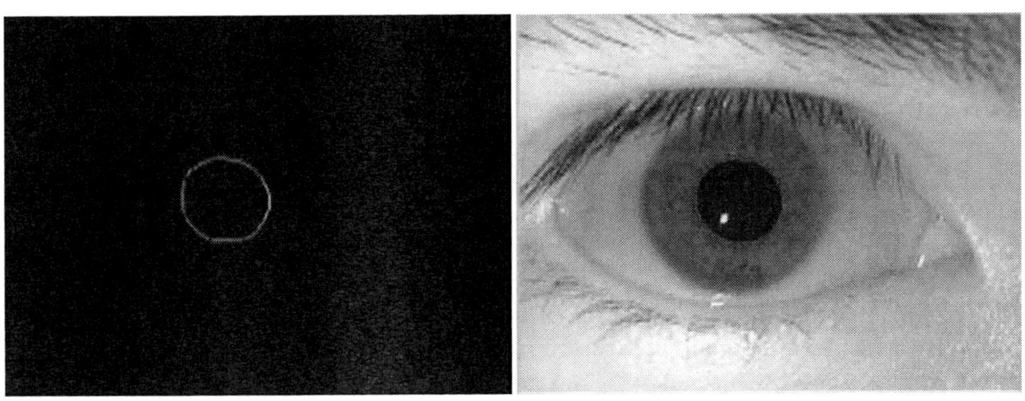

(a) (b)

Figure 8. Image output (a) outline morphology image (b) pupil segmentation

IOP Conf. Series: Materials Science and Engineering **210** (2017) 012031 doi:10.1088/1757-899X/210/1/012031

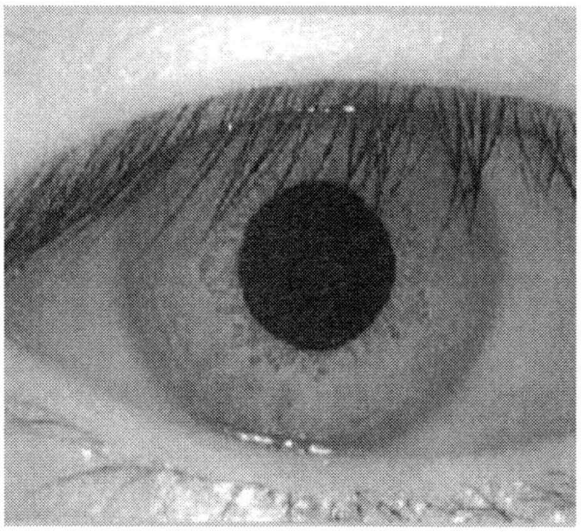

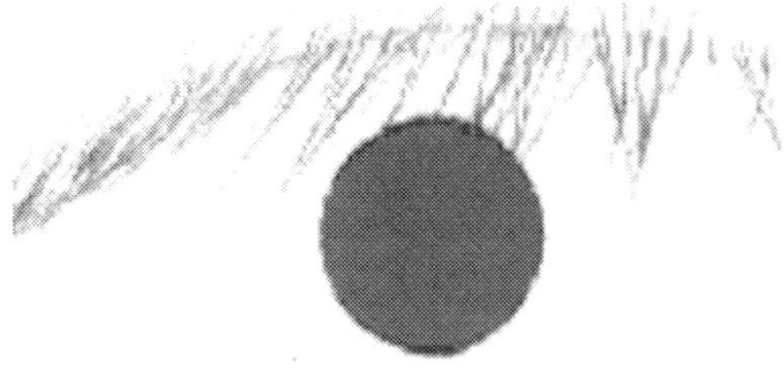

(a) (b)

Figure 9. Image (a) original image (b) Logarithm operator output

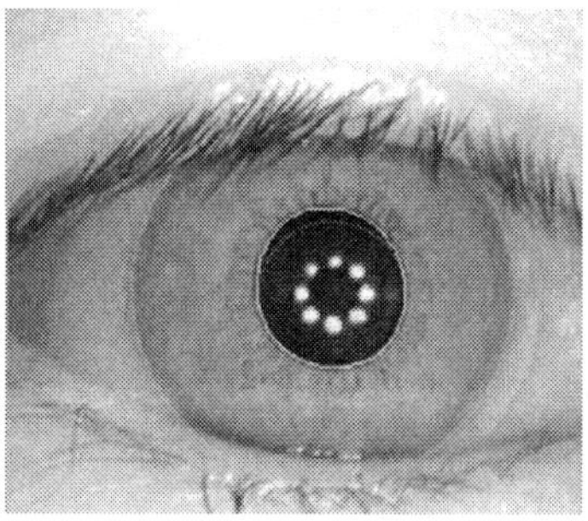

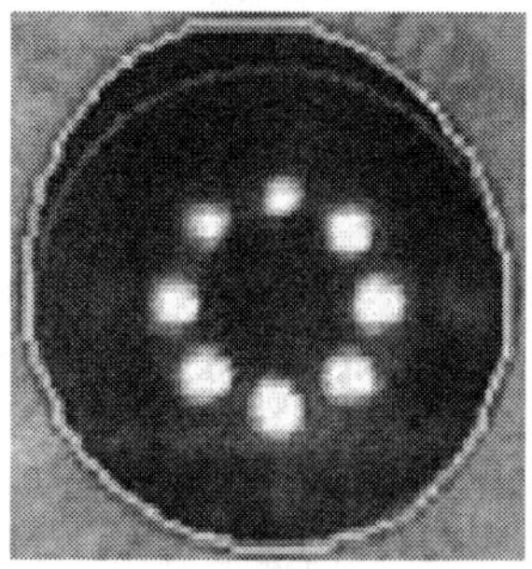

(a) (b)

Figure 10. Image (a) Comparison of segmentation method using Anirudh's algorithm [13] and our proposed algorithm (b) Zooming of segmentation pupil area; red outline using Anirudh's algorithm and white outline using proposed algorithm.

4. Result and Discussion

Experiments carried out in this project used CASIA and MMU iris image database and for synechia eye the images taken from online website iritis with synechia. In Figure 10, the segmentation result of the pupil is shown based on the proposed algorithm and the algorithm by [13]. The red outline is the segmentation result [13], while the white outline segmentation is the suggested algorithm. It demonstrated that our proposed algorithm able to segment fully, the region of the pupil. However, the algorithm by [13] failed to do so, because their method based on the circular segmentation.

In Figure 11, demonstrated some form of segmentation of abnormal pupil who has succeeded in segmenting correctly according to the shape of the pupil. Figure 11 (a) and (b), is a condition known as a synechia pupil. A synechia is a condition where the iris structure attached to the adjacent openings causes the pupil to be abnormal. Meanwhile, the Figure 11 (c) is the condition of the pupil which is a non-symmetrical shape.

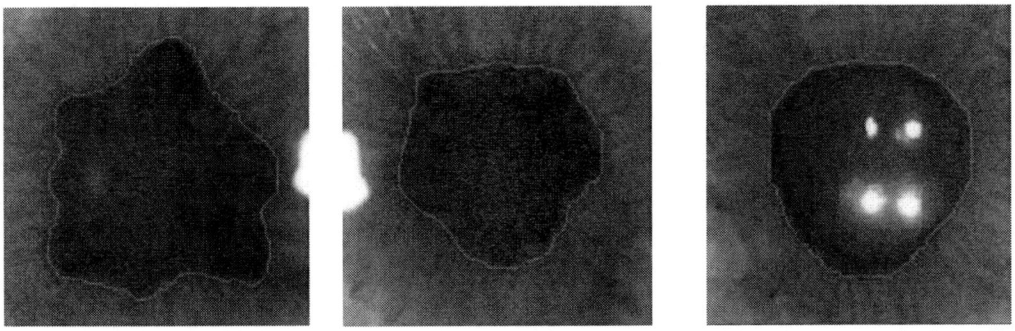

(a) star s synechia (b) synechia (c) non-symmetrical pupil

Figure 11. Images show several shape of non-circular pupil are segmented successfully using proposed segmentation system.

Figure 12 is the segmentation using CHT method and our proposed algorithm. It has obviously seen that using the CHT technique, the area covered by this technique fails to segment properly the region of the pupil as shown by figure (a) to (c). This failure occurs in two situations; whether segmentation exceeded the ROI or segmentation obtained is not the proper covering of the pupil area. Meantime the proposed algorithm successfully segmenting the pupil region correctly such as in the picture (d) to (e).

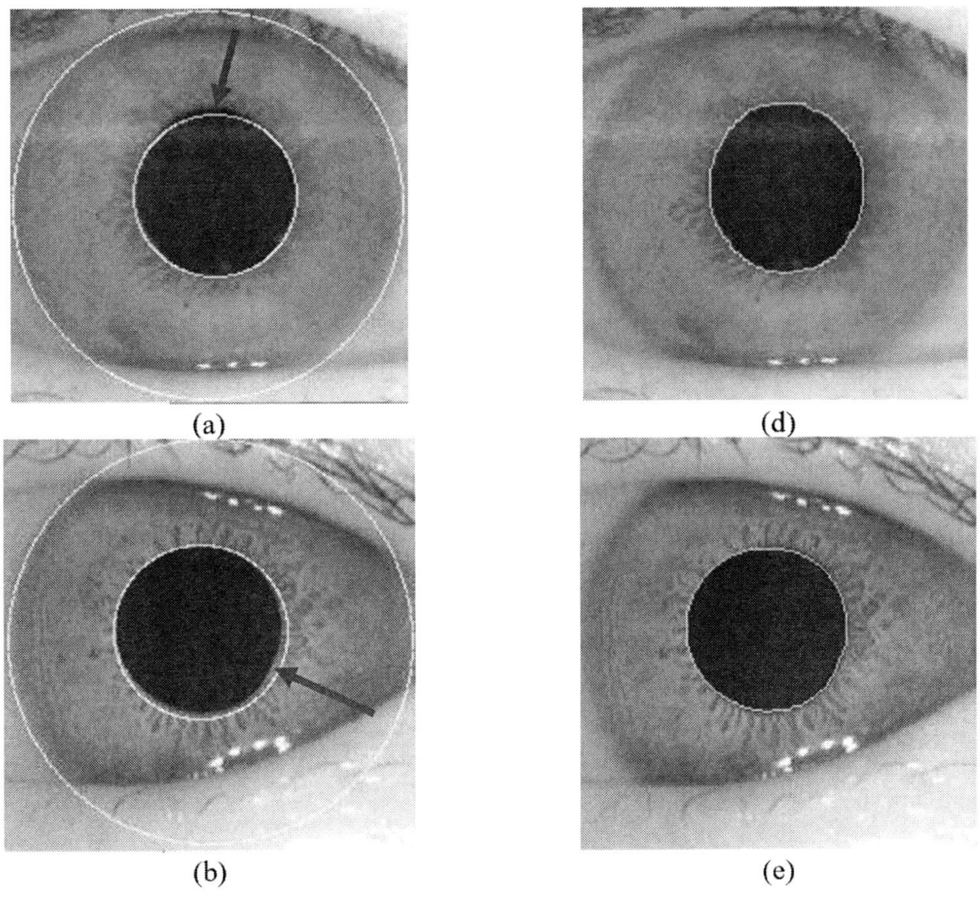

(a) (d)

(b) (e)

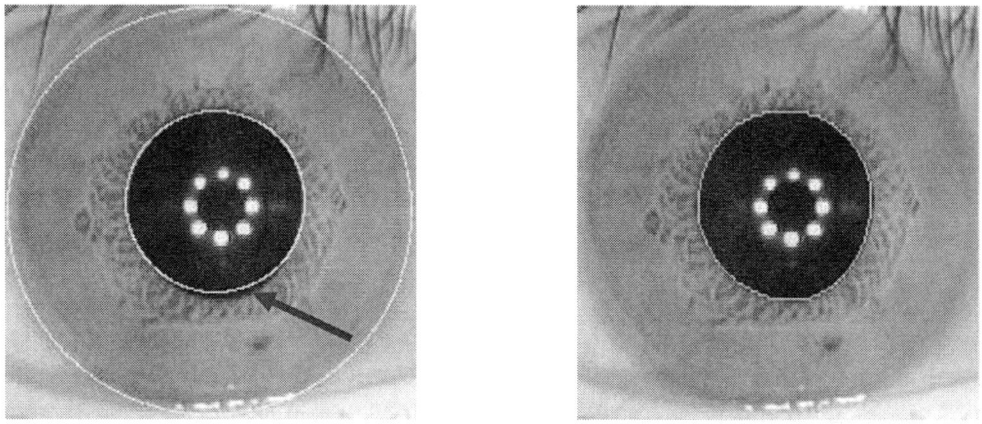

(c) (f)

Figure 12. The experiment results, figures (a), (b), and (c) using the CHT method for segmentation the pupil region, while figures (e), (f), and (g) using the proposed algorithm.

5. Conclusions

Segmentation is one of the primary operations in image processing. The accurate segmentation is important, especially in the iris recognition system in order to reduce mistakes of selection the ROI. In this paper, we propose an exact segmentation system for non-circular pupil (NCP). The NCP normally occurs in the abnormal eye, such as the synechia. To do segmentation process for synechia condition, we use the combination of LT and PLT to get the apparent area, of the pupil region. For synechia condition, the segmentation based on a circle equation is won't work accurately. There is also an eye condition that is not in the circumstances, such as synechia, but has a shape of pupil that is not symmetrical rounded. This will lead to the segmentation using the CHT or DIDO not managed to get the exact section. From the experiments conducted using the proposed algorithm, can produce a more accurate segmentation quality.

In future, we consider doing segmentation based on the other cases of eye diseases such the pterygium, since these diseases related to the part of the pupil, hence our method seen suitable for that matter.

Acknowledgment

The authors would like to thank Universiti Putra Malaysia (UPM) and Universiti Teknikal Malaysia melaka (UTeM) for the support and guidance. The authors would also like to acknowledge the Malaysia of Higher Education (MOHE) for funding this project.

References

[1] J. Daugman 2004 *IEEE Trans. Circuits Syst. Video Technol* **14** 21–30
[2] R. O. Duda and P. E. Hart 1972 *Commun. ACM* **15** 11–15
[3] M. Poe, 2014 [Online]. Available: http://www.ophthalmicphotography.info/website/slit lamp/iritis.html
[4] M. A. M. Abdullah, S. S. Dlay, and W. L. Woo," 2014
[5] M. Abdullah 2014 *Int. J. Inf. Electron. Eng.* **4**
[6] A. a. Jarjes, K. Wang, and G. J. Mohammed 2010 *2010 2nd Int. Asia Conf. Informatics Control. Autom. Robot.* **1** 349–352
[7] L. V. Romaguera, F. P. Romero, C. R. V. Seisdedos, and J. E. Neto 2015 *In VI Latin American Congress on Biomedical Engineering CLAIB 2014, Paraná, Argentina 29, 30 & 31 October 2014* **49** 429–432

[8] A. De Santis and D. Iacoviello 2006 *Comput. Methods Programs Biomed.* **84**, 174–187
[9] H. Mehrabian and P. Hashemi-Tari 2007 *Conf. Image*, no. December 77–82
[10] CASIA, 2003, http://www.sinobiometrics.com.
[11] MMU, 2005, http://pesona.mmu.edu.my/~ccteo/.
[12] MATLAB, 1994, http://www.mathworks.com/products/matlab/.
[13] Anirudh Sivaraman, 2007, http://www.mathworks.com/matlabcentral/fileexchange/15652-iris-segmentation-using-daugman-s-integrodifferential operator/content/Integrodifferential_operator.zip.

Bismuth (III) Telluride (Bi_2Te_3) Based Topological Insulator Embedded in PVA as Passive Saturable Absorber in Erbium-Doped Fiber Laser

N H M Apandi[1*], F Ahmad[1], S Ambran[1], M Yamada[1] and S W Harun[2]

[1]Dept. of Electronic Systems Engineering, Malaysia-Japan International Institute of Technology, Universiti Teknologi Malaysia, Jalan Sultan Yahya Petra, 54100, Kuala Lumpur, Malaysia
[2]Photonic Research Center, University of Malaya, 50603 Kuala Lumpur, Malaysia

Email: *nurhidayahapandi@gmail.com

Abstract. We demonstrate a passive Q-switched by integrating a Bismuth (III) Telluride (Bi_2Te_3) dispersed in Polyvinyl Alcohol (PVA) as passive saturable absorber. The experimental works show that the proposed passive saturable absorber operated at input power ranging from 21.69 mW to 126.89 mW with central operating wavelength of 1531 nm. We observe the tunable repetition rate from 40 kHz to 166 kHz with the shortest pulse width of 1.32 μs. The laser produced maximum instantaneous output peak power and pulse energy of 1.62 mW and 11.2 nJ, respectively. The signal to noise ratio was measured at 49 dB which indicates the stability of the generated pulse.

1. Introduction

In the past years, the study in using real saturable absorbers have become crucial for researchers to generate pulsed laser. Fiber lasers has been regarded as the most promising laser configuration as it is proven to be superior than other type of lasers such as dye, gas, chemical, solid state and semiconductor lasers. The first laser was successfully demonstrated by Maiman in 1960. A few years later, the Q-switched laser was discovered by using colored glass filter and dyes onto ruby laser configuration [1]. Since then, the materials used has evolved from semiconductor saturable absorber mirror (SESAM) to nanomaterials such as quantum dots, carbon nanotubes, graphene, transition metal dichalcogenides (TMDs), black phosphorous (BP) and topological insulators (TIs) [2]. Topological insulators in bulk state have an insulating energy gaps whether in two or three dimensions but the sample boundary were protected by time-reversal symmetry at the surface state or gapless edge [3]. These TIs are robust and have a simple surface states consisting of a single Dirac cone and these stoichiometric crystals are Bismuth (III) Telluride (Bi_2Te_3), Bismuth (III) Selenide (Bi_2Se_3) and Antimony (III) Telluride (Sb_2Te_3). Basically, Bi_2Te_3 share the same rhombohedral crystals structure with five atoms in one unit cell thus the material is known as quintuple layers along the z-axis [3]. Each of the quintuple layer consists of five atoms with two equivalent Te atoms. The atoms was covalently bonded with stacks of Te-Bi-Te-Bi-Te within each layer where the layers (quintuple layer) were held together by van der Waals forces as shown in Figure 1 [4].

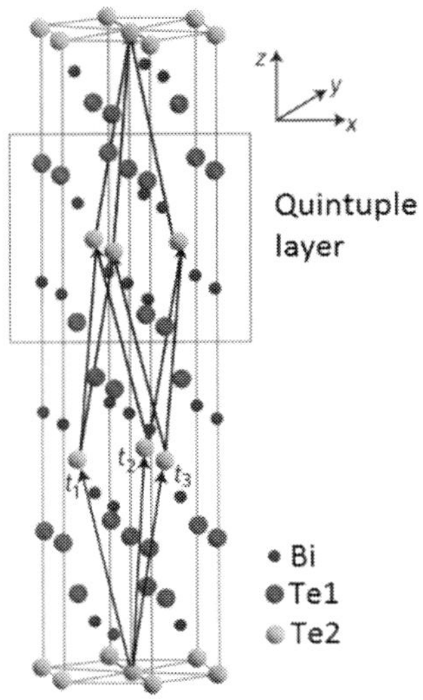

Figure 1. Crystal structure of Bi_2Te_3 with three primitive lattice vectors denoted as $t_{1,2,3}$ and a quintuple layer as indicated by the red square [3]

Experimentally, the first demonstration of passive saturable absorber (SA) of Bi_2Te_3 for ultrafast fiber laser has been done by drop casting techique on a quartz plate [5]. Since that, the integration of the SA is approached either by evanescent wave interaction [6-9], sandwiching [7-8] or free space coupling [10] with a wider operating wavelength from 1 μm to 3 μm [6-10]. The TI based passive SA application has been reported by Chen *et al.* [7] and Wu et *al.* [8] by using Bi_2Te_3. Chen *et al.* [7] synthesized a layer of Bi_2Te_3 nano-sheets by liquid phase exfoliation (LPE) and then dissolved in isopropyl alcohol. The Bi_2Te_3 based SA was attached at the fiber end-facet after optical deposition method with central wavelength of 975 nm and maximum power of 500 mW. The central wavelength tunable from 1510.9 nm to 1589.1nm, with a large energy of 1.5 μJ, range of repetition rate from 2.2 kHz to 12.8 kHz, and the shortest pulse width 13 μs is reported. The signal-to-noise ratio is 36.4dB. Wu *et al.* [8] also synthesized the TI nanomaterial by using liquid exfoliation approach before it is dispersed in isopropyl alcohol. The Bi_2Te_3 nano-sheets were then deposited onto the fiber end-facet with a pump power of 80 mW and central wavelength of 975 nm. The generated pulse recorded maximum pulse energy of 7.5 nJ, range of repetition rate from 12.6 kHz to 177.7 kHz, shortest pulse width of 0.2 μs, and signal-to-noise ratio of 47 dB. The use of optical deposition method is not efficient for scalable production of SA, the process is time consuming and the thickness is uncontrollable. On the contrary, this work proposes a simpler approach in developing Bi_2Te_3-PVA film as passive SA at 1.5 μm. The generated Q-switched pulse operated at a central wavelength of 1531 nm, tunable repetition rate from 40 kHz to166 kHz with the shortest pulse width of 1.32 μs. The laser produces maximum instantaneous output peak power and pulse energy of 1.62 mW and 11.2 nJ, respectively.

2. Methodology

Bismuth (III) Telluride (Bi_2Te_3) powder (Sigma Aldrich) characterized with -325 mesh and 99.99% trace metals basis with molecular weight of 800.76g/mol was used per received. To prepare the host polymer, 1 g of PVA (Sigma Aldrich) was dissolved in 120 ml de-ionized (DI) water with the aid of a magnetic stirrer at room temperature. TI based passive SA film was prepared by mixing 14 mg of Bi_2Te_3 with 3 ml of PVA suspension and thoroughly mixed with the aid of magnetic stirrer for three hours. Then the Bi_2Te_3-PVA suspension was placed in ultrasonic bath for one hour to make sure that the Bi_2Te_3 powder was fully binded with the binder (PVA). After that, the suspension was carefully poured onto a petri dish to avoid any air bubble and was left at room temperature for 48 hours to develop Bi_2Te_3-PVA composite film. The developed film was characterized using FESEM as shown in Figure 2 and the figure clearly shows that the Bi_2Te_3 powder was thoroughly mixed with PVA host polymer. The transmission characteristics of the developed film were further measured using a UV-VIS-NIR spectrophotometer (Perkin Elmer, Lambda 750) with pure PVA film as reference. Figure 3 shows the transmission spectrum of the fabricated Bi_2Te_3-PVA film and it shows the transmittance is at 21% at 1.5 μm region. The film can also be investigated at 1 μm and 2 μm region showing the broadband properties of the Bi_2Te_3 based passive SA as the structure of the TI is not affected by host polymer [11]. Figure 4 shows the Raman spectrum of the developed Bi_2Te_3-PVA film with low peak at ~62 cm-1(A^1_{1g}), ~95(E^2_g) and 11(A^2_{1g}), respectively. The low peak is maybe due to the low concentration of the Bismuth (III) Telluride powder in PVA and the effect of the film thickness [13].The simple approach in this works to mix the Bi_2Te_3 powder with PVA in room temperature for scalable production of Bi_2Te_3 based SA is very attractive.

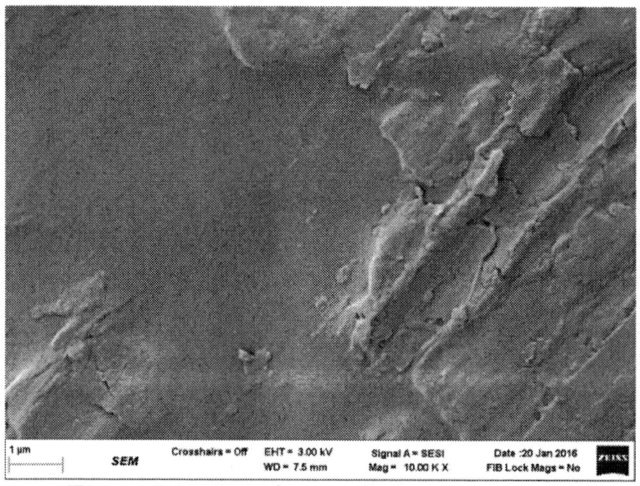

Figure 2. FESEM image of Bi_2Te_3-PVA film

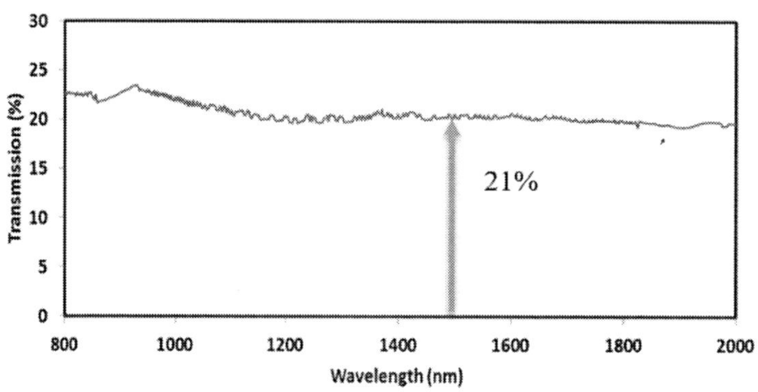

Figure 3. Transmission spectrum of Bi_2Te_3-PVA film

IOP Conf. Series: Materials Science and Engineering **210** (2017) 012032 doi:10.1088/1757-899X/210/1/012032

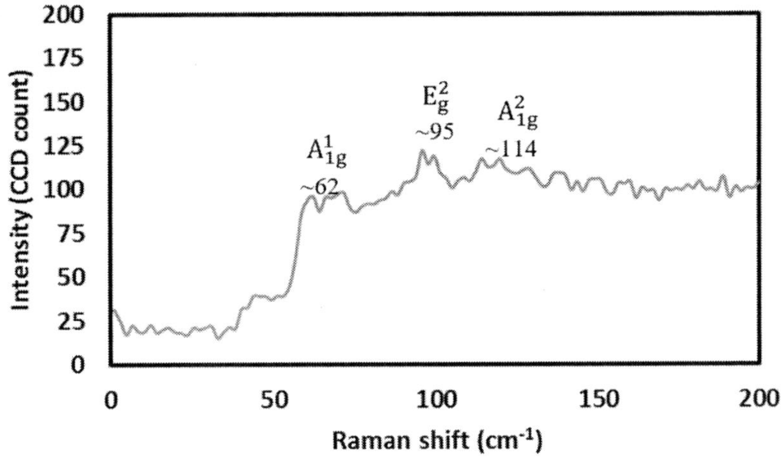

Figure 4. Raman spectrum of free standing Bi_2Te_3-PVA film

3. Experimental setup

In order to investigate the functionality of the prepared of Bi2Te3-PVA as passive saturable absorber, a small portion of the developed film was inserted into the fiber laser cavity. Figure 5 shows the schematic of the Q-switched fiber laser with the bismuth telluride based SA. The pump source is a laser diode (LD) with emission centered at 980 nm. A piece of 3 m long Erbium doped fiber (EDF) was used as the laser gain medium with absorption coefficient of 23 dB/m at 1550 nm. The pump was delivered into EDF via a 980/1550 fused wavelength division multiplexer (WDM). A polarization dependent isolator, placed after the EDF was used to ensure unidirectional operation and eliminate undesired feedback from the output end facet. A fused fiber optical coupler (OC) was used to extract 5% of the propagating light in the laser cavity, and then characterized as output. A saturable absorber was inserted by sandwiching the Bi_2Se_3-PVA sample between two pigtails, and then placed within ring cavity. A 3 dB coupler was placed after the 95/5 OC to investigate the performance of the ring cavity by using an optical spectrum analyser (OSA) which is employed for the spectral analysis of the pulse in Erbium doped fiber laser, whereas the oscilloscope (OSC) is used to observe the output pulse train of the Q-switched operation via photo-detector. Radio frequency spectrum analyzer (RFSA) was used to measure the signal to noise ratio (SNR), while optical power meter was used to measure the average output power.

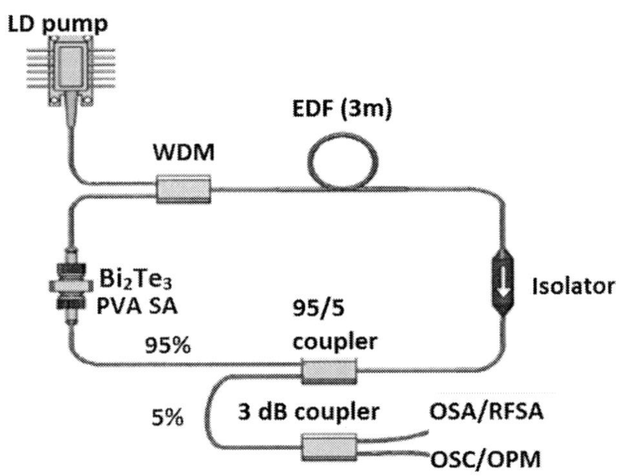

Figure 5. Experimental setup of the proposed passively Q-switched EDFL by residing Bi_2Te_3-PVA based passive saturable absorber

IOP Conf. Series: Materials Science and Engineering **210** (2017) 012032 doi:10.1088/1757-899X/210/1/012032

4. Results and discussion

A stable Q-switched operation was obtained once the pump power exceeded a threshold of 21.69 mW. Figure 6 shows a typical optical spectrum with a central wavelength of 1531 nm and consist of parasitic continuous wave at 1530.64 nm. It has a 3 dB spectral bandwidth of 0.03 nm. The parasitic continuous wave laser is observed due to the reflection light interference from two fiber connectors' end [12]. The stable Q-switched pulse train has a repetition rate of 166 kHz higher than reported work [7,14], corresponding to a time interval between adjacent pulses of 6 μs with maximum input power of 126 mW as shown in Figure 7. Figure 8 shows a zoom-in single pulse profile. The pulse has a symmetric intensity profile with a full width at half maximum (FWHM) of 1.32 μs which is better than previous works [7,14]. Unlike the mode-locking operation where the repetition rate is determined by the cavity length, Q-switching laser has a tunable repetition rate that is dependent on the cavity gain, loss and cavity birefringeance [7]. With the increasing pump power, more electrons can be excited and accumulated in the upper energy level in the laser medium. The rise and fall time of the pulse become simultaneously shorter, leading to the decrease of pulse width and the increase of repetition rate. To further analyze the evolution of the pulse circulating the laser cavity with the changing pump power, the relation of the repetition rate and pulse width as a function to pump power is illustrated in Figure 9. The graph shows that the repetition rate is proportional to the input power where the repetition rate increases from 12.6 kHz to 177.7 kHz with increasing pump power. As observed, the pulse width is inversely proportional to the input power where the pulse width is decreased from 9.33 μs to 1.32 μs. From the measured pulse repetition rate, pulse width and average output power, we calculated the pulse energy and peak power, and the data is tabulated in Figure 10. As shown, the calculated pulse energy decreased proportionally with the input pump power from 11.2 nJ to 0.3 nJ. These values are considered higher as compared to previous reported work [8]. The calculated peak power recorded a different trend with increment from 1.13 mW to the highest of 1.62 mW slightly better than [8] and then started to decrease when the input pump power was tuned from threshold pump power to the maximum pump power. The fluctuation of the pulse energy and peak power is also reported in [12]. One way to reduce the fluctuation is by using higher ratio of output coupler as in their case from 90/10 output coupler to 50/50 output coupler [12].

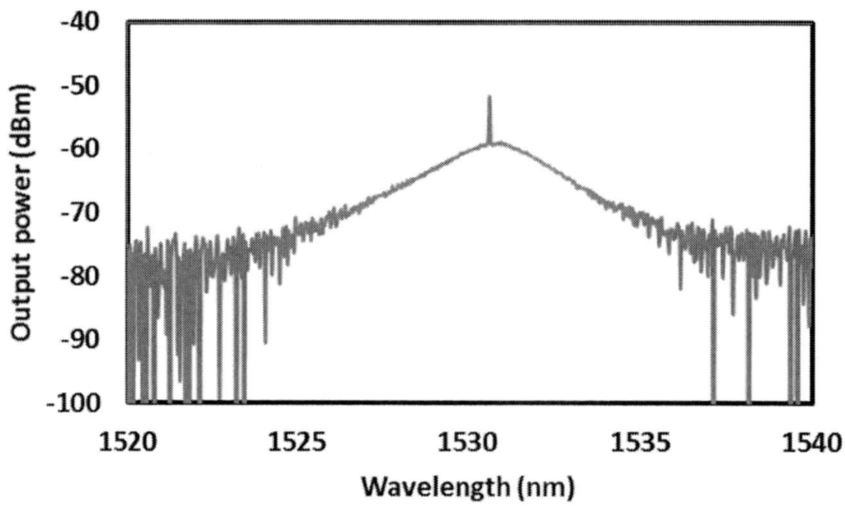

Figure 6. Central wavelength of 1531 nm

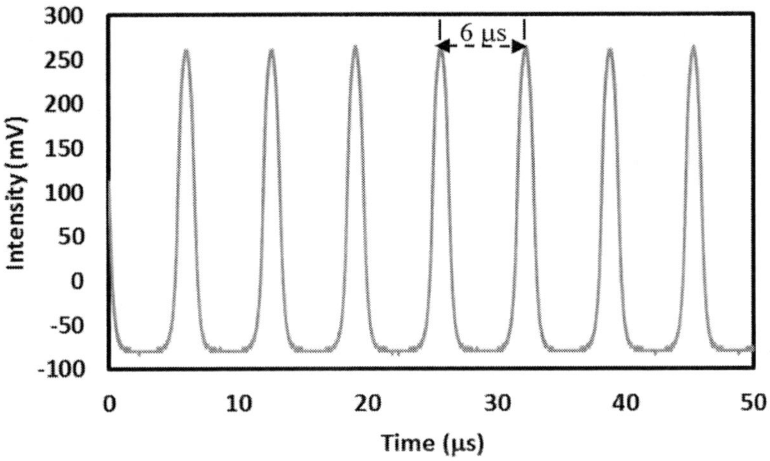

Figure 7. Pulse train with repetition rate of 166 kHz at 126 mW

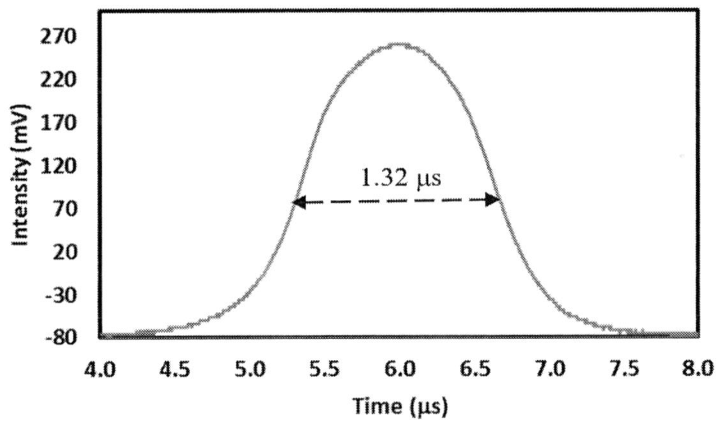

Figure 8. Single pulse envelope with 1.32 μs pulse width at 126 mW

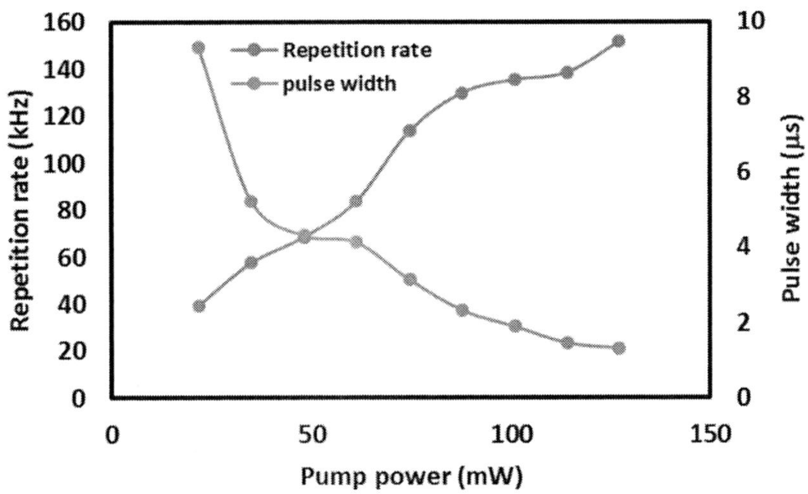

Figure 9. Repetition rate and pulse width versus pump power

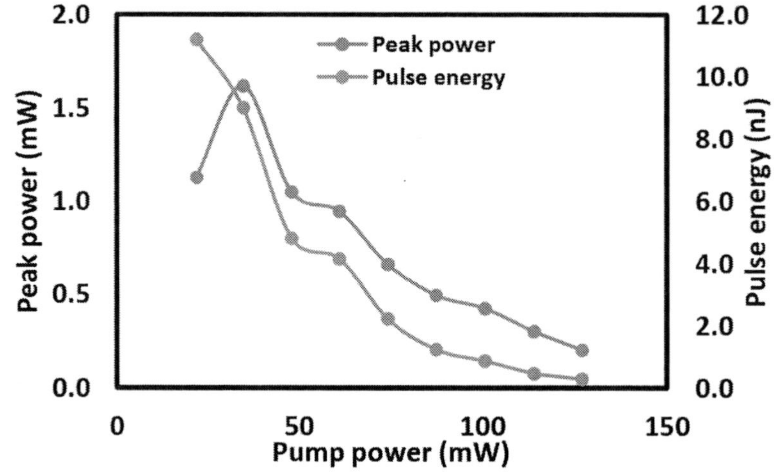

Figure 10. Peak power and pulse energy versus pump power

To investigate the laser stability, we measure its corresponding radio frequency (RF) spectrum. As can be seen in Figure 11, the signal-to-noise ratio (SNR) of our fiber laser is over 49 dB, indicating that the Q-switched pulse operates in a relatively stable regime. Moreover, the SNR recorded in this work is slightly better than the reported Q-switched in previous work [7-8,14] by using the Bismuth (III) Telluride based passive saturable absorber. Table 1 shows some of the previous works that also uses Bi_2Te_3-based passive SA operating in 1.5 μm region.

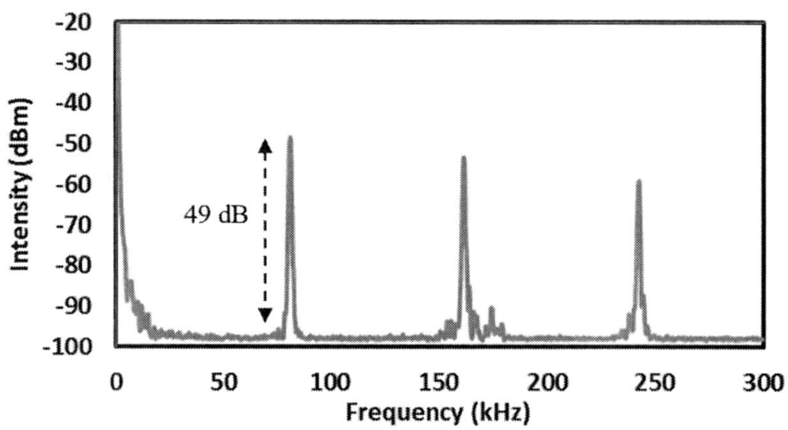

Figure 11. First beat note of 49 dB with frequency of 81.3 kHz

Table 1. Er-doped Q-switched fiber lasers employing Bi_2Te_3 based passive SA.

Repetition rate (kHz)	Pulse width (μs)	Peak power (mW)	Pulse energy (nJ)	SNR (dB)	Refs
2.2-12.8	13	20	1500	36.4	[7]
12.6-177.7	0.2	1.3	7.5	47	[8]
2.6-12	9.5	120	4	40	[14]
40-166	1.32	1.62	11.2	49	This work

5. Conclusion

A simpler approach in the fabrication process of Bi_2Te_3-PVA as passive Q-switcher has been demonstrated in this work. The Bi_2Te_3 was dispersed in PVA by solution casting approach and then dried at ambient temperature to develop a Bi_2Te_3-PVA film. Promisingly, the proposed Q-switcher managed to produce a tunable repetition rate from 40 kHz to 166 kHz with the shortest pulse width of 1.32 μs. Maximum instantaneous output peak power and pulse energy of 1.62 mW and 11.2 nJ respectively, have been recorded throughout the experiment.

References

[1] Soffer B 1964 *Journal of Applied Physics* **35** 2551
[2] Woodward R and Kelleher E 2015 *Applied Sciences* **5** 1440-1456
[3] Zhang H, Liu C, Qi X, Dai X, Fang Z and Zhang S 2009 *Nature Physics* **5** 438-442
[4] Ren L, Qi X, Liu Y, Hao G, Huang Z, Zou X et al. 2012 *Journal of Materials Chemistry* **22** 4921
[5] Zhao C, Lu S, Qi X, Zhang H, Wen S and Tang D 2013 *Applied Physics Letters* **103** 106101
[6] Lee J, Koo J, Chi C and Lee J 2014 *Journal of Optics* **16** 085203
[7] Chen Y, Zhao C, Chen S, Du J, Tang P, Jiang G et al. 2014 *IEEE Journal of Selected Topics in Quantum Electronics* **20** 315-322
[8] Wu M, Chen Y, Zhang H et al. 2014 *IEEE Journal of Quantum Electronics* **50** 393–396
[9] Lee J, Jung M, Koo J et al. 2015 *IEEE Journal of Selected Topics in Quantum Electronics* **21** 31–36
[10] Li J, Luo H, Wang L et al. 2015 *Optics Letters* **40** 3659–3662
[11] Liu H, Zheng X, Liu M, Zhao N, Luo A, Luo Z et al. 2014 *Optics Express* **22** 6868
[12] Xin H, Hang Z, Wei L, Rongfei W, Jianrong Q, Mei Z and Bin H 2015 *Scientific Reports* *5* 15868
[13] Russo V, Bailini A, Zamboni M, Passoni M, Conti C, Casari C S, Li Bassi A and Bottani C E 2008 *Journal of Raman Spectroscopy* **14** 205–210
[14] Chen, S. Q. et al. 2014 Stable *IEEE Photonic. Tech. L.* **26** 987–990

IOP Conf. Series: Materials Science and Engineering **210** (2017) 012033 doi:10.1088/1757-899X/210/1/012033

Performance Configuration of Raman-EDFA Hybrid Optical Amplifier for WDM Applications

N Saidin*, N I A Taib, M S Z Abidin, N F Hasbullah, and A A M Ralib

Department of Electrical and Computer Engineering, International Islamic University Malaysia, Jalan Gombak, 53100 Kuala Lumpur, Malaysia.

Email: **norazlina@iium.edu.my*

Abstract. A hybrid configuration of Raman amplifier and erbium-doped fiber amplifier (EDFA) is proposed to obtain a better performance in term of gain, noise figure and flat gain. It is based on the optimum parameter configuration of a singly-based Raman amplifier and EDFA. The best parameter for both amplification has been analyze in terms of its input signal power, pump power and their fiber length whereas the best erbium ion density has also been analyze in EDFA setup. All the parameters are varied to some values to get the optimum result. The simulation is done by using Optisystem 14.0 software. The hybrid amplifier consists of Raman amplifier with multi-pump power set up and bidirectional pump power of EDFA with the pump wavelength of 980 nm is designed and simulated in order to obtain higher gain and lower noise figure. From the simulation of the hybrid configuration, the optimum output has been achieved. The hybrid configurations exhibit the average gain of 46 dB and average noise figure of 3 dB. The flat gain obtained is between 1530 nm to 1600 nm which include C-Band and L-Band frequency with the gain bandwidth of 70 nm.

1. Introduction

Due to the increasing demand of high transmission capacity and best performance system of optical fiber communication system, hybrid amplifiers become the best option to solve the problem. Asingly based optical amplifier such as semiconductor optical amplifier (SOA), erbium-doped fiber amplifier (EDFA) or Raman amplifier has few limitations to provide high gain amplification as well as broadband operation. These amplifier exhibit its own advantages, for instance SOA uses less electrical power, more compact and fewer components [1], whereas erbium-doped fiber amplifiers (EDFAs) can be develop by using various host and codopant materials, such as silica, alumina, telluride, phosphate, bismuth, and others, to improve the gain performance. Raman amplifier exhibit an ability to provide distributed amplification within the transmission fibre, thereby increasing the length of spans between amplifier and regeneration sites. In contrast with EDFA, the SOA have some drawbacks which makes EDFA has better performance than SOA. SOA has nonlinearities in gain, higher coupling losses, high noise figure, and generation of crosstalk in multichannel systems [2]. EDFA are able to pump the device at different wavelengths, low gain dependance on light polarization and low coupling loss. However, from the previous work reported, most of singly-based amplifier exhibits lower gain compared to hybrid configuration. Combination of Raman amplifier and EDFA performs the best of hybrid amplifier reported previously.

It is necessary to optimize the design parameters such as length of amplifiers, pump power wavelength and input pump power of both amplifiers separately before both are combined. These parameters optimization can be done through simulation using simulation software. In our work, Optisystem software is used to obtain the best performance of optical fiber amplifier for WDM application. As reported by Singh et. al [3], a hybrid two stage Raman-EDFA amplifier can give a flat gain of L-Band. The increasing of input power increase the gain variation over the bandwidth. Flat

IOP Conf. Series: Materials Science and Engineering **210** (2017) 012033 doi:10.1088/1757-899X/210/1/012033

gain of greater than 10 dB is achieved using the input power of 3 mW for the frequency between 187 to 190.975 THz. Without using gain-flattening technique, the gain variation obtained is less than 4.5 dB. Based on the study by Martini et. al [4], the use of recycled multipump lasers for WDM systems in two configurations which are Raman-EDFA and EDFA-Raman is reported. The EDFA-Raman configuration gives the best performances with the global gain of 40 dB, noise figure of 4 dB and gain ripple of 0.8 dB.

Sharma et. al [5] proposed the configuration of hybrid amplifier using 20x50 Gbps WDM system with the range from 1560 nm to 1577 nm wavelength. The input power use is 15 dBm and bit rate of 10 Gbps. The result shows that gain and noise figure variation will increase as the input power, number of pump and pump wavelength increases. The configuration successfully obtained a maximum gain of 22.81 dB and low noise figure of 7 dB. The gain variation is 2.79 dB. Three stage hybrid amplifier has been proposed in a work reported by Kaushik et. al [6] with multiple pump recycling. Using this configuration, the gain bandwidth of 80 nm and gain ripple of 2.5 dB is obtained, while the maximum gain reported is 25 dB. The advantage of this technique is to reduce the cost of equipment in the communication network.

In our proposed design, the EDFA-Raman configuration has been used to perform hybrid amplifier.

2. Design and Simulation

Optisystem software has been used in order to design and simulate the configuration of EDFA, Raman amplifier and hybrid EDFA-Raman amplifier. The simulation started with the configuration and optimization of EDFA and followed by Raman amplifier. Then, using the optimum parameter of both amplifiers, a hybrid optimum Raman-EDFA is configured and simulated.

2.1. Configuration of Erbium Doped Fiber Amplifiers (EDFA)

Figure 1 shows the proposed configuration of EDFA. The parameter involves to be tested in this configuration are pump wavelength, erbium ion density, EDFA length, pump power and input signal power. These parameters will be varied in order to check for the best amplifier performance for the design configuration. The frequency for the WDM channel is from 1530 nm to 1620 nm with 19 channels of 5 nm spacing. The frequency is the range for C-Band and L-Band. The pump wavelength of the EDFA can be 980 nm or 1480 nm with three different configurations which are forward, backward or bidirectional [1,7]. Firstly, the simulation is tested for forward and bidirectional pumping with 980 nm and 1480 nm and the other parameters are set as shown in figure 1. Secondly, for the optimum erbium ion density, the density is varied between 100 ppm-wt to 2500 ppm-wt while the pump configuration is set to optimum which is 980 nm bidirectional pumping. In optimizing the EDFA length, the length is varied between 1 m to 40 m with the previous optimized parameter is set to the configuration. Then, for input signal power optimization, the power is varied between -200 dBm to 0 dBm to obtain the best input signal power.

IOP Conf. Series: Materials Science and Engineering **210** (2017) 012033 doi:10.1088/1757-899X/210/1/012033

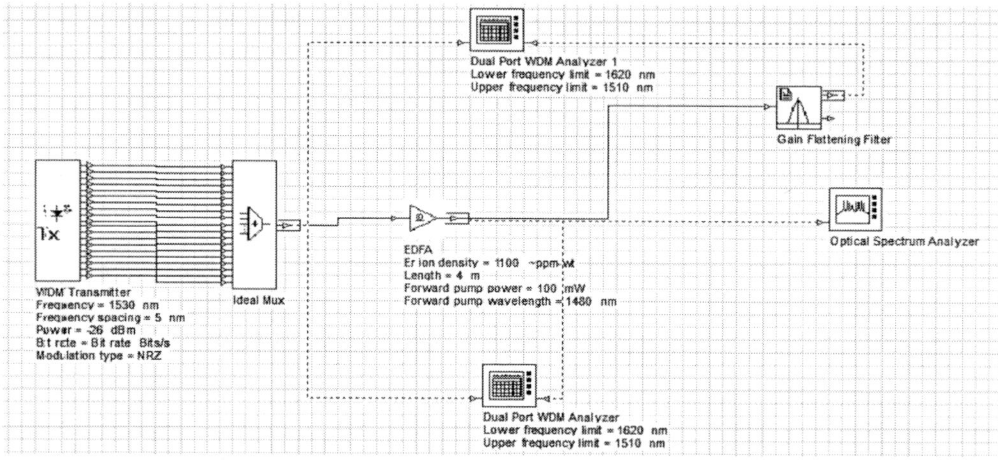

Figure 1. Configuration of EDFA.

2.2. Configuration of Raman Amplifier

Figure 2 shows the proposed configuration of Raman amplifier. The WDM transmitter uses the frequency range between 1520 nm to 1620 nm with 10 nm spacing. The input signal power of -40 dBm is set to constant throughout this simulation. In this simulation, the optimum Raman length and pump power have been observed. There are 11 pump laser arrays of various frequencies ranging from 1420 nm to 1520 nm with increment of 10 nm. The difference between pump wavelength and signal wavelength should be 100 nm to obtain high Raman gain coefficient [8]. The optimization of Raman length is done by varying the length from 10 km to 100 km. Besides, in optimizing the pump power, the amplifier length is set to the optimum length obtained previously which is 45 km with variation of pump power from 50 mW to 250 mW. As the optimum length and pump power has been achieved, the value has been used in the configuration and the result obtained is shown in figure 5.

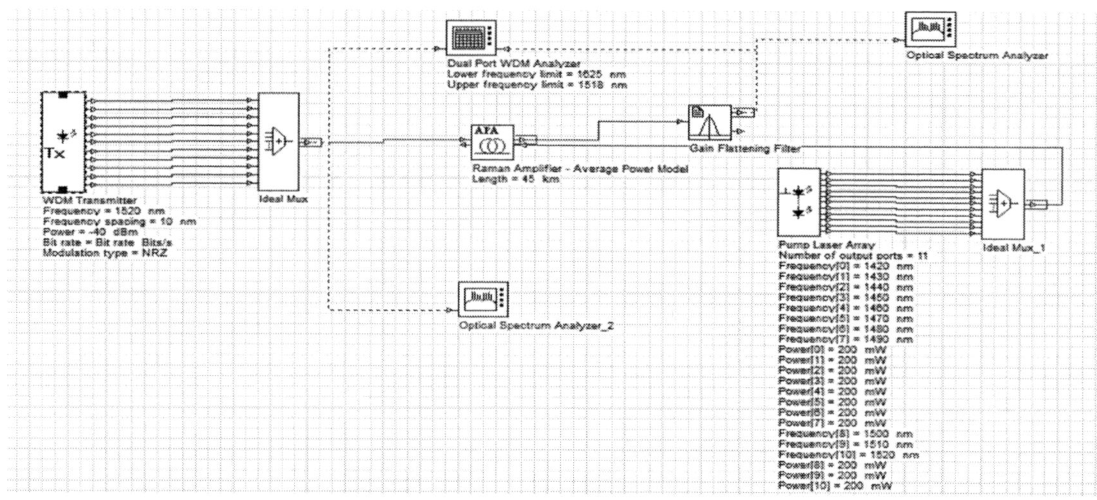

Figure 2. Configuration of Raman Amplifier.

2.3. Configuration of Hybrid EDFA-Raman amplifier

Figure 3 shows the proposed configurations of Hybrid EDFA-Raman Amplifier with the optimum parameter of EDFA and Raman amplifer tested previously. The WDM transmitter is set to 11 channels with 1530 nm to 1620 nm wavelength and 10 nm spacing. The input signal power is set to -40 dBm. The EDFA parameter used in this simulation is 300 mW bidirectional pump power with 980 nm pump wavelength, 1000 ppm-wt of erbium ion density and EDFA length of 5 m. The Raman parameter used in this configuration is the Raman length of 45 km and 150 mW pump power of each pump laser. Then, the configuration is changed with the combination of Raman amplifier followed by EDFA.

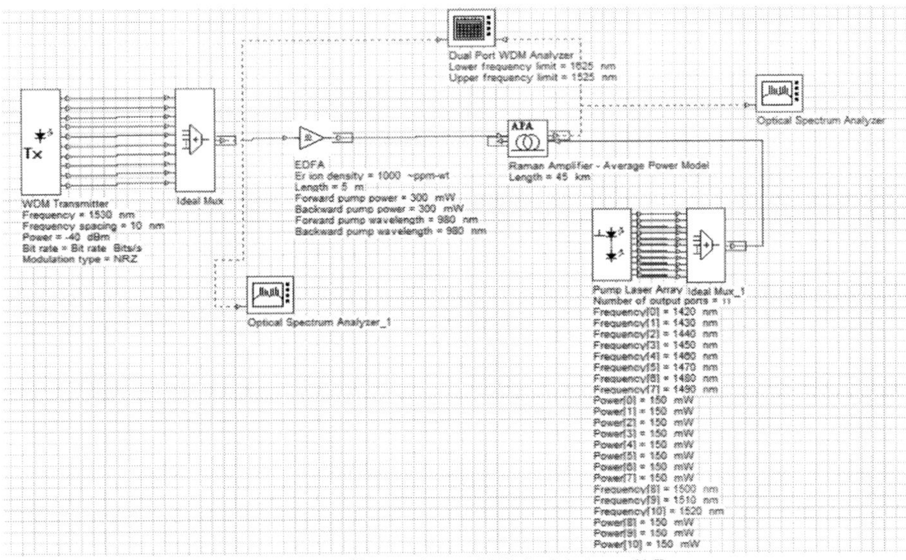

Figure 3. Configuration of Hybrid Amplifier.

3. Result and discussion

3.1. Optimization of EDFA

Figure 4 shows the gain and noise figure performances for EDFA. It is shown that the flat gain can be obtained from 1540 nm to 1560 nm with the average gain around 41 dB and average noise figure of 4 dB. The achievable flat gain bandwidth is 20 nm. This optimum EDFA is obtained based on the 300 mW bidirectional pump power with 980 nm pump wavelength. Pumping EDFA with 980 nm can give less noise and larger population inversion compared to 1480 nm pump wavelength [1]. In terms of pumping configurations, the best configuration is by using bidirectional pumping as it will provide higher gain and lower noise figure. The optimum erbium ion density used to obtain this result is 1000 ppm-wt. For this configuration, 5 m EDFA length is chosen as the optimum length as it gives highest gain, lowest noise figure and flat gain compared to other length. When longer length is used, the gain will decrease as the pump power does not have sufficient energy to produce complete population inversion. The optimum input signal power used is -40 dBm which is the threshold input signal. If the input signal is higher than the threshold, the gain will decrease.

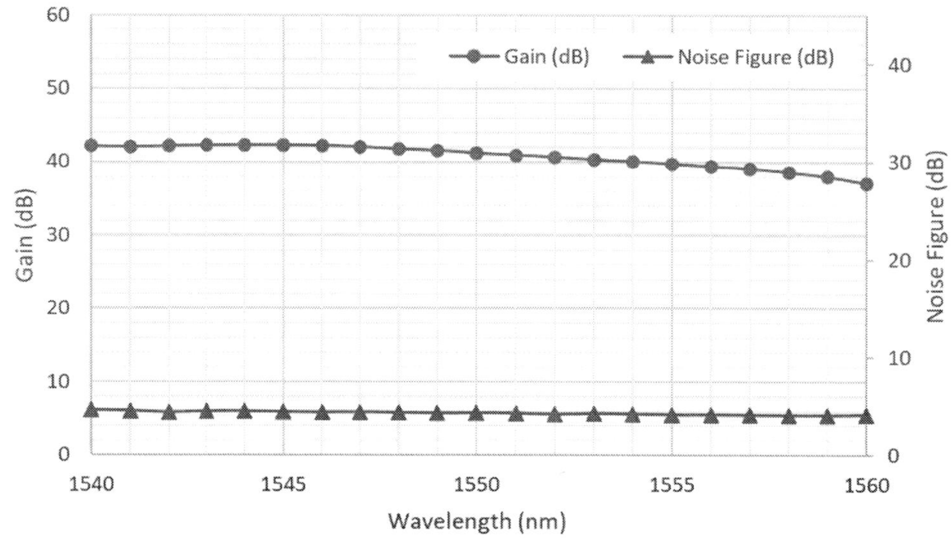

Figure 4. Gain and Noise Figure of Optimum EDFA.

3.2. Optimization of Raman Amplifier

Figure 5 shows the performance of optimum Raman amplifier. As in the figure, the flat gain is obtained from 1604 nm to 1620 nm with the bandwidth of 16 nm. The average gain obtained from the flat gain is 38 dB while the average noise figure is -38 dB. This shows that Raman amplifier will gives better noise figure compared to EDFA. In this configuration, the optimum Raman length used is 45 km as it obtained the highest gain for the signal wavelength of 1600 nm. Gain increases with the increasing Raman length and decreases when the length is longer than 45 km. The optimum pump power for each of the pump laser is 150 mW as it gives less gain variation when compared with 200 mW pump power. This Raman configuration gives better noise figure compared to EDFA.

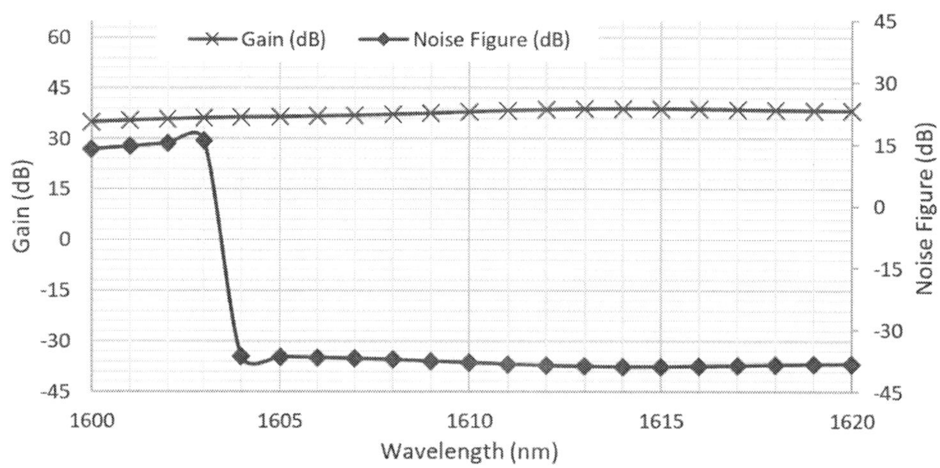

Figure 5. Gain and Noise Figure of Raman Amplifier.

3.3. Optimization of Hybrid Raman Amplifier

Figure 6 shows the performance of Hybrid Raman amplifier. The flat gain can be obtained from 1530 nm to 1600 nm which is 70 nm bandwidth. The best performances of Hybrid Raman amplifier is the configuration of hybrid EDFA-Raman as is gives the highest average gain of 46 dB and lowest noise figure of 3 dB as compared to optimum singly-based EDFA and Raman as well as hybrid Raman-EDFA. Research in [4] also proves that the best performance of hybrid amplifier is the combination of EDFA-Raman amplifier. This hybrid EDFA-Raman amplifier give the higher gain compared to [3] and [4]. In comparing with the latest work in year 2016 which is [5] and [6], the performances of configuration in figure 3 is much better in terms of gain and noise figure even though the bandwidth is reduced in 10 nm compared to [6].

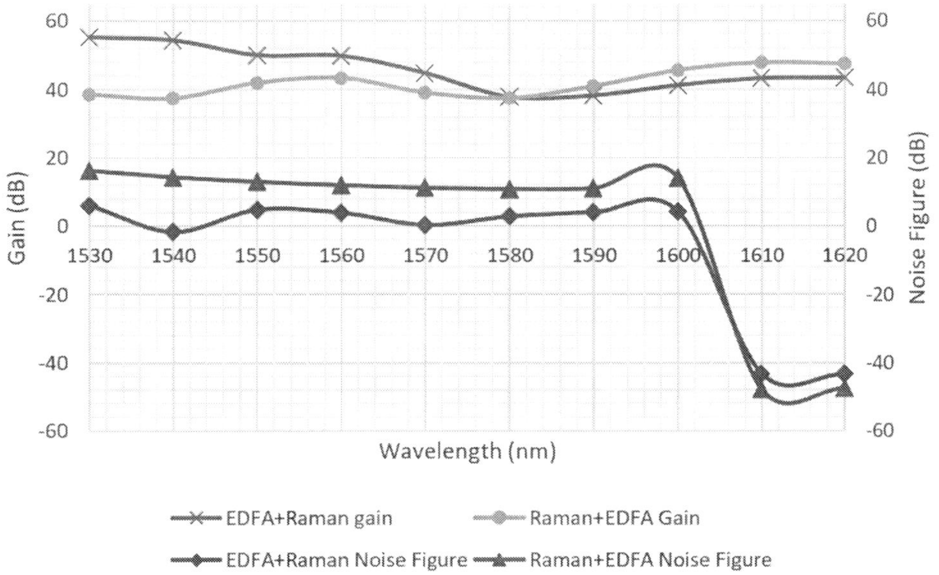

Figure 6. Optimum performances of Hybrid Raman amplifier.

4. Conclusion

In conclusion, the optimization of hybrid EDFA-Raman amplifier performance improved significantly in terms of flat gain bandwidth as well as higher gain of transmission system. The design of optimum hybrid amplifiers is very important in obtaining a broad flat gain and better noise performances. The parameters which can be optimize the EDFA's performance in term of noise figure and gain are erbium ion density, EDFA length, input signal power, pump power and pump wavelength while the parameters to optimize the Raman amplifier are Raman length and pump power. As these parameters are used in the hybrid amplifiers, it gives a higher gain, low noise figure and a better flat gain bandwidth.

Acknowledgement

This work is financially supported by the International Islamic University of Malaysia (Grant No.: RIGS15-145-0145) and (Grant No.: RIGS16-083-0247)

IOP Conf. Series: Materials Science and Engineering **210** (2017) 012033 doi:10.1088/1757-899X/210/1/012033

References

[1] Keiser G. 2010 *Optical Fiber Communications*: McGraw-Hill Education.
[2] Naji AW, Hossain N. 2010 *Modeling of hybrid EDFA/DRA: Theory, Numerical Simulation And Experimental Evaluation*: Lambert Academic Publishing (LAP).
[3] Singh S, Kaler R. 2013 *IEEE Photonics Technology Letters* **25** 250-252
[4] Martini M, Castellani C, Pontes M, Ribeiro M, Kalinowski H, editors. 2015 *Microwave and Optoelectronics Conf. (IMOC), 2015 SBMO/IEEE MTT-S International*: IEEE.
[5] Sharma SR, Sharma VR, editors. *2016 Signal Processing and Integrated Networks (SPIN), 2016 3rd Int. Conf.* on: 2016 IEEE.
[6] Kaushik A, Kapoor V, editors. 2016 *Communication and Signal Processing (ICCSP), 2016 Int. Conf.* on; 2016: IEEE.
[7] Cokrak AC, Altuncu A. 2004 *Journal of electrical & electronics engineering* **4**
[8] Islam MN. 2002 *IEEE Journal of selected topics in Quantum Electronics* **8** 548-59.

IOP Conf. Series: Materials Science and Engineering **210** (2017) 012034 doi:10.1088/1757-899X/210/1/012034

Partial Discharge Behaviour within Palm Oil-based Fe_2O_3 Nanofluids under AC Voltage

M Z H Makmud[1,2], H A Illias[1] and C Y Chee[3]

[1]UM High Voltage Laboratory (UMHVL), Department of Electrical Engineering, Faculty of Engineering, University of Malaya, 50603 Kuala Lumpur, Malaysia

[2]Complex of Science and Technology, Faculty of Science and Natural Resources, Universiti Malaysia Sabah, 88400 Kota Kinabalu, Malaysia

[3]Department of Chemical Engineering, Faculty of Engineering, University of Malaya, 50603 Kuala Lumpur, Malaysia

E-mail: mzhilmey@ums.edu.my and h.illias@um.edu.my

Abstract. In this work, comparison studies of partial discharge (PD) behaviours within palm oil-based Fe2O3 nanofluids under AC voltage were performed. The samples were prepared with different weight percentage of Fe2O3 nanoparticles; LC (Low Concentration), MC (Medium Concentration) and HC (High Concentration) in order to investigate the „electron trapping" effectiveness, their discharge withstand capability, discharge numbers, magnitude and Phase Resolved Partial Discharge (PRPD) patterns under high voltage stress. The results indicate that the presence of Fe2O3 nanoparticles with LC enhances the withstand characteristics of palm oil against PD.

1. Introduction

Nowadays, nanofluids have attracted a wide interest due to their superior performance and engineering impact. This leads to the trend of researches to conduct the studies on the characteristics and suitability of nanofluids in various fields especially in high voltage engineering for power transformer application. The combination of natural ester and nanotechnology for example, may produce dielectric nanofluids with higher breakdown strength, prone to ageing process and biodegradable. This could replace the usage of mineral oil in power transformer in future.

Generally, natural ester such as palm oil can be synthesized to several derivatives with different molecular structure and physiochemical properties. In fact, the differences would provide different performance and characteristics as insulation purpose. For example, increasing the length of the side branches of chemically modified alkyl ester of palm kernel oil possess an excellent breakdown strength compared to mineral oil [1]. Partial discharge activities are also hindered in the larger molecular structure with double bonds in hydrocarbon and extra energy is needed to break the chemical bonds under the same electric field [2].

Partial discharge (PD) is defined as a localized electrical discharge, which partially bridges between electrodes inside the insulation system. The measurement of PD behaviour is merely important for assessment on insulation quality or as an early diagnostic before the insulation failure. Therefore, the objective of this work is to compare the PD behaviours of palm oil-based Fe_2O_3 nanofluids with different range of concentration level. The experimental results present the partial discharge inception voltage (PDIV), partial discharge numbers, phase resolved partial discharge (PRPD) patterns and morphology analysis.

2. Experimental Method

2.1. Material selection
The natural ester used in this study is Refined Bleached Deodorized Palm Olien (RBDPO), which is high in dielectric strength [3]. The nanoparticles of Fe_2O_3 were obtained commercially from Sigma Aldrich in the form of nanopowder with physiochemical properties as shown in Table 1. All the materials were dried through vacuum chamber to remove the impurities and moisture.

Table 1. Physiochemical properties of Fe_2O_3 nanoparticles

	Ferum Oxide, Fe_2O_3
Type	Conductive
Size	50 – 100 (nm)
Density	5.17 (g/cm^3)
Electric Conductivity	1×10^4 - 1×10^5 (S/m)
Relative Dielectric Constant	80
Relaxation time constant	7.47×10^{-14} (s)

2.2. Nanofluids preparation
The Fe_2O_3 nanoparticles in all formulation samples were dispersed by ultrasonication in two-step method. The weight percentage and amount of Fe_2O_3 in each nanofluid samples are tabulated in the Table 2.

Table 2. Palm oil-based Fe_2O_3 nanofluids formulation samples

Unit	Concentration level			
	Null (Pure)	Low (LC)	Medium (MC)	High (HC)
Weight percent (%)	0.0	0.1	10	100
Amount (g/L)	0.0	0.01	0.1	1.0

After the sonication, all samples were dried again in the vacuum chamber to remove the micro bubbles during the mixing process and then stored inside the bottle for experimental purpose as shown in Figure 1. The variations of colours indicate the concentration of Fe_2O_3 nanoparticles in the sample.

Figure 1. Snapshot of palm oil-based Fe_2O_3 nanofluids sample

2.3. Partial discharge measurement setup

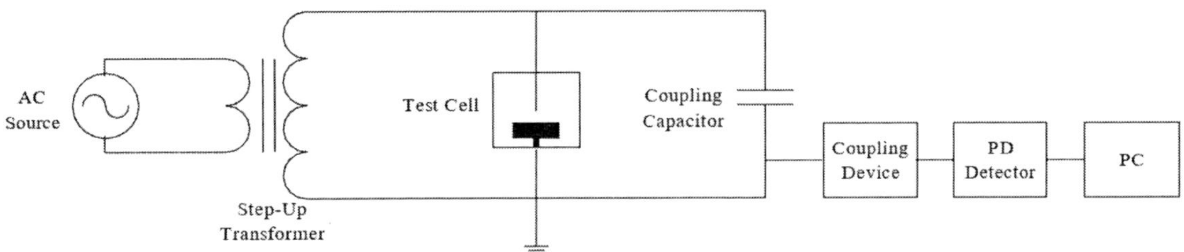

Figure 2. Experimental setup for PD measurement

Figure 2 shows the experimental setup used for the PD measurement. The test setup includes a HVAC supply (100/0.22 kV, voltage rating), measuring impedance (coupling device which is equivalent to RLC circuit), a coupling capacitor (1 nF), PD detector and a test cell which contains the sample. In order to initiate PD, a high voltage AC supply was connected to the needle electrode with curvature radius of 10 μm and 10 mm gap distance to the plane electrode (grounded). The voltage rise was controlled to have a rate of rise around 1 kV/s. Every single PD event captured was then transmitted to a personal computer via optic cable for further analysis.

3. Result and discussion

This section presents how the concentration of Fe_2O_3 nanoparticles influences the PD activities within the samples.

3.1. Partial discharge inception voltage

The results of PDIV are shown in Figure 3 and it clearly indicates the increment of inception voltage of LC sample compared to the pure sample. The result also shows slightly reduction of inception voltage for MC and HC samples respectively. PDs were initiated when the electric field at the tip of needle exceeds the dielectric breakdown strength of the samples due to ionization of surrounding at tangential needle to the ground electrode. Thus, for LC sample, the electron trapping effect due to the existence of nanoparticle prevents the charge mitigation and consequently increases the inception voltage. This was also reported by several researchers when low concentration of conductive nanoparticles was dispersed in mineral oil [4-5]. However, the inception voltage decreases for MC and HC due to the electron trapping effect is deteriorated when the Fe_2O_3 nanoparticle concentration increases. Similar trends were observed in PD histograms and PRPD patterns.

IOP Conf. Series: Materials Science and Engineering **210** (2017) 012034 doi:10.1088/1757-899X/210/1/012034

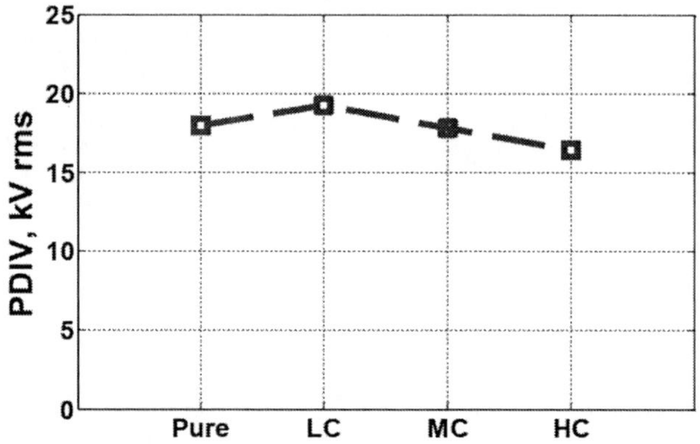

Figure 3. Partial discharge inception voltage (PDIV) result

3.2. *Partial discharge numbers*

In order to compare PD behaviours from various samples, the applied voltage were constantly set at 20 kV and higher than their PDIV level. Figure 4 shows the histogram of PD at 0 – 360 phase degree within positive and negative cycles of the applied voltage. At positive cycle, LC sample shows the lowest number of PD and increases with the increasing Fe_2O_3 concentration. This is due to the "electron trapping" effect that is associated with the applied electric field when Fe_2O_3 nanoparticle was dispersed in the palm oil. The nanoparticles induce the trapping and de-trapping process of charge carrier due to the differences of dielectric constant in a medium [6-7]. However, in high concentration of Fe_2O_3 nanoparticles, the trapping effect deteriorates due to the agglomeration of nanoparticles. At negative cycle the LC still dominant to maintain the lowest number of PD events. From the PD activity at negative cycle, it is also revealed that in larger surface area of negative electrode permits higher ionization process and produces more deterioration of liquid dispersion, which consequently produces more PD activities. In overall, the total number of PDs is higher in HC due to the 'bridging electron' effect is more dominant than 'electron trapping'.

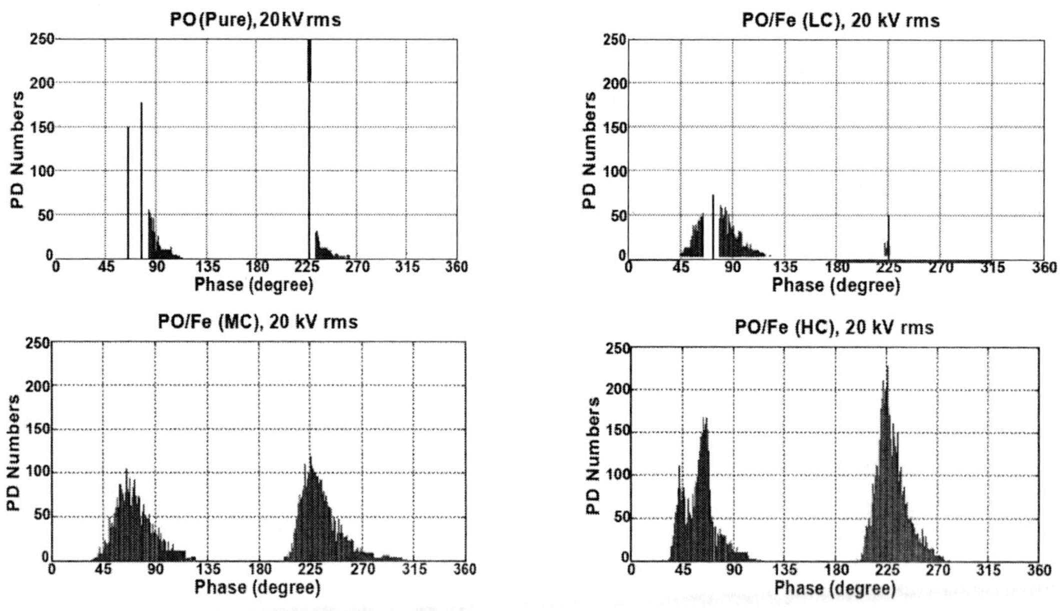

Figure 4. PD histogram with phases measured from samples

3.3. Phase resolved partial discharge pattern

The Phase Resolved Partial Discharge (PRPD) patterns have been widely used in high voltage system diagnostic to analyse the insulation condition [8]. Typical PRPD patterns obtained for pure, LC, MC, and HC with 3000 cycle of 20 kV rms of applied voltage are presented in Figure 5. The PRPD patterns show the PD charge magnitude and phase of all obtained pulses during the measurement. Referring to Figure 5, the PRPD pattern for pure sample is not symmetrical while the PD charge magnitude at positive cycle is much higher than negative cycle. The unbalanced geometry of the positive and negative electrode contributes to the asymmetrical shape of PD corona generation in the pure sample [9]. On the other hand, the PRPD patterns for nanofluids samples are generally equal. When the electron avalanche escapes from the trapping position, their velocity is higher, resulting in higher PD magnitude.

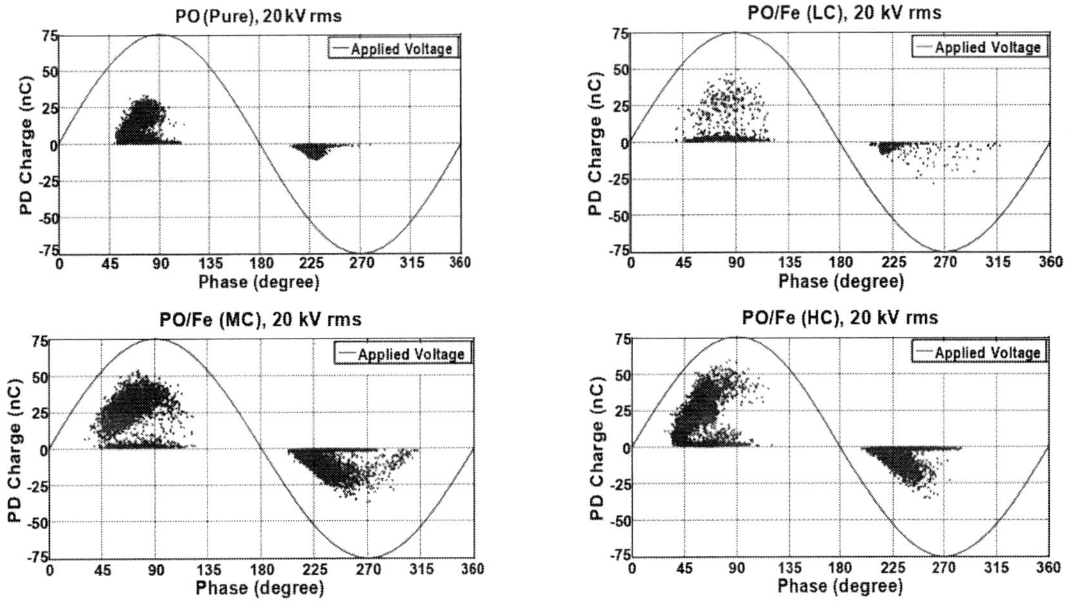

Figure 5. PRPD patterns measured for various samples

3.4. Morphology analysis

The morphology analysis was used to reveal extra information on how PD activity is affecting the nanofluids sample after 3000 cycles of applied voltage. Within the cycles, only in HC sample shows black layer developed within the surface area of the ground electrode. This is due to the bombardment of high velocity electron on the negative electrode, which could degrade the dispersion stability of palm oil-based Fe_2O_3 nanofluids. Hence, the Van der Walls force between particles deteriorates and leads to the sedimentation process that was obviously observed in HC sample. In fact, identical process happens on solid insulation when it was exposed to corona discharge [10]. Figure 6 shows the ground electrode condition after PD activity in PO/Fe (HC) sample. Figure 6 shows ground electrode condition after PD activity in PO/Fe (HC) sample.

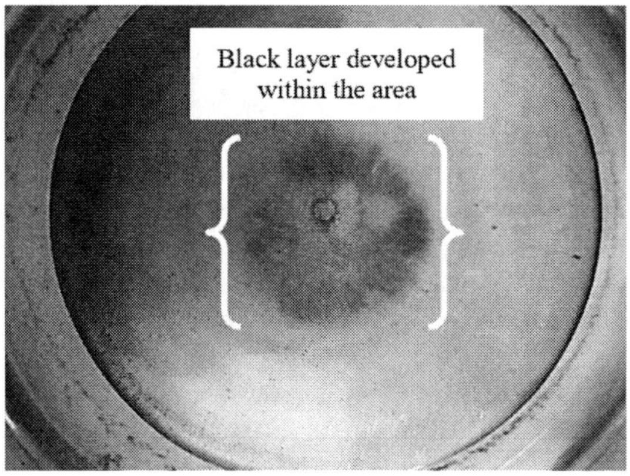

Figure 6. Ground electrode condition after PD activity in PO/Fe (HC) sample.

4. Conclusions

In conclusion, the PDIV withstand characteristic of palm oil is enhanced when the low concentration (LC) of Fe_2O_3 nanoparticles was dispersed in the liquid-based and performs the 'trapping' effect. However, it will deteriorate if the concentration exceeds certain limit. The PD charge magnitude is higher in the nanofluid samples compared to the pure sample. This is due to the accelerated velocity of the electron after escaping from the 'trapping' effect. The bombardment of high velocity electron avalanche will produce electrical ageing and therefore degraded the nanofluids stability. This effect can be investigated by performing the morphology analysis on the ground electrode after PD activities.

5. Acknowledgments

The authors thank to the University of Malaya and Malaysian Ministry of Higher Education for supporting this work through Postgraduate Research Fund (PPP) (grant no: PG236-2016A) and scholarship under SLAB.

6. References

[1] A. A. Abdelmalik 2014 *Sustain. Mater. Tech* **1–2** 42–51

[2] M. Hikita, J. Tokiyoshi, M. Tsuchie, M. Kozako, T. Suzuki, A. Kanetani, and T. Kano 2012 *Electrical Insulation (ISEI), Conference Record of the 2012 IEEE International Symposium on.* 26–29

[3] S. S. Sinan, S. N. Shawaludin, J. Jasni, N. Azis, M. Z. A. A. Kadir, and M. N. Mohtar *2014 IEEE Innovative Smart Grid Technologies - Asia (ISGT ASIA)* 760–763

[4] A. Cavallini, R. Karthik, and F. Negri 2015 *IEEE Trans. Dielectr. Electr. Insul.* **22** 2592–2600

[5] H. Jin, P. H. F. Morshuis, A. R. Mor, and T. Andritsch 2014 *IEEE 18th International Conference on Dielectric Liquids (ICDL).* 1–4

[6] Y. Hayase, Y. Tanaka, T. Takada, Y. Murata, Y. Sekiguchi, and C. C. Reddy 2009 *J. Phys. Conf. Ser.* **183** 12004.

[7] J. Li, B. Du, F. Wang, W. Yao, and S. Yao 2016 *Phys. Lett. A* **380** 604–608.

[8] T. Tanaka, "Partial discharge pulse distribution pattern analysis," *Sci. Meas. Technol. IEE Proc.* -, vol. 142, no. 1, pp. 46–50, 1995.
[9] H. Illias, T. S. Yuan, A. H. A. Bakar, H. Mokhlis, G. Chen, and P. L. Lewin, "Partial discharge patterns in high voltage insulation," *Power and Energy (PECon), 2012 IEEE International Conference on*. pp. 750–755, 2012.
[10] Y. Zhu, K. Haji, M. Otsubo, and C. Honda, "Surface Degradation of Silicone Rubber Exposed to Corona Discharge," *IEEE Transactions on Plasma Science*, vol. 34, no. 4. pp. 1094– 1098, 2006.

IOP Conf. Series: Materials Science and Engineering **210** (2017) 012035 doi:10.1088/1757-899X/210/1/012035

Analytical Investigation of Step-Square Wave in Time and Frequency Domain Using Transform Equations

F E Fajingbesi[1,2*], N S Midi[1], S Khan[1]

[1]Department of Electrical and Computer Engineering, International Islamic University Malaysia
[2]Department of Physics Electronics and Earth Science, Fountain University Osogbo Nigeria

E-mail: *fawwazfajingbesi@yahoo.com

Abstract. A unit step function can be used for generating a pulse of given width and height. The same pulse can be used for generating a stepped waveform of known levels and a given width in time-domain which when translated to frequency or complex S-domain provides the fundamental constituent frequencies. This paper is about studying the frequency of the above-generated functions in time as well in the frequency domain using transform equations. The result of this simulations are indicative of the correlation between step square waving and reductions in total harmonic distortion (THD) during DC to AC conversion. A decrease in the pulse width difference between each step shows better conversion with reduced harmonics when subjected to simple passive filter of low transfer function.

1. Introduction

In today's broad field of wave applications for theoretical and physical interest, transforms are used in understanding the fundamental component and simplification of the complex equations at different domain [1]. Such commonly used transform include the Fourier and Laplace transform which when carried out on a function f(t) in time domain produces the signal in the frequency domain F(ω) and complex frequency s-domain F(s) respectively. On a general term, the Fourier Transform decomposes any function into a sum of sinusoidal basis functions where each of these basis functions is a complex exponential of a different frequency [1-2]. The Fourier transform therefore offers a unique way of viewing any function - as the sum of simple sinusoids. Through this, noise reduction and harmonics cancelation system could be achieved for any signal. For a square like waveform, described as a pulse train, or pulse wave, is a periodic waveform consisting of instantaneous transitions between two distinct levels easily generated by the switching High and low state of a DC voltage source has found much application in today worlds such as timings circuits, PWM, H-bridge, DC-DC & DC-AC conversion, and generation etc. [1-13]. In most application where independent source are required to combine in steps or linearly every individual source offers additional input to either the stability of the output or extra source of noise hence, This work is about studying step square wave forms analytically using MATLAB R2016a, deriving their expressions and behavior at different domain by analyzing their transform. The effect of amplitude and pulse width change would be considered at this domain, and also the effect of combining the signals before and after subjection to filter circuit of known transfer function $H(s||j\omega)$ would be explored since in the real world circuits behavior with different sources are mostly linear—adding or subtracting based on the combination [2, 7, 8, 10].

2. Theory and transform equations

The Fourier transform $\mathcal{F}(t)$ is intimately related with the Laplace transform $\mathcal{L}(s)$ which is also used for the solution of differential equations and the analysis of filters systems. The Laplace and the Fourier transforms for a causal function are the same, provided that the real part of s is zero [3, 14] therefore more focus is on the Fourier transform. For any given function $f(t)$ the Fourier transform and the inverse transform can be written respectively as

$$\mathcal{F}\{f(t)\} = F(f) = \int_{-\infty}^{\infty} f(t)e^{-2\pi i f t}dt \quad (1)$$

$$\mathcal{F}^{-1}\{F(f)\} = \int_{-\infty}^{\infty} F(f)e^{2\pi i f t}df = f(t) \quad (2)$$

The Laplace transform on the other hand for a function $f(t)$ takes it to a function of a complex variable $F(s)$ in the frequency domain. Where $s = \sigma + \omega$. The transform and inverse operator is given as

$$\mathcal{L}\{f(t)\} = F(s) = \int_{0^-}^{\infty} f(t)e^{-st}dt \quad (3)$$

$$\mathcal{L}^{-1}\{F(s)\} = \frac{1}{2\pi i}\lim_{T\to\infty}\int_{\gamma-iT}^{\gamma+iT} F(s)e^{st}ds = f(t) \quad (4)$$

If we consider a unit square wave in time domain $f(t)$ generated by a sequence of close-open-close of a light switch. The resulting wave can be described as

$$f(t) = \begin{cases} A \ if -\tau \leq t \leq \tau \\ 0 \ otherwise \end{cases} \quad (5)$$

Where $\tau = On\ Time\ \&\ A = DC\ Amplitude$. Since a square wave is the sum of sinusoid with different fundamental frequency it can also be expressed as

$$f(t) = u(t+\tau) - u(t-\tau) \quad (6)$$

Where $u(t)$ is called the step function defined as

$$u(t) = \begin{cases} A \ t > 0 \\ 0 \ t < 0 \end{cases} \quad (7)$$

Where $f(t)$ is a linear sum of two unit step function which has been time delayed and inversed and the other time advanced. Hence The Fourier transform on this square pulse should give the $sinc(t)$ function described mathematically as

$$\mathcal{F}\{f(t)\} = F(f) = \int_{-\infty}^{\infty} f(t)e^{-2\pi i f t}dt$$

$$= \int_{-\tau}^{\tau} Ae^{-2\pi ift}\, dt = \frac{A}{-2\pi if}\left[e^{-2\pi ift}\right]_{-\tau}^{\tau}$$

$$= \frac{A}{-2\pi if}\left[e^{-2\pi if\tau} - e^{2\pi if\tau}\right] = \frac{A\tau}{\pi f\tau}\left[\frac{e^{2\pi if\tau} - e^{-2\pi if\tau}}{2i}\right]$$

$$= \frac{A\tau}{\pi f\tau}\sin(\pi f\tau) = A\tau[sinc(f\tau)] \quad (8)$$

The transfer function $H(s||j\omega)$ of a known filter can then be linearly applied to condition the wave. Where

$$H(s||j\omega) = \frac{1}{s+1} = \frac{1}{j\omega + 1} \quad (9)$$

$$where\ R\ and\ C\ are\ unity\ otherwise$$

$$= \frac{1}{sc+1} = \frac{1}{\omega\tau + 1} \quad (10)$$

3. Analytical plots from transform equations

Considering the theoretical equations in section II, the equivalent model is created using MATLAB R2016a to obtain various analytical plots of a square pulse derived from constituent unit step functions in the time domain. The acquired pulse is then further combined to obtain waves of different width, height and steps. The resulting pulse train is then subjected to various transform from time domain to frequency domain and complex frequency S-domain shown in figure 1-12.

4. Discussion and application of results

In this work we have modeled a DC to AC conversion using analytical approach in MATLAB and studied the effect of different changes along the conversion process. The design is based on Unit step function for creating square pulse and step square wave. In Figure 1. An alternating (sinusoidal) wave obtained analytically from a unit step function — square pulse — step square wave + transform + filter. It was observed that the higher the step number the lower the transfer function of the filter required for the conversion and hence an overall reduction in the total harmonic distortion (THD) for the system. In figure 2, an impulse function was created in time domain and transformed, the result in the frequency and complex frequency domain are a DC with zero frequency component likewise the unit step function in time domain produces an impulse function in the frequency domain and other fundamental frequency peak attenuated. From figure 3, the unit step function have been used to derive a unit square pulse. Its transform into the frequency domain at each step is shown in figure 4 and the unit step square wave function in the frequency domain is described as a sinc (ω) function shown at the bottom right corner of Figure 3 and in the bottom row of figure 4 the single side double sided band of the Fourier transform is show with their imaginary components. In figure 5 the effect of change the width and height respectively is illustrated in both time and frequency domain using transform. For change in width the fundamental frequency is observed to start converging and as the pulse width tends to infinity in the time domain the frequency converges to an impulse while a change is amplitude or pulse height only increase the power spectrum energy and no clear difference in frequency shift was observed at different heights. In figure 6 the unit step square wave of varying width are added and transformed, the effect of adding this unit step is linear. The result of addition is increase in the number of step and in the frequency domain the fundamental frequency is found to start converging and as the step increases the lesser the harmonics present in the transformed wave. In figure 7, the step wave have been inverted and low pass filtered at different transfer function. In figure 8, the delayed and inverted step square wave have been linearly added to its original wave and the resulting wave is full cycle step wave. Figure 9 and 10 shows the effect of applying filter to the resulting wave at different cut off frequency in the transfer function of the low pass filter. It is seen that a very nice AC signal was produced and is shown in figure 11 to figure 12. In figure 12 (a) and 12 (b) respectively, a single cycle and multiple cycle filtered step square wave is then transformed into frequency domain.

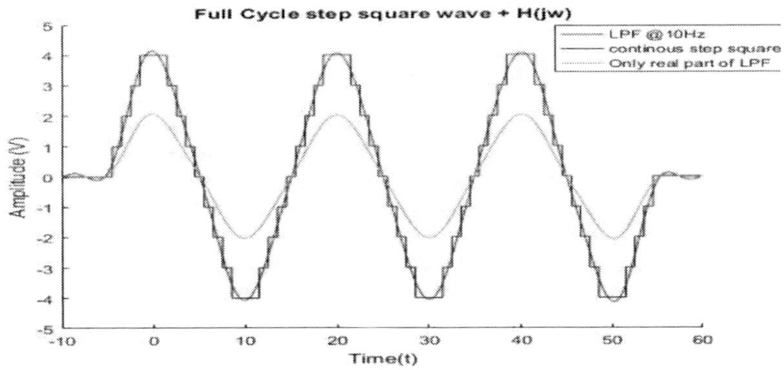

Figure 1. A step square wave is F(ω) transform and low pass Filtered at various f_c then inverse F(ω) producing a sinusoid f (t)

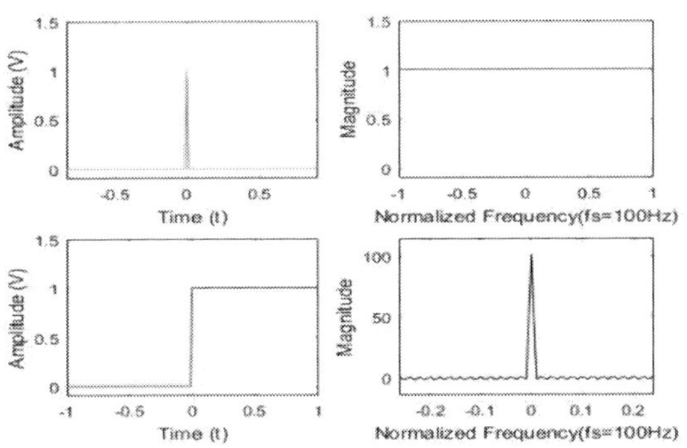

Figure 2. δ (t) and u(t) transformed to δ(ω) and u(ω)

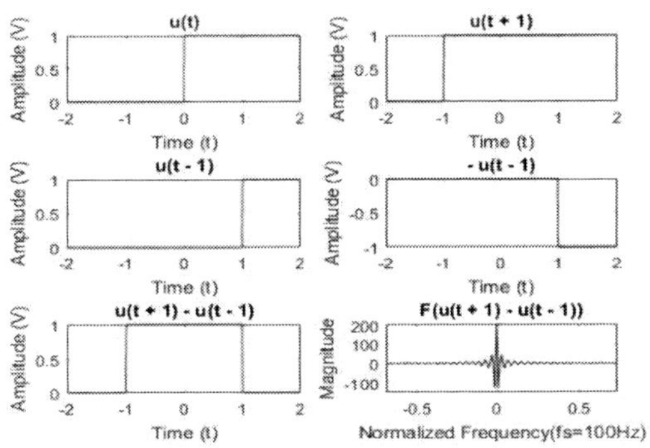

Figure 3. Converting Unit pulse into square pulse then fourier transform

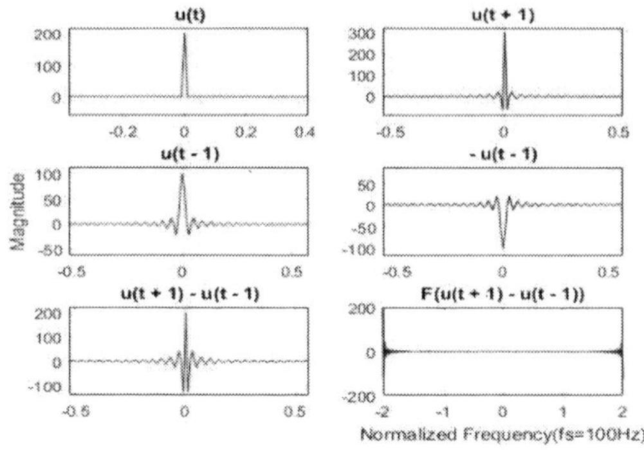

Figure 4. Fourier transform of all function in figure 3.

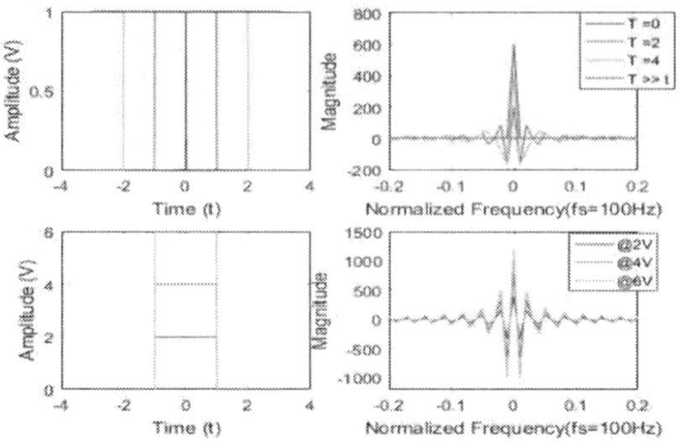

Figure 5. Effect of width and height change in t and ω domain

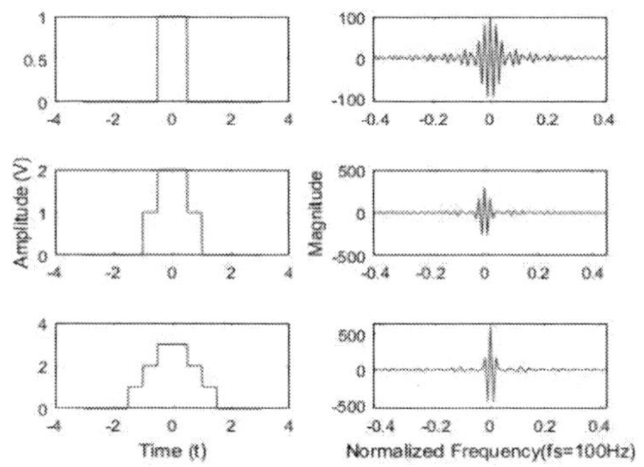

Figure 6. Effect of adding signal (step pulse) in t and ω domain

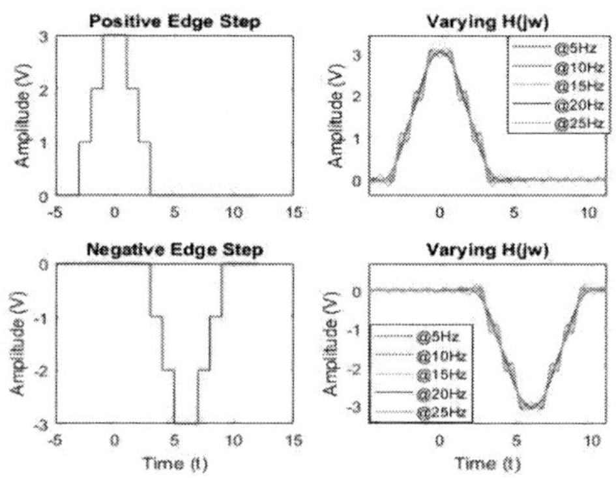

Figure 7. Effect of filtering on step pulse in t and ω domain

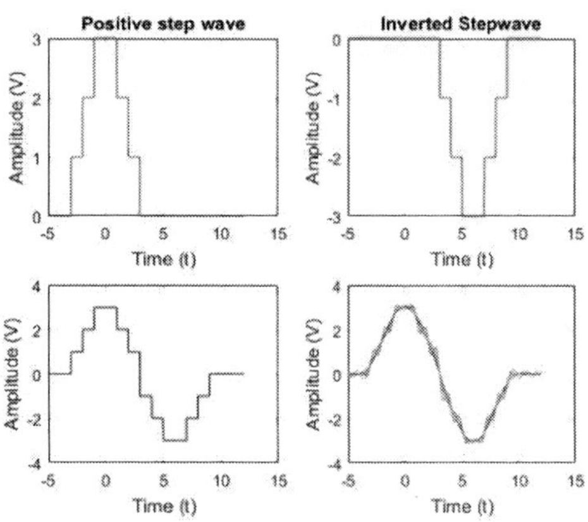

Figure 8. Effect of adding step signal in time domain

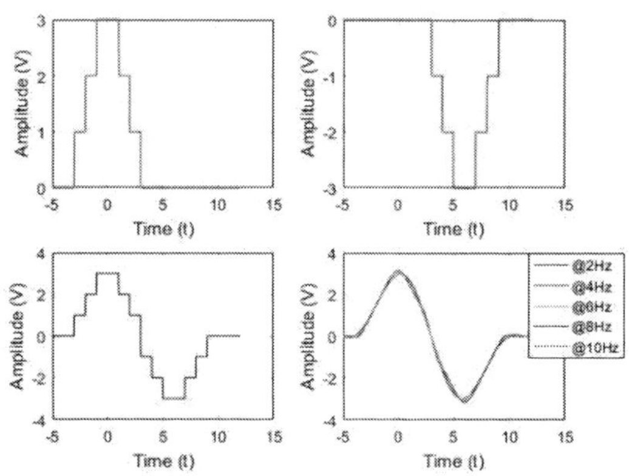

Figure 9. Filtering at different H (s||jω)

IOP Conf. Series: Materials Science and Engineering **210** (2017) 012035 doi:10.1088/1757-899X/210/1/012035

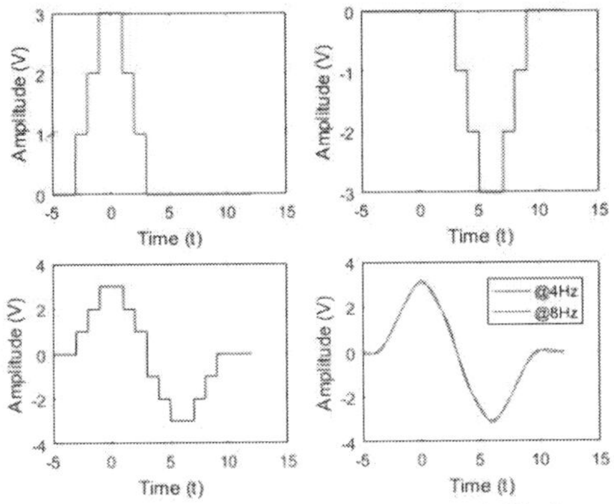

Figure 10. Filtering at different H(s||jω)

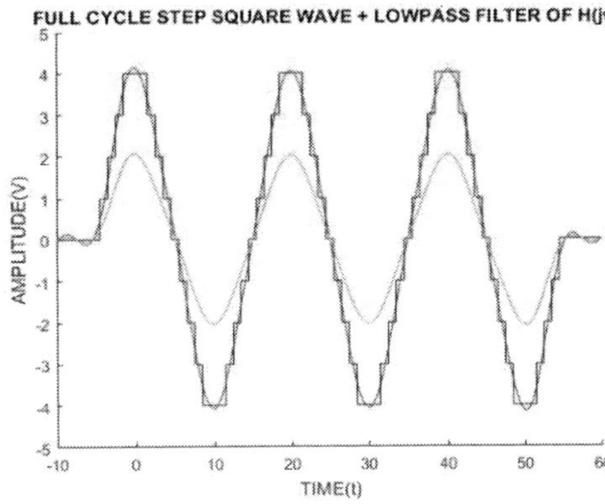

Figure 11. Filtering step square wave to obtain sinusoid

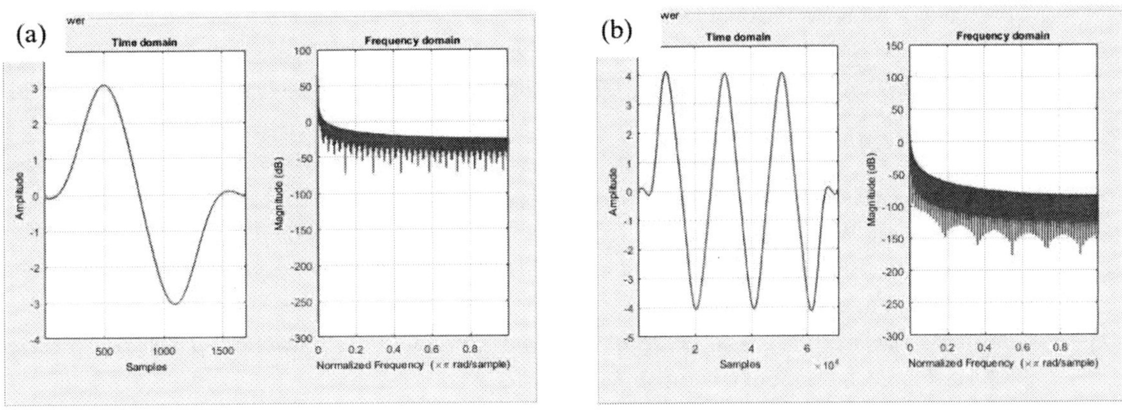

Figure 12. Single cycle (a) and multiple cycle (b) filtered step square wave in t and ω domain

5. Conclusion

In this research, we have generated a step square wave function for analytical investigation of the square pulse from a unit step function in the time domain using MATLAB R2016a. The square wave has been subjected to analytical transform in frequency and time domain where the effect of width and height change, time delay and advance, time reversal and inversion and filtering where considered. By increasing the step number the transfer function required to generate a perfect sinusoid also reduce making it possible for DC to AC conversion with simple low pass filter—preferable single pole low passive filter. Also by increasing the width of the pulse in the time domain, the fundamental frequency present tend to converge reducing the number of harmonics present in the system. When the height was change and the power spectral density was plotted, the power intensity of the spectrum also increased. Combining the width change and the height change a step square wave was generated which when subjected to the transform produced a much lower harmonics distortion. This is very much important in real world application where reduction of total harmonic distortion (THD) in the system is necessary. In conclusion, we could generalize that when the width of a pulse in the time domain tend towards a DC produces an Impulse in the frequency domain where all the fundamental frequency has converged and center at zero. Subsequently, an impulse in the time domain would also produce DC in the frequency domain.

Acknowledgment
This research has been supported by Malaysian Ministry of Higher Education Research Grant, RAGS15-063-0126.

References

[1] A. M. Shaarawi and S. M. Riad 1986 *IEEE Transactions on Instrumentation and Measurement,* **IM-35**91-92

[2] T. Mori, H. Funato, S. Ogasawara, F. Okazaki and Y. Hirota 2012 *2012 International Conference on Renewable Energy Research and Applications (ICRERA)*, Nagasaki 1-6

[3] S. Ohnuki, S. Watanabe and K. Nagasawa *2016 Progress in Electromagnetic Research Symposium (PIERS),* Shanghai 1786-1786

[4] L. P. Zha, Z. H. Fan, D. Z. Ding and R. S. Chen 2014 *IEEE Transactions on Antennas and Propagation,* **62** 5846-5851

[5] M. J. Corinthios, *2006 14th European Signal Processing Conference,* Florence pp. 1-5.

[6] L. Venkataramanan, T. M. Habashy and D. E. Freed *2010 18th European Signal Processing Conference,* Aalborg 1150-1153

[7] S. Kishimoto, S. Ohnuki, Y. Ashizawa, K. Nakagawa, S. Y. Huang and W. C. Chew *2013 USNC-URSI Radio Science Meeting (Joint with AP-S Symposium),* Lake Buena Vista, FL 47-47

[8] C. H. Noh, G. H. Gwon, Y. S. Oh, J. Han and C. H. Kim 2014 *12th IET International Conference on Developments in Power System Protection (DPSP 2014),* Copenhagen 1-5.

[9] Weisstein, Eric W. http://mathworld.wolfram.com/SquareWave.html

[10] S. Bruscoli, F. Berizzi, M. Martorella and M. Bernabò *2006 14th European Signal Processing Conference,* Florence 1-5.

[11] Chatfield, Chris (2004), *The Analysis of Time Series: An Introduction, Texts in Statistical Science* (6th ed.), London: Chapman & Hall/CRC.

[12] H. Chen and Y. Zhang *2015 3rd International Conference on Computer and Computing Science (COMCOMS)*, Hanoi 32-36

[13] Bracewell, Ronald N. 1978 *The Fourier Transform and its Applications* (2nd ed.), McGraw-Hill Kogakusha.

IOP Conf. Series: Materials Science and Engineering **210** (2017) 012036 doi:10.1088/1757-899X/210/1/012036

A Novel Multilevel DC – AC Converter from Green Energy Power Generators Using Step-Square Waving and PWM Technique

F E Fajingbesi[1,2*], N S Midi[1] and S Khan[1]

[1]Department of Electrical and Computer Engineering, International Islamic University Malaysia
[2]Department of Physics Electronics and Earth Science, Fountain University Osogbo Nigeria

E-mail: *fawwazfajingbesi@yahoo.com

Abstract. Green energy sources or renewable energy system generally utilize modular approach in their design. This sort of power sources are generally in DC form or in single cases AC. Due to high fluctuation in the natural origin of this energy (wind & solar) source they are stored as DC. DC power however are difficult to transfer over long distances hence DC to AC converters and storage system are very important in green energy system design. In this work we have designed a novel multilevel DC to AC converter that takes into account the modular design of green energy systems. A power conversion efficiency of 99% with reduced total harmonic distortion (THD) was recorded from our simulated system design.

1. Introduction

The need and use of green (renewable) energy systems are rapidly increasing across the globe occupying an increasing share in the global power generation market in other to attain an emission free environment and hopefully restore the grave extent of damage caused by the old systems to nature and environment. However the downsides of this source are majorly the instability in the natural sources (Wind & Solar). Green energy generators are mostly DC and are modular in nature. In other to be compatible into grid systems this modular sources have to be efficiently synthesized into one single source hence the research on multilevel modular converters [1-11]. Multilevel Modular converter are now a common place in both industrial and general grid electronics since the inception in 2001 and major research achievements have been recorded [1-17]. They offer high voltages with low harmonics mostly without any need for major reactive devices such as transformers and inductors. Many designs topologies have since been proposed each having its merits and downside with the H-Bridge been the bases for most design.

2. System overview

This work has presented a novel multilevel DC converter from modular energy sources seen in green energy system consisting of less reactive element to offer grater reduction in size, power loss and total harmonic distortion (THD). In figure 1, the block diagram illustrating the cascading concept of adapting multiple modular source to achieve higher output voltage is illustrated using our multilevel DC converter design shown in figure 2. The novel design seen in figure 2 consist of a multilevel DC converter and an Inverted multilevel DC converter. Each single multilevel DC converter consist of four MOSFET switch and two Capacitor with a power source.

IOP Conf. Series: Materials Science and Engineering **210** (2017) 012036 doi:10.1088/1757-899X/210/1/012036

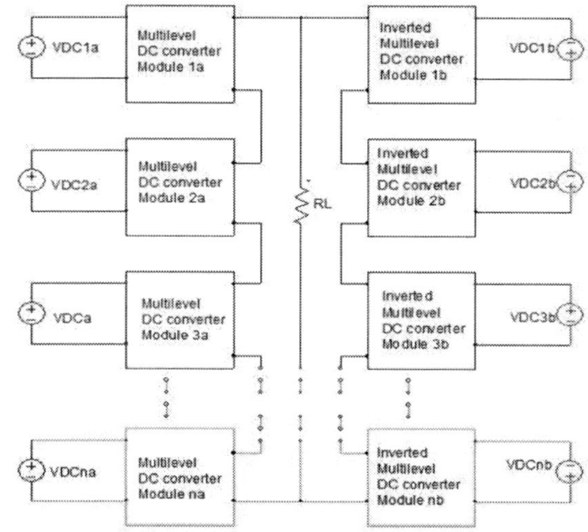

Figure 1. Multilevel DC power converter module system configuration

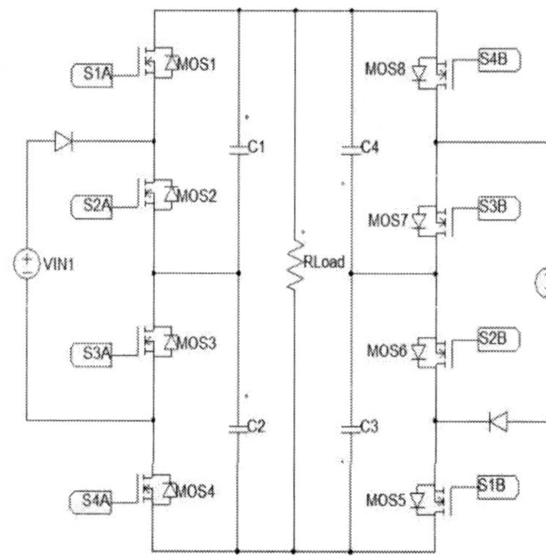

Figure 2. Multilevel DC converter circuit

The two separate block of this novel multilevel converter shown in figure 2 are switch intelligently in a complementary fashion to produce a positive and negative cycle multilevel DC output as seen in Figure 10. There are several possible output from this system based on the switching configuration. In our work we have chosen a 3 state system where the output moves from off state to V_{IN} then to $2V_{IN}$ and back to V_{IN} then off producing the positive half cycle. The transition for producing the negative half cycle is also achieved using by the inverted DC converter module section similar to the positive cycle process. For the cascaded input parallel output series shown in fig 1. The output is dependent on the PWM timing configuration govern by the novel PWM switching Output algorithm below

$$if\ (\mathrm{PWM}_{1A\,\&\,1B} == \mathrm{PWM}_{2A\,\&\,2B})$$

$$Then\ R_{L(P-P)} = 4 \times V_{IN}\ ;$$

$$No.of\ Steps = 2\ ;$$

$$Else\ if\ (\mathrm{PWM}_{1A\,\&\,1B} \gg \mathrm{PWM}_{2A\,\&\,2B})$$

$$Then\ R_{L(P-P)} = 4 \times V_{IN}$$

$$No.of\ Steps = 4\ ;$$

Therefore one can conclude for any N^th^ number of cascade with PWM signal time scaled by a factor $E = E\ \pm 1$, then the number of steps is given by $2N$ and the peak to peak voltage is given by $2N \times V_{IN}$. The advantage of higher step have been seen in several literature as they offer not only higher stable voltage but also extreme reduction in THD [1–11], [15]. Therefore by combining cascades with equal PWM and those with time scaled PWM the novel converter can achieve higher steps and amplitude for modular sources.

3. PWM Switching States

Each module of this multilevel converter consist of four MOSFET switch whose gating signal are systematically design using PWM technique. The switching state of each transistor in achieving the three fundamental operating mode earlier discussed in section II is summarized in the table 1. S_{1A}

IOP Conf. Series: Materials Science and Engineering **210** (2017) 012036 doi:10.1088/1757-899X/210/1/012036

represents the first switch in Module 1A and S_{1B} represent the last switch in the inverted Multilevel converter Module 1B (*labeling from top to bottom)*

Table 1. Converter switching state

V_{OUT}	S_{1A}	S_{2A}	S_{3A}	S_{4A}	S_{1B}	S_{2B}	S_{3B}	S_{4B}
V_{IN}	1	0	0	1	0	0	0	0
$2V_{IN}$	1	0	1	0	0	0	0	0
	0	1	0	1	0	0	0	0
V_{IN}	1	0	0	1	0	0	0	0
$-V_{IN}$	0	0	0	0	1	0	0	1
$-2V_{IN}$	0	0	0	0	1	0	1	0
	0	0	0	0	0	1	0	1
$-V_{IN}$	0	0	0	0	1	0	0	1

When the converter is in switching state for mode 1, (only switches S_{1A} and S_{4A} are on and all others are off), then

$$V_{OUT} = V_{IN1} \tag{1}$$

After a time τ the PWM frequency change and S_{1A} S_{3A} and S_{2A} S_{4A} are switched on as pairs complimentarily with 50% duty cycle for a time $\tau + 1$ and hence

$$V_{C1} = V_{IN1} \ and \ V_{C2} = V_{IN1} \tag{2}$$

$$V_{OUT} = 2V_{IN1} \tag{3}$$

After this switching state, the PWM frequency is returned back to its original state for another time τ where S_{1A} and S_{4A} are only on hence the output voltage falls back to

$$V_{OUT} = V_{IN1} \tag{4}$$

At the end of this cycle the inverted multilevel switch are then activated and the process is repeated producing the sequence from

$$-V_{IN2} \ to \ -2V_{IN2} \ to - V_{IN2} \tag{5}$$

The result of this is a positive and negative cycle of multilevel DC signal in time domain. The graphical illustration of the switching is illustrated in the next section

4. Result and Discussions

We have designed a novel multilevel DC converter for integrating modular energy sources (green renewable energy). The system is based on CMOS PWM gated switch. The switching states are given graphically in figure 3 to 11 below.

IOP Conf. Series: Materials Science and Engineering **210** (2017) 012036 doi:10.1088/1757-899X/210/1/012036

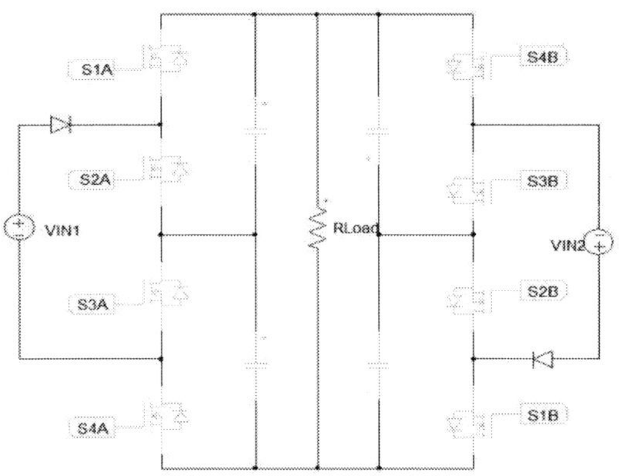

Figure 3. Converter at a t < 0 – all switch off

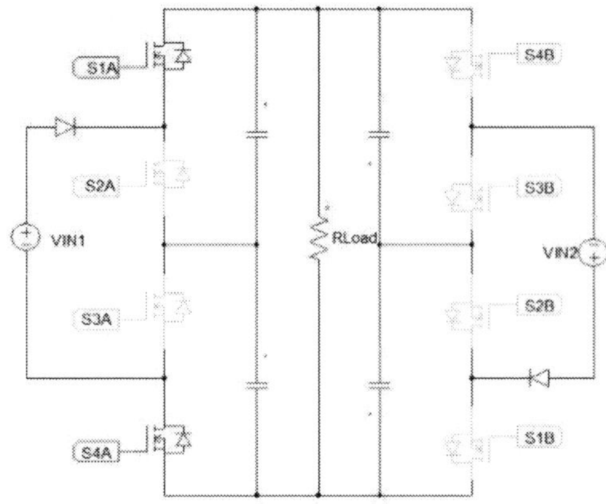

Figure 4. $S1_A$ and $S4_A$ receive PWM signal

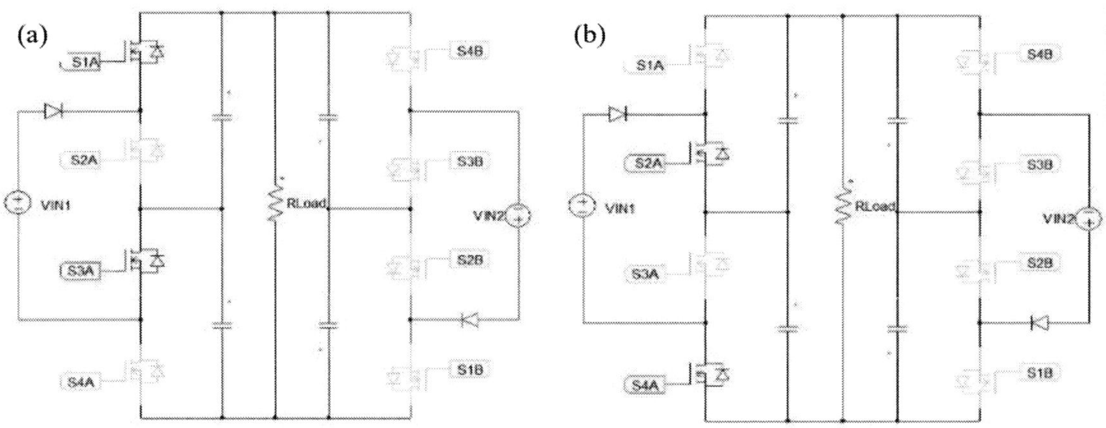

Figure 5. Transition state to twice V_{IN} : PWM operate at High f.(a) $S1_A$ and $S3_A$ receive PWM signal for half time period. (b)) $S2_A$ and $S4_A$ receive PWM signal for the second half time period

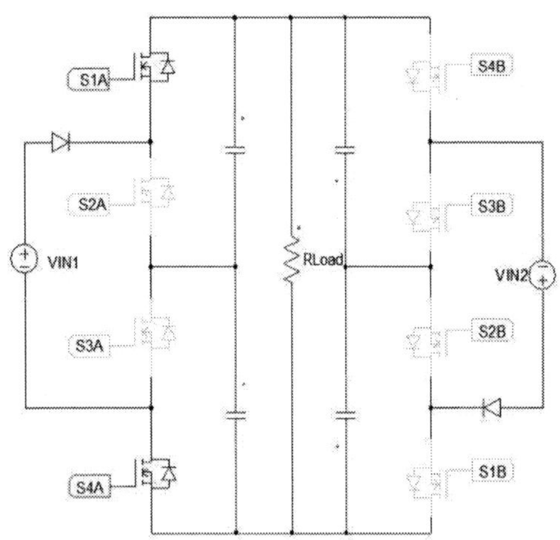

Figure 6. Transition state back to V_{IN}

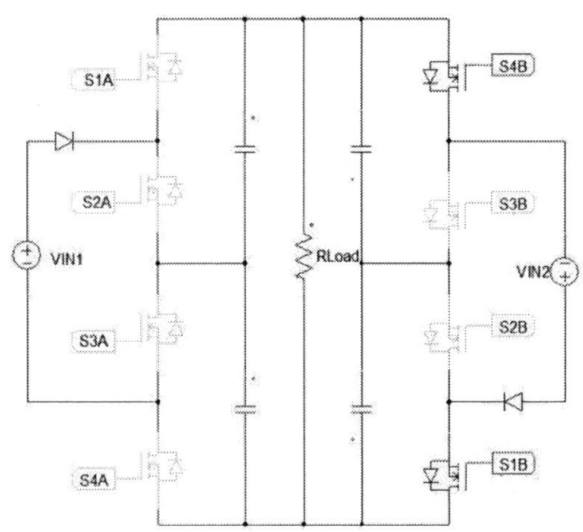

Figure 7. Transition state from V_{IN} to $-V_{IN}$

IOP Conf. Series: Materials Science and Engineering **210** (2017) 012036 doi:10.1088/1757-899X/210/1/012036

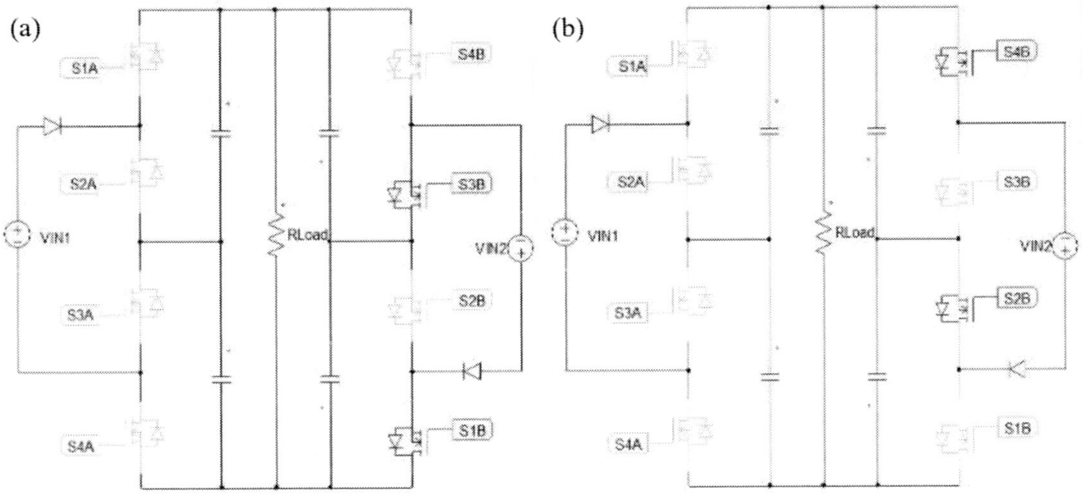

Figure 8. Transition state to twice $-V_{IN}$: PWM operate at High f.(a) $S1_B$ and $S3_B$ receive PWM signal for half time period. (b)) $S2_B$ and $S4_B$ receive PWM signal for the second half time period

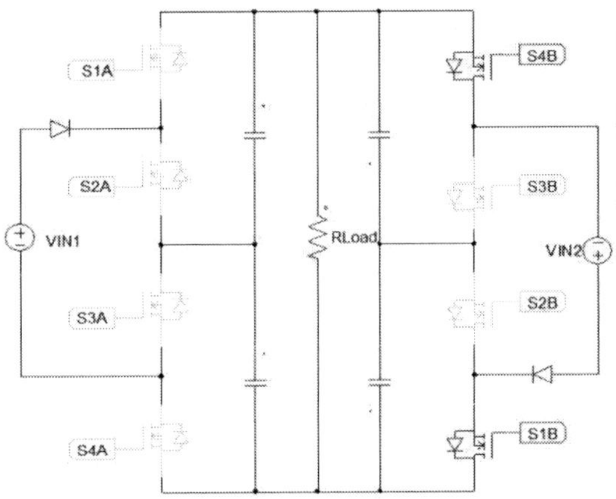

Figure 9. Transition state back to $-V_{IN}$

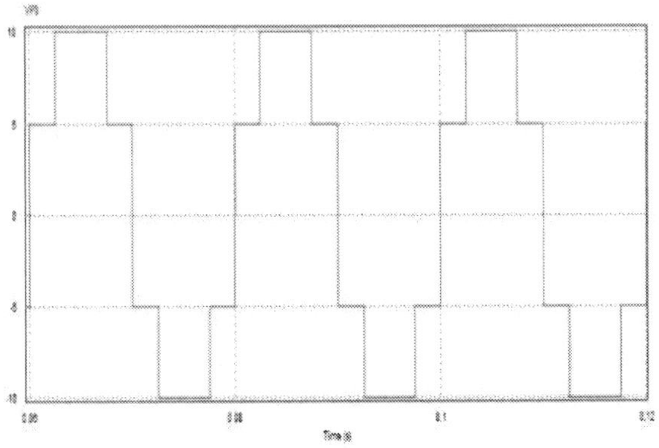

Figure 10. Multilevel full cycle output wave

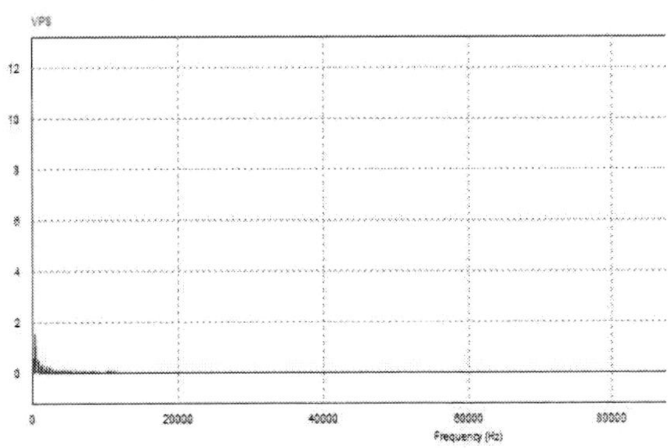

Figure 11. FFT of output wave

5. Conclusion

In this work, we have studied past designs of multilevel converters and now proposed a novel converter that take into account the modular structure of today's rising green renewable energy sources. The simulated efficiency of our converter system is 99% and with high total harmonic distortion (THD) reduction. The design is so simple yet efficient and the core of it is in the PWM switching, timing and delays. The base module can be cascaded to obtain higher steps and peak amplitudes. The future of this converter includes prototyping, adding simple first order low pass filters for real time testing and evaluation.

Acknowledgment

This research has been supported by Malaysian Ministry of Higher Education Research Grant, RAGS15-063-0126.

References

[1] S. Busquets-Monge, A. Filba-Martinez, S. Alepuz, and A. Calle-Prado 2016 *IEEE Trans. Power Electron.* **8993** 1–1.
[2] W. Cai and F. Yi *2015 IEEE Energy Conversion Congress and Exposition (ECCE)* 6667–6674
[3] H. Abu-Rub and Mariusz Malinowski, Kamal Al-Haddad 2014 *Power Electronics for Renewable Energy Systems, Transportation and Industrial Applications*, Chichester, UK: John Wiley & Sons, Ltd 591–607
[4] B. Curuvija, X. Lyu, Y. Li, and D. Cao 2016 *2016 IEEE 4th Workshop on Wide Bandgap Power Devices and Applications* 328–333
[5] H. Dehghani Tafti, A. I. Maswood, G. Konstantinou, and J. Pou 2016 *2016 IEEE Innovative Smart Grid Technologies - Asia* 52–57.
[6] N. Farokhnia, H. Vadizadeh, F. A. Asl, F. Kadkhoda, and A. Vahabzadeh 2010 *The 2010 International Power Electronics Conference* 1244–1251
[7] S. M. Goetz, Z. Li, X. Liang, C. Zhang, S. M. Lukic, and A. V. 2016 *Power Electron* **8993** 1–1
[8] K. Haridas, S. Khandelwal, and A. Das 2016 *2016 IEEE International Conference on Industrial Technology* 257–262
[9] M. F. Moussa, M. O. Abdelslaam, and H. Y. Diab 2016 *2016 UKSim-AMSS 18th International Conference on Computer Modelling and Simulation* 258–263
[10] M. Shen, F. Z. Peng, and L. M. Tolbert 2008 *IEEE Trans. Power Electron.* **23** 420–426
[11] Y. Yu, G. Konstantinou, C. D. Townsend, R. P. Aguilera, and V. G. Agelidis 2016 *IEEE Trans. Ind. Electron.* **46** 1–1
[12] Y. Otani, T. Isobe, and R. Shimada 2011 *2011 IEEE Energy Conversion Congress and Exposition* 2102–2108
[13] N. Vamanan and V. John 2016 *IEEE Trans. Ind. Appl.* **52** 3267–3278
[14] S. Takuma, K. Orikawa, J. Itoh, R. Oshima, and H. Takahashi 2015 *2015 IEEE Energy Conversion Congress and Exposition* 4678–4684
[15] M. Rashad, D. M. Minhas, M. Ashraf, and S. Hussain 2016 *2016 18th Mediterranean Electrotechnical Conference* 1–6
[16] K. Suzuki, W. Kitagawa, and T. Takeshita 2016 *2016 18th European Conference on Power Electronics and Applications* 1–10
[17] F. Irshad, M. Rizwan, H. Khalid, and N. A. Zaffar 2016 *2016 IEEE 8th International Power Electronics and Motion Control Conference* 239–245

Piezoelectric Pre-Stressed Bending Mechanism for Impact-Driven Energy Harvester

A M Abdal* **and K S Leong**
Department of Electronic and computer Engineering Universiti Teknikal Malaysia Melaka

E-mail: *ali.challenger89@yahoo.com

Abstract. This paper experimentally demonstrates and evaluates a piezoelectric power generator bending mechanism based on pre-stressed condition whereby the piezoelectric transducer being bended and remained in the stressed condition before applying a force on the piezoelectric bending structure, which increase the stress on the piezoelectric surface and hence increase the generated electrical charges. An impact force is being exerted onto bending the piezoelectric beam and hence generating electrical power across an external resistive load. The proposed bending mechanism prototype has been manufactured by employing 3D printer technology in order to conduct the evaluation. A free fall test has been conducted as the evaluation method with varying force using a series of different masses and different fall heights. A rectangular piezoelectric harvester beam with the size of 32mm in width, 70mm in length, and 0.55mm in thickness is used to demonstrate the experiment. It can be seen from the experiment that the instantaneous peak to peak AC volt output measured at open-circuit is increasing and saturated at about of 70V when an impact force of about 80N is being applied. It is also found that a maximum power of about 53mW is generated at an impact force of 50N when it is connected to an external resistive load of 0.7KΩ. The reported mechanism is a promising candidate in the application of energy harvesting for powering various wireless sensor nodes (WSN) which is the core of Internet of Things (IoT).

1. Introduction

Recently, Microelectromechanical systems (MEMS), wireless sensors network (WSN), as well as internet-of-thing (IoT) has grabbed the attention of many researchers worldwide, and particularly on microactuators, microsensors, together with microsystems that achieved by merge micromechanical devices with microelectronic for many-sided purposes. Precisely, impact-based energy harvesting (IBEH), which are capable of converting ambient mechanical impact energy into usable electrical energy, are attracting majority of researchers attention for the sake of increasing demand of self-powered WSN for IoT applications [1][2], in which many of this WSN are embedded in the stand alone structure or remote spatial environments whereby there will be a huge numbers of wireless sensors and smart objects surrounding our living space in the "IoT" era, where it will lead to make our living space being smarter, green, and for better quality of lives [3]. Up-to-date IBEHs are facing the trade-off among the impact force, impact velocity, and the output power, whereby the applied impact parameters will affect the stress on the piezoelectric surface which lead to change to output power [4].

As aforementioned, IBEH is one of the attractive approaches that proposed to be used for powering up low power portable autonomous microsystems, due to the fact that, mechanical impact energy resources are available in abundance and ubiquitous surrounding us, starting from human beings activities (walking, dancing, exercising), ending to the industries environment whereby all the machines are generates a noticeable and almost continues amount mechanical impacting energy, whereby all that impacts energy has been dissipated and wasted to the ground or into the impact absorbents. Therefore, this research paper presents a new pre-stressed bending mechanism along with laboratory testing which allocated for harvesting that wasted mechanical impact energy. Moreover the proposed method employs an off the shelf piezoelectric cantilever harvester element due to its high electrical output density contrast with the piezoelectric ceramic disc.

1.1. Piezoelectricity

Piezoelectricity is the electrical charge that accumulates in certain crystal in corresponding to applied mechanical stress. In another word, the word piezoelectricity means electricity that generated from pressure. This piezoelectricity effect is considered a reversible effect, whereby this type of crystal has the ability to behave as an actuator (the shape of the piezoelectric crystal will deform) in according to an AC signal supplied with a certain frequency connects to its terminals [5]. Brothers Pierre, Curie, and Jacques are the researchers who discovered this effect in 1880 [6]. Whereby they found that a certain crystals have the ability of producing dual polarities charges when mechanical stress was applied on it. Subsequent and in 1881, Lippmann has been mathematically deduced the inverse piezoelectricity from fundamental thermodynamic principles, and has been confirmed later on by Curies [7].

The piezoelectric element crystal is addressed by a constitutive equation, which is by merging the electrical induction *D*, electric field *E*, strain *S*, and the stress *T*. Moreover, according to the principle of energy conservation at low frequency we have:

$$D = dT + \varepsilon^T E \tag{1}$$

$$S = S^E T - d'E \tag{2}$$

Where: ε^T: Permittivity at constant stress.
S^E: Compliance at constant electric field.
d: Piezoelectric charge coefficient.

Consequently, and compared to a non-piezoelectric material, there will be always a strain due to the applied electrical field, and vice versa, an electrical charges will be deduced due to the mechanical stress (electrical charges that displaced inside the material are inducing the opposite polarity surface charges on the plates). When the surface area does not deformed under the applied stress (which is impossible in polymers), then ($d=d'$). When solving Equation (1) for *E*, and equation (2) for *T*, we will get:

$$E = \frac{D}{\varepsilon^T} - \frac{Td}{\varepsilon^T} = \frac{D}{\varepsilon^T} - gT \tag{3}$$

$$T = -\frac{d}{S^E}E + \frac{1}{S^E} = C^E S - eE \tag{4}$$

Furthermore, while stressing the piezoelectric materials by applying an external force F_3 in the direction as shown in Figure 1, which bend the structure downward, stressing the piezoelectric materials and building up electric charges $Q_3=D_3A_3$ on the surfaces of area A_3 to yield a voltage $V_3=Q_3/C_3$. The electric charge density is $D_3 = d_{33}T_1$, whereby $T_1=d_{33}F_1/A_1$ and $C_3 = \varepsilon_{33}^T A_3/h$. Hence, the output voltage generated by applying a corresponding amount of force can be calculated from:

$$V_3 = d_{33}F_3\left(\frac{1}{A_3}\frac{h}{\varepsilon_{33}^T}\right) \tag{5}$$

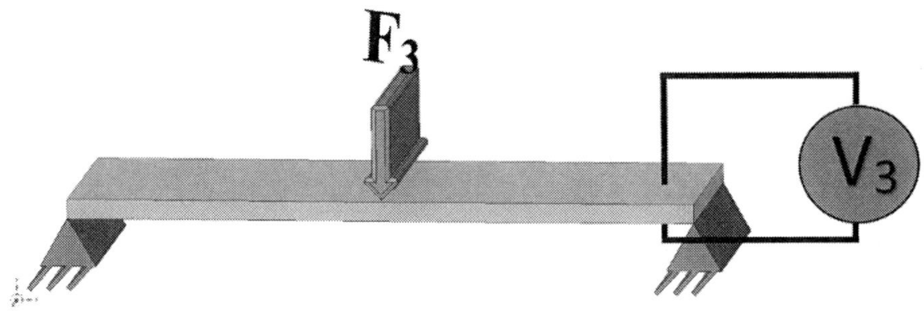

Figure 1. Open circuit voltage measurement of a piezoelectric module while applying external force.

Obviously it can be concluded from Equation (5) that the output voltage *V3* which generated at the piezoelectric transducer terminals is directly proportional to the applied force F_3, by assuming that the other variables are constants. A general piezoelectric energy harvester equivalent circuit can be illustrated as shown in Figure 2, which consisting of mechanical domain clarifies the direct effect of the piezoelectric materials, where m representing the mass of the harvester, d the damping factor, k the mechanical stiffness, α the piezoelectric coupling coefficient and σ_{in} represent the applied force, while the electrical domain consisting of v represents the output voltage, *Cp* and *RL* are representing the inner capacity and external resistive load respectively. Whenever an external force is being applied to deform the piezoelectric structure, the voltage output would be generated according to the responses of the force changes over the time [8][9].

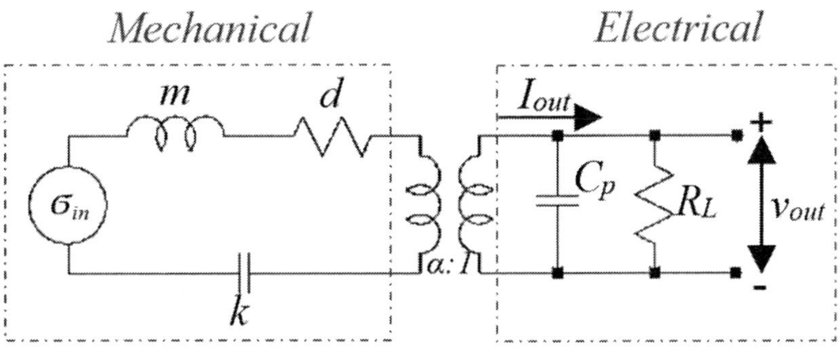

Figure 2. The equivalent circuit of the piezoelectric harvester.

1.2. Impact Force of Object in Free Fall

A free fall is a special type of motion in which the only force acting upon a falling object is the gravity assume that there is no drag force of air. Likewise, when an object with mass equal to m is freely fall from h_1 height, and lets presume its velocity is v upon the impact, therefore its energy equation can be written as the following equation (6):

$$mgh_1 = \frac{1}{2}mv^2 \tag{6}$$

Where g is the gravity acceleration. When the free fall height h_1 is known, subsequently the impact velocity of the falling object can be calculated from:

$$v = \sqrt{2gh_1} \tag{7}$$

The impact force for a falling object is dependent on the impact depth, h_2 before it comes to a totally stop, which can be written as follow:

$$F = \frac{E_k}{h_2} = \frac{mv^2}{2h_2} \tag{8}$$

Where E_k is the kinetic energy. From equation (8) can tell that the falling object force upon the impact is inversely proportional to the penetrated distance h_2 or knowing as bending offset. Therefore and in order to increase the piezoelectric electrical charge generation, and from the stress equation bellow:

$$T = б * S \tag{9}$$

Where: T: Stress
б: Young's Modulus
S: Strain

So from equation (8) and equation (9) can say that h_2 need to be enlarged so that the strain S, and the stress T will be increased accordingly. Consequently, the voltage output is proportional to the stress (Equation 5). Thus a pre-stress bending mechanism has been designed to enlarge the bending offset h_2, and hence maximizing the piezoelectric power generation as mentioned earlier rather than the unstressed bending mechanism, as clear in Figure 3.

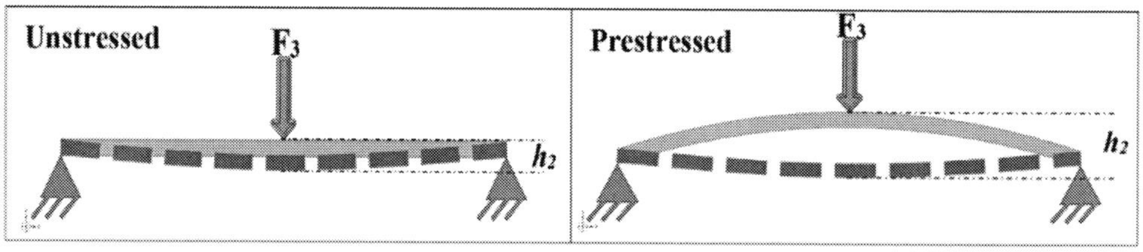

Figure 3. Unstressed vs prestressed bending offset h_2.

This mechanism is designed to house the piezoelectric, pre-stress it, and then stress it according to the applied force. The bending offset h_2 was set to be at a maximum of ≈ 5mm according to the used piezoelectric elasticity properties before it cracked.

A falling object is expected to be bouncing back and forth several times until its momentum reduced to zero and then its impact force can be written as follow:

$$F = \frac{dp}{dt} = m(\frac{dv}{dt}) \tag{10}$$

Where: p: is the momentum.

The maximum electrical power P_{out} generation is fully relying on the impact force F, impact velocity v, as well as the external resistive load R_L, where $P_{out} = V^2/ R_L$.

2. Related Work

Mechanical impact energy to electrical power conversion by employing the piezoelectric effect has received a considerable research attention around the globe in order to develop real-world applications. At present, most studies focus on using a single piezoelectric module. However researchers like M. Renaud and his team [10] have developed a prototype for utilizing the human beings foot strikes to impact piezoelectric beams. Their experimental results showed 600µW harvested from 10Hz frequency and 10cm amplitude linear motion. M. Ferrari et al [11] moved a step forward by presenting a prototype of an autonomous battery-less sensor module. An electrical power of 35mW was harvested from their module and successfully power up the sensor module and send the data within a range of 25m in a laboratory environment. S. Ju and C.-H. Ji in [12] presented a prototype of an impact based piezoelectric cantilever element for harvesting vibration energy by employing free movable metal spheres as a proof mass and an MFC (Macro Fiber Composite) beam as a piezoelectric cantilever, their design have the form factor of a conventional wristwatch in order to harvest energy from low frequency human-body-induced vibrations. When the device was mounted on a wristwatch and shaken manually they achieved a maximum peak-to-peak open circuit voltage of 54V and 621μW of output power have been measured at resistive load 1KΩ. Kyoo N. C. and Hee H. R. [13] presented a novel impact based piezoelectric harvester mechanism. The authors succeeded to harvest a continuous electricity by applying a continuous impact on the piezoelectric surface from wind flow. And they tried to increase efficiency by increasing the impact duty cycle with a spring retention action has been used to increase the frequency. In this prototype they had cascaded five elements of piezoelectric harvesters connected in series and four piezoelectric harvesters connected in parallel and it was able deliver about 16mW with optimum resistive load of 1KΩ.

3. Impact based experiment setup for piezoelectric power generator

For the sake of proving the concept that the prestressed piezoelectric mechanism can further improve and maximize the output wattage generated by the piezoelectric comparing to the unstressed piezoelectric bending mechanism as mentioned in Figure 3 previously. In consequence, a free fall experiment was conducted to evaluate the piezoelectric transducer performance (in term of the electrical power generated via the conversion of impact force) that seated into the prestressed bending mechanism as shown in Figure 4, which is designed to maximize the piezoelectric transducer output from a single impact.

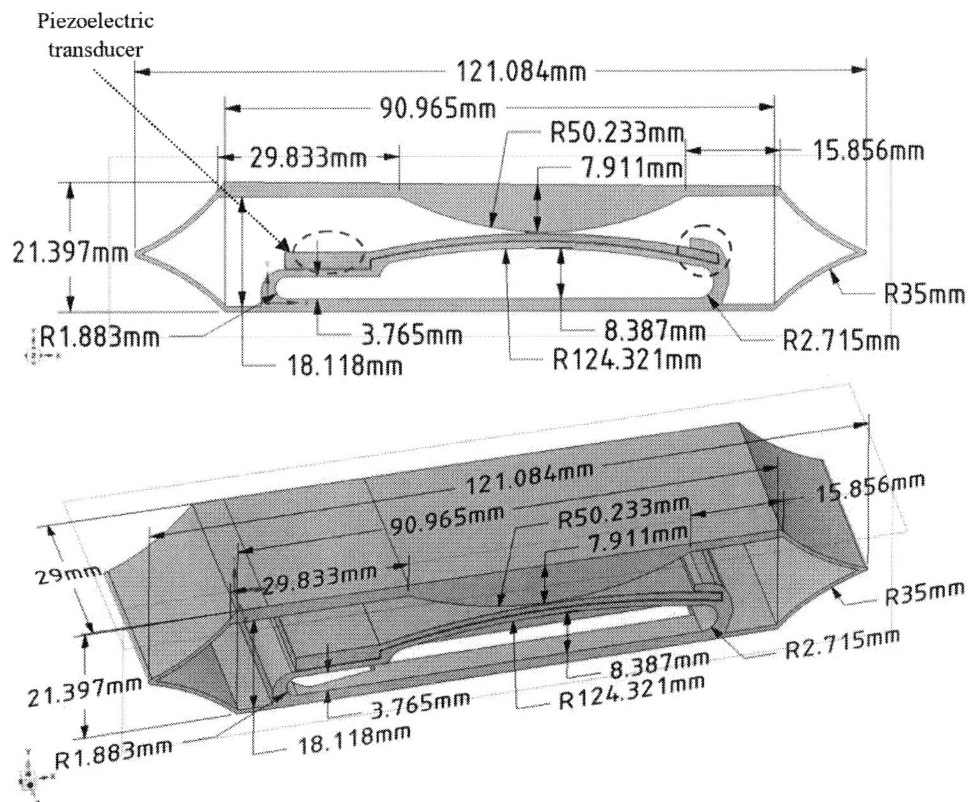

Figure 4. 3D model of the proposed piezoelectric prestressed bending mechanism.

The first task of this piezoelectric housing mechanism is to upward stress (in a concave manner) the piezoelectric by 3mm. The upper lip of the mechanism is convex oppositely to the piezoelectric, where it has the ability to press down on the top of the piezoelectric and strained it. And then when the impact force F_3 being applied on the top of the mechanism (upper lip) the concave piezoelectric will gradually flatten and then becomes convex due to the magnitude of the applied force. The bending offset h_2 for this mechanism was set to be at a maximum of ≈ 5mm according to the used piezoelectric elasticity properties before it cracked. An off-the-shelf rectangle shaped piezoelectric transducer manufactured by "Piezo Systems, Inc" is used to conduct the experiment as shown in Figure 5. Table 1 illustrates the dimensions of the used piezoelectric transducer and the overall dimension of the proposed bending mechanism including the piezoelectric transduce.

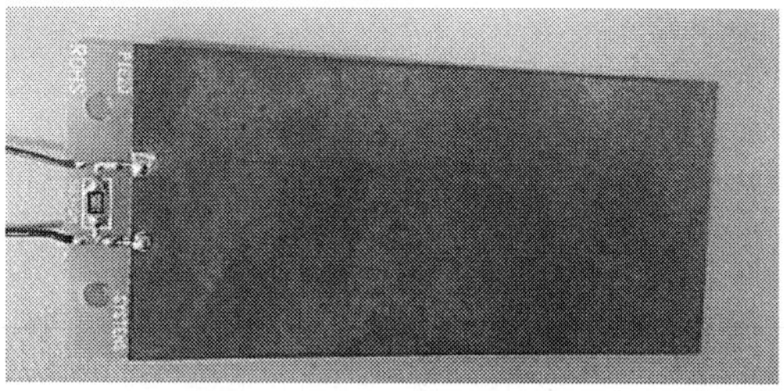

Figure 5. Piezoelectric transducer.

Table 1. Dimensions of the piezoelectric transducer and the proposed mechanical housing.

Parameter	*Piezoelectric transducer*	*Mechanical housing*
Width	29mm	29mm
Length	70mm	121mm
Thickness	0.55mm	21.4mm

The piezoelectric transducer shows in Figure 5 above is being rigidly fixed by two screws at the left end (blue circle in Figure 2), whereas the second right end it was freely lying into a slot (red circle in Figure 2). Since the piezoelectric transducer will be bend by 5mm due to the impact force F_3 so it is obviously the transducer be elongates by about 1mm so here the usefulness of this slot comes to allow the transducer to elongates and shrieks during the bending phase and returned back to the original length after the bending force vanish. The 3D printer used to manufacture the proposed design using ABS material as shown in Figure 6.

Figure 6. 3D printed prototype that used for conducting the experiment.

Next, can start the evaluation experiment by applying different magnitudes of forces on the top of the manufactured prototype by impacting it with different objects have a different mass, freely falls from different heights. And simultaneously measure the maximum harvested AC voltage by using a digital storage oscilloscope. A guiding tube was used so that can produce a consistent result. And also to guarantee that the free falling object is precisely impact on the designated surface of the bending mechanism as shown in Figure 7.

IOP Conf. Series: Materials Science and Engineering **210** (2017) 012037 doi:10.1088/1757-899X/210/1/012037

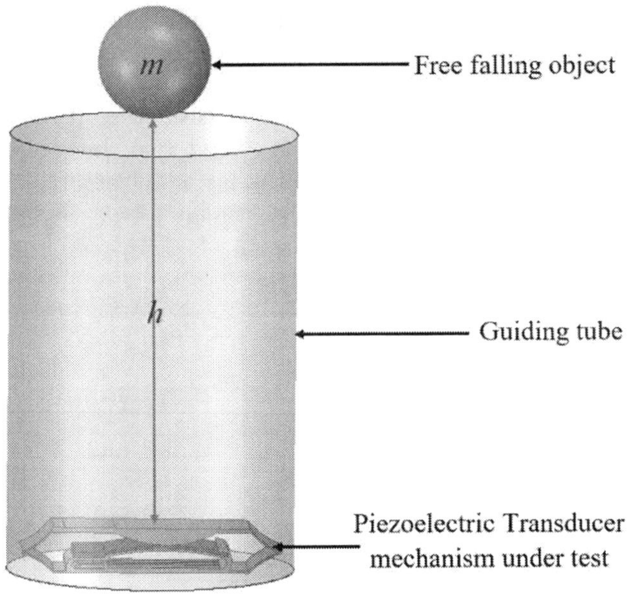

Figure 7. Drop test experiment set-up.

4. Experimental results and discussion

In pursuance of proving the concept of the proposed prestressed bending mechanism, the output AC voltage harvested by the piezoelectric has been measured with the aid of digital storage oscilloscope DSO while the mechanism was impacted by several forces. During the free fall test, and according to the fact of that the manufactured (3D printed) mechanism's surface is very compact (more solid) comparing to the falling object, in consequence, the free fall object will follow the guide tube and precisely impact the mechanism's top surface at its full momentum generating the highest AC peak voltage. And thereafter it will bounce forth and back generating smaller AC peaks until its momentum becomes zero. Figure 8 shows these two peaks.

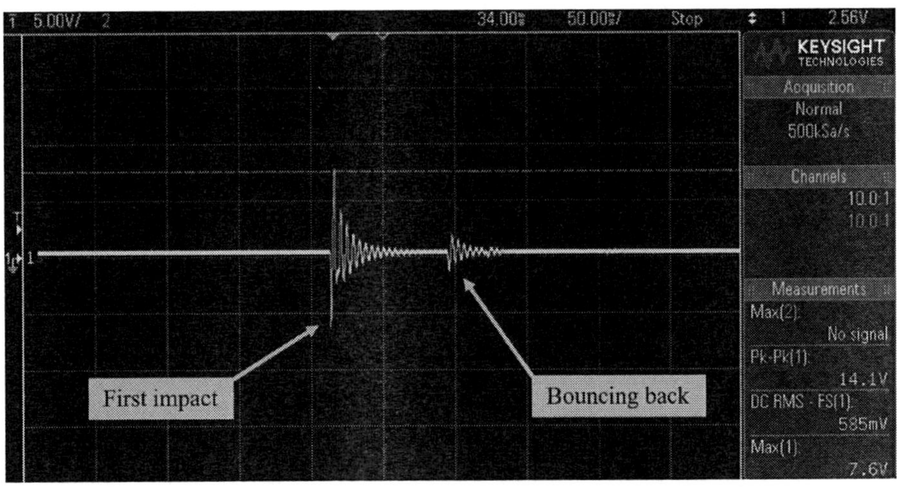

Figure 8. Screenshot of DSO measuring an open-circuit AC output voltage waveform of piezoelectric when it's been impacted at certain force.

From Figure 8, can observe that the first peak is a noticeably higher comparing to the rest of the signal, that's due to the high momentum of the free falling object at the first impact time, where its gradually losing its momentum until reach to the steady state zero. Accordingly with regard to the evaluation aim, only the peak of the first AC pulse will be considered and compared, however the other low signals are still considerably useful harvested signals where it have a small noticeable contribution of charging a storage device in case of energy harvesting powered system. As shown in Figure 8, which illustrates a screenshot of DSO measuring an open-circuit AC output voltage waveform of piezoelectric when it's been impacted at certain force. The AC.V output of this mechanism shows a logarithmic behaver where its starts from about 14.5 AC.V while impacted with a minimum of 3.3N, and gradually increases along with applied force. From Figure 9 can see that the AC.V out was saturated at 70 AC.V with 78.5N onwards, therefore there is no need of applying more amount of forces.

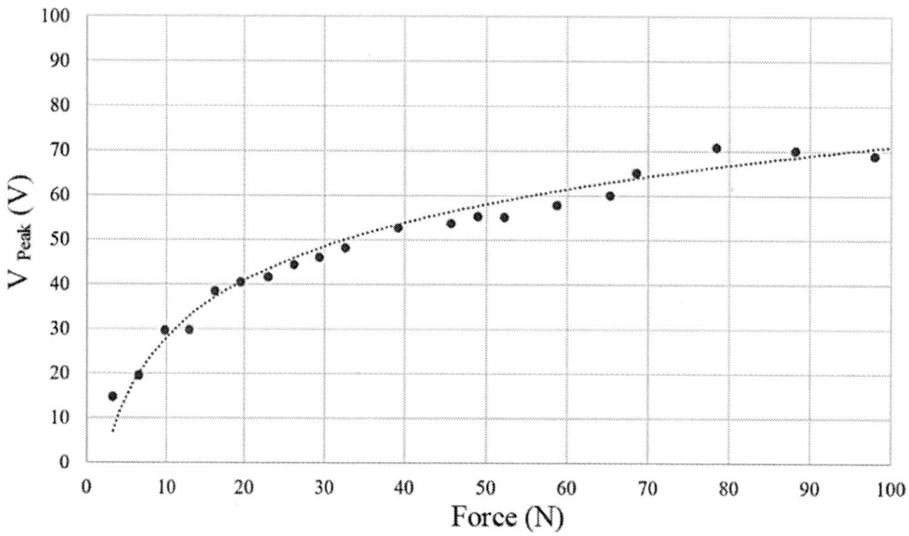

Figure 9. Peak AC. V harvested by piezoelectric transducer that prestressed and been impacted at varying forces.

Even though can get very high amount of harvested voltage from a single piezoelectric harvester, where on the other hand it generates very small amount of alternating current due to its properties and the internal impedance of the piezoelectric itself. Wherefore and in order to maximize the output current of piezoelectric and hence the power, the load impedance has to be matched with the value of the internal impedance where can attain the maximum power output.

Therefore, in order to find the maximum power that can be harvested by the piezoelectric with the aid of this prestressed mechanism, four SMD schottky diodes are preferred to use as a full-wave bridge rectifier due to their low forward voltage drop comparing to a normal diodes. 0.1 μF smoothing filter capacitor used to smoothen the output voltage. A resistor decade box in the magnitude of 0Ω up to 10 KΩ used as a pure resistive load. From Figure 10 can observe the relation of the DC power and the DC voltage output verses the resistive load. The mechanism was impacted at fixed magnitude all the time of force of 80N (the open circuit voltage at this force is about 70AC.V). The load was varied from 0.5 KΩ up to 6 KΩ. Can realize from Figure 10 that the maximum power can be achieved is about 53mW when connecting a resistive load of 700Ω. That elucidates that the load impedance matches the internal impedance of the Piezoelectric therefore produce the maximum power out at this point. At this amount of load the measured DC voltage was about 6V and when applying power law $P = I * V$ will find that the output current is about 9mA.

These values considered more than enough and can play a noticeable role in charging a storage element, powering up, and as a result prolong an autonomous system's lifespan. Where it can be used for portable low-power devices such as wireless sensor network node (WSN) or an internet of thing (IoT) applications to substitute the conventional batteries where frequent changing batteries, economically not a good idea.

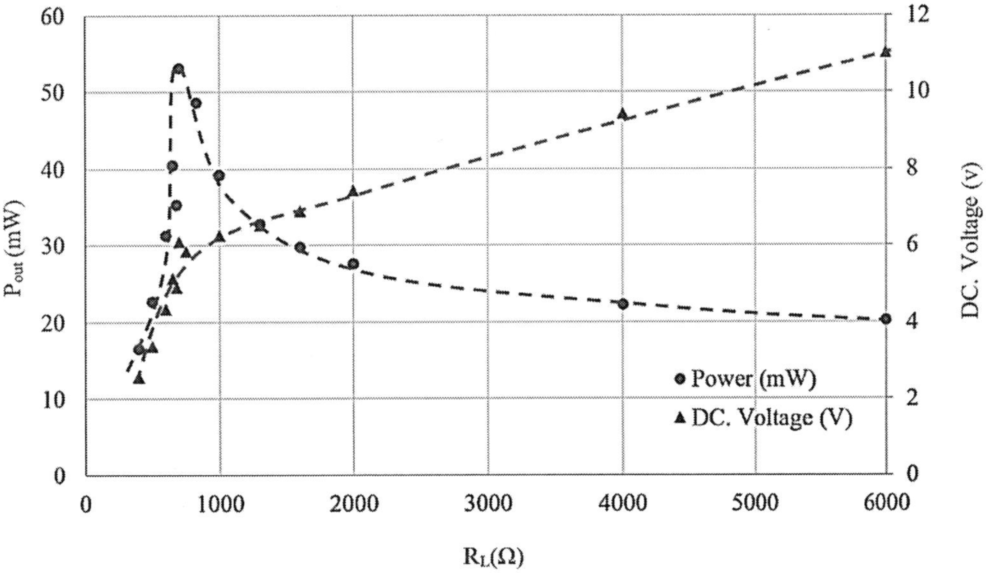

Figure 10. DC output power and voltage at varying external resistive.

5. Conclusion

A piezoelectric prestressed bending mechanism has been presented in this research paper. This mechanism improves the mechanical impact force into bending strain on the piezoelectric harvester and hence generating higher electrical energy compared to non-prestressed piezoelectric structure. Proof of concept along with the evaluation procedure was performed with the aid of a free fall test. The suggested mechanism was capable of harvesting 70 AC.V at open circuit measurement. After rectified it was able to deliver a maximum DC power of 53mW with a load of 700Ω. This amount of electrical power is considerably good enough to be stored and then power up low power portable devices, where this mechanism comes handy to be used for such devices where it's not an easy choice to replace the battery, for instance, wireless sensor network nodes and IoT devices. This prestressed mechanism is meant to be used for harvesting human foot strike impact energy since it can be easily embedded inside the shoes or even built-in beneath the tiles. Furthermore, It also can be used in the industry environment to harvest the wasted impact/bending energy that generates from machines and automobile chassis (car wheel strut assembly) when driving on to undulating roads.

Acknowledgments

The authors would like to acknowledge the support of this work by the Malaysian Ministry of Higher Education under the grant no. FRGS/2/2014/SG02/FKEKK/02/F00244, UTeM Zamalah scholarship, as well as the facility support by the Faculty of Engineering Technology (FTK), Advanced Sensors and Embedded Control Systems (ASECs) Research Group, CeTRI, UTeM.

Reference

[1] Fatemeh M, Loay S I, Tarek M E and Mohamed H A 2017 *IEEE Early Access Articles* 99

[2] Deepti and Sukesha S 2016 *1_{st} IEEE International Conference on Power Electronics. Intelligent Control and Energy Systems*

[3] Parisa R, Diane J C, Lawrence B H and Maureen S E 2011 *IEEE Trans. Knowledge & Data Eng.* **23** 527-539

[4] Ali M A, Kok S L and Kok-Tee L 2016 *Journal of Telecommunication, Electronic and Computer Engineering* **8**

[5] Swee L K, Nile W and Nick H 2011 *Free- standing thick-film piezoelectric cantilever-energy harvesting*

[6] Cady W 1964 *Piezoelectricity*, New York:McGraw-Hill.

[7] Marson W P 1950 *Piezoelectric crystal and their application to ultrasonic,* D. Van nostrand Company Inc.

[8] Yangyiwei Y, Xiang S, Haoran L, Zhao X and Ying D, Yaoze L And Tongqing Y 2015 *APPLIED PHYSICS LETTERS* **106** 173902

[9] Meiling Z, Emma W and James N 2009 *IEEE Transactions on Ultrasonics, Ferroelectrics, and Frequency Control* **56**

[10] Renaud M, Fiorini P, Schaijk and Hoof C 2009 *Transducers* 2009, Denver, CO, USA

[11] Ferrari M, Cerini F and Ferrari V 2013 *Transducers* 2013, Barcelona, SPAIN

[12] Ju S and Ji C-H 2015 *Transducers 2015*, Anchorage, Alaska, USA

[13] Kyoo N C and Hee H R 2015 *IEEE 2nd International Future Energy Electronics Conference*

IOP Conf. Series: Materials Science and Engineering **210** (2017) 012038 doi:10.1088/1757-899X/210/1/012038

Investigation on Phase Shifting Effect on The Voltage Output of Piezoelectric Cantilever Array

B Y Jing and K S Leong*

Faculty of Electronic and Computer Engineering, Universiti Teknikal Malaysia Melaka, Hang Tuah Jaya, 76100 Durian Tunggal, Melaka, Malaysia.

Email: *sweeleong@utem.edu.my

Abstract. This paper analyses the phase shifting of the output waveform produced by piezoelectric cantilevers under a range of vibration frequencies. The phase shift of four piezoelectric cantilevers with different resonant frequency are inspected while it is excited with the vibration from the electrodynamics shaker at a range of frequencies from 100 Hz to 500 Hz with the acceleration level (g-force) fixed at constant magnitude of 1g-level (9.81 m/s2). Time different and Lissajous pattern methods were used in this research to measure the phase shift of the output waveform. Both methods show similar result where the major phase shift happened at the resonant frequency of respective cantilevers. The phase difference remains low around 0 degrees or in other term in phase before the resonant frequency of the cantilever. When the frequency of the vibration source approaches the resonant frequency of respective cantilever, the phase different start to increase rapidly and reach 180 degree which is out of phase after the resonant frequency. This major phase shifting contributes to the significant rise of the gap in between the peaks formed when multiple piezoelectric cantilevers are connected together. As a result, it indirectly improves the output performance of the piezoelectric cantilevers array.

1. Introduction

Piezoelectric based energy harvesting had gained interest over the time due to the advancement in integrated circuit [1]. Improved integrated circuits had its power consumption reduced greatly, making it possible to be powered up by using micro-power energy harvester. Since piezoelectric had shown promising result and potential in harvesting energy to power-up micro-power electronic devices [2-5], it sparked research interest in lot aspects, mainly on maximizing the performance of the piezoelectric material, including technique to optimize the output from the piezoelectric material and widening the operating bandwidth of the piezoelectric energy harvester [1].

Piezoelectric energy harvester especially cantilever based piezoelectric will only produce optimum output within certain excitation frequency range. If the frequency of the vibration shifts, the output of the piezoelectric material will decrease radically. Therefore, many techniques were introduced to broaden the operating frequency bandwidth of the piezoelectric material before its practical deployment. The techniques included resonant frequency tuning [6] which adjust the resonant frequency of the piezoelectric material using mechanical, magnetic, or piezoelectric method; nonlinear energy harvesting configuration that use magnets to adjust the nonlinear stiffness and nonlinear piezoelectric coupling of the material; and multimodal energy harvesting which utilise multiple degree-of-freedom system of the piezoelectric cantilever by using multiple bending modes of a piezoelectric cantilever or by connecting multiple cantilevers into an array.

Researches shown when multiple cantilevers are connected together into an array, the operating frequency bandwidth of the model broadened [7-13]. Ideally, when connecting two or more alternating current (AC) outputs in series connection with same polarity the total output should be the sum of all the connected output, while series connection with alternating polarities will cause the generated output to cancel out each other. However, researches shown different outcome when piezoelectric cantilevers with different resonance frequencies are connected in an array [14]. This implied that the output produced by the piezoelectric cantilevers with different resonance frequencies might not in phase.

Hence, this paper investigate the phase difference of the output waveform from the piezoelectric cantilevers with difference resonance frequencies and also its effect on the output of the piezoelectric cantilever array when connected in series with same polarity and alternating polarities.

2. Experimental Set-up

This research uses piezoelectric standard quick-mount bending generator with pre-mounted and wired at one end (Q220-A4-303YB) from Piezo Systems, Inc [15] as the test subject. The dimension of the piezoelectric generator is shown in Figure 1.

(a) Top View *(b) Side View*

Figure 1. Dimension of Piezoelectric Cantilever (Q220-A4-303YB) [11].

The experiment set-up of this research is as shown in Figure 2, which consists of a G-link wireless sensor and receiver, an electrodynamics shaker, a gain amplifier, a function generator, an oscilloscope, and piezoelectric cantilevers. In order to generate a controllable artificial vibration for test purpose, function generator, gain amplifier, electrodynamics shaker, and G-link wireless accelerometer were used. The function generator was used to supply AC input power to the electrodynamics shaker. Since the power supplied by the function generator alone is not sufficient to generate vibration with high acceleration level (g-force), the gain amplifier was used to amplifier the power before supplying it to the electrodynamics shaker.

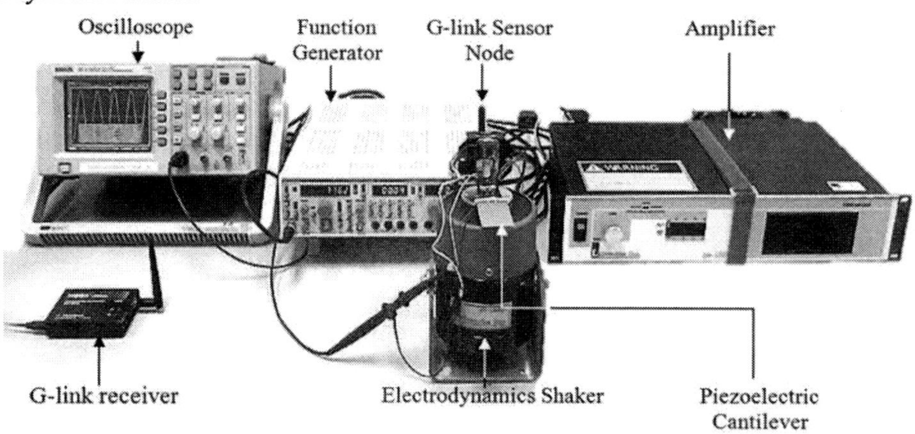

Figure 2. Experimental Set-up

Five units of piezoelectric cantilevers were used in this research. The cantilever is named C1, C2, C3, C4 and CC for reference convenience. In this case, the constant cantilever, CC is a cantilever with its resonant frequency far away from the other cantilevers to acts as a test subject to compare the phase shifting of the other cantilevers. Its resonant frequency is tuned to higher frequency by clamping the cantilever over than its clamping base, towards its flexible beam [16]. While the resonant frequencies of the piezoelectric cantilever C2, C2, and C4 is altered to lower frequency by attaching proof mass to the tip of the cantilevers. After the resonant frequency of each cantilever is tuned to different value of 180 Hz (C4), 220Hz (C3), 270Hz (C2), and 300Hz (C1), the phase difference between cantilever C1, C2, C3, C4 and a constant cantilever CC is recorded and examined. In this research, the phase shift of the cantilevers is measured by using time different method and Lissajous pattern method. The results obtained from both methods are then compared for verification purpose.

After the phase difference of the cantilevers are measured and examined, the research continued with observing the output produced by the cantilevers when connected in series connections. The output is observed when cantilevers are connected in series with same polarity and series with alternating polarities connections to examine the effect of phase shifting of the piezoelectric cantilevers towards it output performance.

2.1. Time Different Method

For time different method, the cantilever CC and C1 are connected individually to the oscilloscope, and the output is viewed in time base setting as shown in Figure 3. The cantilevers were excited with vibration from the electrodynamics shaker at a range of frequency from 10 Hz to 500 Hz with the acceleration level (g-force) fixed at constant magnitude 1-g (9.81 m/s2) and the phase different between the cantilevers was calculated using Equation 1.

$$\theta = \left(\frac{w}{x}\right) \times 360^{\circ} \quad (1)$$

where, θ is the phase shift of cantilever, w represents the phase shift in term of centimeter (cm) and x represents one period in term of cm. The test was then repeated by replacing the cantilever C1 with C2, C3 and C4.

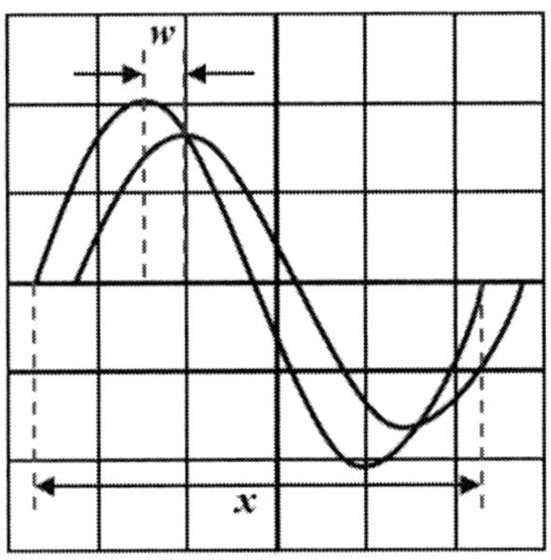

Figure 3. Illustration of phase shift measurement by using Time Different Method

2.2. Lissajous Pattern Method

While for Lissajous pattern method, the cantilever CC and C1 are connected individually to the oscilloscope, and the output is viewed in X-Y mode of the oscilloscope as shown in Figure 4, where the phase different between cantilevers CC and CL can be observed and measured. The cantilevers were again excited with vibration from the electrodynamics shaker at a range of frequency from 10 Hz to 500 Hz with the acceleration level (g-force) fixed at constant magnitude 1-g (9.81 m/s2) and the phase different between the cantilevers was calculated using Equation 2.

$$\theta = 180^\circ - \left(\sin^{-1}\left(\frac{y}{z}\right)\right) \quad (2)$$

where, y represent the intersection at y-axis in cm and z represent the maximum height of the ellipse in cm. The test was then repeated by replacing the cantilever C1 with C2, C3, and C4.

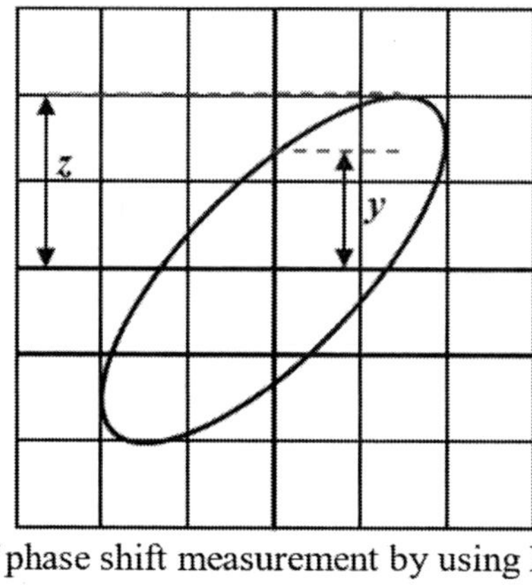

Figure 4. Illustration of phase shift measurement by using Lissajous Pattern Method

3. Experimental Result

The output waveform of C1, C2, C3, and C4 is being observed by comparing to CC. The result of the phase different for the cantilevers is shown in Figure 5 for Time Difference Method and Figure 6 for Lissajous Pattern Method.

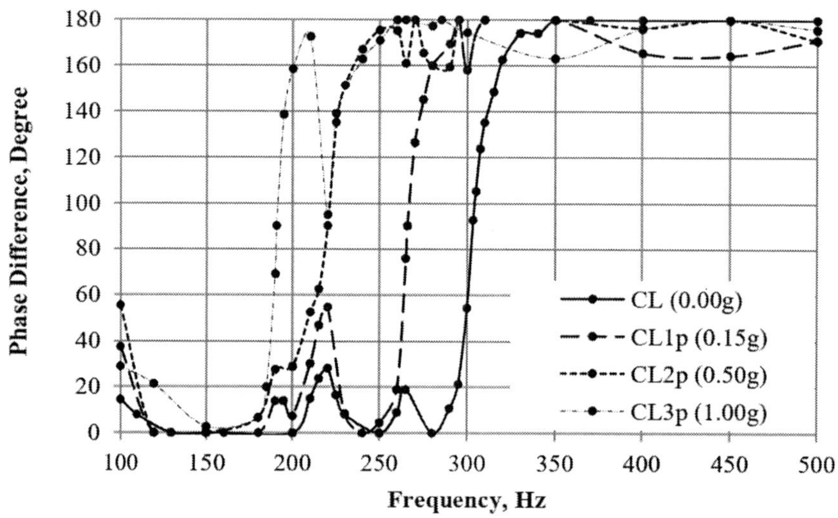

Figure 5. Phase shift of Cantilevers with Proof Mass Attached Obtained using Time Different Method

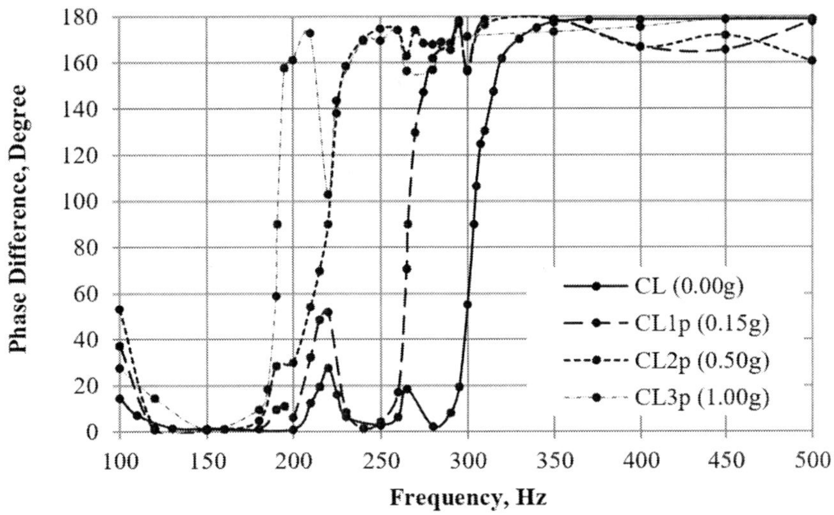

Figure 6. Phase shift of Cantilevers with Proof Mass Attached Obtained using Lissajous Pattern Method

Both methods showed similar result where the major phase shift happened at the resonant frequency of respective cantilevers. The phase difference remains low around 0 degrees before their resonant frequency. When the frequency of the vibration source approaches the resonant frequency of respective cantilever, the phase different between the test cantilever and the constant cantilever start to increase rapidly and reach 180 degree after the resonant frequency. Major phase shifting happen at 300Hz for C1, 270Hz for C2, 220Hz for C3, and 180Hz for C4. Notice that there are also ripples shows in the graph. Cantilevers with heavier proof mass, have larger ripples. Proof mass was attached to the tips of the cantilevers in order to alter its resonant frequency. Frequency responses for individual cantilevers with added proof mass of 0.15g, 0.50g, and 1.00g to C2, C3, and C4 respectively are shown in Figure 7. Notice that the resonant frequency of C1, C2 and C3 is reduced. Resonant frequency for C1, C2, C3, and C4 are 300 Hz, 270 Hz, 220 Hz, and 180 Hz with maximum voltage of 1.99V, 2.24V, 1.88V, and 1.82V respectively.

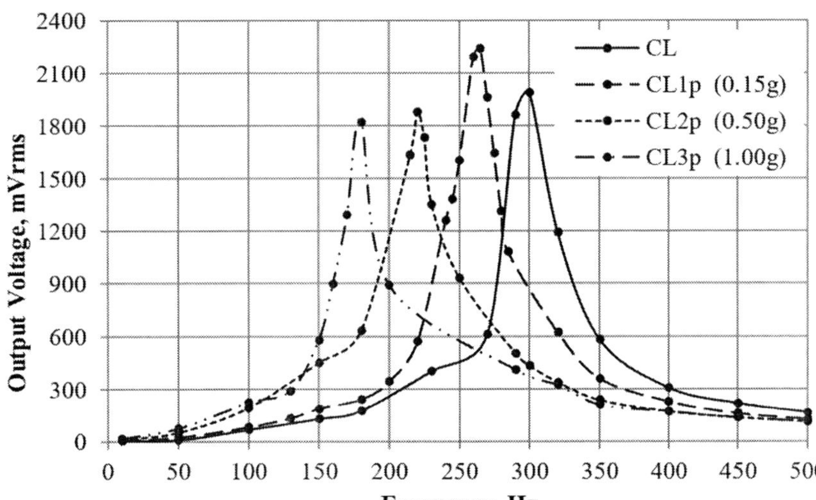

Figure 7. Frequency Response in term of Output Voltage for Multiple Cantilevers with Proof Mass

These altered cantilevers were then connected together electrically in series connections in order to observe the effect of phase shifting of the piezoelectric cantilever toward its output performance. The frequency responses for cantilevers connected in series with same polarity connection is shown in Figure 8, while result for series with alternating polarities connection is shown in Figure 9.

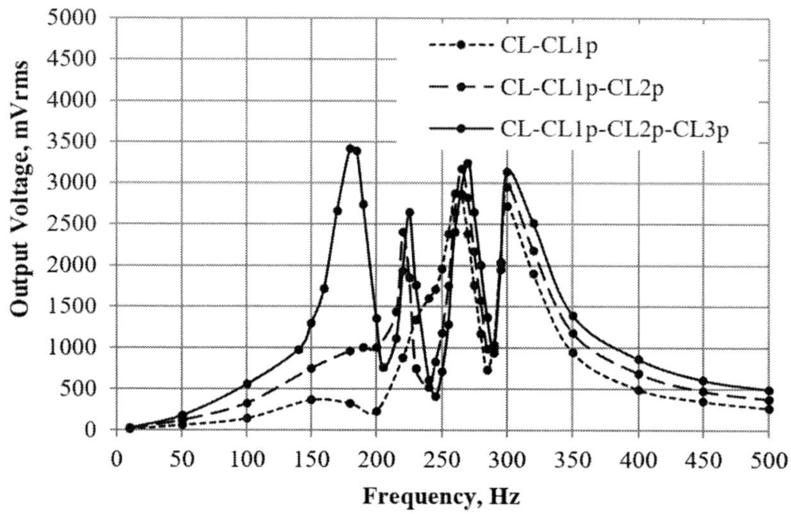

Figure 8. Frequency Responses (Voltage) for Cantilevers Connected in Series with Same Polarity Connection.

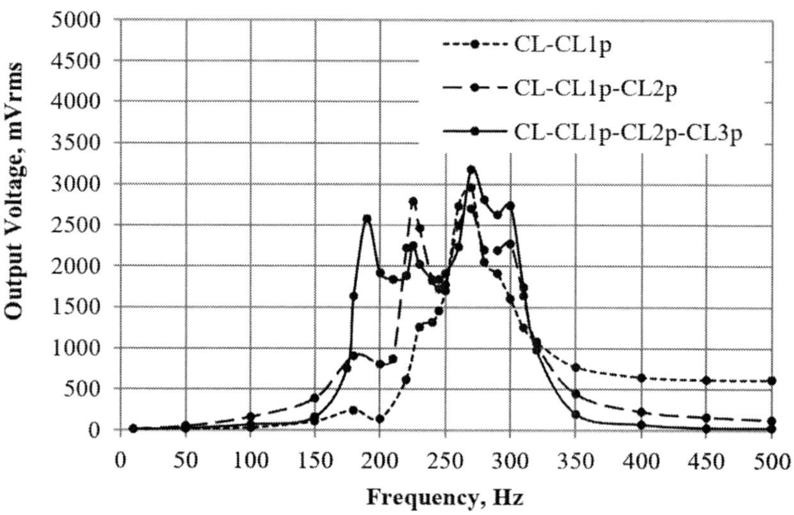

Figure 9. Frequency Responses (Voltage) for Cantilevers Connected in Series with Alternating Polarities Connection.

Notice that another peak is formed whenever another cantilever is added. For example, when two cantilevers are connected, two peaks were formed; while when three cantilevers are connected, three peaks were formed. These peaks are formed around the resonant frequency of each respective cantilevers added, where when C1 and C2 are connected, peaks formed at 300Hz and 270Hz; when C1, C2 and C3 are connected, peaks formed at 300Hz, 270Hz and 220Hz; and when C1, C2, C3 and C4 are connected, peaks formed at 300Hz, 270Hz, 220Hz, and 180Hz. Also notice that the operating bandwidth increases in respect of increasing number of the connected cantilevers. The maximum voltage produced by each peak is also higher than their original voltage output produced. However, there is huge drop of voltage in between the peaks and it is undesirable. For example, note that when connecting four cantilevers all together in series with same polarity, the operating frequency bandwidth widen to between 160Hz to 335Hz for application that need 1.80V voltage input. However, the application will not work in between 145Hz to 220Hz, between 230Hz to 260Hz, and between 280Hz to 295Hz because the voltage drop lowers than 1.80V in between that frequency.

This narrow gap will not form if the output waveforms for each cantilever are in phase. Both result in Figure 5 and 6 proved that the phase of the output waveform shift at the resonant frequency of each cantilever, it explain the reason why when two waveforms are connected in series with same polarity the total output is not the sum of maximum output of both waveforms, but lesser because the waveforms are not in phase.

This report is further proven by the result shown in Figure 9, where the piezoelectric cantilever is connected in series but with alternating polarities. Same as in series with same polarity connection, peaks were formed whenever cantilevers are added. These peaks were also formed around the resonant frequency of each respective cantilevers added with maximum output voltage higher than the original voltage output produced from non-connected cantilevers. However, notice that the waveform formed has clear difference from the waveform formed in by same polarity connection, where the narrow gap between peaks are not as low as in waveform produced by the same polarity connection. For example, for applications that operate with 1.8V voltage input, the operating frequency of four connected cantilevers increased to the range of 180Hz to 310Hz compared to same polarity connecting, this alternating polarities connecting does not have gaps in between the operating frequency.

This major phase shifting is the reason for the significant rise at the gap in between the peaks formed in Figure 9. As the waveform is shifted, when the cantilevers are connected in alternating polarities, the value of the positive peak will not totally cancel out by the value from negative peak of the other cantilever. Hence, the gap in between the peak is not narrow as shown in Figure 8. In fact, the output performance of the piezoelectric cantilever array can be improves when the relation of the phase shifting and resonant frequency is carefully calculated and takes into account.

4. Conclusion

The result showed the output waveform of piezoelectric cantilevers do shift at its respective resonant frequencies. Both time different and Lissajous pattern method showed that the major phase shift happened at 300Hz for C1, 270Hz for C2, 220Hz for C3, and 180Hz for C4. The phase difference remains low around 0 degrees before their resonant frequency. When the frequency of the vibration source approaches the resonant frequency of respective cantilever, the phase different between the test cantilever and the constant cantilever start to increase rapidly and reach 180 degree which is out of phase after the resonant frequency. Since the output waveform is not in phase, it contributed to the significant rise at the gap in between the peaks formed when piezoelectric cantilevers are connected in series with alternating polarities.

In nutshell, it proves that the output waveform of piezoelectric cantilever do experience major phase shifting at its respective resonant frequencies. This finding could take into account in order to maximize the performance of the output from piezoelectric cantilever array in other applications.

Acknowledgment

The authors would like to thank the Malaysian Ministry of Higher Education Research grant, FRGS/2/2014/SG02/FKEKK/02/F00244, UTeM Zamalah Scheme for sponsor this research project as well as the facility support by the Advanced Sensors and Embedded Control System (ASECs) Research Group, Centre for Telecommunication Research and Innovation (CeTRI), UTeM.

References

[1] L, Tang,, Y, Yang., and C.K, Soh., 2013. Chapter 2 Broadband Vibration Energy Harvesting Techniques. Advances in Energy Harvesting Methods. Springer Science+Business Media, pp.17-61.

[2] Yu, H., Zhou, J., Deng, L., and Wen, Z. 2014 Sensors **14** 3323-3341

[3] Gambier, P., Anton, S.R., Kong, N., Erturk, A., and Inman, D.J., 2012 Measurement Science and Technology **23**.

[4] Zhao, J., and You, Z., 2014 Sensor **14** 12497-12510

[5] Lee, H.J., Zhang, S., Yoseph, B-C., and Steward, S., 2014 Sensor **14** 14526-14552

[6] Roundy, S., and Zhang, Y., 2005 Proceedings of the SPIE **5649** 373–384

[7] Yang, Z., and Yang, J., 2009 Journal of Intelligent Material Systems and Structures **20** 569–574

[8] Kim, I-H., Jung, H-J., Lee, B.M., and Jang, S-J., 2011 Applied Physics Letters **98** 214102

[9] Shahruz, SM., 2006 Journal of Sound and Vibration **292** 987–998

[10] Xue, H., Hu, Y., and Wang, Q., 2008. IEEE Transactions on Ultrasonics, Ferroelectrics, and Frequency Control **55** 2104–2108

[11] Ferrari, M., Ferrari, V., Guizzetti, M., Marioli, D., and Taroni, A., 2008. Sensors and Actuators A: Physical **142** 329–335

[12] Liu, J., Fang, H., Xu, Z., Mao, X., Shen, X., Chen, D., Liao, H., and Cai, B., 2008 Microelectronics Journal **39** 802–806

[13] Sari, I., Balkan, T., and Kulah, H., 2008 Sensors and Actuators A: Physical 405–413

[14] Bong, Y.J., Kok, S.L, and Thong, L.W. 2014 Malaysia Technical Universities Conference on Engineering & Technology 10 – 11 Nov, 2014

[15] Datasheet of Q220-A4, Piezo Systems, Inc. 2016 http://www.piezo.com/catalog8.pdf%20files/Cat8.42.pdf (accessed on 2nd December 2016).

[16] Bong, Y.J., and Kok, S.L., 2016. Journal of Telecommunication, Electronic and Computer Engineering **8** 119-123

IOP Conf. Series: Materials Science and Engineering **210** (2017) 012039 doi:10.1088/1757-899X/210/1/012039

Electrical Energy Harvesting from Thermal Energy with Converged Infrared Light

S Y Goh and S L Kok

Faculty of Electronic and Computer Engineering, Universiti Teknikal Malaysia Melaka.

E-mail: m021610014@student.utem.edu.my and siew_yun92@hotmail.com

Abstract. Photovoltaics (PV) cell is a common energy harvester that had been used to harvest solar energy and convert it into electrical energy. However, the vast energy from the spectrum of sunlight is not fully harvested. Therefore, thermoelectric (TE) module that harvest electrical energy from heat is being proposed in this paper. Generally, the part of the sunlight spectrum that induce heat is in the spectrum band of infrared (IR). For the experimental set-up in this paper, infrared (IR) light bulb was being used to simulate the IR spectrum band of the sunlight. In order to maximize the heat energy collection, a convex lens was being used to converge the IR light and therefore focused the heat on an aluminium sheet and heat sink which was placed on top of the hot side of the TE module. The distance between convex lens and IR light bulb is varying in between 10cm and 55cm and the reading was taken at an interval of 5cm. Firstly, the temperature of the IR light and converged IR light were recorded and plotted in graph. The graph showed that the temperature of the converged IR light bulb is higher than the IR light bulb. Lastly, the voltage and power output of the TE module with different heat source was compared. The output voltage and power of the TE module increased inverse proportional to the distance between IR light bulb and TE module.

1. Introduction

Sunlight has a wide spectrum which covers ultraviolet at one extreme to infrared light at the other extreme and in between is the visible light. The solar irradiation on earth are 5% UV, 43% visible and 52% Infrared. In term of wavelength, ultraviolet is in between 0.2μ – 0.39μm, visible is in between 0.39 μ – 0.78μm, whereas infrared is in between 0.78μ - 100μm[4]. The photon energy of the spectrum is inversely proportional to the wavelength. Infrared has lowest photon energy compare to ultraviolet and visible light and therefore is insufficient to break through the bandgap of the photovoltaics cell. This mean that the vast energy of sunlight is not fully utilized by PV cell to generate electrical power.

TE module is suggested to be used for harvesting heat energy from sunlight and convert into useful electrical power. In this paper, an experiment set-up an Infrared light bulb is being used to simulate the sunlight.

TE module is popular for generating electrical energy from industries and machinery which produce relative high heat energy. Toshihiko and Hiroaki proposed heat pipes as heat source for TE module, so electrical energy can be produced from the heat that dissipates from heat pipes [5]. The TE module they used was attached to the heat pipes, and heat sink was used to cold down the cold surface of the TE module. The temperature gradient that they achieved was at 50°C. The output voltage per temperature gradient was about 0.00656v/°C. In another study, Lauri and Matti suggested to used automotive internal combustion engines as heat source [6]. The temperature gradient of combustion engine can achieve up to 200°C which generated 0.0376V/°C. On the other hand, Jie Chen considered gamma radiation as heat source and heat sink as cold surface to generate electrical energy [7]. By

IOP Conf. Series: Materials Science and Engineering **210** (2017) 012039 doi:10.1088/1757-899X/210/1/012039

using gamma radiation, the temperature gradient that he can produce was at about 120°C hence achieved a performance of 0.0508V/°C. In addition, Rajeh and Kiran discovered that chimney from industrial plant was suitable for TE module to generate electrical energy [8]. From their paper, the highest temperature gradient is they can produce was at about 85°C and the output voltage is 0.0198V/°C.

2. Thermoelectric Module (TE module)

The fundamental process of thermoelectric module is based on the Seebeck effect, named by Thomas Seebeck in 1821. The Seebeck effect is the generation of electricity within two dissimilar metals and these metals placed in two different temperature, thus it will generate voltage output. Therefore, it usually used as thermometer and its call thermocouples. However, the dissimilar metal can be improved with semiconductors because semiconductors have higher Seebeck coefficients, so they are more suitable to be used for generating electrical power compared to metal [1,3]. Figure 1 shows the P-N junction that built up TE module. It connected electrically in series and thermally in parallel. Besides, ceramic plate is used to sandwich the N pairs of thermocouple. Using ceramic plates because it is good in heat conduction and resistive of electricity [2].

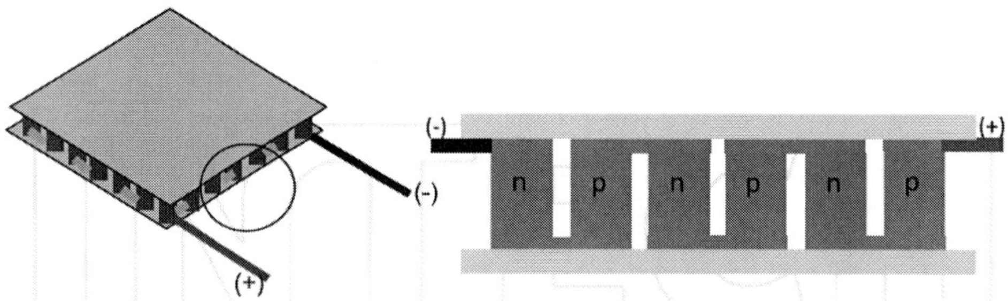

Figure 1. Thermoelectric module [2].

According to the Seebeck effect, the excess of electron will flow from hot junction to cold junction in n-type material. In p-type material, the holes of electron will flow from cold junction toward hot junction. Therefore, this flow produce the current flow of the TE module[1]. In addition, the Seebeck coefficient, α, is defined as the change in voltage per degree of temperature gradient [1,3]:

$$\alpha = \frac{dV}{dT} \text{ volts/K} \quad (1)$$

Seebeck coefficient also included in voltage V_s produces from a TE module while it is applied temperature gradient which is T_h as hot temperature and T_c as cold temperature [2]:

$$V_s = \alpha(T_h - T_c) \quad (2)$$

3. Sunlight spectrum and PV cell

Table 1 shows the bandgap of variety type of solar cell and Table 2 shows the solar irradiation, wavelength and photon energy of spectrum sunlight. Polycrystalline silicon solar cell is most common used nowadays. In addition, the bandgap required is about 1.1eV which means most of the infrared light from sunlight cannot breakthrough the bandgap and generate the electricity. Therefore, sunlight is not fully utilized by using solar cell as energy harvester

IOP Conf. Series: Materials Science and Engineering **210** (2017) 012039 doi:10.1088/1757-899X/210/1/012039

Table 1. Bandgap of material used for manufacture solar cell [4]

Solar Cell Material	Bandgap (eV)
Cadmium Telluride (CdTe)	~1.58
Gallium Diselenide (CIGS)	~1.38
Polycrystalline Silicon	1.1
Germanium	0.477

Table 2. Solar irradiation, Wavelength and Photon energy of spectrum sunlight [4]

Spectrum Sunlight	Solar Irradiation (%)	Wavelength (μm)	Photon Energy (eV)
Ultraviolet	5	0.2 – 0.39	6.2 – 3.2
Violet Blue Green Yellow Orange Red	43	0.39 – 0.78	3.2 – 1.59
Infrared	52	0.78 - 100	1.59 – 0.0124

4. Experimental Set-up

There are three experiments carried out in this paper. All of the experiments were using IR light bulb as the heat source to simulate the heat energy from sunlight. Besides, the convex lenses were used to converge the IR light to focus the heat.

First of all, the temperature of radiated IR light bulb and converged IR light bulb was measured by using thermometer and compared in graph. In order to dissipate the heat evenly, aluminium sheet or heat sink was place on hot surface of TE module. Besides, heat sink was place on the cold surface of the TE module, so the temperature gradient of the TE module can be maintained. Figure 2 shows illustration of these three experiment. Figure 2 (a) shows measuring temperature of radiated IR light and converged IR light, Figure 2 (b) shows converged IR light on TE module with aluminium sheet on top and Figure 2 (c) shows converged IR light on TE module with heat sink on top.

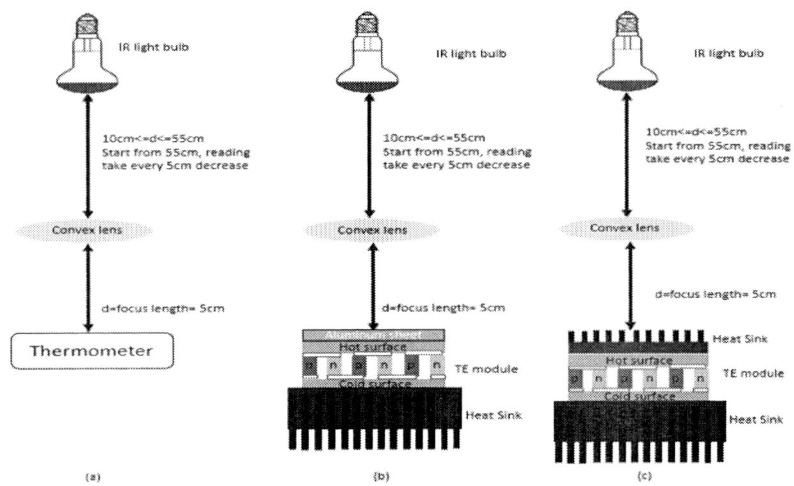

Figure 2. Illustration of (a) measuring temperature of radiated IR light and converged IR light (b) Converged IR light on TE module with aluminium sheet on top (c) Converged IR light on TE module with heat sink on top.

IOP Conf. Series: Materials Science and Engineering **210** (2017) 012039 doi:10.1088/1757-899X/210/1/012039

Figure 3 shows the experiment set-up in lab. The focus length of convex lens was fixed as 5cm. In addition, the distances between IR light bulb and convex lens are 55cm, the voltage and current output were taken every 5cm IR light bulb near to convex lens.

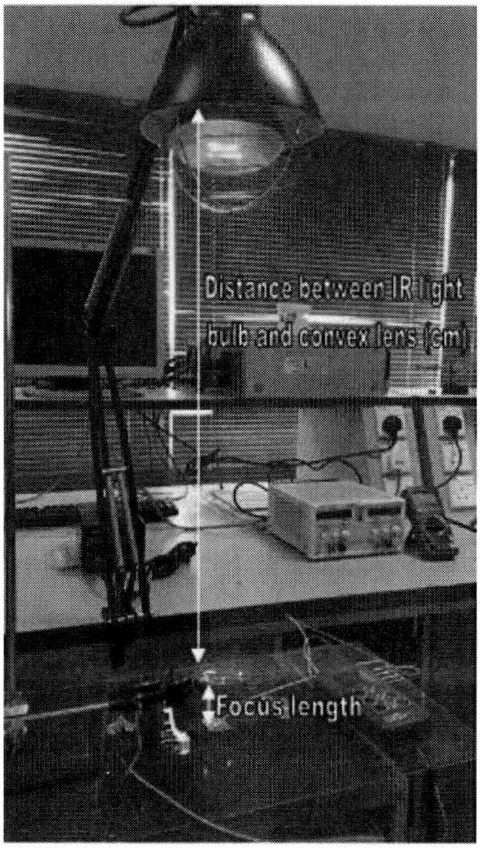

Figure 3. Using IR light bulb as experimental set-up

Figure 4 shows the first lab experiment which is being used to measure the temperature from IR light bulb. Two temperatures were taken in the experiment, which are IR light and converged IR light. One of the thermocouple was placed on the bottom of convex, another thermocouple was placed outside the focus area of a convex lens. After that, both of the temperature readings were plotted in graph and compared.

Next, the converged IR light were used as a heat source for TE module to generate an electrical energy from the heat energy. Firstly, an aluminium sheet with a dimension of 8cm length and 8cm width was put on the top of TE module. This was followed by placing a 19cm in diameter and 9cm height heat sink to replace the aluminium sheet.

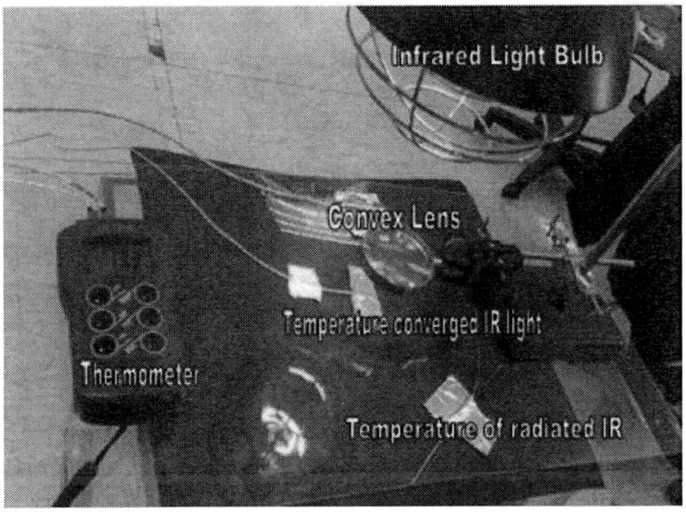

Figure 4. Measuring temperature of radiated IR light and converged IR light

Figure 5 shows the experimental set-up of converged IR light on the aluminium sheet. TE module that used in the paper is from Laird Technology UltraTEC series. The TE module was placed between the aluminium sheet and heat sink, so that the aluminium sheet can dissipate the heat on TE module evenly. Besides, the heat sink can release the heat from the TE module to maintain the cold surface of the TE module. Additionally, an aluminium foil with PE foam was used to reflect and resist the heat energy from IR light radiate to heat sink.

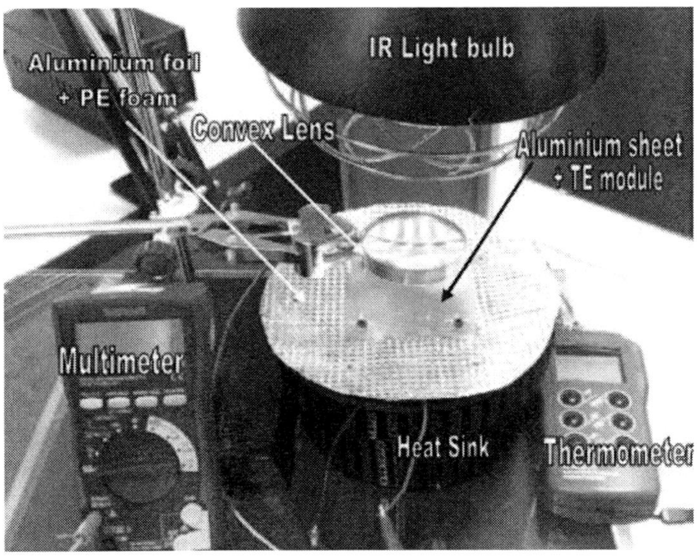

Figure 5. Converged IR light is focused on aluminium plate.

Figure 6 shows the experimental set-up of converged IR light on heat sink. The TE module was placed between two heat sinks, the small heat sink has the same function with aluminium sheet in former experimental set-up which is heat dissipation on TE module uniformly. Lastly, the voltage output and power of the TE module in different experimental set-up is plotted and compared.

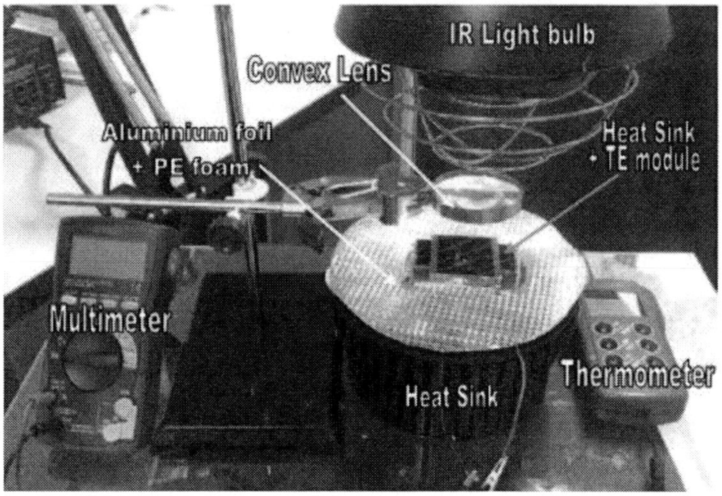

Figure 6. Converged IR light is focused on heat sink

5. Result and Discussion

Figure 7 shows temperature of the IR temperature, the converged IR temperature, and the temperature different at different distance between IR light bulb and convex lens. From the graph above, it shows that the temperature between IR light and converged IR light have not much different from 55cm until 40cm. However, temperature changed dramatically from 45cm until 10cm. There is a 1.3°C temperature gain from the converged IR light when the distance between the IR light bulb and the convex lens is at 40cm. Moreover, the temperature different between the IR light and the converged IR light is 30.9°C when the distance is at 10cm. Therefore, it shows that the convex lens has the potential to double up the temperature that was produce by the IR light bulb. It clearly shows that a convex len is suitable to be used for focusing IR light for better heat energy production from the same heat source.

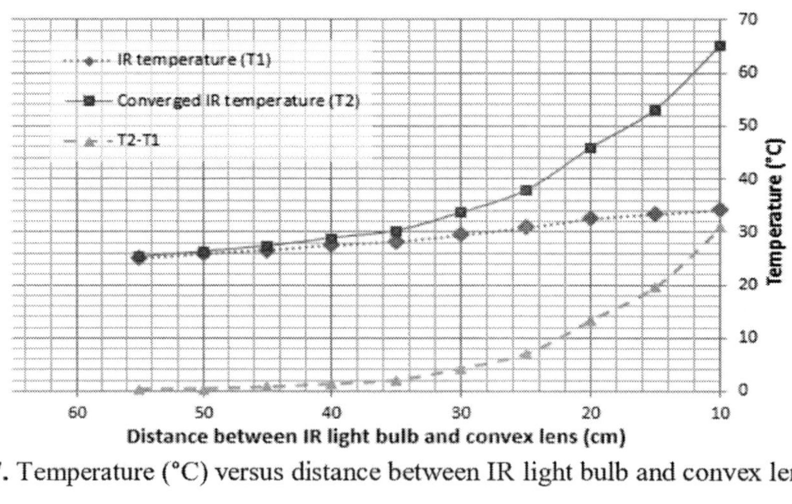

Figure 7. Temperature (°C) versus distance between IR light bulb and convex lens (cm).

Figure 8 shows the voltage output of TE module while different temperature gradient was applied on it. Besides, there are two different materials that was put on top of TE module, which are a aluminium sheet and a heat sink. When the distance between IR light bulb and convex lens is at 10cm, the temperature of the aluminium plate and the heat sink obtained the highest temperature which are at 10.7°C and 12.05°C respectively. Although the highest temperature showed a different of only about 1°C, but the outputs voltage have a significant different between them. When the distance between IR light bulb and convex lens is at 10cm, the output voltage of TE module with aluminium plate on top is 0.1059V while the voltage output of the TE module with the heat sink on top is 0.341V. From Figure 7, the increment of the voltage per temperature of the TE module with aluminium plate and heat sink are 0.00978V/°C and 0.0286V/°C respectively. It is about a factor of 3 times different between these two different experimental set-ups.

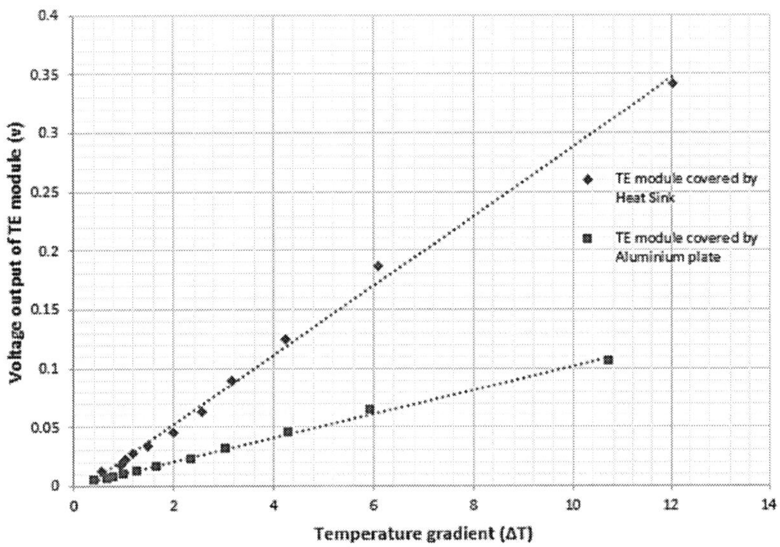

Figure 8. Voltage output of TE module from variance temperature gradient.

Figure 9 compares the power output of TE module from different perspective, including variance resister load, variance distance between IR light bulb with convex lens, and different material cover TE module. The power increase from origin until peak power, the peak power output occurs between 80Ω and 140Ω resister load.

There are three different distances between IR light bulb and convex lens included in output power measuring which are 30cm, 35cm, and 40cm. The temperature applied on the TE module increased, while the distance, *d* between the IR light bulb and the convex lens decreased. In comparison to the same material that used to cover TE module, the power output is at its highest when distance is at 30cm. When aluminium was used to cover the TE module, the peak power that can be obtained is about 1.23μW at separation distance of 30cm. Besides, the peak power is 0.6μW and 0.31μW when the distance is at 35cm and 40cm respectively. This scenario is same when the TE module covered by the heat sink, the highest power is obtained when the distance is at 30cm. The peak power when distance equal to 30cm is about two times of the peak power when distance is at 35cm.

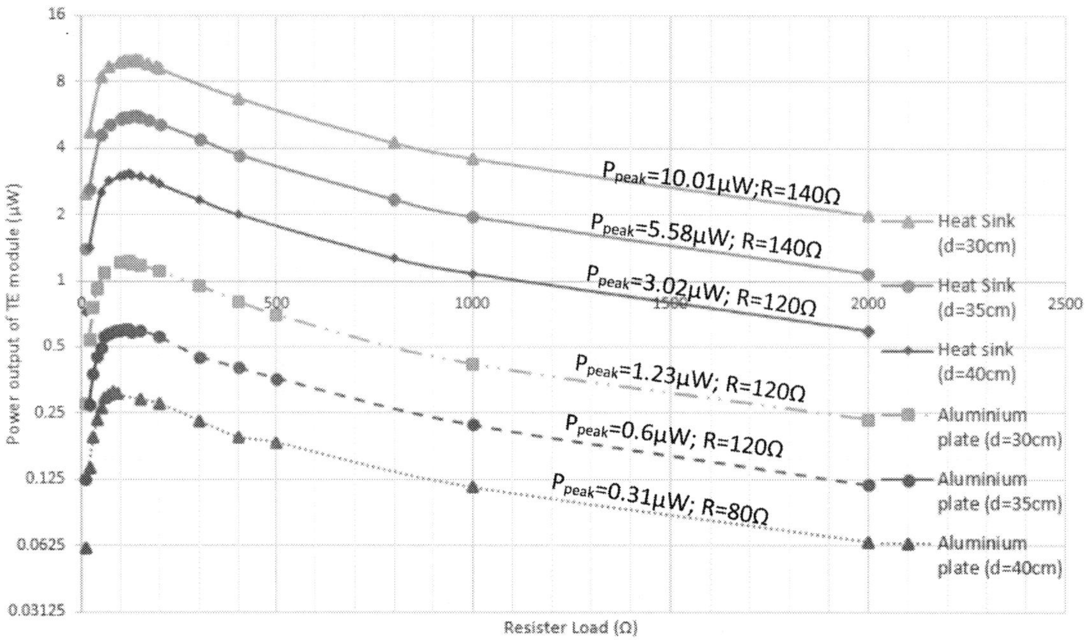

Figure 9. Power output of TE module from variance resister load

According to Figure 8, the power output of using heat sink to cover the TE module is at its higher than using the aluminium plate to cover it. When using the same distance between IR light bulb and convex lens, the power output of using heat sink as the cover is 8 to 10 times higher than that using aluminium plate as the cover. Therefore, it can be concluded that the heat sink is more suitable to be used as the cover to dissipate the heat onto the TE module uniformly.

6. Conclusion

The above results showed that, the heat energy collection is greater for a converged IR light than a radiated IR light. Therefore, it can be applied to produce higher temperature to act on a TE module and hence generating more electrical energy. According to the voltage output and power output of the TE module, a heat sink is more suitable to be used as a cover of a TE module instead of using an aluminium plate as a cover. Since, a TE module can produce more voltage and power output when using heat sink as a cover. Therefore, using a converged IR light as heat source generation of electrical energy from the sunlight as a potential heat source for TE module.

Acknowledgments

The authors would like to thank the Ministry of Higher Education of Malaysia for the financial support under the research grants no. FRGS/2//2014/SG02/FKEKK/02, RAGS/1/12014/SG02/FTK/B0080 and UTeM Zamalah as well as the facility support by the Advanced Sensors and Embedded Control System (ASECs) Research Group, Centre of Excellence for Telecommunication Research and Innovation (CeTRI), Universiti Teknikal Malaysia Melaka (UTeM).

References

[1] Hyder A K, Wiley R L, Halpert G, Flood D J, and Sabripour S 2000 Static Energy Conversion *Spacecraft Power Technologies* (Singapore: World Scientific) chapter 6 288-294
[2] Piotr Dziurdzia 2011 *Modeling and Simulation of Thermoelectric energy harvesting processes* Sustainable Energy Harvesting Technologies-Past, Present and Future (Cracow: AGH University of Science and Technology) chapter 4 109-116

[3] Basshuysen R v and Schafer F 2016 Energy Management in the engine and vehicle *Internal Combustion engine handbook- basics, components, system, and perspectives* (UK:SAE Internation) ed 2 chapter 31 1056-1060

[4] Anwar S, Efstathiadis H, and Qazi S 2013 Principal of photovoltaic cell *Handbook of Research on Solar Energy systems and technologies* (USA:IGI Global) chapter 6 163-191

[5] I.Toshihiko and Y.Hiroaki 2012 *Renewable Energy Research and Applications (ICRERA), 2012 International Conference*. Nov 11-14 2012 1-4

[6] L.Kutt and M.Lehtonen 2015 *IEEE 5th International Conference on Power Engineering , Energy and ELectrical Drives (POWERING) Sept. 2015* 55-62

[7] J.Chen, J.Klein, Y.J.Wu, S.X.Xing, R.Flammang, M.Heibel, and L.Zuo 2016 *IEEE Transaction on Nuclear Science* **5** 2738-2746

[8] Rajeh.B.K and Kiran.B. 2015 *Nascrent Technologies in the Engineering Field (ICNTE), 2015 International Conferenc,* Jan 9-10 2015 1-6

High Voltage Discharge Profile on Soil Breakdown Using Impulse Discharge

F E Fajingbesi[1,2*], N S Midi[1], E M A Elsheikh[1] and S H Yusoff[1]

[1]Department of Electrical and Computer Engineering, International Islamic University Malaysia
[2]Department of Physics Electronics and Earth Science, Fountain University Osogbo Nigeria

E-mail: *fawwazfajingbesi@yahoo.com

Abstract. Grounding terminals are mandatory in electrical appliance design as they provide safety route during overvoltage faults. The soil (earth) been the universal ground is assumed to be at zero electric potential. However, due to properties like moisture, pH and available nutrients; the electric potential may fluctuate between positive and negative values that could be harmful for internally connected circuits on the grounding terminal. Fluctuations in soil properties may also lead to current crowding effect similar to those seen at the emitters of semiconductor transistors. In this work, soil samples are subjected to high impulse voltage discharge and the breakdown characteristics was profiled. The results from profiling discharge characteristics of soil in this work will contribute to the optimization of grounding protection system design in terms of electrode placement. This would also contribute to avoiding grounding electrode current crowding, ground potential rise fault and electromagnetic coupling faults.

1. Introduction

Experiments on electrical conductivity of soil have long been investigated for numerous reasons ranging from its use in agricultural purpose to use in electrical grounding systems. The soil electrical conductivity vary with time and place due to factors like pH, moisture and available nutrients present[1, 2]. The naturally existing variation in such soil property makes it difficult in designing efficient grounding system for electrical terminal. Furthermore, this soil properties could also vary due to electrical activities in its vicinity such as lightning strikes, over voltage power dissipation and other electrical activity leading to soil ionization and breakdown [3, 4].

In several literature, this adverse effect of soil conductivity variation at grounding sites have majorly been attributed to ground potential rise (GPR) that can directly fail connected systems or increasing the risk and occurrence of step and touch voltage harmful to humans and other living organism [5 – 7]. Protection designs against the detrimental effects from the enormous electrical discharge of lightning phenomena strongly rely on a total dissipation through the soil whose impedance isn't constant [8]. With designs of more sensitive technological devices that are susceptible to effect resulting from earth impedance fluctuations in widespread profiling the breakdown would provide better information for optimization of grounding protection systems. One way to make clear of soil characteristics influence on current distribution and breakdown properties is by investigating the discharge propagation in soil when exposed to high voltage discharge.

This research aims to profile the discharge breakdown phenomena of soil when subjected to high voltage discharge. In this work, high voltage impulse discharge was injected on a soil surface. The

IOP Conf. Series: Materials Science and Engineering **210** (2017) 012040 doi:10.1088/1757-899X/210/1/012040

discharge propagation resulting in ionization and breakdown was then observed and discussed using the discharge current distribution during the process on a laboratory scale.

2. Experimental Setup and Procedures

The experimental setup design shown in figure 1 consist of a point to plane electrode system. The stratified point to plane electrode system is designed to offer a dielectric two-phase gas-solid experimental environment. The point electrode is an 8 mm diameter pointed stainless steel electrode which serve as the discharge electrode. The discharge electrode tip is separated from the soil surface by a 5mm airgap. There are nine grounding electrodes subdivided into five horizontal (E1 – E5) and four vertical (E6 – E9) electrodes placed inside a plastic tub. Electrode 1 is a 1 mm thick, 40 mm diameter copper electrode. Electrodes 2 until 5 are ring-shaped copper electrodes with 1 mm thickness each. Their outer radius is 40 mm longer than their inner radius. Electrodes (E1 – E5) were attached to an acrylic sheet placed at the bottom of a 480 mm diameter plastic tub. Electrodes (E6 – E9) are 0.1 mm thick and 40 mm width copper tape attached to the inner wall of the plastic tub with a 15 mm separation between each horizontal and vertical electrode. Electrode (E9) is placed leveled with the soil surface and electrodes (E8 – E1) were set as under soil electrodes respectively. The soil surface level was set 220 mm from the bottom electrodes. The discharge electrode is connected to a 280 kV impulse voltage generator (TERCO) to generate high voltage spark discharge on the soil surface.

The experiment was conducted by continuously injecting 25 – 28 kV of standard impulse voltage (1.2/50μs) to the discharge electrode at a distance of 5 mm from the soil surface. The discharge current due to the soil ionization and breakdown in the soil were observed at each of the 9 grounding electrodes using a current probe (Pearson 6585) connected to a digital oscilloscope (LeCroy Waverunner 625Zi) and the discharge breakdown condition on the surface was recorded using a digital camera (RICOH CX3) at 5fps shooting speed. In this experiment, the variation in soil pH was achieved by varying the soil moisture content and available nutrients resulting in an increase at the soil electrical conductivity as shown in table 1. The available nutrients was varied adding various weight of fertilizer (NPK and $Ca(OH)_2$) also shown in table 1. The soil electrical conductivity were also measured using earth resistance meter (Megger DET4TCR2) prior to and after the discharge process.

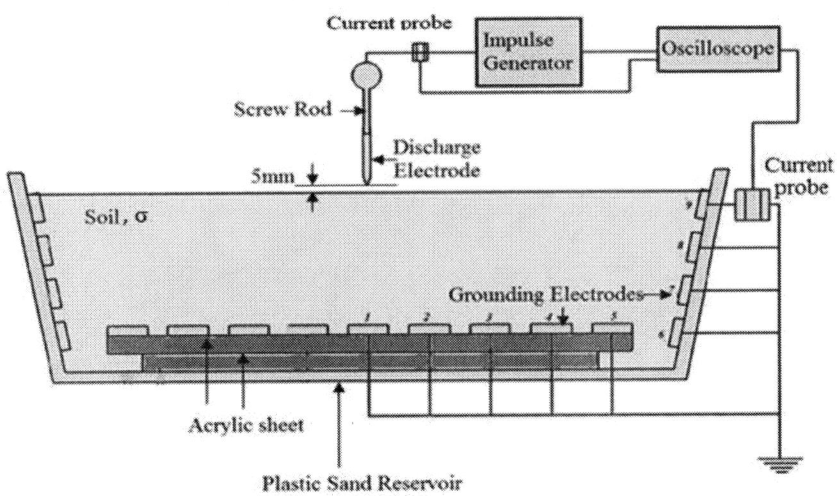

Figure 1. Experimental setup for observing lightning discharge phenomena

IOP Conf. Series: Materials Science and Engineering **210** (2017) 012040 doi:10.1088/1757-899X/210/1/012040

Table 1. Soil electrical conductivity variation through physical properties alteration

Constituents						
Soil(kg)	Water(mL)	NPK 11-11-11 (g)	NPK 4-16-16 (g)	$Ca(OH)_2$ (g)	Moisture[%]	Conductivity [mS/m]
30	3300	100	0	0	23.98	29.31
30	3600	100	100	0	24.78	43.53
30	3900	100	100	100	25.57	57.89

3. Result and Discussion

3.1. Voltage and current waveform

Figure 2, 3 and 4 shows the voltage and current waveforms at σ = 29.31, 43.53 and 57.89 [mS/m] respectively. Figure (a) shows the typical injected voltage and current waveform while figure (b) shows the discharge current waveform relative to the breakdown currents at the soil layer around the grounding electrodes resulting from the injected discharge. From the waveforms, it can be seen clearly that the discharge process between soils of different conductivity are distinct as the breakdown phenomena are clearly evident with increase in conductivity from each other. The presence of soil ionization breakdown is confirmed with the sudden drop in input voltage and current waveform which becomes more stepper with increase in conductivity shown in figure 2, 3 and 4 respectively. The evolvement of the discharge current in this figures is somewhat synchronized with that of the injected input current where the increase time pattern and decrease time pattern are almost matching each other. The observed secondary peaks further down the line as time elapsed also suggest a multiple secondary ionization breakdown or partial discharge which could be due to micro-voids in soil as the electric potential traverse the soil [9]. Also the input current and the discharge current waveforms are not in sync with the input voltage waveform which further suggest that further breakdown occurs in the soil as conductivity increases [10]. Looking at the variation in the peak discharge current at grounding electrodes and the injected voltage relationship as a function of soil conductivity, it shows that the discharge current increase linearly with the conductivity as the required voltage for the dielectric breakdown and discharge to occur decreased with increase in soil conductivity[11, 12].

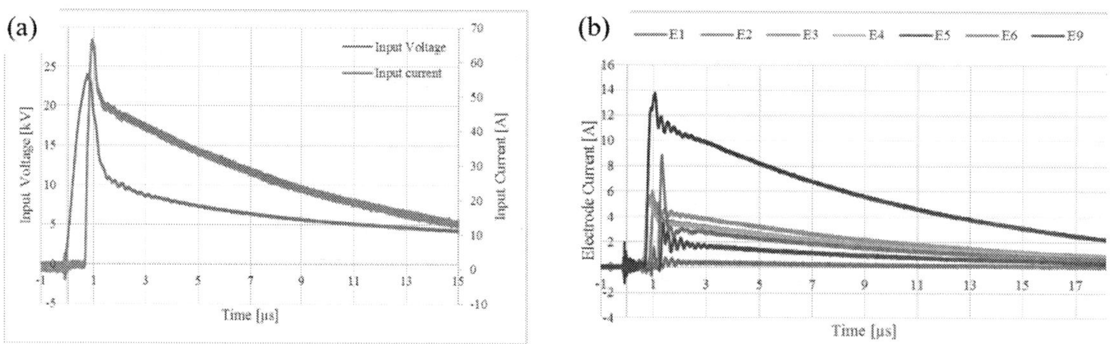

Figure 2. (a) Input voltage and current waveform at σ = 29.31 [mS/m]. (b) Electrode discharge current waveforms at σ = 29.31 [mS/m]

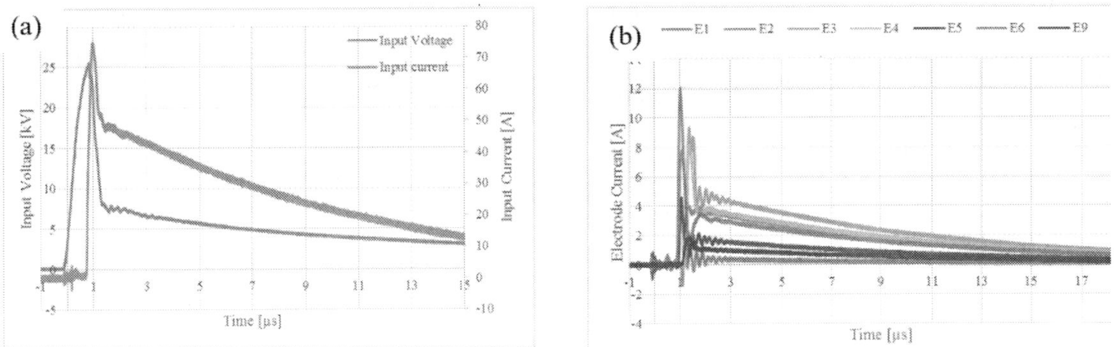

Figure 3. (a) Input voltage and current waveform at σ = 43.53 [mS/m]. (b) Electrode discharge current waveforms at σ = 43.53 [mS/m]

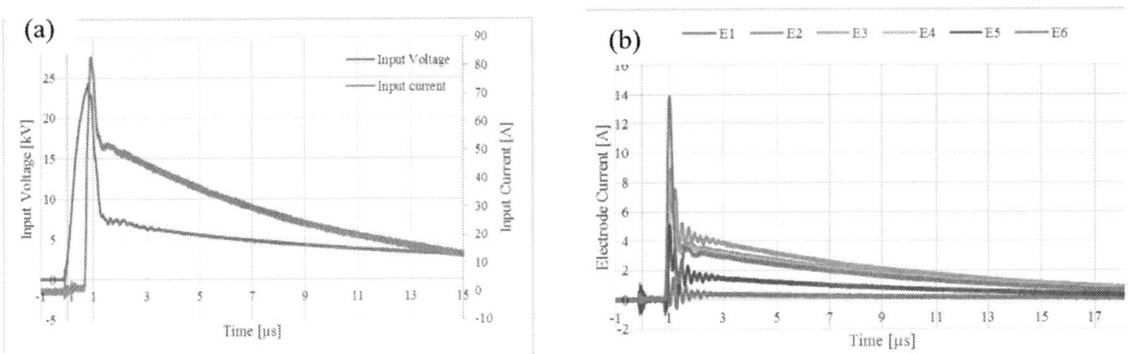

Figure 4. (a) Input voltage and current waveform at σ = 57.89 [mS/m]. (b) Electrode discharge current waveforms at σ = 57.89 [mS/m]

3.2. High voltage breakdown discharge

Following the analysis of figure 2 – 4, the waveforms strongly support the occurrence of soil ionization and breakdown phenomena. The discharge emission profile captured at 4 different times using the RICOH CX3 digital camera during the multiple injection discharge on soil surface with 3 conductivity σ = 29.31, 43.53 and 57.89 [mS/m] is shown in figure 5 – 7. The discharge conditions labeled (a) – (d) are captured at different intervals during the multiple injection process on each of the three conductivity.

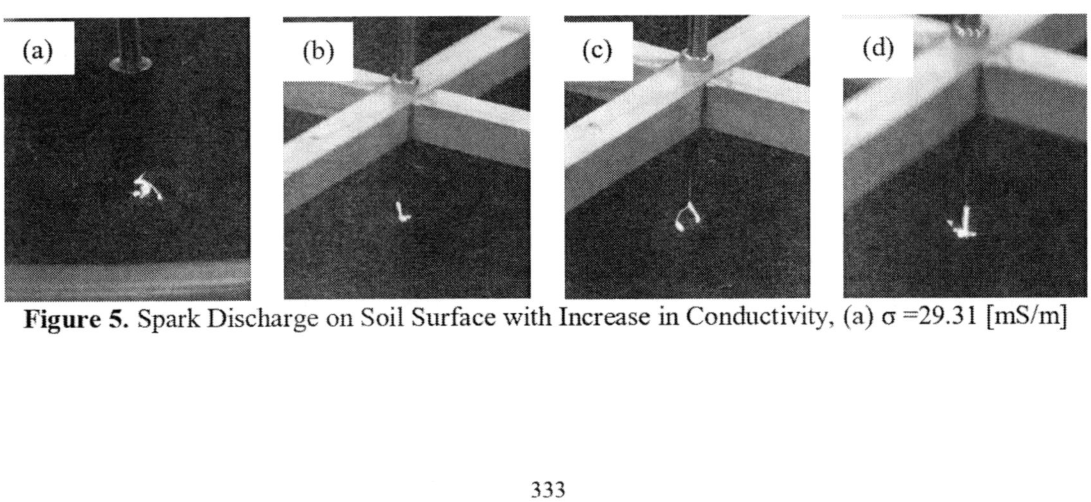

Figure 5. Spark Discharge on Soil Surface with Increase in Conductivity, (a) σ =29.31 [mS/m]

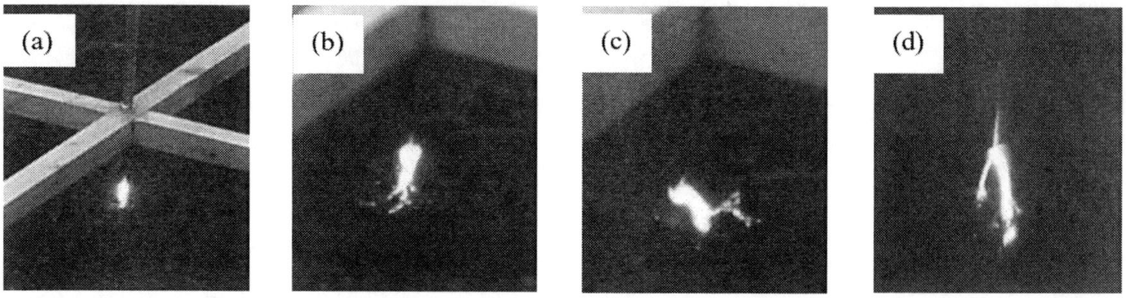

Figure 6. Spark Discharge on Soil Surface with Increase in Conductivity, (a) $\sigma = 43.53$ [mS/m]

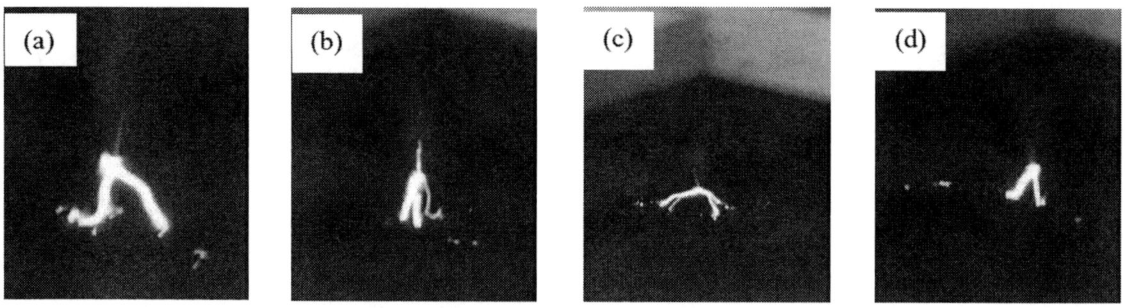

Figure 7. Spark Discharge on Soil Surface with Increase in Conductivity, (a) $\sigma = 57.89$ [mS/m]

From figure 5, it can be seen that there were no visible spark discharge propagation directly on the soil surface. The discharge to the soil surface were thin and localized at a point not far from the discharge electrode tip with minimal difference over the time period as shown in figure 5(b) – (c). In figure 5(d), the discharge channel increased signifying further breakdown would occur with more discharge injection. The discharge channel to soil surface in figure 6 is a lot wider than that seen in the lower $\sigma = 29.31$ [mS/m]. Also the width and the impact zone on the soil surface shows an increase with time as shown by figure 6 (a) – (d). From figure 7, one would see that not only does the discharge channel grow wider but also become branched occupying larger localization area on the soil surface as seen in figure 7 (a) – (d). This is because the increase in soil conductivity resulted in larger area for current propagation easily. Also this increase in electric conductivity resulted in lowering the air-soil phase dielectric breakdown voltage. From all these observations, it can be concluded that the discharge emission on the surface and soil ionization breakdown is strongly associated with the voltage and current waveforms.

3.3. Soil electrical conductivity

The shallow soil conductivity measured before (pre-ECs) and after (post-ECs) along with the deep soil conductivity measured before (pre-ECd) and after (post-ECd) the high impulse spark discharge on the soil surface is shown in figure 8. This graph shows a significant difference in the conductivity of the soil before and after imitated lightning discharge. This difference in the conductivity measured shows the reason for increase in the discharge channel as observed by the images of discharge condition shown in figure 5 – 7. This also validates the occurrence of permanent soil ionization breakdown during the discharge process.

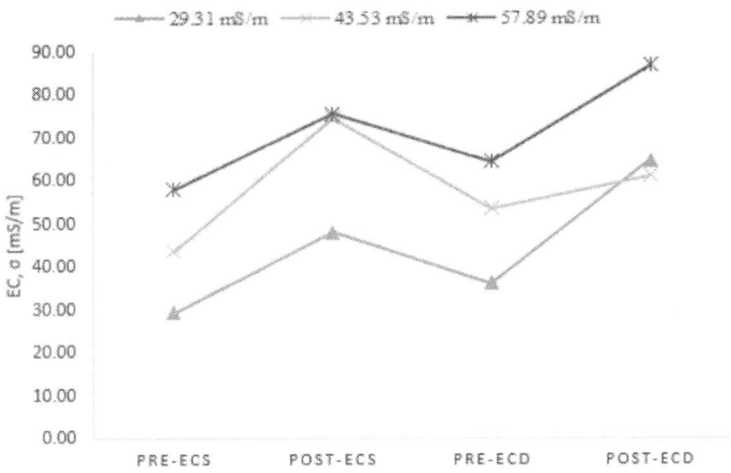

Figure 8. Soil σ variation due to impulse discharge

4. Conclusion

In this work, high voltage discharge breakdown of soil was investigated using impulse discharge applied to the surface of soil placed in a point to plane electrode system. The soil sample at different conductivity obtained from soil with different moisture content, available nutrient and pH was investigated. From there, breakdown discharge characteristics were observed. From the experiment it can be concluded that complete discharge leading to multiple soil ionization and breakdown occurred depending on conductivity of the soil. Furthermore, the breakdown that occur result in varying the electrical conductivity of the soil. The soil ionization is seen to occur readily in higher conductivity and the permanently alters the electrical conductivity of the soil after multiple injection.

Acknowledgment

This research has been supported by Malaysian Ministry of Higher Education Research Grant, RAGS15-063-0126.

References

[1] S. Visacro and F. H. Silveira 2015 *IEEE Trans. Electromagn. Compat.* **57** 434–441
[2] I. N. Aini, M. H. Ezrin, and W. Aimrun 2014 *Agric. Agric. Sci. Procedia* **2** 199–206
[3] W. Sima, S. Liu, T. Yuan, D. Luo, P. Wu, and B. Zhu 2015 *IEEE Trans. Ind. Appl.* **51** 5343–5351
[4] N. Harid, H. Griffiths, S. Mousa, D. Clark, S. Robson, and A. Haddad 2015 *IEEE Trans. Ind. Appl.* **51** 5324–5334
[5] M. Nassereddine, J. Rizk, and G. Nasserddine 2013 *Int. J. Electr. Comput. Energ. Electron. Commun. Eng.* **7** 35–40
[6] N. Harid, H. Griffiths, N. Ullah, M. Ahmeda, and A. Haddad 2012 *J. Light. Res.* 36–44
[7] A. Kalair, N. Abas, and N. Khan 2013 *J. Light. Res.* **5** 11–2
[8] W. Bai, L. Kong, and A. Guo 2013 *J. Rock Mech. Geotech. Eng.* **5** 406–411
[9] S. M. Kim, Y. W. Yoo, and B. H. Lee 2016 *J. Electr. Eng. Technol.* **11** 446–454
[10] N. S. Midi and R. I. Ohyama 2011 *Annu. Rep. - Conf. Electr. Insul. Dielectr. Phenomena, CEIDP* 223–226
[11] A. Elzowawi, A. Haddad, H. Griffiths, and D. Clark 2015 *2015 50th International Universities Power Engineering Conference (UPEC)* **2015–Novem** 1–5 [12] J. He and B. Zhang 2015 *IEEE Trans. Ind. Appl.* **51** 4924–4933

IOP Conf. Series: Materials Science and Engineering **210** (2017) 012041 doi:10.1088/1757-899X/210/1/012041

Fabricate Optical Microfiber by Using Flame Brushing Technique and Coated with Polymer Polyaniline for Sensing Application

N A Razak[1], B A Hamida[1*], N Irawati[2] and M H Habaebi[1]
[1]Department Electrical and Computer Engineering, International Islamic University Malaysia 53100 Kuala Lumpur, Malaysia
[2]Photonics Research Center, Department of Physics, University of Malaya, 50603 Kuala Lumpur, Malaysia

E-mail: *belal@iium.edu.my

Abstract. Adiabaticity is one of the essential criteria in producing good fabricated tapered fibers. Good tapered fibers can be use in sensor application such as humidity sensor, temperature sensor and refractive index sensor. In this paper, good tapering silica fiber is produced by using flame brushing technique and then, the microfiber is coated with polymer Polyaniline (PAni) to sense different type of alcohols with different concentrations. The outcome of this experiment gives excellent repeatability in the detection of alcohol sensing with a sensitivity of 0.1332 μW/% and a resolution of 3.764%. In conclusion, conducting polymer coated optical microfiber sensor for alcohol detection with low cost, effective and simple set-up was successfully achieved in this study.

1. Introduction

Recently, tapered fibers have attracted more researchers around the globe to focus in the optical sensor area due to its extensively outstanding performances and powerful in numerous sensing applications [1]. Fiber optics sensor is immune to the electromagnetic interference which make more desirable in sensing applications. Fiber optic sensing has been in market demands for quite some time alongside with the rapid growth in micro and nanotechnology nowadays. The tapered fiber is called microfiber when the diameter is in range of micrometer (μm) and called nanofiber when the diameter of the tapered fiber is below 1 μm [2]. The small in diameter made the fiber optics more sensitive, highly responsive, large dynamic range, low loss of attenuation, tight optical confinement and strong evanescent fields [3-7]. With such characteristics, it makes the tapered optical fiber an ideal application in remote sensing such as in humidity [8,9], refractive index sensing [10-12], vapour sensing, biomedicines, chemical analysis, environmental engineering and in an automotive industry [3].

Due to the smaller in diameter of the microfiber, the evanescent field surrounding the area is large and make it sensitive to the ambient changes. The evanescent part will have effect on the propagating mode inside the core and the output transmission power of this sensor changes with its optical properties, depending on the ambient medium or solutions. Therefore, with the modification of the optical fiber cladding it brings many attractive advantages due to its fast responsive, high sensitivity and larger integration with other fiber systems.

Conducting polymer polyaniline (PAni) as a cladding optical fiber sensor result in quick and large responses to chemical solutions. The advantages of using the polymer is due to easy fabrication, excellent stability, low power consumption, rapid and reversible adsorption [13]. To add, PAni characteristics is independent on detecting air humidity and temperature compared to traditional sensing material such as metal oxide which is highly dependent on detecting environment. Thus, polymer PAni make more favourable use as sensing applications compare with traditional material. Chiam et al. demonstrated polymer PAni coated optical microfiber in detection of alcohols. The outcome of this study gives red shift in the output spectrum with increasing steric effect. The presence of PAni result in an increasing of dihedral angle and band gap energy [14].

Both microfiber and nanofiber (MNF) are usually manufactured from the regular sized optical fiber by heating and stretching it [2]. The process of reducing the diameter of the fiber is called fabrication technique. There are many techniques to fabricate MNF. The most commonly use to fabricate microfiber is flame brushing technique because of their uniformity to maintain heat source, controlling the flame movement, stretching the fiber length and speed [15]. Flame brushing technique gives many advantages as it will determine the adiabatic criterion and to produce good quality of tapered fibers.

In this study, an optical microfiber is fabricated by using flame brushing technique and the microfiber produced is coated with polymer Polyaniline (PAni) to sense different type of alcohols with different concentrations. The polymer PAni is coated onto the tapered fiber by using drop coating method. The coated microfiber is sensitive to selective chemical or biological species. Thus, good is sensing applications.

2. Fabrication of microfiber

Fabrication process of microfiber is where the core diameter of the structured fiber is reduced by heating and stretching the single-mode fiber (SMF). The smallest diameter of the tapered fiber is called waist and between the waist and the unstretched SMF is transition regions. Transition regions is where the diameter of the core and cladding are decrease exponentially to range micrometer or nanometer, depending on how long the stretching of the SMF took place. When there is changing of size in the core and cladding diameter along the transition regions, the field distribution of waveguide will vary. Due to the rate of diameter change, the propagating wave may experience energy transfer from the fundamental mode to some closest few higher order modes. Most of the time loss will be appear in the system. To minimize the excess loss, adiabatic criteria should be taken into consideration throughout the tapering process [15].

3. Experimental Method

The experiment is started with the preparation of the optical fiber by connecting the fiber with the pigtails. The preparation of the optical fiber includes stripping, cleaving and splicing. 0.01 dB transmission loss has achieved in the splicing process. The experiment is continued with the fabrication of the microfiber by using (a) flame brushing technique and conducting (b) coated polymer Polyaniline for alcohol detection.

3.1 Flame Brushing Technique

Figure 1, shows a schematic diagram for fabrication of tapered fiber by using flame brushing technique. There are two stepper motors, one motor controller board, and two fiber holders on a linear stage, a sliding stage with an oxy-butane torch being fixed to it. The control board acts as a brain of the system which will give instruction to the stepper motors such as speed, moving direction and controlling the stages position. The two fiber holders are on a linear stage and one of them were controlled by precision stepper motors. One of the clamps will hold the fiber while the other one that have stepper motor will stretch the fiber during the tapering process. The linear stage travels at slow speed in order to have low loss and smooth process of the tapered fiber. At this time, the other motorized stage which acts as a brushing flame which will heats and brush the uncoated segment of the fiber. The torch is on sliding stage and move at a very high speed than a linear stage. The reason is to provide uniform heat along the tapered fiber. The flame is produced from oxygen and butane or called oxy-butane torch. The flame also plays an important role in tapering process. To obtain good flame, the convection air flow from the flame need to be kept at an acceptably low level and the temperature of the flame need to be high in order to heat enough and soften the silica fiber. The distance between burner's tip and silica fiber also need to be concern in order to prevent excessive heat that can break the tapered fiber during the fabrication process. The length of the tapered fiber is successfully fabricated at 3 cm.

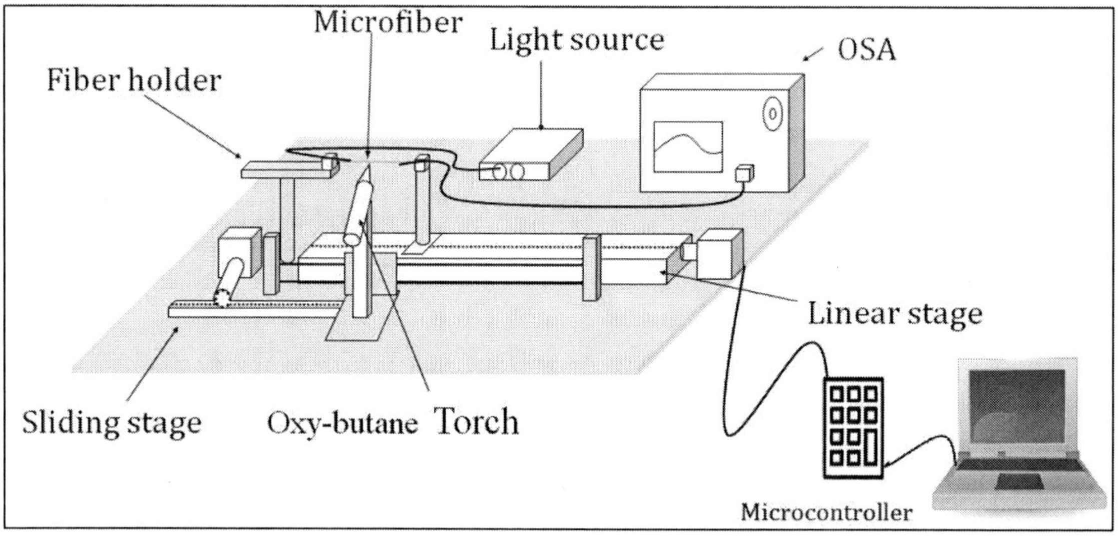

Figure 1. Flame Brushing Machine.

3.2 Coated Polymer Polyaniline

After successfully fabricate microfiber by using flame brushing technique, the microfiber is then coated with polymer Polyaniline (PAni) by using drop coating method. 3 drops of PAni solution is drop onto microfiber with interval of 10 seconds each and the structure is left to dry for 30 minutes. Figure 2 shows the coated PAni microfiber at a diameter of 4 µm. Figure 3 shows the optical set-up for microfiber sensor coated with PAni solution to sense different type of alcohols such as Methanol, Ethanol and Propanol with different concentrations. In the experiment, the sensing device is located between the light source, amplified spontaneous emission (ASE) and output wave, optical spectrum analyser (OSA). Weighing dish is placed below the tapered coated fiber for sensing different kind of solutions. The coated PAni microfiber was immersed by different kind of alcohols for sensing application and their outcome was analysed and recorded by OSA. Figure 4 shows a picture of the experiment in this project where the end of the fiber is connected to OSA and ASE.

IOP Conf. Series: Materials Science and Engineering **210** (2017) 012041 doi:10.1088/1757-899X/210/1/012041

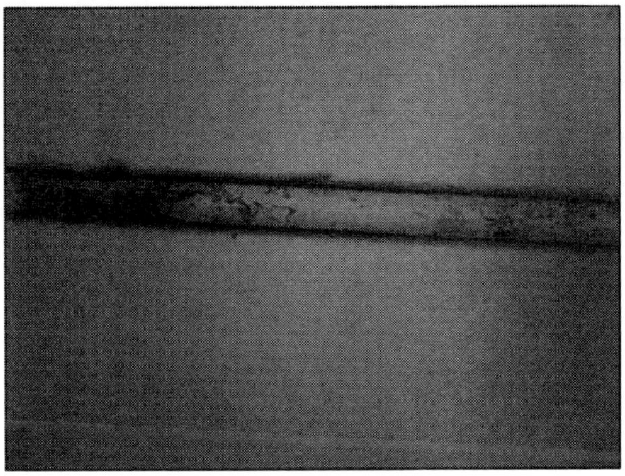

Figure 2. Microscopic image of microfiber coated with Polyaniline.

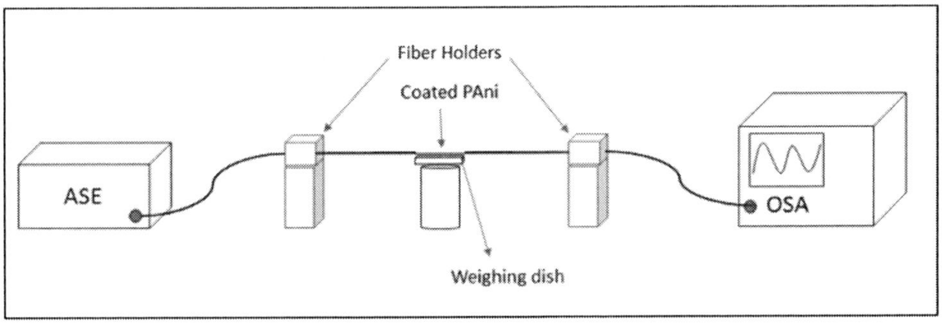

Figure 3. Experiment setup for conducting PAni coated optical microfiber sensor.

Figure 4. Picture of the overall experiment assembled in the laboratory.

IOP Conf. Series: Materials Science and Engineering **210** (2017) 012041 doi:10.1088/1757-899X/210/1/012041

4. Result and Discussion

Figure 5(a) shows the transmission power of different alcohols with different concentrations. The uncoated microfiber was immersed in alcohol solutions. Based on the plotted graph, Propanol have the highest transmission power for all the concentrations compared to Ethanol and Methanol. As the concentration of propanol increases from 100 ppm to 500 ppm, the output spectra reduce from 797.1 µW to 761.7 µW. The uncoated microfiber sensor give a sensitivity of 0.0327 µW/% with a linearity of 95.56 % and a limit detection of 13.76 %. Whereas in figure 5(b) the graph is plotted for coated microfiber with polymer PAni and the graph illustrated that as the concentration of propanol increases from 100 ppm to 500 ppm, the output spectra reduce from 375.3 µW to 320.38 µW. The coated microfiber sensor give a sensitivity of 0.126 µW/% with a linearity of 96.54 % and a limit detection of 5.723 %. As the number of carbon atoms in the alkyl group of alcohol increasing the transmission power also increasing. However, the increasing of concentration of the solvent reduces the transmission power. The pattern is repeatability by different concentration of alcohols.

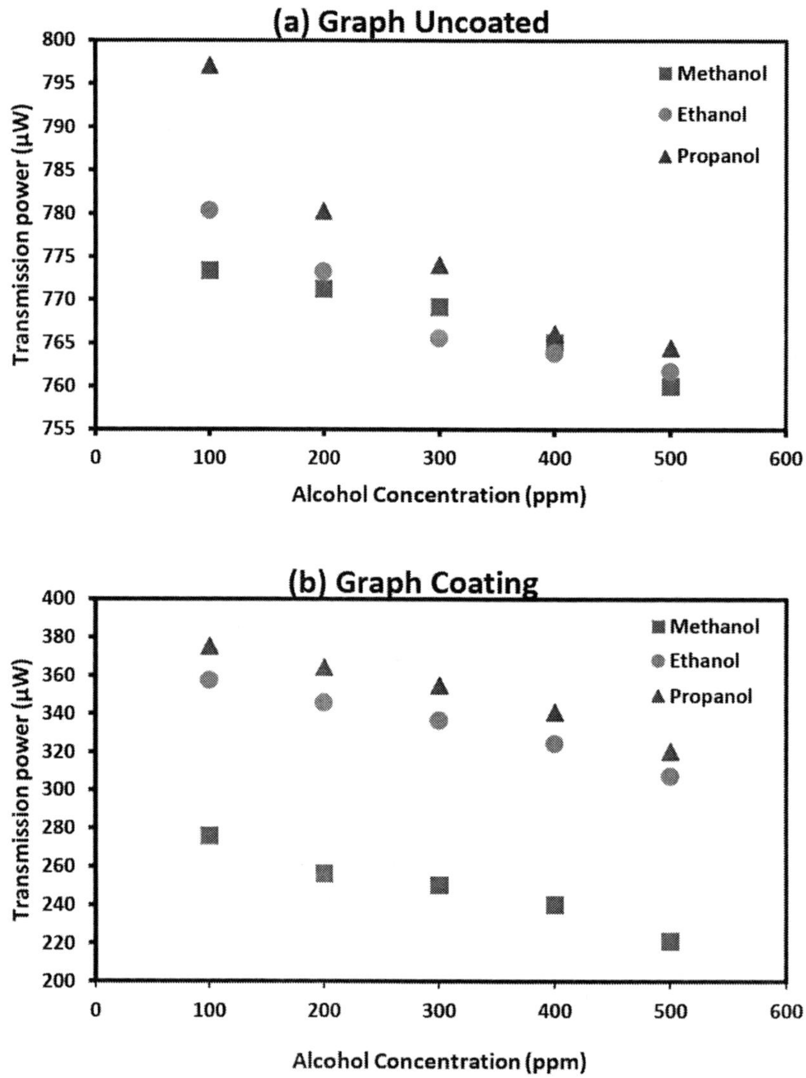

Figure 5. Power spectrum in detection of various kind of alcohols at different concentrations (a) uncoated microfiber and (b) coated microfiber with Polymer PAni.

Figure 6 shows the output spectra of the transmitted light from the microfiber with and without coating of Polyaniline. As illustrated in the figure, without coating of PAni in the tapered fiber the spectrum output power is -23.60 dBm and after the coating of PAni the spectrum output power decrease to -49.63 dBm. The transmitted light intensity drops by around 26 dBm at 1532 nm after the microfiber is coated with the polymer. This indicate that by coating polymer PAni to the microfiber make the fiber more sensitive to the surrounding because there is difference in the refractive index of the inner cladding. The evanescent fields surrounding the coated fiber and air is high which is very sensitive to the index change when there is presence of other medium or solution around the tapered area. The resonant peaks in the transmission spectrum will change with the variation in the refractive index of the environment. Microfiber becomes more sensitive when coated with polymer Polyaniline. Hence, the increasing in the sensitivity gives a good sensing properties.

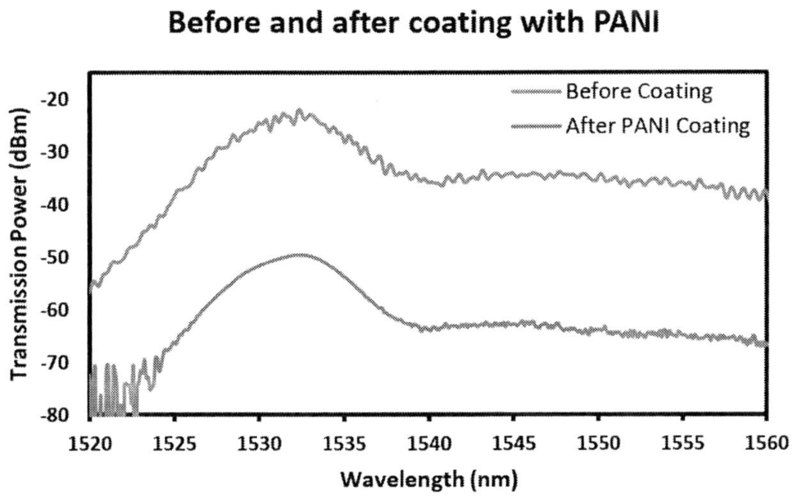

Figure 6. Output spectra of the microfiber before and after coating of PAni.

Figure 7 shows the spectral output response of coated polymer PAni microfiber in Methanol with different concentrations. As illustrated in the figure, different concentrations of alcohols have different transmission power due to their different refractive index of the solution. The refractive index increases with the increment of the alcohol concentrations. Higher concentration allows a strong enhancement of the evanescent field intensity and a great reduction in the transmitted power of the microfiber due to enhanced interactions between the evanescent field and the alcohol solution. Therefore, more leakage from the light that propagates inside the tapered region to the surrounding, which result in decreasing of output power. Thus, when the concentrations of methanol increase from 100 ppm to 400 ppm the output voltage drops from -55.97 dBm to -60.37 dBm.

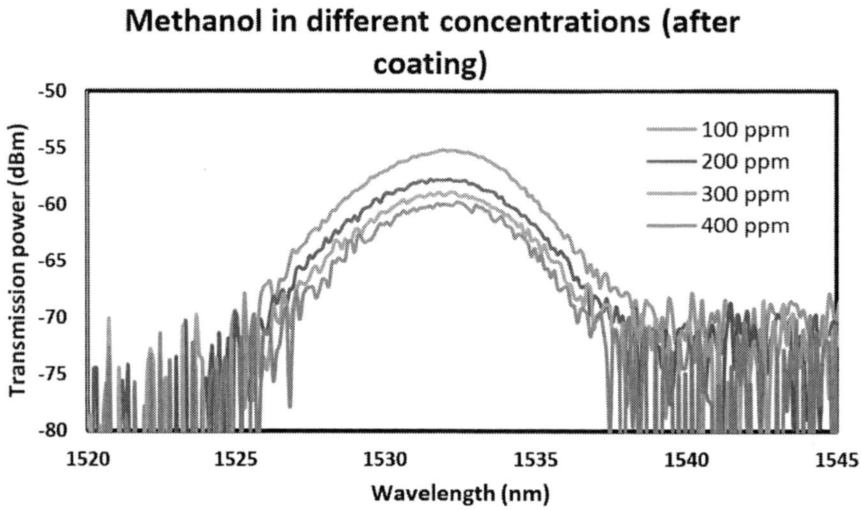

Figure 7. Transmission power (dBm) reduced when the concentration of methanol increasing.

In Figure 8, the graph has been plotted for transmission power of coating PAni in different alcohol between Methanol, Ethanol and Propanol. 100 ppm concentration of each alcohols is taken to compare their transmission spectrum. Methanol has the highest transmission power of -55.28 dBm, whereas Propanol has the lowest transmission power of -62.07 dBm. It is observed that with the increasing of carbon atoms in the alkyl group of alcohol molecules, it increases the reduction in the transmitted power because of the strong interactions between the evanescent field and the alcohol solutions. Furthermore, with the increasing of carbon atoms in the alkyl group also increases the refractive index and lead to increases in reduction in the transmitted power. With different kind of alcohol, the output power will differ from one another and results in capability to sense different kind of alcohols.

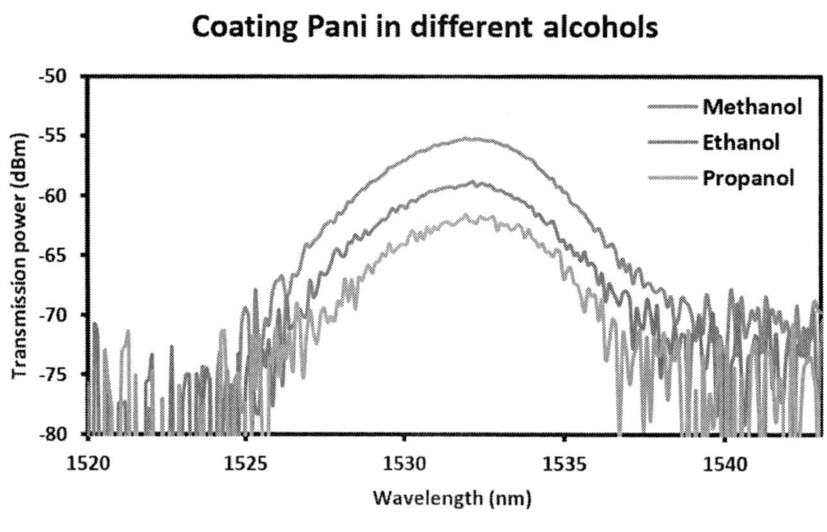

Figure 8. Different type of alcohols lead to different output spectra (dBm).

The overall performance of the refractive index sensor is summarized in Table 1. The microfiber coated with PAni display a higher sensitivity than the uncoated microfiber when the alcohol solutions is immersed. The microfiber coated with PAni shows a sensitivity of 0.1332 μW/% with a linearity of 97.52% when it immersed with Methanol. Its limit of detection is calculated at 5.6%. As for the uncoated microfiber with Methanol solution, the sensitivity display 0.0794 μW/% with a linearity of 90.43%. The detection limit is 9.08%. With the comparison between coated and uncoated microfiber, the coated microfiber has low limit of detection compared to with uncoated microfiber. The lower limit of detection gives higher system efficiency and performance. To conclude, based on the Table 1, coated PAni microfiber provides more sensitivity in term of sensing applications.

Table 1. Performance characteristic.

	Uncoated Microfiber		**Coated Microfiber**	
	Methanol	**Propanol**	**Methanol**	**Propanol**
Sensitivity (μW/%)	0.0794	0.0327	0.1332	0.126
Linearity (%)	90.43	95.56	97.52	96.54
Standard Deviation (μW)	0.721	0.450	0.746	0.721
Limit of Detection (%)	9.08	13.76	5.60	5.72

5. Conclusion

A simple refractive index sensor is proposed and demonstrated using a tapered silica optical fiber coated with a polymer composite for measurement of different kind of alcohols with various concentrations. The microfiber is fabricated by using flame brushing technique with tapering length of 3 cm. The microfiber coated with polymer PAni gives a sensitivity of 0.1332 μW/% with a linearity more than 97.52%. The presence of coated polymer give more sensitivity to the system due to the effective refractive index of the deposited coated that allows more light to be transmitted through the tapered fiber. Therefore, the proposed sensor provides outstanding advantages such as low cost, simple, and effective setup for sensing applications. This refractive index sensor is very convenience in monitoring and controlling the quality of alcohol concentration in the food processing and medical industry.

Acknowledgements

This work financial support from IIUM under grant RIGS (RIGS 15-148-0148). We also extend our thanks and gratitude to the photonic research center to allow us to use their lab to do practical experiments.

References

[1] Udd, E., Spillman, W.B., Eds. 2011 *Fiber Optic Sensors: An Introduction for Engineers and Scientists*, 2rd ed. John Wiley & Sons Inc.: Hoboken, NJ, USA
[2] Brambilla, G. 2010 *J. Opt.* **12** 043001
[3] Wu, X.; Tong, L. 2013 *Nanophotonics* **2**, 407–428
[4] G. Brambilla, V. Finazzi, and D. J. Richardson 2004 *Opt. Exp.* **12** 2258
[5] Tong, L.; Sumetsky, M. 2009 *Subwavelength and Nanometer Diameter Optical Fibers*; Zhejiang University Press: Hangzhou, China
[6] S.W. Harun, K.S. Lim, A.A. Jasim and H. Ahmad 2010 *Laser Physics* **20** 1629–1631
[7] Harun, S. W., Lim, K. S., Tio, C. K., Dimyati, K., & Ahmad, H. 2013 *Optik-International Journal for Light and Electron Optics* **124** 538-543

[8] Irawati, N., Rahman, H. A., Ahmad, H., & Harun, S. W. 2016 *Measurement* **99** 128-133
[9] Rahman, H. A., Irawati, N., Abdullah, T. N. R., & Harun, S. W. 2015 *IOP Conference Series: Materials Science and Engineering* **99** 012025
[10] John, V., Muhammad, M. Z., Irawati, N., Ali, N. M., & Harun, S. W. 2015 *JOURNAL OF OPTOELECTRONICS AND ADVANCED MATERIALS* **17** 1656-1660
[11] Irawati, N., Harun, S. W., Adwan, S., Alnowami, M., & Ahmad, H. 2015 *Microwave and Optical Technology Letters* **57** 2455-2457
[12] Irawati, N., Yusuf, N. A. M., Rahman, H. A., Yasin, M., Ahmad, H., & Harun, S. W. Published by., Yasin, M., Soelistiono, S., Yhuwana, Y. G., Khasanah, M., Arof, H., Irawati, N., & Harun, S. W. 2015 *OPTOELECTRONICS AND ADVANCED MATERIALS-RAPID COMMUNICATIONS* **9** 1185-1189
[13] Yuan, J., & El-Sherif, M. A. 2003 *IEEE sensors journal* **3** 5-12
[14] Chiam, Y. S., Lim, K. S., Harun, S. W., Gan, S. N., & Phang, S. W. 2014 *Sensors and Actuators A: Physical* **205** 58-62
[15] Lim, K. S., Ahmad, H., Arof, H., & Harun, S. W. 2012 *Selected Topics on Optical Fiber Technology* INTECH Open Access Publisher.

IOP Conf. Series: Materials Science and Engineering **210** (2017) 012042 doi:10.1088/1757-899X/210/1/012042

Graphene Oxide saturable absorber for generating eye-safe Q-switched fiber laser

A. H. A. Rosol[1], Z. Jusoh[3*], H. A. Rahman[1], M. F. M. Rusdi[2], S. W. Harun[2] and A. A. Latiff[4,5]

[1]Faculty of Electrical Engineering, Universiti Teknologi Mara, 40450 Shah Alam, Selangor

[2]Department of Electrical Engineering, Faculty of Engineering, University of Malaya 50603 Kuala Lumpur, Malaysia.

[3]Faculty of Electrical Engineering, University Teknologi Mara (Terengganu), 23000 Dungun, Terengganu, Malaysia

[4]Photonic Research Centre, University of Malaya, 50603 Kuala Lumpur, Malaysia.

[5]Faculty of Electronic and Computer Engineering (FKEKK), Universiti Teknikal Malaysia Melaka, Hang Tuah Jaya, 76100 Durian Tunggal, Melaka, Malaysia.

E-mail: *zulzi5352@tganu.uitm.edu.my

Abstract. This paper reports the generation of Q-switched fiber laser using thulium doped fiber (TDF) as a gain medium and graphene oxide (GO) as a saturable absorber (SA). The GO powder is embedded into polyvinyl alcohol (PVA) to form an SA film based on a drop-casting technique. GO-SA film is sandwiched between two fiber connectors and tighten by FC adapter before it is incorporated into an TDF laser cavity for Q-switching pulse generation. At 344 mW pump level, a stable Q-switching regime presence at 1943 nm with a 3-dB spectral bandwidth of 9 nm. The maximum repetition rate, pulse width, and pulse energy are at 25 kHz, 4.2 μs, and 0.68 μJ, respectively. All finding results are comparable with other reported pulse fiber lasers.

1. Introduction

Q-switched thulium doped fiber laser (TDFL) operating at 2 μm region have gained many research interests in recent years due to their wide range of potential applications in the fields such as medicine, remote sensing and LIDAR. The TDFL operates at wavelengths near atmospheric transparency window of 2 μm with a broad emission spectrum ranging from 1.8 μm to 2.1 μm. Basically Q-switched laser can be obtained by either active or passive methods [1-3]. The active method normally requires the insertion of bulk modulator such as acoustic-optic and electro-optic modulator into the laser cavity, thus make the laser configuration complex and expensive. Passively Q-switched fiber lasers serve many advantages over the active one including low cost, simplicity and flexibility. Previously, several saturable absorbers (SAs) have been reported to operate at the 2 μm wavelength including crystal of Cr^{2+}: ZnSe, semiconductor saturable absorber mirror (SESAM). However, the crystal SAs required an additional component to couple the light into the fiber, which conceding the key benefit of the compactness of the fiber. Even though SESAMs are widely commercialized and matured in the field, however it is considered to be expensive, narrow bandwidth operation and complex in device fabrication especially for 2 μm wavelength operation. Recently, carbon materials such as carbon nanotubes (CNTs) and graphene show promising performances as saturable absorber (SA). For CNTs, it is required to control the diameter of nanotubes for obtaining saturable absorption in the desired wavelength [4]. Graphene is a dirac two-dimensional nanomaterial which offers high performance saturable absorption characteristics with wide band operation from visible to near infrared region, ultrafast recovery time and low non-saturable loss [5]. The chemical synthesis of graphene oxide (GO) is cheaper than that of the pure graphene and thus it can serve as an alternative SA material with ultra-fast relaxation and strong nonlinear saturable absorption. GO also has inherent advantages such as good solubility in water and easy in fabrication [6].

Thulium doped fibers (TDFs) have a very broad emission spectrum, which spanning the wavelength range from 1700 nm to 2100 nm [7, 8]. It can provide great flexibility in the operating wavelength and serve as an excellent pump wavelength for nonlinear frequency conversion to the mid-infrared wavelength [9]. Recently, there are also several reports on generating of Q-switching pulse using graphene based saturable absorber [10-12]. In this paper, a Q-switched TDFL is demonstrated by using GO SA as a Q-switcher. The Q-switched laser produces pulse train, which repetition rate can be tuned from 15.76 kHz to 25.08 kHz as the pump power increases from 344 to 491 mW.

2. Fabrication and Characterization of GO SA

For constructing the GO based SA, the first step is to produce graphene flakes by using electrochemical exfoliation process. During the exfoliation process, a pair of electrodes was placed in electrolyte. When electrical potential was applied between the electrodes, the negative ion move towards anode while the positive ion to the cathode. This results in the loosening of the graphene layers from the graphite rod. The exfoliation process continues for two hours to obtain a stable graphene suspension in the sodium dodecyl sulfate (SDS) solution. The stable graphene suspension was centrifuged at 3000 rpm for 30 minutes to remove large agglomerates. Afterwards, the supernatant portion of the suspension was decanted. The graphene obtain from the electrochemical exfoliation was then mix with PVA solution at the ratio of 1:4 with drop casting technique for SA film. Then the solution was dried in the petri dish at room temperature to obtain free standing film with thickness of around 50 μm. The nonlinear absorption properties of GO SA was investigated by two level SA model. Figure 1 shows the absorption profile of GO SA. These data are fitted based on a simple two-level SA model [13];

$$\alpha(I) = \frac{\alpha_s}{1 + I/I_{sat}} + \alpha_{ns}$$

where $\alpha(I)$ is the absorption, α_s is the modulation depth, α_{ns} is the non-saturable absorption, I_{sat} is the saturation intensity, and I is the input intensity. From the analysed graph, the value of modulation depth, non-saturable absorption, and saturation intensity obtained to be 11 %, 9 % and 1.3 MW/cm^2, respectively.

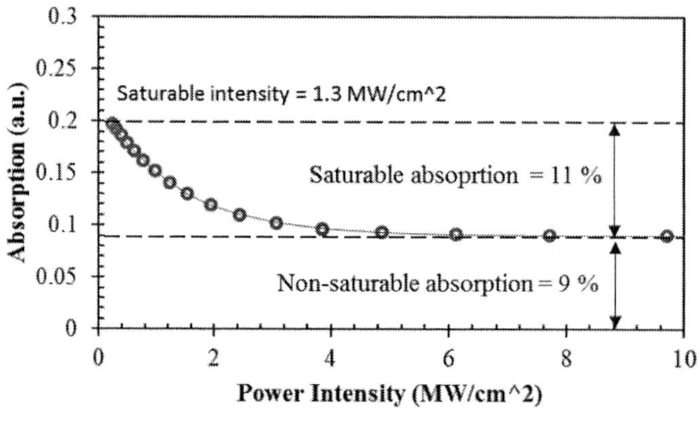

Figure 1. The nonlinear absorption profile of GO.

3. TDFL Configuration

The Q-switching pulse is generated using a TDFL ring cavity as shown in Figure 2. It uses a 5-m-long TDF as a gain medium, which is forward pumped by1550 nm erbium–ytterbium doped fiber laser via 1550/2000 nm wavelength division multiplexer. The TDF used has numerical aperture of 0.15, core and cladding diameters of 9 and 125 μm, respectively, loss of less than 0.2 dB/km at 1900 nm, and peak core absorption at 1180 and 793 nm are 9.3 and 27 dB/m, respectively. The isolator is used for unidirectional laser propagation and maintain it direction. The Q-switched laser was measured on the 10% output coupler, and the remaining 90% of output coupler used as a feedback for laser oscillated in the ring cavity.

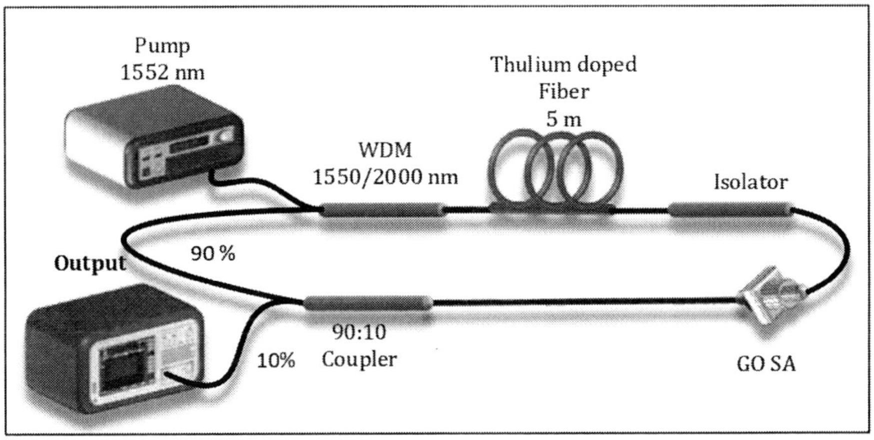

Figure 2. The experiment setup of Q-switched TDFL using GO SA.

The 1 mm x 1 mm GO film was sandwiched with a pair of fiber ferrule. Index matching gel was used as an adhesive and also as an element to reduce the back-scattering effect between the ferrules. As for confirmation of generation Q-switched laser by GO film, we remove the film from the ferule and there is no Q-switched lasing was observed. The total cavity length of the ring cavity approximately 11 meters. The output spectrum was observed by using optical spectrum analyser (OSA, Yokogawa: AQ6375) with resolution of 0.005 nm. Power meter was measured by using optical power meter (OPM, OPHIR: AN-2) connected with thermal sensor (OPHIR: D30A-SA_V1). The temporal characteristic in time domain of the Q-switched laser was observed by using 500 MHz oscilloscope (OSC, LeCroy: WAVEJET 352 A) through 7 GHz InGaAs photodetector with rise and fall time less than 50 ps. The result from oscilloscope will be confirm with the radio frequency spectrum analyser (RFSA, Anritsu: MS2683A) for measure in frequency domain of Q-switched laser.

4. Result and Discussion

At first, we sandwiched the 1mm x1 mm of GO film in between two fiber ferules and lock them by a fiber adapter. The Q-switched operation starts at the pump power of 344 mW and continues stable up to pump power of 491 mW. However, it disappears after the pump power is increased beyond 491 mW. Figure 3 shows the output characteristics of the Q-switched laser. Figure 3(a) shows the output spectrum at the maximum pump power of 491 mW, which indicates the operation of laser at the wavelength of 1943 nm. The figure also compares the output spectrum with and without the GO SA in the cavity. It is observed that the spectrum shifts to the shorter wavelength with the incorporation of the SA. This is attributed to an additional loss in the cavity created by the GO SA. The output power and pulsed energy characteristics of the Q-switched laser against the pump power are depicted in Figure 3(b). In the experiment, the pump power is varied from 344 mW to 491 mW. It is shown that the output power is decreased as the pump power is increased before it increases steadily as the pump power is

IOP Conf. Series: Materials Science and Engineering **210** (2017) 012042 doi:10.1088/1757-899X/210/1/012042

increased from 380 mW to 499 mW. The maximum output power is 17 mW at the pump power of 344 mW. On the graph, the slope efficiency of output power is 4.68% when maximum pulse energy was calculated at 1.66 nJ. The pulse energy shows the increment steadily with the increment of pump power.

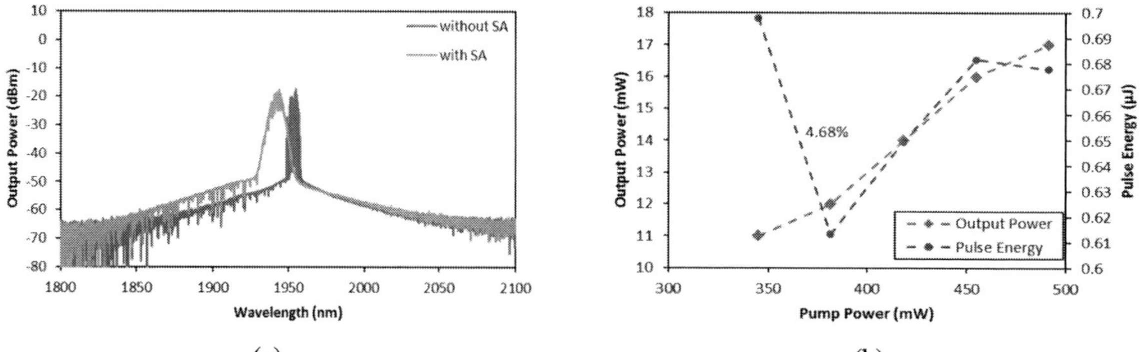

(a) (b)

Figure 3. Performance of the Q-switched TDFL (a) Output spectrum with and without SA at pump power of 344 mW. (b) Output power and pulse energy against the pump power.

Figure 4(a) shows the repetition rate and pulse width of the Q-switched laser against the pump power. As shown in the graph, by tuning pump power from 344 mW to 499 mW the repetition rate increase from 15.76 – 25.08 kHz. Also, the pulse width is narrowed from 5.08 μs to 4.2 μs with the increasing of repetition rate. Figure 4(b) shows the radio frequency (RF) spectrum of the TDFL Q-switched laser at the maximum pump power of 499 mW with the span of 380 kHz. Signal to noise ratio (SNR) of the RF spectrum is 47 dB for the fundamental frequency.

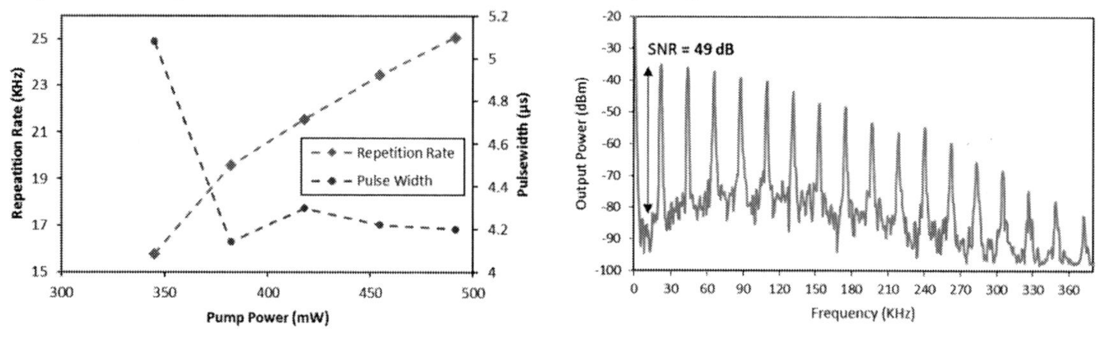

(a) (b)

Figure 4. Temporal characteristic of the laser (a) Repetition rate and pulse width against the pump power (b) RF spectrum within 380 kHz span.

5. Conclusion

The Q-switched generation by the GO SA was successfully demonstrated with the repetition rate tunable from 15.76 – 25.08 kHz when the pump power is varied from 344 mW to 491 mW. The laser has a slope efficiency of 4.58%. The homemade GO SA was fabricated by embedding the GO into PVA to form a film by using a drop-casting technique. The pulse width decreases from 5.08 μs to 4.2 μs with the increment of pump power. At the maximum pump power, the pulse energy of 1.66 nJ was achieved.

Acknowledgment

This research work is financially supported by Ministry of Higher Education Grant Scheme (FRGS/1/2015/SG02/UITM/03/3)

Reference

[1] Wang F, Torrisi F, Jiang Z, Popa D, Hasan T, Sun Z, Cho W and Ferrari A C 2012 *CLEO: Science and Innovations* Optical Society of America JW2A-72
[2] Eichhorn M and Jackson S D 2007 *Optics Letters* **32** 2780-2
[3] Latiff A, Rusdi M, Jusoh Z, Yasin M, Ahmad H and Harun S 2016 *OPTOELECTRONICS AND ADVANCED MATERIALS-RAPID COMMUNICATIONS* **10** 801-6
[4] Luo Z, Liu C, Huang Y, Wu D, Wu J, Xu H, Cai Z, Lin Z, Sun L and Weng J 2014 *IEEE Journal of Selected Topics in Quantum Electronics* **20** 1-8
[5] Bao Q, Zhang H, Wang Y, Ni Z, Yan Y, Shen Z X, Loh K P and Tang D Y 2009 *Advanced Functional Materials* **19** 3077-83
[6] Eda G and Chhowalla M 2010 *Advanced materials* **22** 2392-415
[7] Stoneman R and Esterowitz L 1990 *Optics letters* **15** 486-8
[8] Paul M, Dhar A, Das S, Latiff A, Ahmad M and Harun S 2015 *IEEE Photonics Journal* **7** 1-8
[9] Lippert E, Nicolas S, Arisholm G, Stenersen K and Rustad G 2006 *Applied Optics* **45** 3839-45
[10] Sharif N M, Yusoff N M, Ahmad F, Harun S and Ahmad H 2015 *2015 1st International Conference on*: IEEE *Telematics and Future Generation Networks (TAFGEN)* 103-6
[11] Luo Z, Zhou M, Weng J, Huang G, Xu H, Ye C and Cai Z 2010 *Optics letters* **35** 3709-11
[12] Latiff A, Shamsudin H, Aziz N, Hashim A, Irawati N, Ahmad H and Harun S 2016 *Optik-International Journal for Light and Electron Optics* **127** 11119-23
[13] Sobon G 2015 *Photonics Research* **3** A56-A63

IOP Conf. Series: Materials Science and Engineering **210** (2017) 012043 doi:10.1088/1757-899X/210/1/012043

The generation of Q-switched erbium-doped fiber laser using black phosphorus saturable absorber with 8% modulation depth

C. M. Fauziah[1, 2*], A. H. A. Rosol[3], A. A. Latiff [4, 5] and S. W. Harun[2]
[1]Politeknik Sultan Azlan Shah, 35950 Behrang, Perak, Malaysia
[2]Department of Electrical Engineering, University of Malaya, 50603 Kuala Lumpur, Malaysia
[3]Faculty of Electrical Engineering, Universiti Teknologi Mara, 40450, Shah Alam, Selangor
[4]Photonics Research Centre, University of Malaya, 50603 Kuala Lumpur, Malaysia
[5]Faculty of Electronic and Computer Engineering (FKEKK), Universiti Teknikal Malaysia Melaka, Hang Tuah Jaya, 76100 Durian Tunggal, Melaka, Malaysia.

E-mail: *cmfauziah@gmail.com

Abstract. We report a generation of the Q-switched laser operating in 1.55-micron region by using black phosphorus (BP) as a saturable absorber (SA). A 980-nm laser diode was pumped into Erbium-doped fiber (EDF) gain medium in ring cavity configuration. The BP-based SA was prepared by mechanically exfoliating the BP crystal using scotch tape. The obtained BP-tape SA has a modulation depth of 8 %. To realize a Q-switching operation, a small piece of the tape is then integrated into between two fiber ferrule tips. A stable Q-switching operation started at 40 mW. The maximum repetition rate obtainable at 28.57 kHz, with pulse width of 5.35 μs. This finding shows the BP is one of the promising material to work as an SA for pulsed laser generation.

1. Introduction

Pulse fiber lasers have undergone tremendous development in the past few decades since they have quite different characteristics compared with continuous wave lasers. High peak power, controllable repetition rate, and pulse width enable the pulse fiber laser to contribute significantly to laser development for practical applications, including material processing, LIDAR, laser communications, environmental detection, medical care, nonlinear frequency conversion, and laser acceleration [1-4]. Q-switching is one type of pulse fiber lasers, which is of great interest in recent years for the above-mentioned applications. Also, Q-switched all-fiber lasers have advantages in term of their flexibility, large accumulated one-trip gain, high beam quality and intense power confined in mode field diameters of only a few micrometers. Q-switching in all-fiber lasers has been achieved by applying a mechanical force on highly dispersive fiber Bragg gratings (FBG). The mechanical force or vibration could be provided by piezoelectric actuators [5], magnetostrictive transducers [6], or acoustooptics modulators [7]. Modulation of the Q-factor can also be realized in passive approaches using semiconductor saturable-absorber mirrors [8] and solid-state saturable-absorber fibers [9]. Recently, the search for new saturable absorber (SA) materials, essential for passive Q-switching, has intensified, as the alternative to the traditional SAs. For instance, single wall carbon nanotubes (SWNTs) and two-dimensional (2D)

IOP Conf. Series: Materials Science and Engineering **210** (2017) 012043 doi:10.1088/1757-899X/210/1/012043

materials such as graphene and transition metal di-chalcogenides (TMDs) have emerged as new SA material with superior performance, such as wide-band operation, mechanical and environmental robustness [10].

Recently, black phosphorus (BP), a newly emerged 2D material, has also gained wide attention in recent years for potential application in developing the next-generation optoelectronics devices such as a sensor, field-effect transistors and solar cells [11-13]. Multi-layer BP has a similar structure with bulk graphite. In a single layer, each phosphorus atom is covalently bonded to three adjacent phosphorus atoms to form a stable honeycomb structure, and different layers are stacked together by van der Waals interaction [14]. A multi-layer BP has a direct energy bandgap structure, with bandgap from 0.3 eV to 2 eV depending on the number of layer [15]. Naturally, BP has the common properties of 2D materials such as wideband absorption, ultrafast carrier dynamics and planar characteristic [16]. Note that BP comprises only the elemental "phosphorus". Hence it could be easily peeled off by using scotch tape. Stimulated by the similarity between graphene and BP regarding single elemental component and direct band-gap, it is natural to find out whether BP could be used as an SA for the Q-switching and mode-locking applications. The Q-switched of BP SA has been reported by [17, 18] with repetition rate from 31 kHz to 82.85 kHz at the wavelength of 1550 nm and also by using Ytterbium dope fiber for the mode-locked laser.

In this paper, we demonstrate a Q-switched Erbium-doped fiber laser (EDFL) using a mechanically exfoliated BP as SA. The BP-based SA tape was prepared by peeling bulk-structured BP 2D thin layer using a scotch tape. By inserting a small piece of the BP -SA tape into the EDFL cavity, Q-switched pulses with maximum pulse energy of 28.3 nJ is achieved, the shortest pulse width is 5.35 μs. To the best of our knowledge, this is the first report on BP -SA tape based passively Q-switched fiber lasers.

2. Preparation and characterization of the BP based SA

In this work, the BP based SA was prepared by mechanical exfoliation method, which has been widely used in graphene-based ultra-fast fiber laser applications [19, 20]. The nonlinear optical response property for the multilayer BP on the scotch tape is then investigated to confirm its saturable absorption by using the balance twin-detector measurement technique. A self-constructed mode-locked fiber laser (1562 nm wavelength, 1.05 ps pulse width, 16.3 MHz repetition rate, 3.2 MW/cm^2 maximum intensity before sample) is used as the input pulse source. The transmitted power and also a reference power for normalization are recorded as a function of incident intensity on the tape by varying the input laser power. With increasing peak intensity, the material absorption decreases as shown in Figure 1, confirming saturable absorption. The experimental data for absorption are fitted according to $\alpha(I) = \alpha_s/(1 + I/I_{sat}) + \alpha_{ns}$, where $\alpha(I)$ is the absorption, α_s is the modulation depth, I is the input intensity, I_{sat} is the saturation intensity, and α_{ns} is the non-saturable absorption. As shown in the figure, a modulation depth, non-saturable intensity, and saturation intensity were obtained to be 8 %, 57 % and 0.35 MW/cm^2, respectively. Taking into account its nonlinear optical response leading to absorption saturation at relatively low fluence, the mechanically exfoliated BP meets basic criteria of a passive SA for fiber lasers.

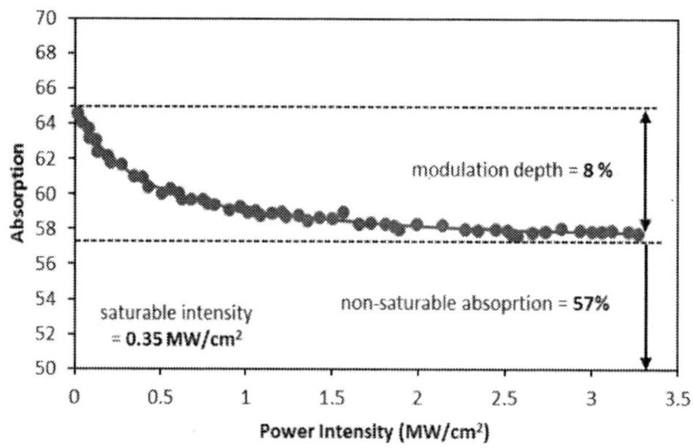

Figure 1. Nonlinear absorption profile.

3. Experimental setup

The setup of the Q-switched fiber laser is depicted in Figure 2. A 3 m long piece of erbium-doped fiber (EDF) was used as an active fiber. The EDF has core and cladding diameters of 4 μm and 125 μm respectively, a numerical aperture of 0.16 and Erbium ion absorption of 23 dB/m at 980 nm. It was co-directionally pumped via a fused 980/1550 nm wavelength division multiplexer (WDM) by a 980-nm laser diode. The signal was coupled out using 95/5 output couplers (OC) which channeled out 5 % of the light oscillating in the ring cavity for both spectral and temporal diagnostics. The fiber isolator forced a unidirectional signal propagation. As mentioned above, we chose a BP based SA to obtain the Q-switched pulse. The prepared multi-layered BP on the scotch tape was integrated into the fiber laser cavity by sandwiching a ~1 mm × 1 mm piece of the composite tape between two fiber connectors. The SA was adhered to the ferrule using index matching gel. The cavity length was measured to be approximately 12 meter. The optical spectrum analyzer (OSA) with a spectral resolution of 0.07 nm was used for the spectral analysis while an oscilloscope (OSC) was used to analyze the output pulse train of the Q-switching operation via a photo-detector.

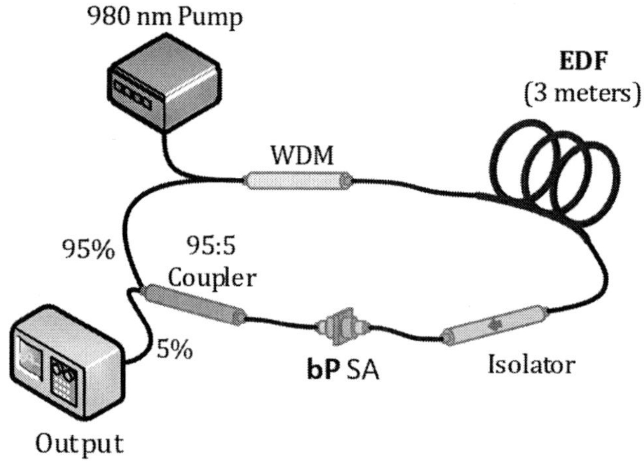

Figure 2. Experimental setup of the proposed Q-switched EDFL employing a mechanically exfoliated BP.

4. Experimental results

Continuous wave operation begins at a pump power of 25 mW, and stable Q-switched pulses were achieved when the pump power was increased to 40 mW. The Q-switching operation was maintained up to the maximum pump power of 170 mW. Typical characteristics of the Q-switched pulses emitted from the fiber laser at a pump power of 170 mW are presented in Figure 3. The optical spectrum of the Q-switched pulses is centered at 1550.9 nm, and the 3-dB spectral width is approximately 0.4 nm, as shown in Figure 3 (a). Figure 3(b) displays a typical oscilloscope trace of the Q-switched pulse train, which shows a pulse period and pulse width of 35.0 μs and 5.35 μs, respectively. There is no distinct amplitude modulation is the entire Q-switched envelope in the spectrum, which leads to knowledge that the self-mode locking effect on the Q-switching is weak.

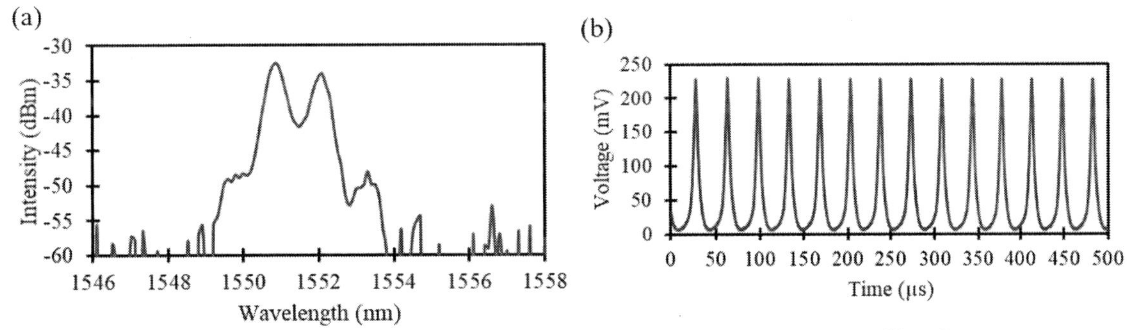

Figure 3. Typical characteristics of the Q-switched pulses emitted from the fiber laser at a pump power of 170 mW, where (a) Optical spectrum and (b) oscilloscope trace.

To verify that the passive Q-switching was attributed to the BP based SA, the FC fiber ferrule filled with BP tape was replaced with a common clean ferrule. In this case, no Q-switched pulses were observed on the oscilloscope even when the pump power was adjusted over a wide range. This finding has confirmed that the BP-based SA was responsible for the passively Q-switched operation of the laser. Unlike the fixed repetition rate of a mode-locked fiber laser the pulse repetition rate in our laser increased with the pump power, which is a typical feature of passive Q-switching operation. Figure 4 shows the pulse repetition rate and pulse width of the Q-switched fiber laser as functions of the incident pump power. By increasing the pump power from 40 to 170 mW, the pulse repetition rate could be varied over a wide range of frequencies, from 7.46 to 28.57 kHz. On the other hand, the pulse width decreased from 22.34 μs near the pump threshold of 40 mW to 5.35 μs at a pump power of 170 mW. The minimum pulse width obtained in our experiment could be narrowed by shortening the cavity length and improving the modulation depth of the few-layer BP.

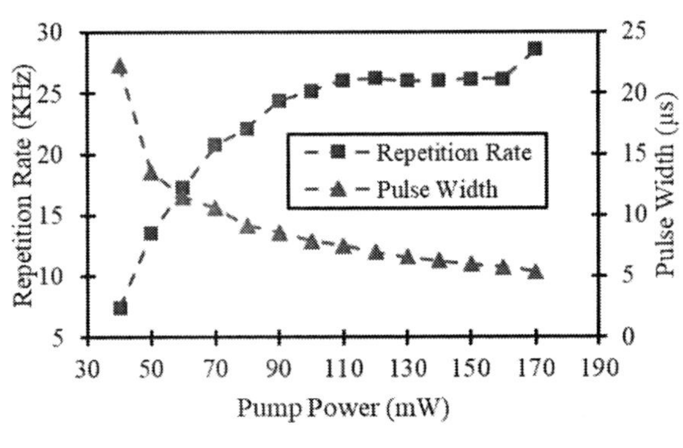

Figure 4. Pulse repetition rate and pulse width at different pump powers.

Figure 5 shows the dependence of the average output power and pulse energy of the Q-switched fiber laser versus the pump power. The pulse energy was calculated based on the measured average output power and the repetition rate. When the pump power exceeded the threshold, both average output power and pulse energy increased almost linearly with the pump power. The maximum average output power was 0.18 mW at a pump power of 170 mW, corresponding to a pulse energy of 6.4 nJ. We believe that the performance of Q-switched pulses produced by the laser could be further improved by optimizing the SA parameters of BP tape and the cavity design.

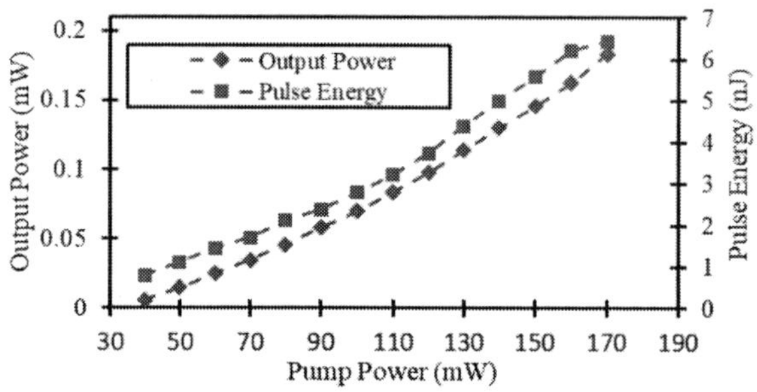

Figure 5. Average output power and pulse energy versus incident pump power.

5. Conclusion

We experimentally demonstrated a passively Q-switched ring EDFL using a few-layer bP flakes onto a scotch tape as a SA. The BP tape was prepared by the mechanical exfoliation method and sandwiched between two fiber connectors with a fiber adapter to form a fiber-compatible BP -based SA. Stable Q-switched pulses at 1550.9 nm were successfully obtained. The laser showed a pump threshold of 40 mW and a minimum pulse duration of 5.35 μs. The pulse repetition rate could be varied over a wide range of frequencies, from 7.46 to 28.57 kHz, by adjusting the pump power. Our experimental results suggest that few-layer BP is a promising material for pulsed laser applications.

References

[[1] Fermann M E and Hartl I 2013 *Nature photonics* **7** 868-74
[2] Nishizawa N 2014 *Japanese Journal of Applied Physics* **53** 090101
[3] Scholle K, Fuhrberg P, Koopmann P and Lamrini S 2010 INTECH Open Access Publisher
[4] Clowes J 2008 *Optik & Photonik* **3** 36-8
[5] Russo N, Duchowicz R, Mora J, Cruz J and Andrés M 2002 *Optics communications* **210** 361-6
[6] Pérez-Millán P, Díez A, Andrés M, Zalvidea D and Duchowicz R 2005 *Optics express* **13** 5046-51
[7] Delgado-Pinar M, Zalvidea D, Diez A, Pérez-Millán P and Andrés M 2006 *Optics express* **14** 1106-12
[8] Wang H-Y, Xu W-C, Luo A-P, Dong J-L, Cao W-J and Wang L-Y 2012 *Optics Communications* **285** 1905-7
[9] Tsai T-Y, Fang Y-C and Hung S-H 2010 *Optics express* **18** 10049-54
[10] Martinez A and Sun Z 2013 *Nature Photonics* **7** 842-5
[11] Abbas A N, Liu B, Chen L, Ma Y, Cong S, Aroonyadet N, Köpf M, Nilges T and Zhou C 2015 *ACS nano* **9** 5618-24
[12] Li L, Yu Y, Ye G J, Ge Q, Ou X, Wu H, Feng D, Chen X H and Zhang Y 2014 *Nature nanotechnology* **9** 372-7
[13] Buscema M, Groenendijk D J, Steele G A, Van Der Zant H S and Castellanos-Gomez A 2014 *Nature Communications* **5** 4651
[14] Geim A K and Grigorieva I V 2013 *Nature* **499** 419-25
[15] Tran V, Soklaski R, Liang Y and Yang L 2014 *Physical Review B* **89** 235319

[16] Qiao J, Kong X, Hu Z-X, Yang F and Ji W 2014 *Nature communications* **5**
[17] Ismail E, Kadir N, Latiff A, Ahmad H and Harun S 2016 *RSC Advances* **6** 72692-7
[18] Hisyam M, Rusdi M, Latiff A and Harun S 2017 *IEEE Journal of Selected Topics in Quantum Electronics* **23** 1100205
[19] Novoselov K 2011 *Reviews of Modern Physics* **83** 837
[20] Martinez A, Fuse K and Yamashita S 2011 *Applied Physics Letters* **99** 121107

Multispectral image segmentation using localized spectral binarization

Nur Salma Mohd Mokhtar[1], Khairuddin Osman[1*], Khairul Muzzammil Saipullah[1], Muhammad Haziq Faris Hasnol[1],
[1]Faculty of Electronic and Computer Engineering (FKEKK), Universiti Teknikal Malaysia Melaka, Hang Tuah Jaya, 76100 Durian Tunggal, Melaka, Malaysia
[2]RETAK Lab, REKA Axis PLT

E-mail: *kharuddin.osman@utem.edu.my

Abstract. This paper proposes a new feature extraction method for multispectral image segmentation based in Localized Spectral Binarization (LSB). In contrast with the standard image operation, which is applied with traditional image, LSB is computed on a single pixel with numerous bands. The proposed algorithm calculates differences of spectra locally on same pixel's coordinate in different bands of the multispectral image. The difference value is converted into binary by breaking the difference values into two directions, which are the positive and negative value then the differences are thresholded to form a binary codeword. A binomial factor is assigned to these codewords to form another unique value. These values are then grouped to construct the LSB feature image where is used in the image segmentation. LANDSAT multispectral images are used in the experiment to evaluate the segmentation and classification accuracy of the proposed LSB in terms of pixel-wise image segmentation. The result shows that LSB feature outperforms the spectral feature.

1. Introduction

Segmentation is process of dividing an image space into some none overlying meaningful homogeneous regions [1]. Multispectral images cover more information than single band images. One of the vital approaches of information extraction, is the multispectral image segmentation. Object-based image analysis secret to success in accurateness is dependent on the quality of the image segmentation. Commonly, multispectral has three to seven bands. To achieve a higher accurate image segmentation and classification result, the extra information from multispectral will significantly help with the result. Meanwhile, mishandling of the extra band information can cause to low performance of image segmentation. Grey value thresholding and pixel classification are the two popular approaches for image segmentation in remote sensing process. In image thresholding [2], a set of thresholds T is searched so that all pixels with gray values in range constitute the ith region type. Homogeneous regions are determined by clustering the feature space of multiple image bands in pixel allocation. Pixel classification and thresholding methods can be either local or global. Pixel classification-based segmentation is frequently applied because of each pixel in multispectral image is represented by a set of values.

Histogram-based and cluster based methods are divided into two based on pixel classification-based segmentation. The histogram-based method assumes that homogeneous regions in the image correspond to modes of image histogram. They are based on the spectral feature space [3]. Cluster-based segmentation methods assume that interesting structures in the image form clusters in the band domain.

The basic idea behind histogram and cluster-based segmentation is that image pixels are processed as general data samples in the feature space. The performance of cluster-based segmentation method depends on the feature of the each pixel. Standard cluster-based method for multispectral image applies the spectral information as the feature of the each pixel. However this information is not enough and often leads to misclassification. In order to increase the feature information, extra information needs to be extracted from the spectral.

In this paper, we propose a novel feature extraction of spectral information based on Localized Spectral Binarization (LSB) [4]. LSB is based on the local neighborhood difference between the surrounding and the central pixels in a all bands. The remainder of this paper is organized as follows: in Section 2, related work is presented; in Section 3, detailed algorithm to implement LSB in multispectral image segmentation is presented; experimental evaluations are described in Section 4 and finally conclusions and future works are given in Section 5.

2. Related Work

2.1. Local Binary Pattern

The detail explanation about LBP [5] algorithm using 3x3 neighborhoods is shown in Figure 1. First, the neighboring pixels of the 3x3 neighborhood are extracted. For each 3x3 window of an image, neighborhood differences which are the centralized difference v(n) are calculated. Then the v(n) are thresholded against zero in order to convert the difference values into 8-bit binary codeword. A binomial factor of 2 is assigned for each binary code to transform the codeword into a unique LBP number that represents the texture unit of the 3x3 neighborhood. This LBP value is a decimal value between 0 and 255 resulting from the 8-bit binary code. Next, a histogram is constructed with 256 dimensions using the LBP values and the histogram denotes the distribution of the CND values. Finally, the texture descriptor is obtained from the histogram.

The extremely simple computations for constructing the LBP make it very fast in extracting image feature. The performance of LBP is also similar with other local texture descriptors such as combined neighborhood difference (CND) [9] and local neighbor differences (LND) [6].

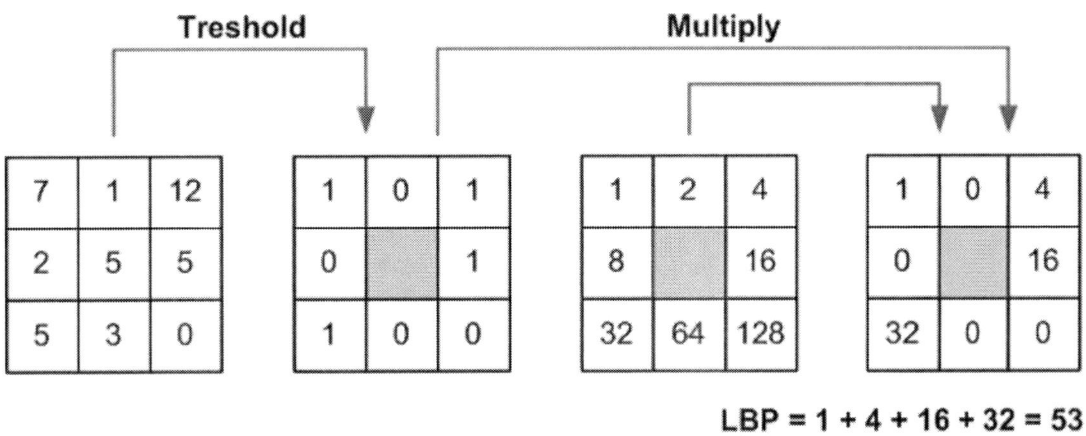

Figure 1. Figure shows a LBP example.

IOP Conf. Series: Materials Science and Engineering **210** (2017) 012044 doi:10.1088/1757-899X/210/1/012044

3. Proposed Method

3.1. The Localized Spectral Binarization Histogram

Multispectral image uses more than one band to represent an image. Each band acquires one digital image in a small band of visible spectra, ranging 0.4 µm to 0.7 µm, called red-green-blue (RGB) region, and going to infra-red wavelengths of 0.7 µm to 10 µm or more, classified as Near InfraRed (NIR), Middle InfraRed (MIR) and Far InfraRed (FIR) or Thermal [7].

Each pixels of a multispectral image contains several values depending on the number of the band. To increase the significant information, LSB feature is extracted from each band for each pixel. The multispectral images can be converted into the feature space F(x,y,c) using LSB using several steps. First, the pixel value from each band of the multispectral image is gathered to form a 1D data p(a), a=1, 2,…, n where n is the number of band in the multispectral image. Then, p(a) in divided into the center group c(a), a=1, 2,…, n and the neighborhood group x(a,b), a=1, 2, …, n, b=1, 2,…, n-1 as follows:

$$\begin{aligned} c(a) &= p(a) & a &= 1,2,\ldots,n \\ x(a,b) &= p\{1+(a+b-1)\bmod n\}, & b &= 1,2,\ldots,n-1 \end{aligned} \quad (1)$$

Then the differences between x(a,b) and c(a) is calculated and the values are thresholded against 0 as follows:

$$T(a,b) = \begin{cases} 1, \text{ if } \{x(a,b)-c(a)\}>0 \\ 0, \quad \text{otherwise} \end{cases}, \quad (2)$$

$$a = 1,2,\ldots,n, \; b = 1,2,\ldots,n-1$$

T(a,b) can be seen as an 2D array of binary code. By handling each row in T(a,b) as a n-1 bit of binary codeword, it is possible to transform (3) into a unique LSB number for each band. It is done by assigning a binomial factor H for each T(a,b), given by

$$f(a) = \sum_{b=1}^{n-1} T(a,b)H^{b-1}, \; a = 1,2,\ldots,n \quad (3)$$

f(*a*) is then saved in the feature image F(x,y,a). If the size of the multispectral image is MxN, x=1, 2, …, M, y=1, 2,…, N and a=1, 2,…, n. This process is repeated to each pixel of the multispectral image. F(x,y,a) is then used as the feature in the segmentation of the multispectral image. In the feature image F(x,y,a), each pixels have a set of LSB values. The number of LSB code each pixel has is the same as the number of band of the multispectral image.

3.2. The Segmentation

The K-Means algorithm [8] is used as the clustering method in the unsupervised segmentation of the multispectral image. K-Means algorithm classifies the input data points into multiple classes based on their inherent distance from each other. The algorithm assumes that the data features form a vector space and tries to find natural clustering in them. Given a set of points x=( ), where each point is a n-dimensional real vector. The points are clustered around centroids which are obtained by minimizing the within-cluster sum of squares

IOP Conf. Series: Materials Science and Engineering **210** (2017) 012044 doi:10.1088/1757-899X/210/1/012044

$$V=\sum_{i=1}^{k}\sum_{x_j\in S_i}\left(x_j-\mu_i\right)^2 \quad (4)$$

Where there are k clusters, , i=1,2, …, k and is the centroids or mean point of all of the points . To segment the multispectral image, pixel values in F(x,y,a) feature image are grouped into a group of points x=( ), where each LSB values of each pixel in F(x,y,a) is assigned as the n-dimensional vector of the points.

4. Experimental Results

To evaluate the performance of the LSB in segmentation of multispectral images, the classification and segmentation of two LANDSAT 7 multispectral images of Rio Janeiro. The images of Rio Janeiro in the RGB band can be seen in Figure 2. We can judge the segmentation performance by scrutinizing the clarity of the segmentation result. The precision of the segmentation can also be assessed using classification accuracy, which can be calculated as follows:

$$\text{accuracy}(\%)=\frac{\text{no. of correctly classified pixels}}{\text{no. of pixels in each class}} \quad (5)$$

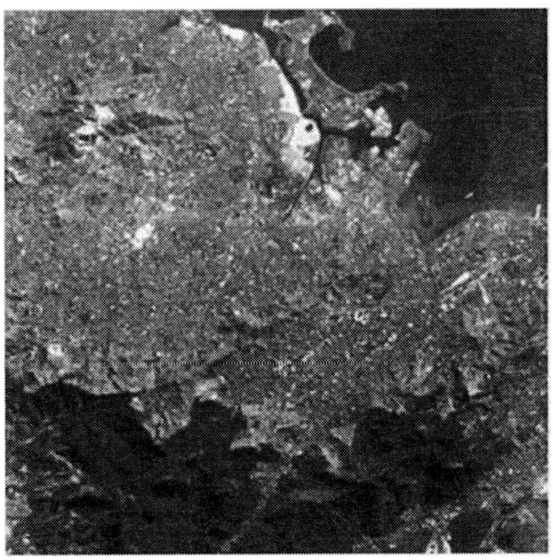

Figure 2. Figure shows Pixel Rio Janeiro

The Rio Janeiro image contains seven bands and the binomial factor H assigned for Rio Janeiro image is 2. The pixels are grouped into a set of points in order to perform the K-mean clustering with Euclidean distance. The pixels in the image are clustered into three main classes (K=3) namely, water, build-up and vegetation. The segmentation performance using LSB feature is compared with that of segmentation using spectral feature. The results of the segmentation of Rio Janeiro multispectral image using LSB feature and spectral feature are shown in Figure 3. The classification accuracy result can be seen in Table 1.

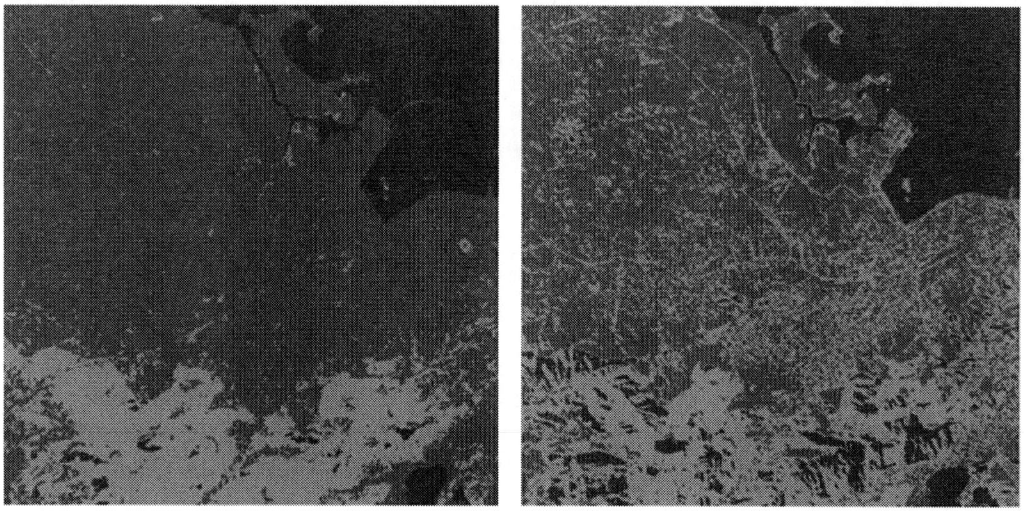

Figure 3. The segmentation result of Rio Janeiro images

Table 1. The classification accuracies

Method	Water	Build-up	Vegetation	Average
LSB	95.72	89.10	85.12	89.98
Spectral	95.17	70.03	2.22	55.81

From the segmentation results, one can see that the segmentation using LSB feature distinctly clustered each pixel into three classes compared to that of spectral feature. The bridges are always misclassified as water. However, LSB manages to perfectly classify all of the bridges. The color of the vegetation area is similar to that of water area. From the result, we can see that the segmentation using spectral feature misclassified these two areas. On the contrary, LSB manages to classify these two areas almost perfectly. From the classification accuracy results, we can see that LSB perform consistently in the classification with average accuracy of 89.98%. The classification accuracy using spectral feature is unstable since a lot of vegetation areas are misclassified as water areas. This misclassification causes low average classification accuracy, which is 55.81%.

5. Conclusion

In this paper, we propose a novel feature extraction method for multispectral image based on the LSB. The LSB is calculated on single pixel with various bands. The result of segmentation using LSB feature shows greater result compared to that of using spectral feature. LSB feature also achieves high segmentation accuracy compared to those of spectral feature. For the future works, we would like to implement the LSB algorithm in the hyperspectral image analysis such as target detection. Hyperspectral image contains more than 200 dimensions, and the feature extraction of extremely high dimensional image is not an easy task.

Acknowledgments

The authors would like to thank Universiti Teknikal Malaysia Melaka (UTeM) and Ministry of Education (MOE) Malaysia under the Research Acculturation Grant Scheme (RAGS) No. RAGS/1/2014/TK03/FKEKK/B00063 for their support.

References

[1] Pal, S.K. and Mitra, P. 2002 *IEEE Transactions on Geoscience and Remote Sensing* **40** 2495-2501
[2] Pal, S.K., Ghosh, A. and Shankar, B.U. 2000 *International Journal of Remote Sensing* **21** 2269-2300
[3] Paclík, P., Duin, R.P. and van Kempen, G.M. 2001 In *PROCEEDINGS OF THE SCANDINAVIAN CONFERENCE ON IMAGE ANALYSIS* 230-235
[4] Khairul Muzzammil, Shao-hu Peng, and Deok-Hwan Kim 2010 *Journal of the Institute of the Electronics Engineers of Korea* **47**
[5] Ojala, T., Pietikäinen, M. and Harwood, D.1996 *Pattern recognition* **29** 51-59
[6] Khairul Muzzammil, Shao-hu Peng, Min-Wook Park and Deok-Hwan Kim 2010 *Proc. KIPS Spring Conf.*
[7] R.A. Schowengerdt 1997 *CA:Academic*
[8] Lloyd, S. 1982 *IEEE transactions on information theory* **28** 129-137
[9] Khairul Muzzammil Saipullah 2011 *International Journal of Computer Technology and Electronics Engineering* **1** 1-6

Tunable microwave generation using dual-wavelength Brillouin O-band fiber laser

S N Aidit[1*], S I Ooi[1], N A Hassan[2], H Ahmad[1] and Z C Tiu[1]
[1]Photonics Research Centre, University of Malaya, 50603 Kuala Lumpur, Malaysia
[2]Advanced Devices and Materials Engineering iKohza, Malaysia-Japan International Institute of Technology, Universiti Teknologi Malaysia International Campus, 54100 Kuala Lumpur, Malaysia

E-mail: *nabilaaidit@siswa.um.edu.my

Abstract: A dual-wavelength fiber laser with a tunable output for generating microwave signals is proposed and demonstrated. The system utilizes stimulated Brillouin scattering inside a dispersion compensating fiber to generate the desired dual-wavelength output, which then beats to form the desired microwave frequency in the O-band region. A constant channel spacing of 0.072 nm is obtained for wavelengths ranging from 1260.0 nm to 1330.0 nm. The resulting microwave signal can also be tuned from 13.0 GHz to 12.3 GHz is achieved by controlling the Brillouin pump wavelength, and has a stable signal-to-noise ratio of 30.0 dB.

1. Introduction

Microwave and sub-millimeter wave carrier signal generation through optical sources has been the focus of significant research efforts due to their wide range of applications that range from broadband wireless access networks to applications that involve phased-array communications, radio astronomy and electronic warfare [1]. Traditionally, these signals were generated electronically, but incur high levels of complexity and significant cost. Furthermore, most applications require the generation of the microwave signal on-site, as distribution through electrical lines is not suitable due to high loss [2]. These factors make the electrical generation of microwave and sub-millimeter wave carrier signals highly unpractical, and only used when no other options abound.

In this regard optical fibers, with their inherently low loss and significantly broad bandwidth, make the generation and distribution of microwave signals over optical fibers a very desirable option. Numerous techniques have been proposed and demonstrated for the generation of microwave signals using optical fibers, including optical injection locking [3], optical frequency or phase locked loops [4], external modulation and semiconductor lasers that have been resonantly enhanced [5]. Optical heterodyning has been proven to be an effective and cost-friendly solution towards generating optically based microwave signals, simply by having two lasing signals at different wavelengths mixing at a photodetector to produce an electrical beat signal [6]. The difference between the wavelengths corresponds to the desired microwave frequency.

As a result of this, dual-wavelength fiber lasers (DWFL) have come to the forefront of research as these systems are capable of producing two phase coherent signals whose wavelength difference corresponds to the desired microwave frequency in a compact, cost-effective platform. In addition to this, DWFLs, and by that extension multi-wavelength fiber lasers (MWFLs), can be used to generate a tunable microwave output. Pan and Yao et. al. reported a frequency-switchable microwave generation using high-finesse ring filter [7], while Kim et. al. demonstrated optical beat frequency generation in an erbium based DWFL utilizing the nonlinear polarization rotation effect [8]. However, a significant concentration of research has been focused on the generation of multi-wavelength and microwave signals at the C-band region, and extending into the S- and L-band regions [7-11], but not much into the O-band region. This provides a promising avenue for research, especially to cater to broader bandwidth as to allow for switchable microwave frequencies [7-8, 12-15].

In this work, a DWFL with a tunable wavelength output operating in the O-band is proposed and demonstrated. The proposed system exploits Brillouin scattering in order to achieve the desired dual-wavelength output, which is used in turn to generate a microwave signal through optical beating. Tunability is achieved by controlling the center wavelength of the Brillouin pump (BP).

2. Experimental Setup

The setup of the proposed O-band DWFL is shown in Figure 1. The system consists of a Santec TLS-550 tunable laser source (TLS), which has a tuning range from 1260.0 nm to 1365.0 nm and acts as the BP for the Brillouin gain medium (BGM). The signal from the TLS is directed through the input port of the first optical circulator (CIR1), and the output from the common port of CIR1 is fed into the BGM. The BGM in this work is a 6.90 km long dispersion shifted fiber (DSF), and its output is connected to a second optical circulator (CIR2), which is configured to act as a reflector by having its output port linked to its input port, thus looping the signal back towards the common port. The output port of CIR1 is connected to a 50:50 coupler, which splits the output signal evenly for analysis by an AQ6370B optical spectrum analyzer (OSA) from Yokogawa, which has a resolution of 0.02 nm, as well as an Advantest U3771 radio frequency spectrum analyzer connected to a Newport 818-BB-51F photodetector.

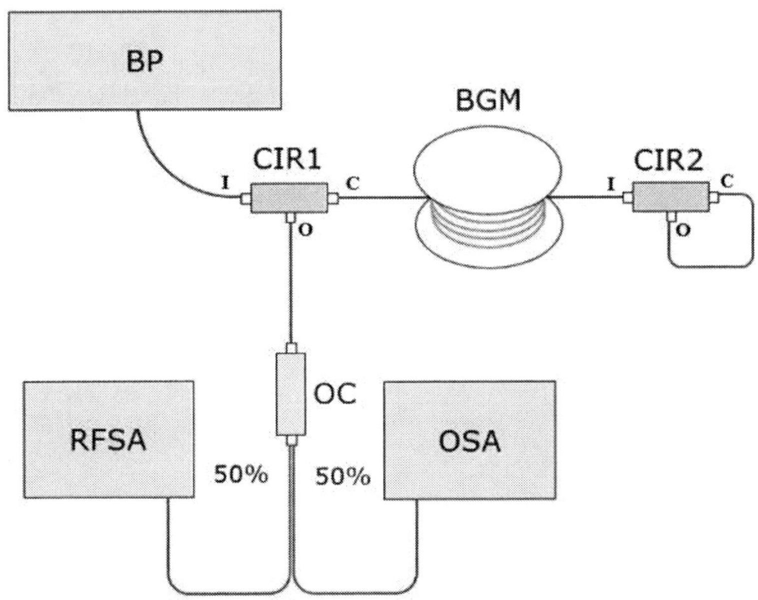

Figure 1. Experimental setup of dual-wavelength Brillouin O-band fiber laser.

3. Results and Discussion

The desired dual-wavelength output is obtained from the system by exploiting the Brillouin effect in the BGM. In this regard, the TLS serves as the BP. The BP signal then travels into the cavity where it encounters the BGM. The 1st Stokes is generated once the intensity of the BP signal is strong enough to excite stimulated Brillouin scattering (SBS) in the BGM. The Stokes generated is red-shifted by 0.072 nm with respect to the BP wavelength and propagates in the opposite direction. The BP signal continues to travel along the linear cavity where it is reflected by CIR2 and now propagates in the opposite direction, parallel to the 1st Stokes. Thus, two lasing wavelengths are obtained; at 1310.00 nm and 1310.07 nm, with intensities of -7.4 dBm and -5.7 dBm respectively. Due to the structure of the cavity, it is only possible to obtain two lasing wavelengths at any given time, as the structure does not support the generation of the 2nd, 3rd and subsequent Stokes wavelengths. The spectrum of the lasing wavelengths is given in Figure 2.

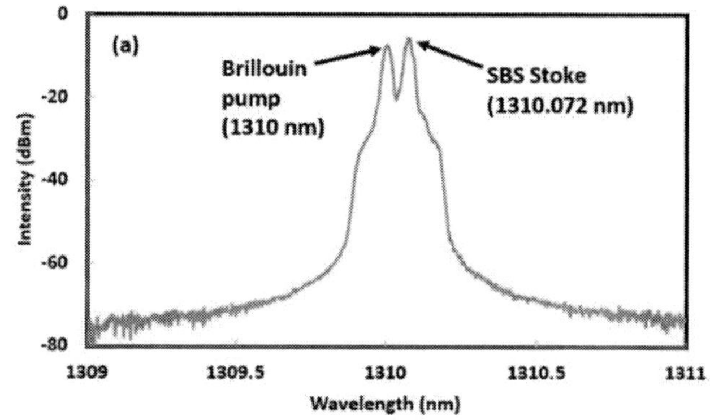

Figure 2. Dual wavelength output spectrum, showing the BP and 1st Stokes wavelength.

The total output power measured at the output port of CIR1, before reaching the 50:50 coupler, is approximately 1.6 mW, with a laser efficiency of about 16%. The heterodyning of the signal at the photodetector induces a beat frequency at about 12.5 GHz and corresponds to the spacing of 0.072 nm between the two lasing wavelengths. The intensity of the beat signal is about -32.0 dBm, with a signal-to-noise ratio (SNR) of about 30 dB as well. The spectrum of the beating frequency, as captured by the RFSA, is shown in Figure 3.

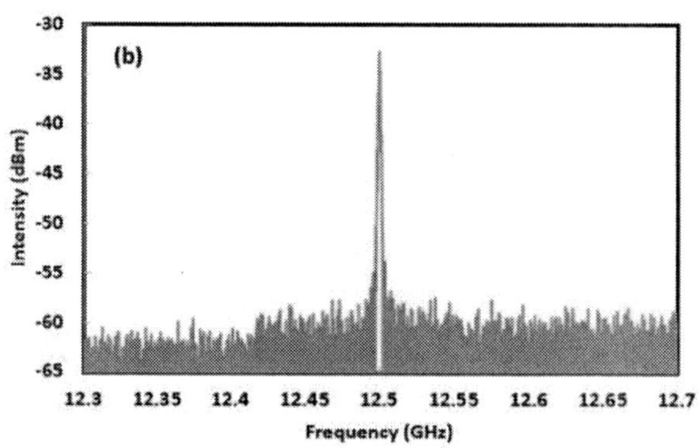

Figure 3. Beat frequency induced by dual-wavelength Brillouin spectrum.

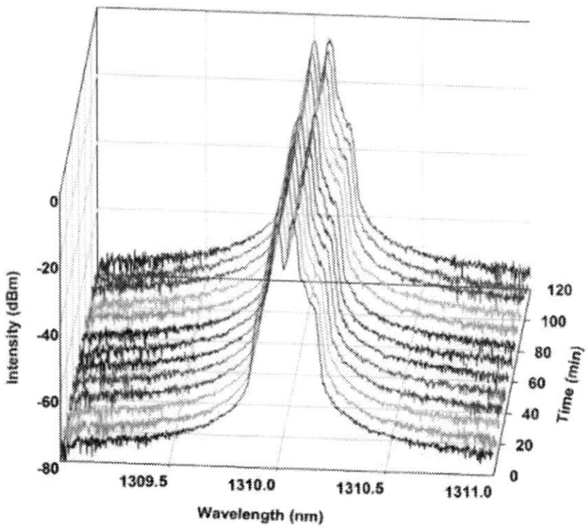

Figure 4. Time series of the dual-wavelength optical output over a period of 120 minutes.

The stability of the dual-wavelength output and by extension, the beat frequency is shown in Figure 4 and Figure 5 respectively. These measurements were taken over a period of 120 minutes. It can be seen from Figure 4 that the lasing wavelengths are highly stable in the optical domain, with the BP having fluctuations in power of less than 0.5 dB and the 1st Stokes having a power fluctuation of less than 1.4 dB. This is attributed to the fact that the SBS process is highly sensitive to multiple internal and external factors, and can be affected by environmental conditions such as vibrations and temperature. As such, the stability of the 1st Stokes would naturally be slightly less than that of the BP. The beat frequency obtained however is highly stable, with fluctuations of less than 0.01 GHz as given in Figure 5. This indicates that the DWFL, and subsequently the generated microwave signal, is highly stable.

The tunability of the generated frequency is achieved by changing the BP wavelength over a tuning range of 1260.0 nm to 1330.0 nm in increments of 10.0 nm. Since the dual-wavelength output is obtained by the SBS process, thus the spacing between the BP and the 1st Stokes for any BP wavelength will remain unchanged.

IOP Conf. Series: Materials Science and Engineering **210** (2017) 012045 doi:10.1088/1757-899X/210/1/012045

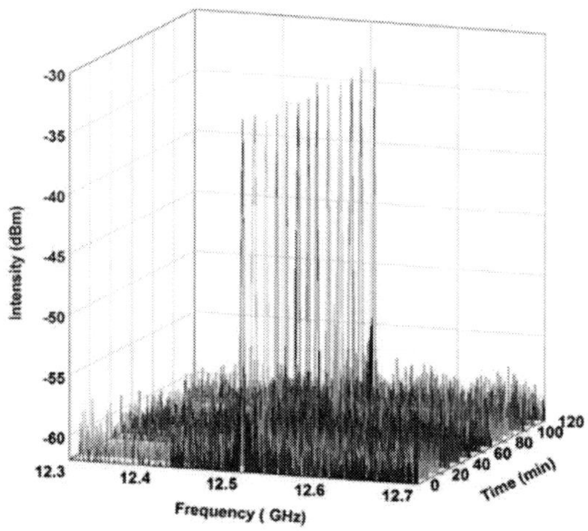

Figure 5. Time series of the dual-wavelength fiber laser electric output over a period of 120 minutes.

Figure 6(a) and 6(b) shows the obtained dual-wavelength output against BP wavelength. It can be seen that the dual-wavelength output responds almost linearly to the BP wavelength, which is as predicted. A dual-wavelength output is observed throughout the tuning range of 1260.0 nm to 1330.0 nm, with fluctuations of less than 1.4 dB in the peak powers.

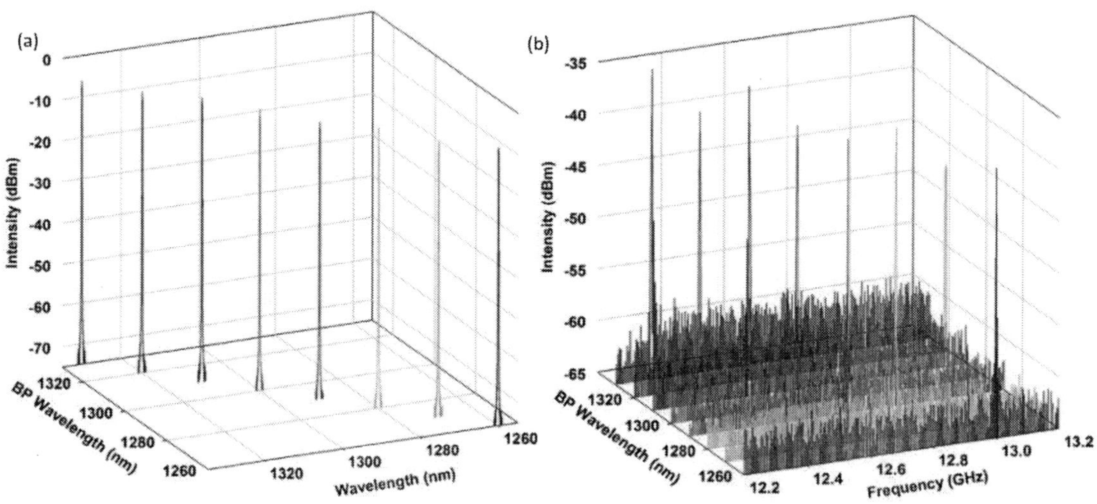

Figure 6. (a) Response of dual-wavelength output against different Brillouin pump wavelengths and (b) Beat frequency shifting corresponded to Brillouin pump wavelength tuning.

Tuning the BP to different wavelengths also results in the tuning of the generated beat frequency, as given in figure 7. By tuning the BP to 1260.0, 1270.0, 1280.0, 1290.0, 1300.0, 1310.0, 1320.0 and 1330.0 nm, a beat frequency of 13.0, 12.9, 12.8, 12.7, 12.6, 12.5, 12.4, and 12.3 GHz respectively was obtained. The output frequencies were observed to have fluctuations of 4.6 dB throughout the tuning range. The fluctuations are attributed to uneven wavelengths and the corresponding frequency response at photodetector. The proposed system has high potential for use in the O-band region, covering the portion of the spectrum that is still crucial but often overlooked in research efforts. Microwave signal generation corresponding to the 1310.0 nm region is crucial in supporting various telecommunication and scientific applications.

4. Conclusions

A tunable DWFL for microwave signal generation in the O-band region is proposed and demonstrated. The system uses a TLS to launch a BP signal into a 6.9 km long DSF to induce SBS in the system. A dual-wavelength spectrum, tunable from 1260.0 nm to 1330.0 nm with a constant channel spacing of 0.072 nm is obtained, and a beat frequency of 12.5 GHz with an SNR of approximately 30.0 dB is achieved. The output is highly stable in both the optical and electrical regimes, and thus have multiple highly potential applications in a multitude of real-world applications.

Acknowledgement
Funding for this research was made possible by the University of Malaya through the UMRG (RP029A-15AFR)

References

[1] K Zhao, Z Ying and S He 2016 *IEEE Antennas and Wireless Propagation Letters* **15** 1132
[2] J Yao 2010 *International Journal of Microwave and Optical Technology* **5** 16-21
[3] P Bouyer, T L Gustavson, K G Haritos and M A Kasevich 1996 *Optics Letters* **21** 1502
[4] L A Johansson and A J Seeds 2000 *IEEE Photonics Technology Letters* **12** 690
[5] R Nagarajan, S Levy, A Mar and J E Bowers 1993 *IEEE Photonics Technology Letters* **5** 4
[6] J L Zhou, L Xia, X P Cheng, X P Dong and P Shum 2008 *Applied Physics B* **91** 99
[7] S Pan and J Yao 2009 *Optics Express* **17** 12167
[8] R K Kim, S Chu and Y G Han 2012 *IEEE Photonics Technology Letters* **24** 521
[9] S W Harun, X S Cheng, N K Saat and H Ahmad 2005 *Electronics Letters* **41** 174-176
[10] S W Harun and M R Shirazi 2007 *Laser Physics Letters* **4** 678-680
[11] S W Harun, S Shahi and H Ahmad 2009 *Optics Letters* **34** 46
[12] S Feng, S Lu, W Peng, Q Li, C Qi, T Feng, S Jian 2013 *Optics and Lasers Technology* **45** 32
[13] X Chen, Z Deng and J Yao 2006 *IEEE Transactions on Microwave Theory and Technique* **54** 804
[14] H Zou, S Lou, G Yin and W Su, 2013 *IEEE Photonics Technology Letters* **25** 1003
[15] G Carpintero, K Balakier, Z Yang, R C Guzman, A Corradi, A Jimenez, G Kervella, M J Fice, M Lamponi, M Chitoui, F van Dijk, C C Renaud, A Wonfor, E A J M Bente, R V Penty, I H White and A J Seeds 2014 *Journal of Lightwave Technology* **32** 3496

IOP Conf. Series: Materials Science and Engineering **210** (2017) 012046 doi:10.1088/1757-899X/210/1/012046

Mode-locked Erbium-doped fiber laser generation using hybrid ZnO/GO saturable absorber

H. Hassan*, M. M. Ariannejad, R. Safaei, I. S. Amiri, H. Ahmad
Photonics Research Centre, University of Malaya, 50603 Kuala Lumpur, Malaysia

Email: *hazirah2711@siswa.um.edu.my

Abstract. Mode-locked generation of erbium-doped fiber laser (EDFL) with hybrid zinc oxide/graphene oxide (ZnO/GO) thin film as saturable absorber (SA) is proposed and practically demonstrated. The SA shows the modulation depth of 18.69% and it has been sandwiched between the fiber ferrules. Mode-locked pulse occurred at pump power of 14.8 mW and by varying the pump power to maximum threshold 27.43 mW, the repetition rate of the pulse fixed at 19.98 MHz at 1563 nm of central wavelength. The pulse width is estimated as 0.90 ps, whereas the pulse energy is calculated as 27.0 nJ.

1. Introduction

Nowadays, ultrafast fiber laser are of great interest among scientists and researchers due to their multifaceted applications for example in optical telecommucation, medicine and material processing. Although there are many types of ultrafast laser, passively mode-locked pulse laser is point out for the low cost, accuracy and compactness [1]. In order to generate passively mode-locked fiber laser, saturable absorber (SA) is governing the startup characteristics of the mode-locking process [2].

Many SAs had been used to produce mode-locked such as nonlinear polarization rotation (NPR), carbon nanotubes (CNTs), graphene oxide and many other 2D materials [3-6]. Compare to artificial saturable absorber approaches, film based SAs give a better performance in enviromental stability and independent to the polarization in the cavity. What is more interesting, when comparing between single material and hybrid material, hybrid material shows compelling technique to upgrade the graphene application by exhibiting the variety characteristics with high performance compare to single material [7]. Oxygen is reactive functional group that contain in the graphene oxide to enhance the chemical functionalization of the material. Besides that, various inorganic nanoparticle used to hybridize with GO or graphene for example Ag, Au, TiO_2 and ZnO.

In this article, mode-locked EDFL using hybrid ZnO/GO is proposed and demonstrated. Hybridization of the ZnO/GO is a new material use as SA to generate mode-locked fiber laser. ZnO/GO film is sandwiched between the fiber ferrules and the spectrum is observed. Addition to that, there is no mode-locked pulse construct using hybrid ZnO/GO SA.

2. Nonlinear Optical Absorption Characteristics of Hybrid ZnO/GO SA

Twin detector technique was used to characterize the nonlinear absorption characteristic [8] of hybrid ZnO/GO SA. A passively mode-locked fiber laser source is employed with pulse width of 0.60 ps, repetition rate of 28.01 MHz and central wavelength (λ_c) of 1563 nm. The output power from both detectors were recorded as the value of attenuation is gradually decreased. Low-dispersion amplifier and variable optical attenuator are used to control the signal power, then splitted into two equal part via 50/50 optical coupler. First output is connect directly to optical power meter (OPM), whereas the other output port is layered with ZnO/GO thin film before connected to another similar OPM.

From the result, the nonlinear optical absorption is calculated and fitted using the saturation model formula [9]:

$$\alpha(I) = \frac{\alpha_s}{1+\frac{I}{I_{sat}}} + \alpha_{ns} \quad (1)$$

where α is optical absorption, I is intensity, α_s is saturable absorption, α_{ns} is non-saturable absorption and I_{sat} is saturation intensity. Figure 1 depicts the absorption against power density graph of the SA.

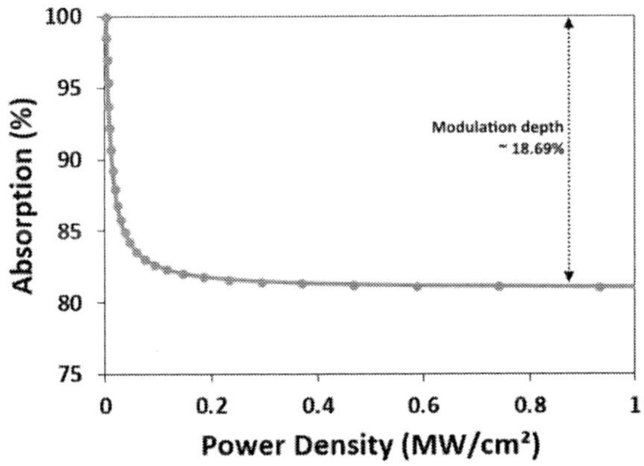

Figure 1. Measured saturable absorption of hybrid ZnO/GO SA.

3. Experimental Arrangement

The schematic drawing of the fiber laser is presented in Figure 2. The laser resonator is consist of 3-m EDF, wavelength division multiplexer (WDM), isolator, 3-dB output coupler and hybrid ZnO/GO SA. The EDF have absorption coefficient of 11.3 dB/m at 979 nm wavelength, mode field diameter of 6.6 μm and numerical aperture (NA) of 0.21. The EDF having backward-pumped by 1480 nm Fitel laser diode (LD) through WDM. The EDF is connected to the common port of the WDM and again to another similar WDM on its other end. The purpose of the second WDM is to remove any excess pump power from the laser cavity in order to optimize the performance of the system. When the EDF is pumped by the LD, amplified spontaneous emission (ASE) spectrum is generated. The ZnO/GO SA is insert between a pair of pigtails surface and integrated into the laser resonator. The cavity has total length around 9.5 m with the rest of the cavity is constructed with single mode-fiber SMF-28. Twenty percent of the light is removed from the ring resonator while another 80% is continues to pass through the resonator, via an optical isolator which ensure that the light propagate in one direction only. The spectrum of the laser is captured by the. An oscilloscope is used with a 1.2 GHz bandwidth photodetector to capture the mode-locked emission.

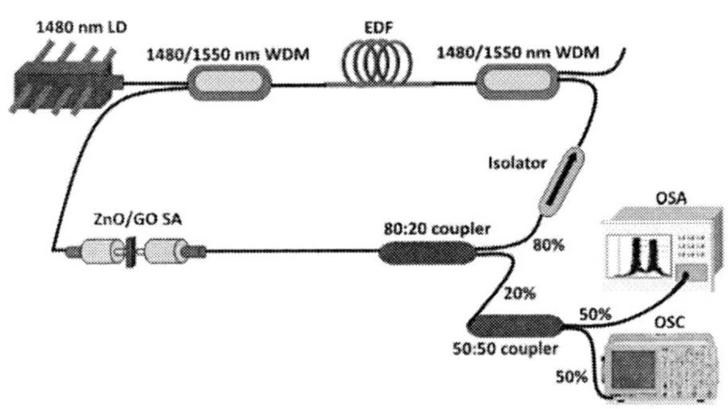

Figure 2. Experimental setup of the purposed mode-locked EDFL.

4. Results and Discussion

An Anritsu MS9740A optical spevtrum analyzer (OSA) with span 200 nm and resolution 0.03 nm is used to observe the output lasing spectrum of the EDFL via 50% of the cavity output. The initial threshold pump of the mode-locked operation started at 14.8 mW. But, when the pump power increases to until 27.43 mA, the mode-locked pulse starts to alter and diminish due to strong non-linear phase shift. A reliable mode-locked pulse train is observed at 14.8 mW pump power as shown in Figure 3 (a). During the operation of the mode-locking, the pulse repetition rate fixed at 19.98 MHz frequency which proved that the mode-locked formed at fundamental frequency. Figure 3 (b) depicts the mode-locked spectrum at central wavelength of 1563 nm and FWHM spectral bandwidth of 5.6 nm.

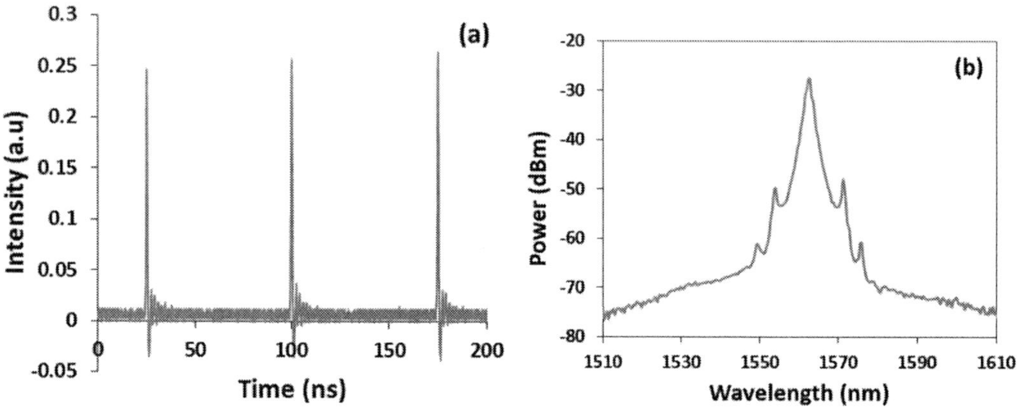

Figure 3. (a) mode-locked pulse train at 14.8 mW. (b) output spectrum from OSA

Besides that, RF spectrum analyzer is employed to check the stability of the mode-locked spectrum as shown in Figure 4 (a). The pulse width is analyze using autocorrelator. Figure 4 (b) depicts single pulse measured at 0.90 ps and resolution of 50 fs.

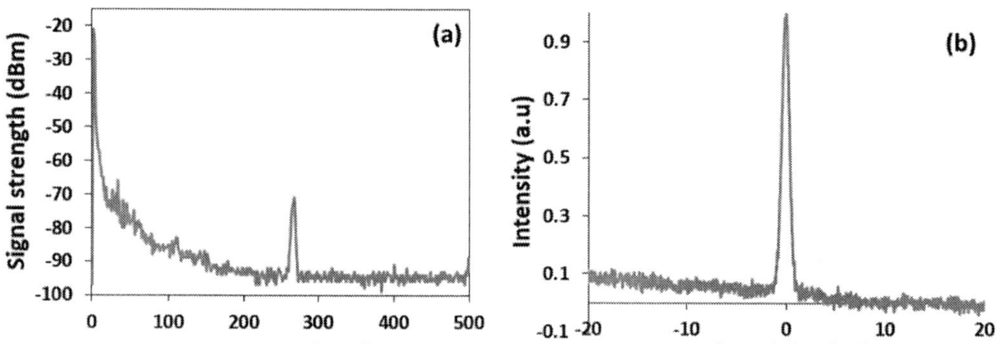

Figure 4. (a) Radio-frequency optical spectrum of the mode-locked EDFL. (b) pulse width captured by autocorrelator

Throughout the mode-locked operation, pulse energy increase when the pump power increase. The highest pulse energy is 37.5 nJ achieved when the pump power is 27.43 mW as shown in Figure 5.

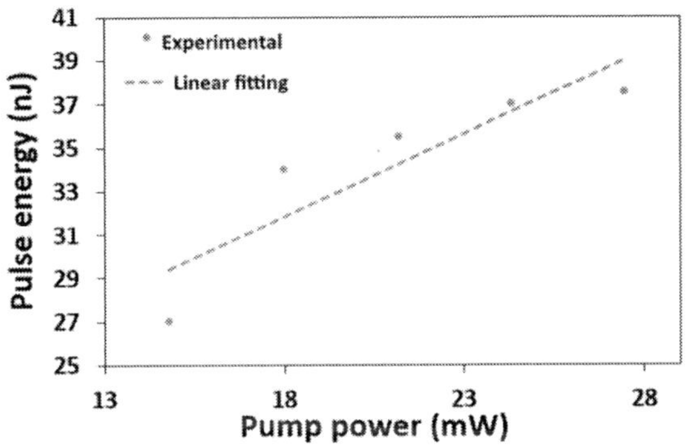

Figure 5. Pulse energy trend when increasing pump power

5. Conclusion

Passively mode-locked EDF laser have been proposed and practically demonstrated operating at 1563 nm with incorporation of hybrid ZnO/GO saturable absorber. The EDFL generates mode-locking pulse at pump power of 14.8 mW. By varying the pump power until 27.43 mW, the pulse repetition rate remain constant at 19.98 MHz with pulse width of 0.90 ps and having pulse energy of 27.0 nJ.

Acknowledgement

Funding for this research was made possible by the University of Malaya through the UMRG (RP029A-15AFR)

References

[1] Latiff A, Shamsudin H, Aziz N, Hashim A, Irawati N, Ahmad H, et al. 2016 *Optik-International Journal for Light and Electron Optics* **23** 11119-11123
[2] Fermann M, Minelly J, Vienne G, Harter D. 1996 *Optics letters* **13** 967-969
[3] Latiff A, Shamsudin H, Tiu Z, Ahmad H and Harun S 2016 *Journal of Nonlinear Optical Physics & Materials* **25** 1650034
[4] Yamashita S, Martinez A, Xu B. 2014 *Optical Fiber Technology* **6** 702-713
[5] Zhao J, Wang Y, Yan P, Ruan S, Zhang G, Li H, et al. 2013 *Laser Physics* **7** 075105
[6] Li X, Yu X, Sun Z, Yan Z, Sun B, Cheng Y, et al. 2015 *Scientific reports* **5** 16624
[7] Yang Y, Liu T. 2011 *Applied Surface Science* **21** 8950-8954
[8] Nonlinear Saturable Absorption of Liquid-Exfoliated Molybdenum/Tungsten Ditelluride Nanosheets Mao D, Du B, Yang D, Zhang S, Wang Y, Zhang W, et al. 2016 *Small*
[9] Garmire E. 2000 *IEEE Journal of Selected Topics in Quantum Electronics* **6** 1094-1110

Continues-wave Brillouin-Raman fiber ring laser using 7.7 km long dispersion compensating fiber at 1563 nm wavelength

Z. I. Rizman[1,2*], U. N. Zakaria[2], A. A. Latiff[3,4], S. W. Harun[2] and Z. Jusoh[1]
[1] Faculty of Electrical Engineering, Universiti Teknologi MARA, Terengganu Campus, Dungun Branch, 23000 Dungun, Terengganu, Malaysia
[2] Department of Electrical Engineering, Faculty of Engineering, University of Malaya, 50603 Kuala Lumpur, Malaysia
[3] Photonics Research Centre, University of Malaya, 50603 Kuala Lumpur, Malaysia
[4] Faculty of Electronic and Computer Engineering (FKEKK), Universiti Teknikal Malaysia Melaka, 76100 Melaka, Malaysia

E-mail:* zairi576@tganu.uitm.edu.my

Abstract. We demonstrate the generation of continues-wave (CW) laser based on Brillouin-Raman fiber laser (BRFL) by incorporating a 7.7 km long dispersion compensating fiber (DCF) as a nonlinear gain medium. The 1455 nm Raman pump (RP) was launched into the DCF via a wavelength division multiplexing (WDM). At 645 mW pump power, stable CW laser presence at 1563.5 nm (1st stokes), 1563.4 (2nd stokes) nm and 1563.3 nm (3rd stokes). Those peak wavelengths have produced most identical output power. This finding proves our ring laser cavity has the capability to generate stable CW BRFL. Thus, it potentially to generate pulse laser operation in the future work.

1. Introduction

Brillouin-Raman fiber laser (BRFL) scattering effects have provided high attention in several significant types of research recently. It offers high benefits and remarkable potentials with wide applications. Brillouin stokes lines have been achieved based on the mixed effect of Rayleigh scattering, Raman amplification and Brillouin shift in a linear laser cavity and single-pass Raman amplification configuration. However, the stability of this wavelength lasing outputs was still inefficient and inaccurate that can clearly observe from the given spectra in their experiments. So, it needs to be improved before those reported BRFL sources are practically used. Also, the stability of the lasing oscillations in their configuration was not analyzed [1].

Moreover, in the previous studies have described vital findings and importance role in this hybrid BRFL. For example, in [1], a Stable room-temperature multi-wavelength lasing oscillations in a Brillouin-Raman fiber ring laser has been demonstrated. In [2] investigate 20 GHz spacing multi-wavelength generation of a Brillouin-Raman fiber laser in a hybrid linear cavity 150-km-range distributed temperature sensor based on coherent detection of spontaneous Brillouin backscatter and in-line Raman amplification. A study of multiwavelength Brillouin-Raman ring-cavity fiber laser with 22-GHz Spacing was also be measured [3]. The research of contribution of Rayleigh scattering on Brillouin comb line generation in Raman fiber laser was reported in [4]. The 17-channels S-band multi-wavelength Brillouin/erbium fiber laser co-pump with Raman source characteristics was also being studied [5]. Distributed Brillouin fiber sensor assisted by first-order Raman amplification features is explained in [6]. In [7] describe S-band multi-wavelength Brillouin-Raman fiber laser. Brillouin optical time-domain analysis assisted by second-order Raman amplification. For [8], Brillouin optical time-domain analysis assisted by second-order Raman amplification was obtained. The study of dynamics of cascaded Brillouin-Rayleigh scattering in a distributed fiber Raman amplifier was also determined [9]. The analysis of Brillouin-Raman comb fiber laser with cooperative Rayleigh scattering in a linear cavity was shown in [10].

In this paper, it indicates the generation of continues-wave (CW) laser with BRFL by using a 7.7 km long dispersion compensating fiber (DCF) as a nonlinear gain medium. The 1455 nm Raman pump (RP) was provided into the DCF through a wavelength division multiplexing (WDM). At 645 pump power, stable CW laser presence at 1563.5 nm (1st stokes), 1563.4 (2nd stokes) nm and 1563.3 nm (3rd stokes). The peak wavelengths have produced most identical output power. This finding proves that the ring laser cavity can generate stable CW BRFL.

2. Methodology

The ring cavity configuration for BRFL is shown in **Figure 1**. The proposed BRFL consists of a 1455-nm Raman pump (RP), wavelength division multiplexer (WDM), 7.7 km long DCF, 10-dB coupler, Brillouin pump (BP), and circulator. The RP has core pumped a nonlinear gain medium of 7.7 km long DCF through WDM. Via a circulator, the narrow linewidth Tunable Light Source (TLS) is integrated into the laser cavity to provide the BP source. This BP signal will generate the CW laser in the 7.7 km DCF, which is also injected with a 1455 nm RP signal at 645 mW and 892 mW. The RP provides Raman amplification to the BP, thus reducing the BP loss and provides higher powered Stokes wavelengths. This will help to generate more subsequent Stokes. The amplification of the generated Stokes wavelengths is accomplished by the use of the DCF based amplifier.

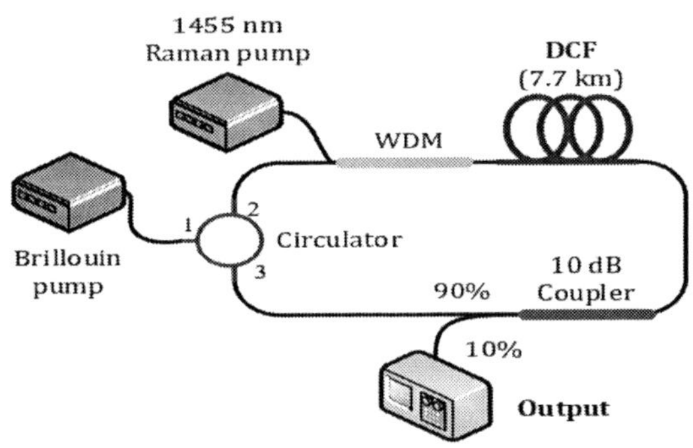

Figure 1. Experimental setup for the proposed DCF based BRFL.

3. Results and discussion

The spectrum generated by the proposed setup is shown in **Figure 2**. The output spectra of Raman fiber laser is obtained at 645 mW and 892 mW. This is for comparison purpose to show its significant results of output power (dBm) versus wavelength. The tuning range of the proposed BRFL is approximately 20 nm from 1550 to 1570 nm respectively. This output spectra will be used to determine whether Stokes wavelengths can be produced.

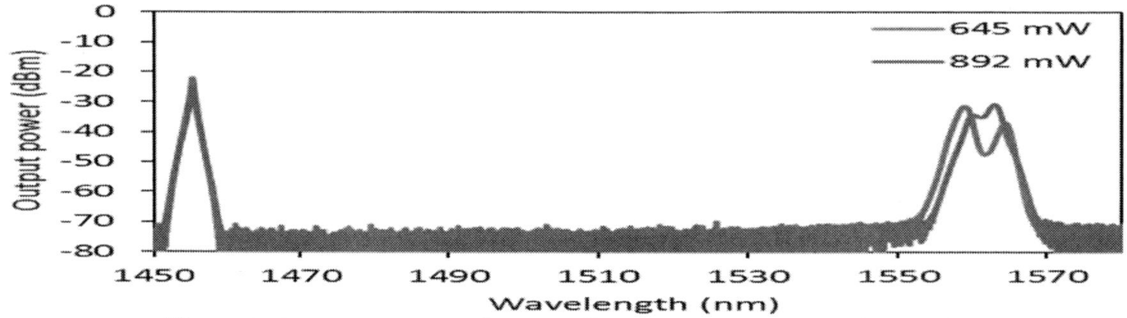

Figure 2. Output spectrum of Raman fiber laser at 645 mW and 892 mW.

In Figure 3, it shows the output spectrum at 892 mW for Brillouin fiber laser and Raman pump. This vital finding indicates the pattern characteristics of the spectrum between both Brillouin and Raman pumps. This is important to lead in studying the relationship of it.

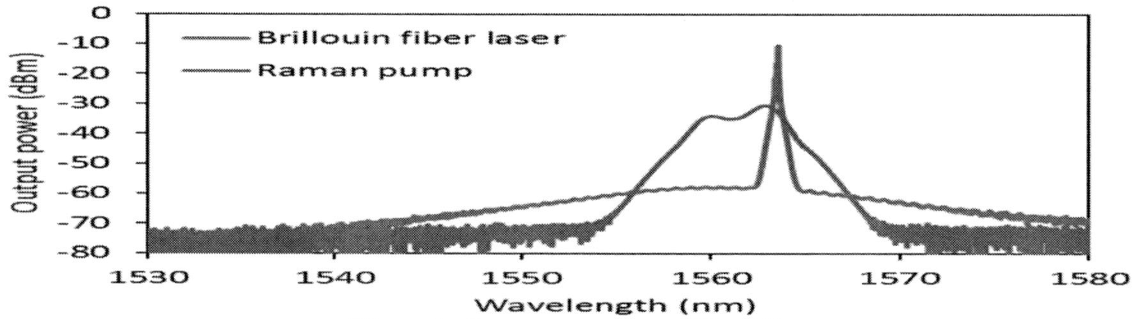

Figure 3. Output spectrum at 892 mW.

To obtain the range for generating the number of Stokes wavelengths and also to determine the tuning range of the proposed BRFL, BP signals at different wavelengths are injected into the BRFL cavity. Figure 4 shows the output spectrum of the Brillouin fiber laser at 892 mW pump power. This comparison of both spectrums offers better understanding in producing results of Stokes wavelengths. A stable CW laser presence at 1563.5 nm (1st stokes), 1563.4 (2nd stokes) nm and 1563.3 nm (3rd stokes) with a spacing of ~ 0.08 nm.

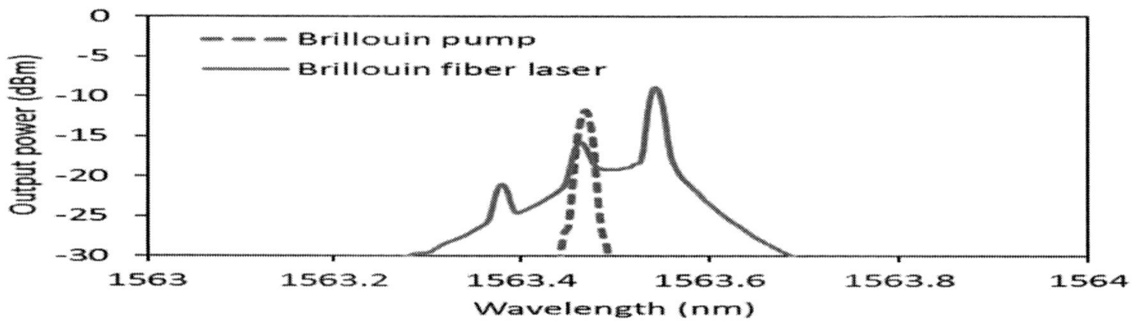

Figure 4. Output spectrum of the Brillouin fiber laser at 892 mW pump power.

Additionally, the output power of Brillouin pump and Raman pump at 1563 nm are compared to Figure 5 and Figure 6. The measured of Stokes power in both pumps indicate good results to show its correlation with the power of Brillouin and Raman pumps.

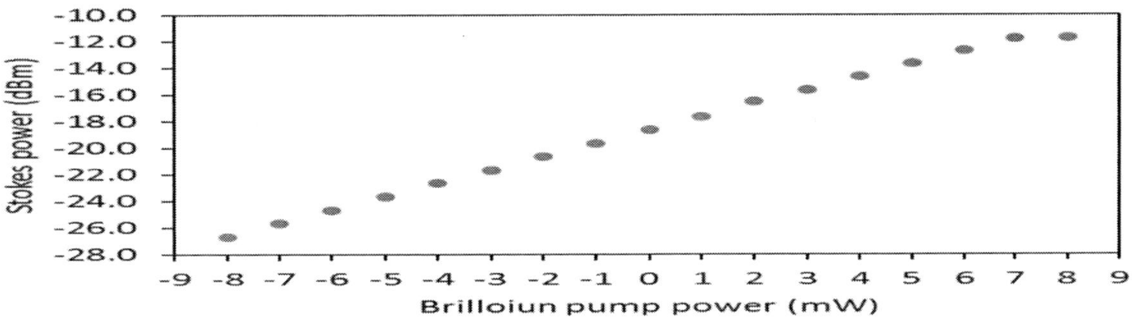

Figure 5. The output power of Brillouin pump at 1563 nm.

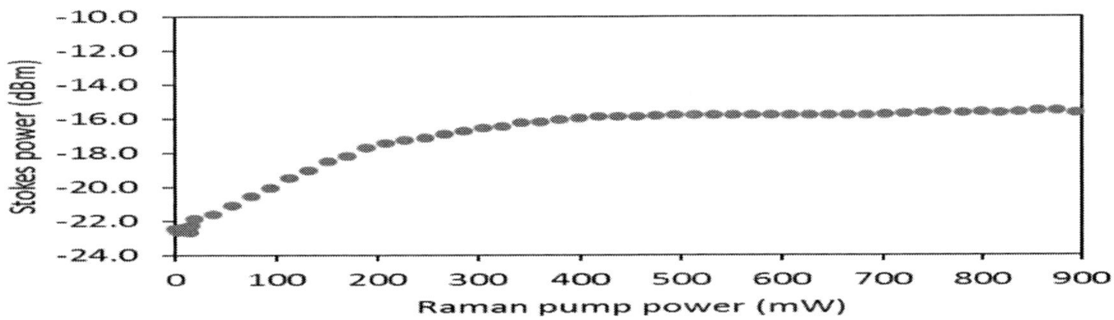

Figure 6. The output power of Raman pump at 1563 nm.

As shown in Figure 7, the injected BP only generates three Stokes wavelengths. It indicates the output power at 1563.5 nm (1st), 1563.4 (2nd) nm and 1563.3 nm (3rd) via a power meter. All these three stokes show the same pattern of uniform and stable power curve.

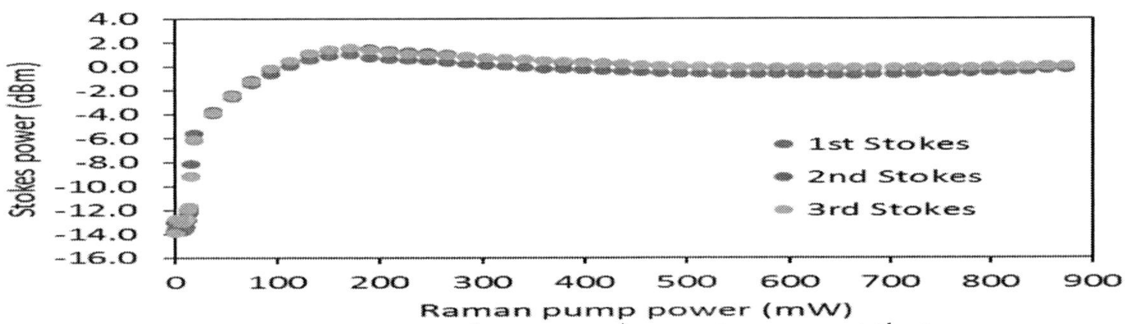

Figure 7. Output power at 1563.5 nm (1st), 1563.4 (2nd) nm and 1563.3 nm (3rd) via a power meter.

4. Conclusion

This paper has successfully demonstrated a CW laser via BRFL through 7.7 km long DCF as a nonlinear gain medium. The 1455 nm RP was injected into the DCF via a WDM. At 645 mW pump power, stable CW laser presence at 1563.5 nm (1st stokes), 1563.4 (2nd stokes) nm and 1563.3 nm (3rd stokes). The peak wavelengths have produced most uniform output power. This result shows that this ring laser cavity capable of generating stable CW BRFL. Thus, it has high potential to produce pulse laser operation in the next extended research of future work.

Acknowledgements

This work is financially supported by Ministry of Higher Education Grant Scheme (FRGS/1/2015/SG02/UITM/03/3).

IOP Conf. Series: Materials Science and Engineering **210** (2017) 012047 doi:10.1088/1757-899X/210/1/012047

References

[1] Liu Y G, Wang D and Dong X 2008 *Opt. Commun.* **281** 5400-5404

[2] Alahbabi M N, Cho Y T and Newson T P 2005 *J. Opt. Soc. Am. B.* **22** 1321-1324

[3] Hambali N A, Al-Mansoori M H, Ajiya M, Bakar A A, Hitam S and Mahdi M A 2011 *Laser Phys.* **21** 1656-1660

[4] Zamzuri A K, Al-Mansoori M H, Samsuri N M and Mahdi M A 2010 *Appl. Optics.* **49** 3506-3510

[5] Ahmad H, Zulkifli M Z, Latif A A, Thambiratnam K and Harun S W 2009 *Laser Phys.* **19** 2188

[6] Rodríguez-Barrios F, Martín-López S, Carrasco-Sanz A, Corredera P, Ania-Castañón J D, Thévenaz L and González-Herráez M 2010 *J. Lightwave Technol.* **28** 2162-2172

[7] Ahmad H, Zulkifli M Z, Norizan S F, Jemangin M H and Harun S W 2011 *Opt. Commun.* **284** 4971-4974

[8] Martin-Lopez S, Alcon-Camas M, Rodriguez F, Corredera P, Ania-Castañon J D, Thévenaz L and Gonzalez-Herraez M 2010 *Opt. Express.* **18** 18769-18778

[9] Park K D, Min B, Kim P, Park N, Lee J H and Chang J S 2002 *Opt. Lett.* **27** 155-157

[10] Zamzuri A K, Ali M M, Ahmad A and Mohamad R 2006 *Opt. Lett.* **31** 918-920

Investigation of Brillouin Raman fiber laser operating at 1558 nm using THDF saturable absorber

Z. I. Rizman[1,2*], U. N. Zakaria[2], Z. Jusoh[1], A. A. Latiff[3,4] and S. W. Harun[2]
[1] Faculty of Electrical Engineering, Universiti Teknologi MARA, Terengganu Campus, Dungun Branch, 23000 Dungun, Terengganu, Malaysia
[2] Department of Electrical Engineering, Faculty of Engineering, University of Malaya, 50603 Kuala Lumpur, Malaysia
[3] Photonics Research Centre, University of Malaya, 50603 Kuala Lumpur, Malaysia
[4] Faculty of Electronic and Computer Engineering, Universiti Teknikal Malaysia Melaka, 76100 Melaka, Malaysia

E-mail: *zairi576@tganu.uitm.edu.my

Abstract. We report a generation of Brillouin-Raman fiber laser (BRFL) which operating at 1558 nm wavelength. The proposed configuration utilizes a nonlinear gain medium of 7.7 km long dispersion compensating fiber (DCF) and using thulium holmium-doped fiber (THDF) as a fiber saturable absorber (SA). The self-started Q-switching operation starts at 569 mW pump power and keep stable until 721 mW pump level. The stable repetition rate of 22 kHz and pulse width of 1.791 μs are obtained with a signal to noise ratio (SNR) of 32 dB. This work shows the fiber SA can be used promisingly in a BRFL for pulse laser generation.

1. Introduction

Recently, Raman Brillouin fiber lasers (BRFL) researches have tremendously been reported. It offers wide interesting opportunity to be explored in studying its additional unique characteristics [1]. For example, fiber Bragg grating interrogation technique for remote sensing (100-km) using a hybrid Brillouin-Raman fiber laser was reported in [2]. S-band dual cavity multi-wavelength Brillouin Raman fiber laser employing dual Raman amplifier characteristics was also being studied [3]. S-band multi-wavelength ring cavity Brillouin Raman fiber laser exploiting different FBG reflectivity features is explained in [4]. In [5] describe about L-band single wavelength ring cavity configured Brillouin/Raman fiber laser. For [6], multi-wavelength Brillouin-Raman fiber laser utilizing a partially reflective fiber Bragg grating in L-band was obtained. A comparative study of multi-wavelength Brillouin-Raman fiber laser utilizing linear and ring cavity structures was also determined [7]. The analysis of Raman amplification effects on stimulated Brillouin scattering threshold in multi-wavelength Brillouin-Raman fiber laser was shown in [8].

Thulium holmium-doped fiber (THDF) is convenient to be as one of best option to be operated around 1.5-micron region. Many investigations have been carried out in the field of 1.5-micron lasers based on THDF lasers. It has gained considerable attention previously [1]. The pulsed output from a fiber laser cavity can be determined either by Q-switching [9] or mode locking [10-11], and the related working principles are well known.

This paper demonstrates a study of BRFL that operating at 1558 nm wavelength. The experimental set-up uses a 7.7 km long nonlinear gain medium of dispersion compensating fiber (DCF) and utilizing THDF saturable absorber (SA). The self-starting function was at 569 mW pump power and keep constant until it disappear when it reaches 721 mW pump level onwards. Consistent of measured 22 kHz repetition rate, 32 dB signal to noise ratio (SNR) and 1.791 μs pulse width was determined.

2. Methodology

Figure 1 shows the proposed setup of the experimental for BRFL which operating at 1558 nm using 0.18 m THDF as a SA. It incorporates a 7.7 km long dispersion compensating fiber (DCF) that act as a nonlinear gain medium and Brillouin pump (BP). An optical circulator (OC) is placed between Raman pump (RP) and BP to function it as a linear cavity. The 1455 nm RP was launched into the DCF via a wavelength division multiplexing (WDM). A narrow linewidth of Tunable Light Source (TLS) produces BP signal which injected into the linear cavity. It will generate the continues-wave (CW) laser in the DCF that integrated with a 1455 nm RP at 569 mW and 721 mW. The 10% output of 90:10 coupler was monitored by an optical spectrum analyser (OSA) and an oscilloscope.

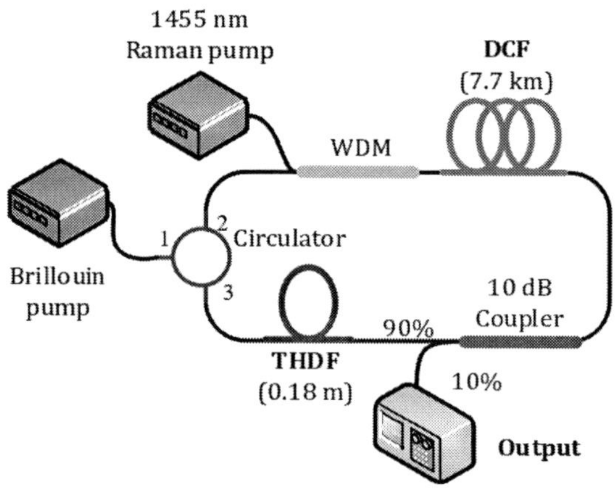

Figure 1. Configuration of proposed experimental setup for the DCF-BRFL with THDF-SA.

3. Results and discussion

As we inetgrate the THDF inside the cavity, stable Q-switching operation started from 569 mW to 721 mW pump power. Figure 2 shows the ouput spectrum without THDF and with THDF. More ripples present in spectrum with THDF. Output power and pulse energy versus pump power is shown in Figure 3. As pump power increased, the output power and pulse energy also increased. This correlation proves that the pulse energy can be increased by inserting a dispersion compensator inside the cavity [12]. Therefore, higher output pulse energy is promising but limited by the damage threshold of the fiber component [13].

IOP Conf. Series: Materials Science and Engineering **210** (2017) 012048 doi:10.1088/1757-899X/210/1/012048

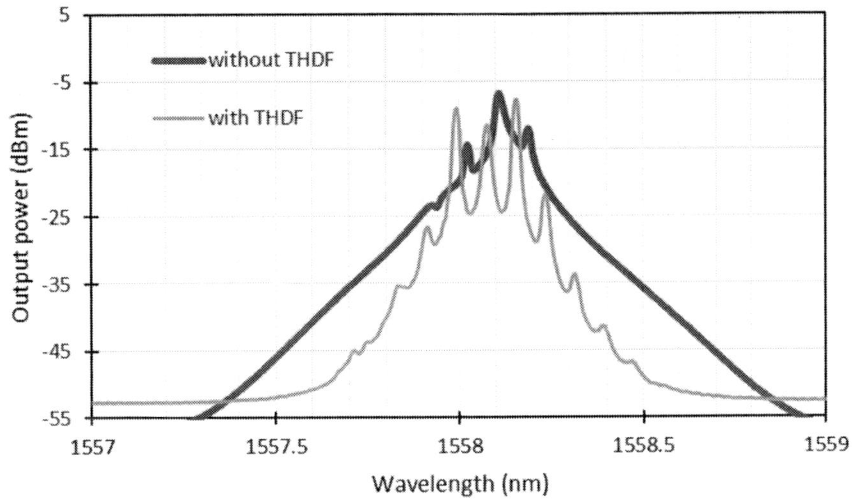

Figure 2. Ouptut spectrum of Q-swithcing operation.

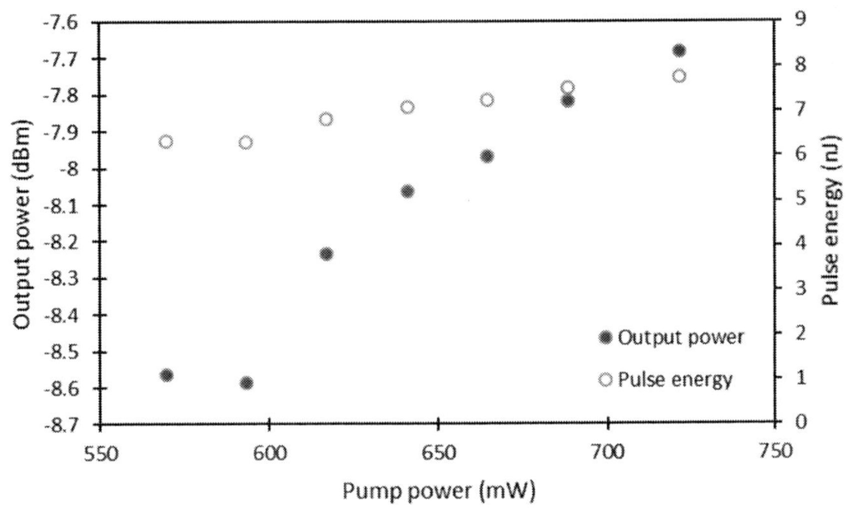

Figure 3. Output power and pulse energy between 569 to 721 mW.

In Figure 4, it shows the relationship between voltage and response time. The pulse train displayed on the oscilloscope is constant between 569 mW to 721 mW pump power. The laser produces a stable pulse train [14]. When the amplitude of control pulse was optimized, the pulse train was firmly switched [15].

IOP Conf. Series: Materials Science and Engineering **210** (2017) 012048 doi:10.1088/1757-899X/210/1/012048

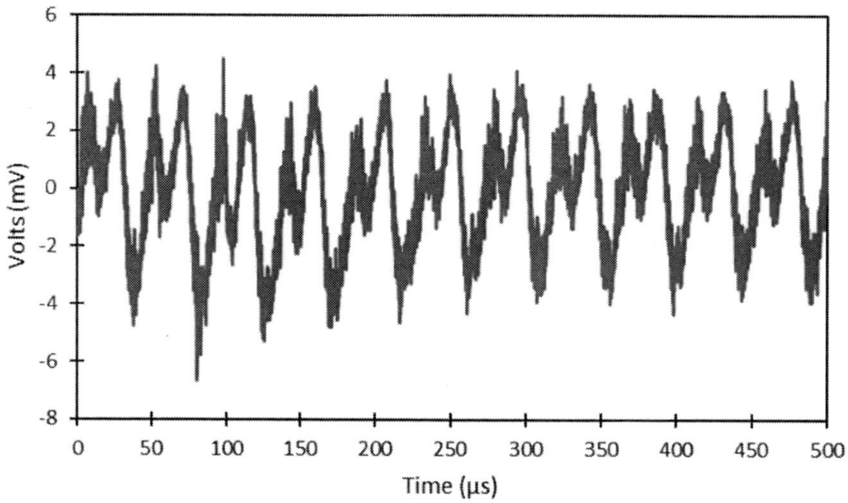

Figure 4. Oscilloscope train constant between 569 mW to 721 mW pump power.

Additionally, Figure 5 illustrates the output power versus pulse repetition frequency. The RF spectrum generates 150 kHz span at 1558 nm. The results produce a fixed repetition rate of 22 kHz (fix for all pump level), 1.791 μs pulse width with signal to noise ratio (SNR) of 32 dB. The low and decent repetition rate is limited by the optical damage threshold and the response time [16].

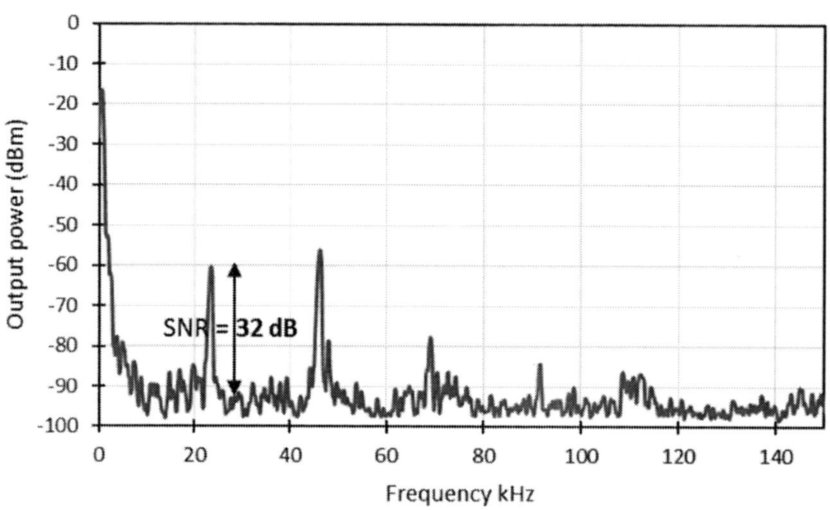

Figure 5. RF spectrum at 1558 nm with 150 kHz span.

4. Conclusion

In conclusion, we have reported an important findings via this experimental measurement. It indicates a free-running regime at pump power from 569 mW to 721 mW. The significant results show obtained repetition rate, pulse width, and SNR are 22 kHz, 1.791 μs, and 32 dB, respectively. Thus, it comparable to a current passively Q-switched fiber laser. Further investigation is needed to aim Q-switched and mode-locked pulsed output.

Acknowledgements
This work is financially supported by Ministry of Higher Education Grant Scheme (FRGS/1/2015/SG02/UITM/03/3).

References

[1] Gumenyuk R, Vartiainen I, Tuovinen H and Okhotnikov O G 2011 *Opt. Lett.* **36** 609-611
[2] Fernandez-Vallejo M, Leandro D, Loayssa A and Lopez-Amo M 2011 *21st International Conference on Optical Fibre Sensors (OFS21)* 77537I1-77537I4
[3] Roshidah N, Hambali N A, Murtadha M M, Wahid M H and Azidin M A 2016 *3rd IEEE International Conference on Electronic Design (ICED)* 391-394
[4] Sohaimi M S, Hambali N A, Roshidah N, Wahid M H, Shahimin M M, Harun A and Isa SS 2016 *3rd IEEE International Conference on Electronic Design (ICED)* 438-441
[5] Ajiya M, Hambali N M and Wahab M S 2014 6th IEEE International Conference on Adaptive Science and Technology (ICAST) 1-3
[6] Toor M A, Hambali N A, Abdul H A and Yusoff Z 2014 *5th IEEE International Conference on Photonics (ICP)* 90-92
[7] Abass A K, Al-Mansoori M H, Jamaludin M Z, Abdullah F, Al-Mashhadani T F and Ali M H 2013 *4th IEEE International Conference on Photonics (ICP*) 138-140
[8] Abass A K, Al-Mansoori M H, Jamaludin M Z, Abdullah F and Al-Mashhadani T F 2012 *3rd IEEE International Conference on Photonics (ICP*) 171-174
[9] Kivistö S, Hakulinen T, Guina M, Rößner K, Forchel A and Okhotnikov O 2008 *Photonics Europe* 1-8
[10] Wang Q, Geng J, Jiang Z, Luo T and Jiang S 2011 *CLEO: Science and Innovations* 1-2
[11] Kadel R, Jones A M and Washburn B R 2011 *Frontiers in Optics* paper FTuI3
[12] Engelbrecht M, Haxsen F, Ruehl A, Wandt D and Kracht D 2008 *Opt. Lett.* **33** 690-692
[13] Fu B, Gui L, Li X, Xiao X, Zhu H and Yang C 2013 *IEEE Photonic. Technol. Lett.* **25** 1447-1449
[14] Kadel R and Washburn B R 2012 *Appl. Optics.* **51** 6465-6470
[15] Kivisto S and Okhotnikov O G 2011 *IEEE Photonic. Technol. Lett.* **23** 477-479
[16] Jung M and Lee J H 2013 *Appl. Optics.* **52** 2706-2710

IOP Conf. Series: Materials Science and Engineering **210** (2017) 012049 doi:10.1088/1757-899X/210/1/012049

Silhouette-based approach of 3D image reconstruction for automated image acquisition using robotic arm

N. Azhar, W. H. M. Saad*, N. A. Manap, N. M. Saad and A. R. Syafeeza
Faculty of Electronic and Computer Engineering, Universiti Teknikal Malaysia Melaka, Hang Tuah Jaya, 76100 Durian Tunggal, Melaka, Malaysia

E-mail: *wira_yugi@utem.edu.my

Abstract. This study presents the approach of 3D image reconstruction using an autonomous robotic arm for the image acquisition process. A low cost of the automated imaging platform is created using a pair of G15 servo motor connected in series to an Arduino UNO as a main microcontroller. Two sets of sequential images were obtained using different projection angle of the camera. The silhouette-based approach is used in this study for 3D reconstruction from the sequential images captured from several different angles of the object. Other than that, an analysis based on the effect of different number of sequential images on the accuracy of 3D model reconstruction was also carried out with a fixed projection angle of the camera. The effecting elements in the 3D reconstruction are discussed and the overall result of the analysis is concluded according to the prototype of imaging platform.

1. Introduction

Three-dimensional (3D) reconstruction is a process to build the 3D image by capturing the surface shape of real object [1]. It has different modalities in order to create 3D models for example, using computer vision, radio waves and magnetic field, high frequency sound waves and gamma rays. The 3D reconstruction process can be accomplished either by active techniques or passive techniques [2]. It can be further classified as uni-directional and multi-directional. With active techniques the light sources are specially controlled as part of the strategy to arrive at the 3D information. Unlike active techniques, passive techniques don't need specially controlled light and can work with ambient light available and also require simple instrumentation. From the computational point of view, active techniques are less demanding compared to the passive techniques as their applicability is restricted to the environment where the special illumination techniques can be applied. In this study the 3D reconstruction using computer vision is done by using passive techniques for multi-directional points [3]. A few sets of sequential images of objects are taken by using a robotic arm attached to a camera surrounding the object.

There is a lot of methods have been proposed in the robotic arm control system. A simple way to design the robotic arm, which can pick up, the object is using a freedom servo motor [4]. Microcontroller (ATMEGA-328) is used as the hardware interface platform to control the motion of robotic arm [5]. The potentiometer is fitted with the remote in the design while the electric pulse is produced by remote motion of the potentiometer. Servo motor will receive the digital pulse after the signal is converted by the Arduino board. In this study, the robotic arm that capable to handle the medium weight of the object is implemented. DC servo motor is an important device which is widely used in many applications. In robotic application, servo motor normally has priority to use in moving the robotic arm to a relevant position in the application which require high-speed control accuracy and high performance dynamic respond [6].

IOP Conf. Series: Materials Science and Engineering **210** (2017) 012049 doi:10.1088/1757-899X/210/1/012049

In this study, the shape from silhouette technique is used for the reconstruction of 3D shape. The silhouettes are computed for every 2D image captured surrounding the object. The computed silhouettes of every image along the center of the corresponding camera are used to define a volume. A back projected volume in 3D space can be assumed to bind the object. The intersection of these volumes associated with the set of acquired images yields a reasonable approximation of the real object. The intersection volume is known as visual hull and described as the maximal object that gives the same silhouette with the real object from any possible viewpoint [7].

2. Research Methodology

The methodology in this study is divided into three parts, image acquisition, camera calibration and 3D reconstruction.

2.1. Image Acquisition

In the block diagram of Figure 1 Arduino UNO microcontroller and G15 Shield are used to control the servo motor in order to rotate the camera. For motion system, two Cytron servo motors and two U-shaped holders are connected in series, which can rotate accordingly in a time. A USB webcam with the resolution of 0.3 megapixels act as the recognition system which is used to capture the image surrounding the object (Figure 2(a)).

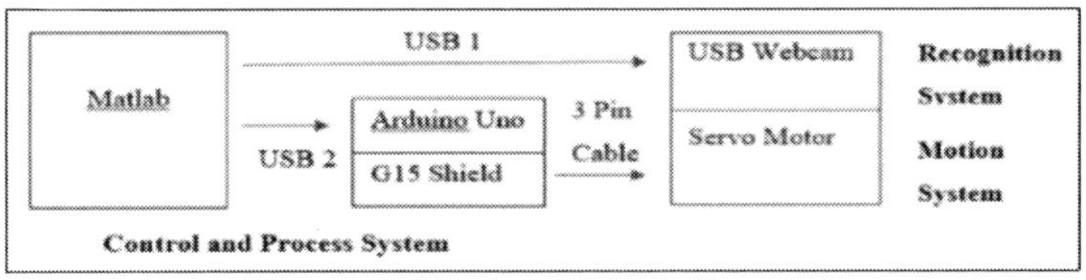

Figure 1. Hardware block diagram.

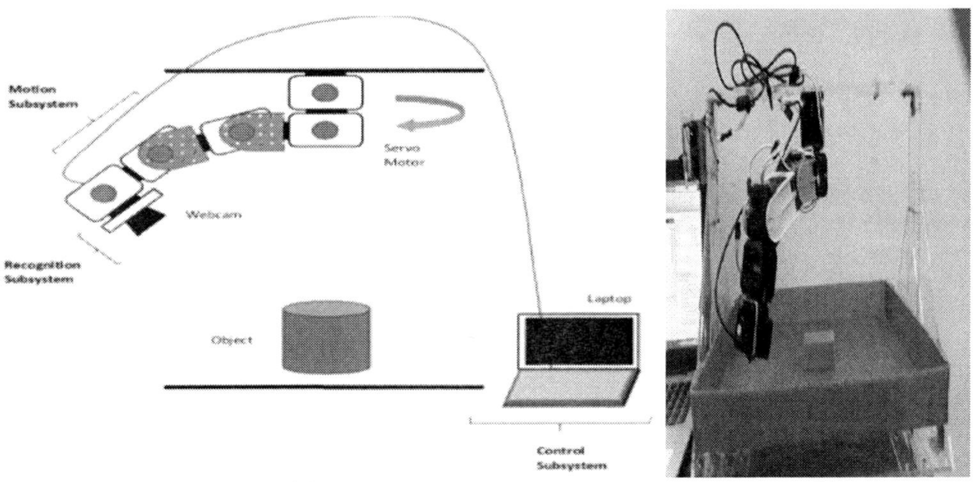

(a) (b)

Figure 2. (a) Project subsystem and (b) The robotic arm.

The sequence of images of the checkerboard pattern and the object is taken by using the autonomous robotic arm attached with a USB webcam. The checkerboard pattern and the object are placed on the platform and the autonomous robotic arm will capture the image surrounding it (Figure 2 (b)). The angular position of the robotic arm is initialized to zero degree and angle interval required which is 5^0 and 10^0.

2.2. Camera Calibration

Camera calibration is a process to recover the camera extrinsic and intrinsic parameters that can affect the imaging process. The checkerboard pattern (Figure 3) is used for camera calibration due to its robustness and easy to implement. The alternately bright and dark grids as well as the grid corner features can be a very strong feature to detect and recognize. The Matlab Camera Calibrator app is used to estimate camera extrinsic, intrinsic and lens distortion parameters for a single camera. These camera parameters will be used to remove lens distortion effect from an image, measure the planar objects, reconstruct 3D scenes from multiple cameras, and perform other computer vision application.

Figure 3. Checkerboard pattern used for camera calibration.

2.3. 3D Reconstruction

Space carving is a simple and a fast technique to recover the 3D structure from multiple images. The 3D shape is reconstructed by removing the background scene which are not consistent with the input images [8]. This process can be summaries of the following steps: load the camera and image data, convert the image into silhouette, create a voxel array, and carve the voxels using the camera images. In this method, a structure of voxels is projected into the 2D views of the scene and carved if the voxels lie outside the silhouette [8].

The sequential images of objects are converted into binary images before proceeding to the voxel creation. The process is done using a threshold conversion technique based on the color threshold values as shown in Figure 4. However, this technique faced a problem when the color of the object and platform are almost similar. This problem affects the accuracy of the silhouette image. Thus the RGB images are converted into HSV images before applying the threshold technique. Since R, G, and B components in digital image are correlated with the amount of the light hitting the object, image descriptions in terms of hue/lightness/saturation or hue/lightness/chroma are more relevant.

Silhouette image will be projected onto a regular grid in 3-dimensional space called a voxel. Along with the camera viewing parameters, the silhouette defines a back-projected cone that contains the actual object. This cone is called silhouette cones. The intersection of the cones is called a visual hull which is a bounding geometry of the actual 3D object [7].

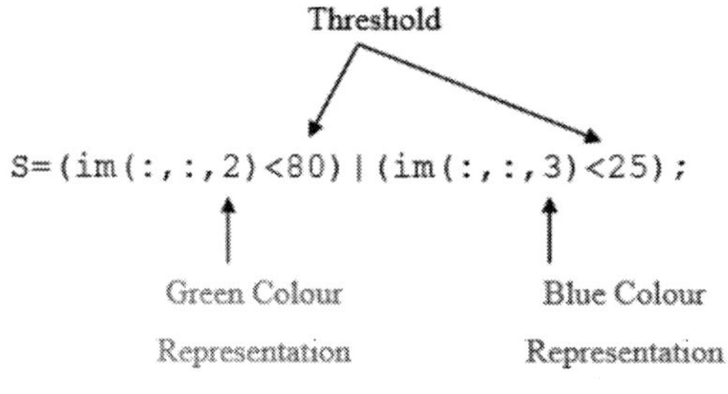

Figure 4. Threshold value of colour.

3. 3D Reconstruction Result and Analysis

3.1. Different image sequence

Figure 5 shows the final 3D model by using an increasing number of silhouette images, 36 and 72 images. In terms of shape produced, there is not much different with a higher or lower number of images used. However, the surface of the 3D model in figure 5 (b) is much smoother compared to the 3D model in Figure 5 (a). According to [9], the higher number of image sequence holds more information for the 3D reconstruction thus it will produce more accurate final 3D model.

(a) (b)

Figure 5. Final 3D model using (a) 36 image sequence, (b) 72 image sequence using an autonomous robotic arm.

3.2. Different camera position

The position of the camera in Position 1 is closer to the object compared in Position 2. In terms of shape, the final 3D model shape in Figure 6 (b) is closer to the real object compared to 3D model Position 1 in Figure 6 (a). The deformation cone at the bottom edges of the final 3D model in Figure 6 (b) is smaller compared to the result in Figure 6 (a).

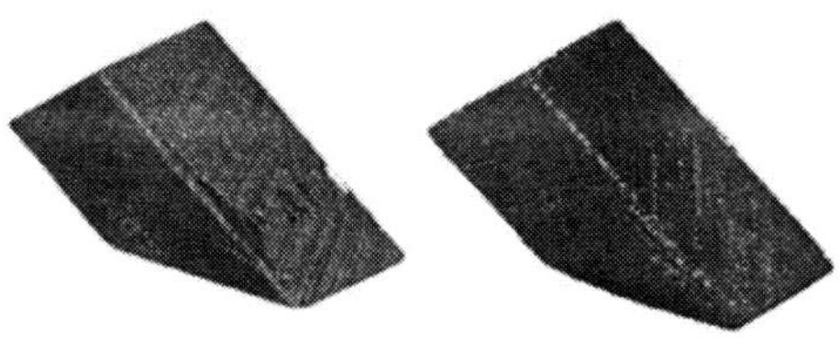

(a) (b)

Figure 6. Final 3D model of (a) Position 1, (b) Position 2.

Based on the analysis of camera calibration, it can be concluded that the higher the projection angle of the camera, the smaller the reprojection error. However, both Position 1 and Position 2 still produce a deformation cone like shape at the bottom of the 3D model produced. This is because the higher the camera position of the object causing the silhouette cones projected through the silhouette higher as well. The shape of the hidden parts of the camera will be estimated thus the intersection of the silhouette cones result in poor 3D reconstruction model.

4. Conclusion

The silhouette-based approach has been successfully tested for the sequence of object image taken by using the autonomous robotic arm. Other than that, the constructed autonomous robotic arm is successfully working to capture the sequence of object image surrounding the object for a different angle interval as well as different camera position. However, the 3D models produced are not accurately constructed as the original object as the models having deformation shape formed at the bottom of the 3D models. The positioning of the camera, the number of 2D image sequence, background colour and calibration technique can significantly influence the computed model. Apart from that, the technique of 3D reconstruction using Shape-from-Silhouette is prone to error. Other

IOP Conf. Series: Materials Science and Engineering **210** (2017) 012049 doi:10.1088/1757-899X/210/1/012049

than time consuming steps, the silhouette calculation is relatively sensitive for error like noise or wrong calibration.

Acknowledgement
The authors would like to thank to the Rehabilitation Engineering & Assistive Technology (REAT) research group under Center of Robotics & Industrial Automation (CeRIA) and Machine Learning & Signal Processing (MLSP) research group under Center for Telecommunication Research and Innovation (CeTRI) of Universiti Teknikal Malaysia Melaka (UTeM), Faculty of Electronics and Computer Engineering (FKEKK), Universiti Teknikal Malaysia Melaka (UTeM) for the funding provided under 'UTeM Zamalah Scheme' and Ministry of Higher Education (MOHE), Malaysia for sponsoring this work under project RAGS/1/2014/ICT06/FKEKK/B00065 and the use of the existing facilities to complete this project.

References

[1] Notano L G and Minghim R 2001 *Computer Graphic Forum* **20** 161-174
[2] Bianco G, Gallo A 2011 *3D-ARCH 2011*
[3] Moons T and Gool L V 2009 *Found. and Trends in Comp. Graphic and Vision* **4** 287-404
[4] Cytron Technologies, G15 Servo Motor User's manual. Retrieved from www.cytron.com.my
[5] Krishna A R and Bala G S 2012 *Int. J. of Eng. Res. and Applic.* **2** 3098-3103
[6] Perera K N D, Fernando S R M. et al 2013 *Computer controlled DC servo motor, Working paper* 1-3
[7] Laurentini A 2014 *IEEE Transactions on pattern analysis and machine intelligence* **16** 150-162
[8] Perez M C and Casamitjana M 2009 *Lect. Notes in Computer Science (LNCS)* **5524** 112-119
[9] Fredriksson L 2011 *Bachelor Thesis*, University of Gavle, Sweden

IOP Conf. Series: Materials Science and Engineering **210** (2017) 012050 doi:10.1088/1757-899X/210/1/012050

Development of sleep monitoring system for observing the effect of the room ambient toward the quality of sleep

W. H. M. Saad*, C. W. Khoo, S. I. Ab Rahman, M. M. Ibrahim, N. H. M. Saad
Faculty of Electronic and Computer Engineering (FKEKK), Universiti Teknikal Malaysia Melaka, 76100 Melaka, Malaysia

E-mail: *wira_yugi@utem.edu.my

Abstract. Getting enough sleep at the right times can help in improving quality of life and protect mental and physical health. This study proposes a portable sleep monitoring device to determine the relationship between the room ambient and quality of sleep. Body condition parameter such as heart rate, body temperature and body movement was used to determine quality of sleep and Audio/video-based monitoring system. The functionality test on all sensors is carried out to make sure that all sensors is working properly. The functionality of the overall system is designed for a better experience with a very minimal intervention to the user. The simple test on the body condition (body temperature and heart rate) while asleep with several different ambient parameters (humidity, brightness and temperature) are varied and the result shows that someone has a better sleep in a dark and colder ambient. This can prove by lower body temperature and lower heart rate.

1. Introduction

Based on the general definition, sleep is a type of "brain activity" and the purpose of this activity is to recover from brain fatigue [1]. Good sleep quality can contribute in many positive outcomes including in having better health in general, reducing daytime sleepiness and a better psychological health [2]. According to [3], there are three major factors that cause the sleep abnormalities which are psychological, physiological, and environmental. Psychological factors of sleeping disorder are usually short-term and it is involving in mental issue such as stress that being the number one cause. Physiological factor is involving in changes of physiological system that affect the quality of sleep. This factor can occur because of discrepancies in the nervous system, brain activity, immune system, metabolic functions and cardiovascular system. Lastly, environmental factors involving in environmental parameter such as temperature, light, humidity and other ambient factors including the comfortability of the room, the quietness or perhaps sound that make one feel calm to asleep.

When someone is asleep, heart rate and respiration rate will be varied. The paper [4] stated that the change in heart rate and respiration rate can be determined by Electrocardiogram (ECG)-derived respiration (EDR) during apnoea. By using real respiratory signal, accuracy up to 85 % of sleep apnoea can be detected. Other than that, movement of the chest can be used to detect the breathing phase. This movement shows the changes of ontogenetic [5]. The duration of without interrupt periods of sleep can be used to investigate the sleep continuity. In order to measure the movement of body during sleep stage, movement time (MT) and movement events (ME) are two important parameters to be measured. Brain is always in a very active phase but it will keep on reducing in term of regulatory response when someone is in REM sleep stage. Low ambient temperature can cause REM sleep to decrease while more heat may cause slow-wave sleep (SWS) to decrease. Regulation of temperature and sweating response will be decreased during sleep stage if only if the body's thermoregulatory responses are dependent [6].

Typical healthy person will experience up to five complete sleep cycle per night. It comprises of one stage for Rapid Eye Movement (REM) and four stages of Non–Rapid Eye Movement (NREM) sleep stages from NREM1 to NREM4 [7][8]. NREM3 and NREM4 are called as deep sleep. Each phase of NREM sleep can be last from 5 to 15 minutes. NREM1 is the beginning of sleep which named as transition stage [9]. When someone is in this stage, his or her eyes will move slowly, the muscle activity will be slowing down and people can be awakened easily. During stage two, the eyes movement will stop and the physical parameter like heart rate and body temperature will be decreased. During stage 3 and 4, he or she will be more difficult to be awaken. Dream sleep is the state of REM where people fall asleep and dream is taking part. In this stage, brain and body need a lot of energy and it will supply by high heart rate and body temperature [10]. The most common measurement to differentiate each of NREM sleep stages is by using electroencephalogram (EEG) reading, while the most commonly use in differentiating REM sleep from other sleep stages are electromyogram (EMG) and electrooculogram (EOG). Muscle tone is normally measured by using electromyography (EMG) method. Electrical signals are transferred by the electrodes and cause muscles to contract [11]. Needle electrodes are placed into the muscle to record electrical activity and the signal will be translated into graphs or numerical values for a specialist to interpret. EEG is a very powerful tool in the neurological field as it can capture normal electrical activity of brain as well as abnormal activity. It is measured directly from the electrocardiogram which is the cortical surface of the brain. Although these measurement is powerful but there also have a major limitation which is only can be performed by trained professionals and only in clinical environment. They also limit the monitoring to one or two night.

Nowadays, there are many existing procedures and methods to measure the quality of sleep that use several different types of sensor and from the measurement, it can be relate to the sleep stages. Polysomnography (PSG) for an example is used to measure the physiological variables including brain electrical activity, muscle movement respiratory effort and oxygen saturation [12]. Through this technique, researchers can identify sleep-onset and wake-up time [13]. A device must be worn on the wrist of user during sleep to measure their activity. The device must be placed on the user's wrist because it is believed that there is a strong connection between wrist movement and sleep status of the user [14]. The other method that is available is the use of video and audio recording together with Passive Infra-Red (PIR) sensor to detect the user sleep status [15]. The paper in [16] presented a research on ambient sleep monitoring which using sensors that are installed inside living apartments. The type of sensors that had been used in this study include passive infrared (PIR) sensors as motion detection, contact sensors that had been attached to the windows and doors, temperature sensors to detect room temperature and sensors that can detect heat application usage and electricity.

Mobile device is now a very useful and synonym to our daily life. It can do a lot of work including attached with sleep monitoring system. A system which named as ubiquitous architecture is a system with heart rate, accelerometer and sound signals. The idea is joining the sleep monitoring system with smart watch and smartphone. Framework with API need to be developed in order to able the data exchange between the smart watch and smart device. Other than that, this system also includes the novel algorithms for signal treatment. Ubiquitous architecture can be considered to be very suitable and feasible for sleep monitoring system.

In this study, we would prefer to use PC instead of smartphone because while we are in sleep, we did not require too much of mobility ergo PC with higher processing capability is more suitable than smartphone that allowing user to do a lot of processing and video recording. The developed sleep monitoring system is a combination of sensors and BLUNO microcontroller board to monitor and record the ambient parameters and body condition while user is sleeping. Our goals is to develop a sleep monitoring system that can be used at home and can do the monitoring in more than one night. Other

IOP Conf. Series: Materials Science and Engineering **210** (2017) 012050 doi:10.1088/1757-899X/210/1/012050

than that, this monitoring system can help to determine the relationship of room ambient and quality of sleep.

2. Methodology

The monitoring system is developed by using two different type of BLUNO microcontroller board, light dependent resistor (LDR) and DHT22 as an ambience sensor and pulse sensor, thermistor and accelerometer as a body condition sensor. The audio video recording is also a part of this project to monitor the behavior of user during sleep. The obtained value of light intensity, temperature, surrounding humidity, heartrate, body temperature and the movement of the user will be displayed on the Windows Apps interface.

The value obtained which is light intensity, temperature, humidity of the room surrounding, heart rate, body temperature and the movements of the user will be displayed on the Window surface apps.

2.1. System Overview

The whole developed system consists of two separate systems which are ambient monitoring system and portable body monitoring system. Figure 1 shows an illustration on how the system work.

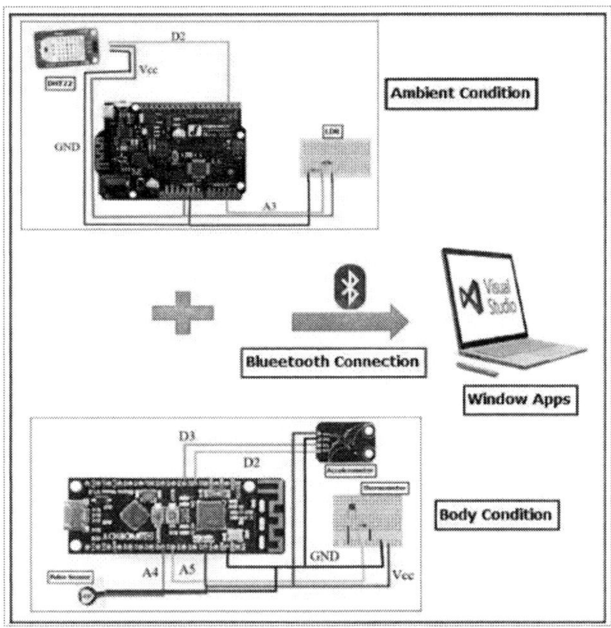

Figure 1. System Overview

Ambient monitoring system included LDR and DHT22 while portable body monitoring system included heart pulse, thermistor and accelerometer. The ambient sensor and body condition sensor are connected to BLUNO and BLUNO NANO respectively to capture the input data. In this system, the room temperature, light intensity and humidity is the input for ambient sensor hence to measure heart rate, the pulse sensor is simply clipped anywhere near to the blood vessel. To measure body temperature, the thermistor need to be clipped on any of fingertip for a few second to get a consistent reading.

This sleep monitoring system involves Arduino IDE software and Visual Studio 2015. After done with designing interface in Visual Studio software, the values obtained are displayed in the Window apps interface. Arduino IDE is programmed that consist of sensor algorithms to enable those sensors and to read the value that has been captured from room ambience and body condition. A Window

Application was programed by Visual Studio to display the value of those parameters as can be seen in Figure 2. The sensor value is transferred between hardware part and software part via Bluetooth communication. The video recording using a webcam is also included in this project.

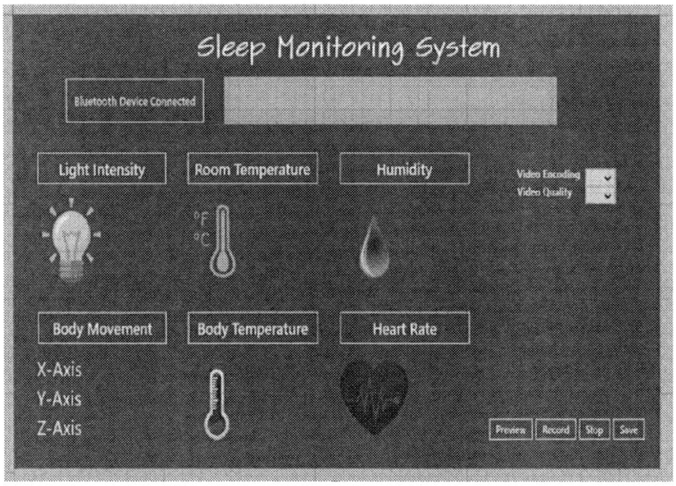

Figure 2. Window apps display in *MainPage.xaml* file.

2.2. Room Ambient Monitoring System

DHT22 sensor is a combination of a thermistor and capacitive humidity sensor to read the analogue value of temperature and humidity of the surrounding. It is also comprised with basic chip that convert the analogue signal to digital signal of both temperatures in degree Celsius and humidity in percentage of relative humidity (%RH) [17]. To measure the intensity of light in the surrounding, Light Dependent Resistor (LDR) is used which act as light sensors. BLUNO is an Arduino UNO module that is integrated with Bluetooth BT 4.0 (BLE). Connection between sensors and BLUNO is shown as in Figure 3.

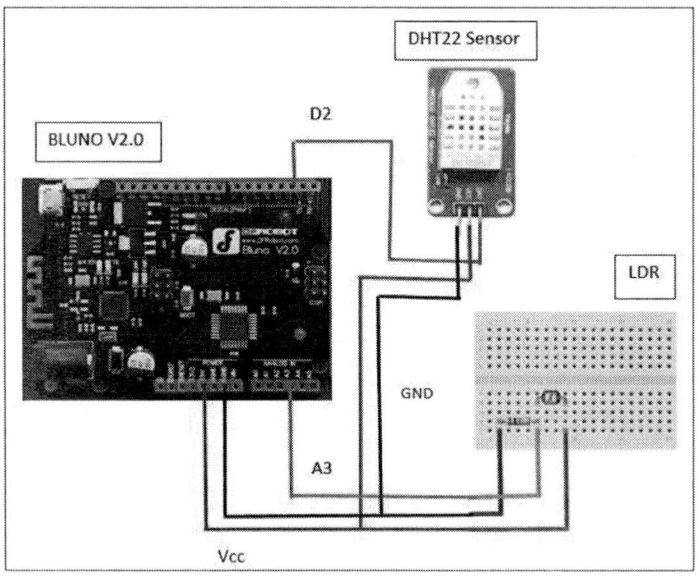

Figure 3. Hardware connection for Ambience monitoring device

2.3. Portable Body Monitoring System

Pulse Sensor is used to measure heart rate of user while thermistor is used to measure body temperature. To monitor the movement of user in x, y and z axis during sleep, accelerometer is used as well. BLUNO NANO is the Arduino NANO module that is integrated with Bluetooth BT 4.0 (BLE). The data is transferred via wireless communication that makes the system is comfortable to be used. Connection between these components is shown in Figure 4.

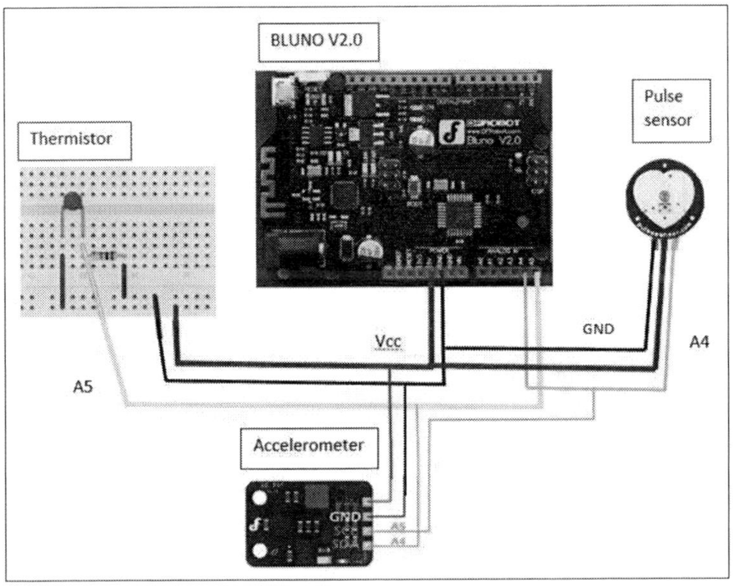

Figure 4. Hardware connection Portable Body monitoring device.

2.4. Functionality test of ambient monitoring system

The experiments on testing the functionality of ambient sensors is carried out to ensure each sensor working properly. The experiment that has been done to measure room temperature, light intensity and room humidity. The reading of light intensity was measured using LDR sensor. Ten readings were taken for each of five set of conditions which are very dark, dark, dim, bright and very bright. The value of light intensity is obtained based on the digital value with a range between 0 and 1023. For the condition 1, the value of light intensity is recorded between the ranges of 1 to 9 which indicate very dark surrounding such as during the night and the lamp inside the house is turned off. For condition 2, the value is between the ranges of 10 to 20 which indicate dark surrounding conditions such as in the early morning and the lights is turned off. For the condition 3, which refer to dim surrounding such as in the morning or late in the evening, the value of light intensity is recorded between the ranges of 21 to 45. Next, the condition 4 recorded the value of light intensity between the ranges of 70 to 487 which indicate bright surrounding in the afternoon in the house. Lastly, the condition 5 is recording the light intensity between the ranges of 501 to 1021 which indicate very bright surrounding such as in the afternoon outside the house. The reading is shown in Figure 5.

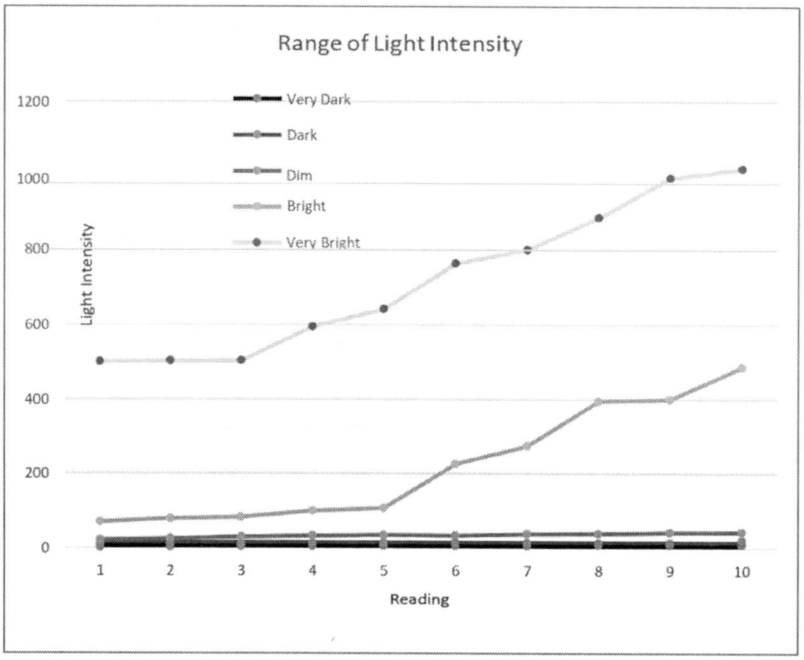

Figure 5. The value of light intensity in 10 readings

The reading of room humidity was measured using DHT 22 sensor. There are 10 reading taken in few conditions which are dry, medium or moist condition. Based on the data measured, the humidity reading for dry condition is ranging between 1 to 2% RH that indicates a dry condition during sunny day such as in the afternoon and in the evening. For the moderate condition, the measured reading is between 3 to 8% RH. The reading indicates medium humidity in several surrounding conditions, such as during raining. The reading of humidity is ranging between 13 and 32% RH which is referred to moist condition when there is raining for quite a long time and right after the raining when the temperature is low. The reading of all condition is shown in Figure 6.

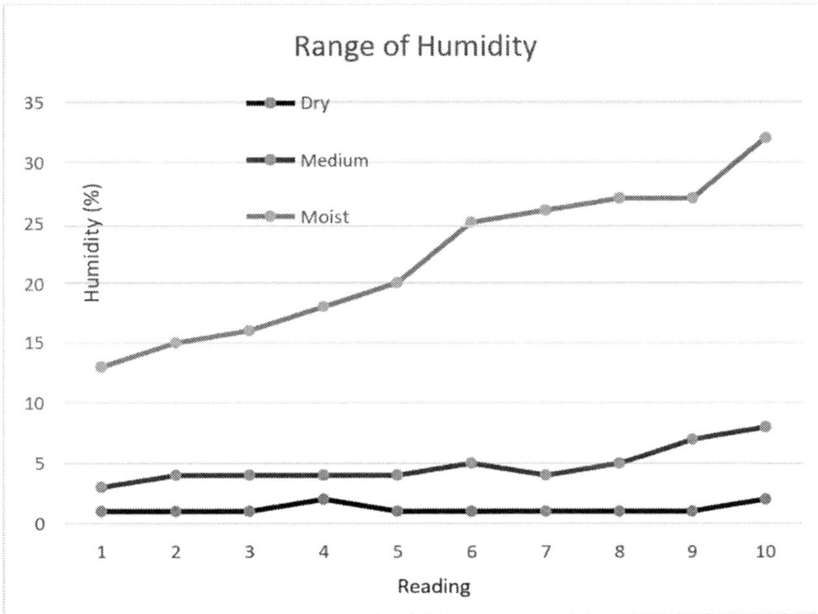

Figure 6. The value of humidity in 10 readings.

Figure 7 shows the reading room temperature in few different conditions. The reading of temperature is compared between sensor reading and thermometer reading. According to Figure 7, the reading obtained from a thermometer and a sensor is almost the same for every measured reading. This proved that sensor that used in this monitoring system can detect the room temperature with high accuracy. The temperature is divided into five different conditions which are very cold, cold, moderate, hot and very hot. Very cold conditions such as inside the room with air conditioner or during heavy rain in the midnight. The measured reading is recorded between 18 and 22 degree Celsius. Cold condition have a range of temperature between 23 and 27 degree Celsius. The reading for cold condition is obtained in early in the morning or right after a rain. The moderate condition has a range of temperature between 29 and 33 degree Celsius. The reading is taken in the bedroom with a speed of fan 3 during the day or night. The temperature for very hot condition is between 37 and 40 degree Celsius. The reading can be obtained outside the house in the afternoon or in the evening.

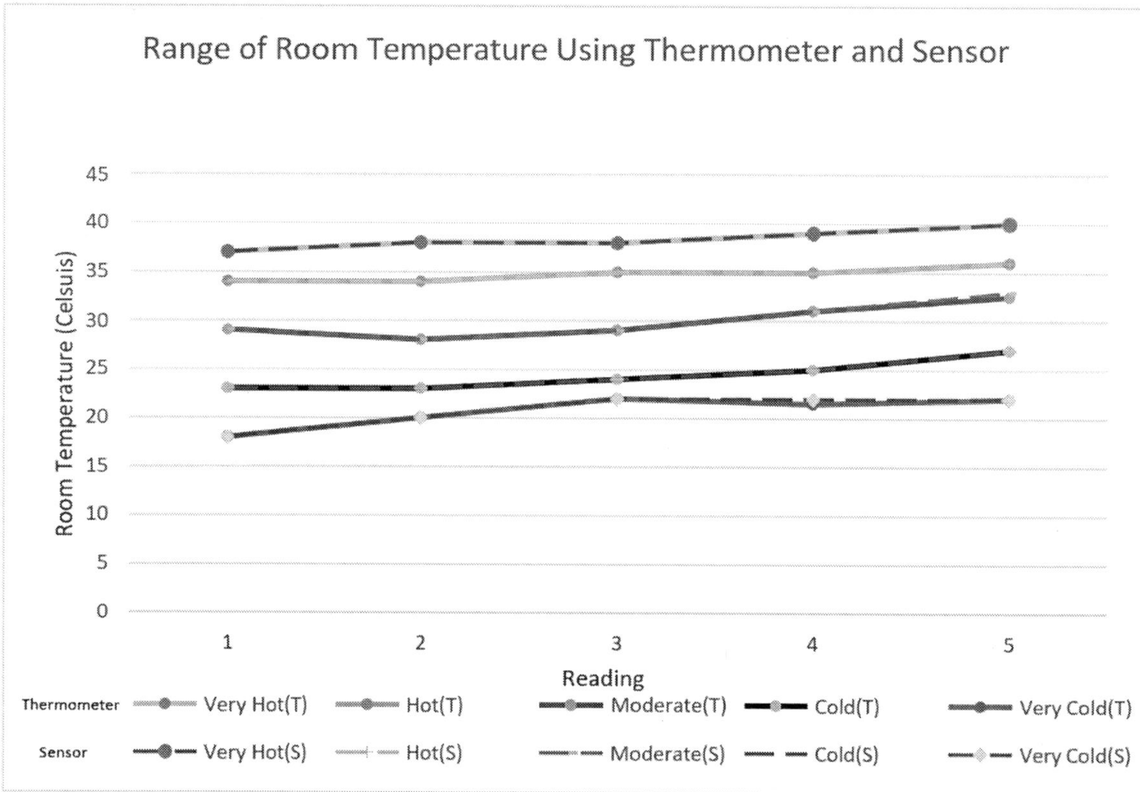

Figure 7. Compare the value of room temperature in 5 readings using thermometer and sensor.

3. Result and Discussion

There are three sets of experiments that have been carried out to monitor the body condition of user based on different ambient. The author in [18] state that a normal heart rate is ranged from 50-100 as most people's hearts beat is between 60 to 80 bpm. At resting time such as when sleeping, the heart rate reading become slows and tends to be lower if the person are very fit. Usually, heart rate reading during sleep will be slower. It faster when doing daily activities or when exercising and the heart rate will be recover quickly back to a resting rate a few minutes after exercise. The normal body temperature is usually 37℃, and can be slightly decreased during sleep [19]. According to [20], temperature sensor able to measure body temperature with high accuracy but it need about 35 second to read a stable reading.

Experiments with different ambient condition are carried out and the value is compared. Heart rate also recorded by using manual method (count the heart beat) and using sensor. Reading of heart rate is also compared. The ambient condition for the experiment set is summarized as in Table 1.

Table 1. Ambient condition of experiment set.

Experiment	Light Intensity	Room Temperature (Celsius)	Humidity (%)
Set 1	Very Dark	32	1
Set 2	Bright	32	1
Set 3	Dark	35	1

The values of body temperature on different ambient condition are measured by using sensor and thermometer. Both values taken have been compared. It shows that the reading was almost the same with not much different. The results are summarized as in Figure 8 and Figure 9.

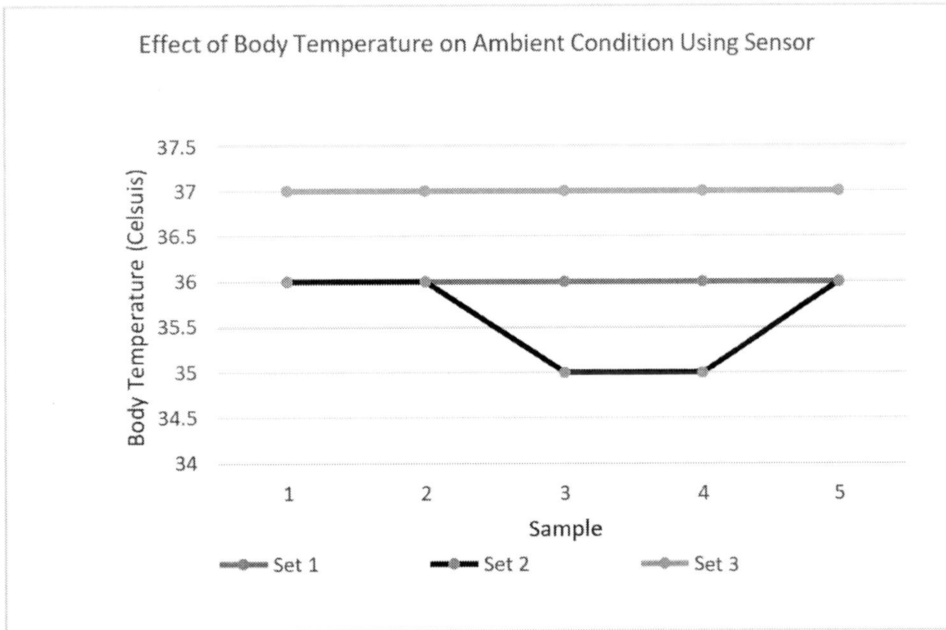

Figure 8. Effect of body temperature on different ambient condition (using sensor)

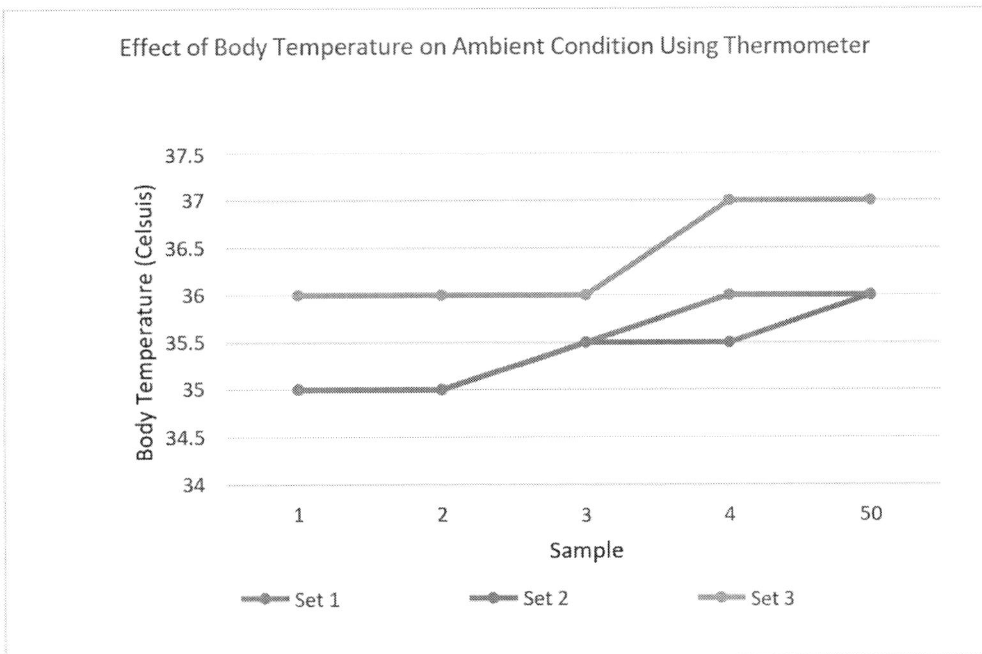

Figure 9. Effect of body temperature on different ambient condition (using thermometer)

According Figure 8 and 9, when the experiment is carried out in set 1, the sensor show average body temperature is 36°C while the thermometer shows 35.4 °C. For set 2, the sensor show average body temperature is 35.6°C while the thermometer shows 35.5 °C. Lastly, the sensor show average body temperature is 37°C while the thermometer shows 36.4 °C for set 3.

3.1. Heart Rate

Heart rate also recorded by using manual method (count the heart beat) and using sensor. Both values taken have been compared. It shows that the reading was slight different. The results are summarized as in Figure 11 and Figure 12.

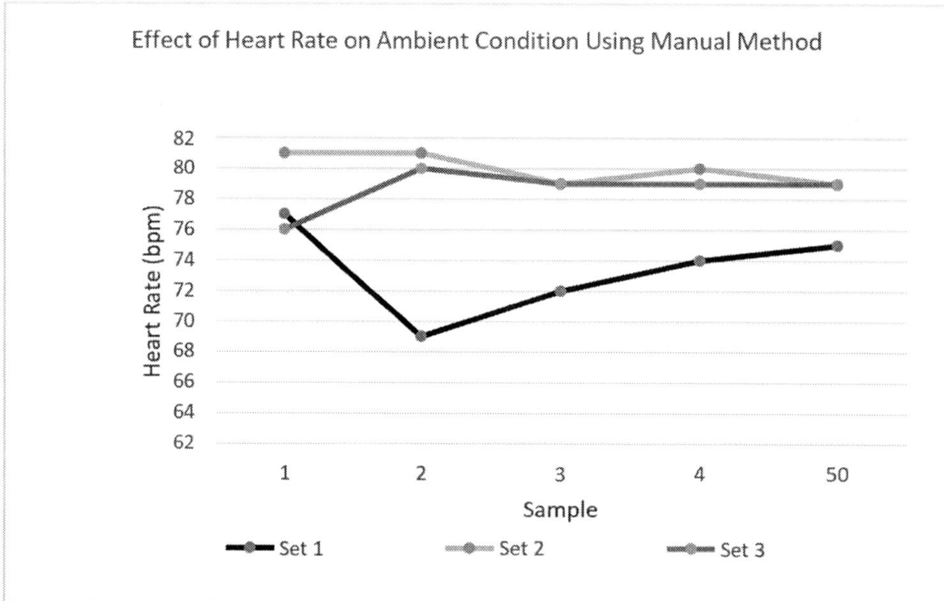

Figure 10. Effect of heart rate on different ambient condition (using manual method)

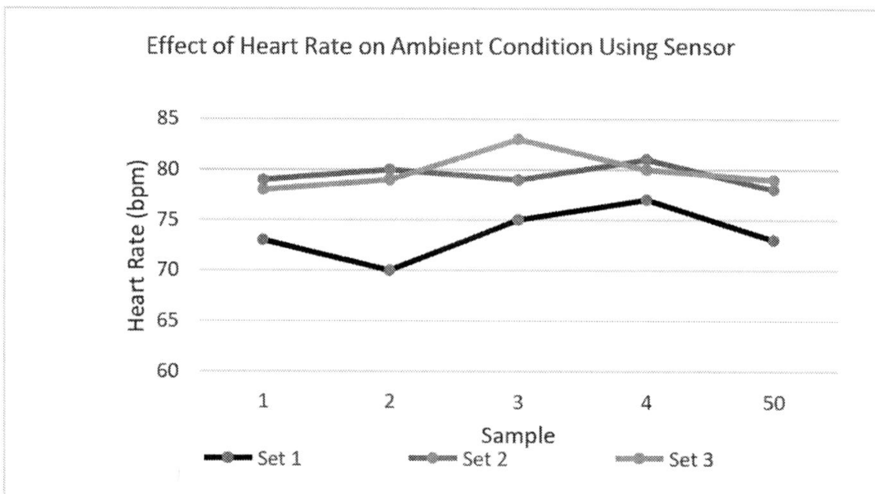

Figure 11. Effect of heart rate on different ambient condition (using sensor)

According Figure 10 and 11, when the experiment is carried out in set 1, the manual method show average heart rate is 74 bpm while the sensor shows 73 bpm. For set 2, the manual method show average heart rate is 80 bpm while the sensor shows 79 bpm. Lastly, the manual method shows average heart rate is 79 bpm while the sensor shows 80 bpm for set 3.

4. Conclusion

At the end of this project, a portable sleep monitoring system for ambience and body condition have been developed. At the same time, camera and microphone application inside the laptop is used to record the person's behavior while sleeping. The ambience and body condition monitoring is successfully displayed on the Window App. Functionality of all the sensors used is proven using functionality test. The simple experiment is carried out to monitor the body condition on several differences ambiences.

Acknowledgement

The authors would like to thank to the Machine Learning & Signal Processing (MLSP) research group under the Center for Telecommunication Research and Innovation (CeTRI) of Universiti Teknikal Malaysia Melaka (UTeM), Faculty of Electronics and Computer Engineering (FKEKK) and Ministry of Higher Education (MOHE), Malaysia for sponsoring this work under project RAGS/1/2014/ICT06/FKEKK/B00065.

References

[1] Suzuki, T., Ouchi, K., Kameyama, K.I. and Takahashi, M. 2009 *In BIODEVICES* 326-331
[2] Harvey, A.G., Stinson, K., Whitaker, K.L., Moskovitz, D. and Virk, H. 2008 *SLEEP-NEW YORK THEN WESTCHESTER* **31** 383-393
[3] *Sleep Disorders* 2016 http://www.sleepfoundation.ord/
[4] V.Carolina, F. Lieven, T. Dries, B. Bertien and Van Huffel 2016 *Proc. of the 43st Annual Computing in Cardiology* 1-4
[5] Coussens, S., Baumert, M., Kohler, M., Martin, J., Kennedy, D., Lushington, K., Saint, D. and Pamula, Y. 2014 *Sleep* **37** 2025
[6] David Gaynor and Kelsey Breseman 2014 *ChiliTech in Sleep Science*
[7] *Stages of Sleep* 2016 https://web.mst.edu
[8] Zhao, W., Wang, X. and Wang, Y. 2010 *Proceedings of the 18th ACM international conference on Multimedia* 1079-1082
[9] S.J. Casey, L.C. Solomons, J.Steier, N.Kabra, A. Bumside, M.F.Pengo, J. Moxham, L.H. Goldstein and M.D. Kopelman 2016 Neuropsychology **30** 931-945
[10] Carskadon, M.A. and Dement, W.C. 2005 *Principles and practice of sleep medicine* **4** 13-23
[11] Şen, B., Peker, M., Çavuşoğlu, A. and Çelebi, F.V. 2014 *Journal of medical systems* **38** 18
[12] Haakma, R. and Beun, R. 2012 *Measuring Behavior* **2012** 122
[13] Ancoli-Israel, S., Cole, R., Alessi, C., Chambers, M., Moorcroft, W. and Pollak, C. *American Academy of Sleep Medicine Review Paper* **26** 342-392
[14] Sadeh, A., Sharkey, K.M. and Carskadon, M.A. 1994 *Sleep* **17** 201-207
[15] Peng, Y.T., Lin, C.Y., Sun, M.T. and Landis, C.A. 2007 *IEEE Transactions on Biomedical Circuits and Systems* **1** 217-227
[16] Andrea Kealy, K. M. 2013 *International Conference on Pervasive Computing Technologies for Healthcare and Workshops* 33-40.
[17] DHTxx Sensors 2015 https://learn.adafruit.com
[18] Your Heart Beat and Your Health 2010 http://www.berkeleywellness.com/
[19] Phillips, K. 2015 *Alaska Sleep Education Center* http://www.alaskasleep.com/
[20] W. H. bin Mohd Saad, N. A. B. A. Salam and T.K. Leong 2014 *International Journal of Applied Engineering Research* **9** 24369-24378

IOP Conf. Series: Materials Science and Engineering **210** (2017) 012051 doi:10.1088/1757-899X/210/1/012051

Ultrafast soliton mode-locked Zirconia-based Erbium-doped fiber laser with carbon nanotubes saturable absorber

Arni Munira Markom[1,2]*, Mah Wey Sen-Winson[2], Mukul Chandra Paul[3]*and Sulaiman Wadi Harun[4]

[1]Faculty of Engineering and Computing, First City University College, No.1, Persiaran Bukit Utama, Bandar Utama Damansara, 47800 Petaling Jaya, Selangor, Malaysia.
[2]Faculty of Electrical Engineering, Universiti Teknologi MARA (UiTM), 40450 Shah Alam, Malaysia.
[3]Fiber Optics and Photonics Division, CSIR-Central Glass and Ceramic Research Institute, 196 Raja S.C. Mullick Road, Kolkata 700032, India.
[4]Department of Electrical Engineering, University of Malaya 50603 Kuala Lumpur, Malaysia.

E-mail: *arnimunira@gmail.com and *paulmukul@hotmail.com

Abstract. Ultrafast soliton mode-locked fiber laser was successfully generated in zirconia-yttria-alumina (Zr-Y-Al) co-doped erbium-doped fiber laser cavity using a single-wall carbon nanotubes (SWCNTs) as saturable absorber. The laser cavity was 11.5 m long with the group delay dispersion of -0.04 ps^2. The laser generates soliton pulse train with a center wavelength and 3 dB bandwidth of 1564.2 nm and 3.8 nm, respectively at pump power of 92 mW. Meanwhile, the repetition rate, pulse duration and pulse energy were 17.7 MHz, 770 fs and 51.4 nJ.

1. Introduction

In recent years, mode-locked erbium-doped fiber lasers (EDFLs) are being extremely explored due to their advantages such as flexibility, compact design and cost effective. They have potential scientific applications in many areas such as industry, military, medicine, remote sensing and optical communication [1-2]. Mode locked EDFLs are most commonly achieved by using passive modulation techniques for instance nonlinear polarization rotation (NPR) [3-4] and semiconductor saturable absorbers (SESAMs) [5-6], but either are costly or required some delicacy to implement, thus rendering the development of a compact ultrafast fiber laser impracticable. The use of carbon nanotubes (CNTs) and graphene material as passive saturable absorbers (SAs) has captured the interest of many researchers working on Q-switched and mode-locked fiber lasers [7-9]. This is due to a number of inherent advantages that they possess, including good compatibility with optical fibers, low saturation intensity, fast recovery time, and wide operating bandwidth [10-11].

CNT-based SAs are still popular as they are easier to fabricate [12]. Furthermore, CNT-based SAs may also have a lower non-saturable to saturable absorption ratio as compared to graphene [13]. With the outstanding quality and controllable physical parameters, CNT absorber fabrication technique is very practical for large, cost effective and high speed industrial production. A simple and cost effective fabrication technique of this SA was successfully demonstrated in a polymer matrix with implanted composite SWCNTs [14]. This technique easily adapted to the real application with the used of thin film formation due to outstanding homogenous dispersion of SWCNTs [15-16].

This paper demonstrates a soliton mode locked fiber laser in zirconia-yttria-alumina (Zr-Y-Al) co-doped erbium-doped fiber laser cavity using a simple, compact and cheap SWCNTs based SA. The SWCNTs having a large diameter precisely selected to operate at longer wavelength region of 1.5 µm and prepared using Polyvinyl alcohol (PVA) composite. The SA is designed by sandwiched together the composite film between two fiber connectors to generate mode-locked pulses with a pulse width of 770 fs as well as a repetition rate, average output power and pulse energy of 17.7 MHz, 0.91 mW and 51.4 nJ, respectively. The proposed mode locked operates at 1564 nm region which is desirable for many applications, such as optical communications by using all-fiber configuration.

2. Fabrication and Characterization of SWCNT SA

The main component of the proposed mode-locked laser is the SWCNTs-based SA, which is obtained by embedding SWCNTs into a poly (ethylene) oxide (PEO) film. It uses SWCNTs powder, which was synthesized by catalytic chemical vapour deposition (CCVD) process as raw material in the preparation. The SWCNTs grains were 99 % pure with a diameter of 1-2 nm and length of 3-30 µm and thus they required no purification process. In the experiment, 250 mg SWCNTs were added to 400 ml sodium dodecyl sulphate (SDS) solution in deionized water at 1 % concentration before being sonicated for 30 min at 50 W. The dispersion of single walled CNTs in the solution was achieved ultrasonically with the aid of SDS, which induced the dispersion of the grains as separate tubes in aqueous solution. The solution was centrifuged at 1000 rpm to remove large particles of undispersed SWCNT to obtain dispersed suspension that is stable for weeks.

The 1.8 ml of dispersed SWCNT suspension containing 1.125 mg solid SWCNT was then added into a polymer solution, which was obtained by mixing 1 g of polyethylene oxide (PEO, average molecular weight of 1 x 106 g/mol) with deionized water. The mixture was thoroughly mixed through a centrifuging process to form a SWCNTs-PEO composite precursor solution. Then the precursor solution was casted onto a glass petri dish and kept in a vacuum oven at 60 °C for 48 hours to form a thin film. The free standing film had a peak absorption at 1.5 µm with a thickness of around 50µm and was then used to form a fiber compatible SA device for mode-locking pulse generation in the proposed EDFL setup. The fabrication processes of the SWCNTs film is summarized in Figure 1. Figures 2 (a) and (b) show the 'real' and Field Emission Scanning Electron Microscopy (FESEM) images of the SA film, respectively. Figure 2 (b) indicates that the SWCNTs are randomly distributed in the polymer composite.

IOP Conf. Series: Materials Science and Engineering **210** (2017) 012051 doi:10.1088/1757-899X/210/1/012051

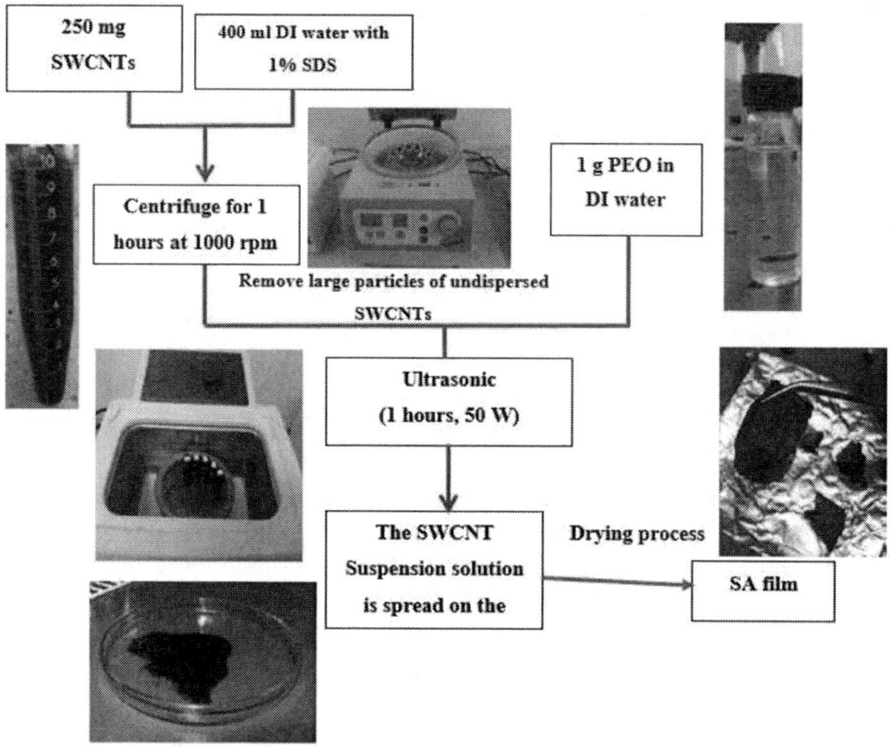

Figure 1. Fabrication procedures of SWCNTs-PEO.

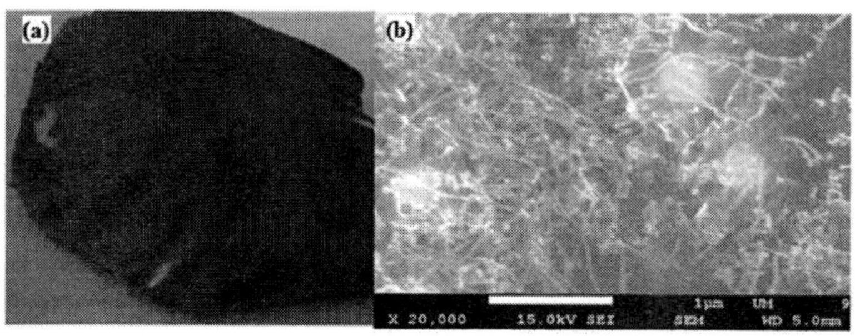

Figure 2. (a) An actual size and (b) FESEM images of the SWCNTs-PEO composite film.

Figure 3 shows the Raman spectrum, which was obtained using a Raman spectrometer where the small piece of the SA film was excited by a 532 nm laser. The spectrum can be used to confirm the presence of SWCNTs in the composite film and to determine the nanotube diameter distribution. As shown in the spectrum, the so-called G peak, which originates from the tangential vibrations of the carbon atoms, is prominent at 1584 cm^{-1}. In the low frequency region, the RBMs are observed, for which their energy is inversely related to SWCNT diameter. Since the distinct peaks are observed at around 214 cm^{-1}, the SWCNT diameter is estimated to be approximately 0.9 nm. This translates to the operating wavelength of around 1550 nm region. The carbon and G′ peaks are also observed at 1363 cm^{-1} and 2670 cm^{-1}, respectively.

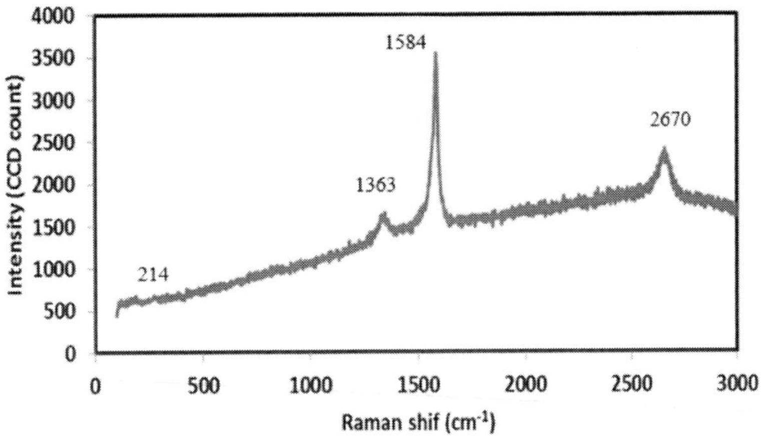

Figure 3. Raman spectrum of the prepared SWCNTs-PEO composite thin film.

3. Experimental Setup

The fiber laser setup used in the experiment is schematically shown in Figure 4. It is based on a ring configuration with the prepared SWCNTs-PEO SA was used as a mode-locker. The SA was fabricated by cutting a small part of the prepared composite film and sandwiching it between two FC/PC fiber connectors, after depositing index-matching gel onto the fiber ends. The insertion loss of the SA is measured at around 2 dB at 1550 nm. The gain medium is a 2 m long Zr-EDF with the absorption coefficient of approximately 220 dB/m and group velocity dispersion (GVD) of -56 ps^2/km at 1550 nm. A 980 nm laser diode is used to pump the Zr-EDF through a WDM. The WDM comprises 3 m long pigtailed fiber with a GVD of -38 ps^2/km at 1550 nm. The single mode fiber that makes up the rest of the ring cavity is 6.5 m in length with GVD of -21 ps^2/km. The estimated net dispersion and group delay dispersion (GDD) of the cavity, ignoring the GVD of the SA are 0.279 ps/nm and -0.04 ps^2, thus the cavity operates at the anomalous dispersion region. A PC is used to adjust the state of polarization in the cavity while a polarization independent isolator is incorporated in the laser cavity to force unidirectional propagation of light. The output of the laser is tapped from the cavity via a 95/5 output coupler, which keeps 98 % of the light oscillating in the cavity. An OSA with wavelength resolution of 0.02 nm is used to capture the output laser spectrum while a 350 MHz oscilloscope in conjunction with 1.2 GHz bandwidth photo-detector is used to detect the pulse train. The total length of the cavity is around 11.5 m.

IOP Conf. Series: Materials Science and Engineering **210** (2017) 012051 doi:10.1088/1757-899X/210/1/012051

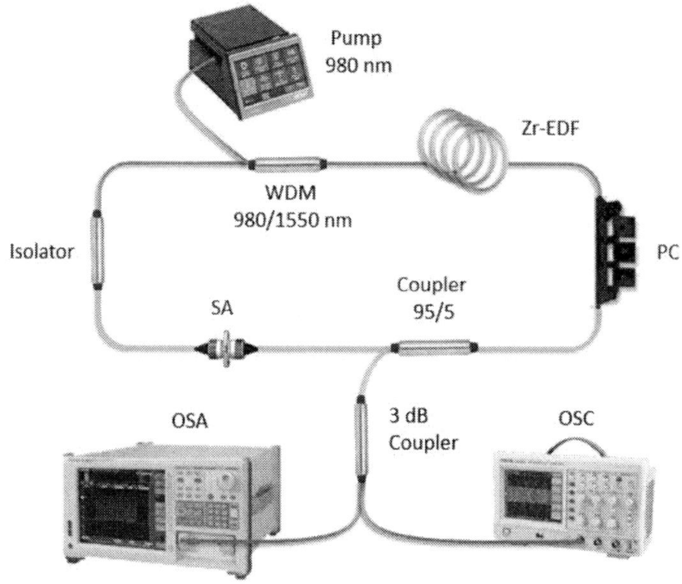

Figure 4. Schematic configuration of Zr-EDFL passively mode-locked by SWCNTs-PEO film-based SA.

4. Results and Discussion

It is observed that the laser always works in the CW operation with a threshold pump power of 45 mW in the absence of the SA in the cavity. As the SWCNT-based SA is incorporated in the laser cavity, the mode-locked laser self-started at pump power of 67 mW with appropriate setting of PC. Figure 5 shows the optical spectrum of the passively mode-locked Zr-EDFL at three different pump powers, analyzed by an OSA limited by a resolution of 0.02 nm. It operates at the central wavelength of 1564.2 nm and the 3 dB bandwidth is approximately 3.8 nm with strong Kelly sidebands at the maximum pump power of 92 mW. The spectrum is free from continuous wave (CW) parasitic lasing. The presence of Kelly sidebands confirms that this mode-lock fiber laser is operating in anomalous dispersion regime.

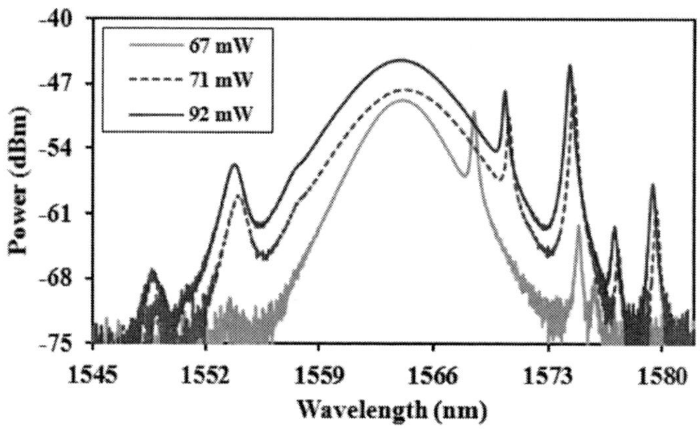

Figure 5. Output spectrum of the proposed mode-locked fiber laser at various pump powers.

Figure 6 (a) shows the pulse train of the passive mode-locked fiber laser obtained at pump power of 67 mW. It has a cavity round trip time of 56 ns, corresponding to a pulse repetition rate of 17.7 MHz and a cavity length of 11.5 m. Figure 6 (b) shows the corresponding second harmonic generation (SHG) autocorrelation trace, with the estimated pulse duration of 770 fs at its full-width half maximum (FWHM). A time-bandwidth product (TBP) calculated from the 3 dB bandwidth of the optical spectrum and the FWHM of the pulse is around 0.677, which indicates that the pulse is slightly chirped. The radio-frequency (RF) spectrum is shown in Figure 7. One can observe that the fundamental repetition rate of soliton is ~17.7 MHz. The peak-to-background ratio of the RF spectrum is 37.4 dB, implying a good mode-locking stability. At the maximum pump power of 92 mW, the output power for this fiber laser is 0.91 mW. Consequently, the resultant pulse energy is estimated to be around 51.4 nJ. We believe that the output power and pulse energy can be improved by reducing of the cavity loss and optimizing the output coupler. Based on the experiment results, we conclude that the fiber laser operates at a stable soliton mode-locking state.

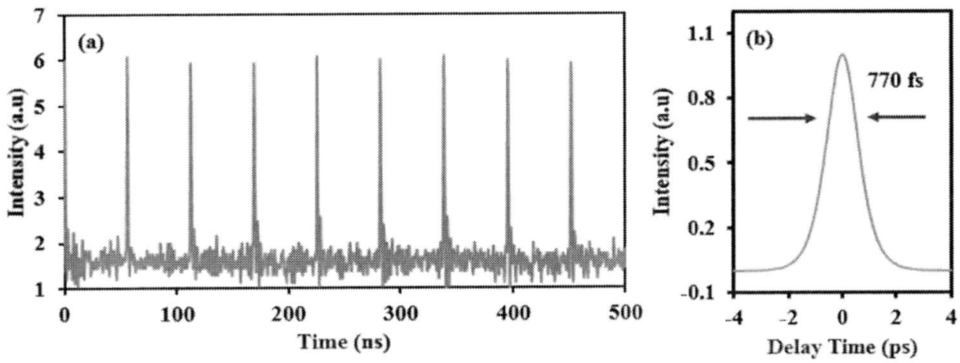

Figure 6. The temporal characteristics of the soliton laser (a) oscilloscope trace (b) autocorrelation trace.

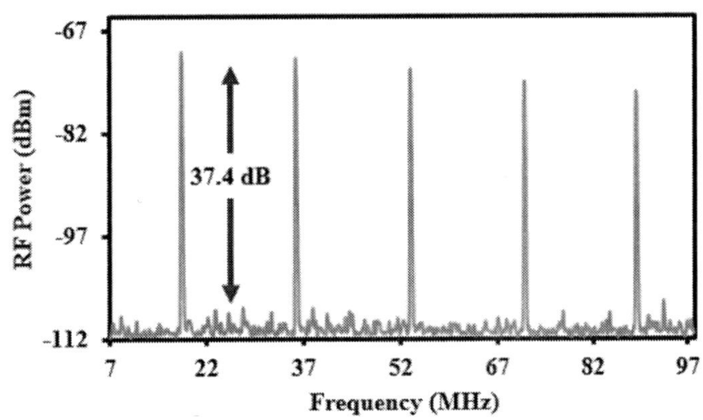

Figure 7. Radio-frequency spectrum of the soliton laser.

5. Conclusion

A mode-locked Zr-EDFL was demonstrated using a SWCNTs based SA to generate a soliton pulse operating at fundamental frequency of 17.7 MHz. The SA is obtained by sandwiching a piece of SWCNT-PEO film between two fiber connectors for passive mode-locking. The cavity operates at the anomalous dispersion region with a length of 11.5 m with estimated net group delay dispersion of -0.04 ps^2. The pulses are characterized by a slightly chirped pulse width of 770 fs, RF signal peak to noise ratio of 37.4 and pulse energy of 51.4 nJ at pump power of 92 mW.

References

[1] Grattan, L. S. and Meggitt, B. T. 2013 *Springer Science & Business Media* Optical fiber sensor technology: advanced applications-Bragg gratings and distributed sensors

[2] Saeedkia, D. 2013 *Elsevier Handbook of terahertz technology for imaging, sensing and communications*

[3] Tan, S. J., Harun, S. W., Arof, H., and Ahmad, H. 2013 *Chinese Optics Letters* **11** 073201

[4] Hamzah, A., Paul, M. C., Awang, N. A., Ahmad, H., Pal, M., Das, S., and Harun, S. W. 2013 *Optics & Laser Technology* **47** 22-25

[5] Major, A. 2016 *IEEE International Conference in Laser Optics* R3-27

[6] Li, J. F., Luo, H. Y., He, Y. L., Liu, Y., Zhang, L., Zhou, K. M., and Turistyn, S. K. 2014 *Laser Physics Letters* **11** 065102

[7] Ahmed, M. H. M., Ali, N. M., Salleh, Z. S., Rahman, A. A., Harun, S. W., Manaf, M., and Arof, H. 2014 *Optics & Laser Technology* **62** 40-43

[8] Ahmad, F., Harun, S. W., Nor, R. M., Zulkepely, N. R., Muhammad, F. D., Arof, H., and Ahmad, H. 2014 *Journal of Modern Optics* **61** 541-545

[9] Harun, S. W., Saidin, N., Zen, D. I. M., Ali, N. M., Ahmad, H., Ahmad, F., & Dimyati, K. 2013 *Chinese Physics Letters* **30** 094204

[10] Liu, X., Han, D., Sun, Z., Zeng, C., Lu, H., Mao, D., and Wang, F. 2013 *Scientific reports* **3** 2718

[11] Cui, Y., & Liu, X. 2013 *Optics express* **21** 18969-18974

[12] Della Valle, G., Osellame, R., Galzerano, G., Chiodo, N., Cerullo, G., Laporta, P., and Ferrari, A. C. 2006 *Applied physics letters* **89** 231115

[13] Martinez, A., Fuse, K., Xu, B., and Yamashita, S. 2010 *Optics express* **18** 23054-23061

[14] Xie, X. L., Mai, Y. W., and Zhou, X. P. 2005 *Materials Science and Engineering: R: Reports* ***49*** 89-112

[15] Ahmad, F., Harun, S. W., Nor, R. M., Zulkepely, N. R., Ahmad, H., and Shum, P. 2013 *Chinese Physics Letters* **30** 054210

[16] Rosdin, R. Z. R. R., Ahmad, F., Ali, N. M., Nor, R. M., Zulkepely, N. R., Harun, S. W., and Arof, H. 2014 *Chinese Physics Letters* **31** 094202

IOP Conf. Series: Materials Science and Engineering **210** (2017) 012052 doi:10.1088/1757-899X/210/1/012052

Analysis performance of proton exchange membrane fuel cell (PEMFC)

A. N. A. Mubin[1*], M. H. Bahrom[1], M. Azri[1], Z. Ibrahim[1], N. A. Rahim[2] and S. R. S. Raihan[2]
[1]Department of Power Electronics and Drives, Faculty of Electrical Engineering, Universiti Teknikal Malaysia Melaka, Malaysia
[2]UMPEDAC, Level 4, Wisma R&D UM, University of Malaya, Malaysia

E-mail: *ayunurfatika@yahoo.com.my

Abstract. Recently, the proton exchange membrane fuel cell (PEMFC) has gained much attention to the technology of renewable energy due to its mechanically ideal and zero emission power source. PEMFC performance reflects from the surroundings such as temperature and pressure. This paper presents an analysis of the performance of the PEMFC by developing the mathematical thermodynamic modelling using Matlab/Simulink. Apart from that, the differential equation of the thermodynamic model of the PEMFC is used to explain the contribution of heat to the performance of the output voltage of the PEMFC. On the other hand, the partial pressure equation of the hydrogen is included in the PEMFC mathematical modeling to study the PEMFC voltage behaviour related to the input variable input hydrogen pressure. The efficiency of the model is 33.8% which calculated by applying the energy conversion device equations on the thermal efficiency. PEMFC's voltage output performance is increased by increasing the hydrogen input pressure and temperature.

1. Introduction

Due to the awareness of the environmental issues, many of the researcher explores for clean energy source technology to replace the conventional energy [1]. In that case, hydrogen is found as one of the alternative sustainable renewable energy source for future [1][3][4]. Regarding to the electrical energy consumption, a system named fuel cell is introduced as a zero-pollutant electrical energy source. The fuel cell uses hydrogen gas and oxygen taken from the surrounding to produce electricity by the chemical reaction that is oxidation reduction. The process of a fuel cell is similar to the process of electrolysis where the usage of two electrodes to separate the electron to produce electricity and water as a result at the anode.

The reaction will also produce heat from the separation of electron in the oxygen atom where the separation of the electron will need a certain amount of heat to successfully remove the electron from the atom and combine with the hydrogen to produces water [5]. The PEM fuel cell is a fuel cell that used hydrogen as an electrolyte and a fast emerging in the green technology to replace the current renewable energy sources problems. The current renewable energy sources are too dependent on the surroundings and didn't produce much energy that will fulfil the requirements of a certain process. The needs in controlling the pollution of using fuel that releasing carbon monoxide that are polluting the air massively.

The research for fuel cell research and applications has been increasing rapidly from the last ten years and reached over 1.5 billion USD in the year 2007 where the fuel cell has been successfully applied on the areas that requires independent power supplies and a primary energy sources. In Malaysia, numbers of economic corridors planned in the future and for the present growth energy consumption where the needs for an independent power supplies would be a heavy demand especially when the development today are more on alternative energy resources [6]. Hydrogen could be an important energy carrier in the future where an energy carrier moves and delivers energy in a useable form towards the consumers. Producing energy using hydrogen which is can be stored and transport to other locations when needed will results in a high demand for hydrogen in the future technology.

The technology can be applied to the industrial production by changing the current machinery supplies and also opens up new market of a renewable energy which leads towards more job opportunity and more experts in the future. The PEM fuel cell can operate up to 200°C temperature and normally operates at 80°C and this was caused by the electric current productions where the higher amount of power produced will lead to a higher amount of heat [7]. Compared with the other types of fuel cells, a proton exchange membrane fuel cell (PEMFC) shows promising results with its advantages such as low temperature, high power density, fast response, and zero emission if it is run with pure hydrogen, and it is suitable for use in portable power supply, vehicles, and residential and distributed power plants [8][9][10].

The difficulties in obtaining an accurate PEMFC dynamic model is getting the parameters of the model to be used in modelling model [11]. The parameters can affect the voltage, power, temperature characteristics of the simulated stack. The parameters value used are primarily based on experiments and manufacturing data. There are some aspects on the operation of PEMFC are still hard to accurately model and some processes are property of the manufacturers. The importance of the physical and electrochemical processes occurs in a fuel cell can be evaluated using a generalized multi-parametric sensitivity analysis (MPSA) which includes the entire model parameter space [12].

The Figure 1 shows the electrical model for PEMFC illustrates the losses of the fuel cell by using resistor and capacitor that important consideration to include in the modelling. The application of the charge double layer capacitance (CDL) where it is a phenomenon of a jumping charge between two contacted surfaces as in this case it is a reaction in the electrode and the electrolyte by connecting capacitor parallel to Rconc and Ract. Ract is known as the cathode and the Rconc known as anode of the fuel cell operation. The open circuit voltage (OCV) is the value of cell voltage exact to an operation under 100°C. The OCV is expressed in Gibbs free energy release, $\Delta\bar{g}f$. The equation of the gibbs free energy released can be expressed in equation (1).

$$E = \frac{-\Delta\bar{g}f}{2F} \tag{1}$$

The purpose of this paper is to analyze the performance of Proton exchange membrane (PEMFC) by using a thermodynamic model. The developing of the mathematical thermodynamic model is by using Matlab/Simulink simulation software. This paper will also discuss the contribution of heat from the differential equation used to develop the thermodynamic mathematical model and the effects on the amount of heat and pressure produced towards the output voltage of the PEMFC.

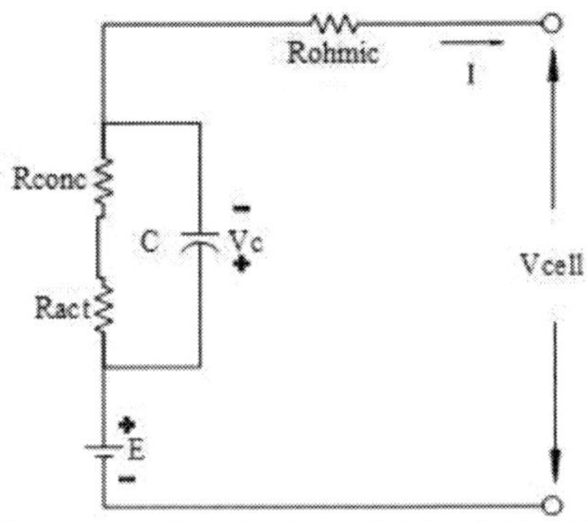

Figure 1. Electrical model for PEMFC.

2. Model specification

This section describes the simulation model for PEMFC. The modeling is using a few mathematical equations [13] and the parameter specifications of Horizon H-500 Fuel Cell System [14] in order to simulate the model as a real-time fuel cell stack system. Firstly, the mathematical modeling of the PEMFC is built in Matlab/Simulink environment using steady-state and thermodynamic model. The area of the cell is important requirement needed to simulate the PEMFC mathematical modelling due to its potential to produce much potential in a wider area cell. Besides, other than the physical characteristics, thermal and pressure behavior are also considered in the simulation. The current demand from the load are used as the input parameter for the simulation PEMFC mathematical model which can expect the power needed for the usage.

Modelling of the fuel cell considered the steady state requirements first before proceeding to the thermodynamic. In the steady-state consideration, there are four conditions that needed to be considered which is an area of cell, temperature of cell, input pressure and the current demand based on the Horizon- 500 parameter. The input pressure of the fuel cell is coming from two types of gases which are hydrogen gas and oxygen gas. The input pressure for hydrogen is set to a certain value which represents the hydrogen gas from a tank while for the oxygen gas is taken from the surrounding of the system.

$$E = 1.229 - 8.5\times10^{-4}(T_{cell} - 298.15) + 4.308\times10^{-5}(\ln PH_2 + \tfrac{1}{2}PO_2) \tag{2}$$

$$P_{H2} = 0.5PH_2O^{sat}[\exp\left(-\frac{4.192J}{T_{cell}{}^{1.334}}\right)\frac{P_a}{PH_2O^{sat}} - 1] \tag{3}$$

$$PO_2 = PH_2O^{sat}[\exp\left(-\frac{4.192J}{T_{cell}{}^{1.334}}\right)\frac{P_a}{PH_2O^{sat}} - 1] \tag{4}$$

$$\log_{10} PH_2O^{sat} = -2.18 + 2.95\times10^{-2}T_c - 9.18\times10^{-5}T_c^2 + 1.44\times10^{-7}T_c^3 \tag{5}$$

In the mathematical model of PEM fuel cell the partial pressure of these two gases gives effect to the ideal equilibrium potential, E as shown in equation (2). Both partial pressure of hydrogen and oxygen can be obtained by using the equation (3) and (4). Pa in equation (3) is the partial pressure for anode while J is the current density (A/cm^2). Besides, Pc in equation (4) is the partial pressure for cathode and P_{H2Osat} is the saturation pressure for water. The hydrogen and oxygen pressure from the surroundings are important in order to get the partial pressure of the hydrogen and partial pressure of the oxygen in the PEM fuel cell.

Thermodynamic is also important to include in the modelling. It is a coupled of a complex system which is the flow of fluid-solid-heat and electrochemistry where it is considered as the process of the electrochemical reaction. In addition, it is the operation for both membrane level and the auxiliary systems [13][15]. In dynamic, there is charge double layer capacitance (CDL) and the open circuit voltage. Charge double layer capacitance is the factor that causing the behavior of the fuel cell where CDL is the jumping or transfer of charge between to metal plate contacted. In the system, CDL occurs between the cathode and the anode and the potential of CDL labelled as V_c can be expressed as in equation (6).

$$V_c = V_{acto} + V_{conc} \tag{6}$$

In equation (6), the V_{conc} is the results for the change of the concentration of the reactants as they consumed in the reaction and the V_{acto} were taken from the steady-state model. The output of the dynamic is the cell voltage and the cell power where taken from the open circuit voltage (OCV) where OCV is the value of an open circuit voltage exacted to an operation of the fuel cell under 100°C. The output cell voltage can be expressed in equation (7) where E is the reversible voltage, V_{conc} is the concentration voltage, V_{act} is the activation voltage and V_{ohmic} is the ohmic voltage.

$$V_{cell} = E - V_{conc} - V_{act} - V_{ohmic} \tag{7}$$

The most common way of expressing the efficiency of the fuel cell is based on the change in standard free energy for the cell reaction which can be expressed in equation (8) where the reaction produced water in form of liquid [16]. The thermal efficiency on energy conversion is defined as the amount of useful energy produced towards the change in chemical energy stored that is released when a fuel cell reacts with an oxidant (oxygen). The ideal thermal efficiency of fuel cell on pure hydrogen and oxygen at standard condition is 0.83 where the actual efficiency of the fuel cell can be expressed in terms of the ratio of the operating cell voltage to the ideal cell voltage as in equation (9). The ideal voltage of a fuel cell is reversibly with pure hydrogen and oxygen in standard condition is 1.229V. Therefore, an actual fuel cell thermal efficiency of the operation at a voltage of V_{cell}, based on greater value of heating hydrogen is expressed as in equation (10).

$$\eta_e = \frac{\Delta G}{\Delta H} \tag{8}$$

$$\eta_e = \frac{0.83 V_{cell}}{V_{ideal}} \tag{9}$$

$$\eta_e = \frac{0.83 V_{cell}}{1.229} \tag{10}$$

A fuel cell may operate in different values of current densities where corresponds of the cell voltage then determines the fuel cell efficiency. A decrease in current density is resulting in an increase in cell voltage and thus increase the efficiency of the fuel cell. It can be described as the current density decreased the active cell area must be increased in order to obtain the desired amount of power.

3. Results and discussion

In this section, the results obtained from the PEMFC modelling are shown and discussed. The model is simulated by using a real-time specification from Horizon H-500 Fuel Cell Systems. Thirty six of cells are used and the area of cells is taken from the dimension stated in the data sheet which is 250mm x 190mm x 75mm are inserted into the simulation. Besides that, the value of pressure and temperature is set manually in order to observe the performance of the fuel cell system depending on that two physical quantities. The maximum temperature of the system is set to 338.15K.

3.1. Performance effects

At standard condition, 293.15K of temperature and 1.00 atm, gas pressure is used in the model [17]. The standard value of temperature is used as the input temperature for the model while the standard value of gas pressure is used at the cathode side which represent input oxygen pressure that obtained from the air the surroundings. Besides, the input pressure of hydrogen gas is set to 1.45 atm in the modelling. This section briefly explains the thermodynamic voltage and power output performance in a single cell. The stack performance of the PEMFC can be obtained from [13].

Figure 2 illustrates the single cell voltage performance for PEMFC versus current density. The single cell voltage is obtained from equation (7). Current density is the current demand per area of the cell in a PEMFC system. The voltage starts at 1.229V then decline rapidly at first due to the activation reaction in the PEMFC, then it's slowly decrease in the middle of time due to the ohmic losses. Finally, the voltage concentrates at 0.3V. Besides, Figure 3 shows the output power deliver in each cell of the PEMFC system. The power is increased by increasing current demand on the system until at one time the power is rapidly dropping before it reaches in a steady state condition due to the dryness of the cell [18][19][20]. Thus, the performance of the PEMFC is depending on the characteristics of the membrane and its limitation [21][22][23].

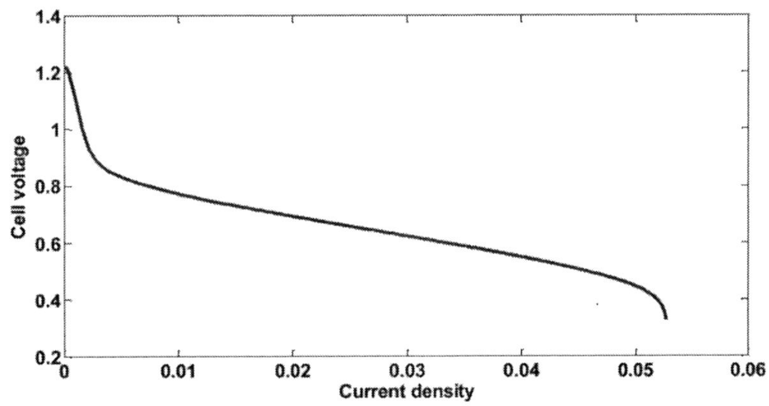

Figure 2. Single cell voltage performance.

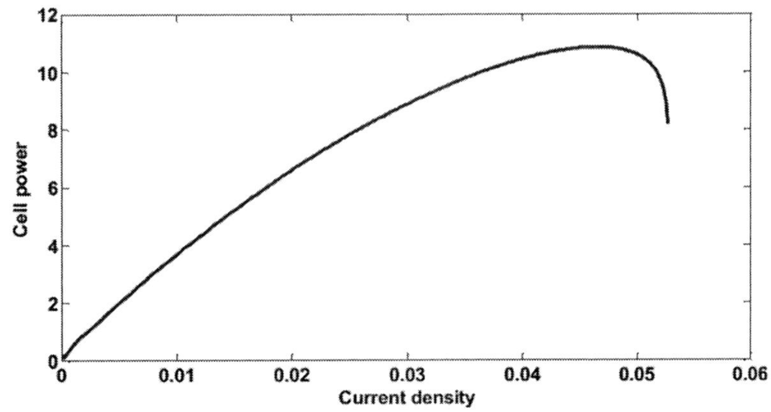

Figure 3. Single cell power performance.

For thermodynamic behavior, equation (6) as discussed in section 2 will affect in the voltage performance. Figure 4 illustrates that the variation of the temperature makes sense of the performance of the fuel cell voltage output in a single cell at a certain duration of current demand. The temperature of the system is varied from 293.15K to 313.15K. As the results, 293.15K of temperature has the lowest of output voltage performance compared to 298.15K, 303.15K, 308.15K and 313.15K. Therefore, from the figure below, it clearly shows that the higher the temperature of the system, the higher the voltage deliver from the PEMFC.

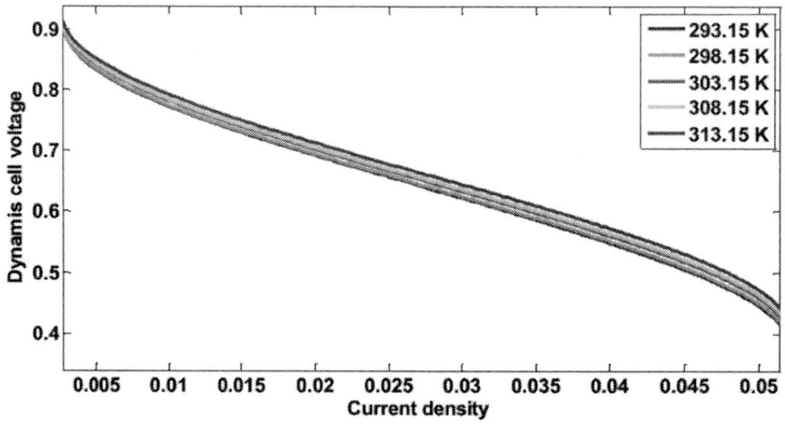

Figure 4. Performance on a single cell of the PEMFC depends on temperature effect.

On the other hand, input pressure also gives affects towards the performance of the PEMFC as stated in equation (2). Input hydrogen pressure and input oxygen pressure are two types of pressure included in the PEMFC model. In this paper, the input hydrogen pressure is varying from 1.00 atm until 1.60 atm while the input oxygen pressure is remaining the same as 1.00 atm which represents as air standard pressure. Figure 5 shows that the output voltage performance in a single cell is slightly different with the variation of the pressure. In the range of 1.00 atm to 1.60 atm, the higher level of the input pressure gives the higher output voltage [24]. However, it might be a large difference in a higher pressure level and in stack performance.

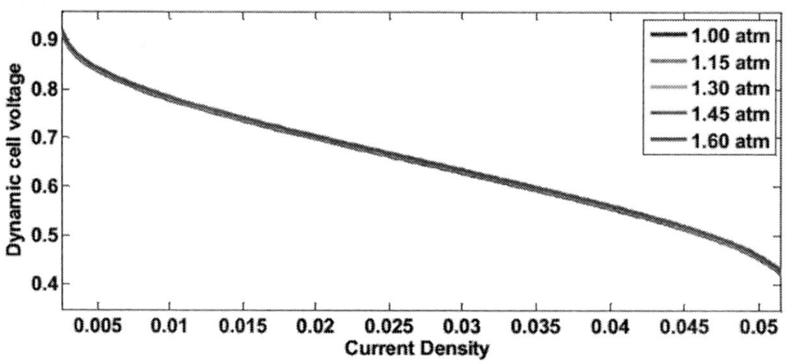

Figure 5. Performance on a single cell of the PEMFC depends on pressure effect.

3.2 Model efficiency

Figure 6 explained that the ideal and the actual voltage deliver from a single cell of the PEMFC is slightly different. The ideal voltage is mean that the voltage of PEMFC which is before the cell considers all the internal losses of the system which is represented in the actual voltage. The efficiency for the modelling of the PEMFC is calculated by using equation (10). Thus, from the formula, the efficiency of the PEMFC modelling is 33.8%. It is 15.5% percent difference with the Horizon H-500 series FC system which deliver 40% efficiency [14].

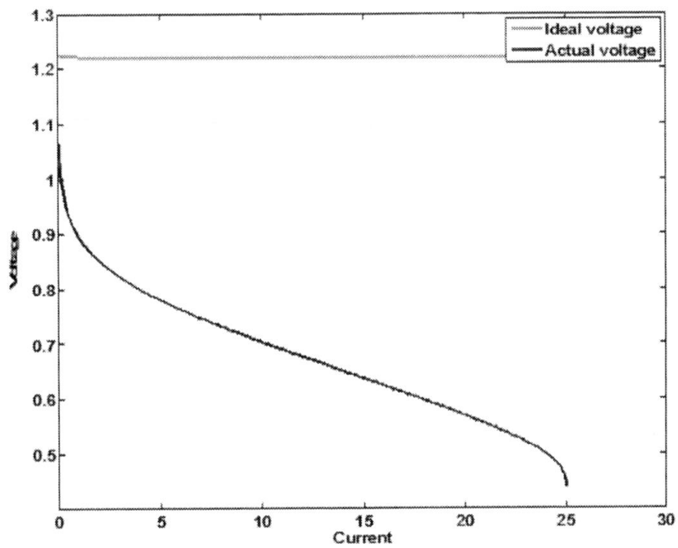

Figure 6. Ideal and actual voltage on a single cell of PEMFC.

4. Conclusion

As a conclusion, the output performance of the PEMFC is affected by varying the two parameters in this paper which are temperature and pressure. The higher the temperature and pressure inserted into the system, the better the performance. However, the over supplied input pressure and temperature will damage the membrane of the PEMFC. For further research, the limitation of the input supply and other parameters should consider in the model.

Acknowledgement

We are honored for the UTeM scholarship for author 1 and 2 that has open up a new opportunity to establish in the electrical engineering field (LRGS/2014/FKE/TK01/02/R00004).

References

[1] Azri, M., Rahim, N. A., & Halim, W. A. 2015 *Electric Power Components and Systems* **43** 928-938
[2] Ajanovic, A. 2008) *International journal of hydrogen energy* **33** 4223-4234
[3] Kruger, P. 2006 *Alternative energy resources: the quest for sustainable energy*
[4] Islam, M. R., Saidur, R., Rahim, N. A., and Solangi, K. H. 2009 *Engineering e-Transaction* **4** 69-72
[5] Yuan, J., Faghri, M., & Sundén, B. 2005 *DEVELOPMENTS IN HEAT TRANSFER* **19** 133
[6] Mohamed, W. A. N. W., Atan, R., & Yiap, T. S. 2009 *Proc. of the Int. Conf. on Advances in Mechanical Engineering*

[7] LOs, F. C. 2000 *Fuel cell basics*
[8] Li, Q., Chen, W., Wang, Y., Jia, J., & Han, M. 2009 *Journal of Power Sources* **194** 338-348
[9] Faghri, A., & Guo, Z. 2005 *International Journal of Heat and Mass Transfer* **48** 3891-3920
[10] Zhang, J. 2008 *PEM fuel cell electrocatalysts and catalyst layers: fundamentals and applications*
[11] Correa, J. M., Farret, F. A., Popov, V. A., & Simoes, M. G. 2005 *IEEE Transactions on energy conversion* **20** 211-218
[12] Reatrived at The World's #1 *Renewable Energy Network for News, Information, and Companies*.
[13] Azri, M., Mubin, A. N. A., Ibrahim, Z., Rahim, N. A., & Raihan, S. R. S. 2016 *Journal of Theoretical and Applied Information Technology* **86** 409
[14] TW, "*Horizon Fuel Cell Technologies*," ed, 2013. Retrieved from http://www.horizonfuelcell.com/
[15] Lee, J. M., & Cho, B. H. 2009 *Applied Power Electronics Conference and Exposition* 720-724
[16] Pilatowsky, I., Romero, R. J., Isaza, C. A., Gamboa, S. A., Sebastian, P. J., & Rivera, W. 2011 *Cogeneration fuel cell-sorption air conditioning systems*
[17] Olabisi, O., & Simha, R. 1975 *Experimental. Macromolecules* **8** 206-210
[18] Ye, D. H., & Zhan, Z. G. 2013 *Journal of power sources* **231** 285-292
[19] Bose, S., Kuila, T., Nguyen, T. X. H., Kim, N. H., Lau, K. T., & Lee, J. H. 2011 *Progress in Polymer Science* **36** 813-843
[20] Sharma, S., & Pollet, B. G. 2012 *Journal of Power Sources* **208** 96-119
[21] Vengatesan, S., Kim, H. J., Lee, S. Y., Cho, E., Ha, H. Y., Oh, I. H., and Lim, T. H. 2008 *International Journal of Hydrogen Energy* **33** 171-178
[22] Chao, W. K., Lee, C. M., Tsai, D. C., Chou, C. C., Hsueh, K. L., & Shieu, F. S. 2008 *Journal of Power Sources* **185** 136-142
[23] Jung, U. H., Jeong, S. U., Park, K. T., Lee, H. M., Chun, K., Choi, D. W., & Kim, S. H. 2007 *International Journal of Hydrogen Energy* **32** 4459-4465
[24] Rayment, C., & Sherwin, S. 2003 Introduction to fuel cell technology **46556** 11-12.

IOP Conf. Series: Materials Science and Engineering **210** (2017) 012053 doi:10.1088/1757-899X/210/1/012053

Contouring control of ball screw mechanism using a practical control method

Norhaslinda Hasim[1,2], Shin Horng Chong[1,2]*, Wai Keat Hee[2], and Zulkifilie Ibrahim[2]
[1]Motion Control Research Laboratory, Universiti Teknikal Malaysia Melaka, Malaysia
[2]Faculty of Electrical Engineering, Universiti Teknikal malaysia Melaka, Malaysia

E-mail: *horng@utem.edu.my

Abstract. This paper presents an improved practical controller for enhancing the contouring motion of a ball screw mechanism. Practically, a high motion control performance and ease of controller design are desired. A continuous motion nominal characteristic trajectory following (CM NCTF) control has been considered to fulfill the desired performance. The NCTF controller comprises of a nominal characteristic trajectory (NCT) and a PI compensator where the controller parameters are easily determined and it is free from exact modeling. In the present paper, the CM NCTF controller has been proposed in order to enhance the continuous motion performance such as tracking and contouring accuracies of the system. In order to justify the advantages, the CM NCTF controller was examined in tracking motion performances using an AC driven X-Y ball screw mechanism. The bandwidth of the CM NCTF controller is larger compared to the PID controller, therefore, it proved that the CM NCTF controller has fast response as compared with the PID controller. The experimental results proved that the CM NCTF controller achieves better contouring motion performances than the PID controller by showing a two times smaller motion error.

1. Introduction

AC servo mechanisms are commonly used in industrial application due to its advantages in greater efficiency, fewer stability problems, less maintenance and smoother operation as compared to the DC servo mechanism. But, the AC servo mechanism is tougher to be controlled in positioning applications and it frequently suffers due to its inherently nonlinear characteristics.

Recently, several researchers had studied the effectiveness of conventional PID control versus advance PID control system in ball screw mechanism. The advance PID control system used such as nonlinear PID control, P/PI control, nonlinear PID control cascade with feedforward [1, 2], and fuzzy PID control [3]. These researchers prove that, the advance PID control has better positioning performance as compared to the conventional PID control when the control system is applied to a nonlinear mechanism. In terms of contouring control, [4] had proposed a PID controller with friction feedforward for the ball screw system, but the results do not able to provide a sufficient friction compensation in the system. A position domain PID controller had been introduced by [5] to a 2-DOF Cartesian robot, where the results demonstrated that the proposed control is stable and can obtain very good contour tracking performances. However, the design procedures required an inverse kinematics analysis with mathematical derivations.

Advanced controllers have been suggested to achieve better positioning performance in ball screw mechanism. In [6], a disturbance observer with variable structure controller has overcome the problems due to friction and disturbances. A back-stepping sliding mode controller was proposed by [6,7] to improve the performance of the system used. Besides that, a Reinforcement Learning methods by learning the feedback controller of a ball screw feed drive is used in [8] can be an alternative to typical feedback control approaches such as PID controllers. Although these advanced controllers are able to show good performance in the robustness of precision positioning, but it involves with model parameter determination and the performance of these controllers is based on how accurate the known model and known parameters are used in their design procedures, which affect the design processes in terms of time consuming and more work is acquired to identify the accurate parameters, which create hurdles to their practical use.

A control method using NCTF control was first proposed by [9] as a practical controller for point-to-point positioning system and was examined with a rotary mechanism. After that, the continuous motion nominal characteristic trajectory following (CM NCTF) controller has been proposed by [10] as a high performance and simple control structure to a DC ball screw mechanism in 2009. The best part of this controller is it has a simple control structure as compared with the other advanced controllers. Besides that, the design procedure of a NCTF controller is much simpler and easy to understand. It does not require an exact model or parameter identification of the plant which sometimes troublesome the researchers. It applied a general pulse input to drive the mechanism and stop smoothly in open-loop response.

In year 2010, the CM NCTF controller was validated with a non-contact mechanism [11]. It proved that the procedure of the NCTF controller is independent of the friction characteristic. The effectiveness of the CM NCTF controller was evaluated by [12] using the 1-DOF rotary system, which is driven using DC servo motor in 2013. The third generation, Acceleration Reference-Continuous Motion NCTF (AR-CM NCTF) controller was proposed in year 2010 [13].

The main goal of this paper is to examine and clarify the usefulness of the CM NCTF control to a ball screw mechanism that driven using AC servo motor. AC servo motor produces higher acceleration as compared to DC motor. A suitable input signal is used in this paper which is able to produce sufficient rapid and smooth response during deceleration motion in open-loop experiment. The open-loop responses are then used to construct the nominal characteristic trajectory (NCT). This procedure is simple yet important because it has significant influences to the reference following characteristic of the control system. The detail procedures in designing the CM NCTF controller and PID controller can be referred to [14]. Experiments in tracking motion in difference axis will be conducted in order to validate the effectiveness of the CM NCTF control system in comparison to the PID controller.

The rest of this paper is written as follows: Section 2 describes the experimental setup of the ball screw mechanism that used in this research. In section 3, the control structure of the CM NCTF controller and PID controller control system with their design procedure are discussed. The performance of both controller are validated and discussed in section 4. Lastly, the remark conclusion and future studies are made.

2. Experimental setup

The contouring control performance in this research is examined using a ball screw mechanism shown in Figure 1. The mechanism is driven by an AC servo motor (model: MSMD022P1U) through ball

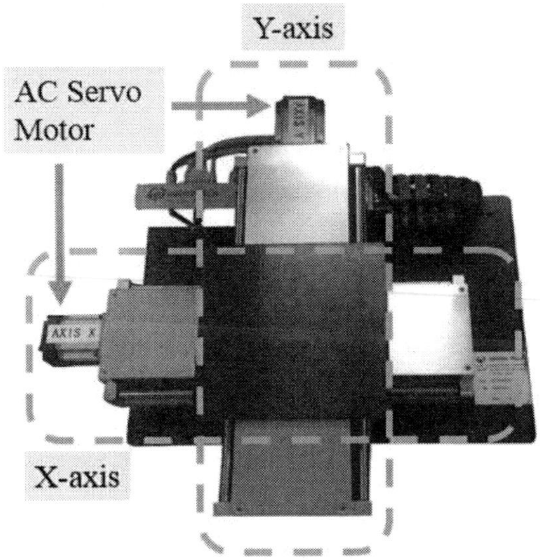

Figure 1. X-Y ball screw mechanism.

screws with a 5 mm pitch and the lead of the ball screw is 1.25 mm/rev. Each motor is controlled independently by the same servo driver. The ball screw movement is measured by an incremental encoder with a resolution of 0.5 m/pulse. The maximum travel of the table is 175 mm in X and Y directions. The velocity in this research is obtained using a backward differential algorithm from the measured displacement. Basically, the material for both mechanism (X-axis and Y-axis) are similar except for their weight. As X-axis is mounted at the bottom of Y-axis, therefore, the weight for X-axis is heavier (total mass ± 16.71 kg) compared to Y-axis is only about 3.11 kg since it only sustain the mass of the moving table.

3. Controller Design

A controller with a similar control structure is chosen to compare with the CM NCTF controller in tracking to see their contouring motion performance. The design procedure of the Continuous Motion Nominal Characteristic Trajectory Following (CM NCTF) controller and the PID controller will be explained in this section.

3.1. Continuous Motion Nominal Characteristic Trajectory Following (CM NCTF) Control Design

The CM NCTF controller is a NCTF based controller where it has a simple structure and easy to design. The NCTF controller is adjusted to improve the contouring motion performance. The CM NCTF controller comprises of a nominal characteristic trajectory (NCT) element and a Proportional and Integrator (PI) compensator as illustrates in Figure 2. The NCT plays an important role as a virtual error rate for the object to follow and it is expressed on the phase plane. The NCT is designed using the object responses in open loop experiment and the inclination of the NCT near origin is denoted as β. Instead, the PI compensator controls the velocity of the ball screw mechanism and make it follows the NCT and end at origin. When the object stops at the origin, it means the end of the positioning motion.

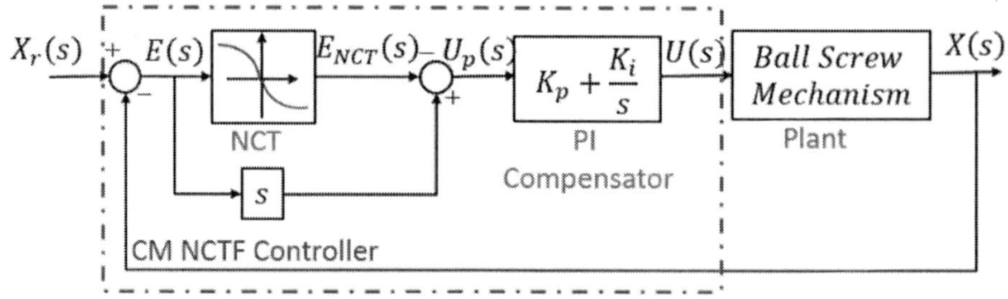

Figure 2. Control Structure of Continuous Motion NCTF (CM NCTF) controller.

The design procedure of the CM NCTF controller is consisted of three simple steps:

I. **Open-loop response,** where the ball screw mechanism is driven in open loop and the displacement and velocity responses are measured.

II. **Construction of NCT**, where the NCT is constructed on the phase plane using the displacement and velocity of the mechanism during deceleration motion.

III. **Design of PI compensator**, where it is designed based on the open-loop responses and NCT information.

The constructed NCT with inclination near origin, β equals to $87\,s^{-1}$ is used in this research. The values of K_p and K_i of the PI compensator are 0.08 Vs/mm and 1.34 V/mm respectively. The details design procedures for the CM NCTF controller and PID controller are discussed in [14].

3.2. PID Control Design

PID controller is chosen to be compared with the CM NCTF controller because it has a similar control structure with the CM NCTF controller. Furthermore, the design procedure of the PID controller and the CM NCTF controller are simple and easy. Figure 3 shows the control structure of the PID controller used in this research. The step to design a PID controller is simple where the values of K_p , K_i and K_d are chosen using trial error method. Next, all the compensator gains are fine-tuned until the transient response of the PID controller similar with the CM NCTF controller. The final values of K_p , K_i and K_d are 3.00 V/mm, 0.80/mms, and 0.03 Vs/mm respectively.

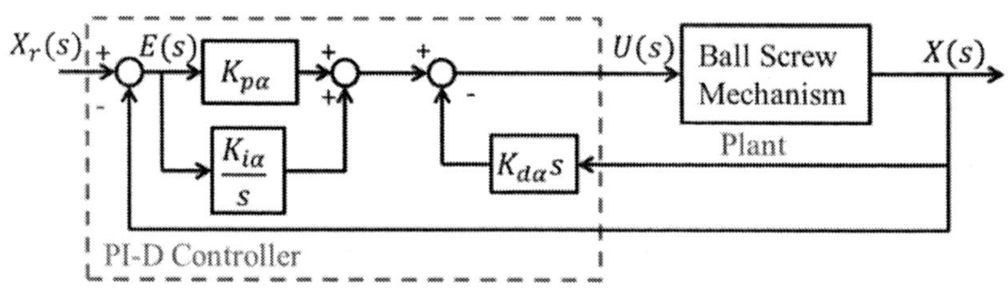

Figure 3. Control structure of PID controller.

4. Performance Evaluation

In this section, the contouring performance for both controllers are experimentally examined and the effectiveness of the CM NCTF controller is evaluated in comparison with PID controller. The performance of both controller is evaluated based on the continuous motion for X-axis and Y-axis. For circular motions, sinusoidal reference inputs are applied to the ball screw mechanism. Figure 4 shows the continuous motion performances of the PID controller at frequency of 0.1 Hz and the reference radius is 10 mm, while the performance of the CM NCTF controller is shown in Figure 5. The experiments conducted in such a way that the Y-axis is mounted on top of X-axis. Then, a cosine wave reference is used for X-axis and a sine wave for Y-axis mechanism, thus the resulting contour is a circle. The CM NCTF controller generates a much smaller tracking errors than the PID controller. The reference used in the experiments is a circle with 10 mm radius and the frequency is 0.1 Hz. The radial error is reduced by more than ten times with the CM NCTF controller, and it guarantees the proposed controller is suitable for continuous motion.

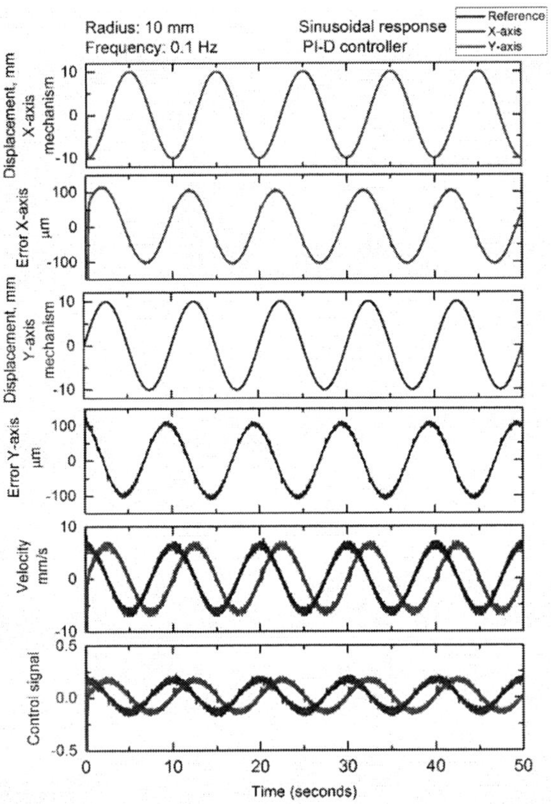

Figure 4. Continuous motion performances with the PID controller at frequency 0.1 Hz.

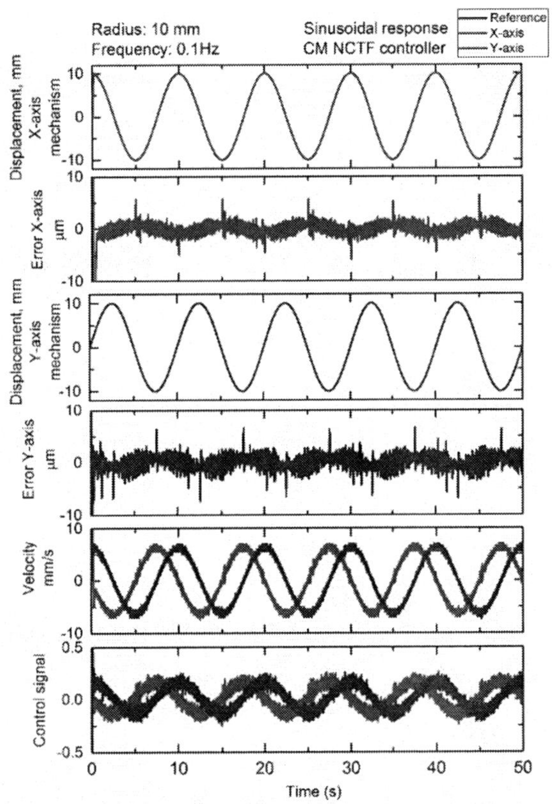

Figure 5. Continuous motion performances with the CM NCTF controller at frequency of 0.1 Hz.

4.1. Contouring control performances

Figure 6 shows the circular motion of both CM NCTF controller and PID controller at frequency of 0.1 Hz with radius 10 mm and 1 mm. At lower frequency, both controller managed to perform with less error compared to higher frequency, however, PID still outperforms by more than twice those obtained using the CM NCTF controller. The circular motion at 5 Hz with 0.1 mm and 1 mm are shown in Figure 7 and 8 respectively. The outer and inner radial errors are summarized in Table 1. Based from Table 1,

IOP Conf. Series: Materials Science and Engineering **210** (2017) 012053 doi:10.1088/1757-899X/210/1/012053

it can be clearly observed that the experimental results proved that the CM NCTF controller achieves better contouring motion performances than the PID controller by showing at least two times smaller motion error.

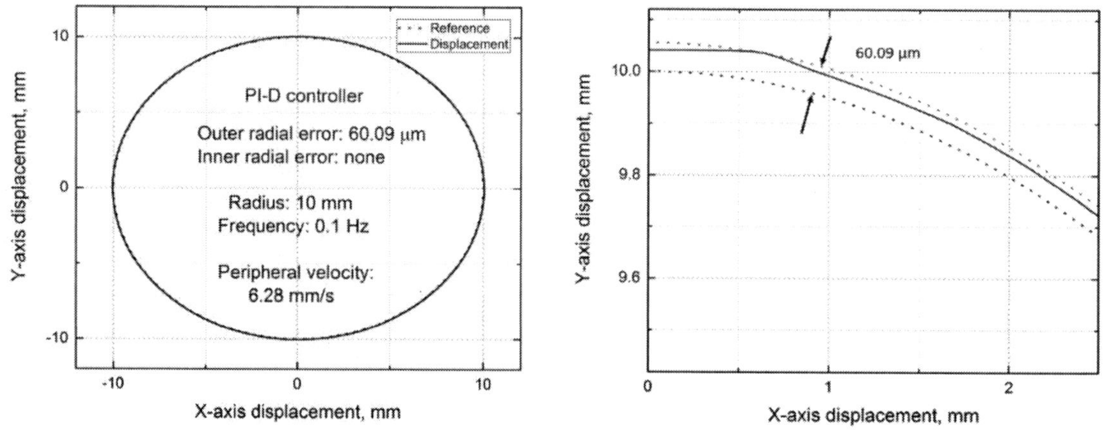

(a) PID controller

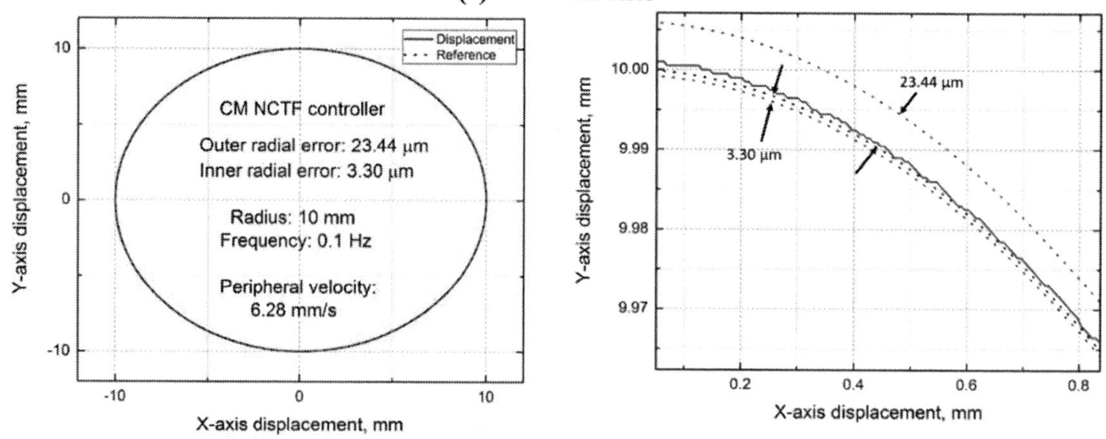

(b) CM NCTF controller

Figure 6. Circular motion with reference input of 10 mm and frequency 0.1 Hz.

Table 1. Contour control performances of the CM NCTF and PID controllers.

Frequency (Hz)	Amplitude (mm)	Radial error (μm)					
		PID controller			CM NCTF controller		
		Outer	Inner	Total	Outer	Inner	Total
0.1	10.0	60.09	none	**60.09**	23.44	3.30	**26.74**
	0.1	13.54	17.51	**31.05**	11.80	none	**11.80**
5.0	1.0	none	175.55	**175.55**	37.00	25.05	**62.05**

IOP Conf. Series: Materials Science and Engineering **210** (2017) 012053 doi:10.1088/1757-899X/210/1/012053

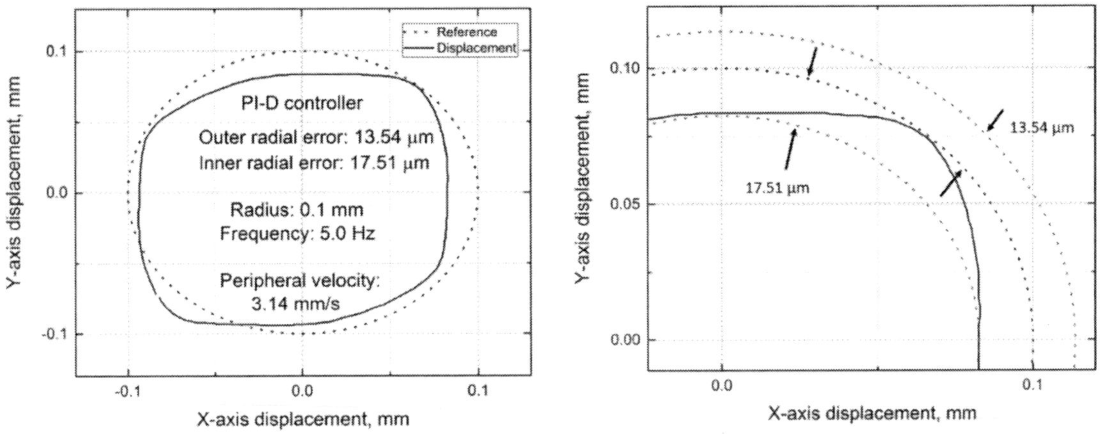

(a) PID controller

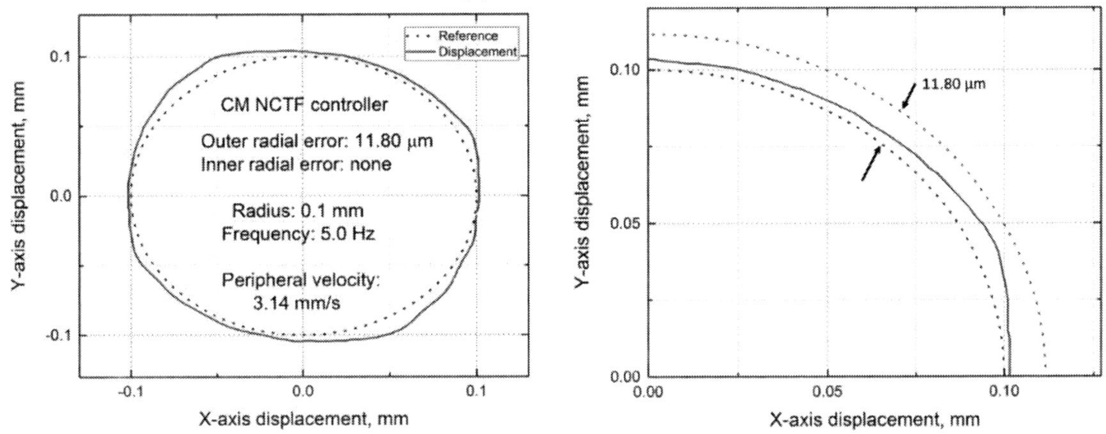

(b) CM NCTF controller

Figure 7. Circular motion with reference input of 0.1 mm and frequency 5.0 Hz.

5. Conclusion

In this paper, the CM NCTF controller was suggested as a high-performance with a practical design method controller. The designed controller was then realized to a ball screw mechanism to evaluate its performances. The CM NCTF has almost similar control structure with PID controller and their design procedure are simple and easy. The experimental results proved that the CM NCTF controller achieves better contouring motion performances than the PID controller by showing a two times smaller motion error.

Acknowledgments

The authors would like to be obliged to Motion Control Research Laboratory, Center for Robotics and Industrial Automation (CeRIA), and Faculty of Electrical Engineering (FKE) in Universiti Teknikal Malaysia Melaka for providing the laboratory facilities and equipment support. The Centre for Research and Innovation Management (CRIM) from UTeM is gratefully acknowledged.

References

[1] Abdullah L, Jamaludin Z, Maslan M N, Jamaludin J, Halim I, Rafan N a and Chiew T H 2015 *Procedia CIRP* **26** 212-6
[2] Xu J, Qiao M, Wang W and Miao Y 2011 *The 6th Int. Forum on Strategic Technology* 706-711
[3] Zhang C and Chen Y 2016 *IEEE Transactions on Industrial Electronics* **46** 7682-7692
[4] Rafan N A, Jamaludin Z, Chiew T H, Abdullah L and Maslan M N 2015 Contour error analysis of precise positioning for ball screw driven stage using friction model feedforward *Procedia CIRP* **26** pp 712-717
[5] Ouyang P R, Pano V and Dam T 2016 *IEEE International Symposium on Industrial Electronics* 1297-1302
[6] Lin C J and Lee C Y 2011 *Mechatronics* **21** 185-203
[7] Chen J, Li Z, Gan M G and Zhang G 2011 *IET Control Theory and Applications* **5** 1895-905
[8] Fernandez-Gauna B, Ansoategui I, Etxeberria-Agiriano I and Grana M 2014 *Engineering Applications of Artificial Intelligence* **30** 107-117
[9] Wahyudi, Sato K and Shimokohbe A 2003 *Precision Engineering* **27** 157-169
[10] Sato K and Maeda G J 2009 *Precision Engineering* **33** pp 175-186
[11] Chong S H and Sato K 2010 *Precision Engineering* **34** 286-300
[12] Rozilawati M N and Chong S H 2013 *IEEE Int. Conf. on Control System, Computing and Engineering* 381-6
[13] Chong S H and Sato K 2010 *In Proceedings of ASPE Spring Topical Meeting on Control of Precision Systems*
[14] Chong S H, Hee W K, Norhaslinda Hasim and Sato K 2017 *Int. J. of Control and Automation* **10** Accepted

Q-switched double-clad Ytterbium-doped fiber laser using MoS_2 flakes saturable absorber

K. L. Tey[1*], M. F. A. Rahman[2, 5], A. H. A. Rosol[4], M. F. M. Rusdi[2], M. B. H. Mahyuddin[2], S. W. Harun[2*] and A. A. Latiff[3]
[1]Kumpulan Abex Sdn Bhd (167376-M), 40400 Shah Alam, Selangor, Malaysia.
[2]Department of Electrical Engineering, Faculty of Engineering, University of Malaya 50603 Kuala Lumpur, Malaysia.
[3]Photonic Research Centre, University of Malaya, 50603 Kuala Lumpur, Malaysia.
[4]Faculty of Electrical Engineering, Universiti Teknologi Mara, 40450 Shah Alam, Selangor.
[5]Faculty of Engineering Technology (FTK), Kampus Teknologi, Universiti Teknikal Malaysia Melaka, Hang Tuah Jaya, 76100 Durian Tunggal, Melaka, Malaysia.

E-mail: *kl_tey@kabex.com.my

Abstract. By sandwiching MoS_2 thin flakes between two fiber ferrule connectors in the ring cavity, we successfully demonstrate the passive Q-switched fiber laser operation. The cavity consists of all-fiber double-clad ytterbium-doped fiber (YDF) as a gain medium and MoS_2 tape as a saturable absorber (SA). The SA was prepared from a commercial MoS_2 crystal (~ 99.8 % purity) through mechanical exfoliation process. A piece of 1 mm x 1 mm tape contains MoS_2 thin flakes was sandwiched between two fiber connectors. By controlling loss and gain in the cavity, stable Q-switching operation were generated. The repetition rate was tunable from 4.8 kHz to 17.2 kHz as the pump power is varied from 753 mW to 1133 mW. The obtained pulse width is also narrowed from 21.3 μs to 10.44 μs with the same variation of pump power.

Keywords: Fiber lasers; pulsed laser; passive saturable absorber and mechanical exfoliation

1. Introduction

In the past decades, various passive saturable absorber (SA) have been developed and reported owing to its promising features for generating pulse laser [1]. For instance, semiconductor saturable absorber mirror (SESAM) was introduced in 1990s and it is widely used in commercial lasers [2]. However, SESAMs have disadvantages associated with narrow operational bandwidth and expensive in fabrication. Furthermore, its operation often limited to several picoseconds pulse width. Therefore, high motivation to find for better SA materials arises which lead to discovery of semiconductor nanocrystal quantum dots (QDs) zero-dimensional material (0D) in 1997 [3], followed by the invention of carbon nanotube (CNTs); one-dimensional material (1D) [4, 5, 6]. Later, graphene [7], a two-dimensional (2D) material sparked a new era of 2D SA for its zero bandgap, that essential for wide operational broadband. Discovery of topological insulators (TIs) [8], transition metal dichalcogenides (TMDs) and black phosphorus (BP) [9, 10] under 2D materials later on, have shown remarkable result for SA since its several layers combination form strong van der Waals heterostructures [11]. However, certain 2D materials have their deficiency. The TIs has limitation on its application in microelectronics since its carrier density is less tuneable due to its Fermi level that not sit well at the Dirac point. The graphene moreover, have very low modulation depth (typically <1% per layer), which associated with poor saturation absorption ability [12, 13]. BP is environmentally unstable and vulnerable when exposed to oxygen and water molecules [14]. The TMDs consist of 40-layered materials with a notable chemical configuration of MX_2. MoSe2, WS_2 and MoS_2 are useful in photonics and electronic application including photodetectors, logic devices and light emitters [11]. The advantage of TMDs is the bandgap can be tuneable from indirect 1.29 eV of bulk MoS_2 to direct 1.80 eV of monolayer MoS_2 [15]. This direct bandgap results in photoluminescence from monolayer MoS_2, which is good for optoelectronic applications [16]. In this paper, we demonstrate the generation of Q-switched pulse fiber laser using MoS_2 as SA.

2. Fabrication and Characterization of MoS_2 SA

The MoS_2-SA was formed by mechanical exfoliation process. This method is simpler and cheaper, where a very small piece MoS2-SA was peeled off from a bulk MoS_2 crystal (purity of 99.995 %). The piece was placed on a clear scotch tape and moderately pressed using an acrylic plate. Then, the 1 mm x 1 mm MoS_2 film was sandwiched between FC/PC fiber ferrules, adhered with index matching gel. The nonlinear optical respond property for MoS_2 was then investigated by a balance twin-detector measurement technique to determine the MoS_2-SA saturation absorption. A self-standing mode-locked fiber laser was used as input laser source. The transmitted power and also a reference power for normalization were recorded as a function of incident intensity on the tape by varying the input laser power. The data obtained as shown in Figure 1 where the MoS_2 absorption decreases with the increment of intensity. These data are fitted based on a simple two-level SA model;

$$\alpha(I) = \frac{\alpha_s}{1 + I/I_{sat}} + \alpha_{ns}$$

where $\alpha(I)$ is the absorption, α_s is the modulation depth, α_{ns} is the non-saturable absorption, I_{sat} is the saturation intensity, and I is the input intensity. From the analysed graph, the value of modulation depth, non-saturable absorption, and saturation intensity are obtained to be 10 %, 8 % and 0.001 MW/cm^2, respectively.

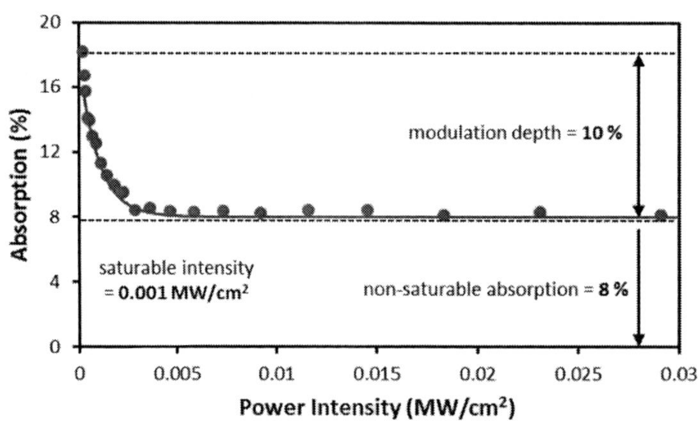

Figure 1. The nonlinear absorption profile of MoS_2

3. Experiment Setup

The ytterbium doped fiber laser (YDFL) setup is shown in Figure 2. The laser cavity comprises 10 m double-clad ytterbium doped fiber (YbDF) with core radius of 5.5 μm, cladding radius of 65 μm, numerical aperture of 0.46 and cladding absorption of 3.9 dB/m at 975 nm. The YbDF was pumped by 980 nm laser diode via a multimode coupler (MMC). A polarization controller (PC) was utilized to adjust the cavity birefringence and polarization state of the laser beam. The MoS2-SA was sandwiched between two FC/PC fiber ferrules and incorporated into the laser cavity. A 10 dB coupler was used to extract out 10% of intra-cavity power for measurement purpose while the rest 90% was retained in the cavity for spectral and temporal analysis. The total cavity length was measured to be 14.8 m. The optical spectra of the laser output was measured by an optical spectrum analyzer (Yokogawa AQ6370B) while the pulse train was recorded by an oscilloscope (LeCroy Wavejet 352A) via a photodetector (EOT ET-5010F). A radio frequency (RF) spectrum analyzer (Anritsu MS2683A) was used to measure the RF spectra.

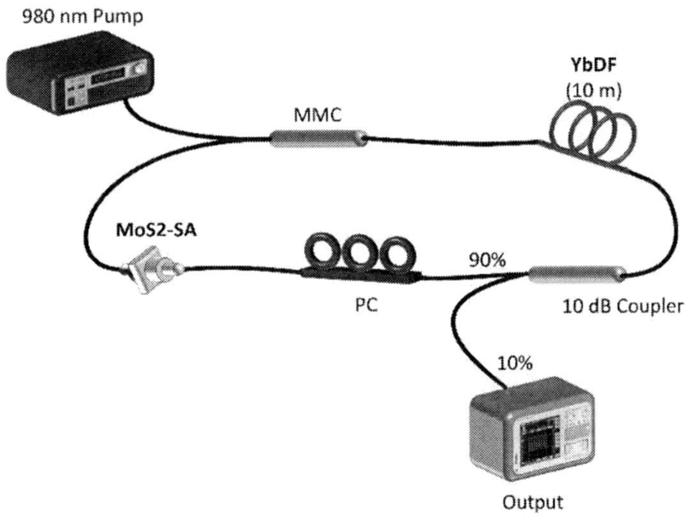

Figure 2. The experiment setup of Q-switched YDFL using MoS_2 SA.

4. Result and Discussion

Figure 3 shows the performance of Q-switched YDFL using MoS_2 saturable absorber. The stable Q-switched pulsed was achieved when the pump power is increased after the threshold pump power of 753 mW. The laser generates dual-wavelength output at 1086 nm and 1093 nm as shown in Figure 3(a). The Q-switched operation establishes a tuneable pulse repetition rate from 4.8 kHz to 17.2 kHz, following to the increment of pump power as shown in Figure 3(b). In addition, the pulse width became narrower from 21.3 μs to 10.44 μs with the increment of pump power. The typical pulse train of the Q-switched pulse as shown in Figure 3(c) represents pulse repetition rate of 17.2 kHz and pulse duration of 58 μs. No pulse train detected when the MoS_2-SA was removed, which confirmed the pulse generation was due to MoS_2-SA. Figure 3(d) shows the average output power that increases linearly with the increment of input pump power with slope efficiency of 9.48 %. The graph also shows the value of pulse energy obtained (4μJ) which is nearly constant regardless of the input pump power value.

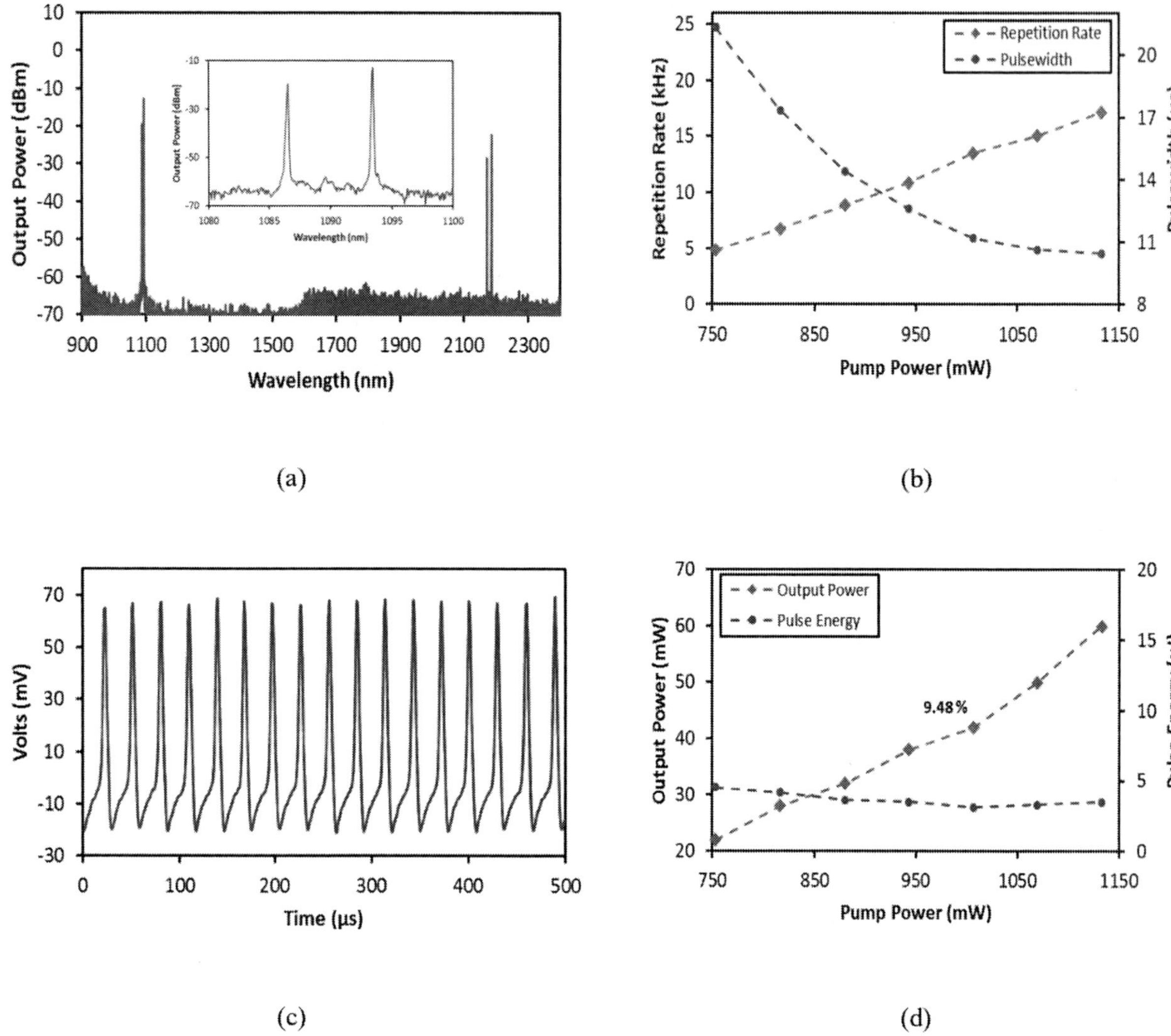

Figure 3. Laser performance. (a) Output spectrum of Q-switched double-clad YbDFL at pump power of 879 mW. (b) Repetition rate and pulse width of the Q-switched double-clad YbDFL. (c) Pulse train of the Q-switched double-clad YDFL. Insert is enlarge pulse train. (d) Output power and pulse energy of the Q-switched double-clad YDFL.

5. Conclusion

We successfully demonstrated a stable Q-switched pulse laser with dual-wavelength output operating at 1086 nm and 1093 nm in YDFL cavity using MoS2 as SA. The achieved tuneable repetition rate was measured to be between 4.8 kHz to 17.2 kHz at various pump power from 753 mW to 1133 mW. The obtained pulse width narrowed from 21.3 µs to 10.44 µs with the same variation of pump power.

Reference

[1] R. I. Woodward and E. J. Kelleher 2015 *Applied Sciences* **5** 1440-1456
[2] U. Keller, K. J. Weingarten, F. X. Kärtner, D. Kopf, B. Braun, I. D. Jung 1996 *IEEE Journal of Selected Topics in Quantum Electronics* **2** 435-453
[3] P. Guerreiro, S. Ten, N. Borrelli, J. Butty, G. Jabbour, and N. Peyghambarian 1997 *Applied Physics Letters* **7** 11595-1597
[4] R. H. Baughman, A. A. Zakhidov, and W. A. de Heer 2002 Science **297** 787-792
[5] S. Iijima and T. Ichihashi 1993 *Nature* **363** 603-305
[6] S. W. Harun, M. A. Ismail, F. Ahmad, M. F. Ismail, R. M. Nor, N. R. Zulkepely and H. Ahmad 2012 *Chin Phys. Lett.* **29** 114202
[7] M. A. Ismail, F. Ahmad, S. W. Harun, H. Arof and H. Ahmad 2013 Laser Phys. Lett. **10** 025102
[8] L. Fu and C. L. Kane 2008 *Physical review letters* **100** 096407
[9] Y. Chen, G. Jiang, S. Chen, Z. Guo, X. Yu, C. Zhao, et al. 2015 *Optics express* **23** 12823-12833
[10] M. Hisyam, M. Rusdi, A. Latiff, and S. Harun 2017 *IEEE Journal of Selected Topics in Quantum Electronics* **23** 1100205
[11] Q. H. Wang, K. Kalantar-Zadeh, A. Kis, J. N. Coleman, and M. S. Strano *Nature nanotechnology* **7** 699-712
[12] S. D. D. D. Cafiso, E. Ugolotti, A. Schmidt, V. Petrov, U. Griebner, A. Agnesi, et al. 2013 *Optics letters* **38** 1745-1747
[13] K. Janulewicz, A. Hapiddin, D. Joseph, K. Geckeler, J. Sung, and P. Nickles 2014 *Applied Physics A* **117** 1811-1819
[14] J. O. Island, G. A. Steele, H. S. van der Zant, and A. Castellanos-Gomez 2015 *2D Materials* **2** 011002
[15] K. F. Mak, C. Lee, J. Hone, J. Shan, and T. F. Heinz 2010 *Review Letters* **105** 136805
[16] A. Kuc, N. Zibouche, and T. Heine 2011 *Physical Review B* **83** 245213

IOP Conf. Series: Materials Science and Engineering **210** (2017) 012055 doi:10.1088/1757-899X/210/1/012055

1563 nm Q-Switched Brillouin-Raman fiber laser using Graphene as a saturable absorber

U. N. Zakaria[1,3*], Z. I. Rizman[1,2], A. H. A. Rosol[4], S. W. Harun[1*] and Z. Jusoh[2]
[1]Department of Electrical Engineering, Faculty of Engineering, University of Malaya, 50603 Kuala Lumpur, Malaysia
[2]Faculty of Electrical Engineering, Universiti Teknologi MARA, Terengganu Campus, Dungun Branch, 23000 Dungun, Terengganu, Malaysia
[3]Photonics Research Centre, University of Malaya, 50603 Kuala Lumpur, Malaysia
[4]Faculty of Electrical Engineering, Universiti Teknologi Mara, 40450 Shah Alam, Selangor, Malaysia

E-mail: *umi.namirah@yahoo.com

Abstract. This paper demonstrates a Q-switched Brillouin-Raman fiber laser (BRFL) operating at 1563nm. The BRFL uses a 7.7 km dispersion compensating fiber (DCF) as nonlinear gain medium and graphene as a saturable absorber. As the pump power is increased, the repetition rate increases from 20.80 kHz to 21.99 kHz while the pulse width decreased from 15.45 μs to 12.34 μs. The highest pulse energy is 1407.913 nJ. This result shows that the graphene SA has a high potential to be used for pulsed BRFL generation.

1. Introduction

Brillouin-Raman fiber lasers (BRFLs) have gained a tremendous interest in recent years, especially for operation at wavelengths that are difficult to access directly to conventional gain media. They are developed based on Stimulated Raman Scattering (SRS) effect. In the process, the amount of the Stokes frequency shift is intrinsically determined by the irradiated medium in which a quantum conversion increase proportionally with the complex part of the third-order nonlinear permittivity [1]. To date, many Brillouin or Raman fiber lasers have been fabricated and demonstrated using various gain media such as germane-silicate or phosphor silicate glass fiber. Most of the proposed lasers are operating in a CW mode with high output power [2-4]. On the other hand, Q-switched fiber lasers were also intensively investigated for many potential applications such as LIDAR, remote sensing, communication and medicine [5-6]. Compared with the active technique, the passive approach is better regarding simplicity, compactness, and flexibility of implementation [7]. The passive Q-switching generation can be realized by using either nonlinear polarization rotation or passive saturable absorber (SA) techniques.

To date, various SAs have been implemented such as semiconductor saturable absorber mirrors (SESAMs), two-dimensional (2D) materials (graphene, topological insulators, and transitions-metal dichalcogenides) and doped-fiber SA for Q-switching generation in various fiber lasers and cavities [8-11]. However, the applications of these SAs are restricted due to their drawbacks such as complex optical alignments, environmentally sensitive, limited operating bandwidth and complicated in fabrication. Therefore, there are many new attempts in recent years to develop new SAs with better performance for Q-switching laser operation.

In this paper, we have successfully demonstrated a Q-switched BRFL using a Graphene as a passive SA. The Graphene was sandwiched between two fiber ferrules via a fiber connector to form a fiber-compatible SA, which is then incorporated in BRFL cavity to generate Q-switched pulses at 1563 nm. As the pump power is increased, the repetition rate increases from 20.80 kHz to 21.99 kHz while the pulse width decreased from 15.45 μs to 12.34 μs. The highest pulse energy is 1407.913 nJ. This result shows that this ring laser cavity capable of generating stable Q-Switched BRFL. Hence, it has great potential to produce pulse laser operation in the next extended research of future employment.

2. Methodology

The experimental setup for Q-switched Brillouin-Raman Fiber Laser pump is depicted in Figure 1. The Graphene is inserted in the RFL cavity to act as SA for Q-switching pulse generation as shown in the figure. The fiber laser cavity employs a 7.7 km long dispersion compensating fiber (DCF) with a dispersion of 584 ps/nm-km as a nonlinear gain medium. The DCF was pumped by a 1455 nm laser via a 1455/1550nm wavelength division multiplexer (WDM). The Graphene SA was incorporated into the ring cavity between the 90/10 coupler and circulator which is connected to the Brillouin pump. The output signal was coupled out using a 90/10 output coupler which 10% of laser for spectrum, pulse, and power measurement and other 90% of the light is kept to oscillate in the cavity for both spectral and temporal diagnostics. The output laser characteristic was measured using an optical spectrum analyzer (OSA) with a spectral resolution of 0.02 nm and optical power meter. The temporal characteristics were measured using a 500 MHz oscilloscope via a 7.8 GHz radio-frequency (RF) spectrum analyzer via a 1.2 GHz InGaAs photodetector to convert photon energy to electrical energy from the output laser.

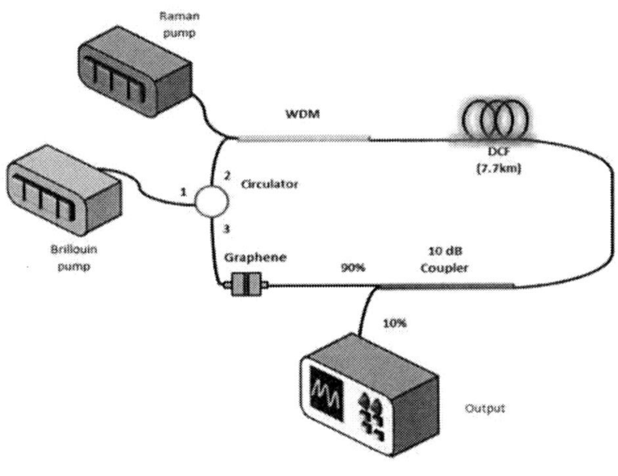

Figure 1. Experimental setup for the Q-Switched BRFL.

3. Results and discussion

The output spectrum generated by the proposed setup is shown in Figure 2(a). The output spectra of Brillouin-Raman fiber laser are obtained between 235 mA to 265 mA pump power. This is to show the results of output power (dBm) versus wavelength.The pulse repetition rate and pulse width of the proposed Q-switched BRFL are investigated as functions of pump power. Figure 2(b) shows the repetition rate of the Q-switching pulse train increased from 20.80 kHz to 21.99 kHz as the pump power is varied from 235 mA to 265 mA. Concurrently, the pulse width also decreased from 15.45 μs to 12.34 μs. The narrower pulse width can be achieved by either shortening the laser cavity length. Also, the average output power and the corresponding pulse energy of the laser are investigated at various pump powers. The Figure 2(c) shows the average output power almost linearly increased with the input pump power up to the maximum pump power. The calculated efficiency is around 0.04% corresponding to the output power. One can observe from the figure that the pulse energy fluctuates with the maximum pulse energy of 1407.913 nJ. The stability of the pulse presence is confirmed by RF spectrum as depicted in Figure 2(d). At maximum pump power, the repetition rate has obtained at 21.99 kHz with signal-to-noise ratio 31.35 dB. Thus, indicates the obtained Q-switching operation is stable and comparable. This Q-switching performance could be improved by further optimizing the cavity design.

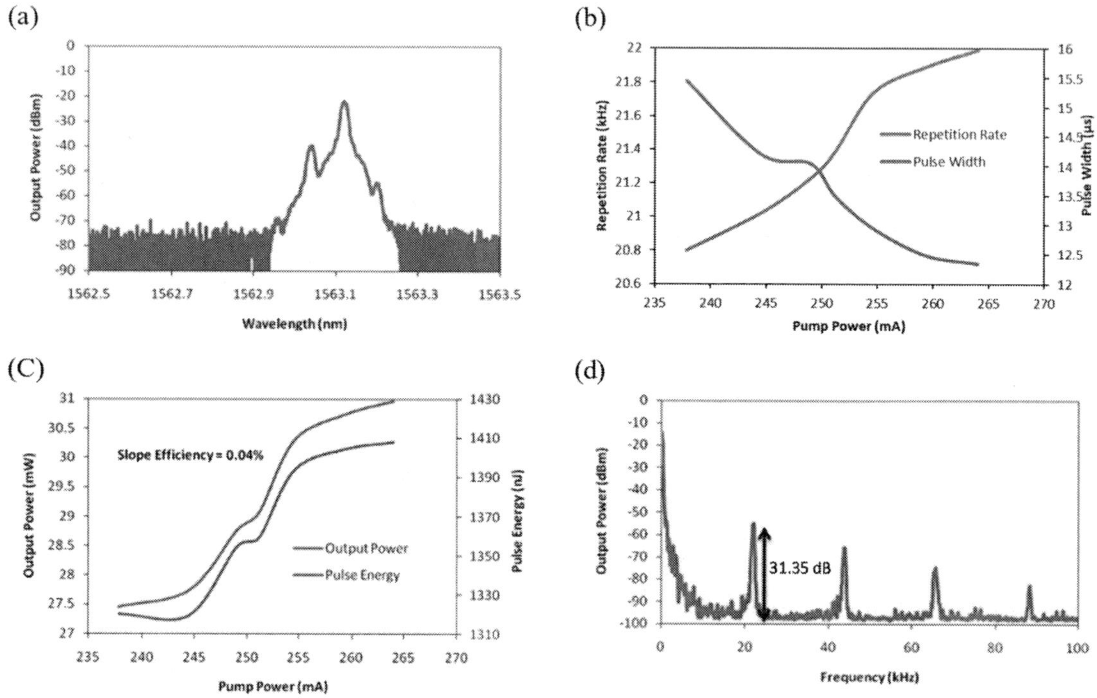

Figure 2. Q-switching performance of the BRFL (a) Output spectrum. (b) Repetition rate and pulse width against pump power (d) Output power and calculated pulse energy against pump power (d) RF spectrum with 100 kHz span.

4. Conclusion

We have successfully demonstrated a Q-switched BRFL using a graphene embedded in polymer film as a passive SA. The graphene film was sandwiched between two fiber ferrules via a fiber connector to form a fiber-compatible SA, which is then incorporated in BRFL cavity to generate Q-switched pulses at 1563 nm. As the pump power is increased, the repetition rate increases from 20.80 kHz to 21.99 kHz while the pulse width decreased from 15.45 µs to 12.34 µs. The highest pulse energy is 1407.913 nJ. This result shows that this ring laser cavity capable of generating stable Q-Switched BRFL with comparable performances.

Acknowledgements
This work is financially supported by Ministry of Higher Education Grant Scheme (FRGS/1/2015/SG02/UITM/03/3).

References

[1] Y. Zhang, Y. Tian, W. Wang, and B. Yao 2010 *Laser Phys. Lett.* **7** 225
[2] M. R. Shirazi, S. W. Harun, M. Biglary and H. Ahmad 2008 *Opt. Lett.* 33 770
[3] R. Parvizi, H. Arof, N. M. Ali, H. Ahmad and S. W. Harun 2011 *Opt & Laser Technol.* 43 866
[4] H. Ahmad, M. Z. Zulkifli, M. H. Jemangin and S. W. Harun 2013 *Laser Phys. Lett.* 10 055102
[5] N. M. Fried and K. E. Murray 2005 *J. Endourol.* **19** 25–31

[6] R. Szlauer, R. Götschl, A. Razmaria, L. Paras, and N. T. Schmeller 2009 *Eur. Urol.* **55** 368–375
[7] S. W. Harun, M. A. Ismail, F. Ahmad, M. F. Ismail, R. M. Nor, N. R. Zulkepely and H. Ahmad 2012 *Chin Phys. Lett.* **29** 114202
[8] M. A. Ismail, F. Ahmad, S. W. Harun, H. Arof and H. Ahmad 2013 Laser Phys. Lett. **10** 025102
[9] H. Haris, S. W. Harun, A. R. Muhammad, C. L. Anyi, S. J. Tan, F. Ahmad, R. M. Nor, N. R. Zulkepely and H. Arof 2017 *Opt & Laser Technol.* 88 121
[10] H. Hisamuddin, U. N. Zakaria, M. Z. Zulkifli, A. A. Latiff, H. Ahmad and S. W. Harun 2016 *Chin Phys. Lett.* **33** 074208
[11] Rahman, M. F. A., Rusdi, M. F. M., Lokman, M. Q., Mahyuddin, M. B. H., Latiff, A. A., Rosol, A. H. A., Dimyati, K. & Harun, S. W. 2017 *Chinese Physics Letters* **34** 054201

Practical controller design for ultra-precision positioning of stages with a pneumatic artificial muscle actuator

T F Tang and S H Chong
Centre for Robotics and Industrial Automation, Faculty of Electrical Engineering, Universiti Teknikal Malaysia Melaka, Hang Tuah Jaya, 76100 Durian Tunggal, Melaka, Malaysia.

E-mail: horng@utem.edu.my

Abstract. This paper presents a practical controller design method for ultra-precision positioning of pneumatic artificial muscle actuator stages. Pneumatic artificial muscle (PAM) actuators are safe to use and have numerous advantages which have brought these actuators to wide applications. However, PAM exhibits strong non-linear characteristics, and these limitations lead to low controllability and limit its application. In practice, the non-linear characteristics of PAM mechanism are difficult to be precisely modeled, and time consuming to model them accurately. The purpose of the present study is to clarify a practical controller design method that emphasizes a simple design procedure that does not acquire plants parameters modeling, and yet is able to demonstrate ultra-precision positioning performance for a PAM driven stage. The practical control approach adopts continuous motion nominal characteristic trajectory following (CM NCTF) control as the feedback controller. The constructed PAM driven stage is in low damping characteristic and causes severe residual vibration that deteriorates motion accuracy of the system. Therefore, the idea to increase the damping characteristic by having an acceleration feedback compensation to the plant has been proposed. The effectiveness of the proposed controller was verified experimentally and compared with a classical PI controller in point-to-point motion. The experiment results proved that the CM NCTF controller demonstrates better positioning performance in smaller motion error than the PI controller. Overall, the CM NCTF controller has successfully to reduce motion error to 3μm, which is 88.7% smaller than the PI controller.

1. Introduction
Nowadays, the human sciences and technology are getting more advanced and widespread. The usage and number of robotic and automation applications are increasing due to the development of advanced technologies. Linear positioning systems are the fundamental mechanisms that play important roles in the industrial machines such as machine tools, semiconductor manufacturing system and robotics. For robotic applications, the system performance is characterized by constraints such as power to weight ratios, response rate, physical size, speed of motion, realiability, controllability, compliance, and cost [1]. Those applications have required an actuation system as function in providing force, torque or mechanical motion, in order to drive the motion of the system. Typically, a robot system is designed extremely stiff and built with strong joint actuators, in order to overcome threshold friction, position-dependent gravity and payloads [2]. However, the robot systems are high cost and high energy consumption due to their heavy construction. Besides, their large strength are too dangerous to use in environment, where they are operating by human.

Pneumatic actuators are traditional actuators and have numerous advantages, such as simple structure, low component cost, high power-to-weight ratio, ease of maintenance, high safety and many more. Therefore, these advantages have led to the development of a novel actuator, namely McKibben pneumatic artificial muscle (PAM), which was originally inverted in the 1950s by the American physician, J. L. McKibben. PAM is a mechanical apparatus that duplicates the behaviour of skeletal muscle, where it contracts and generates a pulling force in a non-linear way once it is activated [3]. This PAM consists of an inflatable rubber tube sheathed by a braided mesh. When pressurized, the rubber tube tends to increase in volume. Due to the non-extensibility of the thread in the braided mesh, it restricts the rubber tube to expand in radius but shortens axially instead [4]. The pneumatic energy in the pressurized air is thus transformed into mechanical work. Since PAM only has one active action which is contraction, thus it must always be operated in antagonistic actions. These antagonistic actions can either involve two active PAMs or a single active PAM coupled to a spring.

The physical configuration of PAM has variable-stiffness spring like characteristics, and it is very light weight compared to other types of actuators [5]. It is a promising actuator in robotics since its driving force is large despite its light weight. The power-to-weight ratio of PAM is about five times higher than the ratio of electric motors or hydraulic actuators. Besides, PAM can exert the force about ten times of a comparably sized pneumatic cylinder [6]. Thus, PAM actuators are safe to use and have numerous advantages which have brought these actuators to wide applications, especially in rehabilitation and welfare devices. For example, they have been applied in the power assist devices [7–10], robotics [11], medical applications [12] and industrial machinery [13]. However, the PAM is exhibits strong non-linear characteristics [14], low damping ability, hysteresis [15] and creep phenomenon [16]. These limitations lead to a low controllability and a high difficulty in achieving the precision system control and limit its application. Furthermore, the air compressibility and lack of damping ability of PAM mechanism causes dynamic delay to the pressure response, which will results in oscillatory motion. As a result, it is not easy to realize the motion with high accuracy and high speed.

An enhancement of the PAM in positioning accuracy is desirable due to its advantages, and then extend its applications in the precision positioning system. For example, the precise motion of PAM allows nursing care robot to provide disabled people with delicate supports such as injection, rehabilitation support and transfer assist. In the industrial, the precision control of PAM allows to create a precise assembly robot which is powerful and lightweight as well as safer to assemblers who work with the robots. As compared with hydraulic and pneumatic cylinders, PAM has superior characteristics when applied to welfare devices, thus the development of PAM with precision motion could be a promising advancement for future neurosurgical robots [17].

In recent years, numerous control methods have been proposed to control the motion of the PAM mechanisms. These methods can be classified in three categories which are classical proportional-integral-derivative (PID) control, non-linear model-based control and intelligent control. Classical PID control is the most popular ones and often used for positioning systems. The PID control is easy to design, but it is not robust to the change of parameters and insufficiently to compensate for the non-linearity of the PAM mechanism which lead to poor accuracy. Caldwell *et al.* [18] proposed a feed-forward PID controller for controlling a pneumatic muscle. The result showed that the performance was quite sensitive to errors in the feed-forward term, due to fluctuations of supply pressure, temperature and muscle characteristics. In [19–21], the authors reported that the motion accuracy produced by classical PID controller is 0.1° for rotational positioning. For linear PAM mechanism, the highest positioning accuracy produced by a PID controller is 200 μm [19]. The classical PID controller does not solve the hysteresis problem.

Non-linear model-based control is a popular approach that applied to the PAM mechanisms. Due to inaccurate modeling, the positioning accuracy for the PAM mechanisms is on the order of a micrometer [22]. Sliding mode controller (SMC) had been applied to the PAM mechanisms in [23, 24]. However, the simulation results were presented only for the effectiveness of the strategy. Tondu and Lopez proposed a sliding mode controller using a non-linear length-tension model of the PAMs [25]. Besides, Li *et al.* [26]

proposed a non-linear dynamic force model that generated by a PAM, which was achieved 6 μm as the highest positioning accuracy. Different modified non-linear model-based controllers have been proposed in the literature, but the controllers were managed to compensate the positioning error on the order of a micrometer due to the inaccurate modeling. In practice, the non-linear characteristics of the PAM mechanism are difficult to be precisely modeled, and time consuming to model them accurately. As a result, it is difficult to design and impracticable for the engineers who are unfamiliar and insufficient knowledge of the control theory with the controller design that requires the mechanism modeling. Intelligent control has also been widely used for the PAM mechanism. Hesselroth *et al.* [27] proposed a Kohonen-type neural network learning algorithm to control a five-joint pneumatic robot arm and gripper through feedback from two video cameras as visual observation of real and desired end-effector position. However,the Kohonen neural network algorithm is compressing a large data set, and the learning sessions are very time consuming. Therefore, the system has limited practical applicability. Chan *et al.* [28] proposed a fuzzy PD+I learning control that has good results with accurate positioning of the pneumatic muscle after a few seconds of operation. However, some limitations still exist because the necessary time for learning is quite long and the controller output functions properly after about 30 to 45 seconds according to the input signal. Furthermore, Ahn *et al.* [29] proposed an intelligent switching control method to adjust the gains of PID controller using neural networks. This method shows the most accurate positioning result with a rotational positioning error of 0.05° for the intelligent control of the PAM mechanism. The neural network controller with learning capabilities provide an efficient method to satisfy the positioning requirements. However, intelligent control has a complex controller design and no general systematic approach to choose the neural control structure, which require sufficient knowledge of the intelligent algorithm when it applied for the PAM mechanisms. Regardless the mechanical complexity, the control design method should still be easy and straightforward. Thus, a practical controller, which has a simple design procedure and capable to demonstrate high robustness with the changes of plant parameters and positioning accuracy performance is desired to overcome the stated problems. This is always the solution that needed in the industry. As mentioned before, the PAM need to be operated in antagonistic actions. In this research, a PAM mechanism is constructed with a mover that is located in the middle of two PAMs in an opposing pair configuration. The purpose of the present study is to clarify a practical controller design method that emphasizes a simple design procedure that does not acquire plants parameters modeling, and yet is able to demonstrate ultra-precision positioning performance for a PAM driven stage. The effectiveness of the proposed controller is verified experimentally and compared with a classical PI controller in point-to-point (PTP) motion. The rest of this paper is organized as follows. Section 2 describes the experimental setup and provides a brief review of the characteristic of the PAM driven stage. Section 3 introduces the CM NCTF control concept and the proposed design procedure. In Section 4, the PTP positioning control performances of the CM NCTF controller are evaluated and compared with the PI controller. The conclusion of the present study is remarked in Section 5.

2. Characteristics of PAM driven stage

2.1. Experimental setup

An experimental setup is designed and constructed as a linear antagonistic structure using two PAMs and a mover is located in between the two PAMs in a horizontal motion, as shown in Figure 1. The two PAMs (FESTO DMSP-10-150N-RM-CM) generate pulling forces to push and pull the 2 kg mover along the horizontal moving direction in a maximum working range of ±3.2 mm. As an input source, air is injected from the pressure supply with a pressure of 0.5 MPa and controlled by a 5-port 3-way proportional servo valve (FESTO MPYE-5-1/8LF-010-B). The pressures in the two PAMs are measured using two pressure sensors (SMC PSE540A-01) with the resolution of 0.0012 MPa, which are not used as feedback sensors for control but are instead used to observe the air pressures. A linear encoder with the resolution of 0.1 μm is used as a single feedback sensor in this mechanism to measure the displacement of the mover. For the signals processing, a data acquisition unit is used to interface with a host computer that installed MATLAB/Simulink software. The sampling time, T_S is 0.1 ms.

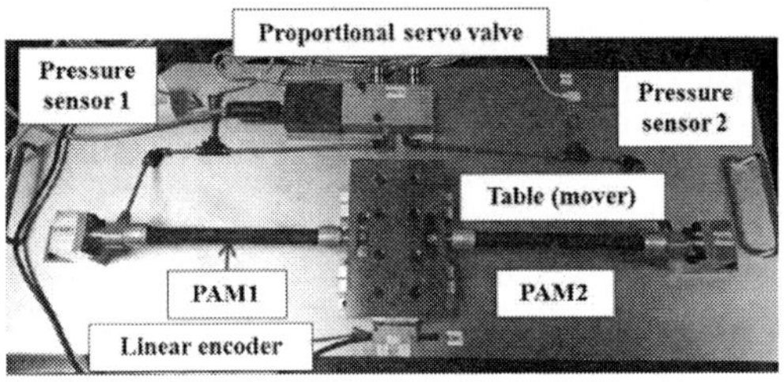

Figure 1. PAM driven stage.

2.2. Static relationships

Based on the experimental open-loop responses, the static relationships between input voltage, pressures of PAM1 and PAM2, and displacement of the mover are summarized. Figure 2 presents the static relationship between input voltage and pressures of PAM1 and PAM2; the static relationship between input voltage and displacement. The result shows that the system is high non-linearities.

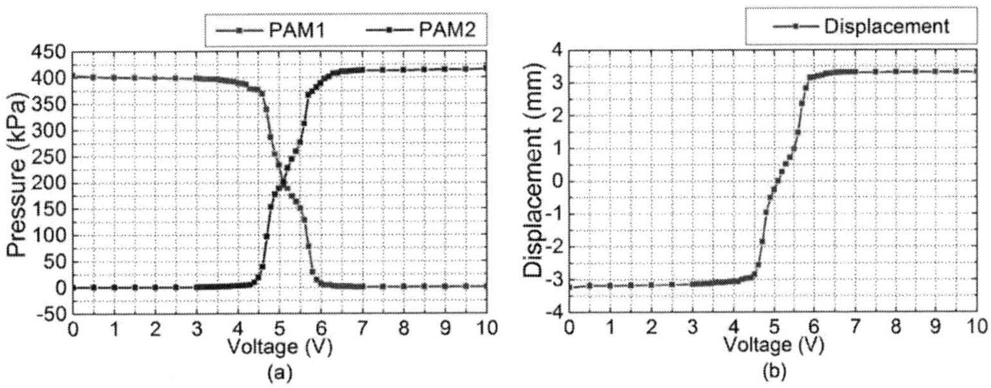

Figure 2. The static relationships between (a) input voltage and pressures of PAM1 and PAM2, and (b) input voltage and displacement.

As shown in Figure 2(a), a small change of an input voltage within 4.5 V to 5.8 V, it gives a large different in the pressure of both PAMs. However, the pressures from 0 V to 3 V and 7 V to 10 V have slight changes with respect to the large changes of input voltage. The ratio of output pressure over input voltage is very small. Also, the working range is saturated around ±3.2 mm as can be found in Figure 2(b). These are due to the antagonistic structure of PAM mechanism. The relaxed PAM becomes stiff when it has been pulled over the elasticity limit, and the contractile force could not against it with the supplied pressure. Besides, the result of the displacement is non-linear between the contraction of the PAM1 and PAM2 due to the hysteresis and non-linear friction force. The maximum pressure of the contracted PAM is about 390 kPa rather than 500 kPa (supply pressure), due to the losses caused by the friction inside the tubing and leakage from the servo valve when the pressurized air transferred from the compressor to the PAM.

2.3. Dynamic model

Figure 3 shows the free body diagram of PAM mechanism. The initial position of the mover is in the middle position, 0 mm. When the contraction of PAM1, the mover is moving to the left direction with negative value of displacement. The displacement of the mover is then considered as positive value when PAM2 is contracting. The PAM is stated as variable-stiffness spring-like characteristics due to its physical structure. The motion equation of the PAM mechanism can be written as Equation 1.

$$M\ddot{x} + B\dot{x} + Kx = F_2 - F_1 - F_{friction} \tag{1}$$

Where M is the mass of the stage, B is the damping coefficient, and K is the spring coefficient. The generated forces of PAM1 and PAM2 are stated as F_2 and F_1, respectively. $F_{friction}$ represents the friction force between the stage and the linear guide.

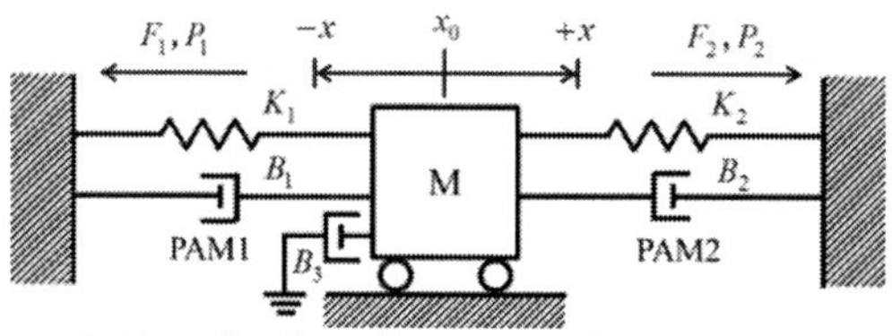

Figure 3. Free body diagram of PAM mechanism.

3. Controller design

3.1. CM NCTF control structure

Continuous Motion Nominal Characteristic Trajectory Following (CM NCTF) controller is an improved control strategy of Nominal Characteristic Trajectory Following (NCTF). The CM NCTF controller has been modified to improve the motion control performances in tracking motion and contouring motion, while it is maintaining point-to-point (PTP) positioning that offered by the NCTF controller [30]. The main contribution of the CM NCTF controller is emphasizes a simple and easy design procedure, in order to achieve the promising results in both positioning motion and continuous motion controls. As an advantage, this controller does not require an exact model of the plant and its parameters, which makes it easy to design, understand and adjust. This makes it is user-friendly to engineers who are lack of control theory knowledge to handle.

Figure 4 shows the structure of the CM NCTF control system. The CM NCTF controller consists of two main parts which are a Nominal Characteristic Trajectory (NCT) and PIcompensator. The NCT is the reference motion trajectory of the control system that used to restrict the motion of the mechanism along the trajectory, and it is expressed on phase plane. The NCT is constructed from the actual responses such as displacement and velocity responses of a mechanism in open-loop condition, which the characteristic of the mechanism that influenced by the friction and saturation effects are captured. The determined NCT represents the deceleration motion of the mechanism in positioning, which it is significantly affect the positioning performance. Subsequently, the PI compensator controls the object motion to follow the determined NCT, and further end the motion at the origin of the phase plane. The PI compensator works for reduction of the difference between the NCT and the actual motion when the difference is increased by the disturbance forces and mechanism characteristic changes. The PI compensator is designed based on the stability of the system.

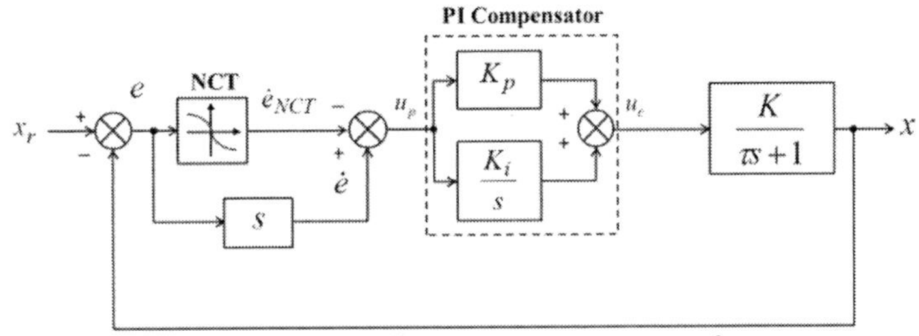

Figure 4. Structure of the CM NCTF control system.

The conventional CM NCTF controller has demonstrated a higher positioning performances for electric-motor driven mechanism [31], non-contact mechanism and contact mechanism [32]. However, the response of the control system exhibits oscillation in the steady-state response when applied in the PAM mechanism. This is due to low damping characteristic of the constructed PAM driven stage causes severe residual vibration that deteriorates motion accuracy of the system. Therefore, the idea to increase the damping characteristic by having an acceleration feedback compensation to the plant has been proposed. An acceleration feedback is then added to increase the damping effect of the PAM mechanism, in order to eliminate the oscillation and achieve better accuracy. Figure 5 presents the structure of CM NCTF control system with the acceleration feedback.

$$G_{po}(s) = \frac{K}{KK_a s^2 + \tau s + 1}$$

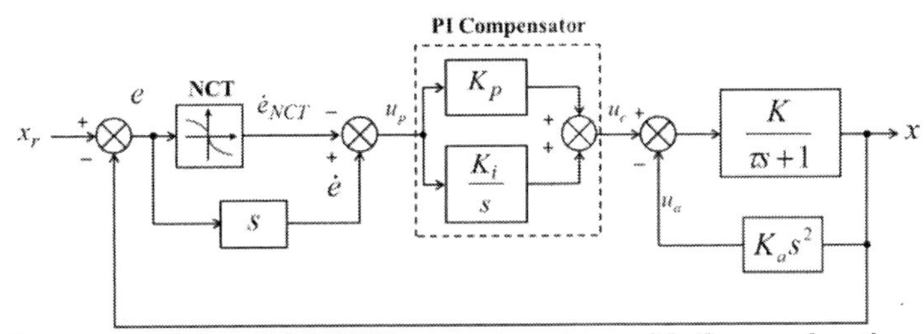

Figure 5. Structure of the CM NCTF control system with the acceleration feedback.

The linearized open-loop transfer function of the PAM mechanism, G_{pon}can be obtained as shown in Equation 2. The plant model is considered as second-order system. Thus, the acceleration feedback is expected to be increasing the damping effect and enhanced the accuracy effectively.

3.2. Design procedure

The main consideration of the CM NCTF control design procedure is simple and easy to design, which the detailed dynamic model and its exact parameters are not required. In this paper, the CM NCTF with an acceleration feedback is designed and applied in the PAM driven stage. Generally, the design procedure of the controller is comprised four major steps:

(1) Determination of open-loop responses.

The mechanism is executed with an open-loop step input, and its displacement and velocity are then measured. Wahyudi *et al.* [33] stated that the amplitude of input is chosen 80% of the rated input to the actuator, which the overall system dynamics are included in the measured responses. Thus, a step input of 4 V is used to measure the PAM mechanism. Figure 6 presents the open-loop responses of the PAM mechanism.

IOP Conf. Series: Materials Science and Engineering **210** (2017) 012056 doi:10.1088/1757-899X/210/1/012056

(2) Construction of NCT.

The NCT is constructed on the phase plane using the displacement and velocity of the mechanism during deceleration. The obtained responses in the open-loop condition during the deceleration are directly selected to construct the NCT, and without further do the mathematics derivation for the dynamics of the mechanism. The NCT works as virtual reference error rate, with the inclination at origin is referred as β. Based on the obtained responses, the NCT can be constructed as shown in Figure 7. The inclination at origin, β in this case is 208 s^{-1}.

(3) Design of PI compensator.

The PI compensator is designed based on the open-loop responses and NCT information. In order to consider the real characteristic of the mechanism in designing the PI compensator, a practical stability limit is determined as the margin in selecting the PI gains that bounded under the stable region of the actual mechanism. The discussion about the determination of practical stability limit and choice of the design parameters could be found in Section 3.3.

(4) Determination of acceleration gain.

The acceleration gain, Ka is determined experimentally for sufficient damping characteristics using a mechanism that embedded the CM NCTF controller with an optimum PI compensator. The fined-tuned Ka is 6×10^{-6} Vs^2/mm.

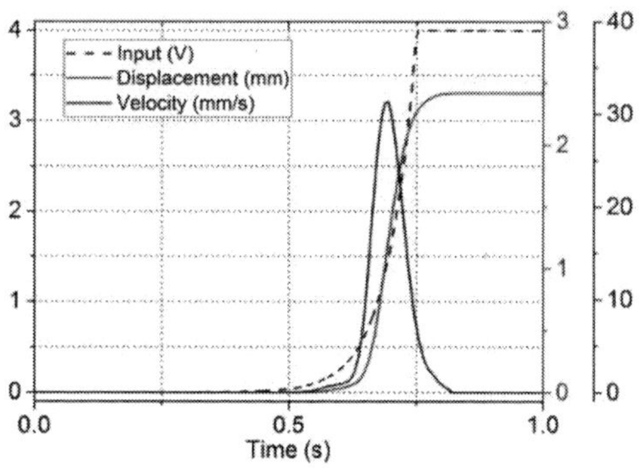

Figure 6. Open-loop responses of the PAM mechanism.

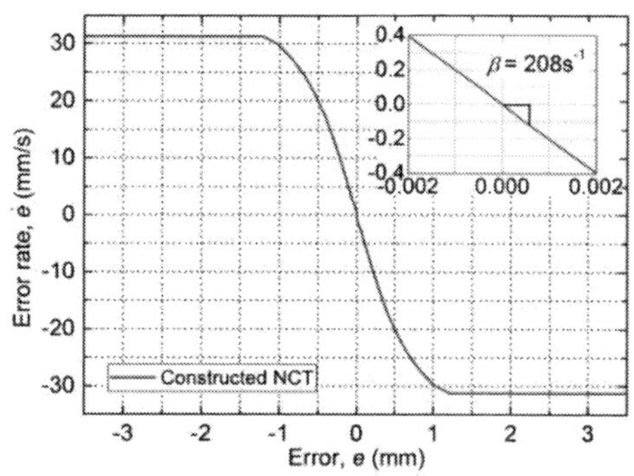

Figure 7. Constructed NCT from open-loop responses.

The practical stability limit is used to choose the PI gains within the stable operation region which can be easily calculated independently of the actual mechanism characteristic. Few parameters have to determine first before find the practical stability limit. In order to estimate the parameters of plant model, the normalized of the open-loop step responses of the PAM mechanism under different step input voltages are measured. Figure 8 and Figure 9 show normalized open-loop step responses and its average step response, and an estimated model respectively.

$$G_{closedloop}(s) = \frac{2\zeta\omega_n s + \omega_n^2}{s^2 + 2\zeta\omega_n s + \omega_n^2} = \frac{KK_p s^2 + (\beta KK_p + KK_i)s + \beta KK_i}{(KK_p + \tau)\left[s^2 + \left(\frac{\beta KK_p + KK_i + 1}{KK_p + \tau}\right)s + \left(\frac{\beta KK_i}{KK_p + \tau}\right)\right]} \quad (2)$$

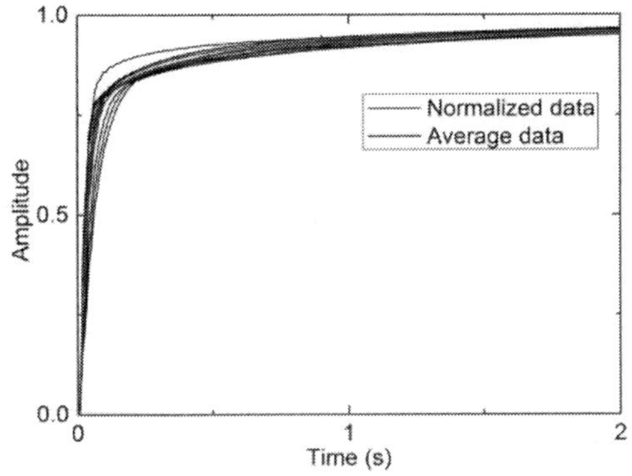

Figure 8. Normalized open-loop step responses and an average response.

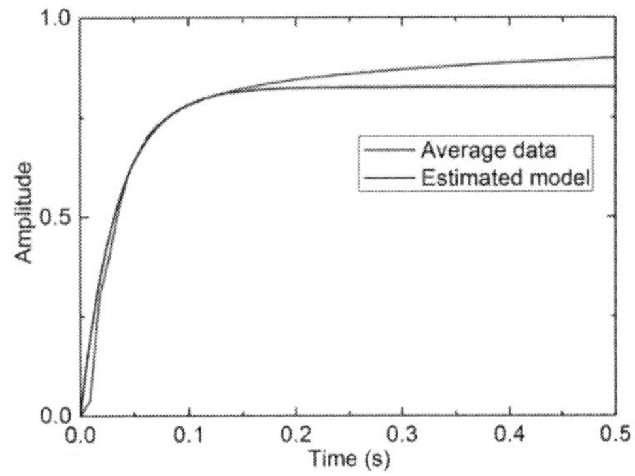

Figure 9. The average response and an estimated first-order model.

An estimated first-order model of the mechanism is based on a linear macrodynamic model expressed as:

$$\zeta_{practical} = \frac{\beta^2 K K_{pu} + K K_{pu}\omega_n^2 + \tau\omega_n^2 + \beta}{2\omega_n K K_{pu}\beta + 2\omega_n \tau \beta} \quad (3)$$

$$K_p = \frac{\tau\omega_n^2 + \beta - 2\zeta\omega_n\beta\tau}{2\zeta\omega_n\beta K - \beta^2 K - K\omega_n^2} \quad (4)$$

$$K_i = \frac{K K_p \omega_n^2 + \tau\omega_n^2}{\beta K} \quad (5)$$

where K is the system gain and τ is the time constant. The creep phenomenon is happened after the final state, which the estimated linear model could not model it. However, the estimated model is sufficient to represent the characteristic of PAM driven stage. Based on the CM NCTF control system as shown in Figure 4, the closed-loop transfer function is expressed as second- order system in Equation 4. The determined inclination at origin, β is 208 s^{-1} from the NCT.

The practical stability limit of the actual mechanism is found by driving the mechanism with the CM NCTF controller using only the proportional element. The value of proportional gain is increased until continuous oscillations which indicates the existence of instability. However, the continuous oscillations is not applicable to the PAM mechanism, due to it is a first-order system. Thus, the maximum proportional gain is decided when the PAM mechanism is vibrating. The maximum proportional gain is referred as an actual ultimate proportional gain, K_{pu} (0.0018Vs/mm in the case of the PAM mechanism). Based on Equation 4, the practical stability limit, $\zeta_{practical}$, proportional gain, K_p and integral gain K_i are derived as follows:

$$G_{po}(s) = \frac{K}{\tau s + 1} = \frac{0.826}{0.034s + 1}$$

For the maximum natural frequency, ω_n of the PAM mechanism, it is determined through the open-loop frequency response and the break frequency is 17.7 rad/s at -3 dB. Therefore, the range of ω_n in the practical stability limit is set from 1 rad/s to 16 rad/s, which the limit is slightly lower than the break frequency. By varying the ω_n, the practical stability limit is constructed using Equation 6. The points are defined by ζ and $\omega_n T_s$ (T_s = 0.0001 s) as presented in Figure 10. The PI compensator is selected within the stable region. From the stability limit (safety margin), the PI compensator is choose 90% of the value of $\zeta_{practical}$, which is 10% of safety margin.

Four points are selected to verify the positioning performances as shown in Figure 11. Table 1 details the parameters of the selected PI compensator. Step input of 2 mm is used as reference input for the PAM mechanism. Figure 12 shows the positioning performances of P1, P2, P3 and P4. P3 is the optimum PI parameters (K_p = 0.0142 Vs/mm and K_i = 0.0384 Vs^2/mm) which demonstrates faster transient response compared to P1 and P2. However, P1 encounter vibration during the transient because it is too fast for the PAM mechanism to react the 2 mm step amplitude. The problem is not occurred in the smaller working range. This is the reason of choosing 2 mm as the reference input rather than 1 mm or smaller working range. Overall, the design procedure of the CM NCTF controller has shown the simplicity and straightforward in determine the required parameters which can be measured directly from the mechanism.

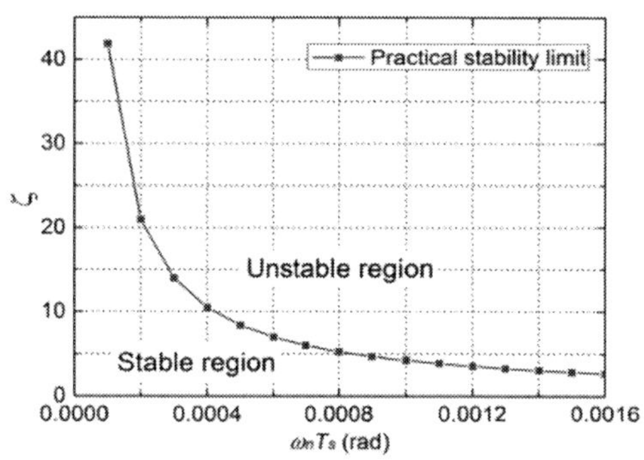

Figure 10. Relationship between stability, ζ and ω_n, and practical stability limit.

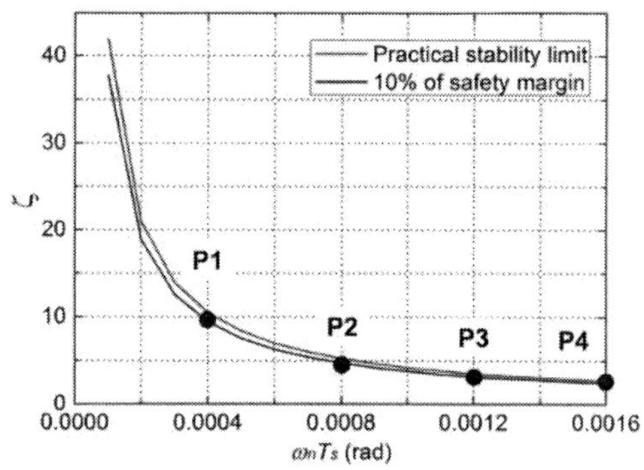

Figure 11. Four different points of PI compensator on 10% of safety

Table 1. Parameters of the selected PI compensator.

	ω_n (rad/s)	$\omega_n T_s$ (rad)	ζ	90% of ζ	K_p (Vs/mm)	K_i
P1	4	0.0004	10.478	9.4302	0.0143	0.0043
P2	8	0.0008	5.2534	4.7281	0.0143	0.0171
P3	12	0.0012	3.5183	3.1665	0.0142	0.0384
P4	16	0.0016	2.6556	2.3900	0.0142	0.0682

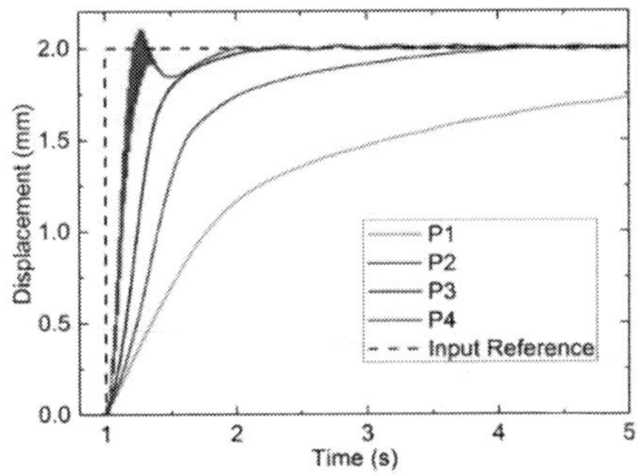

Figure 12. Positioning performances of P1, P2, P3 and P4.

4. Experimental result

In this section, different positioning performances are experimentally examined. The effectiveness of the CM NCTF controller is evaluated in comparison with the PI controller. The PID controller is not chosen instead of the PI controller because the PID controller does not show a significant improvement in the positioning performance for the PAM mechanism. The CM NCTF controller is designed through the procedure as shown in Section 3.2. The NCT shown in Figure 7 and controller parameters designed in Section 3.3 are used. The PI controller which is used to compare the CM NCTF controller is designed to have an optimum performance. The fine-tuned PI controller gains are K_P = 0.4 V/mm and K_i = 3.5 Vs/mm under a reference input of 2 mm. Figure 13, Figure 14 and Figure 15 present the PTP positioning performances comparison between the CM NCTF controller and the PI controller with 0.1 mm, 1 mm and 2 mm respectively.

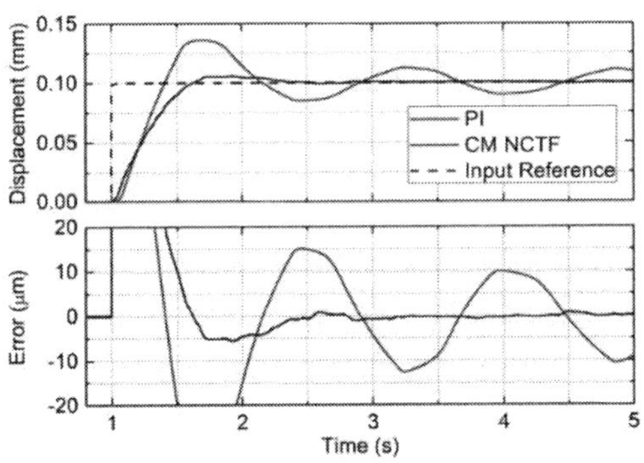

Figure 13. PTP positioning performance comparison in 0.1 mm between the CM NCTF controller and the PI controller.

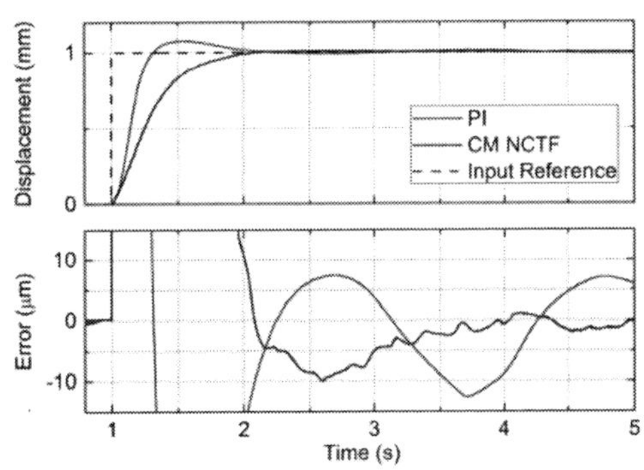

Figure 14. PTP positioning performance comparison in 1 mm between the CM NCTF controller and the PI controller.

The positioning result of the CM NCTF controller differ considerably to the PI controller. The result of both controllers is summarized in Table 2. The CM NCTF controller demonstrates a better positioning performance compared to PI controller. In overall, the CM NCTF controller has achieved the positioning accuracy within 3 μm of steady-state error. The CM NCTF controller also demonstrates the consistency of positioning accuracy in different step height.

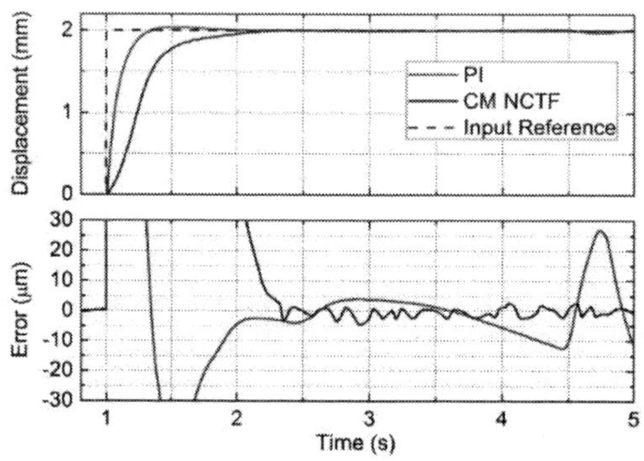

Figure 15. PTP positioning performance comparison in 2 mm between the CM NCTF controller and the PI controller.

In contrast to the PI controller, the CM NCTF controller shows 88.7% smaller steady-state error as an average. As observed from Table 2, the huge overshoot from the PI controller is increasing when the reference step input is reduced. By contrast, the CM NCTF controller exhibits smaller and acceptable overshoot than the PI controller in all the working range. However, the CM NCTF controller offers longer rise time and settling time as compared to the PI controller. This is due to the reference motion trajectory from the NCT and the characteristic of the CM NCTF controller. As aforementioned, the NCT is used to restrict the motion of the mechanism. Thus, it will help in reducing the overshoot, which is found lack in the PI controller. Overall, it can be concluded that, the PI controller is lack of adaptability to precision motion control in comparison with the CM NCTF controller.

Table 2. PTP positioning performances of the CM NCTF controller and the PI controller.

Step height	Performance index	PI controller	CM NCTF controller	Improvement
0.1 mm	Rise time	0.27 s	0.44 s	-
	Settling time	1.13 s	1.29 s	-
	Overshoot	35.90%	5.60%	84.40%
	Steady-state error	15.0 μm	1.1 μm	92.67%
1 mm	Rise time	0.19 s	0.56 s	-
	Settling time	0.98 s	1.08 s	-
	Overshoot	7.53%	1.00%	86.72%
	Steady-state error	12.7 μm	2.1 μm	83.46%
2 mm	Rise time	0.19 s	0.44 s	-

Settling time	0.58 s	1.34 s	-
Overshoot	2.04%	0.18%	91.18%
Steady-state error	26.9 μm	2.7 μm	89.96%

5. Conclusion

In the present paper, the CM NCTF controller was proposed as a practical controller design method for ultra-precision positioning. The CM NCTF controller was applied to the PAM driven stage for evaluation of its effectiveness. The controller structure and design procedure are simple and straightforward. The controller parameters were determined using the practical stability limit which makes the tuning procedure more practical than manual tuning. The PTP positioning performances with the CM NCTF controller were evaluated based on the experimental results. The performances with the CM NCTF controller were then compared with the PI controller. Overall, the results prove that the performances obtained using the CM NCTF controller is better than those obtained using the PI controller. The CM NCTF controller has successfully reduced the motion error to 3 μm, which is 88.7% smaller than the PI controller. Besides, the simplicity in design procedure of the CM NCTF controller showed the capability in performing high precision motion and promising results in positioning control. As the future work, the performance of the CM NCTF controller will improve in the transient response and evaluate in the tracking performance.

Acknowledgments

The authors would like to be obliged to Centre for Robotics and Industrial Automation, Faculty of Electrical Engineering, Universiti Teknikal Malaysia Melaka for providing the laboratory facilities and equipment support. This work is financially supported by the Fundamental Research Grant Project (FRGS/1/2016/TK08/FKE-CeRIA/F00308) and the scholarship of Skim Zamalah from Universiti Teknikal Malaysia Melaka.

References

[1] Caldwell D G, Medrano-Cerda G A and Goodwin M 1995 *IEEE Control Systems* **15** 40-8
[2] Smagt P van der, Groen F and Schulten K 1996 *Biological Cybernetics* **75** 433-40
[3] Sakthivelu V and Horng C S 2013 *IEEE Student Conf. on Research and Development* (Putrajaya: IEEE) 60-4
[4] Minh T V, Tjahjowidodo T, Ramon H and Van Brussel H 2010 Cascade position control of a single pneumatic artificial musclemass system with hysteresis compensation *Mechatronics* **20** 402-14
[5] Chou C -P and Hannaford B 1996 *IEEE Transactions on Robotics and Automation* **12** 90-102
[6] Festo 2010 Fluidic muscle DMSP/MAS Festo brochure https://www.festo.com/cat/de_de/data/doc_de/PDF/DE/DMSP-MAS_DE.PDF
[7] Hussain S, Xie S Q and Jamwal P K 2013 *IEEE Transactions Systems, Man, Cybernetics: Systems* **43** 655-65
[8] Noritsugu T, Takaiwa M and Sasaki D 2010 *Next-Generation Actuators Leading Breakthroughs* (London: Springer) 255-66

[9] Sugar T G, He J, Koeneman E J, Koeneman J B, Herman R, Huang H, Schultz R S, Herring D E, Wanberg J, Balasubramanian S, Swenson P and Ward J A 2007 *IEEE Transactions on Neural Systems and Rehabilitation Engineering* **15** 336-46

[10] Wong Z, Teng C and Chong Y Z 2012 *IEEE-EMBS Conf. on Biomedical Engineering and Sciences* (Langkawi: IEEE) 300-4
[11] Hosoda K, Takuma T, Nakamoto A and Hayashi S 2008 *Robotics and Autonomous Systems* **56** 46-53
[12] Chakravarthy S, Aditya K and Ghosal A 2013 *Proc. of the 1st Int. and 16th National Conf. on Machines and Mechanisms* (Roorkee: Indian Institute of Technology Roorkee) 339-45
[13] Zhu X, Tao G, Yao B and Cao J 2009 *IEEE Transactions on Control Systems Technology* **17** 576-88
[14] Wickramatunge K C and Leephakpreeda T 2010 *Int. J. of Engineering Science* **48** 188-98
[15] Vo-Minh T, Tjahjowidodo T, Ramon H, Van Brussel H 2011 *IEEE/ASME Transactions on Mechatronics* **16** 177-86
[16] Minh T V, Kamers B, Ramon H, Van Brussel H 2012 *Mechatronics* **22** pp 923-33
[17] Wang S and Sato K 2016 *Precision Engineering* **43** 448-61
[18] Caldwell D G, Medrano-Cerda G A and Goodwin M J 1993 *Proc. of IEEE Systems Man and Cybernetics Conf.* (Le Touquet: IEEE) 423-8
[19] Andrikopoulos G, Nikolakopoulos G and Manesis S 2013 *21st Mediterranean Conf. on Control and Automation* (Platanias-Chania, Crete: IEEE) 729-34
[20] Li X, He F, Hong X and Guan T 2011 *Applied Mechanics and Materials* **138-9** 273-8
[21] Pujana-Arrese A, Mendizabal A, Arenas J, Prestamero R and Landaluze J 2010 *Mechatronics* **20** 535-52
[22] Xing K, Xu Q, Huang J, Wang Y, He J and Wu J 2010 *IET Control Theory and Applications* **4** 2058-70
[23] Carbonell P, Jiang Z P and Repperger D W 2001 *Proc. of the IEEE Int. Conf. on Control Applications* (Mexico City: IEEE) 167-72
[24] Lilly J H and Liang Y 2005 *IEEE Transactions on Control Systems Technology* **13** 550-8
[25] Tondu B and Lopez P 2000 *IEEE Control Systems Magazine* **20** 15-38
[26] Li H, Kawashima K, Tadano K, Ganguly S and Nakano S 2013 *IEEE/ASME Transactions on Mechatronics* **18** 74-85
[27] Hesselroth T, Sarkar K, van der Smagt P P and Schulten K 1994 *IEEE Transactions on Systems, Man, and Cybernetics* **24** 28-38
[28] Chan S W, Lilly J H, Repperger D W and Berlin J E 2003 *The 12th IEEE Int. Conf. on Fuzzy Systems* 278-83
[29] Ahn K K, Thanh T D C and Ahn Y K 2005 *JSME Int. J. Series C* **48** 657-67
[30] Chong S -H and Sato K 2016 *IECON 2015 - 41st Annual Conf. of the IEEE Industrial Electronics Society* (Yokohama: IEEE) 4790-5
[31] Sato K and Maeda G J 2009 *Precision Engineering* **33** 175-86
[32] Chong S -H and Sato K 2010 *Precision Engineering* **34** 286-300
[33] Wahyudi, Sato K and Shimokohbe A 2001 *IEEE/ASME Int. Conf. on Advanced Intelligent Mechatronics* vol 2 (Como: IEEE) 843-8

Discharge current distribution in stratified soil under impulse discharge

Fawwaz Eniola Fajingbesi[1,2*], Nur Shahida Midi[1], Elsheikh M. A. Elsheikh[1], Siti Hajar Yusoff[1]

[1]Department of Electrical and Computer Engineering, International Islamic University Malaysia
[2]Department of Physics Electronics and Earth Science, Fountain University Osogbo Nigeria

E-mail: fawwazfajingbesi@yahoo.com

Abstract. The mobility of charge particles traversing a material defines its electrical properties. Soil (earth) have long been the universal grounding before and after the inception of active ground systems for electrical appliance purpose due to it semi-conductive properties. The soil can thus be modelled as a single material exhibiting semi-complex inductive-reactive impedance. Under impulse discharge such as lightning strikes to soil this property of soil could result in electric potential level fluctuation ranging from ground potential rise/fall to electromagnetic pulse coupling that could ultimately fail connected electrical appliance. In this work we have experimentally model the soil and lightning discharge using point to plane electrode setup to observe the current distribution characteristics at different soil conductivity [mS/m] range. The result presented from this research indicate above 5% shift in conductivity before and after discharge which is significant for consideration when dealing with grounding designs. The current distribution in soil have also be successfully observed and analysed from experimental result using mean current magnitude in relation to electrode distance and location, current density variation with depth all showing strong correlation with theoretical assumptions of a semi-complex impedance material.

1. Introduction

Electrical discharge from lightning phenomena has long been investigated mostly for the sole purpose of designing better and more efficient lightning protection systems. The general effects from lightning strike to earth are so severe to humans, their livelihood and equipment. In several literature, this effect have been grouped into direct and indirect lightning strike effects [1]. The effects from this strike on a target is expected to vary where the direct is more intense as all the enormous energy is dissipated through the object. The indirect strike could cause harm in the form of electromagnetic pulse coupling or leading to a ground potential rise (GPR) that end up failing systems in its vicinity [2].

A lightning discharge whose pathway is directly between a cloud and the soil surface without going through a high rise object or any man made facility is referred to as an indirect lightning strike just as lightning hitting an object before the ground surface is called a direct strike [2, 3]. There are enormous literature on direct strike and its effect compared to an indirect lightning strike. The detrimental effect that could result from an indirect strike is believed to be significant and require more attention due to the widespread development of new and sensitive technological devices [4].

The soil due to its large impedance have long been the universal ground for all electrical operation including lightning discharge from clouds. This soil impedance however vary in accordance to several factors under the umbrella of soil electrical conductivity. It is believed that the severity from indirect lightning stroke would vary due to this soil characteristics making it difficult to design a universal efficient protection system to combat indirect lightning strike damages. One way to make clear of indirect discharge lightning phenomena and soil characteristics influence is by investigating the discharge propagation in soil when lightning strike on the surface.

This research aims to investigate the discharge phenomena of lightning strike to soil surface. In this work, impulse discharge was applied on a soil surface as an imitation to lightning strike on soil surface. The discharge propagation was then observed and discussed using the discharge current distribution during the process on a laboratory scale.

2. Experimental Setup and Procedures

An experimental setup consisting of a point to plane electrode system as shown in figure 1 was developed to produce an imitation of the lightning discharge phenomena on soil surface. The stratified point to plane electrode system is designed to offer a dielectric two-phase gas-solid experimental environment. The point electrode is an 8 mm diameter pointed stainless steel electrode which serve as the discharge electrode. The discharge electrode tip is separated from the soil surface by a 5mm airgap. There are nine grounding electrodes divided into five horizontal electrodes (E1 – E5) and four vertical electrodes (E6 – E9) inside a plastic tub. Electrode 1 is a 1 mm thick, 40 mm diameter circle-shaped copper electrode. Electrodes 2 until 5 are ring-shaped copper electrodes with 1 mm thickness. The outer radius is 40 mm longer than the inner radius. All the electrodes were attached to an acrylic sheet and were set at the bottom of a 480 mm diameter plastic tub. Electrodes 6 until 9 are 0.1 mm thick and 40 mm width copper tape attached to the inner side wall of the plastic tub. There were 15 mm spaces between each horizontal and vertical electrode. Electrode 1 until 8 were set as under soil electrodes while electrode 9 was set as the surface electrode. The soil surface level was set 220 mm from the bottom electrodes. The discharge electrode was connected to 280 kV impulse voltage generator (TERCO) to generate spark discharge at the air gap leading on the soil surface.

The experiment was conducted by applying 22 – 28 kV of standard lightning impulse voltage (1.2/50μs) to the discharge electrode depending on the conductivity of soil sample under test with a distance of 5 mm between the soil surface and the point electrode. The discharge current due to the soil surface discharge were observed at each of the 9 grounding electrodes using a current probe (Pearson 6585) connected to a digital oscilloscope (LeCroy Waverunner 625Zi). In this experiment, the soil conductivity were altered by increasing the soil moisture content to obtain conductivity (σ = 10.63 and 20.42 [mS/m]) corresponding to a moisture (14.07 and 23.16 [%]) respectively. The soil electrical conductivity were also measured using earth resistance meter (Megger DET4TCR2) prior to and after the discharge process. This offered a more insight to understanding the effect of indirect lightning discharge and soil electrical conductivity.

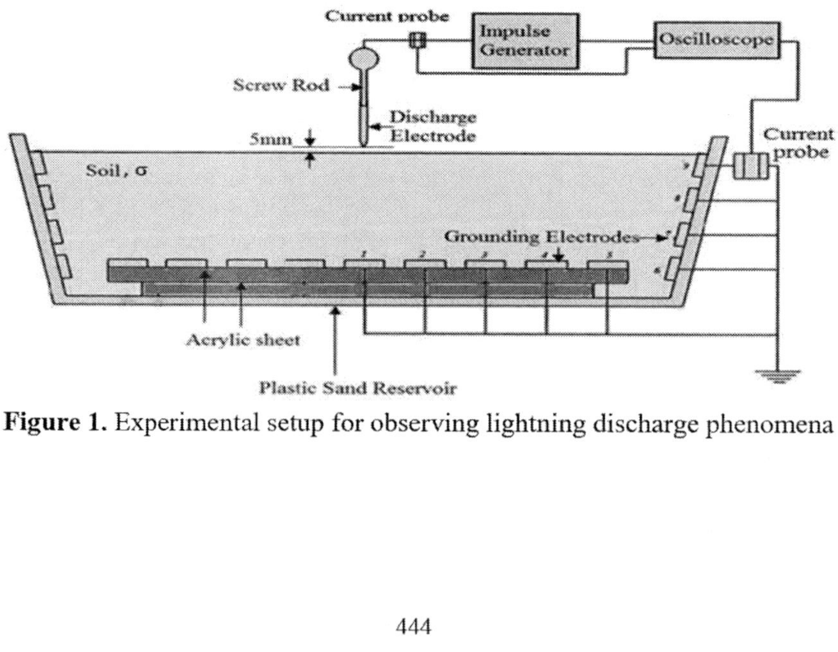

Figure 1. Experimental setup for observing lightning discharge phenomena

3. Result and Discussion

3.1. Voltage and current waveform

Figure 2 and 3 shows the voltage and current waveforms at the minimum (σ = 10.63[mS/m]) and maximum (σ = 20.42[mS/m]) conductivity respectively in the range chosen. From the waveforms, it can be seen clearly that the discharge process between soils of different conductivity are distinct from each other. For σ = 10.64 [mS/m], the input current and the discharge current waveforms are seen to follow similar pattern that is proportional to the input voltage waveform where the increase or decrease in current is seen to follow the increase or decrease in the input voltage. At higher conductivity soil, σ = 20.42 [mS/m] shown in figure 3, a sudden drop in input voltage and current waveform is observed followed by a second peak as time elapsed. The input current and the discharge current waveforms are not exact replica of the input voltage waveform. This suggest that a breakdown occurs in soil as conductivity increase. Also the evident of a second peak shows the characteristics of breakdown in soil [5]. Considering the peak discharge current and voltage in figure 2 (a) and figure 3 (a) measured at the discharge electrode, the discharge current increase with the conductivity. This is because the required voltage for the discharge to occur at the air gap decreased with increase in soil conductivity.

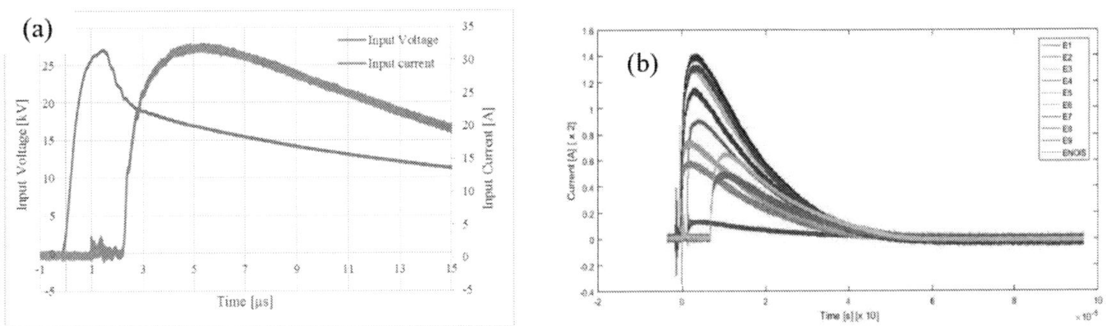

Figure 2. (a) Input voltage and current waveform at σ = 10.64 [mS/m]. (b) Electrode discharge current waveforms at σ = 10.63 [mS/m]

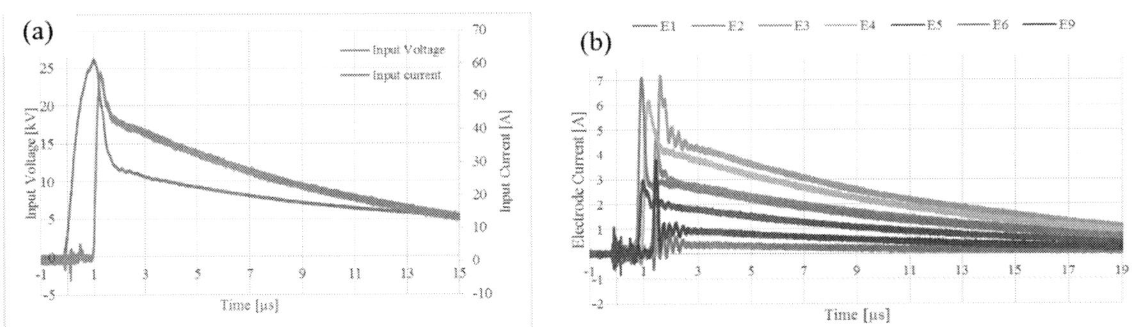

Figure 3. (a) Input voltage and current waveform at σ = 20.42 [mS/m]. (b) Electrode discharge current waveforms at σ = 20.42 [mS/m]

3.2. Discharge current distribution

The distribution of discharge current in soil at two different conductivity is expressed in terms of discharge current magnitude and density. Figure 4(a) show the discharge current peak magnitude [A] measured at the nine underground electrodes. The current magnitude distribution on each electrode

IOP Conf. Series: Materials Science and Engineering **210** (2017) 012057 doi:10.1088/1757-899X/210/1/012057

follows similar pattern across each electrode with increase in conductivity however with greater magnitude at each electrodes. This signifies that current distribution depends on the soil electrical conductivity. Figure 4(b) shows the current density [A/m^2] at each grounding electrode. The current density offers better insight of the current distribution since the electrodes (E1 – E9) area vary from one another. From this data in figure 4(b), current density can be seen to decrease as the electrodes moved further away from the discharge point horizontally for all conductivity. In the vertical electrodes E6 – E9 the current density is seen to rise as the electrode moved closer to the soil surface however started declining after electrode E8. It can be concluded that discharge current is generally distributed into the soil which is to the bottom of the discharge local point (E1) regardless of the soils conductivity. Similarly if the current density in figure 4(b) on each electrode as (E1 – E9) can be expressed as a function of their relative position from the discharge point providing a better insight on the current distribution inside the soil.

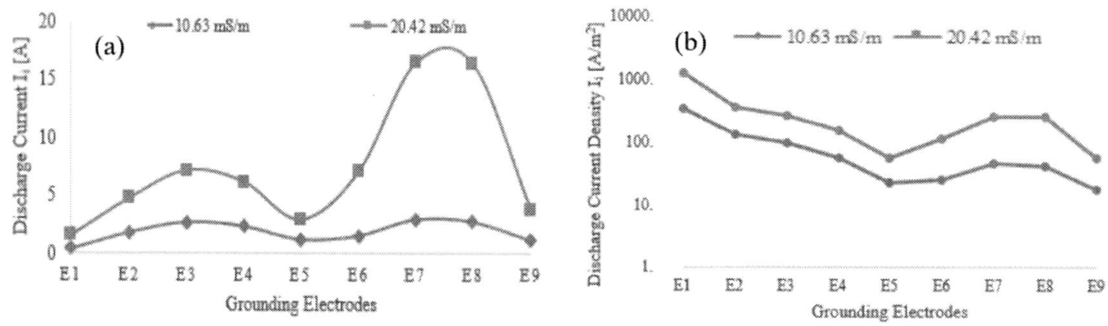

Figure 4. (a) Discharge current magnitude and (b) Discharge current density at grounding electrodes

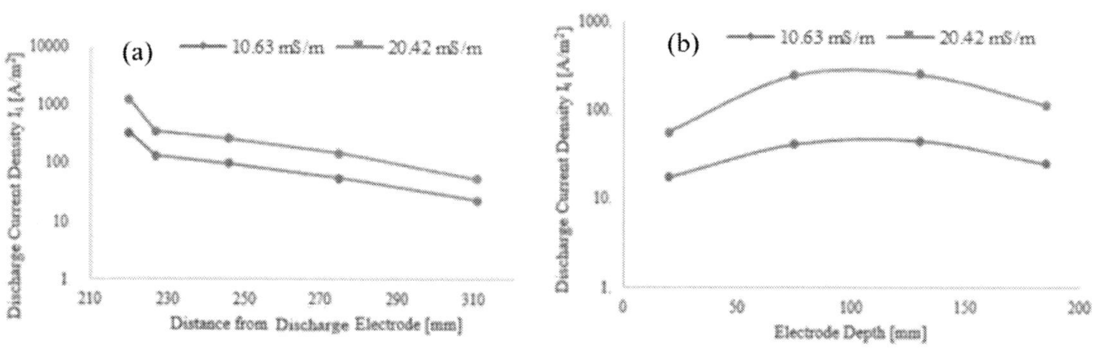

Figure 5. Discharge current density as a function of distance (a) and depth (b)

The discharge current density as a function of electrode distance from the local point of discharge on the surface *d* is shown in figure 5(a). From this graph, it shows the discharge current on the electrode decreased with increase in the distance from the localized discharge point. This almost uniform decrease signifies a uniform drop in the electric potential traversing the soil. Figure 5(b) shows the current density as a function of depth, z for electrode E6 – E9 that share infinitesimal same distance, d. Although an initial increase in current density is observed with increase in soil depth around the surface region the subsequent data shows that as the depth of electrode increase, the discharge current density of the electrodes decreases. This is also likely associated to the decline in electric potential reaching this region. However the initial increase could be associated to the current accumulation on the surface region prior to soil ionization breakdown or partial discharge [6, 7]. In general, this results shows that soil depth and distance is also a factor that affect the discharge distribution in soil.

3.3. Soil electrical conductivity

The shallow soil conductivity measured before (pre-ECs) and after (post-ECs) along with the deep soil conductivity measured before (pre-ECd) and after (post-ECd) spark discharge on the soil surface is shown in figure 6. This graph shows a significant difference in the conductivity of the soil before and after imitated lightning discharge. This also validates the occurrence of permanent soil ionization breakdown during the discharge process. Hence, the design of an efficient protection system should take into account the soil ionization phenomena.

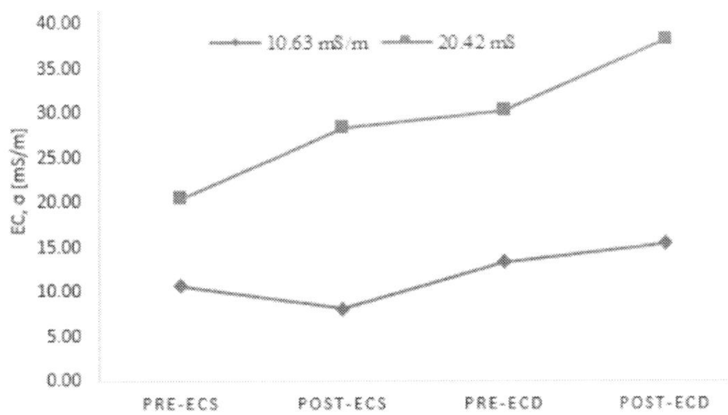

Figure 6. Soil σ variation due to impulse discharge

4. Conclusion

In this work, lightning discharge on soil surface was investigated using impulse discharge applied to the surface of soil placed in a point to plane electrode system. The soil sample at different conductivity obtained from soil with different moisture content was investigated. From there, discharge characteristics were observed. From the experiment it can be concluded that complete discharge occurred depending on conductivity of the soil. Furthermore, on the discharge current distribution, regardless of the soil conductivity, discharge current penetrates into the soil which is to the right below the discharge local point. Also, as conductivity increase, the percentage of discharge current distributed to the soil surface and inside increased. The electric potential propagation in soil with low conductivity is lower hence result in incomplete discharge compared to the propagation in higher conductivity. Finally, soil ionization is seen to occur frequently in higher conductivity.

Acknowledgment

This research has been supported by Malaysian Ministry of Higher Education Research Grant, RAGS15-063-0126.

References

[1] G Punekar and C Kandasamy 2011 *Serbian J. Electr. Eng.* **8** 245–262
[2] A Kalair, N Abas, and N Khan 2013 *J. Light. Res.* **5** 11–28
[3] N Malcolm and R Aggarwal 2014 *2014 IEEE PES General Meeting | Conference & Exposition* 1–5.
[4] R Brooks 2010 *The World Offshore Renewable Energy Report 2004-2008* **2** Department of

Trade and Industry's Renewables 2010
[5] N S Midi and R I Ohyama 2011 *Annu. Rep. - Conf. Electr. Insul. Dielectr. Phenomena, CEIDP* 223–226
[6] A Elzowawi, A Haddad, H Griffiths, and D Clark 2015 *2015 50th International Universities Power Engineering Conference (UPEC)* **2015–Novem** 1–5
[7] J He and B Zhang 2015 *IEEE Trans. Ind. Appl.* **51** 4924–4933

IOP Conf. Series: Materials Science and Engineering **210** (2017) 012058 doi:10.1088/1757-899X/210/1/012058

Influence of MWCNTs addition on mechanical and thermal behaviour of epoxy/kenaf multi-scale nanocomposite

N A M Noor *, **J A Razak** *, **S Ismail, N Mohamad, M Y Yaakob and T H Theng**

Carbon Research Technology Research Group, Engineering Materials Department, Faculty of Manufacturing Engineering, Universiti Teknikal Malaysia Melaka, Hang Tuah Jaya, 76100, Durian Tunggal, Melaka, Malaysia

E-mail: *norainaaliamatnoor@gmail.com

Abstract. This research was conducted to develop kenaf reinforced epoxy/MWCNTs multi-scale composite using kenaf fibre and MWCNTs as the reinforcement in epoxy as the hosted matrix. The composites were produced by using a combination of hand lay-up and vacuum bagging process. The selection of optimum composition of epoxy-MWCNTs is based on the MWCNTs loading and the resulted mixture viscosity. Lower resin viscosity is required to allow good wetting and interaction between matrix and filler, which will yielded superior final performance of the fabricated composites. Therefore, different loading of MWCNTs (0.0 wt. %,
0.5 wt. %, 1.0 wt. %, 3.0 wt. %, 5.0 wt. %, 7.0 wt. %) were used to investigate the mechanical and thermal properties of the composites. As a result, the epoxy/kenaf/MWCNTs multi-scale composite at 1.0 wt. % of MWCNTs addition had yielded substantial improvement by 15.54 % in tensile strength and 90.54 % in fracture toughness. Besides, the fracture surface morphology of the selected samples were analysed via scanning electron microscopy (SEM) observation to further support the reinforcement characteristic of epoxy/kenaf/MWCNTs multi-scale composite.

1. Introduction

The processing technique are improving by the development of technology beginning from macro, micro and current trend to nanoscale inclusion. Theoretically, a dimension that have at least one dimension in the range of 1 to 100 nm are classified as nanoscale inclusion. In the prospectus of the properties produced by the addition of nanofiller, a composite's properties promised a positive tendency to increase but it also depend on the content added in the system. Furthermore, the interfacial interaction between the polymer matrix and reinforcement is also an important parameter which affect the properties of polymer matrix nanocomposite. Thus, a good dispersion between matrix and reinforcement must be thoughtfully considered when producing a nanocomposite [1].

Among all matrix resin that are available in the industries, epoxy resin has its own potential for many high performance application. Epoxy is one of the resin from thermoset family that commonly used as matrix for composite application. In addition, epoxy can be produced from reacting epichlorohydrin with bisphenol [2]. When compared to other polymer, epoxy offers astonishing properties of epoxy such as high corrosion resistance, high mechanical properties and low shrinkage properties [3]. This make epoxy as one of the favorable choice in high performance application especially in aircraft and aerospace industries [4].

However, the extraordinary properties of epoxy discredited by the brittleness behavior of the material in nature of the cross linked molecular structure. Epoxy resin also has low toughness that resulted in weak resistance to crack. Therefore, addition of fine particle is introduced to enhance the properties of epoxy resin [5]. Considering into that the incorporation of nano-size reinforcement of functional filler is advisable to compensate with the limitation of epoxy resin. Functional nanofiller like carbon nanotubes, carbon nanofibre, or graphene were able to modify and improvise the mechanical and thermal properties of polymer matrix resin, like epoxy. In this work, MWCNTs were used as nanofiller in the epoxy matrix.

Carbon nanotubes are outstanding filler for polymer composites. Carbon nanotubes can broadly enhance the electrical, thermal and mechanical properties of the resulting nanocomposites. The Young's modulus of CNTs is nearly to 1 TPa but dependent to the type and diameter of carbon nanotubes. Moreover, the theoretical strength of carbon nanotubes is 100 times higher than steel [6]. Recent advanced in CNTs research, specifically on the synthesis technology and surface modification had allow this nanofiller become compatible with another materials. Therefore, MWCNTs were dispersed in water [7], and due to that the MWCNTs are expected to easily be dispersed in the polymer matrix.

Instead of nanofiller alone, natural fibre also can be used as the reinforcement. According to Akil et al. (2011) [8], natural fibres are commonly used as filler to produce a green composite because it is cheaper, have low density compared to others synthesis fibre and the strength is higher than glass fibre. It is more preferable to produce low cost and high strength composite from many prospective of interest. Among many established natural fibres that were widely exploited for their promising properties, kenaf fibre is one of the kind that capture many interest. In addition, Huda et al. (2006) [9], pointed that kenaf fibre is effectively reinforced the thermoset and thermoplastic composites. By times, kenaf fibre is slowly replacing synthesis material as kenaf fibre has many better properties to offer. In this work, kenaf fibre was used in a non-woven pattern as the natural reinforcement agent for epoxy matrix. The natural fibre based epoxy composite system was further reinforced by multi scaling the epoxy resin by MWCNTs incorporation. It is expected that by hybridizing the MWCNTs nanofiller with kenaf natural fibre through epoxy resin as nanocomposite matrix, the benefit of nanoreinforcement by CNTs will be amplified the potential of kenaf fibre as natural reinforcement agent, for high performance composite application.

2. Experimental

2.1. Materials

The neat epoxy resin used in this work was a standard diglycidylether of bisphenol A (DGEBA) coded CP 360 Part A with an epoxy equivalent viscosity of 8500 mPa.s and the curing agent was amine hardener coded CP 360 Part B that were supplied by Terra Techno Engineering, Malaysia. The multi- walled carbon nanotubes (MWCNTs, Flo Tub 9000 Series, CNano Technology Limited, China) was purchased from CNano Technology Limited, China and were used as it is. The kenaf fiber mat supplied from Lembaga Kenaf & Tembakau Negara (LKTN), Malaysia was conditioned in-house and used as it is.

2.2. Preparation of composites

There are six categories of multi-scale kenaf-MWCNTs/epoxy plate with different MWCNTs loading at 0.00 wt. %, 0.50 wt.%, 1.00 wt.%, 3.00 wt.%, 5.00 wt.% and 7.00 wt. % were prepared by using hand lay-up and vacuum bagging methods. The mixture of epoxy and MWCNTs then stirred by using mechanical overhead stirrer at a speed of 1000 rpm for 60 minutes. Later, the hardener agent CP 360B were poured into the mixture and stirred for 5 minutes prior to integration with kenaf fiber mat.

Meanwhile, a set-up for vacuum bagging was prepared by placing a sheet of plastic on the flat surface before layering with the non-woven kenaf fiber above it. Next, the prepared mixture of

IOP Conf. Series: Materials Science and Engineering **210** (2017) 012058 doi:10.1088/1757-899X/210/1/012058

epoxy/MWCNTs was poured onto the kenaf fibre mat and spread uniformly on the surface of kenaf mat by hand lay-up process. Then, the sample is attach with plastic bagging by using sealant tape around the mould edge. A spherical tube is then install to suck out the excessive resin and bubbles throughout the vacuum bagging process. Finally, the multi-scale kenaf-MWCNTs/epoxy composits were left to cure at room temperature for 24 hours. The resulted composites were fabricated into 300 mm x 300 mm. The after-cured samples were then cut into specific dimension fulfilling the requirements for the tensile, flexural and fracture toughness tests, in accordance to the standard methods of ASTM D3039, ASTM D790 and ASTM D5045, respectively.

2.3. Mechanical properties evaluation

2.3.1. Tensile properties. The tensile properties of the samples were measured by using an Instron 5969 Universal Testing Machine (UTM) equipped with maximum 50 kN load cell. The tensile test was performed at the crosshead speed rate of 2.00 mm/min and gauge length of 60 mm, according to ASTM D3039M-14. The specimens were cut to nominal dimensions of 100 mm x 18 mm and each mix were prepared for three times repetition.

2.3.2. Fracture toughness properties. The fracture toughness evaluation on the multi-scale epoxy/kenaf-MWCNTs composites were performed in accordance to ASTM D5045-14 Standard Test Method for Fracture Toughness and Strain Energy Release Rate of Plastic Materials by closely referred to Bakar et al. (2012) work. A span of 80 mm was specified with crosshead speed of 2 mm/min performed on Autograph UTM 50 kN, Shimadzu, Japan. The fracture toughness was calculated by the formula in Equation 1 and Equation 2 as denoted as below where P is the maximum load, S is the span length, a is the notch length, B is the sample thickness, W is the sample width and y is the geometry correction factor. Three samples were prepared from each category and conditioned at room temperature where the average value were considered as the results.

$$K_{IC} = \left(\frac{PS}{BW^{3/2}}\right) x(y) \quad (1)$$

$$Y = 3(a/w)^{1/2} x\left(\frac{1.99-\left(\frac{a}{w}\right)\left[1-\left(\frac{a}{w}\right)\right]\left[2.15-3.93\left(\frac{a}{w}\right)+2.7\left(\frac{a}{w}\right)^{2}\right]}{2\left[1+2\left(\frac{a}{w}\right)\right]\left[1-\left(\frac{a}{w}\right)\right]^{\frac{3}{2}}}\right) \quad (2)$$

2.4. Thermal analysis

2.4.1. Dynamic Mechanical Analysis (DMA). Dynamic mechanical analysis (DMA) is commonly used to measure temperature dependence properties, such as the storage modulus, E', loss modulus, E", and damping factor, tan σ [11]. The loss modulus, E", was used to determine the glass transition temperature (T_g) of the unfilled epoxy and multi-scale epoxy/MWCNTs-kenaf composites. A Pyris Diamond DMA (Perkin Elmer) with a dual cantilever was used to perform the DMA measurements. Specimens with dimensions of *l x w x t*(40 x 10 x 1) mm were prepared according to ASTM D5418-15. The specimens were evaluated over a temperature range of 30 °C to 180 °C, using a heating rate of 2 °C/min and a fixed frequency of 1.0 Hz.

2.5. Scanning electron microscopy

Morphological structures of the samples were investigated using scanning electron microscope (SEM, Zeiss EVO-50 FESEM) by applying an accelerating voltage of 20 kV. Fractured surfaces from tensile testing were analyzed. All samples were sputter-coated with gold to eliminate electron charging.

3. Results and discussions

3.1. Tensile properties analysis

Figure 1 depicted the ultimate tensile strength of the fabricated composites incorporated with a series of specified content of MWCNTs. It is clearly shows that the ultimate tensile strength were found increases with the increase in the content of MWCNTs inclusion, up to an optimum point at 1.00 wt. %, before drastically dropped after addition of MWCNTs at higher loading. The largest increase of ultimate tensile strength is 25.06 MPa occurred at 1.00 wt. % of MWCNTs with 15.54 % increment from epoxy/kenaf composite acted as control sample for this study. The ultimate tensile strength of the control sample is 21.69 MPa.

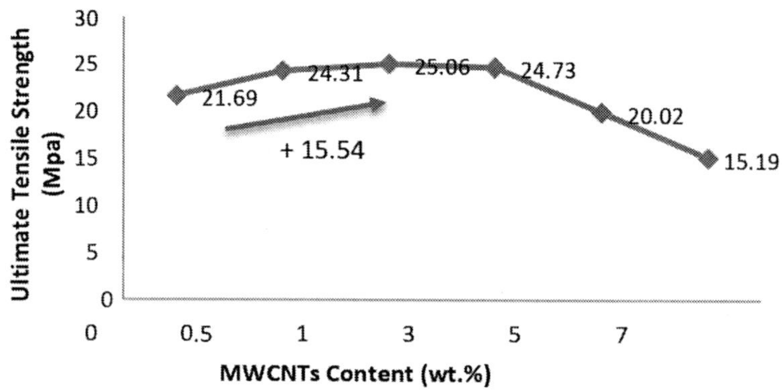

Figure 1. The effect of MWCNTs content on ultimate tensile strength of MWCNTs-kenaf/epoxy multi-scale composite system

An increase in the tensile strength and tensile modulus for multi-scaled hybrid composite system as compared than common micro composites system can be partially explained on the basis of the interaction between the filler and the matrix. It has been found that a greater adhesion between the matrix and nano inclusion had caused less matrix-filler debonding when a stress is been applied. Hence, the elastic modulus and strength are greatly improved due to this beneficial interaction [1]. However, further increased in the percentages of MWCNTs were significantly diminish the strength of produced multi- scale composites. From 1.00 wt. % to 7.00 wt. % addition of MWCNTs added to composite, it was found that the strength was decreases from 25.06 MP to 15.19 MPa which is around 39.39% of reduction. This is because the strength of the nano-reinforcement adhesives begin to degrade due to the agglomeration phenomena that caused by the filler-filler interaction [12]. Furthermore, there is a sharp decreased in the ultimate tensile strength when 7.00 wt. % MWCNTs is added, which is about 24.11 %. This is because the viscosity of epoxy-MWCNTs matrix is getting too high was resulted in a bad dispersion of the nano-reinforcement.

Figure 2 depicts the Young's modulus results of multi-scaled epoxy/kenaf hybrid composites reinforced with MWCNTs nanofiller. The plotted graph shows the effects of various loading system of MWCNTs toward the resulted composites. The addition of MWCNTs yielded an improvement in the Young's modulus up to 1.00 wt. % inclusion, before eventually decrease with more involvement of MWCNTs nanofiller. The trend was similar with the ultimate tensile strength attribute as in Figure 1. The highest Young's modulus of 3.98 GPa was obtained at 1.00 wt. % of MWCNTs loading system. This favourable trend of 48.00 % improvement in MWCNTs-kenaf/epoxy multi-scale composite was noticed when compared to the control sample. The lowest Young's modulus was detected from the sample with the addition of 7.00 wt. % of MWCNTs which is 1.24 GPa. There is almost 53.9 % reduction as compared to the neat epoxy kenaf fibre multi scale composite. The properties dropped might be due to the improper dispersion of MWCNTs at higher concentration which closely related with the agglomeration phenomena.

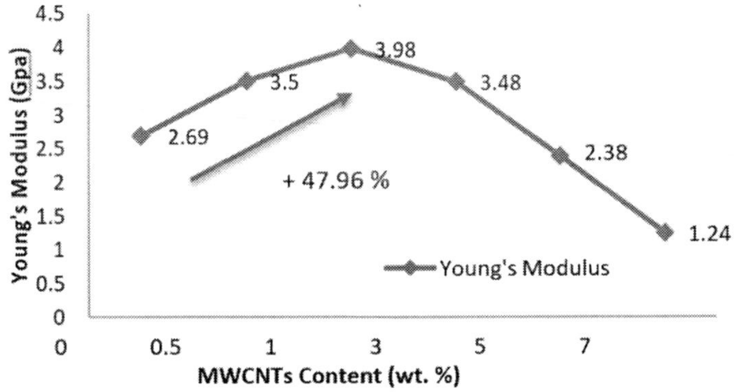

Figure 2. The effect of MWCNTs content on Young's modulus of MWCNT kenaf/epoxy multi- scale composite system.

Figure 3 shows the percentage of elongation at break for the multi-scale composites versus percentages of MWCNTs content. By comparing the graph of Young's modulus and percentage of elongation at break, it can be seen that the elongation attributes was inversely proportional to the Young's modulus results. This is because the stiffer the sample, the lower the percentages of elongation at break of the tested specimen of that particular epoxy system. The percentage of elongation at break was further decreased when the MWCNTs loadings were increased up to 1.00 wt. % of MWCNTs addition. After that, the percentage of elongation at break was unexpectedly increased with the increase of MWCNTs loadings. At 1.00 wt. % of MWCNTs inclusion, the fabricated multi-scaled composite possessed the smallest amount of elongation characteristic which is 1.82 %. This is due to the factor of having good interfacial interaction between the epoxy matrix and hybrid reinforcement. Therefore, the sample is getting stiffer and brittle as representing by the lower percentages of the elongation at break. In contrast, the specimen of 7.00 wt. % of MWCNTs produced the highest percentages of break, which is 3.10 %. The further addition of MWCNTs in the composite caused the viscosity of epoxy-MWCNTs to increase significantly. Thus, serious agglomerations of MWCNTs prevent proper interaction between epoxy resin and hardener. This improper mixing had caused premature curing of the epoxy matrix. As a result, the outer surface of the sample was improperly cured while the internal region was partially uncured. The sample tends to break at the specific region which later caused the testing machine unable to detect the definite elongation properties.

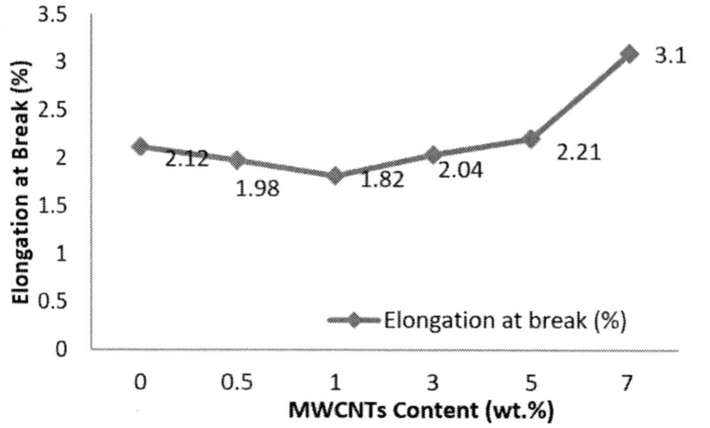

Figure 3. The effect of MWCNTs content on elongation at break (%) of MWCNTs-kenaf/epoxy multi-scale composite system

3.2. Fracture toughness analysis

According to Chandrasekaran et al. [13], the fracture toughness of composite started to decrease beyond certain weight fraction of filler addition which is in parallel with the results found in this work. This phenomenon is due to the MWCNTs fibre that was being pulled out during the impact loading and therefore contributed to the high toughness of produced multi-scaled composites. Figure 4 shows that the fracture toughness was increased when the MWCNTs content increased. The positive increment were noticed to linearly increase up to 1.00 wt. % of MWCNTs addition and later decreased as the content of the nanofiller increased. The maximum fracture toughness is showed by the sample of 1.00 wt. % of MWCNTs content with 90.52 % improvement compared to the unfilled epoxy/kenaf system. At 3.00 wt. % of MWCNTs content, the fracture toughness of the composites are started to diminish by
4.19 % compared to 1.00 wt. MWCNTs-kenaf/epoxy system. This is because when the MWCNTs content increased, the MWCNTs prone to agglomerate and degrade the quality of the resulted composites [12]. All the small fracture were started around the agglomeration area of MWCNTs which later propagate throughout the sample that resulting in fracture or failure of the done hybrid composites [13].

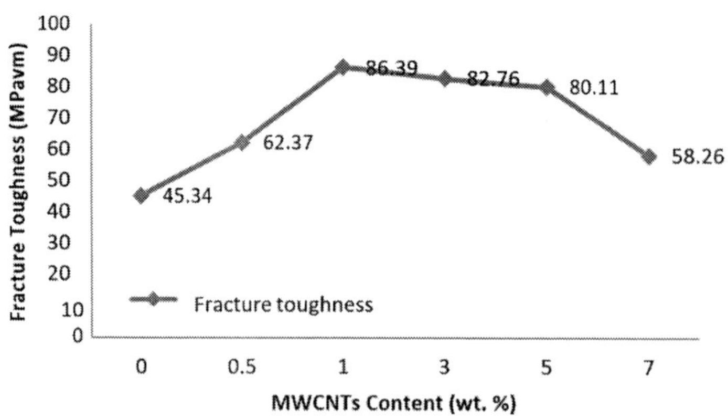

Figure 4. The effect of MWCNTs content on fracture toughness of MWCNTs-kenaf/epoxy multi-scale composite.

3.3. Thermo-mechanical properties evaluation by DMA

Figure 5 shows the storage modulus of multi-scaled hybrid composites with different loading of MWCNTs additions. The graph shows the composition of 1.00 wt. %-MWCNTs addition has the highest storage modulus value of 1700 MPa and decreased with more addition of MWCNTs. This finding proved that the presence of 1.00 wt. %-MWCNTs in the sample introduced a good interfacial bonding between the fibre and matrix. Therefore, the load applied can be transferred efficiently even at the exposure of heat. However, at 0.5 wt. %-MWCNTs addition, the storage modulus was lower than 1.0 wt. %-MWCNTs sample. This is because the small amount of MWCNTs only acted as notches instead of giving reinforcement effect which reducing the load transfer efficiency [14]. While, at 7.0 wt. %- MWCNTs content, the storage modulus are much lower because the higher content of MWCNTs given higher tendency of being affected by agglomeration factor. Thus, the interfacial bonding became weak and load cannot transferred effectively.

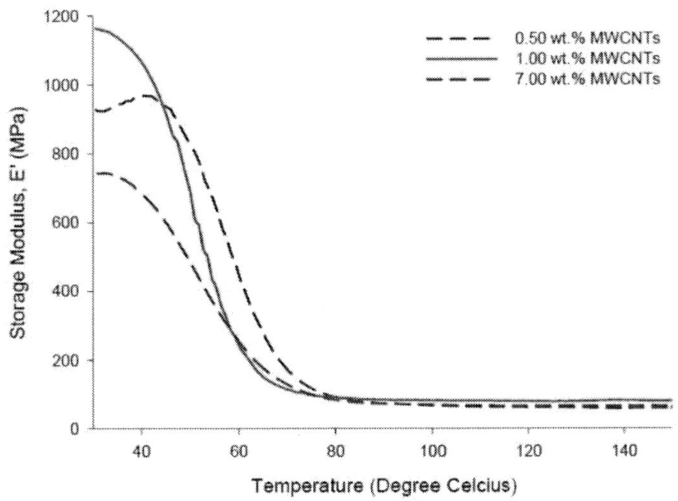

Figure 5. The effect of different MWCNTs addition on storage modulus (E") of epoxy/kenaf multi-scaled composites

Figure 6 depicts the loss modulus of epoxy/kenaf multi-scaled composites with different MWCNTs loading. The loss modulus (E') presented in this work is in parallel with the storage modulus (E") trend as in Figure 6. At 1.0 wt. %-MWCNTs addition, the loss modulus is 150 MPa which is higher than the composition of 0.5 wt. %-MWCNTs composite and 7.0 wt. %-MWCNTs composite. At low addition of MWCNTs, the loss modulus is lower because the energy losses caused by the rearrangement of the molecules and MWCNTs as well as the internal friction between the fibres and epoxy matrix. From the graph, 1.0 wt. %-MWCNTs composite has higher loss modulus showing that the composite has higher potential of decapitated energy as heat. At high addition of MWCNTs, the loss modulus is the lowest because the over-filler-interaction leads to bad dispersion and produced worst agglomeration site. Later, this event caused the bad energy dissipation and affect the mechanical properties.

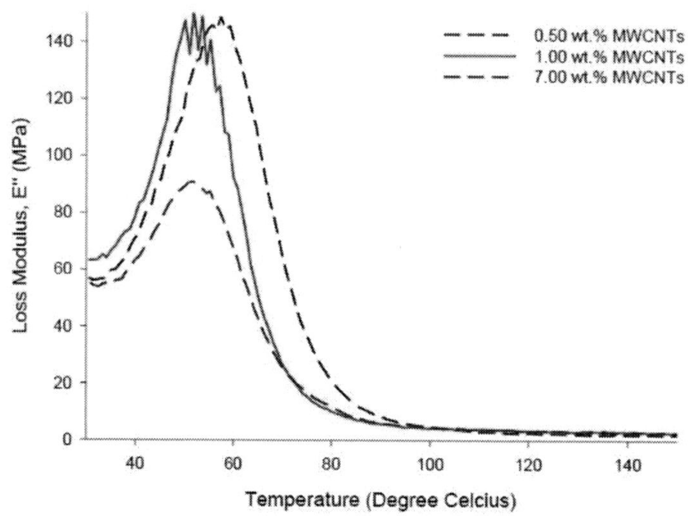

Figure 6. The effect of different MWCNTs addition on loss modulus (E') of epoxy/kenaf multi-scaled composites

Tan delta is the ratio of loss modulus to storage modulus which also known as the damping factor of the produced composites. The increase in tan delta notified that more potential of energy dissipation while a decrease in tan delta means the composite have more elastic properties [14]. At 1.0 wt. %- MWCNTs addition, the sample has the highest damping factor as compared than others. This indicated that the sample with 1.0 wt. % of MWCNTs content has higher potential to dissipate heat and produced a tougher composites system. In contrast, 7.0 wt. %-MWCNTs composite presented the lowest damping factor which is believed due to the decrement of the mobility of polymer molecules chains as hindered by the MWCNTs agglomeration [15].

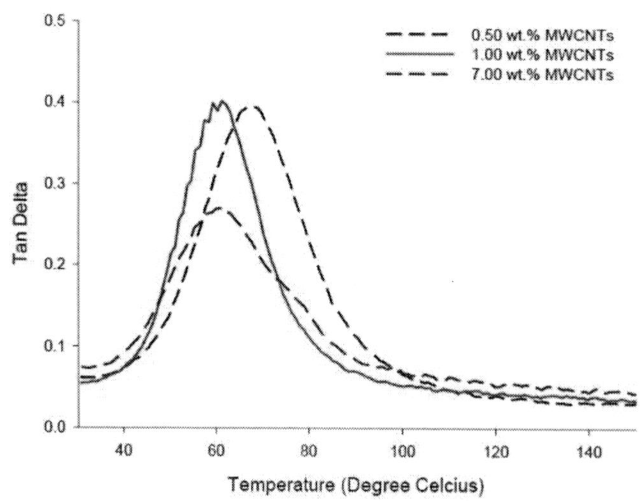

Figure 7. The effect of different MWCNTs addition on tan delta of epoxy/kenaf multi-scaled composites

3.4. Fractured surface observation via SEM

Micrograph at figure 8 shows the fracture surface at the fractured samples with different MWCNTs filler loadings which are 0 wt. % of MWCNTs, 1.00 wt. % of MWCNTs and 7.00 wt. % MWCNTs.

According to Garg et al. [16], the higher the MWCNTs filler loading, the higher the tendency of agglomeration of MWCNTs bundles formation due to the increased of epoxy matrix viscosity. Figure 8
(a) shows the micrograph of 0.0 wt. % of MWCNTs sample. From the micrograph, it was clearly shows that some fibres were being pulled out from the matrix. This is due to the poor interface interaction between the epoxy matrix and the kenaf fibre with the presence of MWCNTs nanofiller. The weak interface was resulting lower strength of the produced hybrid composites. Other than that, there are some internal pores and voids were detected in the kenaf fibres. Those the pores and voids were acted as stress concentration point that becoming a weak point which lowering the resulted strength of the fabricated samples. Micrograph 8 (b) shows that the fibre were break into two rather than pulled out from the epoxy matrix. This proves that the matrix-filler interface are stronger which promoting a better resulted strength. Other than that, a good dispersion of MWCNTs in epoxy was found tends to reduce the stress concentration which further enhanced the uniformity of the stress distribution of the produced hybrid multi-scaled composites [17]. The fracture micrograph is depicted in Figure 8 (c) shows the fibre breakage with non-homogeneous matrix wetting, whereby the agglomeration of MWCNTs was limits the wetting of epoxy to the kenaf fibre surface. Other than that, the micrograph of Figure 8 (c) also shows the poor interfacial of matrix and the kenaf fibre. This totally reduced the strength of produced composites. According to Rahmanian et al. [18], too much of MWCNTs added could lead into a reduction of the resulted strength and modulus of the multi-scaled composites due to the worst agglomeration factor. This is because agglomeration will becoming as a source of crack initiation and propagation [17].

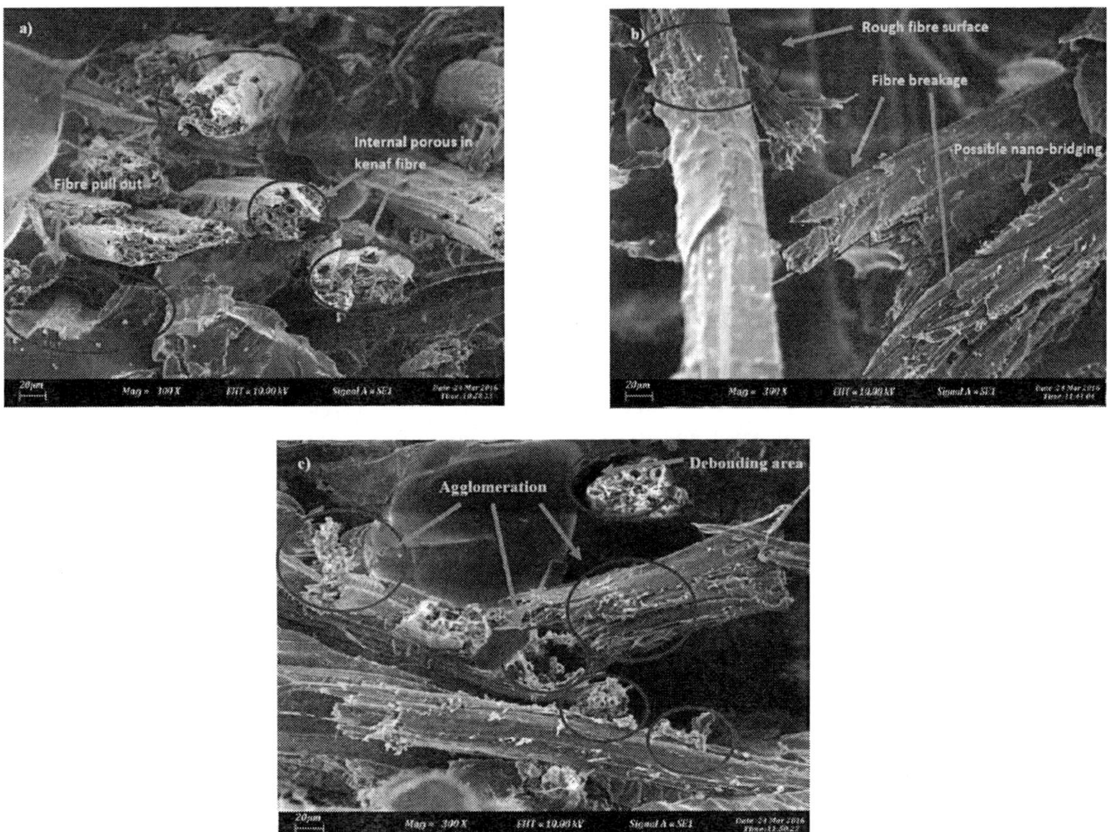

Figure 8. (a) Micrograph of neat epoxy kenaf composite, (b) micrograph of 1.0 wt. % MWCNTs epoxy/kenaf nanocomposite and (c) micrograph of 7.0 wt. % MWCNTs epoxy/kenaf multi-scale composites.

4. Conclusion

Kenaf fibre reinforced epoxy/MWCNTs multi-scaled hybrid composites were successfully produced by using the simplified vacuum bagging technique. The results of tensile, fracture toughness and thermo- mechanical tests showed that at composition of 1.0 wt. % MWCNTs in epoxy/kenaf multi-scaled hybrid composite possess the best tensile, fracture toughness and viscoelastic properties. Fracture surface analysis showed less fibres pulled out for resulted epoxy/kenaf/1.0 wt. % MWCNTs multi-scaled composite.

Acknowledgement

We greatly acknowledge the research funding from Research Acculturation Grants Scheme coded RAGS/1/2014/TK04/FKP/B00072 from the Ministry of Higher Education, Malaysia.

References

[1] J Jordan, K I Jacob, R Tannenbaum, M A Sharaf, and I Jasiuk. 2005 *Materials Science and Engineering A*. **393** 1–11
[2] M Singla and V Chawla 2010 *Mechanical Properties of Epoxy Resin – Fly Ash Composite* **9** 199–210
[3] E J Barbero 2011 *Introduction to Composite Materials Design* (New York: CRC Press)
[4] H He, J Wang, K Li, J Wang, and J Gu 2010 *Materials and Design*. **31** 4631–4637
[5] S C Tjong 2006 *Structural and mechanical properties of polymer nanocomposites* **53** 73–197
[6] K M Liew, Z X Lei, and L W Zhang 2015 *COMPOSITE STRUCTURE*. **120** 90–97
[7] J H Lehman, M Terrones, V Meunier, E Mansfield, and K E Hurst 2011 *Carbon* **49** 2581-2602
[8] Akil, HMl, M. F. Omar, A. A. M. Mazuki, S. Z. A. M. Safiee, ZA Mohd Ishak, and A. Abu Bakar. 2011 *Materials & Design* **32** 4107-4121
[9] Huda, Masud S., Lawrence T. Drzal, Amar K. Mohanty, and Manjusri Misra 2006 *Composites Science and Technology* **66** 1813-1824
[10] A. Bakar, A. S., and K. W. 2012 *Pertanika Journal Science and Technology* **20**129–137
[11] M. J. M. Ridzuan, M. S. A. Majid, M. Afendi, M. N. Mazlee, and A. G. Gibson 2016 *Composite Structures* **152** 850–859
[12] J. M. Wernik and S. A. Meguid. 2014 *Materials & Design* **59** 19–32
[13] S. Chandrasekaran, N. Sato, F. Tölle, R. Mülhaupt, B. Fiedler, and K. Schulte. 2014 *Composites Science and Technology* **97** 90–99
[14] F. Salleh, A. Hassan, R. Yahya, and A. D. Azzahari 2014 *COMPOSITES PART B* **58** 259–266
[15] Loos, M.R., Coelho, L.A.F., Pezzin, S.H. and Amico, S.C., 2008 *Materials Research* **11** 347-352
[16] P. Garg, B. P. Singh, G. Kumar, T. Gupta, I. Pandey, R. K. Seth, R. P. Tandon, and R. B. Mathur. 2011 *Journal of Polymer Research* **18** 1397–1407
[17] A. Montazeri, J. Javadpour, A. Khavandi, A. Tcharkhtchi, and A. Mohajeri 2010 *Materials and Design* **31** 4202–4208
[18] S. Rahmanian, A. R. Suraya, M. A. Shazed, R. Zahari, and E. S. Zainudin 2014 *JOURNAL OF MATERIALS&DESIGN* **60** 34–40

IOP Conf. Series: Materials Science and Engineering **210** (2017) 012059 doi:10.1088/1757-899X/210/1/012059

Prediction of temperature distribution in sericite mica drying with variable temperature and airflow condition

A Mohd Noh[1, 2*], S Mat[1], M H Roslan[1] and E Salleh[1]
[1]Solar Energy Research Institute (SERI), Universiti Kebangsaan Malaysia, 43600 Bangi, Selangor, Malaysia
[2]Engineering Research Center, Ibu Pejabat MARDI, 43400 Serdang, Selangor, Malaysia

E-mail: *arinamn@gmail.com

Abstract. To develop new drying facilities, it was important to know the impact of every input parameter to the drying process. Using a real prototype to carry out the experiment required high cost and time consuming especially for large scale drying. CFD simulation approached was one of the solution. Previous study of drying simulation only focuses on the fix value of the input parameter. This paper presents the result of CFD simulation to predict the heat distribution in sericite mica drying with variable temperature and airflow condition. Variable temperature and airflow was used because the only heat source for the dryer was from the solar energy therefore it's only available in the day time. The analysis was carried out for 24 hours of drying time. The simulation result shows that the temperature inside the sericite mica increase 8 to 10°C when the solar energy is available and it is still increasing about 4 to 7°C for 5 hours after the solar energy is absent. The result also shows that during the drying time the temperature of sericite mica that is closer to the heat source was higher compared to the one that is further away with the maximum difference of 3.8°C.

1. Introduction

Sericite mica is an inert and stable mineral belongs to mica group. It has a sheet or plate like structure with excellent smoothness. The particle size average is 8-9 μm. Sericite mica was used widely in many industries such as cosmetic industry, plastics industry and rubber industry. Currently in the production of sericite mica it needs to be dry to facilitate the handling process and to reduce the transportation cost [1]. Drying process was done using dryer which used electricity or fuel that involve high costs and faced the issue of energy source depletion. For the last four decades, many researchers have successfully developed solar dryer as an alternative dryer for many commodities [2]. Compare to conventional open sun drying, solar dryers produce higher temperatures, lower relative humidity, lower product moisture content and reduced spoilage during the drying process. In addition, they take less space and time, and they are inexpensive compared to artificial mechanical and electrical drying methods. Low capital and operating costs are the primary advantages of solar dryers, as well as the fact that they require little expertise to operate [3]. As far as the authors' knowledge, in the mineral processing industry, there is no report or study on the possibility of using solar dryer. In the development process of solar dryer for sericite mica, especially for large scale it was very important to make sure uniform drying rate throughout the drying chamber. Understanding of the drying process in detail is necessary for the process design optimizations in terms of preserving the quality of the product and energy utilization. Typically full scale design and experiment was used to see the effect of the input parameter during the drying process. This is sometimes not feasible for a large scale drying as it involves high cost and time consuming. In a recent paper by Mujumdar and Wu [4] the authors emphasized the need for cost-effective solutions that can push innovation and creativity in drying equipment's. In the paper they suggest that a computational fluid dynamics (CFD) approach can be one of the solutions. CFD has been proven as an effective computational tool for predicting the flow behaviour and mass transfer phenomena occurring in many processes [5]. CFD based methods are gaining acceptance as a tool to evaluate and design commercial dryers for the many industries [6]. Study of previous researcher [7-9] proved that the performance of new dryer designs can be predicted by simulation work. These studies also proved that the CFD simulation result is comparable with actual conditions.

In the proposed solar dryer, it used only evacuated tube to supply heat to the dryer without the auxiliary heater. Therefore the heat will only be supplied to the system during day time when there is a solar energy from the sun. This was designed to minimize the cost of the dryer. Many past research on CFD simulation of heat distribution in the drying process involved the fix value of input parameter [8-12]. In this study variable input parameter in terms of heat source and air flow was used to represent the actual drying condition. Transient analysis was used to see the effect of the supply heat on sericite mica with time changes. The objective of this study is to predict the heat distribution inside the sericite mica with the changes in the heat source throughout the drying process for 24hours. In the drying process heat distribution was very important not only as its effect the drying rate, but it's also having an impact on the quality of the dried product [13].

2. Methods and Simulation

2.1. Design of the dryer system

The dryer system for sericite mica consist of evacuated tubes, heat exchanger, blower and drying chamber (figure 1). The heat exchanger and blower was placed at one end of the drying chamber. There is an opening at the other end of the drying chamber to remove the evaporated moisture from the product to the environment. The evacuated tube will absorb heat from the sun and heated the water that are flowing in the piping system. The blower and heat exchanger was design to be switched on when the temperature of the water inside the pipe achieved 70°C. The blower will produce an air velocity of 2.5ms^{-1} inside the drying chamber. Both the heat exchanger and blower will turn off once the temperature fall below 65°C. The drying chamber size was 1.2m height x 1.7m width x 17m long. The drying chamber was made of polycarbonate sheet. 10 pallets of sericite mica were placed inside the drying chamber where each pallet consists of 4 stack of sericite mica arrange 14 pieces per stack. The sericite mica have a round shape with approximately 0.63 m diameter and 0.05 m thickness.

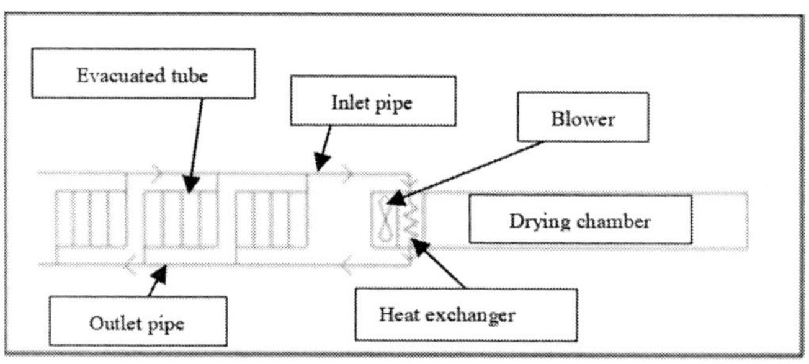

Figure 1. Schematic diagram of the sericite mica solar dryer.

2.2. Basic governing equation.

The governing equations of fluid flow and heat transfer was considered as mathematical formulations of the conservation laws of fluid mechanics and are referred to as the Navier-Stokes equations [5]. By enforcing these conservation laws over discrete spatial volumes in a fluid domain, it is possible to achieve a systematic account of the changes in mass, momentum and energy as the flow crosses the volume boundaries. In Ansys fluent the resulting equations can be written as below [14]

Continuity equation:

$$\frac{\partial \rho}{\partial t} + \nabla.(\rho \vec{v}) = S_M \tag{1}$$

Momentum equation:

$$\frac{\partial}{\partial t}(\rho \vec{v}) + \nabla.(\rho \vec{v}\vec{v}) = -\nabla p + \nabla.(\bar{\bar{\tau}}) + \rho \vec{g} + \vec{F} \tag{2}$$

Energy equation:

$$\frac{\partial}{\partial t}(\rho E) + \nabla.\left(\vec{v}(\rho E + p)\right) = \nabla.\left[k_{eff}\nabla T - \textstyle\sum_j h_j \vec{J}_j + (\bar{\bar{\tau}}_{eff}.\vec{v})\right] + S_h \tag{3}$$

The first three terms on the right-hand side of equation (3) represent an energy transfer due to conduction, species diffusion, and viscous dissipation, respectively. The turbulent kinetic energy, k, and its rate of dissipation, ε, are obtained from the following transport equations:

$$\frac{\partial}{\partial t}(\rho k) + \frac{\partial}{\partial x}(\rho k u_i) = \frac{\partial}{\partial x_j}\left[\left(\mu + \frac{\mu_t}{\sigma_k}\right)\frac{\partial k}{\partial x_j}\right] G_k + G_b - \rho\varepsilon - Y_M - S_k \tag{4}$$

$$\frac{\partial}{\partial t}(\rho \varepsilon) + \frac{\partial}{\partial x}(\rho \varepsilon u_i) = \frac{\partial}{\partial x_j}\left[\left(\mu + \frac{\mu_t}{\sigma_\varepsilon}\right)\frac{\partial \varepsilon}{\partial x_j}\right] + C_{1\varepsilon}\frac{\varepsilon}{k}(G_k + C_{3\varepsilon}G_b) - C_{2\varepsilon}\rho\frac{\varepsilon^2}{k} + S_\varepsilon \tag{5}$$

2.3. Simulation Detail

In this study, commercial CFD software Ansys Fluent v. 14 was used. Ansys Fluent used numerical finite volume methods to solve the equation. Unstructured 3D tetrahedral mesh was used to mesh the solid geometry. The solid geometry of the model is as shown in figure 2. The inlet is where the heat exchanger and blower position and the outlet is at the opening of the drying chamber. A special function inside Ansys Fluent which is UDF (user define function) was used to specify the boundary condition at the inlet because the temperature and air velocity at the inlet varied according to the water temperature inside the pipe. Transient analysis was used in the study. The initial condition of the air and sericite mica temperature used for the simulation was 30°C. To simplify the analysis, for 24 hours drying time the variation of the temperature and air flow was divided into three stages as listed in the table 1. Stage 1 represent the active drying period condition where the heat source from the sun is available, stage 2 start when the sun is set and no heat source available until midnight and stage 3 start from midnight until early morning before the sun rise. In stage 1 only 5 hours was taken into consideration although the actual time where the sun is available is longer because it's the average time where the heat from the sun is enough to heat the water inside the tube to 70°C.

Table 1. Stages in drying time.

Stage	Drying hours	Inlet Temperature (°C)	Air speed (ms^{-1})
1	1st to 5th	70	2.5
2	6th to 15th	40	0
3	15th to 24th	30	0

To simplify the solar radiation effect on the drying chamber, the boundary condition of the top surface of the drying chamber was assumed to receive an average heat flux of 800 w/m^2 [15]. This is applied only at stage 1. The drying chamber and the arrangement of the sericite mica was in symmetrical. Therefore, to save the computing time and simplify the analysis only half of the structure was modelled and symmetry boundary condition was used to represent the other half of the structure. The symmetrical plane is as shown in figure 2.

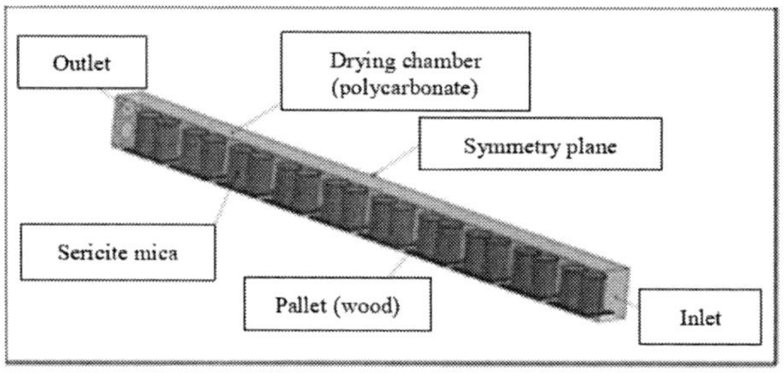

Figure 2. Solid geometry of the simulation

Other set-up of the simulation is as shown in table 2 below:

Table2. Simulation setup and configuration.

Parameter	Value/input/setting
Outlet	Pressure = 0
Side wall	No slip
Bottom wall	No heat loss
Analysis type	Transient analysis
Flow field	Realize k-ε model

The material properties used in the analysis as shown in table 3 below:

Table 3. Material properties.

Properties	Sericite mica	Pallet (wood)	Polycarbonate
Density (kg/m^3)	2820	700	1190
Thermal conductivity(W/m-k)	0.35	0.173	0.2
Specific heat (J/kg-k)	210	2310	1100

3. Results and Discussion

The simulation was carried out and the heat distribution inside the drying chamber was analysed. The temperature distribution at the centre of the sericite mica vertically and horizontally at the end of every stage is as shown in figure 3, figure 4 and figure 5 respectively.

As shown in figure 3, when there is a heat source from the inlet of 70°C and heat flux from the roof of the structure the temperature inside the sericite mica increase significantly. At this stage the blower is switch on to create an air flow of 2.5ms^{-1} flowing from the inlet towards the outlet. This create a uniform heat distribution inside the drying chamber. At the end of stage 2 (figure 4) when the blower is turn off and the heat source was gone as the result the temperature inside the sericite mica was decreasing. At stage 3 (figure 5) the temperature inside the sericite mica is further reduce as the temperature inside the dryer is lower compare to stage 2. To see in detail of the temperature variation inside the sericite mica due to difference in stage condition, a temperature at 5 points inside the sericite mica was plotted versus the drying time. The 5 point location is as shown in figure 6.

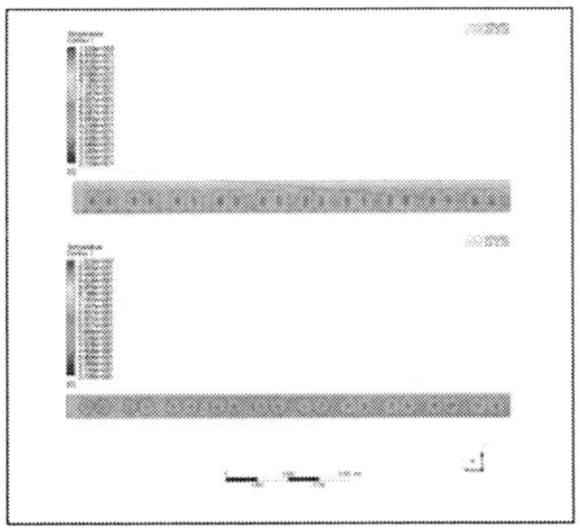

Figure 3. Temperature profile at the centre of the sericite mica (a) vertically and (b) horizontally at the end of stage 1.

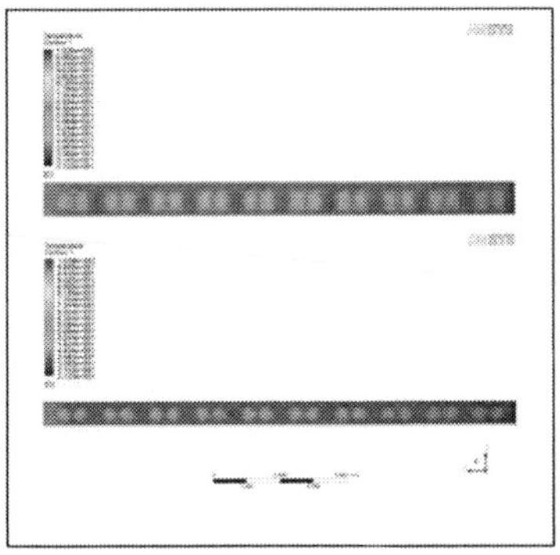

Figure 4. Temperature profile at the centre of the sericite mica (a) vertically and (b) horizontally at the end of stage 2.

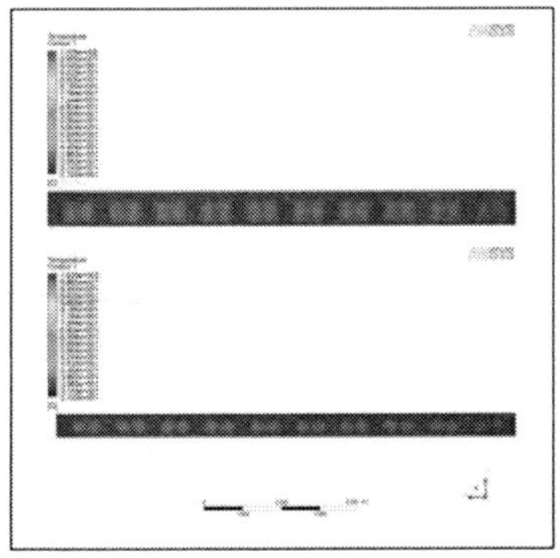

Figure 5. Temperature profile at the centre of the sericite mica (a) vertically and (b) horizontally at the end of stage 3.

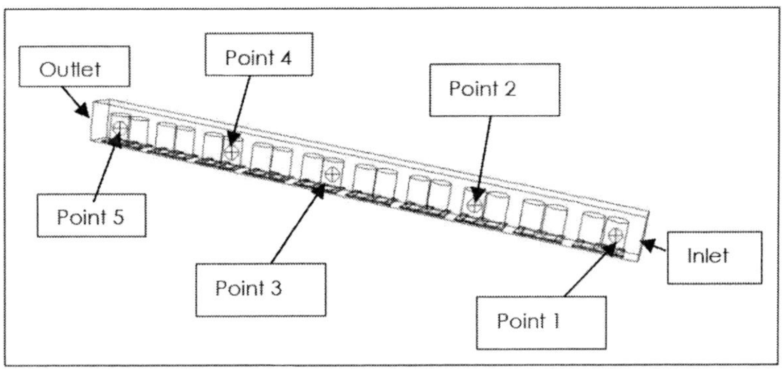

Figure 6. Location of the 5 points inside the sericite mica.

The temperature variation at every point for every hour of drying time is as shown in figure 7. From figure 7 it shows that when the heat source is applied, the temperature at point 1 which is the closest to the inlet increase up to 45.3°C and decrease to 37.3°C at the end of the drying time. The temperature increment of the sericite mica is decreasing in proportion with the distance from the inlet. Therefor point 5 is having the lowest temperature increment, which is maximum at 41.4°C and decrease to 33.3°C at the end of drying time. The temperature variation between point 1 and point 5 was 3.8°C. When the heat source is gone at stage 2 and 3, the temperature drop at point 1 and point 5 is about 8°C. However, for sericite mica located at the centre of the dryer (Point 3, 4 and 5) the temperature drop is smaller, which is about 7°C, this is due to the heat was stored much longer at the centre of the drying chamber compare to the area near the inlet and outlet.

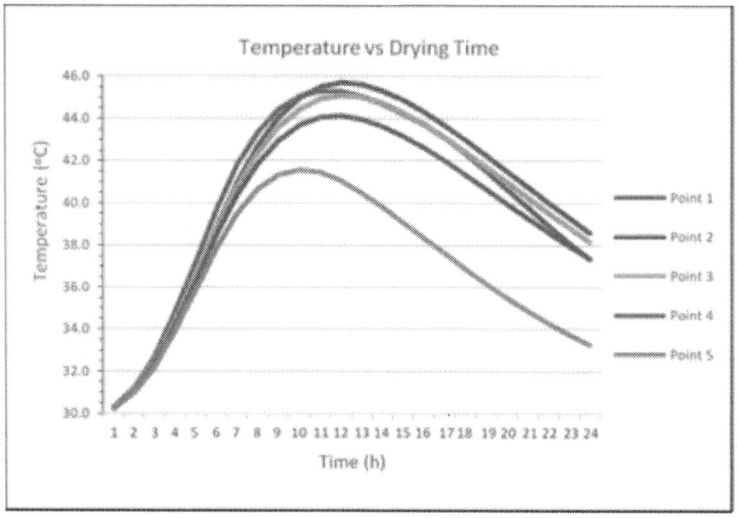

Figure 7. Temperature of every point vs drying time.

As shown in figure 7, although after hour 5 where the heat source was removed, the temperature inside the sericite mica was keep increasing about 6°C until hour 12 for point 1 to point 4. This is due to the heat that was stored inside the drying chamber and the low thermal conductivity properties of sericite mica and polycarbonate that slowing down the heat loss process to the environment. For point 5 the increment of the temperature of sericite mica after the heat source is gone is lower which is only 3°C because it was located near the opening of the outlet and expose to environment condition which is having lower temperature.

4. Conclusion

From this study it can be concluded that CFD simulation was able to predict the temperature distribution inside the sericite mica in a condition where the temperature and air flow is varied during the drying time. The simulation results show that the temperature inside the sericite mica is increasing when the heat source is available. Even after the heat source was gone the temperature of the sericite keep increasing for a few hours before its start to drop. The simulation result also shown that there is a variation in the temperature of the sericite mica depending on the location of the sericite mica from the heat source. The simulation result shows the trend that when the drying process continued the next day the starting temperature of the sericite mica will be higher and higher and this will indirectly reduce the total drying time.

References

[1] Mujumdar A S 2006 Handbook of Industrial Drying. Third Edition. CRC Press
[2] Kumar M, Kumar S S and Khatak P 2016 *Renewable and Sustainable Energy Reviews* **55** 346–360
[3] Yataganbaba A and Kurtbaş I 2016 *Renewable and Sustainable Energy Reviews* **59** 206–224
[4] Mujumdar A S and Wu Z 2008 *Drying Technology* **26** 145–153
[5] Norton T and Sun D W 2006 Trends *in Food Science & Technology* **17** 600–620
[6] Tarek J J and Madhumita B R 2010 *Drying Technology* **28** 120–154
[7] RomeroV M, Cerezo E, Garcia M I and Sanchez M H 2014 *Energy Procedia* **57** 1651 – 1658
[8] Somsila P and Teeboonma U 2014 *Advanced Materials Research* 1238-1242
[9] Yunus Y M and Al-Kayiem H H 2013 *IOP Conf. Series:Earth and Environmental Science* **16** 214 -314
[10] Misha S, Mat S, Ruslan M H, Sopian K and Salleh E 2013 *Latest Trends in Renewable Energy and Environmental Informatics* 59-66

[11] Tayyeb N, Arezou J, Mohammad Hossein K and Mortaza A 2013 *Iran. J. Chem. Chem. Eng* **32** 81-92
[12] Misha S, Mat S, Ruslan M H, Sopian K, and Salleh E 2013 *International Journal of Machine Learning and Computing* **3** 419-423
[13] Mohammad U H Joardder, Karim A and Kumar C 2013 *Journal of Mechanical Engineering and Sciences* **5** 562-568
[14] ANSYS Fluent Theory Guide
[15] Fudholi A, Mohd Yusof Othman M Y, Ruslan M H, Mat S and Sopian K 2015 *Latest Trends in Renewable Energy and Environmental Informatics* 141-149

Nomenclature;

∇	Partial derivative of a quantity with respect to all directions in the chosen coordinate system.
S_M	Mass added to the continuous phase from the dispersed second phase
p	Static pressure
$\bar{\bar{\tau}}$	Stress tensor
ρ	Density of fluid
$\vec{v}$	Velocity vector
u	Dynamic viscosity
$\vec{g}$	Gravitational body force
$\vec{F}$	External body forces
k_{eff}	Effective conductivity
$\vec{J}_j$	Is the diffusion flux of species j
S_h	Includes the heat of chemical reaction, and any other volumetric heat sources
k	Turbulent kinetic energy
ε	Rate of dissipation
G_k	Generation of turbulence kinetic energy due to the mean velocity gradients
G_b	Generation of turbulence kinetic energy due to buoyancy
$C_{1\varepsilon}, C_{2\varepsilon}, C_{3\varepsilon}$	Constant used in turbulent model
σ_k , σ_ε	Turbulent Prandtl numbers for k and ε respectively.
S_k , S_ε	User-defined source terms.
μ	Molecular viscosity
μ_t	Turbulent (or eddy) viscosity

IOP Conf. Series: Materials Science and Engineering **210** (2017) 012060 doi:10.1088/1757-899X/210/1/012060

Chaos-enhanced Stochastic Fractal Search algorithm for Global Optimization with Application to Fault Diagnosis

Tuan A Z Rahman[1], N A Abdul Jalil[1], A As'arry[1] and R K Raja Ahmad[2]

[1]Sound and Vibration Research Group, Department of Mechanical and Manufacturing Engineering, Faculty of Engineering, Universiti Putra Malaysia (UPM) 43400 Serdang, Selangor, Malaysia

[2]Department of Electrical and Electronics Engineering, Faculty of Engineering, Universiti Putra Malaysia (UPM) 43400 Serdang, Selangor, Malaysia

E-mail: tuanzahidi@upm.edu.my

Abstract. Support vector machine (SVM) has been known as one-state-of-the-art pattern recognition method. However, the SVM performance is particularly influenced byits parameter selection. This paper presents the parameter optimization of an SVM classifier using chaos-enhanced stochastic fractal search (SFS) algorithm to classify conditions of a ball bearing. The vibration data for normal and damaged conditions of the ball bearing system obtained from the Case Western Reserve University Bearing Data Centre. Features based on time and frequency domains were generated to characterize the ball bearing conditions. The performance of chaos-enhanced SFS algorithms in comparison to their predecessor algorithm is evaluated. In conclusion, the injection of chaotic maps into SFS algorithm improved its convergence speed and searching accuracy based on the statistical results of CEC 2015 benchmark test suites and their application to ball bearing fault diagnosis.

1. Introduction

Vibration-based condition monitoring (CM) currently plays an important role in manufacturing industries to continuous surveillanceof rotating machinery compared to the conventional methods. Reducing the productivity costs by minimizing themachine's downtime is one of its crucial benefits in order to enhance goods production.As vibration-based condition monitoring area can be treated as pattern recognition, support vector machines (SVMs) approach is commonly used. However, the SVMs performance is really dependable to its kernel and parameters selection. This problem can be formulated as an optimization problem. Thus, many optimization algorithms were applied to overcome this essential problem, systematically.

Zhang *et al.* proposed a hybrid method of the barebones differential evolution (BBDE) algorithm for SVMs parameters tuning [1]. Several others evolutionary algorithms (EAs) have been employed to optimize the SVMs parameters such as genetic algorithm (GA) [2], particle swarm optimization (PSO)[3], ant colony optimization (ACO) [4]and artificialimmunization (AIA) algorithms [5].

In this paper, four chaos-enhanced stochastic fractal search (SFS) algorithms are presented. The performance of these improved variants of SFS algorithms was evaluated using modern benchmark test suites (CEC 2015) and the SVMs parameters tuning for fault diagnosis as its engineering application.

2. Enhanced Stochastic Fractal Search algorithm with chaos

Stochastic fractal search (SFS) optimization algorithm was developed by Salimi to imitate a growth process [6]. The particles in the process try to expand its growing in searching space. This metaheuristic algorithm shows promising results with short computational time. The main advantage of this algorithm is less starting tuning parameters to initiate the searching process.There are two main equations from the original SFS algorithm have been modified with the introduction of a chaotic variable named as α as follow:

$$GW = Gaussian(\mu_{BP}, \sigma) + (\varepsilon \times BP - \alpha \times P_i) \quad (1)$$

and,

$$P'_i(j) = P_r(j) - \alpha \times (P_t(j) - P_i(j)) \quad (2)$$

Noted that, eq. 1 was part of Diffusion Process while eq. 2 in First Updating Process in the original SFS algorithm. This study investigated two different one-dimension non-invertible chaotic maps influence toward the algorithm performance as proposed in previous studies by Saremi*et al.*[7] and Mitic*et al.*[8]. These new chaotic equations will force the particles to move towards the current best optimal solution in a chaotic manner. Table 1 tabulates the mathematical description of the proposed addition of chaotic maps while Figure 1 shows the graphical presentation of these maps over 100 generations. Each of the chaotic maps has the starting point 0.7 and normalized to a range of [0, 1].

Table 1. The mathematical description of chaotic maps

No	Map Name	Equation
1	Chebyshev	$x_{i+1} = \cos(i \cos^{-1}(x_i))$
2	Gauss/Mouse	$x_{i+1} = \begin{cases} 1, for x_i = 0 \\ \frac{1}{\mathrm{mod}(x_i, 1)}, otherwise \end{cases}$

Four chaos-enhanced stochastic fractal search (CFS) algorithms were developed with the implementation of two different chaotic maps at two different parts of SFS algorithm. The initial value of 0.95 was selected for parameter α as suggested in [8]. Table 2 shows the combination of chaotic maps for CFS algorithms. More details regarding the CFS algorithms can be found in [9].

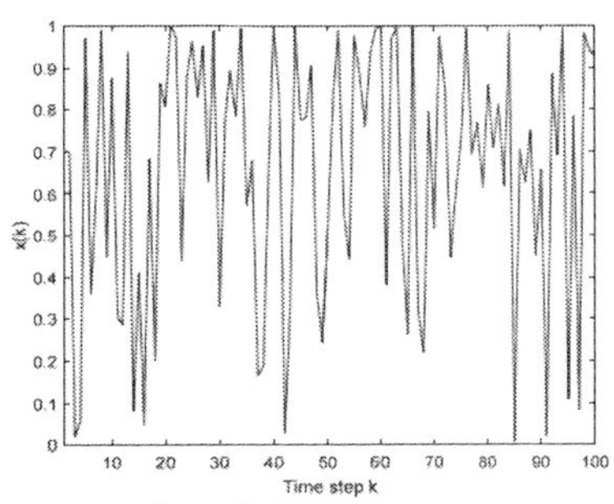

a) Chebyshev map

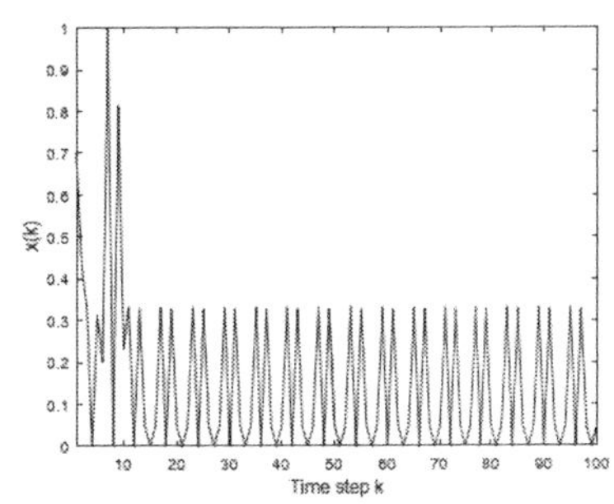

b) Gauss/Mouse map

Figure 1. Visualization of chaotic maps used

Table 2. Chaos-enhanced stochastic fractal search (CFS) algorithms

Algorithm	**Diffusion Process**	**Updating Process**
Original SFS	Random [0, 1]	Random [0, 1]
CFS01	Chebyshev map	Chebyshev map
CFS02	Chebyshev map	Gauss/Mouse map
CFS03	Gauss/Mouse map	Chebyshev map
CFS04	Gauss/Mouse map	Gauss/Mouse map

In this study, 4 modern benchmark test functions from CEC 2015 were used to evaluate the performance of CFS algorithms. Two different levels of problem dimension, D were investigated which are 10 and 30. The number of population (Start Point), *NP* was fixed to 100. Maximum Diffusion Number (*MDN*) equal to 1 with the first Gaussian walk is utilized. Note that, all calculations were performed in MATLAB 2015b software that runs on a desktop PC Intel ® Core ™ i5 of CPU 3.30 GHz with 4GBs RAM and Window 7 (64 bit) operating system. The full results of CFS algorithms evaluation using modern benchmark functions are tabulated in Table 3a and 3b. Non-parametric statistical analysis was conducted based on a total of 50 simulation runs.

3. Support vector machines

Support vector machines (SVMs) can be simplest discussed using a description of linear discriminant analysis. A hyperplane (i.e. a straight line in two dimensions) will be created to separate two classes of data that generalized best (maximum margin). The hyperplane equation as follow,

$$D(x) = \langle w, x \rangle = 0 \tag{3}$$

where$x$ is the input vector and w is the vector of free weights. Each data point x_k in the training set is assigned to a class on the basis of the separating condition given by,

$$x_k \in C_1 \Rightarrow D(x_k) = \langle w, x_k \rangle \geq 1$$

$$x_k \in C_2 \Rightarrow D(x_k) = \langle w, x_k \rangle \geq -1 \tag{4}$$

whereC_1 and C_2 are classes, with respective class labels (1) and (-1), or more concisely as,

$$D(x_k) = y_k \langle w, x_k \rangle \geq 1 \tag{5}$$

wherey_k is the class label. The distance of each point in the training set from the separating hyperplane is,

$$\frac{D(x_k)}{\|w\|} \tag{6}$$

An interval that contains the separating hyperplane but excluding all data is called margin, τif,

$$y_k \frac{D(x_k)}{\|w\|} \geq \tau \tag{7}$$

is satisfied for all k. The optimal margin is illustrated in Figure 2. Note that the parameterization of the hyperplane is currently arbitrary. This can be fixed by specifying,

$$\|w\|_T=1 \tag{8}$$

and this converts eq. (7) into the separation condition of eq. (5).

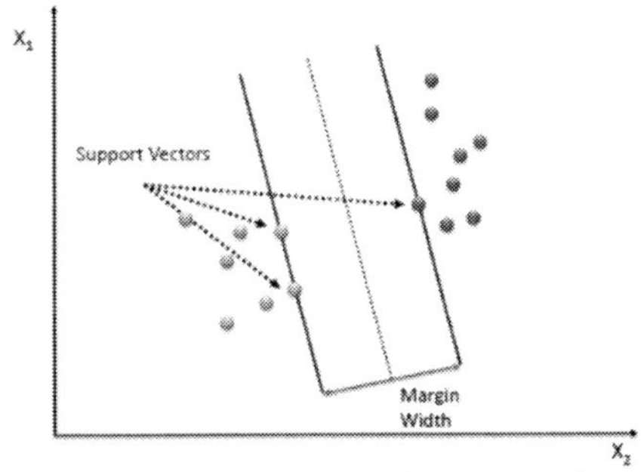

Figure 2. Optimal separating hyperplane

The objective of SVMs is to maximize the margin so that the hyperplane will be placed at the furthest point from data and can be achieved by minimizing $\|w\|$. An appropriate objective function is,

$$Q(w)=\frac{1}{2}\|w\|^2-\sum_{i=1}^{N}\alpha_i y_i\left(\langle w,x_i\rangle-1\right) \tag{9}$$

where the parameters α_i are Lagrange multipliers. Further discussion can be found in [10] regarding convex and the dual formulation of the problem. The analysis so far only considers linear hyperplane. The technique can be made nonlinear by the use of kernel trick which involves quadratic, polynomial and radial basis function (RBF) kernels.

4. Application of CFS algorithms to fault diagnosis

Theapplication of vibration signals is relatively usual in condition monitoring and damage detection area. Signal processing techniques is an important expertise in order to extract diagnosis information from the selected raw vibration data. Features extraction and selection phase are performed to characterize the condition before pattern recognition methods were employed to identify the damages.

Seed faults bearing data from Case Western Reserve University was used in this study [11]. Figure 3 shows the schematic diagram of the experimental setup. Three faults which are outer race fault (OR), inner race (IR) fault and ball (B) fault were introduced to the drive end bearing. The defects sizes are 0.007 inches and 0.021 inches. Each original signalwas divided into 10 signals for each condition of fault types and sizes. Four features (two time-domain and two frequency-domain features) were calculated based on [1].

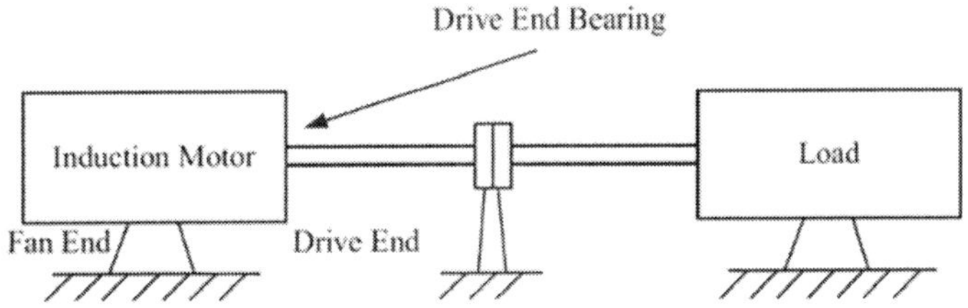

Figure 3. The schematic diagram of the experimental setup (adapted from[1])

Table 3a. Statistical result tested on CEC 2015 benchmark functions (200 Generations)

Function	Dimension		SFS	CFS 01	CFS 02	CFS 03	CFS 04
F1	10	Mean	2.6221e+04	722.9056	**216.5546**	706.9905	316.8680
		SD	1.2390e+04	676.7885	**387.5189**	520.7234	1.1308e+03
	30	Mean	4.3292e+06	**2.9398e+06**	2.9917e+06	2.9521e+06	3.4427e+06
		SD	1.6492e+06	1.6966e+06	1.9840e+06	**1.3451e+06**	1.7882e+06
F3	10	Mean	19.5146	20.2153	**18.8614**	19.0280	20.0611
		SD	3.5328	0.0544	4.8138	4.7194	**0.0471**
	30	Mean	20.7685	20.7608	20.6563	20.7641	**20.6341**
		SD	0.0610	0.0598	0.0648	**0.0566**	0.0806
F6	10	Mean	309.5684	52.8969	**7.0392**	45.0227	8.6524
		SD	77.9991	19.4256	**4.1259**	18.4530	5.5534
	30	Mean	1.3456e+05	1.3207e+04	1.1302e+04	1.1861e+04	**1.0792e+04**
		SD	6.6003e+04	5.9933e+03	7.5751e+03	**5.1988e+03**	7.6252e+03
F10	10	Mean	280.5511	219.9898	**217.5860**	220.0394	217.6341
		SD	20.2745	1.3676	**0.9332**	1.2873	1.0027
	30	Mean	5.8437e+04	5.8325e+03	5.7653e+03	**5.0251e+03**	8.5475e+03
		SD	2.2595e+04	1.8679e+03	7.7974e+03	**1.7204e+03**	1.3749e+04
Algorithm ranking based on average mean value computed via Friedman test			4.5975 (5)	3.2900 (4)	**1.9400 (1)**	3.1950 (3)	1.9775 (2)

Table 3b. Non-Parametric test results based on Table 3a

No. of function	Dimension	Algorithm	Friedman test		Wilcoxon signed rank test			
			Mean rank	*p*	SFS-CFS 01	SFS-CFS 02	SFS-CFS 03	SFS-CFS 04
					p	*p*	*p*	*p*
F1	10	SFS	5.0000	<0.05 7.1986e-31	7.5569e-10	7.5569e-10	7.5569e-10	7.5569e-10
		CFS 01	3.2800					
		CFS 02	1.7200					
		CFS 03	3.2600					
		CFS 04	1.7400					
	30	SFS	4.0000	<0.05 1.2016e-05	2.2093e-05	3.3554e-04	4.0002e-05	**0.0036**
		CFS 01	2.5200					
		CFS 02	2.7400					
		CFS 03	2.6800					

		CFS 04	3.0600					
F3	10	SFS	4.0000	<0.05 1.3843e-26	**0.9040**	1.4705e-07	**0.1463**	2.0113e-07
		CFS 01	3.9600					
		CFS 02	1.6000					
		CFS 03	3.8000					
		CFS 04	1.6400					
	30	SFS	3.8000	<0.05 1.5408e-19	**0.4486**	1.6775e-08	**0.9116**	4.2535e-09
		CFS 01	3.7000					
		CFS 02	2.0000					
		CFS 03	3.8600					
		CFS 04	1.6400					
F6	10	SFS	5.0000	<0.05 5.9277e-38	7.5569e-10	7.5569e-10	7.5569e-10	7.5569e-10
		CFS 01	3.6000					
		CFS 02	1.4600					
		CFS 03	3.4000					
		CFS 04	1.5400					
	30	SFS	5.0000	<0.05 4.1620e-22	7.5569e-10	7.5569e-10	7.5569e-10	7.5569e-10
		CFS 01	2.9400					
		CFS 02	2.2600					
		CFS 03	2.5600					
		CFS 04	2.2400					
F10	10	SFS	5.0000	<0.05 7.4688e-32	7.5569e-10	7.5569e-10	7.5569e-10	7.5569e-10
		CFS 01	3.1800					
		CFS 02	1.7800					
		CFS 03	3.4200					
		CFS 04	1.6200					
	30	SFS	4.9800	<0.05 1.9589e-23	7.5569e-10	7.5569e-10	7.5569e-10	8.0311e-10
		CFS 01	3.1400					
		CFS 02	1.9600					
		CFS 03	2.5800					
		CFS 04	2.3400					

Multiclass SVMs based on one-against-one technique is employed to classify the bearing conditions. One-third of each class fault is used for training the SVM and the remaining data forthe testing purposed. The training and testing data were selected randomly and repeated for 30 times before average classification error is calculated as the objective function. CFS algorithms were evaluated to optimize two SVMs parameters which are the soft margin/penalty parameter, C and the scaling factor for RBF-kernel, γ. Then, the obtained parameters were used to generate an average classification error of 10,000 SVMs runs in verification stage. The performances of each CFS algorithm are compared to their predecessor algorithm. Figure 4plotted the features data based on 1st and 2nd principle components for visualization purpose. All data have seen clearly separated between normal and damaged conditions. For Inner Race (IR021) and Outer Race (OR021) of fault size 0.021 inch, the data can be divided in 3-Dimension view. The original normalized four features data were used to generate the SVMs classifier model.

The initial parameters of SFS and CFS algorithms were set as follow; starting point (*NP*) = 100 particles, number of maximum iterations (*G*) = 50 generations, problem dimension (*D*) = 2, searching range of [0 150] for the soft margin/penalty (*C*) and [0 10] for the scaling factor of RBF-kernel (γ).Maximum Diffusion Number (*MDN*) was set as 1 while 1st Gaussian Walk is selected.

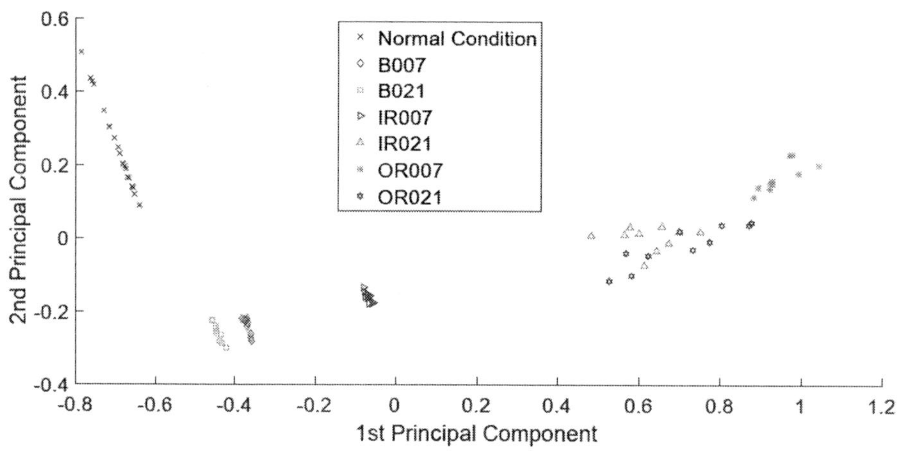

Figure 4. The visualization of features data

5. Results and discussion

Table 4 tabulates the results of bearing fault classification using SVMs-based SFS algorithms.The second and third column shows the obtained scaling factor, γ and soft margin, C respectively. The average of percentage classification accuracy for 10,000 runs is shown in the fourth column with its standard deviation. Based on classification error in the fifth column, CFS 04 algorithm performance was better than other CFS and its predecessor algorithms. The CFS 04 algorithm has achieved 99.992% classification accuracy with the lowest standard deviation of 0.18% in comparison with others. On the other hand, the addition of Chebyshev map in Diffusion and First Updating Processes of SFS algorithm has slightly deteriorated its performance.

The benchmark test suites of CEC 2015 and bearing fault classification results indicated that CFS 04 algorithm shows better searching accuracy compared to the SFS and its other chaos-enhanced algorithms.

Table 4. CFS algorithms performance in comparison to SFS

Algorithm	Scaling Factor (γ)	Soft Margin, (C)	Accuracy (%)	Classification Error
SFS	1.8438	0.3771	99.971±0.340	$2.8750\text{x}10^{-4}$
CFS 01	5.3945	4.0529	99.746±1.240	$2.5375\text{x}10^{-3}$
CFS 02	2.3719	1.0194	99.986±0.240	$1.3750\text{x}10^{-4}$
CFS 03	1.6727	1.1077	99.984±0.260	$1.5833\text{x}10^{-4}$
CFS 04	1.9870	1.1471	**99.992±0.180**	$\mathbf{7.0833x10^{-5}}$

6. Conclusion

In this study, four variants of SFS algorithm enhanced with chaos is introduced. Their searching accuracy and convergence speed performance were evaluated using modern benchmark test suites of CEC 2015 and engineering application to ball bearing fault diagnosis. CFS algorithm with Gauss/Mouse map in Diffusion and First Updating Processes show superiority performance in comparison to its predecessor and other CFS algorithms.

Acknowledgment

The first author would like to acknowledge Universiti Putra Malaysia (UPM) and Ministry of Higher Education (MOHE) Malaysia for the PhD scholarship funding.

References

[1] X. Zhang, D. Qiu, and F. Chen, Support vector machine with parameter optimization by a novel hybrid method and its application to fault diagnosis, *Neurocomputing,* vol. 149, Part B, pp. 641-651, 2/3/ 2015.

[2] A. C. Lorena and A. C. P. L. F. de Carvalho, Evolutionary tuning of SVM parameter values in multiclass problems, *Neurocomputing,* vol. 71, pp. 3326-3334, 10// 2008.

[3] S.-W. Lin, K.-C. Ying, S.-C. Chen, and Z.-J. Lee, Particle swarm optimization for parameter determination and feature selection of support vector machines, *Expert Systems with Applications,* vol. 35, pp. 1817-1824, 11// 2008.

[4] X. Zhang, X. Chen, and Z. He, An ACO-based algorithm for parameter optimization of support vector machines, *Expert Systems with Applications,* vol. 37, pp. 6618-6628, 9// 2010.

[5] I. Aydin, M. Karakose, and E. Akin, A multi-objective artificial immune algorithm for parameter optimization in support vector machine, *Applied Soft Computing,* vol. 11, pp. 120-129, 1// 2011.

[6] H. Salimi, Stochastic Fractal Search: A powerful metaheuristic algorithm, *Knowledge-Based Systems,* vol. 75, pp. 1-18, 2// 2015.

[7] S. Saremi, S. Mirjalili, and A. Lewis, Biogeography-based optimisation with chaos, *Neural Computing and Applications,* vol. 25, pp. 1077-1097, 2014/10/01 2014.

[8] M. Mitić, N. Vuković, M. Petrović, and Z. Miljković, Chaotic fruit fly optimization algorithm, *Knowledge-Based Systems,* vol. 89, pp. 446-458, 11// 2015.

[9] T. A. Z. Rahman and M. O. Tokhi, Enhanced stochastic fractal search algorithm with chaos, in *2016 7th IEEE Control and System Graduate Research Colloquium (ICSGRC)*, 2016, pp. 22-27.

[10] N. Cristianini and J. Shawe-Taylor, *An introduction to support Vector Machines: and other kernel-based learning methods*: Cambridge University Press, 2000.

[11] Case Western Reserve University bearing data centre website, 2017 http://csegroups.case.edu/bearingdatacenter/pages/welcome-case-western-reserve-university-bearing-data-center-website

IOP Conf. Series: Materials Science and Engineering **210** (2017) 012061 doi:10.1088/1757-899X/210/1/012061

Automated platform for designing multiple robot work cells

N S Osman*[1], M A A Rahman[1*], A A Abdul Rahman[1], S H Kamsani[1], B M Bali Mohamad[1], E Mohamad[1], Z A Zaini[2], M F Ab Rahman[2] and M N H Mohamad Hatta[1]

[1]Integrated Manufacturing System (I'Ms), Advanced Manufacturing Centre (AMC), Faculty of Manufacturing Engineering, Universiti Teknikal Malaysia Melaka, Hang Tuah Jaya, 76100 Melaka, Malaysia.
[2]Department of Occupational Safety and Health, Aras 2, 3 & 4, Blok D3, Kompleks D, Pusat Pentadbiran Kerajaan Persekutuan, 62530 W. P. Putrajaya, Malaysia.

E-mail: *norsuriyanti@gmail.com

Abstract. Designing the multiple robot work cells is very knowledge-intensive, intricate, and time-consuming process. This paper elaborates the development process of a computer-aided design program for generating the multiple robot work cells which offer a user-friendly interface. The primary purpose of this work is to provide a fast and easy platform for less cost and human involvement with minimum trial and errors adjustments. The automated platform is constructed based on the variant-shaped configuration concept with its mathematical model. A robot work cell layout, system components, and construction procedure of the automated platform are discussed in this paper where integration of these items will be able to automatically provide the optimum robot work cell design according to the information set by the user. This system is implemented on top of CATIA V5 software and utilises its Part Design, Assembly Design, and Macro tool. The current outcomes of this work provide a basis for future investigation in developing a flexible configuration system for the multiple robot work cells.

1. Introduction

Rapid development of various industrial technologies has stimulated a substantial increase in the use of industrial robot [1]. It is proved through the latest statistic presented by the International Federation of Robotics (IFR) in its 2015 World Robotics report, the utilization of industrial robot in manufacturing industry is shown to be increasing from 2013 to 2017. Also, it is predicted that by 2018, global sales of industrial robots will grow on average by at least 15% yearly [2]. Few researchers claim that industrial robot is used extensively in application such as welding, drilling, assembly, transportation, cutting, spraying, measuring, stacking, and laser processing [3]–[5]. Consequent from the report and claim, configuration work of robot work cell become an intensive challenge among the design engineers in ensuring the developed robot work cell effectively in actual production. Based on the common configuration approach, configuring work requires a high investment cost [6], long commissioning time, high level of expert knowledge and lots of human involvement [1], [7], [8]. Additionally, configuration work is tricky to implement as it needs to account about the safety element [9], [10]. Hence, this work proposes an automated configuration platform of multiple robot work cells, where aiming to provide fast and easy configuration work with minimum cost, human involvement and certain investments such as the trial and error adjustment.

Previous work introduced in [9] on the design safety measure robot work cell are taken into consideration for this work. The illustration of two-dimensional (2D) robot work cell with its safety measure as depict in figure1 is used to model the safety multi-robot work cells.

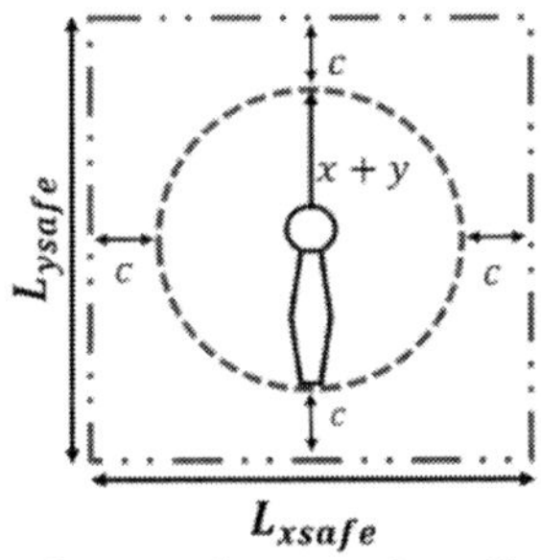

Figure 1. Illustration of 2D Robot Work Cell with Safety Clearance

$$A_{safe} = L_{xsafe} \; x \; L_{ysafe} = 2(X + Y + C) \; x \; 2(X + Y + C) \quad (1)$$

Where,
X: Length of robot arm (mm)
Y: Length of the robot tooling and work piece (mm)
C: Clearance for the worker movement in a work cell taken as 650mm

A concept of variant-shaped configuration introduced in [11], [12] are taken as reference. This concept involves a maximum of ten industrial robots where only two auxiliary equipment are considered which are the robot tooling and the work piece. The variant-shaped configuration was established by joining one or more square robot work cell (figure 1) side by side according to the quantity of robot used. The configuration concept emphasizes only on the simple arrangement. Meanwhile, the complex arrangement is excluded in this work due to the difficulty in tracking the configuration trend. The complex arrangement refers to the corner, half facing, diagonal and mix arrangement. Additionally, this configuration concept excludes the rigid transformation condition such as translation, rotation, reflection or glide reflection. As the results, the data of possible layout of the variant-shaped configuration is presented as shown in table 1. This is the key element to the invention of the next stage.

Table 1. Variant-Shaped Robot Work Cell Configuration Data

Number of Robot, Nr	1	2	3	4	5	6	7	8	9	10
Number of Configuration, Nc	1	1	2	5	12	35	108	369	1285	4655

Afterwards, an optimization concept of configuration work in [11] is utilized. In this work, the earlier variant-shaped configuration (data, pattern and equation) was optimized by grouping the number of configuration into the columns configuration based on the same total number of robot in the references line at the horizontal plane. The columns configuration must avoiding the rigid transformation and the diversity position. Diversity position refers to the layout or configuration which have the same number of robot in every horizontal line but different in position where they can be considered as a same layout. At this stage, the configuration data, pattern and equation have been optimized and the complexity have been reduced. table 2 presents the optimal robot work cell configuration data.

By using the previous concept, the mathematical models of each optimal configuration was developed and the rule of optimal selection was modelled in this work where it provides a framework to create the GUI later on. Few studies [13]–[20] has proved that CATIA is widely used in creating an automatic and intelligent system which able to enhance the development time, minimize the errors and could introduce technologies faster to the market [17], [18]. Therefore, CATIA 3D models, Visual Basic (VB) Interface and macro tools were selected for developing a completed GUI for the automated platform.

Table 2. Optimal Robot Work Cell Configuration Data

Number of Robot, Nr	1	2	3	4	5	6	7	8	9	10
Number of Optimal Configuration, No.c	1	1	2	3	4	6	8	12	16	22

The developed automatic platform may provide the simulation of the robot work cell configuration before the real configuration work where an optimal robot work cell layout was proposed and the workspace area, W_A and manufacturing throughput time, MTT of the robot work cell layout were calculated and presented. The consequence of this work will enhance the current approach of configuring the robot work cell. Also, it could intensify the human-robot interaction as well as alleviate the configuration cost and time in future.

2. Related Works

Configuration work here refers to the laying out the robot work cell according to the manufacturing strategy set by the certain industry. Most of the configuration studies aim to propose a quality configuration approach where it would be very advantageous to the design engineers in handling the configuration work later on. For example, a study in [14] presented the optimal placement for the described duties within the robotic manipulator workspace. This work utilized the response surface approach for the both path translation and rotation. Based on the approach, a robotic optimisation tool as the add-in to the RobotStudio has been developed. The approach was verified properly by optimising the positioning of the industrial robots with its path in 4 varies showcases for attaining a minimal cycle time. Despite, this study only focused on single robot, whereas in the actual application, one or many robots could be present in the work cell. Another study in [15] introduced an optimisation work for laying out multi-robot work cells in a vertical section plane where emphasis to manufacture the outer surface of a large fuselage panel. This work intends to maximise the overlapped workspace among two robots without collision between both robots and work pieces. This work capable to achieve a layout that yielded reasonable positions by tested on an existing layout. Anyhow, this work can only be used for two robot only.

Afterwards, a study in [16] provides a system with its method for optimising of the position of the different workstations in the industrial robot work cell. The developed system and its method involve one or many tasks and the industrial robot for carrying out these tasks. This work intends to enhance the performance of the robot as well as enhance the productivity of the robotic work cell. Nonetheless, the study was not applicable in the cases which used more than one robots. Additionally, an innovative layout approach by [17] was promoted. The developed approach is based on the Differential Evolution (DE) where it is utilized for solving the Facility Layout Planning (FLP). Robotic work cell layout was one FLP example that was described in this work. The mathematical FLP model includes few constraints and optimisation objective was proposed and an automated layout optimisation was developed for accomplishing the 3-D visualisation demonstration for the optimal layout. Yet, this work requires highly skilled and experienced worker for understanding the approach.

Besides, an optimal robot positioning for tasks execution was introduced in [2]. This work aims to optimise the base position of an industrial robot to reach all predefined tasks as well as minimise the cycle time. Apart from that, this work integrated the robot inverse kinematics and collision avoidance with a derivative-free optimisation algorithm. The outcomes of this work had successfully provided a feasible solution in enhancing the cycle time for a robot station by placing the robot in optimal position. However, future work would be focused on the optimization of the placements for several robots, and on the automatic creation of the optimization approach.

The previous studies depicts that various configuration approach was proposed, yet there are still needs in enhancing the current configuration work. Thus, this work aims to settle the raised issue through modelling mathematical models of each optimal robot work cells and the rule of optimal selection where it is the framework of the automated platform in identifying the final optimum layout of robot work cell. The proposed automated platform could simplify the configuration work later on and lessen the configuration cost, time, human involvement with low future investment and as well as satisfy the user necessities.

3. Development Process of the Automated Configuration Platform

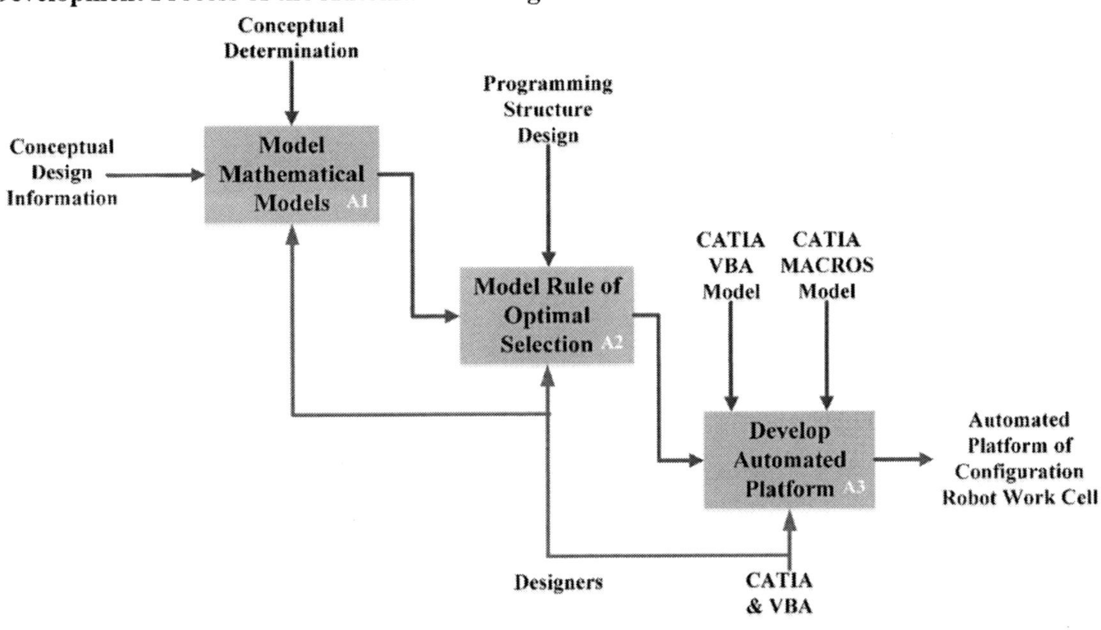

Figure 2. Development Process of the Automated Configuration Platform

Development of the automated configuration platform begins with the determination of previous configuration concept where the probable optimal robot work cells were identified. Later, the mathematical models such as workspace area, A_W and manufacturing throughput time, MTT were derived for each of the probable optimal robot work cell. Next, the rule of optimal selection was modelled in the form of forward chain which it is the framework of the automated configuration platform. Lastly, a completed GUI is generated by using the CATIA VBA and macro. Three levels of user form will be designed where each of the level will running their own specific tasks. The completed GUI of the automated configuration platform is verified by executing a set of user data based on the selection of configuration types (normal configuration, configuration with minimum workspace area, configuration with minimum throughput time and configuration with both minimum workspace area and throughput time). The outcomes of the verification is presented and discussed through this activity.

4. Result and Discussion

4.1 Probable Optimal Robot Work Cell with its Mathematical Models

As the results, 22 probable optimal robot work cell layouts have been finalized for 10 number of robots. Three probable optimal robot work cell configurations were selected to represent the outcome of this work where the workspace are, A_w and the manufacturing throughput time, MTT were derived for each of the configuration. Both workspace area, $A_w$and manufacturing throughput time, MTT equations were modelled based on the basis equations introduced in [21].

4.1.1 Configuration 1

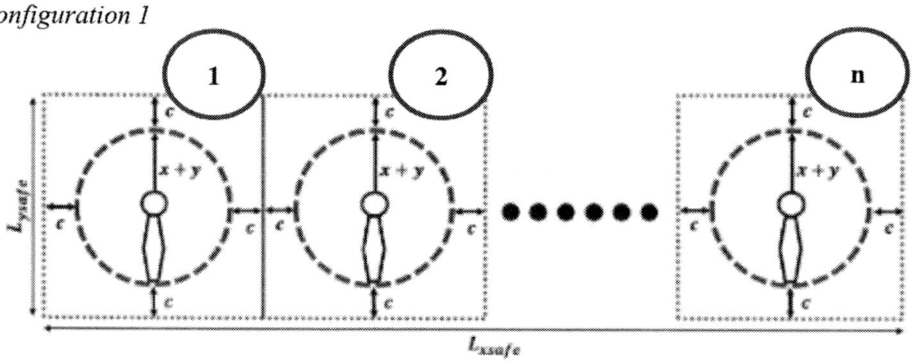

Figure 3. Robot Work Cell Configuration 1

I. Workspace Area, A_w

$$A_W = L_{safe} \times W_{safe}$$

$$W_{safe} = 2 \max_{1 \leq n \leq 10} [x_n + y_n + c]$$

$$L_{safe} = 2 \sum_{i=1}^{n} [(x_i + y_i) + nc]$$

II. Manufacturing Throughput Time, MTT

$$MTT = \sum_{i=1}^{n} (t_p + t_i + t_m + t_q)_i$$

4.1.2 Configuration 2 The manufacturing throughput time for this configuration will be diverse from the configuration 1 where for robot 1 and robot 2, only the maxmimum manufacturing throughput time will be selected to be sum up with the rest of robot's manufacturing throughput time.

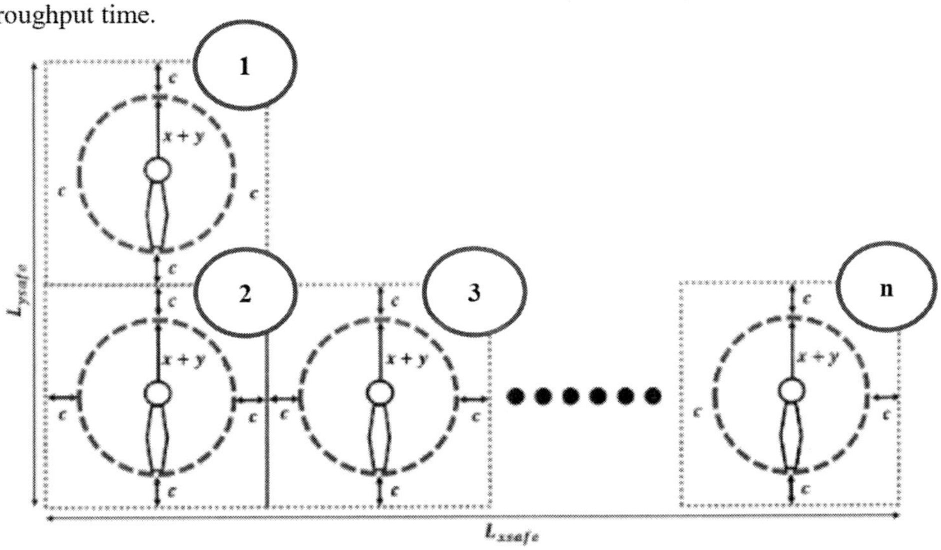

Figure 4. Robot Work Cell Configuration 2

I. Workspace Area, A_W

$$A_W = L_{safe} \times W_{safe}$$

$$L_{safe} = 2\sum_{i=2}^{n}[(x_i + y_i) + (n-1)c]$$

Or

$$= 2[(x_1 + y_1) + c]$$

$$W_{safe} = 2\sum_{i=1}^{2}[(x_i + y_i) + 2c]$$

Or

$$= \max_{3 \leq n \leq 10} 2[(x_n + y_n) + c]$$

II. Manufacturing Throughput Time, MTT

$$MTT = \max \textstyle\sum_{i=1}^{2}(t_p + t_i + t_m + t_q)_i + \sum_{i=3}^{n}(t_p + t_i + t_m + t_q)_i$$

4.1.3 *Configuration 3* The manufacturing throughput time for this configuration is the summation of maxmimum manufacturing throughput time for robot 1 and robot 3 with maxmimum manufacturing throughput time betweeen robot 3 and robot 4 and the rest of robot's manufacturing throughput time.

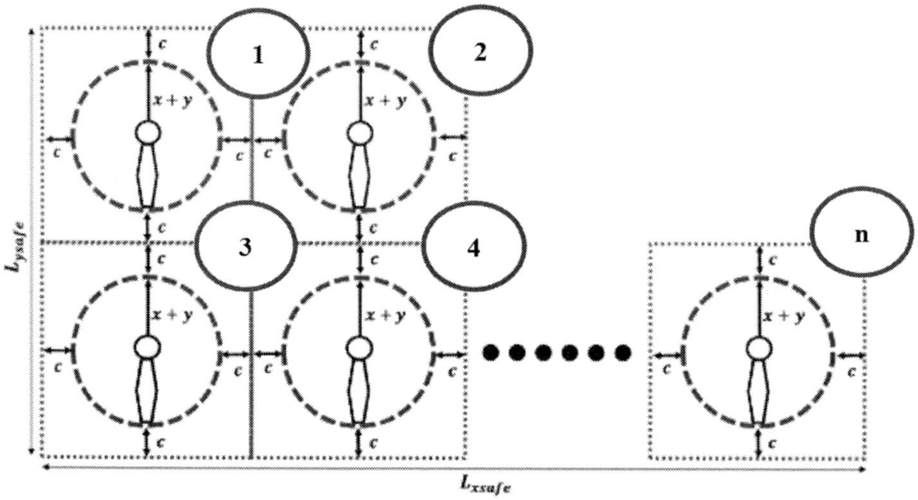

Figure 5. Robot Work Cell Configuration 3

I. Workspace Area, A_w

$$A_W = L_{safe} \times W_{safe}$$

$$L_{safe} = 2\sum_{i=3}^{n}[(x_i + y_i) + (n-2)c]$$

Or

$$= 2\sum_{i=1}^{2}[(x_i + y_i) + 2c]$$

$$W_{safe} = 2\sum_{n=1,3}[(x_n + y_n) + 2c]$$

Or

$$= 2\sum_{n=2,4}[(x_n + y_n) + 2c]$$

Or

$$= \max_{5\le n\le 10} 2[(x_n + y_n) + c]$$

II. Manufacturing Throughput Time, MTT

$$MTT = \max\sum_{i=1,3}(t_p + t_i + t_m + t_q)_i + \max\sum_{i=2,4}(t_p + t_i + t_m + t_q)_i + \sum_{i=5}^{n}(t_p + t_i + t_m + t_q)_i$$

4.2 Rule of Optimal Selection of the Automated Configuration Platform

Figure 6 presents the forward chain of the optimal selection of the robot work cell for up to three number of robots. Basically, 11 forward chains and rules were developed in this work where only 3 forward chains were selected represent the outcome where it is used as a data driven method of deriving a particular goal from a given knowledge base with a set of inference rules.

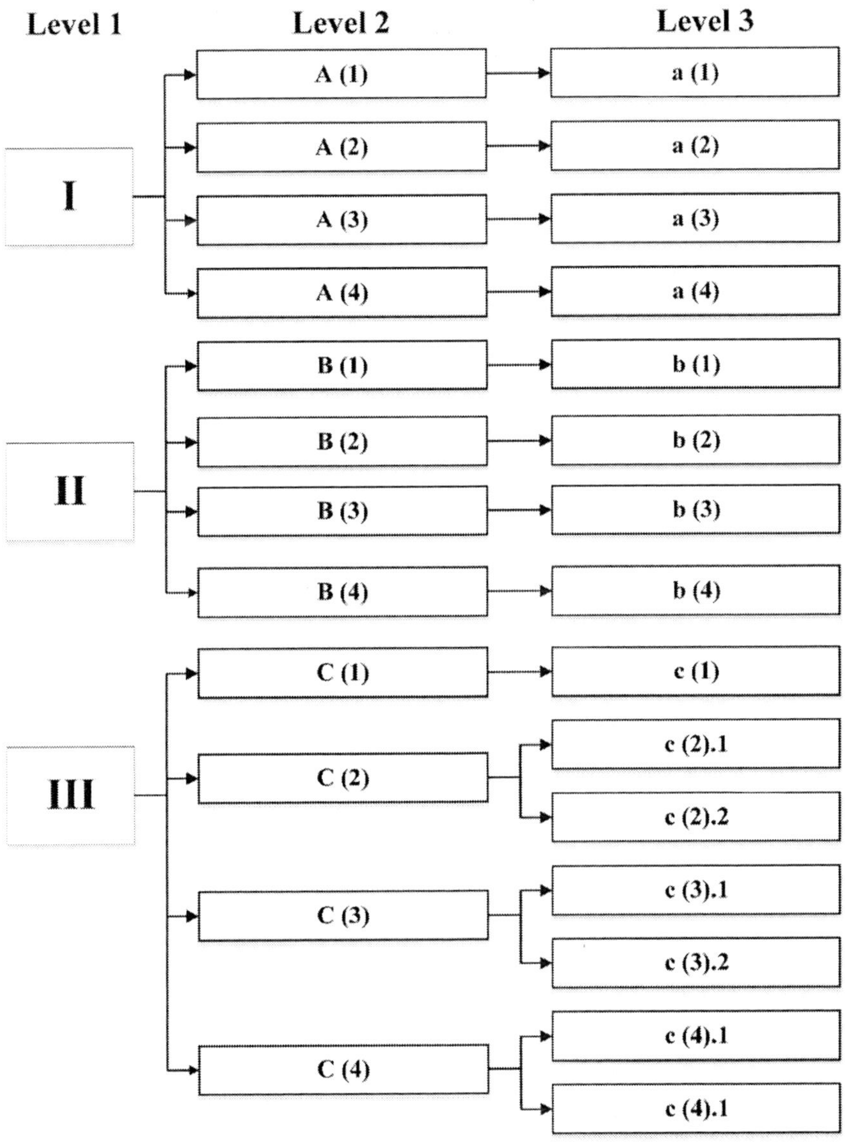

Figure 6. Forward Chain of the Automated Configuration Platform

Basically there are three level of the forward chains which are:

Level 1

I = 1 Robot

II = 2 Robots

III = 3 Robots

Level 2

A(1) / B (1) / C(1) = Normal Configuration

A(2) / B(2) / C(2) = Configuration with Minimum Workspace Area

A(3) / B(3) / C(3) = Configuration with Minimum Manufacturing Throughput Time

A(4) / B(4) / C(4) = Configuration with both Minimum Workspace Area and Manufacturing Throughput Time

Level 3

a(1) / b (1) / c(1).1 = Normal Configuration with Robot Work Cell 1

a(2) / b(2) / c(2).1 = Configuration with Minimum Workspace Area with Robot Work Cell 1

c(2).2 = Configuration with Minimum Workspace Area with Robot Work Cell 2

a(3) / b(3) / c(3).1 = Configuration with Minimum Manufacturing Throughput Time with Robot Work Cell 1

c(3).2 = Configuration with Minimum Manufacturing Throughput Time with Robot Work Cell 2

a(4) / b(4) / c(4).1 = Configuration with both Minimum Workspace Area and Manufacturing Throughput Time with Robot Work Cell 1

c(4).2 = Configuration with both Minimum Workspace Area and Manufacturing Throughput Time with Robot Work Cell 2

For the inference rule, 4 rules elaborated in this paper as shown in table 3. Rule 0 is utilized for identifying the rule number based on the selection of robot quantity. Meanwhile, Rule 1, Rule 2 and Rule 3 were used to determine the final optimal robot work cell with its workspace area and manufacturing throughput time calculation for 1 robot, 2 robots and 3 robots respectively.

Table 3. Inference Rule of Optimal Selection of Robot Work Cell Configuration

Rule 0	**Rule 1**	**Rule 2**	**Rule 3**
If 1 Then Rule 1	If 1 & A(1) Then a(1)	If 2 & B(1) Then b(1)	If 3 & C(1) & c(1).1 Then c(1).1
If 2 Then Rule 2	If 1 & A(2) Then a(2)	If 2 & B(2) Then b(2)	If 3 & C(1) & c(1).2 Then c(1).2
If 3 Then Rule 3	If 1 & A(3) Then a(3)	If 2 & B(3) Then b(3)	If 3 & C(2) & c(2).1 Then c(2).1

If 4 Then Rule 4	If 1 & A(4) Then a(4)	If 2 & B(4) Then b(4)	If 3 & C(2) & c(2).2 Then c(2).2
If 5 Then Rule 5			If 3 & C(3) & c(3).1 Then c(3).1
If 6 Then Rule 6			If 3 & C(3) & c(3).2 Then c(3).2
If 7 Then Rule 7			If 3 & C(4) & c(4).1 Then c(4).1
If 8 Then Rule 8			If 3 & C(4) & c(4).2 Then c(4).2
If 9 Then Rule 9			
If 10 Then Rule 10			

4.3 Automated Configuration Platform of Robot Work Cell

A GUI of the automated configuration platform is invented wherein it comprises of three different levels of user interface. First level user interface is used to begin the configuration process. User is asked to select the "Number of Robot" based on the quantity of robot used and once user click the "Start" button, the next level of user interface will appear automatically. (Refer figure 7).

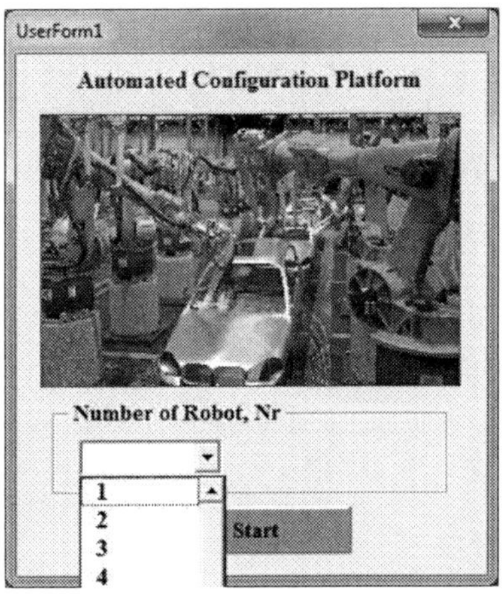

Figure 7. First Level User Interface

Next level user interface (figure 8) is utilized to calculate the workspace area,A_W and manufacturing throughput time,MTT and identified the optimal robot work cell based on the configuration type selection. In this level, user need to input the robot information such as the robot arm length (mm) and the robot tooling and work piece length (mm). Also, the manufacturing throughput time data such as the inspection time, t_i, process time, t_p , move time, t_m and queue time, t_q.

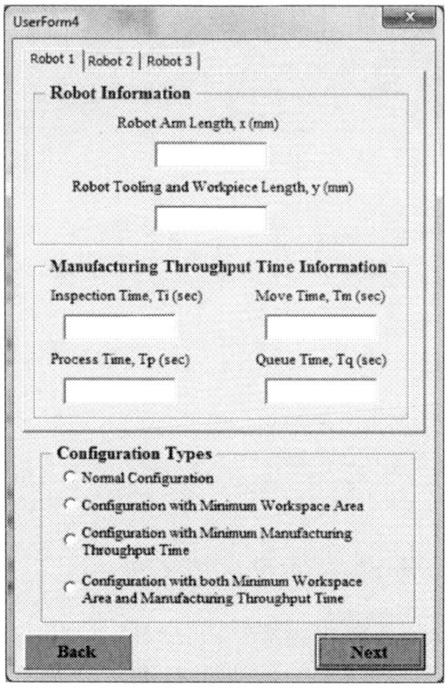

Figure 8. Second Level User Interface

Finally, the last level user interface as in figure 9 is presented where the final robot workspace area, A_W and manufacturing throughput time, MTT with the proposed optimal layout are displayed.

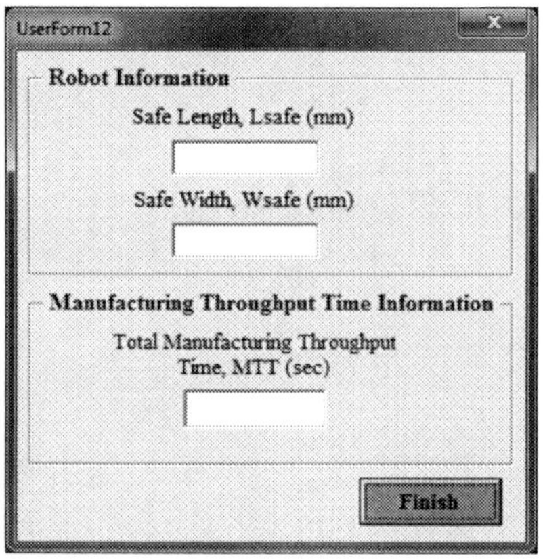

Figure 9. Third Level User Interface

4.4 Verification of the Automated Configuration Platform

For the verification activity, 3 number of robots and a set of robot and manufacturing throughput time data with various configuration types are run to present the result of the developed GUI as shown in the table 4.

IOP Conf. Series: Materials Science and Engineering **210** (2017) 012061 doi:10.1088/1757-899X/210/1/012061

Table 4. Verification Result of the Automated Configuration Platform of Robot Work Cell

Robot 1 Data	**Robot 2 Data**	**Robot 3 Data**	**Configuration Types**	**Workspace Area (mm) & Manufacturing Throughput Time (sec)**	**Proposed Optimal Robot Work Cell**
$x = 100$ $y = 100$ $t_i = 10$ $t_p = 10$ $t_m = 10$ $t_q = 10$	$x = 100$ $y = 100$ $t_i = 10$ $t_p = 10$ $t_m = 10$ $t_q = 10$	$x = 100$ $y = 100$ $t_i = 10$ $t_p = 10$ $t_m = 10$ $t_q = 10$	Normal Configuration	$L_{safe} = 5100$ $W_{safe} = 1700$ $MTT = 120$	
$x = 100$ $y = 25$ $t_i = 10$ $t_p = 10$ $t_m = 10$ $t_q = 10$	$x = 200$ $y = 50$ $t_i = 20$ $t_p = 10$ $t_m = 30$ $t_q = 10$	$x = 300$ $y = 75$ $t_i = 10$ $t_p = 20$ $t_m = 10$ $t_q = 5$	Normal Configuration	$L_{safe} = 5400$ $W_{safe} = 2050$ $MTT = 155$	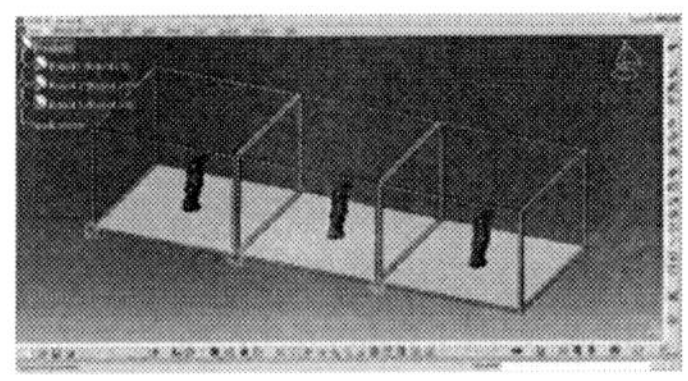
$x = 100$ $y = 100$ $t_i = 10$ $t_p = 10$ $t_m = 10$ $t_q = 10$	$x = 100$ $y = 100$ $t_i = 10$ $t_p = 10$ $t_m = 10$ $t_q = 10$	$x = 100$ $y = 100$ $t_i = 10$ $t_p = 10$ $t_m = 10$ $t_q = 10$	Configuration with Minimum Workspace Area	$L_{safe} = 5100$ $W_{safe} = 1700$ $MTT = 120$	

$x = 100$ $y = 25$ $t_i = 10$ $t_p = 10$ $t_m = 10$ $t_q = 10$	$x = 200$ $y = 50$ $t_i = 20$ $t_p = 10$ $t_m = 30$ $t_q = 10$	$x = 300$ $y = 75$ $t_i = 10$ $t_p = 20$ $t_m = 10$ $t_q = 5$	Configuration with Minimum Workspace Area	$L_{safe} = 5400$ $W_{safe} = 2050$ $MTT = 155$	
$x = 100$ $y = 100$ $t_i = 10$ $t_p = 10$ $t_m = 10$ $t_q = 10$	$x = 100$ $y = 100$ $t_i = 10$ $t_p = 10$ $t_m = 10$ $t_q = 10$	$x = 100$ $y = 100$ $t_i = 10$ $t_p = 10$ $t_m = 10$ $t_q = 10$	Configuration with Minimum Manufacturing Throughput Time	$L_{safe} = 3400$ $W_{safe} = 3400$ $MTT = 80$	
$x = 100$ $y = 25$ $t_i = 10$ $t_p = 10$ $t_m = 10$ $t_q = 10$	$x = 200$ $y = 50$ $t_i = 20$ $t_p = 10$ $t_m = 30$ $t_q = 10$	$x = 300$ $y = 75$ $t_i = 10$ $t_p = 20$ $t_m = 10$ $t_q = 5$	Configuration with Minimum Manufacturing Throughput Time	$L_{safe} = 3850$ $W_{safe} = 3350$ $MTT = 115$	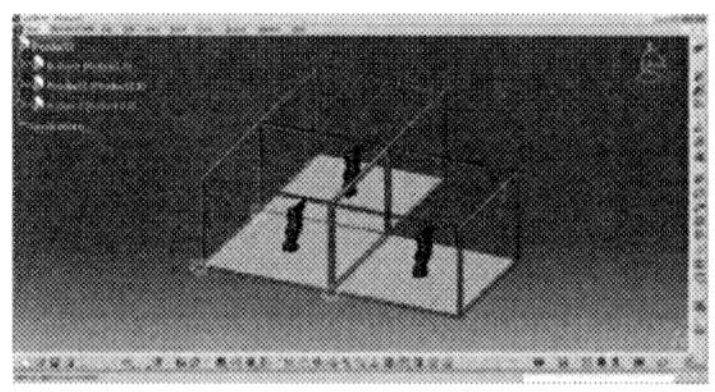
$x = 100$ $y = 100$ $t_i = 10$ $t_p = 10$ $t_m = 10$ $t_q = 10$	$x = 100$ $y = 100$ $t_i = 10$ $t_p = 10$ $t_m = 10$ $t_q = 10$	$x = 100$ $y = 100$ $t_i = 10$ $t_p = 10$ $t_m = 10$ $t_q = 10$	Configuration with both Minimum Workspace Area and Manufacturing Throughput Time	$L_{safe} = 3400$ $W_{safe} = 3400$ $MTT = 80$	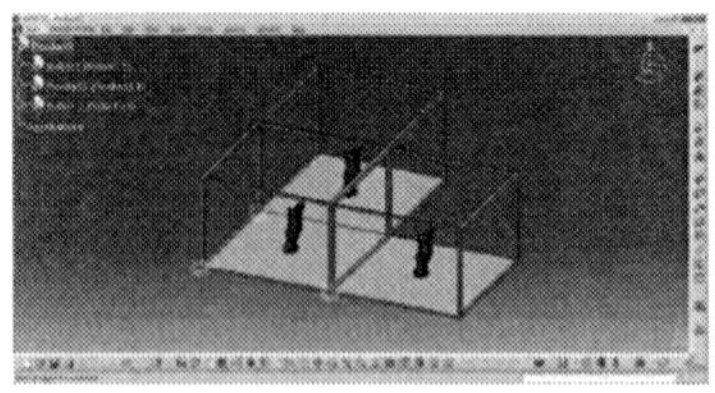

$x = 100$ $y = 25$ $t_i = 10$ $t_p = 10$ $t_m = 10$ $t_q = 10$	$x = 200$ $y = 50$ $t_i = 20$ $t_p = 10$ $t_m = 30$ $t_q = 10$	$x = 300$ $y = 75$ $t_i = 10$ $t_p = 20$ $t_m = 10$ $t_q = 5$	Configuration with both Minimum Workspace Area and Manufacturing Throughput Time	$L_{safe} = 3850$ $W_{safe} = 3350$ $MTT = 115$	

5. Conclusion

An automated configuration platform has been proposed where the configuration concept, mathematical models with its GUI are elaborated in this paper. The essential purpose of this work was providing a better understanding in the designing and developing of the configuration system. Additionally, this work plans to provide a fast and simple platform for design engineers with the present of the significant data in regards to the robot work cell configuration system. Future work will focus on the improvement of the configuration platform which includes other complex user requirements.

Acknowledgement

We are grateful for the support from the Universiti Teknikal Malaysia Melaka (UTeM) and Department of Occupational Safety & Healh (DOSH) Malaysia for providing materials support and useful information in order to complete this work.

References

[1] X Su 2012 *Int. J. Adv. Comput. Technol.* **4** pp 1–9.

[2] I. International Federation of Robotics, "World Robotics 2015 Survey - Executive Summary," 2015.[Online].Available: http://www.worldrobotics.org/uploads/media/Executive_Summary__WR_2015.pdf.

[3] M K Morel 2004 *Custom Robot Solves Palletizing Challenge* **22** pp 14–15.

[4] C C Li 2006 *3rd China-Japan Conference on Mechatronics 2006 Fuzhou* pp 215–219.

[5] Z Xiangquan, L Chao, Y Zhiqiang, Z Xinyi, C Fengyan, N Jiyuan 2008 *International Conference on Computer Science and Information Technology* pp 362–366.

[6] R Kia, F Khaksar-Haghani, N Javadian, and R Tavakkoli-Moghaddam 2014 *J. Manuf. Syst.*, **33** pp 218–232.

[7] M A A Rahman 2014 School of Aerospace Mechanical and Manufacturing Engineering College of Science Engineering and Health RMIT University.

[8] D Spensieri, J S Carlson, R Bohlin, J Kressin, and J Shi 2016 *Procedia CIRP* **44 pp** 395–400.

[9] M A A Rahman 2005 Universiti Tenaga Nasional.

[10] M A A Rahman, N S Osman, C H Boon, G L T Poh, A A A Rahman, B M B Mohamad, S H Kamsani, E Mohamad, Z A Zaini, and M F A Rahman 2016 *J. Adv. Manuf. Technol.* **10**.

[11] N S Osman, M A A Rahman, A A A Rahman, B M B Mohamad, and S H Kamsani 2016 *Innovative Research and Industrial Dialogue, IRID'16.*

[12] N S Osman, M A A Rahman, A A A Rahman, B M B Mohamad, and S H Kamsani, 2017 *4th TMAC Symposium.*

[13] P Zheng, V H Torres, J Ríos, and G Zhao 2012 *Appl. Mech. Mater.* **271–272** pp 974–980.

[14] S Siddesh and B S Suresh 2015 *Automation of Generating CAD Models* **5** pp 55–58.

[15] G Pintzos, C Triantafyllou, N Papakostas, D Mourtzis, and G Chryssolouris 2016 *Int. J. Comput. Integr. Manuf.* **3052** pp 1–13.
[16] B T Lin, K M Huang, K Y Su, and C Y Hsu 2013 *Int. J. Adv. Manuf. Technol.* **68** pp 1887–1899.
[17] Y H Sawant and A Kadam 2016 *Assembly of Horizontal Screw Conveyor in CATIA V5 using VBA* **2** pp 1–6.
[18] S Ćuković, G Devedžić, and I Ghionea 2010 *UPB Sci. Bull. Ser. D Mech. Eng.*, **72** pp 85–960.
[19] Y H Sawant and U M Nimbalkar 2015 *Automated Modeling of Screw Conveyor Components in CATIA* pp 112–116.
[20] P G Student 2014 *Automatic Assembly of Mechanical Joint Based on Extraction of Dimensional Data and Geometric Information From A 3D CAD Model* **3** pp 2192–2195.
[21] N S Osman, M A A Rahman, A A A Rahman, B M B Mohamad, and S H Kamsani 2017 *Proceedings of Mechanical Engineering Research Day 2015.*

IOP Conf. Series: Materials Science and Engineering **210** (2017) 012062 doi:10.1088/1757-899X/210/1/012062

An exploration of polymer adhesion on 3D printer bed

M A Nazan[1*], F R Ramli[1, 2*], M R Alkahari[1, 2], M A Abdullah[1, 2], M N Sudin[1, 2]

[1]Faculty of Mechanical Engineering, Universiti Teknikal Malaysia Melaka, Hang Tuah Jaya, 76100 Durian Tunggal, Melaka, Malaysia

[2]Centre for Advanced Research on Energy, Universiti Teknikal Malaysia Melaka, Hang Tuah Jaya, 76100 Durian Tunggal, Melaka, Malaysia

E-mail: *muhd_afdhal@hotmail.com and faiz@utem.edu.my

Abstract. One of the problems in Fused Deposition Modelling (FDM) 3D Printing process is that the extruded plastic filament tends to shrink and warp from the printing platform. The purpose of this research is to explore the warping deformation problem in four aspects i.e. curling, pincushion effect, trapezoid deformation and blocked shrinkage that usually occur in the process. Epoxy resin based adhesive was applied onto the printing platform to reduce and eliminate the warping deformation. Afterwards, by applying the adhesive, the 3D printed models were measured their curling, pincushion, trapezoid and blocked shrinkage using laser scanner and metrology software. The result shows that the pincushion and trapezoid has low deformation compared to curling and blocked shrinkage. Blocked shrinkage effect shows the highest warping deformation value. In comparison of materials, PLA shows the best geometry result with low warping deformation value and the best surface finish.

1. Introduction

Fused deposition modeling (FDM) is one of the most widely used technique to produce object or part in the field of 3D printer area [1-2]. 3D printer machine use three dimensional CAD data and form layer by layer process to produce prototype part. 3D printer generally use thermoplastic plastic materials such as Polylactic Acid (PLA) and Acrylonitrile Butadiene Styrene (ABS) materials. These materials are melt and extruded by the printer extrusion head and moves correspondingly to the coordinate and position that been set in certain speed [1].

One of the problems in the 3D printing field especially FDM process is that the plastic filament which is extruded from the machine nozzle tends to shrink and warp from the printing bed platform [3]. This has been highlighted by several researchers that an adhesive layer between the first layer and printing platform is required to counter the problem [4-6]. This problem can be reduced by applying heat to the printing bed where the higher the bed temperature is, the less the deformed shape will be [7]. Moreover, the warping deformation has also showing warping reduction when applying epoxy resin based adhesive onto the printing bed [3].

The printing process involved printing onto the printing platform for the first layer of the formation until the last layer. Each of the layer from the molten plastic is solidified either naturally or by the help of external cooling process. Due to the cooling effect, the warpage problems are sometimes occur where it definitely affecting the printed part especially in the geometry size and manufacturing quality.

The warpage problem happens because of the first layer is not stick well and peeled away from the printing platform due to the adhesion of the first layer with the printing platform such in figure 1 and 2. Young et al [7] has stated that in 3D printer it is an undesirable shape error which is occured in the product due to the heat shrinkage phenomenon. The paper also stated that the main reason for inhomogeneous shrinkage of the parts in the 3D printing process are due to time delayed for the solidification of the layers. Moreover, the problem is caused by force transmission between the layers since phase of change from molten plastic to solid during building process. Some of examples of warpage problems occur in the FDM process are pincushion, trapezoid, curling and blocked shrinkages [5].

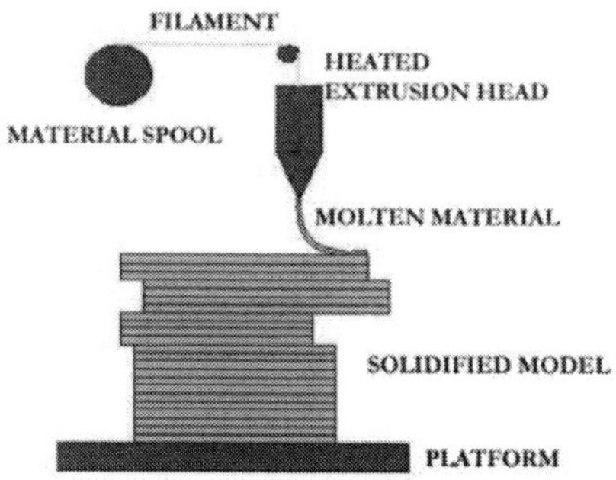

Figure 1. Fused deposition modeling process [2]

Figure 2. Warpage affect during printing process

Epoxy resin adhesive is an adhesive that has been used widely all over the world. The adhesive is proven to be suitable for the repair and restoration of the glass substance in several problem [8-9]. The epoxy has declared containing 5.0% of organic solvents and 56.5% water based that soluble inside the water and have non-toxicity effect to the human and environment [10]. By referring to the safety data sheet, epoxy resin has zero hazard and has harmful effect if it used and handled accordingly to necessary regulation. The epoxy resin based adhesive used in this experiment is shown as in figure 3 [11].

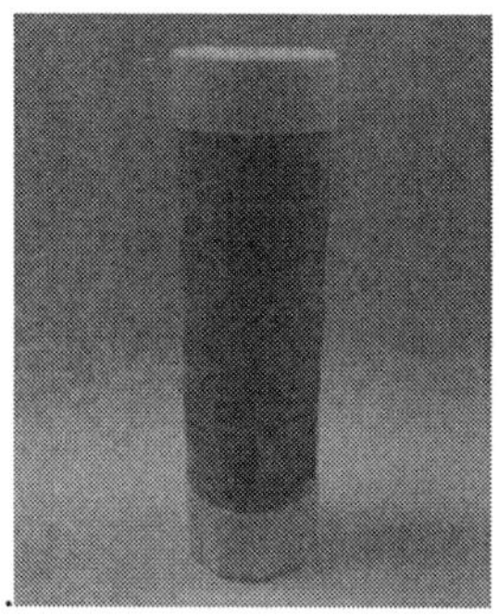

Figure 3. Epoxy resin based adhesive

Thus, the purpose of this paper is to study the relation and the performance of the epoxy resin based adhesive through the surface preparation for the printing platform during the process. In addition, this research also study the warping deformation that occur through pincushion, curling, trapezoid, and blocked shrinkage effects using PLA and ABS material either printing with and without applying heat on bed.

2. Experiment characteristics

2.1. Sample Preparation

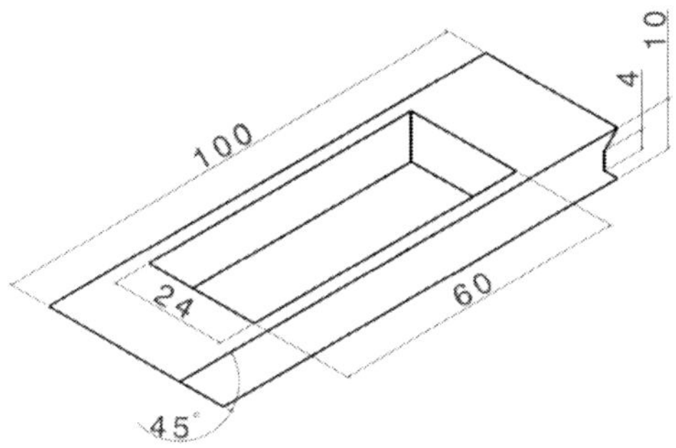

Figure 4. Dimensioned isometric view with angle 45° of specimen

The experiment was started by sketching the solid model of specimens using solid modelling software, CATIA V5 with its dimension shows in figure 4. The solid model was sketched with three different angles of 30°, 45° and 60°, to investigate the warping deformation effect in various angle. Next, the solid model was sliced by using 3D printer slicer software, Repetier Host that was built with Slic3r engines to create machine language, G-Code for 3D printer path direction. This slicing process was required to set up the 3D printer specifications where the infill density, layer temperature, printing speed and layer height need to be defined. Thus, table 1 shows the printing settings that being used for this experiment.

Table 1. Printing setting

Preferences	**Materials**	
	PLA	ABS
Infill density (%)	13	13
First layer temperature, T (°C)	235	235
Other layer temperature, T (°C)	193	230
Printing Speed (mm/s)	40	40
Layer Height (mm)	0.2	0.2

The printing experiment was conducted by using a FDM 3D printer machine, Pursa i3 model 3D printer machine that has printing area of 180mmx180mmx150mm with glass as the material of printing platform. Before the printing process began, the printing platform was prepared by applying the epoxy resin based adhesive and spreading onto the printing area of printing platform. Four specimens type i.e. ABS without the use of heat bed printing platform, ABS with use of heat bed, PLA without heat bed, and PLA with heat bed has been print by using 3D Printer for each angles of 30°, 45° and 60° . Table 2 shows the label for type of specimens used on the experiment.

Table 2. Labelled type of specimens based on material and heat bed

Specimens	**Material**	**Heat bed**
1	ABS	Without
2	ABS	With
3	PLA	Without
4	PLA	with

In addition, the printing setting for the layer temperature was set differently between PLA and ABS material. This is because of the different melting point and different material properties. The bed temperature setting has been set for ABS and PLA are100°C and 60°C respectively.

2.2. Warping Deformation Preparation

Figure 5. Scanning process by Solutionix Rexscan CS2+ laser scanner

After the printing process completed, the next process was scanning the geometry shape that been printed using the 3D printer. The specimens was scanned using laser scanning and metrology software; Solutionix Rexscan CS2+ 3D scanner machine and Geomagic Quality 2013 tools software to compare with the sketched theoretical geometry from drawing. The warping deformation that was measured are shown in figure 6. The geometry of the sample was compared using 3D and 2D compare tools that available in the Geomagic software. Moreover, the result also has been figured using coordinate method which is similar to the conventional method of Coordinate Measuring Machine (CMM).

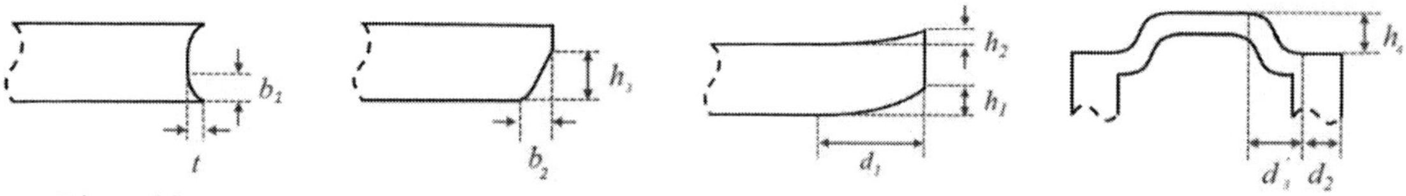

Figure 6. Description of warpage deformation [5]

3. Results and discussions

All the sample of specimen is completely printed using the Pursa i3. However, the printing process of using ABS material without external heat bed had failed where they peeled away from the printing platform. Therefore, the result of the experiment is only presented with 3 type of printing; ABS with heat bed, PLA without heat bed and PLA with heat bed.

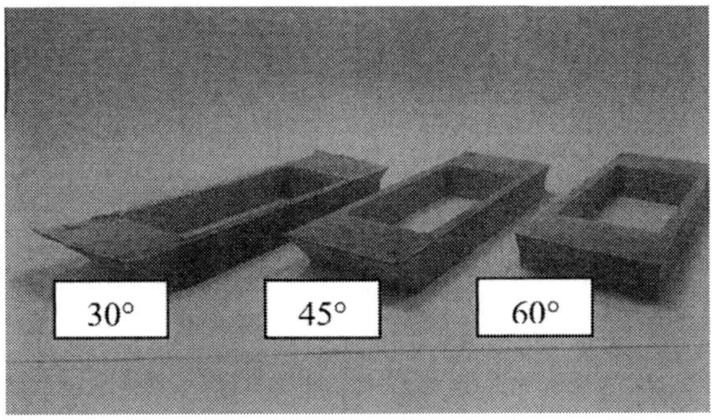

Figure 7a. Photograph of specimens (front view) from left 30°, 45° and 60°

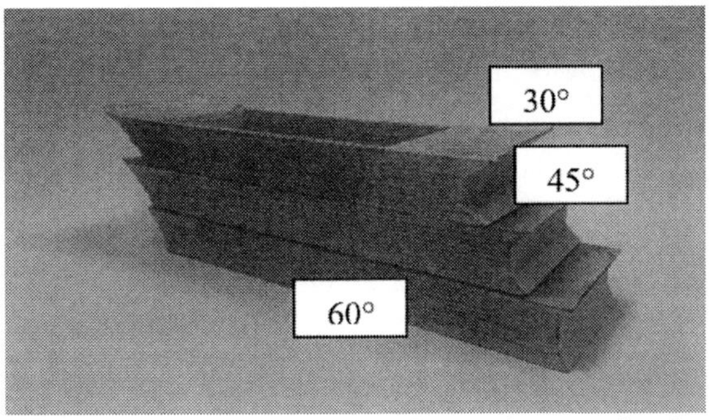

Figure 7b. Photograph of specimens (back view) from top 30°, 45°, and 60°

Figure 7a and 7b is shows the photograph of printed specimen with angle 30°, 45° and 60°. As shown in the photograph, the specimen with 30° of angle shows the fewest shape error especially in the front view of the specimens.

3.1. Measurement results

Data of the measurement has been obtained from the process of laser scanning and metrology process. It has shown that there is some geometry error based on shape.

Figure 8. Photograph of specimen with angle 45° printed using Polylactic Acid (PLA) material

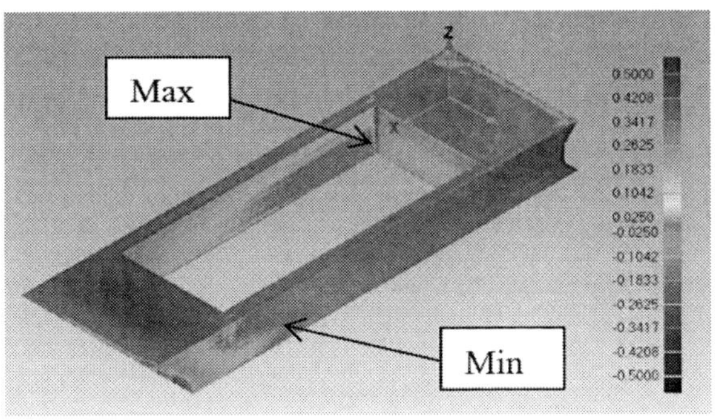

Figure 9. 3D comparison between actual drawing and scanned specimen that has been analyzed using the metrology software with minimum and maximum labeled for the spectrum level

Figure 8 and 9 shows an example from one of the printed specimen that went through the 3D comparison analysis between the actual drawing and the scanned printed specimen. The spectrum colors shows the comparison of the geometry for each of the specimen. As in the figure, the maximum and minimum of deformation spectrum are 0.5 mm and -0.5 mm corresponding to the tolerance of the analysis. The red color shown the positive of deformation spectrum while blue color is the negative of deformation spectrum and then green color is nearly to the dimension of theoretical. Based on the analysis data, the maximum and average of the deformation are 0.4994 mm and 0.1388 mm. As in the figure 9, the rectangular shaped at the center of specimen is shows few of red area while most of top surface shown blue color and green. Table 3, 4 and 5 shows the result of deformation error from measurement of metrology process.

Table 3. Deformation error from measurement from specimens with 30° of angle

Specimens	Pincushion		Trapezoid		Curling		Blocked shrinkages			
	t	b_1	h_3	b_2	h_1	h_2	d_1	d_2	d_3	h_4
1	0.8168	0.2683	0.4080	0.5732	0.1497	0.0864	9.2111	21.3766	3.1463	0.0833
2	2.4176	1.7934	0.7086	1.2975	0.0179	0.0087	1.6314	17.2267	9.8791	0.0936
3	-	-	-	-	-	-	-	-	-	-
4	0.7832	0.2176	0.3149	0.0469	0.0376	0.1213	1.0110	20.0640	12.9380	0.0356

Table 4. Deformation error from measurement from specimens with 45° of angle

Specimens	Pincushion		Trapezoid		Curling		Blocked shrinkage			
	t	b_1	h_3	b_2	h_1	h_2	d_1	d_2	d_3	h_4
1	0.1721	0.1225	0.0076	0.1154	0.0459	0.0219	1.9968	21.8371	0.8987	0.0043
2	0.2115	0.1331	0.0883	0.3917	0.0682	0.0437	1.6837	11.6780	22.9900	0.1041
3	-	-	-	-	-	-	-	-	-	-
4	0.1699	0.5910	0.1518	0.0508	0.0604	0.1357	10.9160	17.8978	14.0234	0.0284

Table 5. Deformation error from measurement of specimens with 60° of angle

Specimens	Pincushion		Trapezoid		Curling		Blocked shrinkage			
	t	b_1	h_3	b_2	h_1	h_2	d_1	d_2	d_3	h_4
1	0.4654	0.3159	0.3007	0.5037	0.0564	0.0077	0.0199	20.5490	2.2459	0.0160
2	1.1335	1.2020	0.6418	0.2787	0.0464	0.0137	0.3373	14.7331	14.9542	0.0927
3	-	-	-	-	-	-	-	-	-	-
4	0.2526	0.4126	0.3976	0.8385	0.2993	0.1957	10.9160	19.8973	3.1310	0.0439

3.2. *Effect of angle*

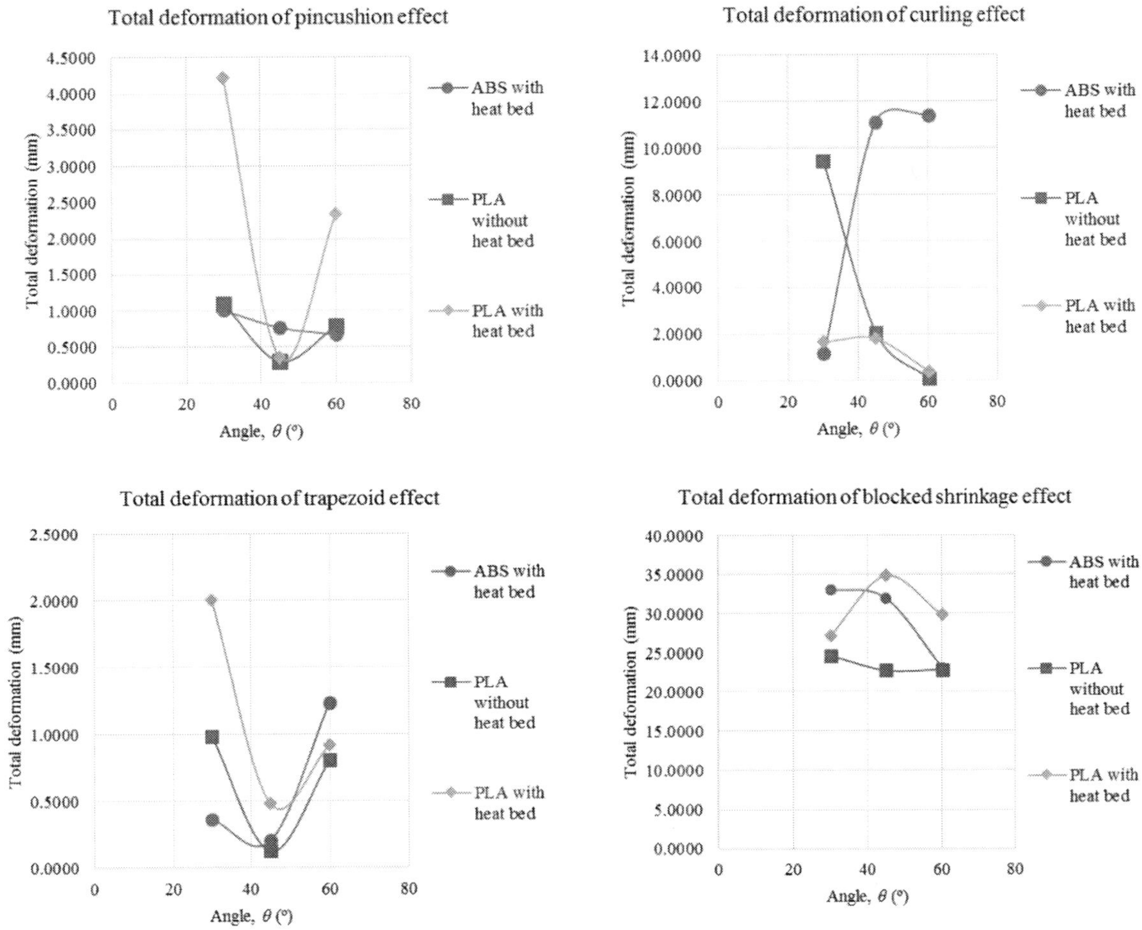

Figure 10. Total deformation of princushion, curling, trapezoid and blocked shrinkage effects.

Figure 10 indicate the results to the pincushion, curling, trapezoid and blocked shrinkage effects that tabulated form the experiments. The graph was plot based on the total deformation for each effects. By referring to the plotted graph, pincushion and trapezoid effects resulted from PLA with heat bed type of specimen has the highest value of deformation especially for the specimens that was built with 30° of angle. Moreover, based on the result for PLA without heat bed, most of graphs shows has lower total deformation than other type of specimens. ABS material that has been printed using heat bed shows that it has lower total deformation compared to PLA with heat bed.

For the curling effect, the result shows that the ABS material that printed using heat bed has high deformation since it has been built with 45° and 60° of angle. This maybe because of the environmental factors that has affecting result to the printing process where the solidification of the material might be different from each other. The ABS material has known issue of solidification in 3D printing.

The effect of the angle shows that the total deformation has lower value when it built with angled 45°. Therefore, it may conclude that best angle of printing with those material is by using 45° of angle while the other angle is resulted has higher value of total deformation.

3.3. *Type of error*

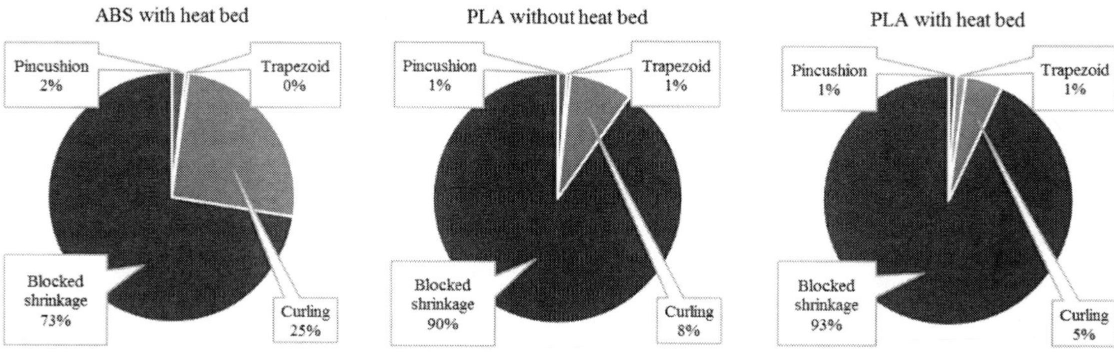

Figure 11. Total of warping deformation by type of error from specimens with 45° of angle

Based on the previous result, the experiment shows that the specimens with 45° of angle are the best compared to the other angle. Thus, figure 11 generally shows the result of total error for the four type of warping deformation. The pie chart shows the total warping deformation that separated by type of material and method of printing. This chart shows that most of experiment has high blocked shrinkages effects. ABS with heat bed shows 2%, 0%, 73% and 25% of pincushion, trapezoid, blocked shrinkage and curling effects. Respectively. PLA material that printed without heat bed and with heat bed shows higher blockade shrinkages where 90% and 93% respectively. Apart from that, the pincushion and trapezoid effects has the lowest deformation among the specimens which is less than 2%. The curling effect shows that ABS material has 25% of deformation while PLA with and without heat bed has 8% and 6% deformation.

3.4. *Effect of material*

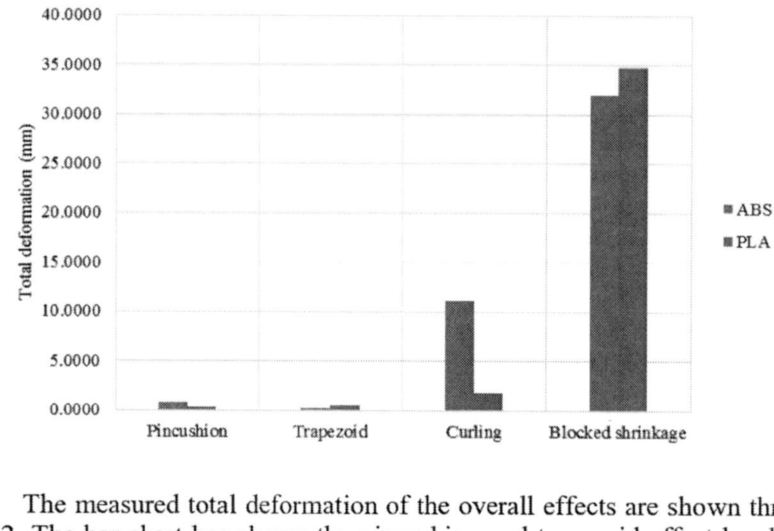

Figure 12. Influence of deformation based on total deformation from material ABS and PLA with heat bed

The measured total deformation of the overall effects are shown through the bar chart as in figure 12. The bar chart has shown the pincushion and trapezoid effect has lower deformation than curling and blocked shrinkage. This result came out similar for both of ABS and PLA material that been print with heat bed. Moreover, the result shows the blocked shrinkage effect have the highest total deformation for the both material. Moreover, the ABS material has less deformation compared to PLA material for the blocked shrinkage effect. However, the ABS exhibit higher deformation for the curling and pincushion effects compare to PLA material. Based on the chart and the observation, PLA material gives the best result where the surface finish and the warping deformation is better than ABS material.

4. Conclusion

Based on the results, it shows that the highest value of deformation is, the lower quality of printing. In the experiment of the geometry shape, the specimen built with 45° of angle that was been print using PLA material without heat bed shows the lowest warping deformation. In addition, the results also shows that blocked shrinkage effect has the highest value of deformation compare to curling, pincushion and trapezoid. In addition, PLA material has the best material where it has low deformation for the curling and pincushion effect compared to ABS material. PLA material printed without heat bed have the lowest total deformation of pincushion, trapezoid and blocked shrinkage effects with 0.2946mm, 0.1230mm and 22.7401mm respectively.

References

[1] Solidwild.wordpress.com, 2017, *Additive Manufacturing | SolidWild.* [online] Available at: https://solidwild.wordpress.com/category/3d-printing-2/additive-manufacturing/ [Accessed 12 Mar. 2017]

[2] Yifan J, Yi W, Bing Z and Zhanqiang L 2017 *Journal of Mechanical Processing Technology,* **240** 233-239

[3] M A Nazan, F R Ramli, M R Alkahari, M N Sudin and M A Abdullah 2016 *Proc. of Mechanical Engineering Research Day 2016* 71-72

[4] Wohlers T, 2012. Wohlers Report 2012: Executive Summary.

[5] W Z Wu, P Geng, J Zhao, Y Zhang, D W Rosen and H B Zhang 2014 *Materials Research Innovations,* **18** 12-16

[6] F R Ramli, M I Jailani, H Unjar, M R Alkahari and M A Abdullah 2015 *Proc. of Mechanical Engineering Research Day 2015* 77-78

[7] Christoph S, Alexander Z, Michael F Z 2016 *48th CIRP Conference on Manufacturing Systems 2015* 1017-1022,

[8] Jane L D 2010 *The Yellowing of Epoxy Resin Adhesives: Report on Natural Dark Aging* **29** 63-76

[9] Davison S, 2010, The problems of restoring glass vessels 3-8

[10] UHU Stic safety data sheet, 2015, **12**

[11] M A Nazan, F R Ramli, M R Alkahari, M N Sudin and M A Abdullah 2017 *Proc. of Mechanical Engineering Research Day 2017* 145-146.

IOP Conf. Series: Materials Science and Engineering **210** (2017) 012063 doi:10.1088/1757-899X/210/1/012063

Analysis of the influence of chemical treatment to the strength and surface roughness of FDM

R. H. Hambali*, K. M. Cheong and N. Azizan*
Faculty of Manufacturing Engineering, Universiti Teknikal Malaysia Melaka, Hang Tuah Jaya, 76100 Durian Tunggal, Melaka, Malaysia

E-mail: *ruzy@utem.edu,my and noorainazizan93@gmail.com

Abstract. The applications of Additive Manufacturing (AM) technology have a greater functionality and wider range of application beyond an intention of prototyping. AM is the process of joining materials to form objects from Computer-Aided Design (CAD) models via layer upon layer process. One of AM technologies is the Fused Deposition Modelling (FDM), which use an extrusion method to create a part. FDM has been applied in many manufacturing applications includes an end-used parts. However, FDM tends to have bad surface quality due to staircase effect and post treatment is required. This chemical treatment is one of a way to improve the surface roughness of FDM fabricated parts. This method is one of economical and faster method. In order to enhance the surface finish of Acrylonitrile-Butadiene-Styrene (ABS) FDM parts by performing chemical treatment in an acetone solution as acetone has very low toxicity, high diffusion and low cost chemical solution. Therefore, the aim of this research is to investigate the influence of chemical treatment to the FDM used part in terms of surface roughness as well as the strength. In this project, ten specimens of standard ASTM D638 dogbone specimens have been fabricated using MOJO 3D printer. Five specimens from the dogbone were tested for surface roughness and tensile testing while another five were immersed in the chemical solution before the same testing. Based on results, the surface roughness of chemically treated dogbone has dramatically improved, compared to untreated dogbone with 97.2% of improvement. However, in term of strength, the tensile strength of dogbone is reduced 42.58% due to the rearrange of material properties and chemical effects to the joining of the filaments. In conclusion, chemical treatment is an economical and sustainable approach to enhance the surface quality of AM parts.

1. Introduction

Additive manufacturing (AM) is a group of techniques used to fabricate quickly a scale model of a physical part using a CAD system. It describes the technology of generative production processes. According to the ASTM definition, AM is the process of joining materials to create a part from the CAD model by layer upon layer. One of famous technologies of AM is Fused Deposition Modelling (FDM). FDM is solid-based manufacturing system that utilizes solids as the primary medium to create parts, which uses extrusion to from parts. FDM can consider as AM by laying down material in layers that use Acrylonitrile-Butadiene-Styrene (ABS) in the form of plastic filament using molten material system [1].

Moreover, FDM technology has expanded not only to be used as prototype, but as well as functional applications as the end-used parts. Therefore, the strength and surface finishing of FDM is important considerations to assure the quality and usability of FDM product with satisfactory physical and mechanical qualities.

In order to attain a good surface finish, researches had been done to recognize the parameters, which can give obvious effect to the surface roughness of the parts produced, include built orientation, slicing height and chemical treatment [2]. In this report, the parameter that will discuss in detail is chemical treatment. It is used to enhance the surface finish of ABS fused modelled part by chemical dipping that yielding significant improvement of the roughness of the treated [3]. Although the implementation of the chemical treatment on the FDM part will gives effect on the tensile strength of the part, the proposed chemical post treatment will be fast, economic and easy to be conducted method to improve the surface finish of the ABS parts [4][5][6].

Therefore, this research focuses on the chemical surface treatment method to eliminate the irregularities on the surface of parts produced via FDM by chemical smoothing reaction observed the improvement of the finishing as well as the behaviour of the finishing. This led in investigating the effect of the chemical treatment to the FDM parts on surface quality and mechanical strength as well maintaining the original dimension without degrading the quality of the parts.

2. Experimental procedure

The experiment works was done to investigate the effect of the chemical treatment on the surface roughness as well as the tensile properties of the FDM used part. Hence, this will include the fabrication of specimens by FDM machine using ABS material. Specimens of dogbone are fabricated in way to carry out analysis of surface roughness and tensile test. Figure 1 show the flow chart of studies.

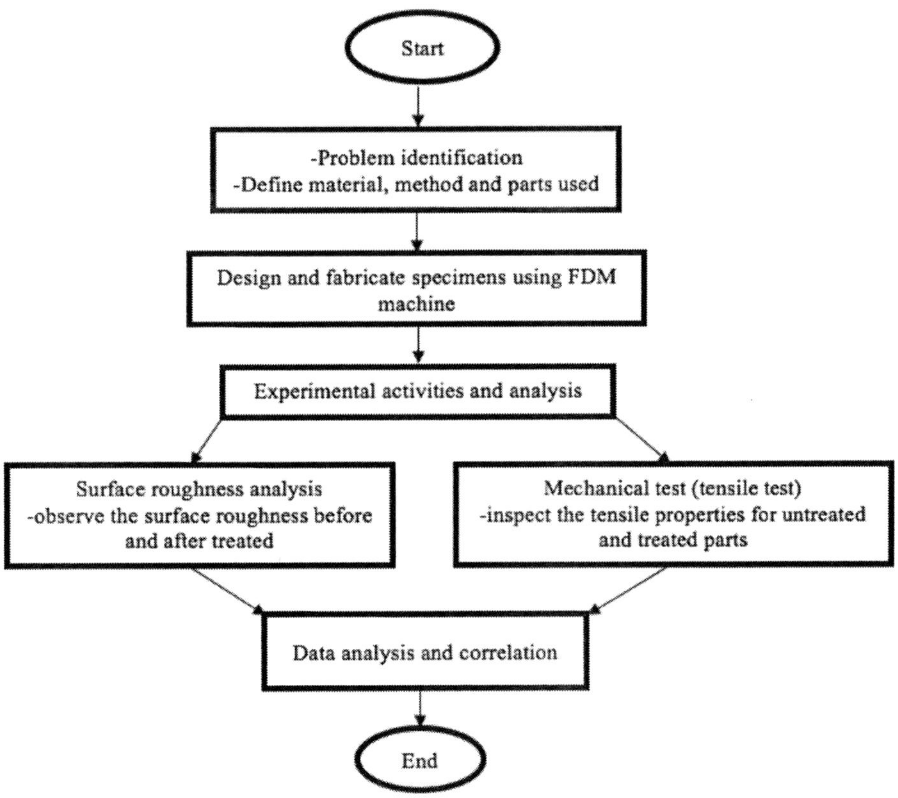

Figure 1. Flow chart of the project

In this study, the dogbone specimens of size and material were created according to Standard Test Method for Tensile Properties of Plastic, Type I ASTM D 638-01. The material of test specimens for this experiment is ABS, which is a rigid plastic. ABS is one of types of thermoplastic that deformable and recyclable. For the mechanical properties, ABS is the toughness and impact resistance. It affects resistance will not fall drastically at low temperature. Figure 2 shows the standard dogbone specimen size and its dimension are presented in table 1.

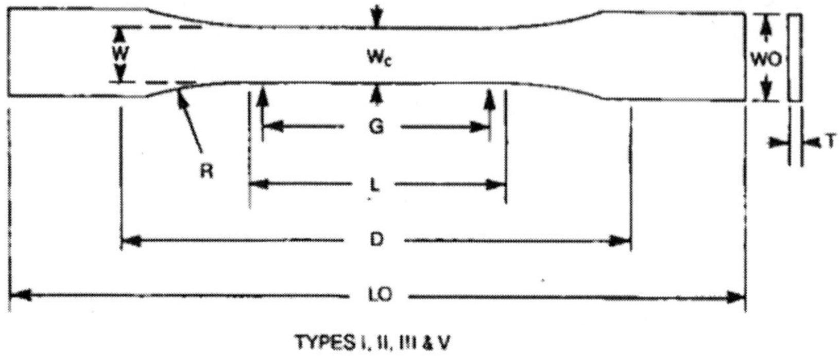

Figure 2. Standard specimen for plastic parts
(Source: ASTM Standard, D638-02)

Table 1. Detail dimension of ASTM 638-01, Type I dogbone test specimen.
(Source: ASTM Standard, D638-02)

Specimen of D638-01 , Type I		**Dimension (mm)**
Lo	Overall length	165
WO	Width overall	19
W	Width of narrow section	13
L	Length of narrow section	57
R	Radius of fillet	19
h	Thickness	3.2
lo	Gauge length	50
D	Distance between grips	115

The fabrications of the parts use the MOJO™ system, which use the FDM technology to build models from CAD STL files. It creates parts layer by layer via extruding engineering grade thermoplastics such as ABSplus model material and SR-30 as support material through print heads. In this machine, it has some control parameters that expected to affect the quality of the parts when produce especially in surface quality. Therefore, in this project, fabrication variables were decided as a constant variable throughout the fabrication. Table 2 show the fabrication variable has been decided.

Table 2. Constant Process Variable.

Process Variable	**Value**
Orientation	Horizontal
Support Style	Sparse

After the dogbone specimen have fabricated, the specimens are subjected to a chemical treatment process. Acetone was chosen as the chemical solution for this experiment because it has very low toxicity, high diffusion and low cost chemical solution. This chemical solution has a low reticulation degree, including Nitrile functionality having a weak reaction with polar solvent on this chemical. This Acetone was selected according to ABS material [7]. Then the surface roughness of specimen been investigated using surface roughness tester while the strength oft has been test using tensile strength analysis using Universal Tensile Test. Figure 3 shows the flow chart of experiment that has been designed as a main method.

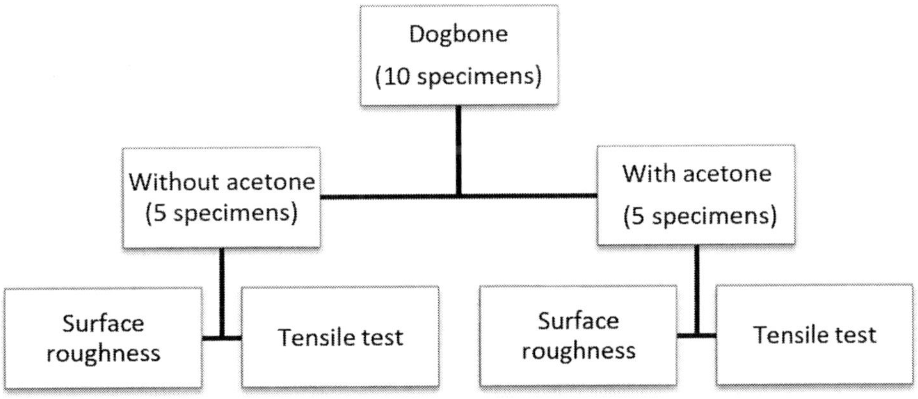

Figure 3. The design of experiment

2.1 Chemical treatment

For the chemical treatment process, Acetone was used as a solution to increase the surface roughness due to it properties towards ABS material. However, there are some difficulties while using acetone, which the chemical is too strong for solvent. Due to that, the chemical solution is added with water with the volume of 90% dimethylketone and 10% water during the process of immersion. All of the 5 dogbone specimens were immersed into the acetone solution for 300 seconds as shown in figure 4. All of the ready specimens were observed and measured in terms of surface roughness values.

Figure 4. An Example of the Specimens Immersing in Acetone Solutions.

2.2 Surface roughness analysis

All the specimens of dogbone were analysed before and after been treated with chemical. The surface roughness of the treated and untreated specimens had been measured using precision measuring instrument, which is Mitutoyo Sueftest SJ 301™ as shown in figure 5. The recorded profile display capability allows surface texture measurements to check instantly with prior to print out.

Figure 5. Mitutoyo Sueftest SJ 301

2.3 Tensile testing

Mechanical properties use in this study to identify the strength of the parts and focus on to the tensile properties. Tensile properties designate the material behaviour will respond to the forces been applied in tension. Therefore, the tensile test is the mechanical test that been done in this research. To conduct a tensile test, the test specimen is gripped at the end and pulled apart until the specimen is broken. In order to minimize the damage on the specimen due to the clamping force, the specimens are designed to have their narrower centre part than the ends as shown in figure 6. The data from the testing includes Young Modulus (E), Maximum Force (F_{max}), strain and tensile strength. Tensile stress is defined as force per unit area of the original cross section within the gauge length. It is usually expressed in megapascal (MPa). The formula is as below

$$\sigma = FA \tag{1}$$

Where;
σ: tensile stress (MPa)
F: Force applied (N)
A: Initial cross sectional area of the specimen (mm2)

Figure 6. Tensile Test Experimental Setup

3. Result and discussion

In this section, the influences of the chemical treatment of acetone on the build parts of FDM will be described. The surface roughness of the parts build is analysed for both treated and untreated parts. The tensile strength of the parts also had been recognizing the effect of chemical treatment to the AM parts. Figure 7 shows the dogbone specimen after printed using MOJO™ FDM machine and before it been treated into acetone. While figure 8 shows the dogbone specimen that has been immersed in acetone in 300 seconds. By visual observation, It shows that the surface finish of the specimens after treated were smooth and shiny compare to untreated specimens.

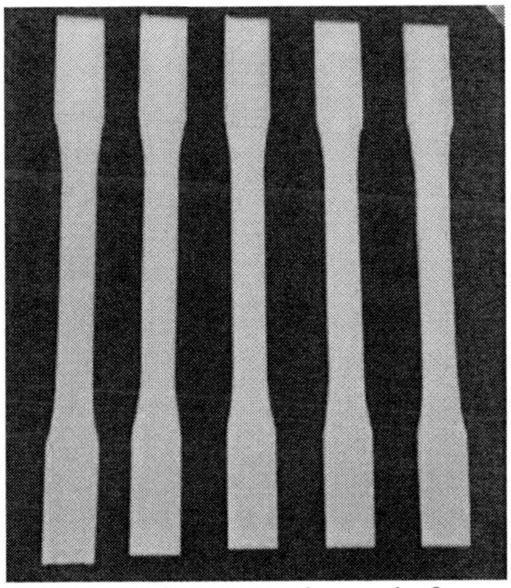

Figure 7. Dogbone Specimens before Immerse in an Acetone

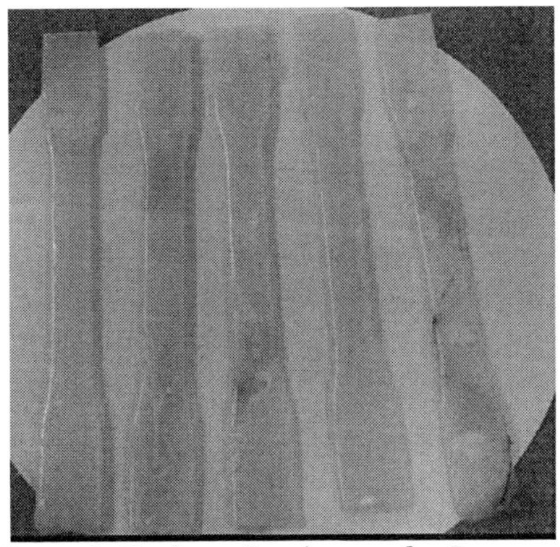

Figure 8. Dogbone Specimens after Immersed in Acetone

Surface roughness is a component of a surface texture which quantified by the normal vector of the real surface in ideal form. To check the surface roughness of the treated and untreated parts of FDM was using Mitutoyo Surftest SJ 301. Table 3 shows the value of surface roughness, R_a for treated and untreated dogbone specimens.

Table 3. The Surface Roughness of the Untreated and Treated Specimens.

Treated /Untreated	Specimen	Surface Roughness, R_a	Mean Roughness Average, R_a
	1	12.84	
	2	12.71	
Untreated	3	12.85	**12.736±0.164**
	4	12.82	
	5	12.46	
	1	0.27	
	2	0.42	
Treated	3	0.37	**0.350±0.054**
	4	0.35	
	5	0.34	

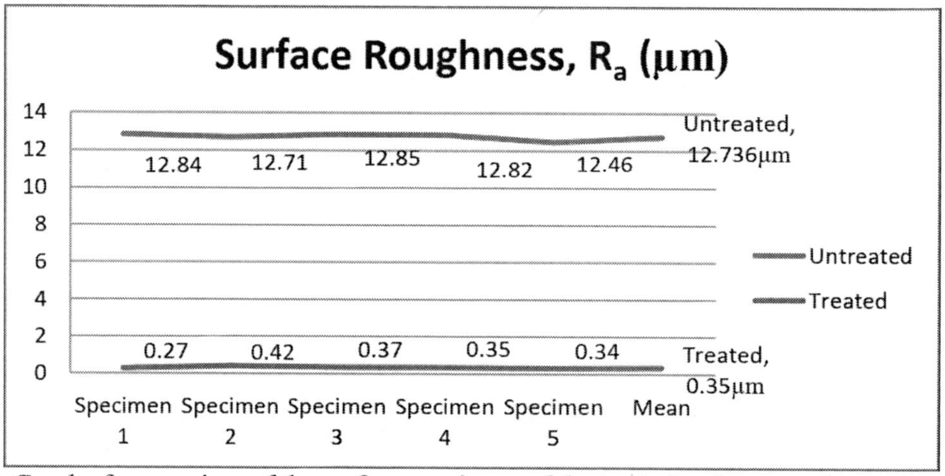

Figure 9. Graph of comparison of the surface roughness of the treated and untreated specimens.

Based on figure 9, the mean surface roughness, R_a for untreated dogbone specimen is 12.736μm while treated is 0.35μm. the surface roughness of specimens treated with acetone has better surface roughness compare to the untreated with the improved 97.25%.

Besides, to know the surface roughness of AM part, the test has been taken for these specimens. The tensile strength of material is the maximum amount of tensile stress that it can take before failure. The table 4 show the tensile strength and its Young Modulus after tensile test experiment.

Table 4. Mechanical Strength Data of Treated and Untreated Specimens

Treated / Untreated	Specimen	Tensile Strength (MPa)	Mean Tensile Strength (MPa)	Young Modulus (MPa)	Mean Young Modulus (MPa)
	1	36.54		7.00	
	2	36.60		6.50	
Untreated	3	35.52	36.012	7.00	**6.5±0.500**
	4	36.12		6.00	**Article I.**
	5	35.28		6.00	
	1	18.84		4.57	
	2	17.28		5.00	
Treated	3	26.30	20.678	5.63	**4.69±0.432**
	4	23.98		5.00	
	5	16.99		4.50	

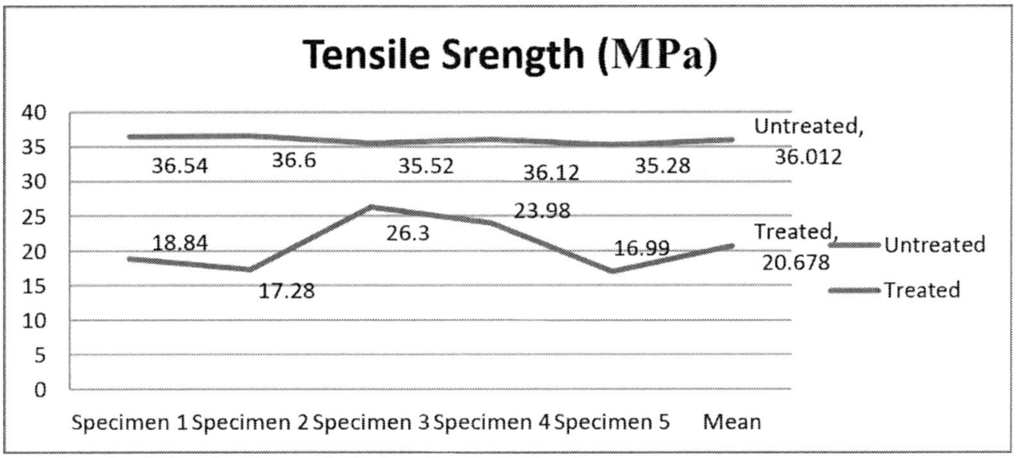

Figure 10. Graph of Tensile Strength Value for Each Specimen.

Based on figure 10, it shows clearly that all tensile strength of the treated specimens has low strength compared to the untreated specimens. The reduction of mean tensile strength was from 36.012MPa to 20.678MPa or 42.5% in percentage value. This can conclude that the specimens that chemically treated with acetone will affects the tensile properties of the parts. Suggested that this may due to the joining of the subsequent filaments affect the arrangement of the atom internally during and after the reaction with the acetone solution.

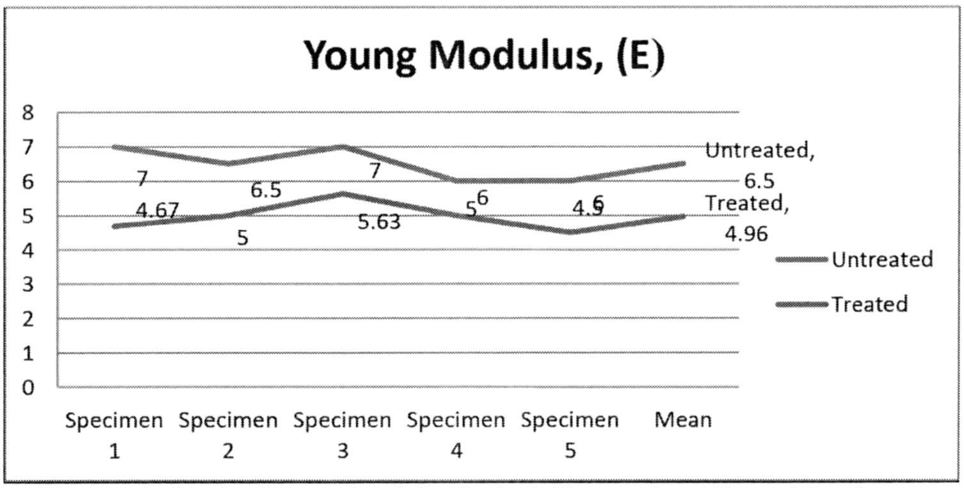

Figure 11. Graph of Young Modulus for both Untreated and Treated Specimens

In addition, the Young Modulus undergoes plastic deformation where the shape of material will not back to original shaper after the force is removed. According to figure 11, all treated specimens also have a lower modulus of elasticity compared to the untreated specimens. This means that there are decreases in term of stiffness after the part immersed into the acetone. In terms of reduction in average, the Young Modulus has dropped from 6.5 to 4.96 or 23.7% in percentage value. This may due to the joining effect of filaments after reacting with acetone.

4. Conclusion

As the conclusion, the main aim of this project is to investigate the influences of chemical treatment to the FDM part in term of surface roughness and mechanical properties has been achieved. The findings of this study generally suggest that chemically treated parts have very good surface finish. However, this will affect the mechanical properties of the parts with reduction of tensile properties. The result from both tensile strength and Modulus of Elasticity, the mechanical properties generally has reduced suggested due to the reaction to the chemical which creating isotropy after chemical treatment. These results can be explained by the joining effect of the filaments after the penetration of Acetone along the structure of the part itself. These show that the treated specimens have improved 97.2% in surface roughness analysis. On the other hand, chemical treatment was able to improve significantly the surface finish of ABS part. Moreover, this chemical treatment is one of the economical and fast methods to enhance the surface quality of the additive manufacturing parts in future manufacturing industry.

Acknowledgments

The authors would like to acknowledge the Ministry of Higher Education (MOHE) for funding this research under Research Acculturation Grant Scheme (RAGS) refers RAGS/1/2015/TKO/UTEM/03/B0010. Faculty of Manufacturing Engineering of UTeM and Centre for Research and Innovation Management (CRIM) for providing facilities and assistance throughout this study.

References

[1] Gibson I, Rosen DW, Stucker B 2010 *Additive Manufacturing Technologies: Rapid Prototyping to Direct Digital Manufacturing* (New York, NY: Springer)
[2] Gurpal Singh Bual, Parlad kumar 2014 *Manufacturing Science and Technology 2* 51-55
[3] L M Galantucci , F Lavecchia, G Percoco 2009 *CIRP Annals-Manufacturing Technology.* **58** 182-192
[4] L M Galantucci, F Lavecchia, G Percoco 2009 *CIRP Annals – Manufacturing Technology* **59** 247-250
[5] Gianluca Percoco, Lavecchia, L M Galantucci 2012 *Research Journal of Applied Sciences, Engineering and Technology* **4** 3838-3842
[6] David Espalin, Danny W Muse, Eric Donald and Ryan B Wicker 2013 *The International Journal of Advanced Manufacturing Technology* **72** 963-978
[7] L M Galantucci , F Lavecchia, G Percoco 2009 *CIRP Annals-Manufacturing Technology* **58** 182-192.

IOP Conf. Series: Materials Science and Engineering **210** (2017) 012064 doi:10.1088/1757-899X/210/1/012064

Enhancing vehicle's engine warm up using integrated mechanical approach

T. M. Ibrahim*[1], A. Z. Syahir[1], N. W. M. Zulkifli[1], H. H. Masjuki[1], A. Osman[2]
[1]Department of Mechanical Engineering, Faculty of Engineering, University of Malaya
[2]Powertrain Department, Engineering Division, Proton Malaysia

E-mail: *ymtmis@um.edu.my

Abstract. Transportation sector covers a large portion of the total energy consumption shares and is highly associated to global warming. Growing concern over the harmful gases being emitted from vehicles and their environmental implications has urged the need for pollutant reduction through more efficient engines. Good engine thermal management especially during cold-start warm up phase has been proven to enhance the engine efficiency in terms of fuel economy and greenhouse emissions specifically. In this study, the viability engine split cooling system was tested in two separate test. The parameters of interest include coolant and transmission temperature as these both parameters indicate the internal engine condition and highly associated with engine efficiency. In the first idle test, coolant temperature within the modified cooling configuration reached the optimum coolant temperature of 60 °C about 41.28% faster when compared to baseline configuration. The modified configuration also heat up the transmission oil around 4 times faster. In the second NEDC test which simulates the real time driving condition, the coolant of the modified vehicle reached the optimum temperature around 28.26% compared to the baseline.

1. Introduction

Energy generation through combustion is expected to remain as the most important method for at least another 30 years [1]. Currently, transportation sector covers most of the total energy consumption shares and accounted to 13.5% of the total global warming [2, 3]. Growing concern over the harmful gases being emitted from vehicles and their environmental implications has urged the need for pollutant reduction through more efficient engines. High demands for efficient engines are also attributed to stringent environmental regulations and high fuel cost. Most of the spark ignition (SI) engines used today are well optimised in terms of combustion efficiency, with approximately 95-98% of the fuel energy content being released during combustion [4, 5]. However, the useful energy leaving the engine (brake power) is only around 40% of the fuel energy [5, 6]. Owing to the above implications, thermal efficiency for internal combustion engines (ICE) has become the main design criteria for academic societies and engine designers.

It is largely accepted that one of the purpose of thermal management is to decrease engine warm up time, consequently reducing the fuel consumption [7]. During engine warmup phase, the engine thermal efficiency is low as more fuel is consumed due to; 1) high friction at engine components exerted by cold lubricant at 2) less efficient combustion at cold condition. Energy loss due to friction within engine during the early stages of warm-up (in the range of 20 °C) was estimated to be 2.5 times higher compared to when in optimum temperature condition [6, 8]. In different studies, Trapy and Damiral [9] and Will and Boretti [8] reported that during cold-start warm up phase, only 9% of energy from the fuel is converted to effective work, a reduction of approximately 30% compared to warm (optimum) engine condition. Figure 1 presents the energy thermal balance during warm up phase. It can be observed from the right-hand chart that only 4% of the transferred heat is used to warm up the engine oil while 52% from the total heat transfers were found to be unused, in other way lost directly to the environment.

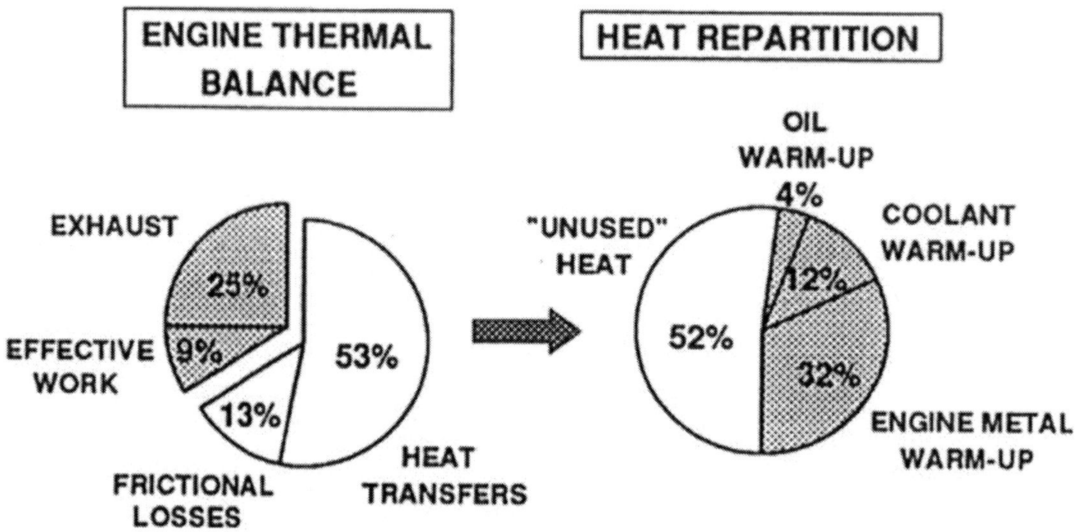

Figure 1. Thermal balance diagram during warm-up phase [9]

Samhaber et al. [10] predicted and increases in fuel consumption can be up to 13.5% during the warm up phase of an engine in a cold-start scenario of around 0 °C. Therefore, a number of strategies and potential solutions have been proposed in the effort of shortening the engine warm up time. By shortening the time in which the engine is operating inefficiently, more fuel can be saved, leading to lower amounts of pollutant emissions will be emitted [11]. The engine components also will experience lower amount of wear corresponding to efficient lubricants performance at optimum temperature, resulting in longer engine life. Previous studies have tried the possibility of controlling the thermal status of the engine (over- and under-cooling) by the implementation of special mechanisms that could manipulate the coolant flow through engine and radiator [12, 13]. Suggested improvements include the utilization of electronically controlled coolant pumps [14-16], fans, thermostats and shutters [17, 18]. In a study, Torregrosa et al. [19] did some modifications on the baseline engine cooling system based on the simulation carried out in their earlier studies. Five different configurations were tested. The results suggested that one of the selected configurations reduced the warm-up period by approximately 159 s when compared with the baseline configuration. Reductions in fuel consumption and pollutant emissions (HC and CO) were also obtained.

In recent years, the feasibility of engine split cooling system as an alternative solution in reducing engine warm up time is gaining in interest. Engine split cooling configuration include the separation of head and block cooling to allow the engine blocks get hot in relatively shorter time while keeping the engine head cool. Engine head operated at a high temperature can experience cracking, subsequently causing catastrophic damage to the engine. Meanwhile, getting the engine blocks to an elevated temperature is important to reduce the warm up time and permitting to increase compression ratio which improves thermodynamic engine efficiency. Using mathematical modelling, Cipollone et al. [7] analysed the reliability of engine split cooling system and proposed two potential solutions. The first solution is that the coolant is let to flow into the engine head first. As it flows out, only a part of it is used to cool the blocks, the remaining part is by passed through and re-mixed together outside the engine. The second solution is that the block cooling is done with a greater coolant flow rate with respect to that crossed the engine head. This can be achieved by re-mixing a part of coolant leaving the engine before re-entering the engine block.

In this study, engine split cooling system will be further explored to prove its scientific viability. Two separate test were conducted. The first test was the idle test to check viability of the split cooling configuration in reducing engine warm up time in a typical cold start condition. In the second test, road tests will be carried out using instrumented vehicle equipped with split cooling system running at different drive cycles as per New European Driving Cycle (NEDC) requirements. The parameters of interest include coolant and transmission temperature as these both parameters indicate the internal engine condition and highly associated with engine efficiency. Test using baseline vehicle was conducted for comparative analysis and therefore to determine whether it is practical and beneficial to replace the existing cooling system with engine split cooling system.

2. Methodology

2.1 Equipment

2.1.1 Chassis dynamometer

Chassis dynamometers is used for complete road simulation in measurement of vehicle emission, fuel consumption, and vehicle performance. A chassis dynamometer can be referred as a rolling road, measures power delivered to the surface of the "drive roller" by the drive wheels. The vehicle is often parked on the roller or rollers, which the car then turns, and the output measured thereby. This machine would be used in conducting the New European Driving Cycle (NEDC) test.

2.1.2 Data collection systems

Integrated Calibration and Acquisition System (INCA), a leading software in automotive industry is used for measurement, calibration and diagnostic. This software is connected to Engine Calibration Unit (ECU) to read and record the data taken from the engine system and built-in sensors of the engine. This software is used on both NEDC and Idle test. Graphtec Data Logger Model GL-820 is used for temperature data logging. The temperature of the targeted point is read and recorded by channelling the thermocouples to this equipment.

2.1.3 Test rig

Two units of vehicles are used for testing the simplified engine split cooling system. One unit is used for baseline (benchmark) with no modification while the other one is installed with the proposed simplified split cooling engine system. Table 1 and table 2 show the specification of the engine and vehicle used in this study. All unit was dedicated with thermocouples installed at certain targeted point as in table 3. By having this test rig, the changes and the characteristic of the enhanced integrated mechanical development of the system, can be evaluated closely based on its original condition (baseline).

Table 1: Engine specifications

Engine Model	Charge Fuel Efficient (CFE) turbo-charged spark ignition internal combustion engine
Displacement Volume	1561 cc
Induction System	Single scroll turbocharger
Fuel Injection System	Gasoline Port Fuel Injection
Stroke	86mm
Bore	76mm
Compression ratio	8.9:1
Coolant	20% Ethylene Glycol and 70% water
Engine Oil	PETRONAS Mach 5 10W30
Maximum Torque	205 Nm at 2000-4000 rpm
Maximum Power	103 kW at 5000 rpm

Table 2: Vehicle specification used for the test rig

Model	**Proton Exora Bold**
Type	7 Seater Multi-purpose vehicle
Kerb Weight	1466 kg
Transmission	Punch CVT
CVT heat exchanger	Oil-to-radiator
Intercooler	Air-to-air
Power Steering	Hydraulic
Maximum Speed	185 km/h

Table 3: Thermocouple points of installation

Channel	**Point of Installation**
1	Ambient
2	Engine Coolant
3	Engine Oil Sump
4	Transmission Oil Sump

2.2 Modifications

Integrated mechanical developments of the engine would be the key of this study. Comparing to other works, this strategy utilizes mechanical approaches rather than complicated electrical and electronic development within the engine compartments. Among the elements of consideration in the modifications are as follows;

2.2.1 Arrangement of engine cooling system, transmission cooling system and turbocharger

Compared to the conventional cooling system, the transmission cooling system and turbocharger cooling system has been arranged in series, which are right after the engine cooling system. This strategy was made to improve heat recovery and thermal distribution during cold start and engine warm up. Heat from the turbocharger and engine cylinder head can heat up the coolant and raise the coolant circuit temperature to optimal temperature. After the coolant reached its optimal temperature, this serial cooling configuration helps the turbocharger and transmission temperature to maintain their optimal temperature and thus avoid the engine, turbocharger and transmission oil from overheat or overcool.

Based on the idea of split cooling, having a separate flow of coolant circuit for the engine cylinder head and cylinder block would contribute in the improvement of engine performance [12]. Believing on every part of the engine have their own needs in temperature, such engine improvement could be achieved by introducing parallel flow which by splitting the normal flow which work in series. By having such "parallel-like" circuit for engine cylindrical head and engine block would improve the internal engine cooling by allowing better flow rate (can reach higher flow rate) and flow rate control (better coolant flow rate range and better feedback control) for the coolant inside the engine itself.

2.2.2 Coolant flow

Coolant flow has been modified as in figure 2. By introducing separate flow of coolant in engine head and engine block, better engine thermal management would be obtained. Some mechanical development has been done along with this coolant flow modification. The proposed system introduced dual exit point (due to dual parallel cooling circuit; cylinder head cooling circuit and cylinder block cooling circuit). This strategy improves the coolant flow rate by reducing pressure drop as well as the internal engine coolant flow restriction is reduced. The permanent bypass is shifted to the cylindrical block. As the cylindrical block has lower flow rate, therefore it is possible to have a bypass line without overcooling the engine block.

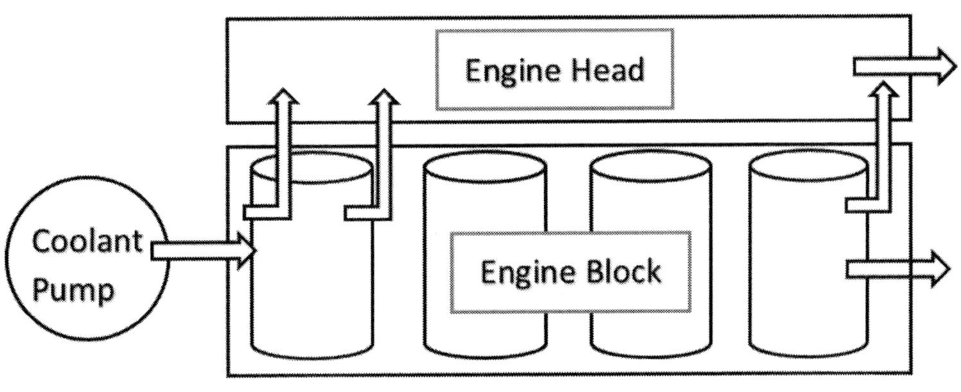

Figure 2. Proposed coolant flow layout

2.2.3 Cooling pack (radiator and fan)

Smaller cooling pack was introduced which is half of the original size. Smaller cooling pack gives lower coolant volume needed for heat exchanging as well as lower thermal inertia. Additionally, it will come with an extra advantage which lower in cost.

2.2.4 Single thermostat opening

The introduction of a mechanically simpler cooling mechanism requires faster response in regulating the coolant temperature. Due to this concern, the opening temperature of the thermostat is modified by lowering the temperature for opening set point. In this study, the thermostat opening temperature is shifted from 72 °C to 67 °C, whereas 60 °C was the lower boundary of the optimum temperature range.

2.3 Test method

The test is separated into two parts which are idle test and NEDC test. The explanation regarding these tests are described below. Temperature changes of two different fluids were measured which are the coolant and transmission oil. Coolant was selected as it represents the relative temperature of the engine blocks and warmed transmission oil was found to contribute in the reduction of the overall fuel consumption [20]. The warm up time is considered as the time taken for the coolant temperature to reach the minimum coolant temperature of a well-functioning engine, which is around 60 °C.

2.3.1 Idle Test

The idle test is conducted in cold-start condition to have extra information on warm up and fuel consumption of a non-moving vehicle. The test is conducted by having "soaking time" with the minimum of 6 hours in control ambient condition (about 25°C). The vehicle's engine is then started with cold-start condition for 600 seconds. The coolant and transmission oil temperature reading is recorded using INCA software that connected to engine calibration unit (ECU) and Graphtec Data Logger.

2.3.2 New European Driving Cycle (NEDC) test

New European Driving Cycle (NEDC) test is used to simulate the driving conditions. NEDC is Driving Cycle pattern consist of idle state, transient state (acceleration and deceleration) and also the steady state (constant velocity). These types of driving pattern will show and trace the behaviour of vehicle performance based on the driving states. NEDC was divided into two sections, Urban Driving Cycle which simulate the relatively common cycle in the city driving, and Extra-Urban Driving Cycle which likely to simulates the common driving style on highway driving styles. During this test, the vehicle act as test subject is equipped with Engine Calibration Unit (ECU) channelled to INCA software and also Graphtec Data Logger are used to determine and record the reading of coolant and transmission oil temperature.

3. Results and Discussion

Part 1 - Idle test

It is posited that the minimum coolant temperature of a well-functioning engine is 60 °C. From figure 3, it can be observed that the baseline (unmodified) engine reached the optimum temperature of 60 °C after 199.96s while the engine with the proposed modification only took around 117.41s for the coolant temperature to reach the optimum point, an improvement of about 41.28%. At the end of the test (after 600s), the coolant in the modified engine reached a higher temperature point when compared to the baseline engine (106.35 °C compared to 80.25 °C). In terms of transmission oil, it can be observed in figure 4 that the modified configuration heat up the transmission oil around 4 times faster, of which the transmission oil in the modified engine reached higher temperature point than the baseline engine (54.40 °C than 30.00 °C) after 600 s.

Looking only at the warm up duration perspective, it can be assumed that the modified engine will perform better than the baseline engine in terms of efficiency. This is because lesser fuel will be burned in order to get the engine to the optimum temperature. The catalytic converter also will reach its optimum temperature condition in shorter time, causing less harmful gases to be emitted. Since the modified engine reached the temperature region of 100 – 110 °C in shorter time, there will be less frictional losses due to better lubrication properties of the engine oil in this temperature range [21]. However, it should be noted that the coolant temperature of 106.65 °C is considered quite high, therefore suggesting that the coolant layout can be further improved for a better cooling system in idle condition.

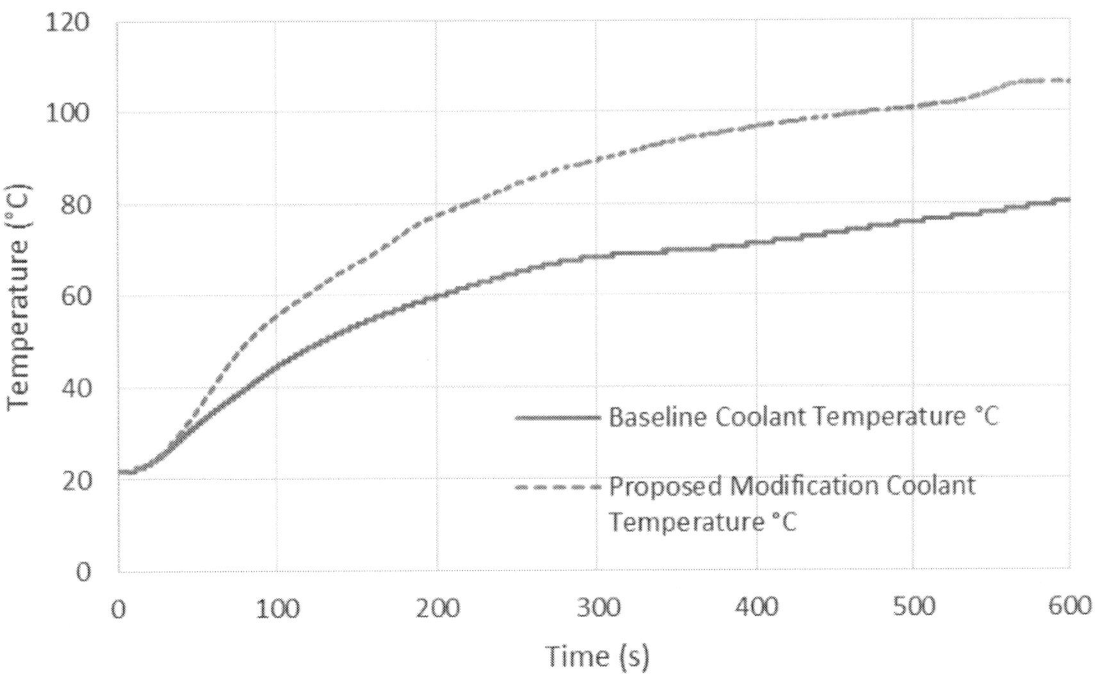

Figure 3. Coolant temperature comparison for idle test

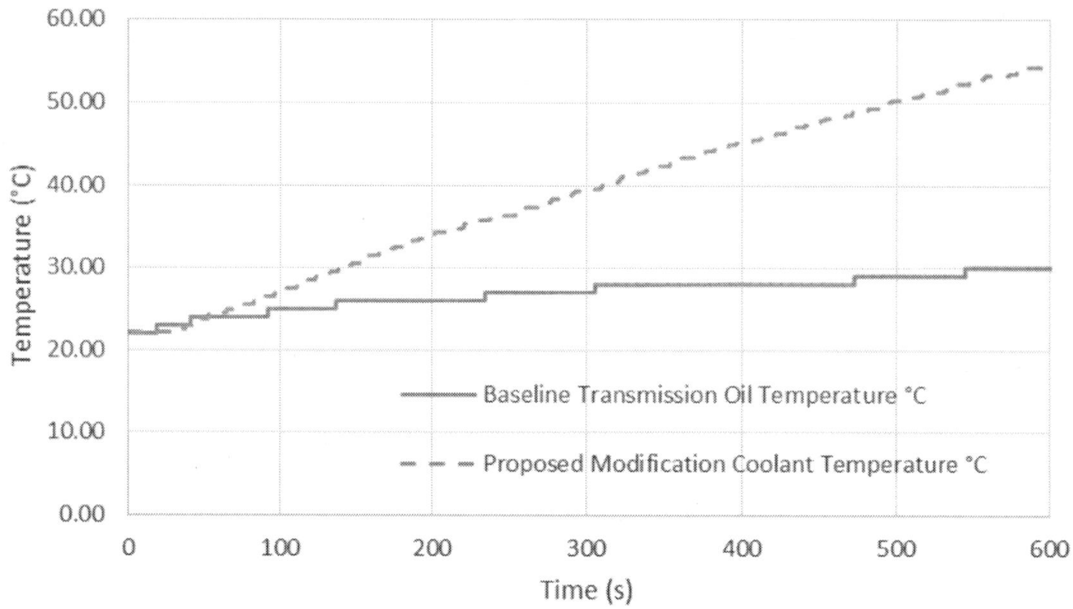

Figure 4. Transmission oil temperature comparison for idle test

Part 2 - NEDC test

Figure 5 shows that the coolant of the modified vehicle reached the temperature point of 60 °C after 114.78 s, which is 28.26% faster when compared to the baseline vehicle (159.99 s). Both vehicles reached 60 °C in the first UDC. After four UDC, the coolant temperature of the baseline and the modified vehicle were 81.00 °C and 84.60 °C respectively. The fluctuations in coolant temperature suggested that the engine of the modified cooling configuration operated at higher temperature compared to the one in baseline vehicle. At the end of the NEDC test, the coolant temperature of the baseline and modified vehicle were 76.50 °C and 83.25 °C respectively.

It should be highlighted that the difference in maximum coolant temperature between idle test (106.65 °C) and NEDC test (95.35 °C) were due to difference in test set up. Idle test was carried out in typical cold-start condition environment, while NEDC test was carried out using the designated test rig. There was no air flow through the heat exchanger (radiator) during idle test as the vehicle was in steady state (non-moving) condition. The presence of air flow into radiator in order to simulate real driving condition at particular speed during NEDC test caused more heat transfer out from the coolant, resulting in lower coolant temperature. From the results obtained, it can be said that the engine using the proposed cooling system operated at higher temperature condition than the one using conventional cooling system almost the entire time, but still within the safe temperature range.

For transmission oil, it can be deduced from the difference in line steepness in figure 6 that the modified configuration heat up the transmission oil by 107.49%, around two times faster. The temperature of the transmission oil within the modified cooling configuration was relatively higher than the baseline configuration almost the entire time. This suggests that the proposed cooling systems was doing a good job increasing the temperature of the transmission oil. Higher temperature of transmission oil is a necessity since its viscosity lower as the temperature increases. Transmission oil with high viscosity has higher intermolecular friction within the oil that can hinder the ability of which the engine can transmit the power through the transmission system efficiently. However, comparing only the line steepness (gradient) for the whole test duration still not good enough due to proposed modification already reached 60 °C (minimum of optimal operating temperature range) after around 500 s. Afterwards, it will increase at a slow rate and then start to maintain its temperature around the optimum range. This results in lower gradient steepness of the linear trend line of proposed cooling configuration. Plus, it was hard to compare the improvement of warm up time in terms of reaching the optimum temperature range as the transmission oil within the baseline configuration did not reached the optimum temperature of 60 °C during the entire NEDC test.

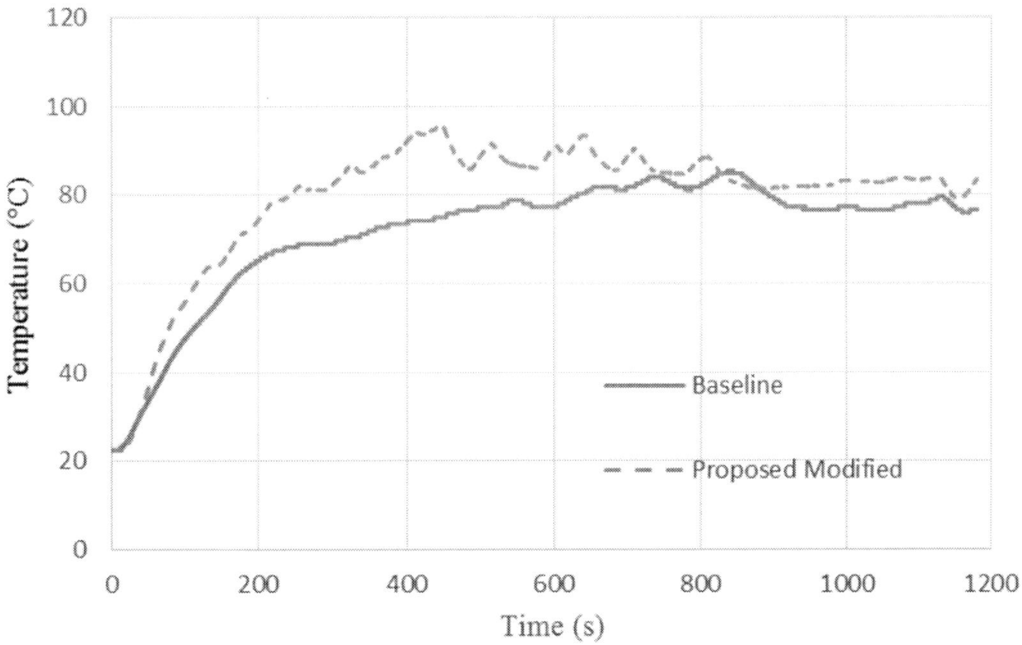

Figure 5. Coolant temperature comparison for NEDC test

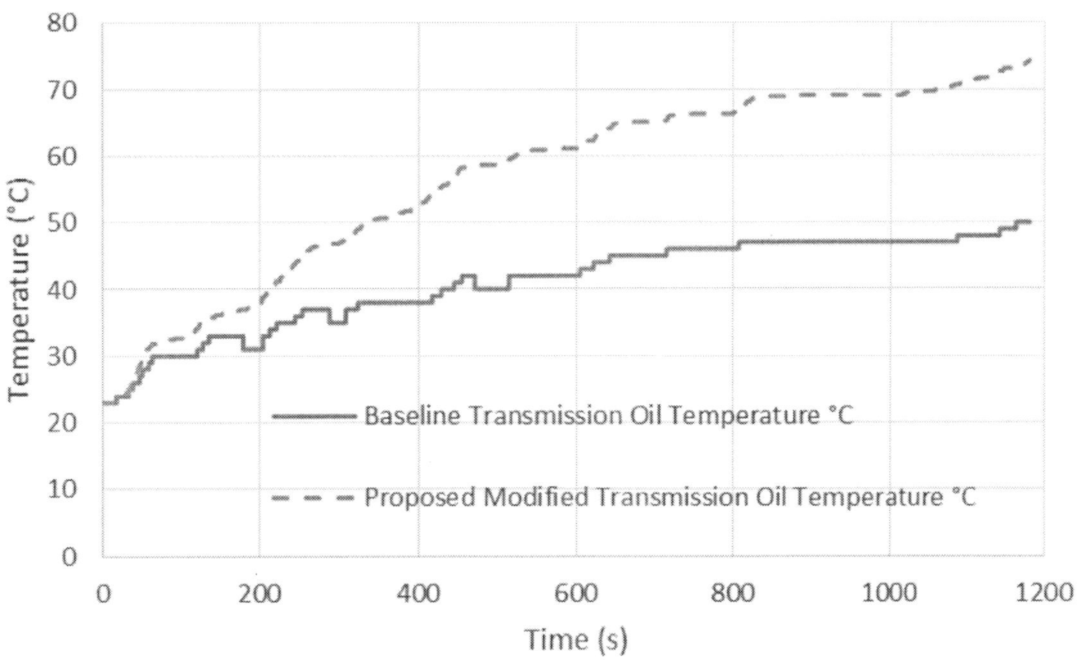

Figure 6. Transmission oil temperature comparison for NEDC test

4. Conclusion

In this study, a new coolant flow layout for replacing the conventional cooling system was proposed. This new cooling system was implemented into the modified vehicle and the performance of the modified vehicle was compared to the performance of the baseline vehicle, which is equipped with the conventional cooling system. The performance evaluation was done in terms of engine warm up time. Both vehicles undergo idle test in cold start condition and also the standard NEDC test using the designated dynamometer. The results suggested that the modified split cooling system performed better than the baseline vehicles. The warm up time of the modified configuration was 82.55 s and 45.21 s faster than the baseline configuration in idle test and standard NEDC test respectively. During the NEDC test, the coolant and the transmission oil of the modified vehicle were at a higher temperature point almost the entire time, implying that the engine operated at higher temperature.

From the test results, the proposed cooling circuit has shown promising attributes of a cost-effective solution, while meeting the future requirement. Nevertheless, the full potential of the proposed configuration is still ambiguous. In particular, the baseline engines which are widely produced today are optimized for the conventional serial cooling system. For future works, it is recommended that more thorough testing for further optimization are carried out which include engine performance and emission testing.

Acknowledgement
The authors wish to extend their greatest appreciation to University of Malaya for providing financial support under Fundamental Research Grant Scheme (Project no.: FP063-2015A. The first author is greatly indebted to Perusahaan Otomobil Nasional (PROTON) for providing the necessary supports under the 'Collaborative Research' programme.

Reference

[1] Shafiee S and Topal E 2009 *Energy policy* **37** 181
[2] Ong H, Mahlia T and Masjuki H 2012 *Renewable and Sustainable Energy Reviews* **16** 532
[3] Baumert K A, Herzog T and Pershing J 2005 *Navigating the numbers: Greenhouse gas data and international climate policy* (California, USA: World Resources Inst)
[4] Roberts A, Brooks R and Shipway P 2014 Energy Conversion and Management **82** 327
[5] Heywood J 1988 *Internal combustion engine fundamentals* (USA: McGraw-Hill Education)
[6] Will F and Boretti A 2011 *SAE Int. J. Engines.* **4** 175-87
[7] Cipollone R, Di Battista D and Gualtieri A 2012 *IFAC Proceedings Volumes* **45** 400
[8] Will F 2012 *Fuel* **102** 247
[9] Trapy J and Damiral P 1990 *SAE Technical Paper 902089*
[10] Samhaber C, Wimmer A and Loibner E 2001 *SAE Technical Paper 2001-01-1697*
[11] Broatch A, Luján J, Ruiz S and Olmeda P 2008 *Int. J. Auto. Tech.* **9** 129
[12] Pang H and Brace C 2004 *Proceedings of the Institution of Mechanical Engineers, Part D: Journal of Automobile Engineering* **218** 1209
[13] Allen D J and Lasecki M P 2001 *SAE International 2001-01-1732*
[14] Cortona E and Onder C H 2000 *SAE Technical Paper 2000-01-0965*
[15] Cortona E, Onder C H and Guzzella L 2002 *Int. J. Engine Res.* **3** 157
[16] Cho H, Jung D and Assanis D 2005 *Int. J. Auto. Tech.* **6** 269
[17] Chanfreau M, Gessier B, Farkh A and Geels P Y 2003 *SAE Technical Paper 2003-01-0274*
[18] Chalgren R D and Allen D J 2005 *SAE Technical Paper 2005-01-2020*

[19] Torregrosa A, Broatch A, Olmeda P and Romero C 2008 *Int. J. Auto. Tech.* **9** 447-58

[20] Hawley J G, Bannister C D, Brace C J, Akehurst S, Pegg I and Avery M 2010 *Proceedings of the Institution of Mechanical Engineers, Part D: Journal of Automobile Engineering* **224** 1213

[21] Zammit J, Shayler P and Pegg I 2011 *Vehicle thermal Management Systems Conference and Exhibition (VTMS10)* (UK: Elsevier) 177

Experimental investigation on momentum and drag reduction of Malaysian crop suspensions in closed conduit flow

S M Ahmed[1*], S N Kazi[1*], G Khan[1], M Dahari[2], M N M Zubir[1], P Ahmad[3], E Montazer[1].

[1] Department of Mechanical Engineering, Faculty of Engineering, University of Malaya, 50603 Kuala Lumpur, Malaysia.
[2] Department of Electrical Engineering, Faculty of Engineering, University of Malaya, 50603 Kuala Lumpur, Malaysia
[3] Department of Physics, Faculty of Science University of Malaya, 50603 Kuala Lumpur,Malaysia.

E-mail: *syed.ma@hotmail.com

Abstract. The study of frictional losses in fiber suspension flow is one of the significant scientific interests as the characteristics of suspension flow considerably changes with shear stress, fiber source, and treatments applied on fibers. Pressure drop measurements were obtained for different Malaysian crop fiber suspensions flowing through a closed conduit. The generated data were gathered over a range of flow rates and suspension concentrations. It was found that the magnitude of the pressure drop of the fiber suspensions is dependent on the concentration, characteristics, and fiber source. Considerable drag reduction is obtained for concentration of 0.6 wt. % at high flow rates. Such a reduction of pressure drop at the particular concentrations and the flow rates is interesting and useful as these data can be used for design and optimization of fiber handling equipment and piping systems. Furthermore, the effect of different fibers, fiber properties and flexibility on pressure drop were studied.

1. Introduction

Fiber suspension flow is different from Two-phase flow occurs in the transportation of oil and gas, mining, petrochemical process, boilers and condensers as fibers are flexible and asymmetric. Usually fibers suspensions flows are occurring in many applications, such as pulp and paper, textile, fiber composites etc. Studies conducted on fiber suspensions flow and their characterization and correlations are mainly wood pulp fibers. The physical characteristics of the fiber suspensions, as well as the properties of the end products, are dependent on the hydrodynamic behavior and structure of the suspensions [3]. Features, such as consistency, flow conditions, interactions and fiber properties affect the hydrodynamic behavior and structure of the suspensions flow. The investigations conducted on fiber suspensions flow and characterization of fiber suspensions are mainly wood pulps and polymeric fibers [4-8].

Due to increasing demand for fibrous materials, limitation of the wood source and environmental concern, the study of non-wood fibers suspensions has become imminent. Crops based fibers attracting the focus of pulp and paper industry as well as researchers, as it is recognized as an alternative source of wood fibers. Kenaf, crop-based fiber reported as an alternative source of wood for the pulp and paper production that have two distinctive stem regions, the outer portion or Bast is about 34 wt.% of the stem and inner, woody core, that is about 66 wt.% [9]. Kenaf bast fiber resembles long softwood fibers which could be used for high-grade pulps production [10-12].

Understanding, the hydrodynamic behavior of pulp suspensions and its correlation with fiber properties, flow conditions and consistency can therefore act as an aid in the design and optimization of processes and products. Kazi, Duffy and Chen [13] have studied drag reduction of wood pulp fiber at different consistencies over a range of flow rates in a pipe. They observed that small amount of fiber in suspension reduces pressure losses than that of water. Duffy, Kazi and Chen [5] considered the findings [13] and correlated the pressure drop with fibers and paper properties. They reported that little modification in fiber properties could be sensed by some little changes in pressure drop at low consistency. They also reported that pressure drop measurements could be correlated and utilized for the prediction of specific fiber and paper properties and these measurements could be significant for retardation of production of low quality or reject papers. Duffy, Kazi and Chen [4] studied pressure drop characteristics of wood (Kraft pine pulps, a spruce pulp, and a eucalypt) pulp fibers and synthetic model polymer fibers and compared with the pipeline studies. They reported the effect of fiber concentrations on pressure drop. The correlation between concentration and fiber properties were generally well versed and they highlighted fiber stiffness and fiber population as key variables. Frictional pressure drop decreases with the increase of fluid velocity, fiber flexibility, and an increase of consistency. Frictional pressure drop was also affected by varying fiber properties, such as fiber length, fiber flexibility, fiber mechanical and chemical treatment as well as different pulping methods used for pulping, furthermore frictional pressure drop can be used to adjust the degree of fiber refining treatment so that the paper made from those fibers are within the product specifications [14, 15].

Crop pulp fiber suspensions flow and their correlation with fiber properties have not been explored as in the case of wood pulp fiber suspensions. Thus, there are extensive requirements for a detailed study of crops pulp fibers on frictional losses and their correlations with fiber properties. The aim of the present work is to generate experimental frictional loss data and their correlation with fiber properties for different crop pulp fibers and their blends for future development of valid models of turbulent fiber suspensions flow and to introduce insight on the advancement of fiber processing, design and optimization of process and product by characterization of the hydrodynamic behavior of fiber suspensions.

2. Experimental

2.1 Material

Kenaf bast and Kenaf core pulp fiber samples were prepared by the mechanical process (refining mechanical pulping RMP) (kbm and kcm). The blend kckb3 of Kenaf core and Kenaf Bast pulps were prepared by mixing Kenaf core and Kenaf bast pulp with a ratio of 9:1. Pulp samples used in the investigation were provided by Forest Research Institute of Malaysia (FRIM).

2.2 *Experimental* set-up

Flow loop of the test section and the equipment used are presented in this section. A schematic diagram of the experimental setup is shown in Figure 1. The flow loop of the test rig consists of stainless steel piping, jacketed stock tank connected with a chiller for cooling of pulp suspensions at the desired bulk temperature, a variable speed pump, a magnetic flow meter, a differential pressure transducer, heated test section and the recycle pipings. A differential pressure transducer (IDP10 FOXBORO) was used to measure the frictional losses across the test section. Ten flat coil heaters, 900 W each were utilized for the heating of the test section. The pulp suspension was pumped from the 100 L capacity stainless steel tank by the LOWARA stock pump driven by a 2.71 KW AC motor which was controlled by a Hoffman MullerTM inverter.

IOP Conf. Series: Materials Science and Engineering **210** (2017) 012065 doi:10.1088/1757-899X/210/1/012065

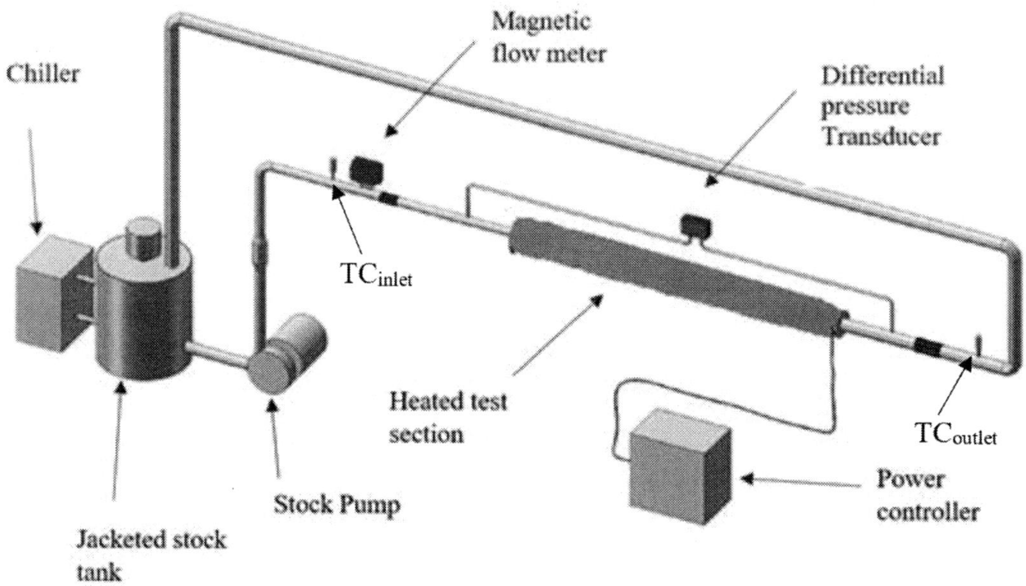

Figure 1. Schematic diagram of experimental setup.

2.3 *Data acquisition*

The inlet temperature data were logged using a data logger (Graphtec, midi logger GL220). The flow rate was measured by using Magnetic flow meter 8000 A SERIES (FOXBORO) and a Differential Pressure cell transmitter IDP10 (FOXBORO) was used to measure the pressure drop. The heater power was adjusted to 3.3 kW by manipulating current and voltage supplied to the heater in order to maintain constant inlet temperature, and the supplied current and voltage data were measured by 1000 A true RMS meter (Agilent). All the data were taken at steady-state conditions at an instantaneous velocity, bulk temperature, and heat flux.

The Moody friction factor (f_M) is calculated by Eq. (3) using the measured data of bulk velocity (u), Pressure drop (ΔP), pipe diameter (D), pipe length (L) and suspension density (ρ).

$$f_M = \frac{D}{L}\frac{\Delta P}{\frac{1}{2}\rho u^2} \qquad (1)$$

The correlations of Petukhov (f_p) [1] and the Blasius relation [2] are presented in equation (2) (3) respectively.

$$f_p = (0.79 \ln Re - 1.64)^{-2} \quad 4000 < Re < 5\times 10^6 \qquad (2)$$

$$f = 4 \times \frac{0.0791}{Re^{0.25}} \qquad (3)$$

2.4 *Experimental* procedures

The pulp fibers were disintegrated in the disintegrator up to 35000 rpm before conducting each experiment. The disintegration process was carried out according to the TAPPI standard where the revolutions were not more than 50,000 (TAPPI, 2002). Then pulp suspensions were prepared in the stock tank, stirred for 1 hour then pumped through the loop for approximately two hours to ensure homogeneous fiber dispersion before any set of data were taken. Once steady state conditions were achieved and inlet temperature maintained at 32°C then pressure drop data were taken in a range of velocity (0.4-2.8 m/s).

3. Results and discussion

3.1 Data Reproducibility and Test rig calibration

To validate the new experimental data, a preliminary set of test were conducted keeping the water as a working fluid. The measured Pressure drop for the new experimental data was correlated with the Petukhov [1] and Blasius [2] correlations and presented in figure 2. It revealed that the measured experimental data are in good agreement with the evaluated values from the empirical correlations and it is observed that with an increase of velocity the deviation between experimental and correlations decreases. In particular, Petukhov [1] empirical correlation are within the agreement of about 7% with the experimental data and Blasius relation [2] are also in good agreement of about 8% with the experimental data, that validates the experimental methodology.

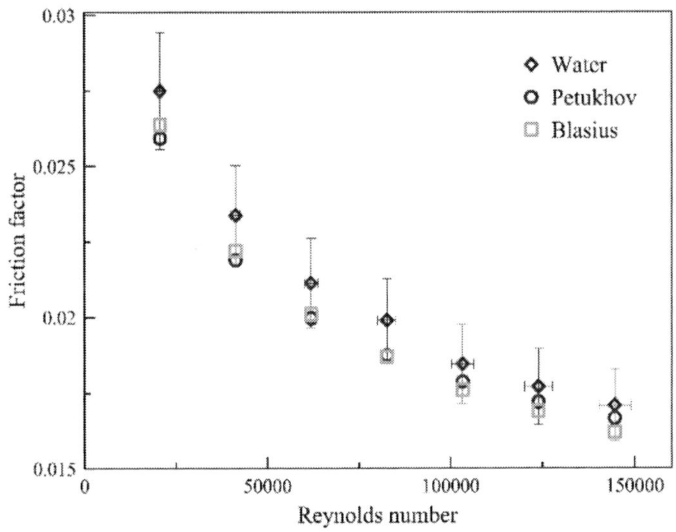

Figure 2. Comparison of the measured experimental friction factor with the Petukhov [1] and the Blasius [2] correlations.

3.2 Concentration effect of fiber on pressure drop

Effect of fiber concentration on drag reduction was studied using kckb3 pulp. Figure 3 presents the ratio of the pressure drop of suspensions and pressure drop of water as the function of velocity for water and kckb3 at different concentrations (0.2, 0.4, and 0.6 wt.%) and a constant inlet temperature of 32 °C and constant heat flux of 23.2 KW/m^2. It is observed that at the lowest velocity 0.4 m/s the drag ratios of suspensions are sufficiently higher than that of water, whereas at the highest concentration of 0.6 wt.% have shown the highest drag ratio. Higher drag ratio at low velocity, substantially owing to the presence of a plug where the turbulent sheared layer between the plug and the pipe wall is very thin and the velocity profile is steep [13]. The drag reduction is not achieved throughout the range of velocity (0.4 – 2.8 m/s) for all consistencies of pulp suspensions. The subsequent decrease in drag ratio with an increase of flow is owing to the dominance of fiber damping the turbulence. It is observed that the magnitude of drag ratio decreases with the increase of velocity and consistency. The minimum drag ratio is observed for highest concentration of 0.6 wt.% at 2.8 m/s and that is slightly higher than the pressure drop of water

3.3 Effect of different fibers on suspension properties

To study the effect fiber source on frictional pressure drop, the blend kckb3 (a mixture of kcm and kbm with mass ratio 9:1) pulp suspension were compared with primary pulp fibers kcm and kbm at a range of velocity 0.4-2.8 m/s and with a concentration of 0.6 wt.% and constant inlet bulk temperature of 32°C along with water presented in figure 4 as a drag ratio. At low velocity of 0.4 m/s, adequate distinction is observed in the drag ratio values of kcm, kbm and kckb3 whereas the the enormous difference between pulp suspensions and water values is marked. Furthermore, kbm showed the highest drag ratio 1.512 followed by kckb3 1.279 and kcm 1.217. The blend increased the drag ratio about 5.09% than that of kcm (the primary pulp in the blend) at a low velocity of 0.4 m/s. High drag

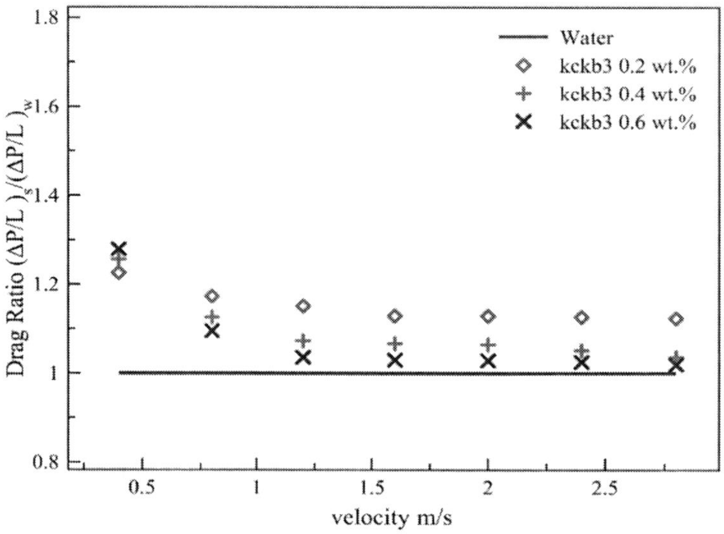

Figure 3. Drag ratio as a function of flow velocity for water and kckb3 pulp suspension at different concentrations.

ratio at low velocity, substantially due to the presence of a plug where the turbulent sheared layer between the plug and pipe wall is very thin and the velocity profile steep. The highest drag ratio marked by kbm at low velocity 0.4 m/s is due to heist aspect ratio that interacts with neighboring eddies, deflect, bend and absorb turbulent energy.

The minimum drag ratios 0.966, 1.02 and 1.035 are observed for kbm, kckb3 and kcm at velocity 1.6, 2.8 and 2.8 respectively. These results indicate that floc and network formation even at same concentration occur at different velocities and in the case of long fiber (kbm) the minimum drag ratio is observed in the mid-range of velocity 1.6 m/s over this point drag ratio successively increases slightly. While short fiber (kcm) and kckb3 showed a successive decrease in drag ratio until the maximum velocity 2.8 m/s investigated in the present study. Whereas, addition (blending) of 10 wt.% long fiber (kbm) in short fiber (kcm) decreased pressure drop about 1.45% than that of kcm.

To magnify the phenomenon of deflecting, bend and absorbing turbulent energy or aid in transmitting momentum with aspect ratio, the blend kckb3 were prepared by mixing pulp fibers with high aspect ratio 146.58 (kbm) with low aspect ratio 25.45 (kcm). The results reveal that with the increase of aspect ratio of fibers in suspension pressure drop decreases. Furthermore, consequent decrease in pressure drop with an increase of velocity is owing to the dominance of shear stress and formation of flocs and network that have special plastic-elastic characteristics

3.4 Fiber characterization

Fiber physical properties could modify turbulent eddies that affect frictional pressure drop. Due to practical and economic reasons, papermaking technologists are interested in flocculation [16]. Flocculation effect paper strength, uniformity, appearance [17-19] and adversely affect subsequent operations, such as coating [20]. Flocculation and consequently the frictional pressured drop would be

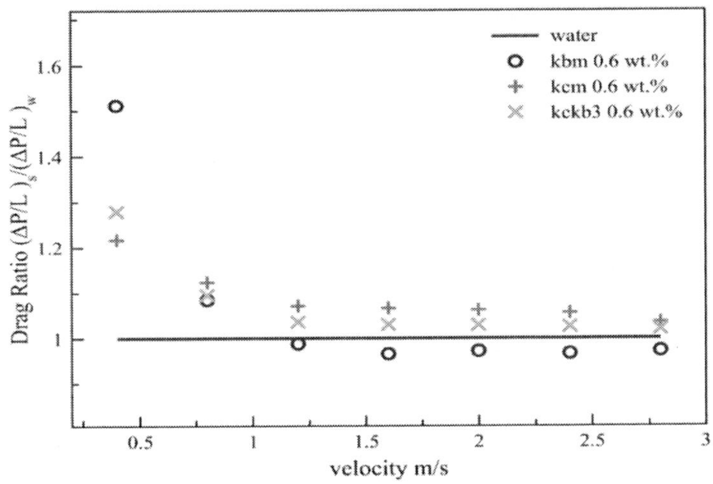

Figure 4. Frictional pressure drop (a,) and Drag ratio (b) as a function of flow velocity for water and different pulp suspensions kbm, kcm and kckb3 at concentration of 0.6 wt.%.

affect by any deviation in fiber properties particularly fiber length and flexibility.Also, many researchers [14, 21] reported that fiber flexibility is dependent on cross-dimensional fiber properties and increases with the decrease of fiber cell wall thickness. Some fiber characteristics could be analyzed by pressure drop measurements [15, 22]. Considering those previous studies which were based on wood and synthetic fibers, there was a need to evaluate pressure drop measurements over a range of flow conditions for crop pulp fibers. To correlate pressure drop of flowing crop pulp suspensions with the fiber length, fiber lumen, and fiber width (other properties are presented in appendix), the consistency of 0.6 wt.% and the velocity of 1.6 m/s were selected in the present study as shown in the Figures 6 (a – c).

The results reveal that the presser drop could be used for monitoring deviations in fiber characteristics and hence pulp and paper quality. The deviations in pressure drop between kcm, kbm

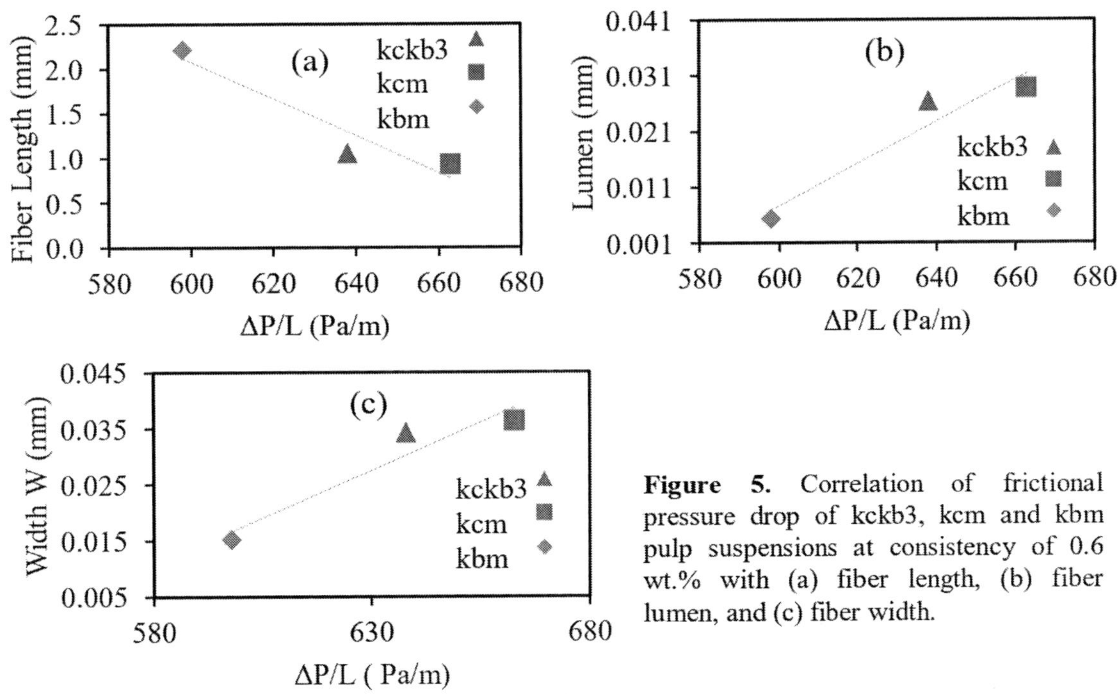

Figure 5. Correlation of frictional pressure drop of kckb3, kcm and kbm pulp suspensions at consistency of 0.6 wt.% with (a) fiber length, (b) fiber lumen, and (c) fiber width.

and kckb3 pulp suspensions are due to fiber flexibility and lignin content and aspect ratios. The findings and data trends are similar with the previous researchers for wood pulp and synthetic fibers [4, 13, 22-24].

4. Conclusions

Suspensions of wood and synthetic fibers flow, properties of different grades of fibers and their correlation with frictional pressure drop have been studied by many researchers. In the present study momentum transfer of fiber suspensions of a different crop pulp fibers (kenaf core and kenaf bast) and their blend has been considered. The fiber consistency, flexibility, aspect ratio of pulp fibers suspensions influence frictional pressure drop, while among these features fiber consistency is an important parameter. It is concluded that with the increase of the concentration of the suspensions, frictional pressure drop reduces at higher flow rates, and drag reduction is observed at a concentration of 0.6 wt. % for kbm pulp suspension. The minimum drag ratio occurs at different flow velocities depending on pulp samples. The drag ratio slightly increases with velocity after the minimum drag ratio point in case of long fibers. Contrarily, drag ratio is successively decreasing with the increase of velocity. The flexible fibers reduces pressure drop by dampening more eddies in comparison to the stiff fibers and hence enhances drag ratio. The results acquired in the present study can elucidate the handling of crop fibers in pulp and paper industries.

Acknowledgements

This research has been financially supported by High Impact Research (MOHE-HIR) grant UM. C/625/1/HIR/MOHE/ENG/45, UMRG-RP012C-13AET, Bantuan kecil penyelidikan BK 009-2016, Faculty of Engineering, University of Malaya and the authors are also grateful to the Forest Research Institute of Malaysia for support to conduct this research work.

References

[1] Petukhov B 1970 *Advances in heat transfer* **6** 503-64
[2] Blasius H 1908 *2. angew. Math*
[3] Jayageeth C, Sharma V I and Singh A 2009 *International Journal of Multiphase Flow* **35** 261-9
[4] Duffy G, Kazi S and Chen X 2011 *Applied Thermal Engineering* **31** 2971-80
[5] Duffy G, Kazi M and Chen X 2002 *Proceedings of the 56th Annual Appita Conference, Rotorua, New Zealand,*
[6] Duffy G G 1972 A study of the flow properties of New Zealand wood pulp suspensions.
[7] Middis J, Muller-Steinhagen H and Duffy G 1994 *Appita* **47** 154-8
[8] Cui H and Grace J R 2007 *International Journal of Multiphase Flow* **33** 921-34
[9] Ashori A 2006 *Polymer-Plastics Technology and Engineering* **45** 1133-6
[10] Villar J, Revilla E, Gómez N, Carbajo J and Simón J 2009 *Industrial crops and products* **29** 301-7
[11] Dutt D, Upadhyay J, Singh B and Tyagi C 2009 *Industrial crops and products* **29** 16-26
[12] Mossello A A, Harun J, Shamsi S R F, Resalati H, Tahir P M, Ibrahim R and Mohmamed A Z 2010 *Agricultural Journal* **5** 131-8
[13] Kazi M S N, Duffy G G and Chen X D 1999 *Chemical Engineering Journal* **73** 247-53
[14] Kazi S, Duffy G and Chen X 2014 *International Journal of Thermal Sciences* **79** 146-60
[15] Kazi S, Duffy G and Chen X 2015 *Applied Thermal Engineering* **78** 172-84

[16] Hubbe M A 2007 *BioResources* **2** 296-331

[17] Rojas O J and Hubbe M A 2005 *Journal of dispersion science and technology* **25** 713-32

[18] Korteoja M, Lukkarinen A, Kaski K and Niskanen K 1997 *Journal of pulp and paper science* **23** J18-J22

[19] Korteoja M, Salminen L, Niskanen K and Alava M 1998 *Journal of pulp and paper science* **24** 1-7

[20] Hua X, Tanguy P A, Li R and Van Wagner J S 1996 *appi journal* **79** 112-6

[21] Paavilainen L 1993 *Paperi ja puu* **75** 689-702

[22] Duffy G, Kazi S and Chen X 2000 *Proceedings of the 54th Annual Appita Conference, Melbourne, Australia, Apr,* pp 3-6

[23] Kazi S, Duffy G and Chen X 2012 *Experimental Thermal and Fluid Science* **38** 210-22

[24] Gharehkhani S, Yarmand H, Goodarzi M S, Shirazi S F S, Amiri A, Zubir M N M, Solangi K, Ibrahim R, Kazi S N and Wongwises S 2017 *International Communications in Heat and Mass Transfer* **80** 60-9

IOP Conf. Series: Materials Science and Engineering **210** (2017) 012066 doi:10.1088/1757-899X/210/1/012066

Influence of electromagnetic field on metal cutting in turning operation of AISI 1018 low carbon steel

J N Nadia*, F Aaron and M Azuddin

Centre of Product Design and Manufacture (CPDM), Department of Mechanical Engineering, Faculty of Engineering, University of Malaya, Kuala Lumpur, Malaysia

E-mail: *[1]nurizaz_haz@siswa.um.edu.my

Abstract. The effects of magnetic field on the machining force and tool wear in turning operation has been investigated. AISI 1018 low carbon steel used during machining as it is generally used in industry. The present and absent of the magnetic field during machining in dry condition was studied, and has been analysed. Taguchi approach has been used when designing the experiment. The results shown in this study related to the machining force and tool wear. The experimental results shown that when magnetic field has been applied, the machining force in turning operation such as radial force (F_r), feed force (F_f), and cutting force (F_c) gave higher results as compared to the absent of magnetic field. As for the tool wear, the results shown that the tool life is getting longer when the magnetic field has been applied which has been proven by the previous studies done before.

1. Introduction

In the present years, manufacturing industry has demand an efficient strategy in order to satisfy market requirements. The researchers are trying to find solutions to reduce the machining costs by having the reduction in force experienced and improvements in tool life when the steel was being cut since earlier time. The machining force is closely related to the tool wear and it can also act as the main feedback of wear level during machining process [1] .Therefore, the application of external electromotive sources such as magnetic field in the machining process has been studied previously.

The investigation has been made on the effects of a magnetic field on the frictional behaviour of steel XC48/graphite couple [2] . The application of the magnetic field has led to a decrease of a mean value of friction coefficient of the sliding contact but the wear rate has not been discussed. The similar studies has been made regarding the wear and friction behaviour of a nickel/XC 48 steel couple which gave the increased value of friction coefficient and has shown the decreased wear rate [3] . This results has agreed to the studies made on the ferromagnetic steel surface in sliding contact under the influence of magnetic field [4] .

As for the machining in turning operation, the effects of heat treatment and magnetic field of turning the mild steel by considering the tool life, surface roughness and chip morphology has been done [5]. The application of magnetic field has later been introduced as one of the external source to provide reduction in tool wear and improving machinability of materials [6-7] . For surface plastic deformation (SPD), the investigation has been done in dry cut of magnetized ferromagnetic AISI 1045 steel which resulted in the strengthens of the cut surface of the chip segmentation in high cutting speed [8].

In this study, the application of magnetic field has been analysed on the turning process of AISI 1018 low carbon steel. The corresponding effect of this application has been obtained to the cutting force and tool life during the machining. The experimental results has been analysed and discussed.

2. Experimental Designs and Procedure

The experimental designs and procedure have been explained below which divided to three subsections which are designation of experiment which explains the tool, material, and parameter used in this experiment. Next subsection would explain regarding the measured tool life and final subsection is the measured machining force during the present and the absence of the magnet.

2.1 Designation of Experiment

In this study, all experiments have been carried out in dry machining condition on the OKUMA LB15 CNC lathe. The range of spindle speed that can be used for this machine is between 75- 4200 RPM. The workpiece material used is AISI 1018 low carbon steel because it has been widely used in industry. The diameter of the workpiece is 100 mm and the length of cut is 180 mm. TNMG 160408N-GU made by Sandvik is used in this experimental work is in the form of triangular insert which is made of coated carbide of grade AC820P. Cutting speed (v), feed rate (f), and depth of cut (d) were selected as the machining parameters to analyze the influence of electromagnetic field on metal cutting which has been shown in The applications of the electromagnets when attached to dynamometer

Table 1. The experiment has used Taguchi method in this study. L9 ($\mathbf{2^4}$) Orthogonal Array has been selected as the chosen parameters has been shown in the Table 2. Figure 1 has shown the sample workpiece of AISI 1018 low carbon steel and Figure 2 shown the DC electromagnet which attached to tool holder.

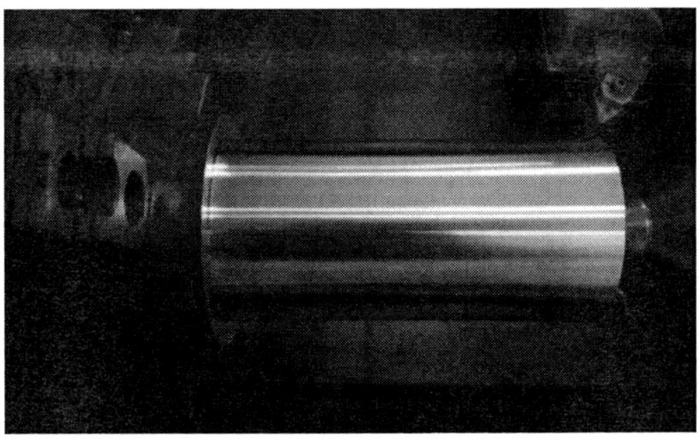

Figure 1. Sample workpiece of AISI 1018 low carbon steel

Figure 2. The applications of the electromagnets when attached to dynamometer

Table 1. The machining parameters and levels.

Symbol	Machining Parameters	Levels		
		Level 1	Level 2	Level 3
v	Cutting speed (m/min)	190	220	250
f	Feed rate (mm/rev)	0.20	0.30	0.40
d	Depth of cut (mm)	0.75	1.00	1.50

Table 2. Machining parameters assigned in a L9 ($\mathbf{2^4}$) orthogonal array.

Experiment No.	**Cutting speed, v (m/min)**	**Feed rate, f (mm/rev)**	**Depth of cut, d (mm)**
1	190	0.20	0.75
2	220	0.30	0.75
3	250	0.40	0.75
4	190	0.30	1.00
5	220	0.40	1.00
6	250	0.20	1.00
7	190	0.40	1.50
8	220	0.20	1.50
9	250	0.30	1.50

2.2 Tool Life

Tool wear measured up to tool failure to analyse the effect of cutting operation under electromagnetic field on tool life. In this study, the tool life measured is in terms of time in minutes. According to International Standard Organization, ISO 3685 (1993), tool wear is achieved when the average flank wear, V_B of the insert has reached 0.30 mm or the maximum flank wear, V_{Bmax} has reached twice of the average which is 0.60 mm.

For every experiment after each pass of length of cut, the insert is taken out and will be observed under microscope to collect the measurement of the flank wear. The average value of the flank wear will be measured after several experiments until the flank wear has reached 0.30mm.

2.3 Machining Forces

The measured machining forces were determined to identify whether the presence of magnets will reduce the machining force from the workpiece. Machining forces were also measured to determine the effects of each machining parameters. The forces of machining were determined using a dynamometer which is based on piezoelectric effect. Then it is connected to a multichannel amplifier to convert those charges into voltages. These voltages are in the form of analogue mode and it will be converted by using a digital storage oscilloscope. These digitized waveforms then will be sent to a computer where the force values are then computed.

Based on the setup specification of the amplifier, three components of forces would be calculated according to the voltage obtained from the oscilloscope. These forces will be compared against forces obtained during different set of parameters and with or without the presence of magnet. The three components of forces which have been obtained are radial force (F_r), feed force (F_f), and cutting force (F_c).

3 Results and Discussion

3.1Tool life and machining force in Normal Experiments

The measured tool life in normal experiments has been determined as shown in Table 3. The mean values of machining forces also have been tabulated shown in Table 4. The longest and shortest time of cut is stated as experiment 1 (83.50 minutes) and experiment 5 (3.08) respectively. Thus, both of these experiments are conducted under electromagnetic field for comparison and further discussion.

Table 3. Tool life in Normal Experiments

Experiment No.	Tool Life (min)
1	83.50
2	19.40
3	5.85
4	22.70
5	3.08
6	41.80
7	4.35
8	15.00
9	4.18

IOP Conf. Series: Materials Science and Engineering **210** (2017) 012066 doi:10.1088/1757-899X/210/1/012066

Table 4. Machining of forces in Normal Experiments

Experiment No.	Radial Force, $\overline{F_r}$ (N)	Feed Force, $\overline{F_f}$ (N)	Cutting Force, $\overline{F_c}$ (N)
1	137.82	156.22	391.83
2	113.07	156.22	416.78
3	210.89	260.37	827.52
4	270.30	327.71	582.52
5	266.34	360.60	894.30
6	84.95	283.07	416.78
7	424.36	555.58	951.42
8	253.66	368.04	477.11
9	345.15	480.01	680.28

3.2 Machining forces between Normal Experiments and Electromagnetic Experiments

Based on Figure 3 below have shown the graph of three component of forces which are radial force (F_r), feed force (F_f), and cutting force (F_c). As shown in Figure 3(b) and Figure 3(d), Experiment 5 has shown both higher values for normal and electromagnet experiment compared to Experiment 1. From these graphs, it has shown that the higher the feed rate, cutting speed, and depth of cut which has been applied, the higher the value of machining force.

As for Figure 3(c), the graph of the feed force versus time has been quite difficult to be compared graphically, as their value often fluctuated and overlapped between each other. Therefore, it has shown in Figure 3(d), the average values of feed forces for Experiments 5 both in normal and electromagnetic experiments have higher feed forces.

Figure 3(e) has been shown that, the cutting force versus time is not easily to be analysed for a comparison of normal and electromagnetic experiment. Based on Figure 3(f) which has shown mean of feed force, there is a little difference which can be seen here. Unlike for other previous two forces which are mean of radial force and feed force have the same pattern value which normal experiment 5 has lower value of force compare to its electromagnetic experiment, the cutting force for normal experiment 5 is a lot higher compared to electromagnetic experiment.

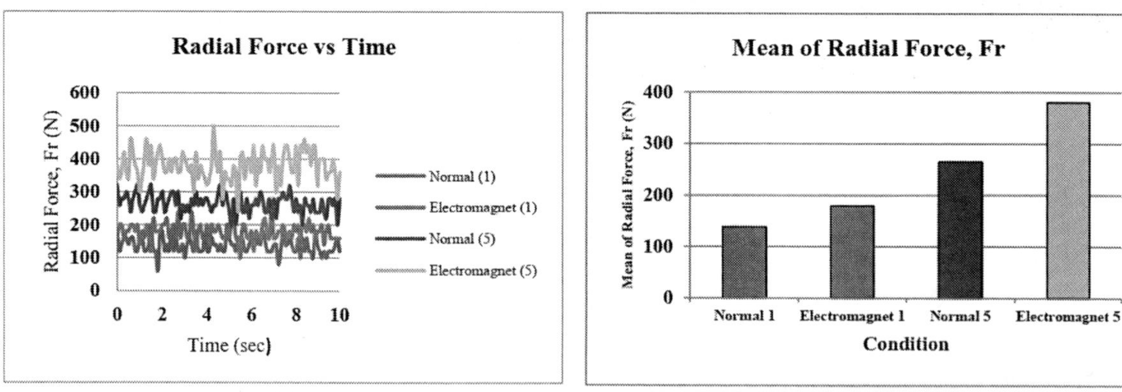

(a) (b)

IOP Conf. Series: Materials Science and Engineering **210** (2017) 012066 doi:10.1088/1757-899X/210/1/012066

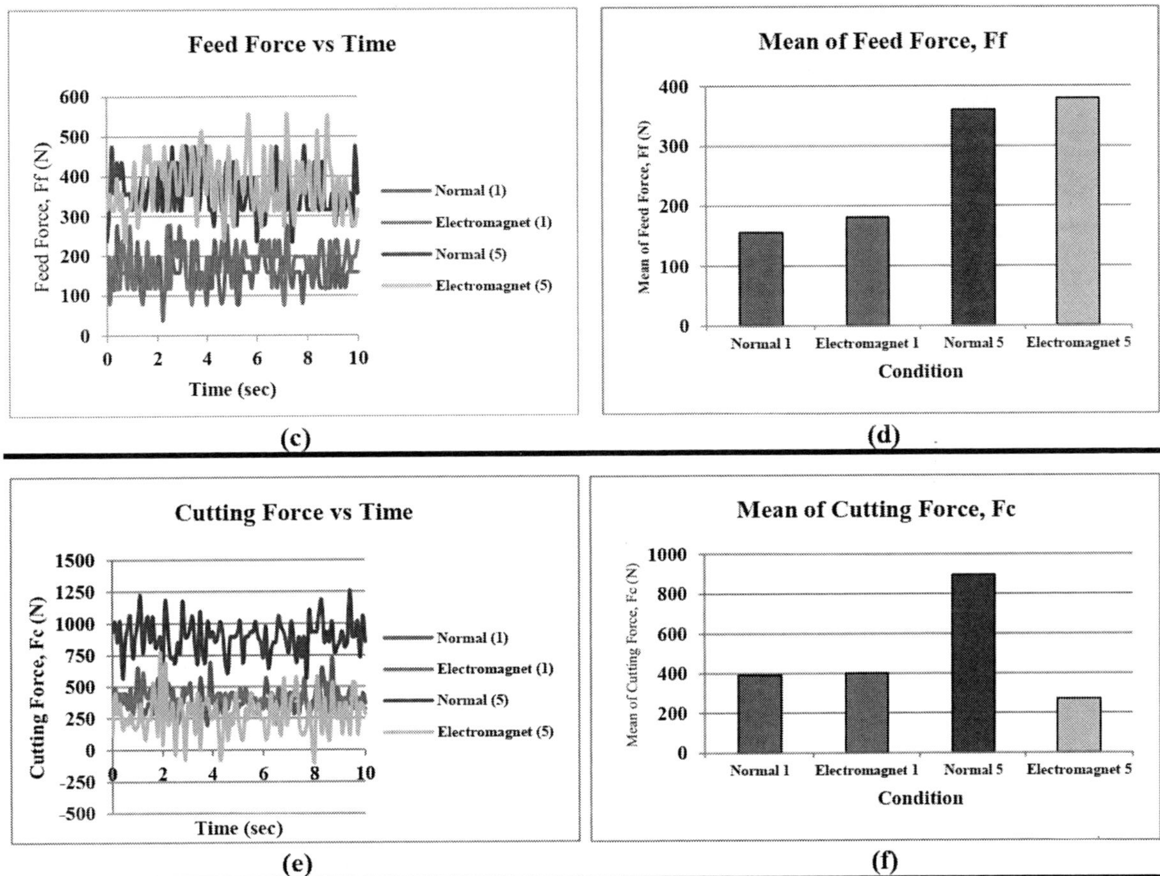

Figure 3. Graph component of forces versus Time and Mean of Component of Forces.

Table 5. Percentage differences of Mean Forces in Normal Experiments and Electromagnetic Experiments

Experiment No.	Condition	Radial Force, $\overline{F_r}$ *(N)*	Feed Force, $\overline{F_f}$ *(N)*	Cutting Force, $\overline{F_c}$ *(N)*
1	Normal	137.82	156.22	391.83
	Electromagnet	179.01	181.28	401.49
Percentage of Difference (%)		+29.89	+16.04	+2.47
5	Normal	266.34	360.60	894.30
	Electromagnet	380.20	379.39	272.85
Percentage of Difference (%)		+42.75	+5.21	-6.95

There are three basic component forces which have been recorded the mean value of forces in both condition which are in electromagnet and normal experiments as shown in Table 5. These forces in electromagnet condition are expected to be lower than the normal experiments. However, based on analyses which have been done apparently shows different result that most of the forces are actually get higher values when machined under electromagnetic field.

As been stated by [1] , the forces measured under electromagnetic field in turning operation were lower compared to corresponding normal experiments. He justified that the presence of electromagnetic field in the form of eddy current during machining is needed to lower the value of forces.

Justification that can be made from this analysis is due to larger part of the tool overhang when machined in electromagnetic condition. Initially, when the experiments have been carried out, normal experiments have been first completed before determining which experiments that would need to be conducted under electromagnetic field. Then, the tool overhang was kept constant throughout the normal experiments. Unfortunately, during setup of electromagnetic experiments, the electromagnets were not able to fit in the space between the tool holder's clamp and the dynamometer. In order to resolve this problem, the position of the tool holder has been shifted for 2 to 3 mm in distance to make sure the electromagnets would attach on the tool holder firmly. Therefore, the tool holder has been shifted outwards. Probably with larger tool overhang and movement during placing of electromagnet, there is higher vibration will occur and these will lead to higher machining forces.

3.3 Tool Life between Normal Experiments and Electromagnetic Experiments

The flank wear plotted against cutting time as shown in Figure 4 and Figure 5 which have shown both Experiment 1 and Experiment 5 respectively. The total cutting time of Experiment 5 has shorter duration compared to Experiment 1. This is due to the increase in the cutting speed, feed rate and depth of cut. In Figure 4, the graph of electromagnet shown longer tool life compared to corresponding normal experiments. This is same as compared to Figure 5 where there has shown longer time of tool life for electromagnet than the normal experiments.

As for further discussion of the analysis, Table 6 has shown the percentage of difference of tool life in normal experiments and electromagnetic experiments. It has shown that for both experiments, the tool life for electromagnetic experiments is higher than corresponding normal experiments. The percentage difference for both experiments has achieved significant improvements since the value is more than 20%. Results obtained were similar to other previous studies. The study has been made of an application on external electromotive force (EMF) such as magnetic field during cutting process [3]. The result has significantly improved the cutting tool life. The previous study have been done for the contribution of better understanding on the influence of magnetic field on the wearing of tools in dry cutting process [4]. The presented result obtained of magnetized ferromagnetic AISI 1045 steel has shown that there was a reduction of wear due to the magnetic applied.

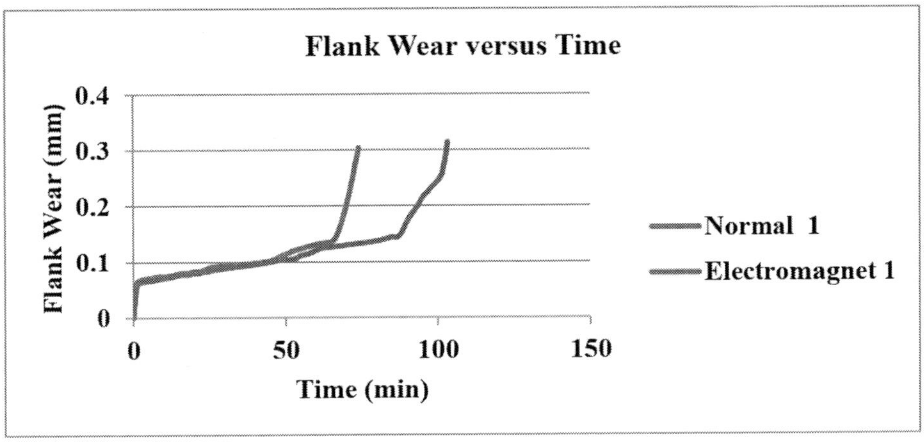

Figure 4. Flank Wear versus Time (Experiment 1)

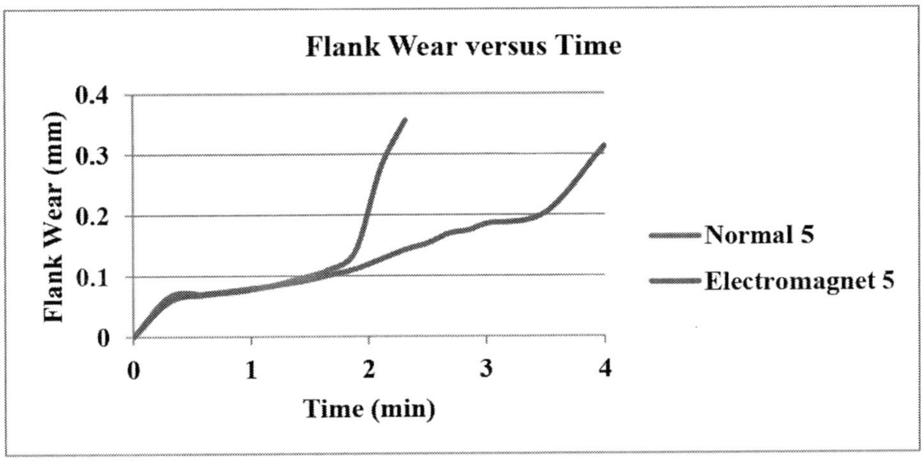

Figure 5. Flank Wear versus Time (Experiment 5)

Table 6. Tool life in Normal Experiments and Electromagnetic Experiments

Exp. No	Tool Life (minutes)		Percentage of Difference (%)
	Normal	Electromagnet	
1	83.50	103.00	+23.25
5	3.08	3.92	+27.27

4. Conclusions

In this study, when direct current (d.c) magnetic field has been applied during machining in turning operation, the analysed results obtained high machining forces and longer tool life. For the tool life it has been proven that when magnetic field is applied, the tool life has achieved significant improvement since the percentage of difference obtained in normal experiments and electromagnet experiments shown more than 20%. The value obtained as for Experiment 1 is +23.25% and Experiment 5 is +27.27%. As for the machining forces, justification has been made due to the larger tool overhang when placing the electromagnetic fields which cause the vibrations to occur frequently. This lead to higher machining forces when machined in electromagnetic condition compared to the corresponding normal experiments. The percentage of difference of experiments for each component of force has shown mostly higher machining forces.
For both experiment when electromagnetic field is absent and present condition, the results obtained will be shown below.

- The percentage of difference obtained in Experiment 1 for mean of component forces for radial force ($\overline{F_r}$), feed force($\overline{F_f}$), and cutting force ($\overline{F_c}$) were +29.89%, +16.04%, and +2.47%.
- The percentage of difference obtained in Experiment 5 for mean of component forces for radial force ($\overline{F_r}$), feed force($\overline{F_f}$), and cutting force ($\overline{F_c}$) were +42.75%, +5.21% and -6.95%.

In future, the further for more number of experiments should have been made and other type of material should have been considered to determine the significant results when the magnetic field been applied. In addition, more research needs to be done in order to determine the extent and limitation of influence of magnetic field on metal cutting.

Acknowledgements
The acknowledgements should have been given to the Ministry of Higher Education and the University of Malaya under the UMRG grant (RP 033A-15AET) for the financial support.

References

[1] Yue X 2007 *Effect of Electromagnetic Field National University of Singapore*
[2] Paulmier D, Mansori M El, Zdidi H 1997 *Wear* **204** 148–54.
[3] Zaidi H, Pan L, Paulmier D, Robert F 1995 *Wear* **181** 799-804.
[4] Paulmier D, Zaidi H, Bedri R, Kadiri EK, Pan L, Jiang Q 1995 *Surf Coatings Technol.* **76** 583-588
[5] Patwari MAU, Mahmood MN, Noor S, Shovon MZH 2013 *Procedia Eng* **56** 713–718
[6] El Mansori M, Pierron F, Paulmier D 2003 *Surf Coatings Technol* **163** 472-477
[7] El Mansori M, Iordache V, Seitier P, Paulmier D 2004 *Surf Coatings Technol* **188** 566-571
[8] El Mansori M, Mkaddem A 2007 *Surf Coatings Technol* **202** 1118–1122

IOP Conf. Series: Materials Science and Engineering **210** (2017) 012067 doi:10.1088/1757-899X/210/1/012067

The effect of heating temperature and methods towards the formability of deep drawn square metal cup

M.A.M. Basril*, H.M. Teng, M. Azuddin, I.A. Choudhury
Department of Mechanical Engineering, Faculty of Engineering, University of Malaya, 50603 Kuala Lumpur, Malaysia

E-mail: *muhammadbasril@siswa.um.edu.my

Abstract. Deep drawing operation is one of the most crucial sheet metal forming processes in industrial applications, but it usually requires expensive multi-step production processes, which is necessary in order to produce complex parts. On top of that, room temperature may cause poor formability or failure due to mechanical properties of the material. The objective of this study is to investigate the effects of heating temperature and the most efficient heating position to perform warm square deep drawing operation without failure. Besides that, this study also aims to compare the thickness distribution of drawn cup's profile obtained from experiment and finite element analysis. A warm formability study of aluminium, mild steel and stainless steel sheet metals are tested by deep drawing experimental methods. The Taguchi approach, which applies L9 orthogonal array, is used to conduct experiments. Different sizes of square blanks are deep drawn at room temperature, 100°C, 150°C and 200°C using three heating techniques, which are heating die only, heating punch only and heating both the die and punch. The results show that warm deep drawing process has more uniform thickness distribution within the square cup profile compared to room temperature condition. Furthermore, maximum thinning condition at the punch corner is noticeably reduced. Lower and upper limit of heating temperature exists, as there is a low and high-temperature failure. Combination of stainless steel with 45 mm blank size, heating temperature of 150°C and die heating technique is optimal in order to obtain uniform thickness distribution in square cup deep drawing process.

1. Introduction

Deep drawing is a forming process where a sheet metal is pushed by a punch into a die cavity and formed into a desired shape such as cup or box like structures. In this operation, a metal blank is placed over a die cavity and held by a blank holder which shown in figure 1. A mechanical action of a punch forces the blank into the forming die involving tensile and compressive forces. It is widely used in numerous industrial fields as it can achieve high production rate. Typical examples of deep drawn components are beverage cans, kitchenware, electrical appliance bodies, aircraft panels and automobile bodies.

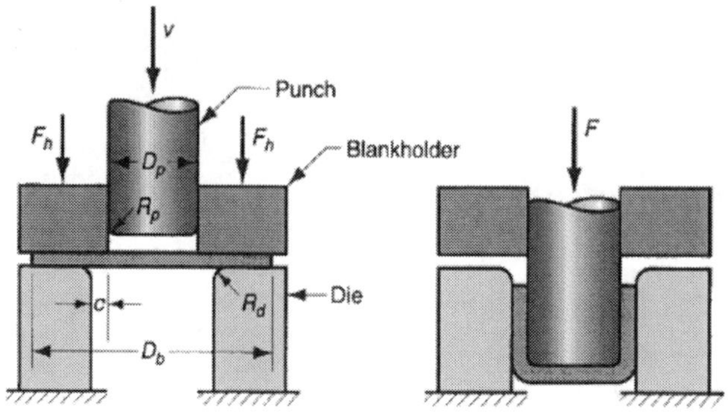

Figure 1. Deep Drawing operation

However, parts with complex geometries can only be achieved by costly multi-step deep drawing process. One of the methods to solve this problem is to conduct deep drawing process at elevated temperature due to its significant effects on formability of a sheet metal.

Formability of a material can be described by limiting drawing ratio (LDR), which is ratio of the maximum diameter of the workpiece that is drawn without tearing and the diameter of the punch [1]. Many factors that can influence the formability of a material such as characteristics of sheet metal die and punch corner radii, blank holder force, thickness of blank, forming temperature and heating technique. According to Reddy, R. et al. [2], the increment of the wrinkle height can be observed when the small die radius was implemented, whilst for the shorter drawn cup, the wrinkles have a greater amplitude as the die radius increased. Based on curiosity, a further investigation on the formability of deep drawn cup from deep drawing process with a proper design of experiment is needed.

Taguchi approach is one of the designs of experiment methods, which is frequently utilized to determine the effect of parameters on performances in a deep drawing process. Taguchi approach is an effective and standardized method to achieve experimental cost and time savings by omitting some of the possible combinations of parameters in a full factorial design of experiment. Quality of a process can be improved and optimized by determining the best combination of the variables. For example, Colgan & Monaghan [3] aims to determine the vital factors that influence deep drawing process through experimental and finite element analysis (FEA) approach. Therefore, Taguchi method is used in order to specify the runs of the experiment conduct.

Besides that, previous scholars found that the formability of the material expanded by deforming it at higher temperatures [4] [5]. According to Zhang et al. [6], the limiting drawing ratio (LDR) increased significantly when conduct in warm condition compare to room temperature condition. Effects of forming temperature and heating technique in deep drawing process are the parameters, which will be further investigated in this study.

2. Methodology

2.1 Designation of Experiment

The set up for the experimental procedures for warm deep drawing has been shown in figure 2. A single acting hydraulic cylinder is attached to a die set and it is controlled by hydraulic pump as shown in figure 3. On top of that, a temperature controller which shown in figure 4, is used to control the die set temperature while the die set is heated up by heater. A set of type K thermocouple is used to sense temperature and the reading is shown on the controller. The punching force generated from single acting hydraulic cylinder and being control by hydraulic pump. It has 101 kN capacity of force with 105 mm stroke which will press sheet metal downward into die cavity when the process takes place. The material, blank size (mm), temperature (°C), and heating technique has been chosen as experimental parameters in order to analyze the formability of the deep drawn cup which can be refer to table 1. In this study, the Taguchi method has been decided as design of experiment method and implementing L9 orthogonal array which shown in table 2. The experimental works will conducted at room temperature first, and then proceed with warm temperatures (100°C, 150°C, and 200°C).

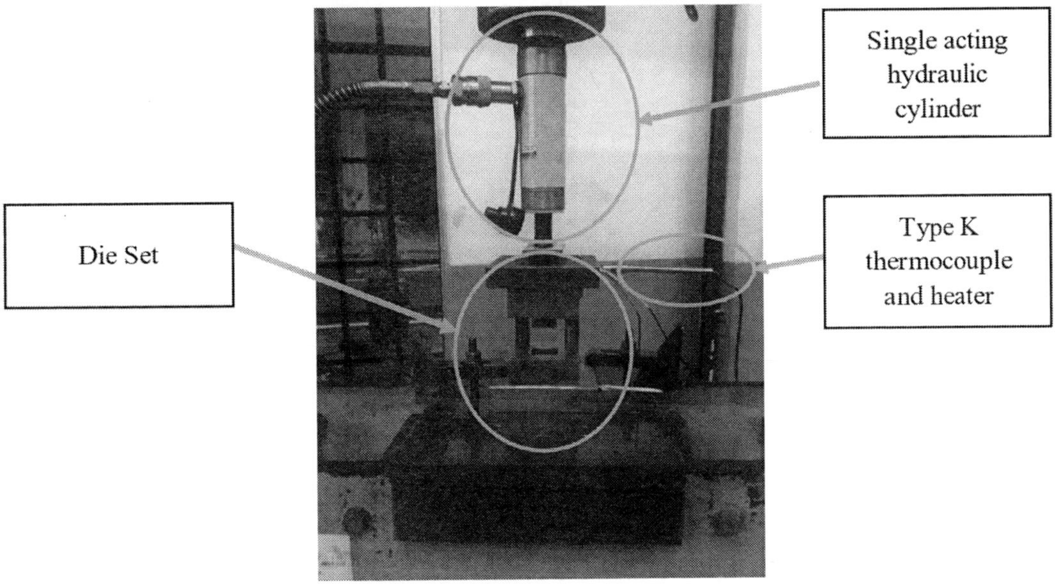

Figure 2. Experimental setup

Figure 3. Hydraulic pump

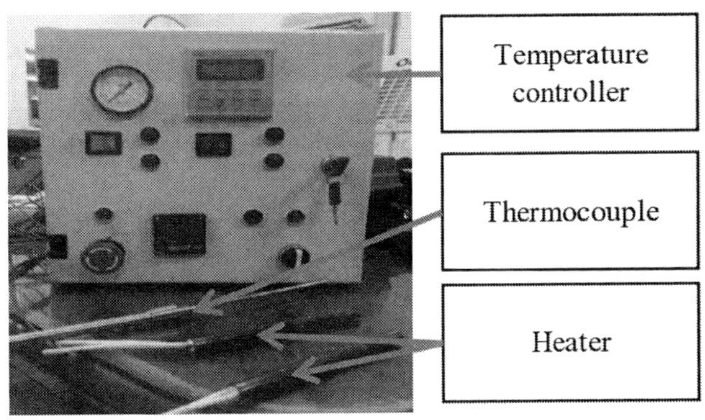

Figure 4. Temperature controller, heater and type K thermocouple

Table 1. Selected factors and levels for square deep drawing

Factors	Level		
	1	2	3
Material	Aluminium	Mild steel	Stainless steel
Blank size (mm)	45	50	55
Temperature (°C)	100	150	200
Heating techniques	Die	Punch	Die and punch

Table 2. L9 Orthogonal array for square deep drawing

Experiment No.	Material	Blank size (mm)	Temperature (°C)	Heating Technique
1	Aluminium	45	100	Die
2	Aluminium	50	150	Punch
3	Aluminium	55	200	Die and punch
4	Mild steel	45	150	Die and punch
5	Mild steel	50	200	Die
6	Mild steel	55	100	Punch
7	Stainless steel	45	200	Punch
8	Stainless steel	50	100	Die and punch
9	Stainless steel	55	150	Die

2.2 Die set and square blanks
Next, the square die set consists of square die and punch as its components. For this particular experiment, the square die has fix height of 35 mm, fix external width and length of 90 mm and 2 mm die corner radius. On the other hand, the square punch has the dimension of 32.5 mm for width and length. Meanwhile, it also has height of 20 mm along with 2 mm punch corner radius. It also equipped with full type blank holder that has thickness of 5 mm along with internal width and length of 34.5 mm. The material used in this research involving aluminium 1050, mild steel and stainless steel 304. Based this material, the square blank is prepared as appeared in figure 5. It has been prepared into three different sizes, which is 45 mm, 50 mm, and 55 mm with 5 mm of blank corner radii.

Figure 5. Samples of squares blank with different sizes

The formability of the drawn cup will be measured in term of thickness distribution and the most influential parameters will be researched by using ANOVA method. For this purpose, the deep drawn square cup will be dissected into half and the thickness of 24 point along the dissected cup profile will be measured by using Dino-Lite Pro Digital Microscope.

3. Results and Discussion

3.1 Condition of deep drawn square cup
The outcomes of proper drawn cup formation for each experiment are recorded in table 3 for room temperature condition while warm temperature condition being recorded at table 4. Based on observation, most of the drawn cups manage to be drawn successfully without defect. Failed outcomes from room temperature condition happened on second and third experiment as the drawn cup suffers tearing condition. On the other hand, for the experiment that conducted in warm temperature conditions, it shows three failures of deep drawn square cup through the outcome of experiment number 2, 3 and 6.

Table 3. Drawability condition for Aluminium, Mild Steel and Stainless Steel in Deep Drawing at Room Temperature

Experiment No.	Material	Blank size (mm)	Temperature (°C)	Heating Technique
1	Aluminium	45	Room Temperature	Successful
2	Aluminium	50	Room Temperature	Failed
3	Aluminium	55	Room Temperature	Failed
4	Mild Steel	45	Room Temperature	Successful
5	Mild Steel	50	Room Temperature	Successful
6	Mild Steel	55	Room Temperature	Successful
7	Stainless Steel	45	Room Temperature	Successful
8	Stainless Steel	50	Room Temperature	Successful
9	Stainless Steel	55	Room Temperature	Successful

Table 4. Drawability condition Aluminium, Mild Steel and Stainless Steel in Warm Deep Drawing

Experiment No.	Material	Blank size (mm)	Temperature (°C)	Heating Technique	Results
1	Aluminium	45	100	Die	Successful
2	Aluminium	50	150	Punch	Failed
3	Aluminium	55	200	Die & Punch	Failed
4	Mild Steel	45	150	Die & Punch	Successful
5	Mild Steel	50	200	Die	Successful
6	Mild Steel	55	100	Punch	Failed
7	Stainless Steel	45	200	Punch	Successful
8	Stainless Steel	50	100	Die & Punch	Successful
9	Stainless Steel	55	150	Die	Successful

3.2 Thickness distribution of deep drawn square cup

Figure 6 illustrates the comparison of thickness measured from cup wall to bottom centre of square aluminium cup with 45 mm blank size in warm deep drawing and deep drawing at room temperature. From the graph, maximum thinning at the punch corner is lower in warm temperature condition compared to deep drawing at room temperature. The numerical descriptive measures are shown in table 5. The results show that the standard deviation of thickness distribution in warm deep drawing is 31.87% lower than deep drawing at room temperature. Maximum thinning is 17.45% less than initial thickness in warm deep drawing and 24.98% less than initial thickness in deep drawing at room temperature.

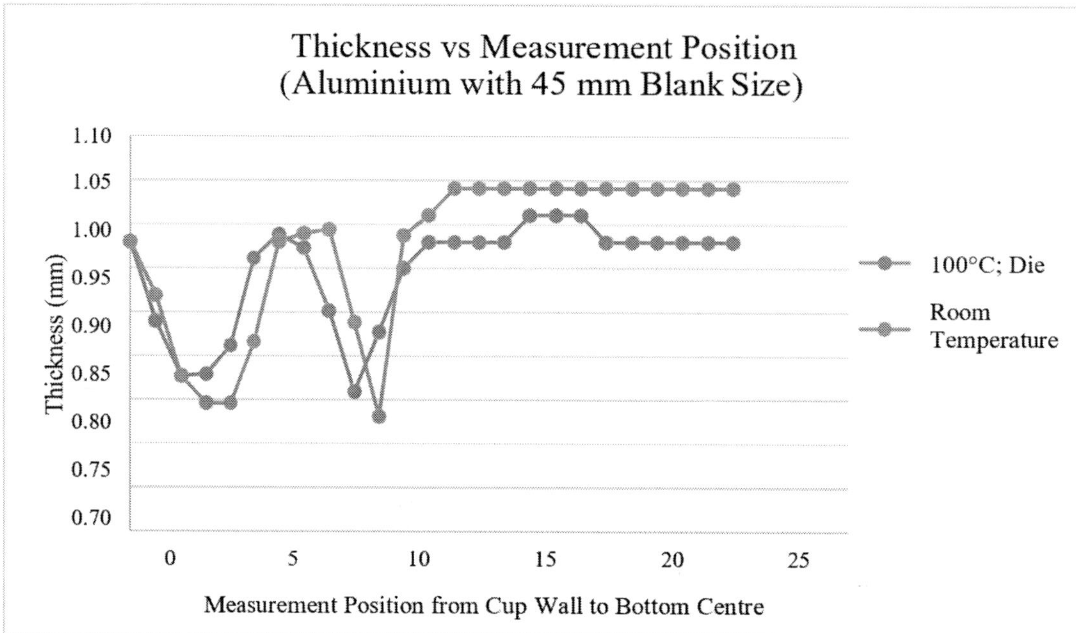

Figure 6. Thickness Distribution for deep drawn aluminium square cup with 45 mm blank size

Table 5. Numerical Descriptive Measures for deep drawn aluminium square cup with 45 mm blank size

Numerical Descriptive Measures	Aluminium with 45 mm Blank Size	
	Warm Deep Drawing (100°C; Die)	Deep Drawing at Room Temperature
Initial Thickness	0.980	1.041
Mean	0.947	0.972
Standard Deviation	0.062	0.091
Maximum	1.011	1.041
Minimum	0.809	0.781
Range	0.202	0.260

Figure 7 illustrates the comparison of thickness measured from cup wall to bottom centre of square mild steel cup with 45 mm blank size in warm deep drawing and deep drawing at room temperature. The graph shows that maximum thinning at the punch corner in warm deep drawing is about the same as deep drawing at room temperature. However, it can be seen that thickness is more evenly distributed in deep drawing at room temperature. The numerical descriptive measure is shown in table 6. The results show that the standard deviation of thickness distribution in warm deep drawing is 12.99% lower than deep drawing at room temperature. Maximum thinning is 19.31% less than initial thickness in warm deep drawing and 21.32% less than initial thickness in deep drawing at room temperature.

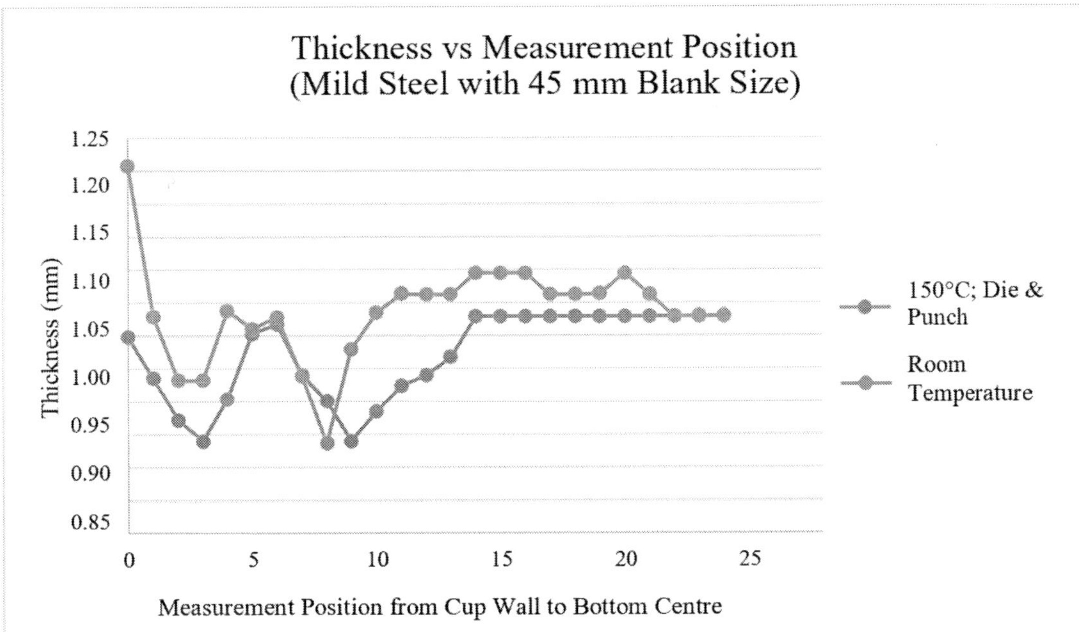

Figure 7. Thickness Distribution for deep drawn square mild steel cup with 45 mm blank size

Table 6. Numerical Descriptive Measures for deep drawn square mild steel cup with 45 mm blank size

Numerical Descriptive Measures	Mild Steel with 45 mm Blank Size	
	Warm Deep Drawing (150°C; Die & Punch)	Deep Drawing at Room Temperature
Initial Thickness	0.979	0.999
Mean	0.921	0.987
Standard Deviation	0.067	0.077
Maximum	0.980	1.208
Minimum	0.790	0.786
Range	0.190	0.422

Figure 8 illustrates the comparison of thickness measured from cup wall to bottom centre of square mild steel with 50 mm blank size in warm deep drawing and deep drawing at room temperature. Both graphs exhibit identical trend of data. The numerical descriptive measure is shown in table 7. The results show that the standard deviation of thickness distribution in warm deep drawing is 8.00% lower than deep drawing at room temperature. Maximum thinning is 28.79% less than initial thickness in warm deep drawing and 28.40% less than initial thickness in deep drawing at room temperature.

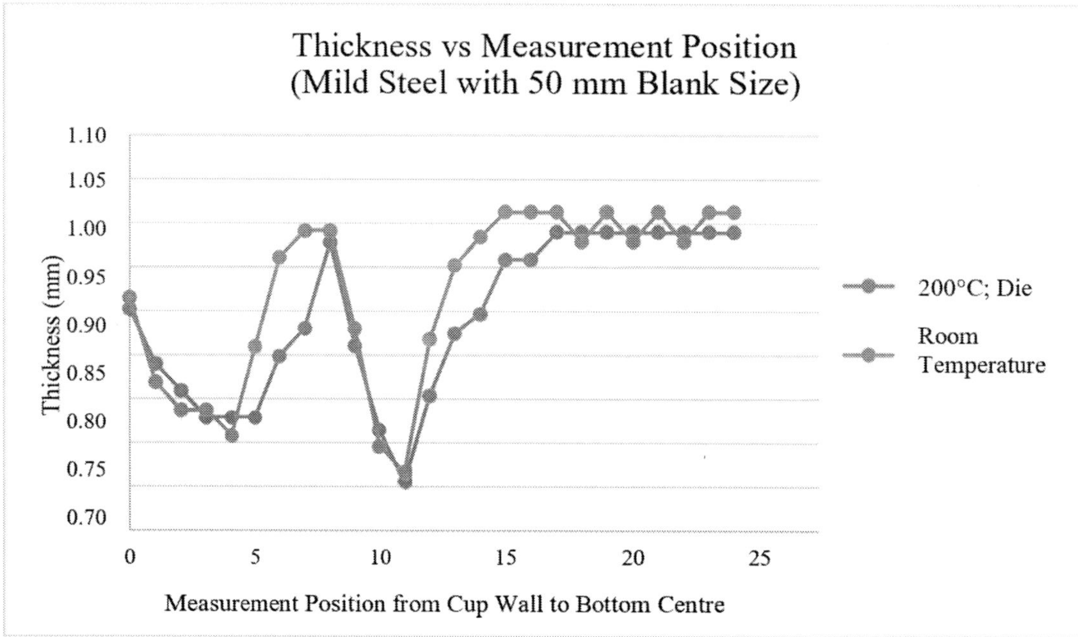

Figure 8. Thickness Distribution for deep drawn square mild steel cup with 50 mm blank size

Table 7. Numerical Descriptive Measures for deep drawn square mild steel cup with 50 mm blank size

Numerical Descriptive Measures	Mild Steel with 50mm Blank Size	
	Warm Deep Drawing (200°C; Die)	Deep Drawing at Room Temperature
Initial Thickness	0.990	1.000
Mean	0.894	0.922
Standard Deviation	0.092	0.100
Maximum	0.990	1.013
Minimum	0.705	0.716
Range	0.285	0.297

Figure 9 illustrates the comparison of thickness measured from cup wall to bottom centre of square stainless steel with 45 mm blank size in warm deep drawing and deep drawing at room temperature. From the graph, it can be seen that maximum thinning at the punch corner is lower in warm deep drawing as compared to deep drawing at room temperature. The numerical descriptive measure is shown in table 8. The results show that the standard deviation of thickness distribution in warm deep drawing is 16.05% lower than deep drawing at room temperature. Maximum thinning is 18.28% less than initial thickness in warm deep drawing and 25.08% less than initial thickness in deep drawing at room temperature.

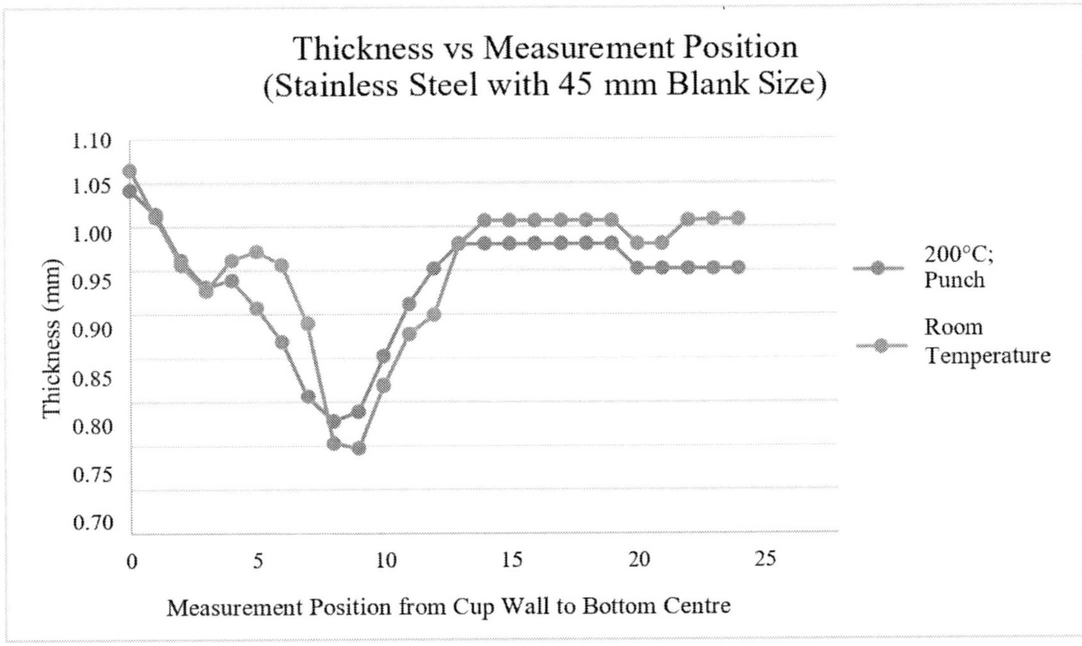

Figure 9: Thickness Distribution for deep drawn square stainless steel cup with 45 mm blank size

Table 8. Numerical Descriptive Measures for deep drawn square stainless steel cup with 45 mm blank size

Numerical Descriptive Measures	Stainless Steel with 45mm Blank Size	
	Warm Deep Drawing (200°C; Punch)	Deep Drawing at Room Temperature
Initial Thickness	0.952	0.997
Mean	0.935	0.954
Standard Deviation	0.068	0.081
Maximum	1.042	1.065
Minimum	0.778	0.747
Range	0.264	0.318

Figure 10 illustrates the comparison of thickness measured from cup wall to bottom centre of square stainless steel with 50 mm blank size in warm deep drawing and deep drawing at room temperature. From the graph, maximum thinning at punch corner is higher in warm deep drawing as compared to deep drawing at room temperature. The numerical descriptive measures are shown in table 9. The results show that the standard deviation of thickness distribution in warm deep drawing is 30.99% higher deep drawing at room temperature. Maximum thinning is 28.86% less than initial thickness in warm deep drawing and 21.84% less than initial thickness in deep drawing at room temperature.

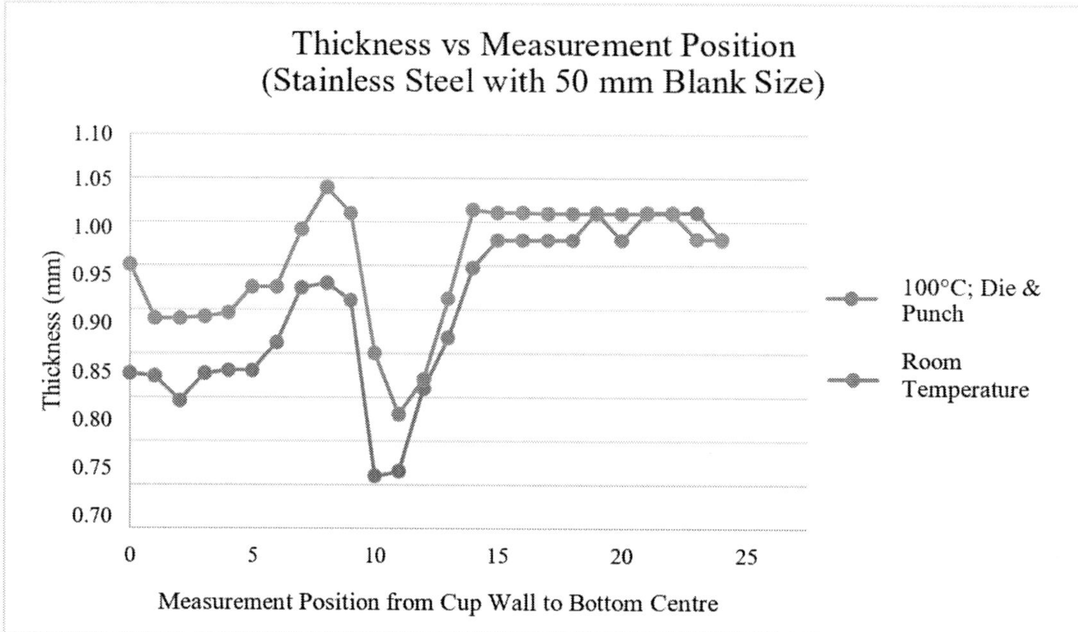

Figure 10. Thickness Distribution for deep drawn square stainless steel cup with 50 mm blank size

Table 9. Numerical Descriptive Measures for deep drawn square stainless steel cup with 50 mm blank size

Numerical Descriptive Measures	Stainless Steel with 50mm Blank Size	
	Warm Deep Drawing (100°C; Die & Punch)	Deep Drawing at Room Temperature
Initial Thickness	0.998	0.998
Mean	0.901	0.953
Standard Deviation	0.093	0.071
Maximum	1.011	1.039
Minimum	0.710	0.780
Range	0.301	0.259

Figure 11 illustrates the comparison of thickness measured from cup wall to bottom centre of square stainless steel with 55mm blank size in warm deep drawing and deep drawing at room temperature. The graph shows that maximum thinning at punch corner is lower in warm deep drawing as compared to deep drawing at room temperature. The numerical descriptive measures are shown in table 10. The results show that the standard deviation of thickness distribution in warm deep drawing is 6.76% lower than deep drawing at room temperature. Maximum thinning is 21.32% less than initial thickness in warm deep drawing and 24.23% less than initial thickness in deep drawing at room temperature.

IOP Conf. Series: Materials Science and Engineering **210** (2017) 012067 doi:10.1088/1757-899X/210/1/012067

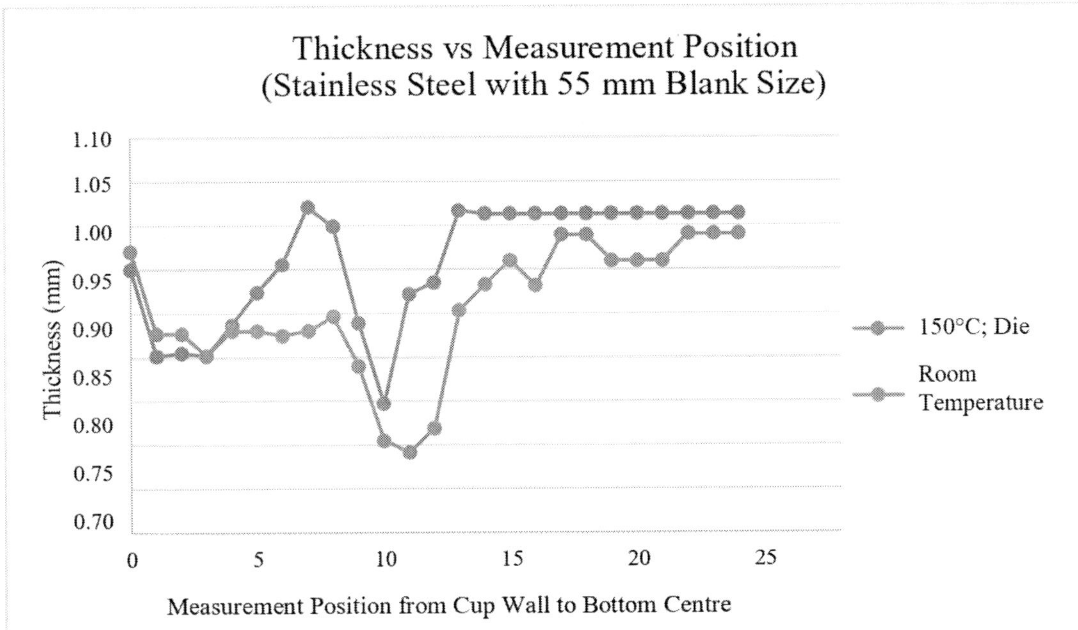

Figure 11. Thickness Distribution for deep drawn square stainless steel cup with 55 mm blank size

Table 10. Numerical Descriptive Measures for deep drawn square stainless steel cup with 55 mm blank size

Numerical Descriptive Measures	Stainless Steel with 55mm Blank Size	
	Warm Deep Drawing (150°C; Die)	Deep Drawing at Room Temperature
Initial Thickness	1.013	0.978
Mean	0.960	0.906
Standard Deviation	0.069	0.074
Maximum	1.021	0.990
Minimum	0.797	0.741
Range	0.224	0.249

Maximum thinning occurs at the punch corner because of high stress concentration in this region. The material at this region will first undergo bending around the die radius, then unbending when it is being drawn against the wall of the die under tensile stress and finally bending around the punch corner in the opposite direction. Therefore, the punch corner is the most critical region as severe thinning can lead to fracture of the cup [7].

Thinning occurs at the centre of the cup wall in the deep drawing of square cup in both warm deep drawing and deep drawing at room temperature. Thinning at this region occurs due to the small die radius. Material is not easily to flow over a small die radius and it is highly stressed. As the die radius goes smaller, the greater the force on the blank, which leads to a thinner wall thickness [8].

On the other hand, the square cups are not fully deformed due to restriction in the height of the punch based on observation. The material cannot be fully drawn into the die cavity which causes incomplete flow of the material. Thereby, thickness is not well distributed in the cup.

Table 11 shows the analysis on thickness distribution in square deep drawing. It shows that, during the warm deep drawing of aluminium with 45 mm blank size, mild steel with 45 mm and 50 mm blank size as well as stainless steel with 45 mm and 55 mm blank size, there is an improvement in the standard deviation when compared to deep drawn cup at room temperature. Thickness is more uniformly distributed in these experiments and the percentage of maximum thinning also reduced compare to deep drawn cup at room temperature. These results prove that thickness distribution and thinning condition at the punch corner can be improved by warm deep drawing under certain heating condition [9].

However, in the warm deep drawing of stainless steel with 50 mm blank size, the value of standard deviation is higher than deep drawing at room temperature. Moreover, the value of percentage of maximum thinning at warm temperature condition is higher compare to room temperature condition. This indicates that thickness is unevenly distributed and the maximum thinning condition at the punch corner is not improved by warm deep drawing. The low heating temperature (100°C) is insufficient to impart uniform temperature distribution within the stainless steel blank which has a low thermal conductivity. On top of that, the both die and punch heating technique also influences this outcome. Based on previous works, it been proven that partial heating in the holder or matrices area was much better when compared with the homogeneously heated tools [10].

Table 11. Analysis on thickness distribution in square deep drawing

Material	Size (mm)	Temperature and Heating Technique	Improvement in Standard Deviation (%)	Maximum thinning in warm deep drawing (%)	Maximum thinning in deep drawing at room temperature (%)
Aluminium	45	100°C; Die	31.87	17.45	24.98
	50	150°C; Punch	N/A	N/A	N/A
	55	200°C; Die & Punch	N/A	N/A	N/A
Mild Steel	45	150°C; Die & Punch	12.99	19.31	21.32
	50	200°C; Die	8.00	28.79	28.40
	55	100°C; Punch	N/A	N/A	N/A
Stainless Steel	45	200°C; Punch	16.05	18.28	25.08
	50	100°C; Die & Punch	-30.99	28.86	21.84
	55	150°C; Die	6.76	21.32	24.23

3.3 Analysis of Variance (ANOVA) and Analysis of Means (ANOM) for deep drawn square cup

The quality characteristic in this experimental study using Taguchi approach is thickness distribution. As the results are desired to be a uniform thickness distribution, quality characteristic is of the nominal-the-best type. The nominal-the-best S/N ratios on thickness distribution are calculated based on equation (1). Table 12 shows the results of ANOVA for square cup. The level mean S/N ratio is the average S/N ratio at each level of the process parameters.

$$\frac{S}{N} = 10log_{10}\left(\frac{\bar{Y}}{\sigma^2}\right) \quad (1)$$

Table 12. ANOVA results for four parameters tested at three levels

Parameter	Level	Experiment No.	S/N	Level Mean S/N
Material	1	1	23.698	5.892
		2	-3.010	
		3	-3.010	
	2	4	22.772	13.167
		5	19.739	
		6	-3.010	
	3	7	22.823	21.802
		8	19.713	
		9	22.872	
Blank Size	1	1	23.698	23.097
		4	22.772	
		7	22.823	
	2	2	-3.010	12.147
		5	19.739	
		8	19.713	
	3	3	-3.010	5.617
		6	-3.010	
		9	22.872	
Heating Temperature	1	1	23.698	13.467
		6	-3.010	
		8	19.713	
	2	2	-3.010	14.211
		4	22.772	
		9	22.872	
	3	3	-3.010	13.184
		5	19.739	
		7	22.823	
Heating Technique	1	1	23.698	22.103
		5	19.739	
		9	22.872	
	2	2	-3.010	5.601
		6	-3.010	
		7	22.823	
	3	3	-3.010	13.158
		4	22.772	
		8	19.713	

Applying equation (2), the sum of squares due to the variation from the average S/N ratio for the yth factor, SSy can be calculated. Applying equation (3) and equation (4), the percentage of contribution of yth factor can be obtained. The contribution of process parameters on warm deep drawing process are shown in table 13.

$$SS_y = \sum_{i=1} \left[\left(\frac{S}{N}\right)_{yz} - \left(\frac{S}{N}\right) \right] \quad (2)$$

$$SS_T = \sum_{i=1}^{n} \left[\left(\frac{S}{N}\right)_{i} - \left(\frac{\bar{S}}{N}\right) \right]^2 \quad (3)$$

$$Percentage\ of\ contribution\ of\ y\ parameter = \frac{SS_y}{SS_T} \times 100\% \quad (4)$$

Table 13. Contribution of Process Parameters on Warm Deep Drawing Process

Process Parameter	Sum of Square, SS_y	% of Contribution
Material	126.874	30.2%
Blank Size (mm)	156.038	37.2%
Heating Temperature (°C)	0.564	0.1%
Heating Technique	136.480	32.5%

Figure 12 illustrates the plots of level average values of material, blank size, heating temperature and heating technique. The optimum levels of the three process parameters are analysed through ANOM and presented in table 14.

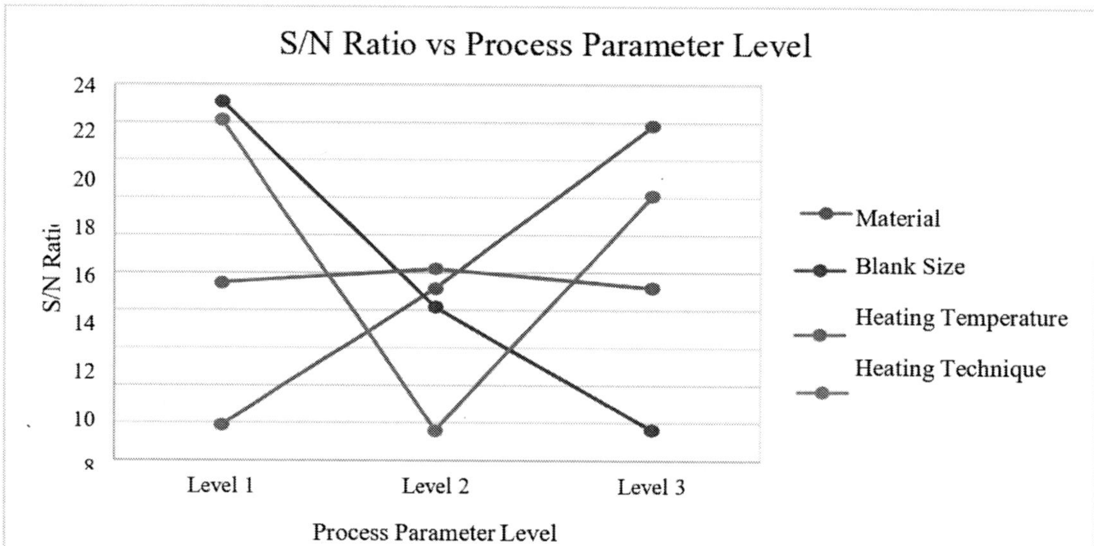

Figure 12: Plots of Level Average Values of Four Process Parameters

Table 14. Optimal Condition for Warm Deep Drawing

Factor	Level Description	Level
Material	Stainless Steel	3
Blank Size (mm)	45	1
Heating Temperature (°C)	150	2
Heating Technique	Die	1

The ANOVA results shown in table 13 indicates the degrees of importance of each process parameter considered, namely, material, blank size, heating temperature and heating technique. It shows how the process parameters affect the thickness of the square cup in warm deep drawing. Blank size has the greatest influence (37.2%) on the thickness of the square cup, followed by heating technique (32.5%), material (30.2%) and heating temperature (0.1%).

Optimum condition for a uniform thickness distribution of square cup can be obtained through ANOM. The level which provides the greatest S/N ratio in process parameters is known as optimal level for that factor. Combination of stainless steel with blank size 45 mm, heating temperature of 150°C and die heating technique is optimal in order to obtain uniform thickness distribution of square cup.

4. Conclusion

In order to investigate the warm formability of aluminium, mild steel and stainless steel sheet metals were going through deep drawing experimental methods. The outcomes of the experiments such as, the relationship between thickness distribution of cup's wall and heating temperature and method were discussed. Therefore, the important conclusions that can be highlighted from the study were as follows:

1. In square deep drawing process, the results of thickness distribution show that in certain heating condition, thickness is more uniformly distributed within the round and square cup and the maximum thinning at the punch corner is reduced when compared to deep drawing at room temperature.
2. For square deep drawn cup, there was no improvement in thickness distribution and thinning condition when low heating temperature is applied shows that lower limit of heating temperature exists. At low heating temperature, uneven distribution of temperature within the blank occurs due to short heating duration and low thermal conductivity of the material such as stainless steel and mild steel, which lead to uneven deformation of the blank.
3. The ANOVA results show that blank size has the greatest influence (37.2%) on the thickness of the square cup, followed by heating technique (32.5%), material (30.2%) and heating temperature (0.1%). Meanwhile, ANOM results show that combination of stainless steel with blank size 45mm, heating temperature of 150°C and die heating technique is optimal in order to obtain uniform thickness distribution of square cup.

Acknowledgements

The acknowledgements should have been given to the Ministry of Higher Education and the University of Malaya under the FRGS grant (FP040-2013B) for the financial support.

References

[1] D Banabic 2010 *Sheet Metal Forming Processes: Constitutive Modelling and Numerical Simulation.*
[2] D G C M R Reddy, R Venkat, Dr T A Janardhan Reddy 2012 *Int. J. Eng. Trends Technol.* **3** 53–58
[3] M Colgan and J Monaghan 2003 *J. Mater. Process. Technol.* **132** 35–41
[4] R K Desu, S K Singh, and A K Gupta 2015 *Int. J. Adv. Manuf. Technol.* 11
[5] Y W Lean and M Azuddin 2016 *IOP Conf. Ser. Mater. Sci. Eng.* **114** 12009
[6] K F Zhang, D L Yin, and D Z Wu 2006 *Int. J. Mach. Tools Manuf.* **46** 1276–1280
[7] Z C Zhang, Y C Xu, and S J Yuan 2016 *Trans. Nonferrous Met. Soc. China (English Ed.)* **26** 1538–1545
[8] F K Chen, T B Huang, and C K Chang 2003 *Int. J. Mach. Tools Manuf.* **43** 1553–1559
[9] Y Moon, Y Kang, J Park, and S Gong 2001 *Int. J. Mach. Tools Manuf.* **41** 1283–1294
[10] J Winklhofer, G Trattnig, C Lind, C Sommitsch, and H Feuerhuber 2010 *AIP Conf. Proc.* **1252** 927–934

IOP Conf. Series: Materials Science and Engineering **210** (2017) 012068 doi:10.1088/1757-899X/210/1/012068

A system dynamics approach to develop a recovery model in the Malaysian automotive industry

N Mohamad-Ali[*1], R A R Ghazilla[*1], S H Abdul-Rashid[1], N Sakundarini[2], A Ahmad-Yazid[1] and L Stephenie[1]

[*1]Centre for Product Design and Manufacturing, Department of Mechanical Engineering, Faculty of Engineering, University of Malaya, 50603 Kuala Lumpur, Malaysia

[2]Department of Mechanical, Materials and Manufacturing Engineering, Jalan Broga, 43500 Semenyih, Selangor, Malaysia

E-mail: [*1]r_ariffin@um.edu.my and [*1]nurhasyimahali@um.edu.my

Abstract. Design strategies play a significant role to enhance recovery effectiveness at the end of product life cycle. By reviewing previous study, there are many factors involved to enhance recovery effectiveness but limited to linking design strategies factors in holistic and dynamics view. Proposed method are explained and an initial model for end-of-life vehicles (ELVs) recovery model illustrated in graphical and numerical data is presented. However this is limited to authors understanding and preliminary data which requires collaboration between designers and other stakeholders to develop a model based on actual situation.

1. Introduction

Industrial growth for instance, Electronic equipment, Transportation, Chemical, Medical and pharmaceutical, Food and beverage, Consumer products, Materials production, Energy production and Services are examples of industries who runs economic activities to meet customers demand and are engaged with common challenges regarding sustainability issues. Statistic shows that manufacturers are the biggest contributor to merchandise trade activities which is USD 18,494 billion compare than commercial services which is USD 4,940 billion [1]. In 2015, total World Motor Vehicle Production is 90,780,583 (cars and commercial vehicles) which is an increase of 1.11% compared to 2014. Additionally, Asia-Ocean countries are leading in produced vehicles compared to other countries followed by America, Europe and Africa. China is the leading country producing vehicles in recent years and vehicles manufacturing in China increased by 3.15% in 2015. Similarly with Malaysia, total vehicles produced increased 3.18% in 2015. Malaysia was the third country leading in vehicles production in Southeast Asia after Thailand and Indonesia.

Malaysia automotive industry dominated by Proton (Perusahaan Otomobil Nasional/National Automobile Company) for 32 years since 1985 followed by 24 years ago for Perodua (Perusahaan Otomobil Kedua Sendirian Berhad /Second Automobile Manufacturer Private Limited). Even though this industry is growing well, they are required to follow the global legislation namely European Union Directive towards End-of-life vehicle (ELV) which is reuse and recovery should be achieved at 85% and 80% reuse and recycling [2]. In 2015, these target should be increase to 10% for reuse and recovery and 5% for reuse and recycling.

Currently, manufacturer are working on end-of-life vehicles issues to fulfill European Union requirement and take responsibility on their product after life span. There are many research on this issues to support automotive local industries towards these requirement. By reviewing previous research, there is a gap that must be fulfilled which links product design strategies towards recovery effectiveness. Thus, this paper is to propose A System Dynamics approach to develop a Recovery Model in the Malaysian Automotive Industry.

The rest of this paper is organized as follows. Malaysia automotive industry is presented in Section 2. System dynamics and initial recovery model are shown in Section 3 and finally recommendation and conclusion are presented.

2. Malaysia automotive industry

This section presents an overview on research and development to enhance ELVs management recovery activities in Malaysia automotive starting on 2009 to 2017 in general.

In recent years, there has been an increasing amount of literature on ELVs management to enhance environmental performance and in line with European Union directive. Malaysia is far behind compared with developed countries where the manufactured vehicles does not implement reusing components & parts into newly manufactured vehicles [3]. Additionally there are a lack of awareness regarding Design for environment (DFE) and Design for disassembly (DFD) among Malaysian stakeholders especially designers [4]. Numerous studies have attempted to explain how DFE is beneficial to the Malaysian automotive industry which will be presented next.

[5] proposed Fuzzy axiomatic design model to reduced recycling and disassembly cost by looking at weight, thickness and manufacturing cost. In the following year, these researchers focus on material section to enhance recyclability towards side mirrors. They proposed ABS, PET and PVC material because these material have good characteristics in terms of Low cost, high recyclability and lighter weight.

In 2012, [6] reported a simple replacement in the product material can lead a manufacturer toward producing more sustainable products. Similarly, Go et al. 2012 demonstrated that, Genetic Algorithm method provide a better disassembly sequences which leads to shorter disassembly time. They found that total disassembly time of the suggested sequence is less than the drafted sequence.

[7] focused on disassembly sequences, whilst [8] concerns on lightweight material selection for doors. These researchers found that for recyclability, aluminum and mild steel are recommended, while for light weight design, a combination of aluminum and carbon fibre are proposed.

In the view of ELVs management, [9] proposed framework for ELVs recycling management by Integrating 6R components in ELV framework. Similarly, [10] identified some ELVs management decision to minimize the environmental impacts and achieving economic benefits. In the same vein, [11] proposed Decision-Making Trial and Evaluation Laboratory (DEMATEL), Analytic hierarchy process (AHP) and fuzzy AHP (FEAHP). All of the studies are concerned on developing ELV management because it is important as initial starting for ELVs recovery activities for Malaysia automotive industry.

In 2014, Malaysia government introduced National Automotive Policy (NAP) as an outline to provide benefit among automotive industry stakeholders [12] and in line with European Union directive. Two years later, Malaysia Automotive Institute (MAI) launched 4R2S (Repair, Reuse, Recycle, Remanufacture, Service and Spare (replacement) parts) transformative plan toward automotive after-market industries [13] and [14].

A recent study by [15] examined factors in ELV management while [16] narrow the study to green supply chain factors and [17] look at how culture influences on choosing green product among Malaysian.

All of the studies reviewed here are summarized in figure 1.

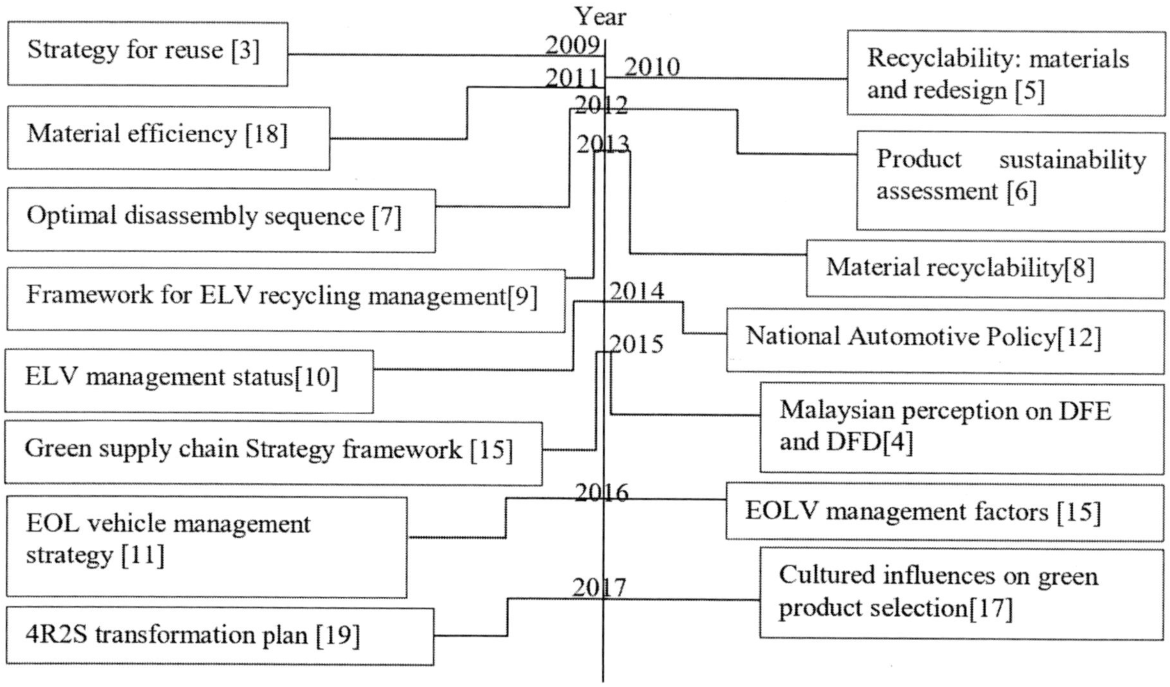

Figure 1. Related work based on Malaysia.

Considering all of these evidences, there are many factors to enhance ELV recovery such as material selection, end-of-life (EOL) option strategies, disassembly sequences, management, supply chain and policy. Taken together, these studies support the notion that ELVs management is an important factors to enhance environmental performance by considering design strategies as a critical role in influencing ELV recovery effectiveness. These studies clearly indicate that there is a complex relationship among these factors and requires a dynamic approach to cater these complexity and dynamic changes such as demand, government policy and global regulation. This makes prediction of ELV outcome by designers to be very challenging to evaluate the best design strategies that should be implemented in order to meet government and European Union recovery target. Thus, this study proposes a holistic approach to evaluate design strategies by linking critical factors on ELV towards recovery effectiveness by gathering information from a variety of stakeholders such as manufacturer, distributors, collectors and policy makers.

3. System dynamics

System dynamics was founded by Jay Wright Forrester, who introduced the causal loop effect. The central idea of this approach is simulating the social behaviour, followed by designing effective policies in order to improve the system performance in reality [20]. There are two types of analysis used in system dynamics: (1) qualitative analysis (i.e. development of a causal loop diagram (see figure 2), and (2) quantitative analysis (i.e. development of stock and flow (see figure 3)). Both positive and negative signs are used in the causal loop diagram [21].

Figure 2 shows the relationships between the main variables of a model for spare parts [22]. We provide a succinct description of this model, as follows:

'Sales of equipment' is the starting point of this causal loop diagram, which is highlighted in bold typeface. The '–' sign indicates that there is a decrease in the dependent variable when there is an increase in the independent variable. For instance, an increase in the delivery rate will decrease the order backlog. In contrast, the '+' sign indicates that there is an increase in the dependent variable when there is an increase in the independent variable. For instance, an increase in the recovery rate will lead to an increase in the stock of serviceable spare parts. An arrow is marked with double lines if the variations of the independent variables will influence the dependent variables with significant delays. This will be the case if the organization increases its expenses for reverse logistics. The owners of the respective equipment may not give their responses immediately and return their obsolete equipment.

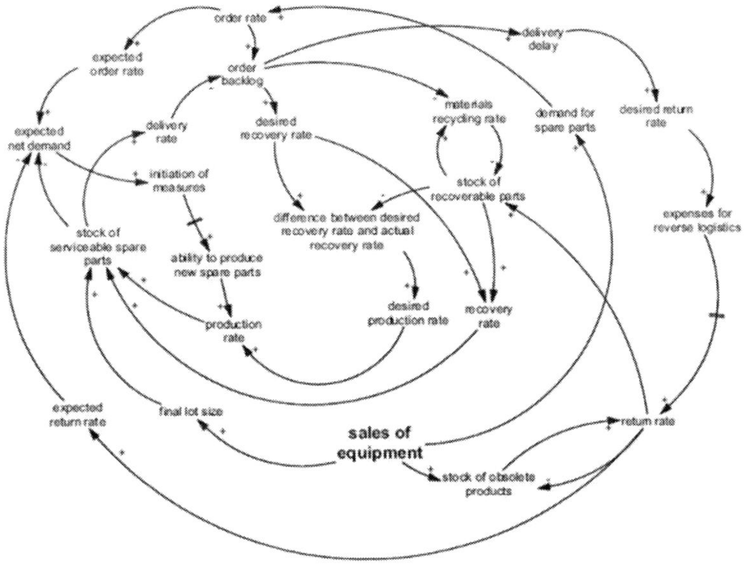

Figure 2. Causal loop diagram for a model of spare parts [22].

3.1. Why system Dynamics?

Achieving sustainability is particularly challenging due to the complexity of the interrelated issues. For this reason, methods such as system dynamics are required to deal with these issues. Additionally there are six reason why these approach is selected and the reason is as follows:

1. System dynamics is an appealing technique since it can be used to analyse the relationships between many variables and improve the practicality of the Triple Bottom Line [23].
2. System dynamics is a powerful tool to analyse the behaviour of complex systems based on non-linear dynamic theory and feedback control.
3. System dynamics can be used to represent human behaviour as well as physical and technical systems and it is a specialized field that draws on cognitive and social psychology, economics and other branches of social sciences [21].
4. System dynamics enable one to identify the causal structure between variables for a given problem [24].
5. The qualitative and quantitative nature of system dynamics facilitates decision-makers during the decision-making process [25].
6. Causal loop diagrams provide clear information [26]; [27] as well as a holistic view of the entire system [28]; [29].

In general, previous study used these approach to cater ELVs recovery issues [30], [31], [32] and [33]. These is evidence that these approach have a high capability to handle complex issues such as product recovery startegies effectiveness.

3.2. Initial recovery model

Graph A represent recovery behavior as a result in this model which different design strategies represent in different values (see B and see in D if prefer in graphical values), how much recovery rate added represent in C by incorporating level, stock and flow diagram (further explanation regarding level, stock and flow diagram refer [34]).

A – Presents a chart comparing recovery rate for design strategy in 4 quarters
B – Represents recovery percentage of 5 different design strategy
C – Shows percentage of recovery added annually and the total percentage of recovery added
D – Outlines the graph chart of the 5 design strategy

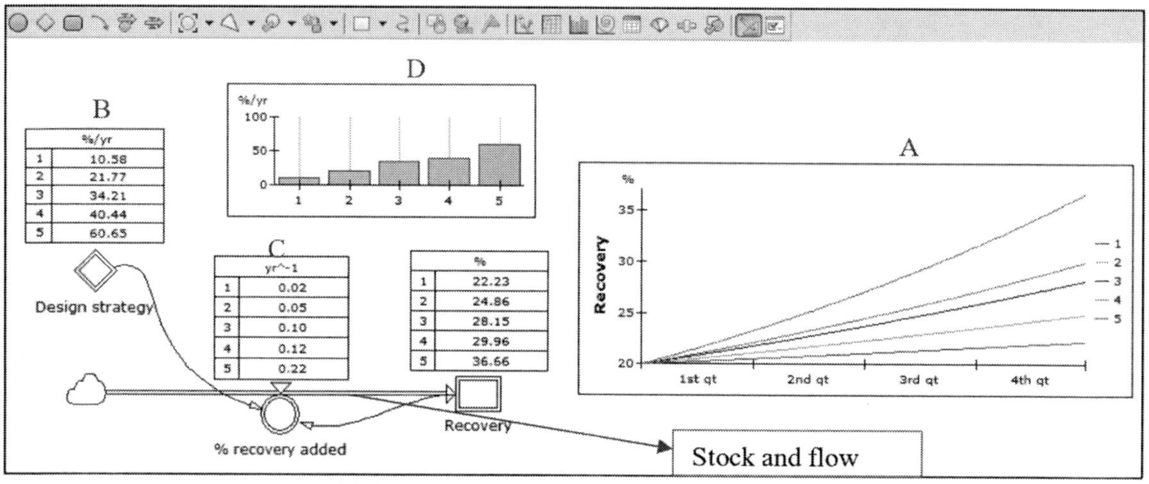

Figure 3. Initial recovery model using system dynamics approach.

4. Conclusion and recommendation

Design for recovery is an important agenda to ensure that the objectives set in the National Automotive Policy are achieved and to alleviate the level of the Malaysian automotive industry to a global standard. This paper proposed system dynamics approach to linking product design strategies with other recovery factors to enhance recovery effectiveness in automotive industries. The technique proposed provides powerful tools to deal with complex issues such as product design strategies recovery effectiveness in evaluating interrelated factors and predicted outcome by decision makers. This is important to assist designers in evaluating the best design strategies and newly proposed model challenges in implementing recovery design to ensure that the process is sustainable and can be economically competitive.

Acknowledgments

We wish to express our greatest appreciation to University of Malaya for funding this work under the University of Malaya Research Grant Scheme (Grant no.: RP033C-15AET)

References

[1] OICA 2016 International organization of motor vehicle manufacturers Retrieved: August 20. 2016 from http://www.oica.net/
[2] Directive 2000/53/EC 2000 Directive 2000/53/EC of the European Parliment and of the Council on end of life vehicles (London: Department of Trade and Industry)
[3] Amelia L, Wahab D, Haron C C, Muhamad N and Azhari C 2009 *J. Clean. Prod.* **17** 1572–79
[4] Ghazilla R A, Sakundarini N, Taha Z, Abdul-Rashid S H and Yusoff S 2015 *J. Clean. Prod.* **108** 331-42
[5] Taha Z, Sakundarini N, Ghazila R A, Gonzales J and Abdul-Rashid S H 2010 *Asia Pacific*

Industrial Engineering and Management Systems Conf (Melaka: APIEM) p 1-8
[6] Ghadimi P, Azadnia A H, Yusof N M and Saman M Z 2012 *J. Clean. Prod.* **33** 10-21
[7] Go T, Wahab D, Rahman M, Ramli R and Hussain A 2012 *Expert Syst. Appl.* **39** 5409–17
[8] Sakundarini N, Taha Z, Abdul-Rashid S H and Ghazila R A 2013 *Mater. Design* **50** 846–57
[9] Azmi M, Saman M Z and Sharif S 2013 *Proc. of the 11th Global Conf. on Sustainable Manufacturing - Innovative Solutions* (Berlin: Universitätsverlag der TU Berlin) 187-93
[10] Ahmed S, Ahmed S, Shumon M R and Quader M A 2014 *J. Appl. Sci. & Agric.* **9** 227-37
[11] Ahmed S, Ahmed S, Shumon M R, Falatoonitoosi E and Quader M A 2016 *Int. J. Sust. Dev. World.* **23** 83-97
[12] NAP 2014 National Automotive Policy Malaysia
[13] MAI 2016 Code of practice for motor vehicle aftermarket: repair, reuse, recycle and remanufacture (4R) for parts and components (Cyberjaya: Malaysia Automotive Institute and SIRIM Berhad)
[14] MAI 2016b Code of practice for motor vehicle aftermarket - service and spare (replacement) parts (2S) (Cyberjaya: Malaysia Automotive Institute and SIRIM Berhad)
[15] Mamat T N, Saman M Z, Sharif, S and Simic V 2016 *J. Clean. Prod.* **135** 1289-97
[16] K S M, Abdul-Rashid S H and E U Olugu, R A R Ghazilla 2015 *I. E.* **22** 46-61
[17] Ghazali I, Abdul-Rashid S H, Dawal S Z, Aoyama H, Tontowi A E and Sakundarini N 2017 Cultural influences on choosing green products: an empirical study in Malaysia *Sustainable Development*
[18] Taha Z, Sakundarini N, Abdul-Rashid S H, Gonzales J and Ghazila R A 2011 *Int. Conf. on IML* 1-7
[19] MAI 2017 4R2S Industry Standards Cyberjaya, Selangor, Malaysia.
[20] Lane D C and Sterman J D 2011 *Profiles in Operations Research: Jay Wright Forrester* in A A Assad and S I Gass, Profiles in Operations Research (New York: Springer) 363-86
[21] Sterman J D 2000 *Business Dynamics System Thinking and Modelling for a Complex World.* (Boston Burr Ridge: McGraw-Hill)
[22] Spengler T and Marcus Schröter 2003 *Interfaces.* **33** 7-17
[23] Lee S, Geum Y, Lee H and Park Y 2012 *J. Clean. Prod.* **32** 173-82
[24] Byggeth S H, Ny H, Wall J, Broman G and Robèrt K H 2007 *Int. Conf. on Engineering Design* (Paris: Cite Des Sciences Et De L'Industrie) p 1-11
[25] Galanakis K 2006 *Technovation.* **26** 1222–32
[26] Sumari S, Ibrahim R, Zakaria N H and Hamid A H *I. J. M. E.* **1** 54-9
[27] Tan B S and Yap E H 2015 *J. E. S. T.* **10** 1150- 61
[28] Hjorth P and Bagheri A 2006 *Futures.* **38** 74–92
[29] Kasperek D, Lindinger M, Maisenbacher S and Maurer M 2014 Paris DSM 14 Proc 234-42
[30] Georgiadis P and Besiou M 2008 *J. Clean. Prod.* **16** 1665–78
[31] Halabi E E and Doolan M 2012 *Int. Society for the Systems Sciences* 1-16
[32] Wang Y, Chang X, Chen Z, Zhong Y and Fan T 2014 *J. Clean. Prod.* **74** 161-71
[33] Kumar S and Yamaoka T 2007 *J. M. T. M.* **18** 115-38
[34] Richardson G P and Pugh A L 1986 *Introduction to System Dynamics Modeling with DYNAMO* (London: The Massachusetts Institute of Technology)

IOP Conf. Series: Materials Science and Engineering **210** (2017) 012069 doi:10.1088/1757-899X/210/1/012069

Design and analysis of wudu' (ablution) workstation for elderly in Malaysia

A Aman*, S Z M Dawal and N I A Rahman

Department of Mechanical Engineering, Faculty of Engineering, University of Malaya, 50603 Kuala Lumpur, Malaysia.

Email: *atikah_aman@yahoo.com, sitizawiahmd@um.edu.my, izzahrahman@um.edu.my

Abstract. Wudu' (Ablution) workstation is one of the facilities used by most Muslims in all categories. At present, there are numbers of design guidelines for praying facilities but still lacking on wudu' (ablution) area specification especially or elderly. Thus, It is timely to develop an ergonomic wudu' workstation for elderly to perform ablution independently and confidently. This study was conducted to design an ergonomic ablution unit for the Muslim's elderly in Malaysia. An ablution workstation was designed based on elderly anthropometric dimensions and was then analyse using CATIA V5R21 for posture investigation using RULAs. The results of the study has identified significant anthropometric dimensions in designing wudu' (ablution) workstation for elderly people. This study can be considered as preliminary study for the development of an ergonomic ablution design for elderly. This effort will become one of the significant social contributions to our elderly population in developing our nation holistically.

1. Introduction

Malaysia population is facing a high rate of demo-graphic rate as in 2016; the population reached 32 million compared to 31 million in 2015 according to Malaysia Department of Statistic: Population Quick Info. High demographic rate not occurs only in Malaysia but this aging issue actually had become a global problem [1]. Malaysian population is projected to increase by 35% in 2040 but the annual growth rate is projected to decrease to 0.6% due to decrease in fertility rate and active international migration based on Department of Statistic at 2015. Elderly people commonly referred as population having one or more illnesses other than having movement or functional problem [2]. Improvement in health-care services results in existence of active elderly people which prefer to live independently [3], demands development of facilities which able toadapt to their capabilities and abilities also increasing to conduct their daily activity, including praying facilities.

As a Muslim, praying and ablution are considered as part of the daily activity that has to be conducted wherever they are. World Muslim population in the world is estimated to be around 20-25% of total world population. Majority of the Muslim residents is approximately in 40 countries including Malaysia and almost every country has small Muslim population [4]. In Malaysia, Selangor, Kuala Lumpur and Putrajaya themselves already had 645 and 144 mosques and musolla, registered with the government bodies. Prior to obligate five times praying every day, Muslims are required to perform ablution as a cleaning procedure for the Muslim [5]. In Malaysia, praying facilities equipped with ablution area can be easily found around public places including shopping complex and highway rest area for the Muslim to pray.

Most of the existing praying facilities design guideline developed by some countries focus solely on the appearance of the praying facilities while ignoring proper design of supporting spaces such as toilet and ablution area [4]. In Malaysia, set of Malaysia Standard has been developed as a guideline for the developer and designer to build public facilities which can be accessible by the vulnerable group such as elderly and disabled population to provide design specification such as height of ramp and handrail, slope of ramp and finishing material. Therefore, it is necessary for praying facilities designer to promote an adequate environment for the elderly and disabled, considering their needs, capabilities, abilities and limitations.

Even though ablutions space is considered as one of the important public facilities especially in country with high Muslim population, there are numbers of design guideline for mosque but still lacking in guideline, research or document specifically on ablution area [6]. Most designer will neglect design of supporting area such as toilets and ablution space since most of them emphasized on aesthetic appearance of their design. As elderly population is increasing, their requirement should be taken into design consideration so that they will function with limited assistance from other people. Ergonomic aspect also should be taken into the consideration for safe and low risk environment. In Malaysia, research on anthropometric data is expanding which can be used to design an ergonomic ablution area at mosques for public use. Therefore, this study focused on designing and analysis of wudu" workstation for elderly in Malaysia.

2. Procedure

2.1 *Subject*

This study had been carried out in Malaysia which covers state of Kuala Lumpur, Selangor, Kelantan and Terengganu who met the following criteria: a) were able to understand or speak basic Bahasa Melayu and English, b) signed informed consent, c) aged between 60 years and above. Ethical approval for the study was obtained from the University of Malaya Medical Centre Ethics Committee. A total of 186 elderly participants involved in the survey.

2.2 *Anthropometry Measurements*

Anthropometry of the subjects was measured using a standard professional anthropometer (TTM Martin's Human Body Measuring Kit, Mentone Educational Centre, Carnegie, Vic., Australia), a sliding calliper, a weighing scale, a plastic measuring tape and an adjustable chair for sitting postures. The seat could be adjusted to different heights and acted as a reference point for the measurements in the sitting position. The measuring kit consisted of instruments for measurements of distances in straight lines, curves, circumferences, and thickness. The sliding calliper and spreading calliper were used to measure small breadths and depths of body segments. The plastic measuring tape was used to measure body circumferences. An adequate description of the human body may require over 300 dimensions [7], but the scope of this study was limited to measurement of body dimensions that were considered important to design the wudu' workstation design. Thus, 11 body dimensions were selected. All dimensions were as defined in the Malaysian Standard MS ISO 3638:1981: Size designation of clothes-definitions and body measurement procedure (Malaysia, Jabatan Standard, 2005).

2.3 *Statistical Analysis*

The IBM Statistical Package for Social Science (SPSS) for Windows version 23.0 (IBM SPSS Statistics for Windows Version 23.0, Armonk, NY: IBM Corp) was used in the following statistical analysis. Data was expressed as means ± standard deviations (SD), and percentiles (5th and 95th) of the above measurements and were calculated for both the male and female subjects. Pearson's correlation coefficients were calculated and used to test the significance of the linear relationship among the variables.

2.4 *Ablution Workstation Design Analysis- Rapid Upper Limb Assessment*

RULA provides a quick analysis of demands on a person's musculoskeletal system when performing a specific task. It involves assessments on neck and upper limb loading in mainly sedentary tasks (repetitive tasks). The outcome of the analysis presents the exposure of individual workers to risks associated with work-related upper limb disorders. It examines risk factors such as number of movements, static muscle work, force, working posture, and time worked without a break. All these factors combine to provide a final score that ranges from 1(Good) to 7(Worse). In this study, the software used for RULA analysis was CATIA V5R21.

Parameters Used for RULA Analysis

1) Posture: Intermittent
2) Repeat Frequency: Fewer than 4 times per minute
3) Worker's posture:
    A) Arms are working across midline: No
    B) Check balance: No
4) Load: 0.
5) Manikin: Japanese (Asian, closely resembling to Malaysians (Not listed in software)), male, weight 69 kg and stature 159cm approximately.

According to above parameters, a manikin is built under human builder module in CATIA as shown in figure 1, figure 2 and figure 3. There is no load exerted because no lifting required.

3. Results and Discussion

3.1 *The important Elderly anthropometric dimensions for wudu design*

Table 1. Anthropometric data for elderly in Malaysia (all dimensions in cm)

		Male (N=91)					*Female (N=96)*				
		Mean	***SD***	***5th***	***50th***	***95th***	***Mean***	***SD***	***5th***	***50th***	***95th***
1	Stature	158.81	7.52	145.39	160.00	169.38	147.88	6.91	135.32	148.00	159.02
2	Shoulder height	132.37	6.09	121.53	133.35	141.94	121.81	5.75	111.64	121.60	129.70
3	Elbow height	100.40	9.74	89.48	98.90	115.10	91.36	5.23	83.00	91.40	99.34
4	Kneecap height	45.51	3.33	40.61	45.55	50.92	44.55	4.97	38.10	44.40	51.58
5	Span	160.02	16.35	135.69	163.05	174.02	148.68	10.87	132.00	150.10	163.90
6	Elbow span	81.96	6.05	69.07	83.05	91.05	73.29	7.47	59.72	74.40	82.30

7	Shoulder breadth	21.08	1.97	18.95	22.05	22.83	22.22	4.88	16.21	24.35	27.09
8	Arm reach forward	78.98	7.09	71.19	78.90	86.73	72.69	4.90	64.08	73.40	79.39
9	Forearm-hand length	43.70	4.13	35.73	44.50	47.92	40.46	5.48	26.02	41.70	45.70
10	Hand length	17.74	1.82	15.30	17.90	19.75	16.66	1.13	15.28	16.70	18.33
11	Foot length	23.83	1.30	21.54	24.00	26.22	22.11	1.33	19.64	22.10	24.30

Table 1 shows the dimensions of anthropometric data of elderly in Malaysia. There are 187 subjects involved in measuring the data which include 91 males and 96 females throughout Malaysia. 11 dimensions are selected which relate to the wudu' workstation out of 93 dimensions that have been taken. Some of the parts are clearly seen that male dominate the data while some parts such as shoulder breadth shows that female is bigger.

3.2 *RULA Scores*

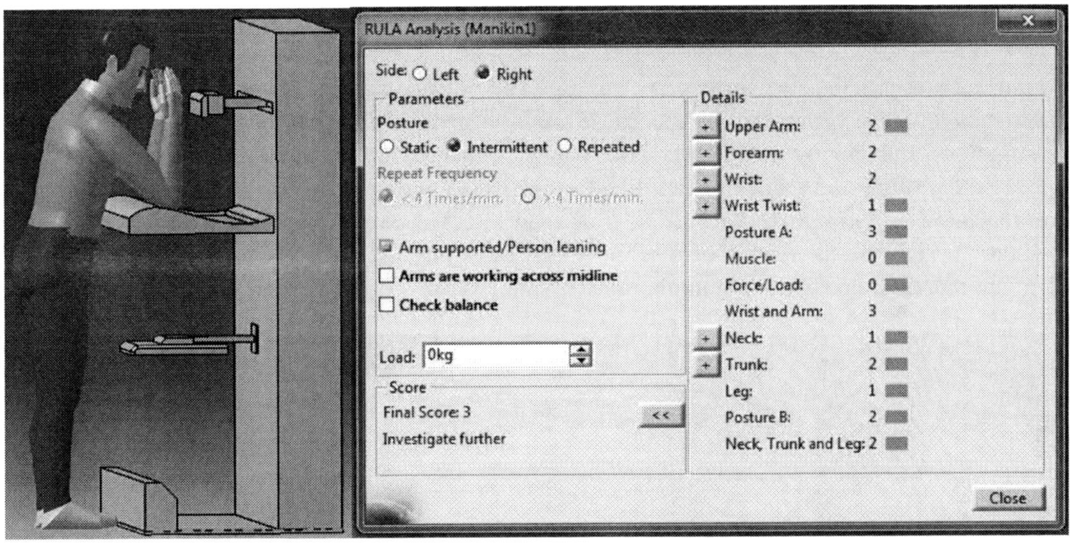

Figure 1. A Model posture attained while face cleaning

Figure 1 shows the posture of the step face cleaning in completing the ablution. The score is 3 and resulted in yellow colour that indicates further investigation is needed and changes may be required. The results show that the posture while performing the wudu' (ablution) is in good condition and further investigation is needed for any changes. The overall analysis indicated that the design is still acceptable but an investigation is needed to improve on the working posture to reduce any discomforts.

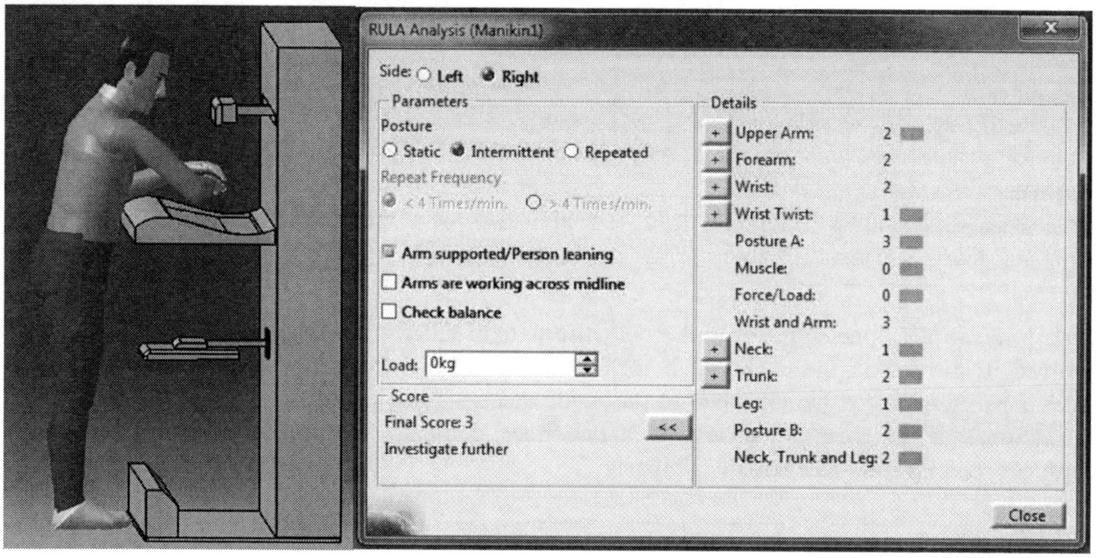

Figure 2. A Model posture attained while hands cleaning

Figure 2 presents the step for hands cleaning. The posture scores the same value, 3 and resulted in yellow colour. The posture will not cause any hazardous and there is no need for immediate investigations and changes. However, the results indicated that further investigation is needed to improve the posture and reduce difficulties for elderly to perform their ablution. The significant body parts that need to be improved are neck, trunk and leg. The results highlighted that the design of the workstation might be the reason of discomfort postures and the anthropometry dimensions should be taken into thorough consideration in planning to improve the design.

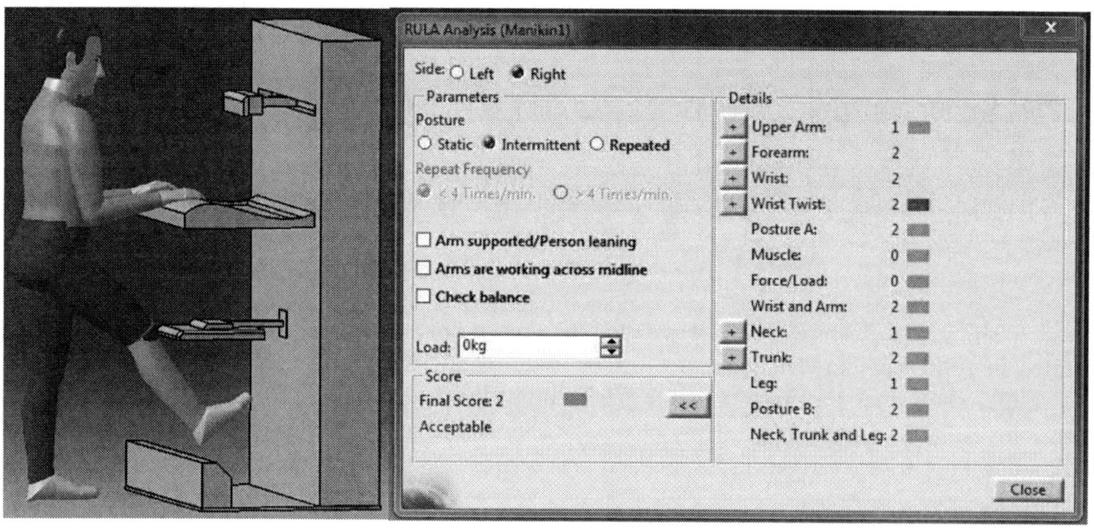

Figure 3. A Model posture attained while foot cleaning

Figure 3 illustrates the posture for foot cleaning in ablution. It scores 2 which resulted in green colour. The green colour indicates that the posture is acceptable. It can be maintained or repeated for long periods of time and no changes are required for improvements. This posture will avoid any discomfort of elderly in order for them to perform their ablution.

4. Conclusion

The results highlighted that the significant factors in designing wudu' (ablution) workstation are the right postures within the limitation and ability of the elderly pertinent to anthropometric dimensions. The significant body parts that related to the design of wudu' (ablution) that need to be improved are forearms, neck, trunk and leg. Therefore, it is hope that this effort will become one of the significant social contributions in developing our nation holistically.

Acknowledgement

This work is financially supported by the University of Malaya under the University Malaya Research Grant (Project title: Development of wudu' ablution workstation for elderly and disabled person, Grant no: RP027-15SBS).

References

[1] Wang Q and Yang A 2012 Soft Computing in Information Communication Technology *Springer Berlin Heidelberg* 501-509
[2] Rodrigues-Barbosa, A., Miranda, L. M. D., Vieira-Guimarães, A., Xavier-Corseuil, H., & Weber-Corseuil, M. 2011 *Revista de salud publica* **13** 54-66
[3] Fitzpatrick R B 2013 *J Natl AssocAdm Law Judiciary* **11** 13-24
[4] Mynatt E D and Rogers W A 2001 *Ageing International* **27** 24
[5] Mokhtar A 2003 *Journal of Architecture Engineering* **2** 55-61
[6] Besari A, Zamri R, Yusaeri A 2009 *Industrial Electronics & Applications (IEEE Symposium Vol 1)* 506-50
[7] Pheasant S T (1986). *Bodyspace: Anthropometry, Ergonomics and Design. Taylor & Francis,* London, UK.

IOP Conf. Series: Materials Science and Engineering **210** (2017) 012070 doi:10.1088/1757-899X/210/1/012070

The application of CAD, CAE & CAM in development of butterfly valve's disc

Muhammad Asiff Razif Shah Ranjit, Nazlin Hanie Abdullah*
Division of Mechanical Engineering, Department of Engineering, Faculty of Engineering and Life Sciences, Universiti Selangor, 45600 Bestari Jaya, Selangor

Email: asiffrazif@gmail.com and *nazlin@unisel.edu.my

Abstract. The improved design of a butterfly valve disc is based on the concept of sandwich theory. Butterfly valves are mostly used in various industries such as oil and gas plant. The primary failure modes for valves are indented disc, keyways and shaft failure and the cavitation damage. Emphasis on the application of CAD, a new model of the butterfly valve's disc structure was designed. The structure analysis was analysed using the finite element analysis. Butterfly valve performance factors can be obtained is by using Computational Fluid Dynamics (CFD) software to simulate the physics of fluid flow in a piping system around a butterfly valve. A comparison analysis was done using the finite element to justify the performance of the structure. The second application of CAE is the computational fluid flow analysis. The upstream pressure and the downstream pressure was analysed to calculate the cavitation index and determine the performance throughout each opening position of the valve. The CAM process was done using 3D printer to produce a prototype and analysed the structure in form of prototype. The structure was downscale fabricated based on the model designed initially through the application of CAD. This study is utilized the application of CAD, CAE and CAM for a better improvement of the butterfly valve's disc components.

1. Introduction

Butterfly valve is widely used in various application especially in industrial sector. A butterfly valve is a simple mechanical operation device preferable for separate or isolate a process and regulate a fluid flow for a certain distance. Industrial sector generally favored butterfly valve due to their lower in cost to other valve designs as well as being lighter in weight, meaning less support is required. Butterfly valve consists of body, disc, stem, and seat. Most of butterfly valve are not applicable for a high pressure fluid flow. The design of the butterfly valve are based on the principle of pipe damper. Element that control the flow which is the disc, acts like a gate and has approximately the same diameter as the inside diameter of the pipe joint [2].

A butterfly valve consists of a circular disc or plate built with a stem through the middle or attached offset. The disc is positioned in the center of the pipe, passing through the disc is a rod connected to an actuator on the outside of the valve. The disc manufactured based on the solid element design. Based on the review that had been done, butterfly valve's disc experience failure due to certain aspects. The first failure is the shaft-disk separation due to pin failure [7]. An incident occur in the 1997 involve the pin connected to the shaft and disc. Another way failure is when the butterfly valve experience disc tear (indented disc). This happen because of the presence of foreign object inside a pipeline flow and the number of valve usage [11].The final problem is due to the presence of cavitation on the low region inside the pipeline flow. High pressure differential between the upstream and downstream of the flow is the main cause of the cavitation [1]. The disc structure of the butterfly valve is based on two consideration of analysis. The first one is the Computational Fluid Dynamics (CFD) and the Structural Analysis using the Finite Element Analysis (FEA). The structure analysis is to analyze the condition of the disc structure and determine the weaker section of the disc and determine the condition of the structure if there is any failure. CFD application is to determine the flow of the fluid in pipeline with the designed model of the disc. This will determine the area of low pressure on the downstream inside the pipeline for cavitation justification.

2. Problem Identification

The most common failures of butterfly valve were described in the following subsections.

2.1. Indented Disc

Foreign particle that travel at high speed inside the pipeline system course the disk from the butterfly valve being indented and as days passed by, the disk will corroded and will experience metal lost. figure 1 shows the corroded disc. As a result, the butterfly valve experience leakage and ultimately it will not block the flow of the gas. The sandwich-structured of composite is a unique and special class of composite materials. Attaching two thin but stiff skins to a thick core but lightweight, the composite materials was fabricated. The core material is normally low strength material, sandwich composite with high bending stiffness due to its higher thickness and overall low density [1].

2.2. Pin and Keyways Failure

Butterfly valve was design with a drive shaft that connects the internal valve disk to an external pneumatic cylinder (actuator). The valve fain when a dowel pin design to fasten the dry shaft to the key shared and key design to transfer torque from the driver shaft from the disk. System pressure was high enough to eject the unrestrained drive shaft from the valve carry with it external counterweight [1].

A number of design may contribute to this hazard. These include the valve has a shaft or stem piece which penetrates the pressure boundary and ends inside the pressurized portion of the valve. This feature results in an unbalanced axial thrust on the shaft which tends to force it (if unconstrained) out of the valve. Figure 2 shows the current disc design of the valve. The second design problem is that valve contains potential internal failure points, such as shaft dowel-pins, keys, or bolts such that shaft-disk separation can occur inside the valve [6].

Figure 1. Corroded Disc

Figure 2. Pin location

The third problem is that the dimensions and manufacturing tolerances of critical internal parts (e.g., keys, keyways, pins, and pin holes) as designed or as fabricated cause these parts to carry abnormally high loads (e.g., in the 1997 accident, the dowel pin rather than the key transmitted torque from the shaft to the disk). The final concern on the design problem is that the valve stem or shaft is not blow-out resistant. Non blow-out resistant design features may include two-piece valve stems that penetrate the pressure boundary (resulting in a differential pressure and unbalanced axial thrust as described above), single-diameter valve shafts (i.e., a shaft not having an internal diameter larger than the diameter of its packing gland) or shafts without thrust retaining devices. A Hobbs triple offset butterfly valves require no pins and keyways on the shaft that will connected to the disc. Replacing the pins with bolts, the disc directly bolted to the shaft and this eliminates the problems of a weaker shaft and the problem of shearing during high velocity applications. The keyways of the shaft was eliminated by using the square shaft, the keyways will not be required anymore [5].

2.3. Cavitation Damage

When a butterfly valve is used for throttling or modulating flow rates, the operating conditions should be evaluated to determine whether significant cavitation will occur. Cavitation can cause objectionable noise, vibration, and decrease the useful life of a valve and nearby piping components [3]. Figure 3 shows the disc was damaged due to cavitation. Cavitation is the vaporization and subsequent violent condensation of a fluid caused by localized areas of low pressure in a piping system. When water flows through a partially open butterfly valve, a localized low-pressure zone may occur immediately downstream of the valve disc because of the sudden changes in flow velocity and flow separation. When the pressure in this zone falls below the vapor pressure of the fluid, the liquid vaporizes, forming a vapor pocket or vapor bubbles. As the bubbles flow downstream and the pipeline pressure recovers, the bubbles violently collapse or implode.

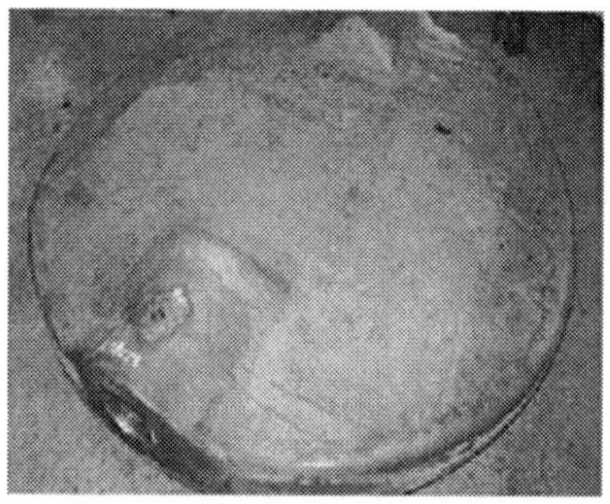

Figure 3. Cavitation Damage

Cavitation can form in a butterfly valve immediately downstream of the valve disc where a low-pressure zone occurs. Cavitation bubbles can implode just downstream of the disc or many times the pipe diameter downstream, depending on where the pressure recover. The process produces an unmistakable noise and vibration that sound like gravel flowing through the pipe. Many simple shutoff valve applications produce cavitation when the valves are near the closed position, because the differential pressure reaches its highest level at that point. However, since a shutoff valve is usually at a near-closed angle for only a short period of time, appreciable damage to the valve or piping usually does not occur. When a valve is exposed to caviting condition continuously, however, such as when it is used for flow modulation or pressure control, significant damage can occur to the metal surfaces of the valve or downstream piping in a short period of time [3]. Figure 3 shows the cavitation effect.

2.4. Predicting Cavitation

Three terms are commonly used to classify cavitation in valves according to the Instrument Society of America namely Incipient cavitation, constant cavitation and choked cavitation. The start of steady cavitation, termed incipient cavitation, can be indicated by an intermittent popping sound in the flow stream. Incipient cavitation typically does not cause damage or objectionably loud noise. If the pressure differential increases, however, the constant cavitation level is reached, which can be indicated by a continuous popping similar to the sound of gravel flowing through the pipe or bacon frying. Continuous flow above the constant cavitation level is often accompanied by objectionable noise and valve or piping damage. The choking cavitation level occurs when the valve is passing the maximum flow possible for a given upstream pressure. The vapor pocket may become extremely long,

IOP Conf. Series: Materials Science and Engineering **210** (2017) 012070 doi:10.1088/1757-899X/210/1/012070

causing damage far down- stream from the valve. Choking cavitation may cause a reduction in noise, but this change is usually preceded by the highest level of noise and vibration. Tests have shown that conditions likely to produce cavitation in a butterfly valve can be predicted and possibly reduced or prevented. The cavitation index is typically used as a predictor of valve damage and is expressed quantitatively at each valve angle as follows:

$$\sigma = (P_u - P_v) / (P_u - P_d) \quad (1)$$

Table 1. Variable and Definition of cavitation index

Variable	Description	Units US Customary (SI-metric)
P_d	Reference downstream pressure for cavitation analysis	psi (kPa)
P_u	Reference upstream pressure for cavitation analysis	psi (kPa)
P_v	Vapor pressure adjusted for temperature and atmospheric pressure. (Example: P, = -14.4 psig [-99.6 kPa] for water at 60" F [16" C], (kPa) measured at sea level).	psi (kPa)
σ	Cavitation index, general form	dimensionless

Cavitation indices for, constant, and choked levels can be determined from flow testing in a laboratory environment. The operating cavitation index can be compared to the cavitation indices for valves to predict what level of cavitation will occur (incipient, constant, or choking). It should be noted that in some earlier texts the constant index is referred to as critical. Later texts change this nomenclature to constant to be more descriptive of the condition without implying a crucial operating condition. Cavitation indices for incipient, constant, and choked levels can be determined from flow testing in a laboratory environment. Cavitation can be observed using a hydrophone or accelerometer during the flow test. The lower the calculated cavitation index, the greater the likelihood of cavitation damage. For example, if a valve is throttled at 45" open with a calculated index of 6.0, then cavitation will likely not occur. If, however, the valve is closed further to 30" open with a calculated index of 2.2, then the cavitation in the range between incipient and constant will occur. Sounds of cavitation will be heard, but serious damage will occur only after a prolonged period of time under those conditions [3]. Figure 4 shows the three different region of the cavitation index.

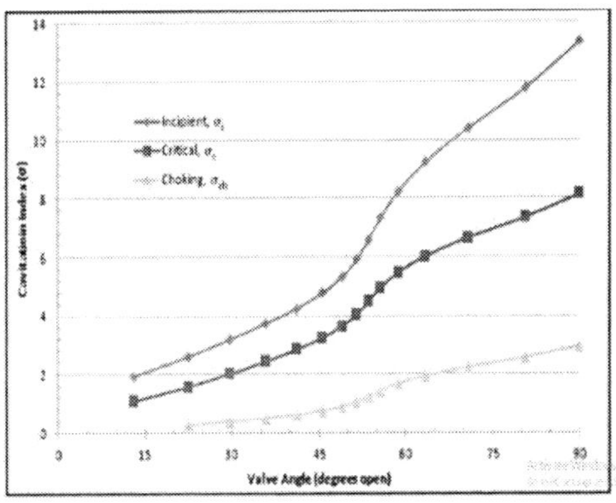

Figure 4. Cavitation Index Boundaries

Design provisions for water systems completely without cavitation are beyond the scope of this manual, but some general recommendations to reduce cavitation can be considered. To reduce cavitation, the value of the cavitation index, a, must be increased above the constant cavitation index for the valve. One way to do this is to increase the downstream pressure, P_d, which will increase the value of the cavitation index. Another strategy is to decrease the differential pressure across the valve

(P_s - P_d). The value of the constant cavitation index, u_c, can also be changed by using the valve at a different opening position or using a different valve model. Finally, air can be introduced to mitigate cavitation.

3. Methodology

3.1 Computer Aided Design (CAD)

A 3D graphical modeling of the butterfly valve's disc components will be created using the computer program software, Computer Aided Design (CAD). In product and industrial design, CAD is used mainly for the creation of detailed 3D solid or surface models, or 2D vector-based drawings of physical components. Graphic representation is considered as one of the most important aspect in designing can be achieved through geometric modeling of mechanical parts in designing the butterfly valve's disc components. CAD software is used to increase the productivity of the designer, improve the quality of design, improve communications through documentation, and to create a database for manufacturing.

3.2 Computer Aided Engineering (CAE)

Computer Aided Engineering (CAE) is the broad usage of computer software to aid in engineering analysis tasks. It includes Finite Element Analysis (FEA) and Computational Fluid Dynamics (CFD). A CAE process comprises of preprocessing, solving, and post-processing steps. In the preprocessing phase, the model of the butterfly valve's disc and the physical properties of the design, as well as the environment in the form of applied loads or constraints. The Finite Element Analysis application is a process of detecting, analyzing and finding the structural performance issues. The strength of the butterfly valve's disc component plays an important role of controlling the flow of fluid in pipeline. Using the application of numerically and mathematically, FEM analyzed complex structural, fluid and variety of physical problems. A fluid flow through a design model in this case a butterfly valve's disc needs to be analyzed using the application of Computational fluid dynamics (CFD) which was compromised with numerical analysis and algorithms for a result. The calculation that required to imitate the interaction of liquid and gases with surfaces defined by boundary conditions. This application will be an ideal method to analyze pressure differential of butterfly valve especially identifying the low pressure zone for each turning angle of butterfly valve. The cavitation index data will be plotted based on CFD analysis.

3.3 Computer Aided Manufacturing (CAM)

The Manufacturing process of the butterfly valve's disc is based on the application of Computer-aided manufacturing (CAM) or commonly refer to the application of numerical control (NC) which used computational software to create the flow or instruction (G-code) that run computer numerical control (CNC) machine tools for manufacturing parts [8]. The creation of a 3D printed object is achieved using additive processes. In an additive process an object is created by laying down successive layers of material until the object is created. Each of these layers can be seen as a thinly sliced horizontal cross-section of the eventual object.

4. Results and Discussion

The butterfly valve was designed using the application of Computer Aided Design. Using CATIA, the detail design of the disc was constructed. The dimensions of the butterfly valve's disc and shaft are based on the standard specified by the American Water Work Association. The butterfly valve with a diameter of 48 inch (1200 mm) has been selected as the reference size for analyzing the performance of the valve using the application of computational aided. The diameter of shaft is also based on the minimum standard that had been specified by the American Water Works Association that is 82.6 mm.

4.1. Computer Aided Design (CAD)

The design of this particular butterfly disc is based on the sandwich theory but with a different "core" structure. Most sandwich materials used honeycomb shape as the core structure to increase the rigidity of the materials in the same time it is lightweight. Considering the drawback of honeycomb structure, which is expensive, a new design will be taken as consideration. An I-beam structure has a promising characteristic especially when considering the rigidity and toughness of the structure. I-beam structure acts as the important influence while designing the disc design [12]. Figure 5 shows how the influence of I-beam in the design. The I-beam sandwich structure will acts as the layer that support the main disc structure. Both I-beam sandwich structure and the disc acts as one component [4].

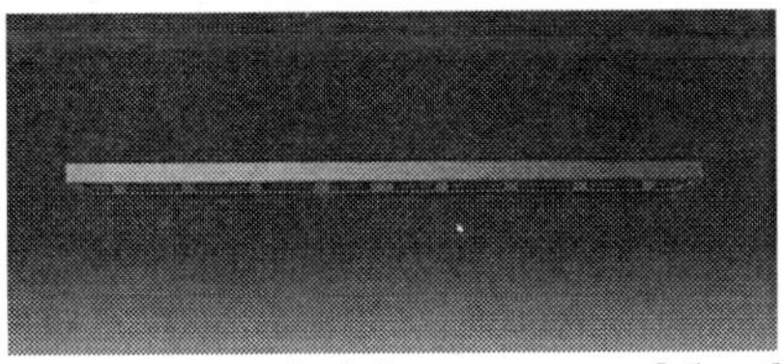

Figure 5. Sandwich structure of the disc

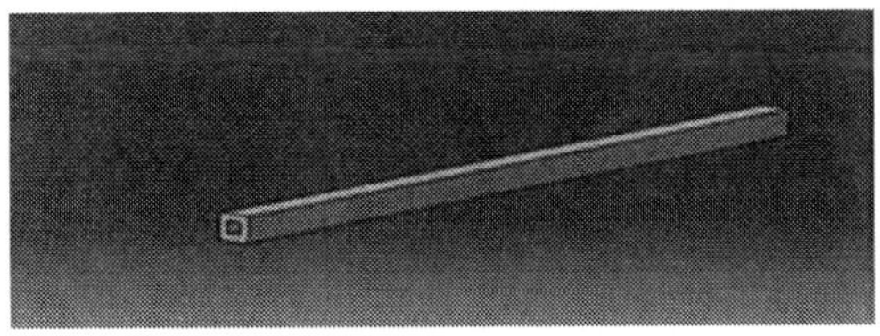

Figure 6. Rectangular shaft

The shaft of the disc was designed based on the rectangular cross-section. The cross length of the rectangle has the same diameter with the round cross-section shaft. The rectangular shaft design is to completely remove the keyways. Figure 6 illustrates the design of a rectangular cross-section shaft that will be connected with the disc. The connector that holds the shaft and the disc was designed to accommodate the structure operation. The connector design will affect the performance of the butterfly valve's disc. One of the consideration that must be included is that the design of the connector must be based on the concept or a shape of an 'aero foil'. The importance of this consideration is reduce the amount of drag force [10]. Figure 7 shows the flow of fluid during the fully open butterfly valve or 90° fully open valve.

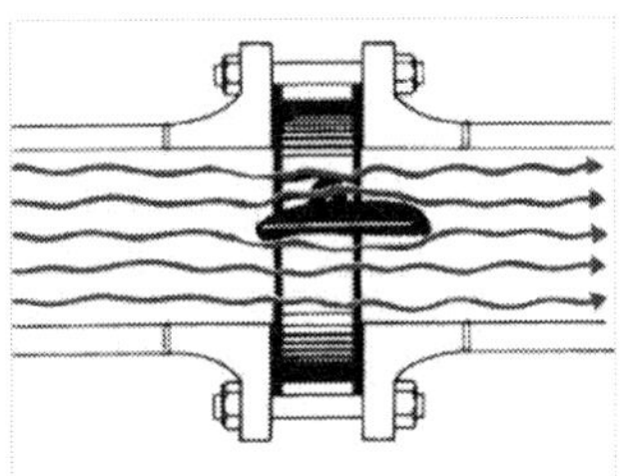

Figure 7. Flow of fluid at fully open valve

Figure 8. Connector

The connector design must reduce the amount of drag force during the opening of a valve. Figure 8 is the detail design of the connector that attached the shaft and the disc. The eight holes appear on the part is to replace the pin on the shaft based on the design that already manufactured on today industry. The pin is replaced with a standard bolt to hold the disc and shaft together. Figure 9 and figure 10 picturized the fully assembled disc of a butterfly valve. The complete assembly was designed to meet the requirement of the standard design performance of perform a much better performance. This fully assembled will be analyzed structurally and fluid flow analysis under the application of computer aided engineering.

IOP Conf. Series: Materials Science and Engineering **210** (2017) 012070 doi:10.1088/1757-899X/210/1/012070

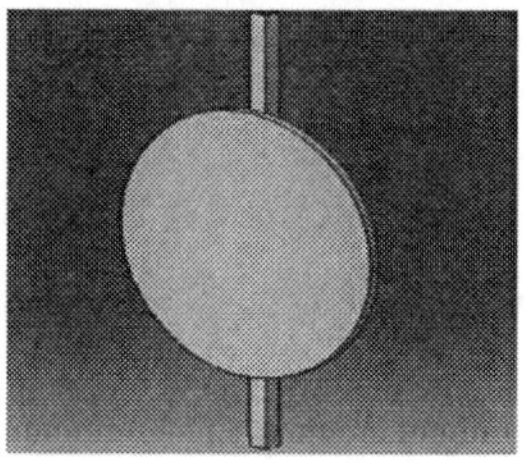

Figure 9. Front view

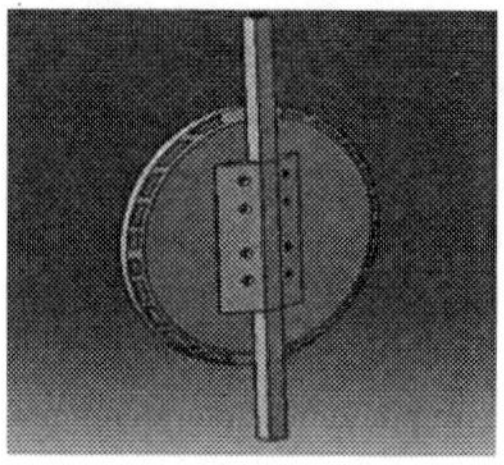

Figure 10. Back view

4.2. Computer Aided Engineering (CAE)

The application of computational aided engineering was applied to analyze the performance of the butterfly valve's disc, shaft and the connector. This is a crucial method to determine the reliability of the design. The application of computational aided engineering is separated into two section. The first section is to analyze the structure analysis using the understanding of finite element. The second section is to analyze the fluid flow of this butterfly valve inside a pipeline to measure the pressure differential for the cavitation index measurement. In this section, the structure analysis was performed to analyze the stress configuration on the structure. The material used for the butterfly valve is ASTM A108 standard based on American Water Works Association. The ASTM A108 materials standard specification for steel bar, carbon and alloy, cold finished. The upstream pressure inside the butterfly valve will create a distribution force acting on the surface of the disc. Based on the standard given by the American Water Works Association (AWWA), modern butterfly-valve designs for water service include cast-body construction in 25-psi (172-kPa), 75-psi (517-kPa), 150-psi (1,034-kPa), and 250-psi (1,723-kPa) pressure classes (Association, 2012). Using 172 kPa as an upstream pressure, the force acting on the surface of the disc of the butterfly valve. The maximum stress focus on the shaft, especially on the support. Based on this observation, the shaft suffer the most stress compare to the disc [13]. Figure 11, figure 12 and figure 13 show the stress and displacement result of the structure analysis.

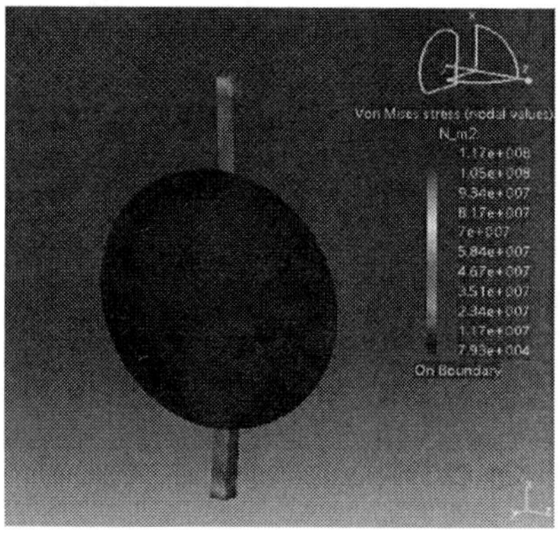

Figure 11. Stress structure analysis (front view)

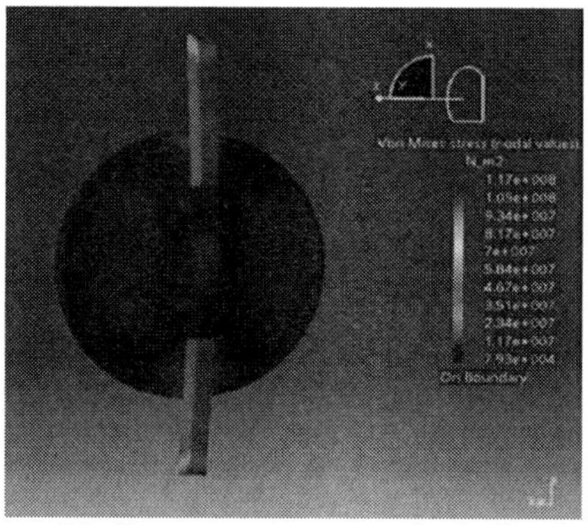

Figure 12. Stress structure analysis (back view)

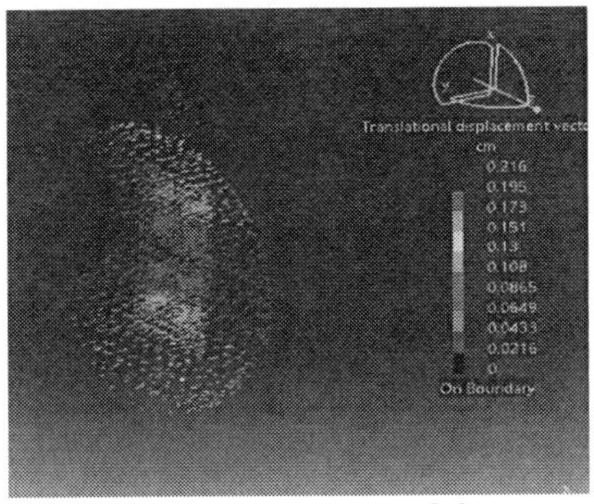

Figure 13. The displacement vector of the structure analysis

The current disc then was analyzed for a comparison. Figure 14 shows the current design of the butterfly valve. The diameter and thickness of the disc is the same as the sandwich plate disc. The diameter of the shaft has the same hypotenuse length. Figure 15 shows the back view of the current disc structure. The one hole attached on the back of the disc is where the pin is positioned .The same distribution force applied on the surface of the disc.

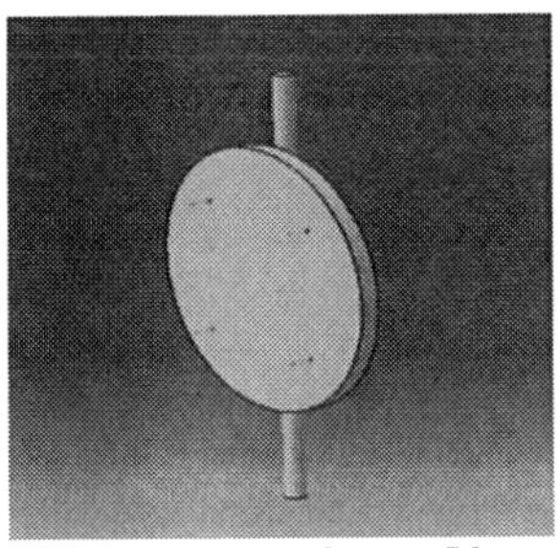

Figure 14. Front view of improved design

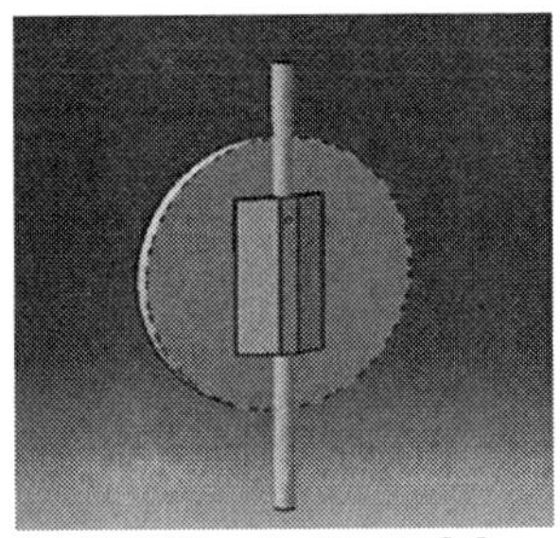

Figure 15. Back view of the current design

The same boundary conditions was applied as the sandwich disc, the current disc was analyzed and the result of the stress configuration and the translational displacement was recorded. Figure 16 and figure 17 is the result of the stress configuration and figure 18 shows the translational displacement of the structure.

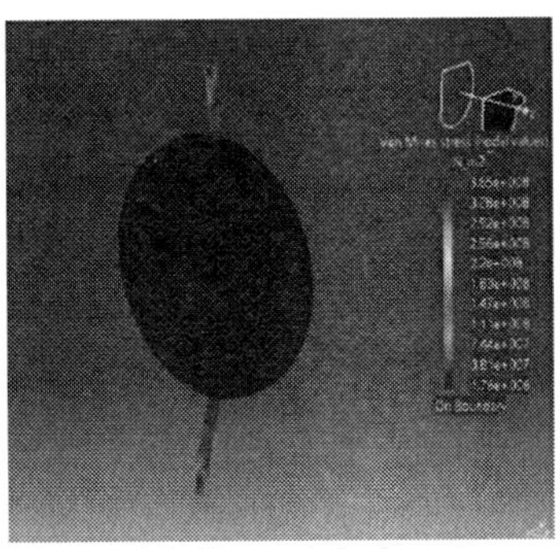

Figure 16. Current design stress analysis (front view)

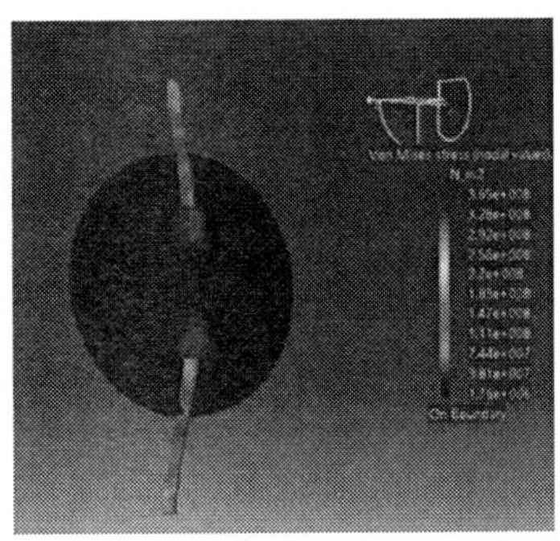

Figure 17. Current design stress analysis (back view)

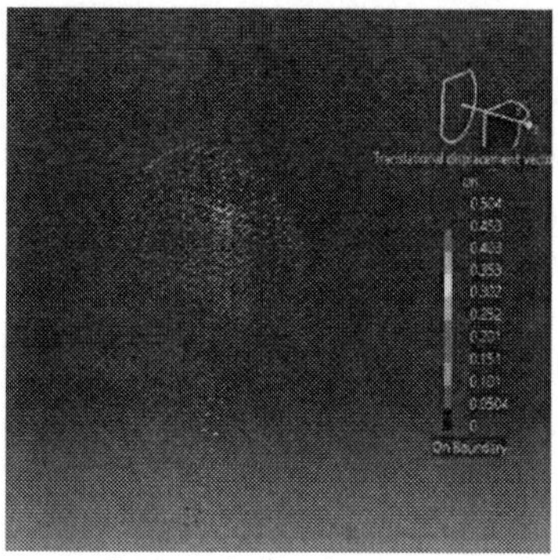

Figure 18. The translational displacement of the current disc

Figure 11, figure 12 and figure 13 show the result of the stress analysis of the sandwich disc butterfly disc. The figures represent the stress configuration and the displacement vector while figure 16, figure 17 and figure 18 represent the stress analysis and displacement vector for the current disc and shaft. From this evaluation, noting that the maximum stress on the current design is higher than the stress on the sandwich butterfly disc structure. The maximum stress concentrate on the shaft due to the fixed support based on the boundary condition that applied on the shaft. The stress configuration on the disc for both designs doesn't suffer a great amount of pressure. However the displacement vector on the current design has the higher deflection due to the distribution force action on the surface of the disc. This can be justify that the disc deflects the most will allow more leakage during fully closed of the valve. From figure 17, the configuration stress on the pin section has a much higher stress concentration. This can be compared from figure 12 which the stress concentration on the bolt section is much lower. Table 2 is the comparison between the current disc design and the sandwich disc design.

Table 2. The stress comparison between the two designs

	Sandwich Design	**Current Design**
Maximum stress (MPa)	117	365
Stress at the Pin –(current)(MPa) Stress at the bolt – (sandwich)(MPa)	39.2	153
Maximum Translational displacement (mm)	2.16	5.04

Fluid analysis is conducted to observe the flow patterns and to measure pressure loss coefficient of this butterfly valve when the butterfly valve works with various opening positions [9]. The boundary condition of a pipeline was created to analyze the flow of fluid through the different opening of butterfly valve. The fluid medium that was used for this simulation is water with a density of 998 kg/m^3. The valve opening is measured between 10° to 90° with an increment of 10° for each simulation. The properties used in this analysis is water with a constant temperature of 30°C. The vapor pressure on 30°C is 4.243 kPa. According to the specification of American Water Works Association, the maximum velocity through this valve cannot be more than 4.8768m/s. Hence, a uniform velocity of 4.8768m/s is imposed at the inlet boundary.

Figure 19. Valve opening at 30 degree

Figure 20. Valve opening at 50 degree

The data of the upstream pressure and downstream pressure was recorded on the table 3. The pressure differential was abruptly decrease based on the increasing of the opening angle between 10° to 40°. The pressure differential the decrease gradually from the valve opening between 50° to 90°. The upstream pressure then decrease whenever the opening of the valve is increase. This can be seen on table 3. The pressure differential that was analyzed based on the graph plotted on figure 21. The graph represent the pressure drop based on the positon of the opening of the valve.

Table 3. Recorded data of each valve opening

Valve opening	Upstream Pressure (pascal)	Downstream Pressure (pascal)	Pressure Differential (pascal)	Cavitation Index
10	623330	232421.31	390908.88	1.58
20	439971	253880.69	186091	2.34
30	337780	220774.97	117005.19	2.85
40	256276	197373.92	58902.39	4.27
50	221572	178831.86	42740.31	5.08
60	201134	182526.92	18607.49	10.58
70	190308	182957.56	7351.21	25.31
80	184636	178690.94	5945.61	30.34
90	184736	182175.39	2561.08	70.47

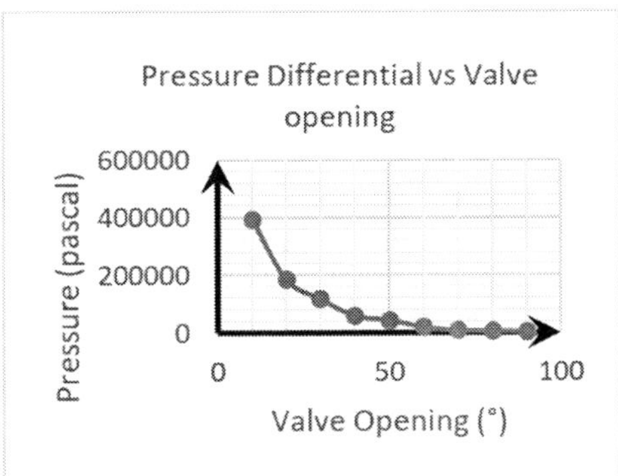

Figure 21. The pressure differential

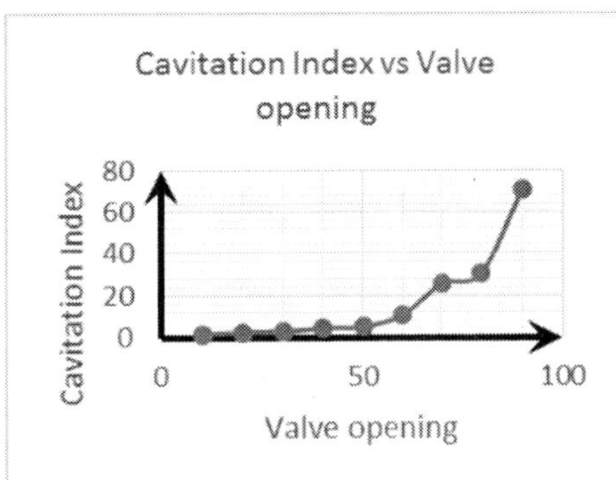

Figure 22. Cavitation index

The upstream and downstream pressure differential was used to analyze the cavitation index of the valve in each of the opening position. Using the reference of figure 4, the cavitation index performance was analyzed through plotting the graph on figure 22. Based on the graph plotted in figure 22 and the data from table 3, the performance of the valve can be estimate with the cavitation index based on the reference from figure 4. From the comparison, the cavitation index at the valve opening of 10°, 20° and 30° are in between the incipient and critical region. However the cavitation index for that three opening positions are much closer to the incipient. Valve opening at 40° through 90° are above the incipient region. It means that the cavitation index is on safe operating zone. The opening at the incipient cavitation region typically does not cause damage or objectionably loud noise.

4.3. Computer Aided Manufacturing (CAM)

The butterfly valve's disc model that was designed using the application of CAD. The file was save "STP" format for a compatibility towards the 3D printer. The purpose of CAM process is to analyze the structure in form of prototyping. The model was downscaled with a ratio of 1:16. Figure 23, figure 24 and figure 25 show the components of the butterfly valve.

Figure 23. Disc prototype

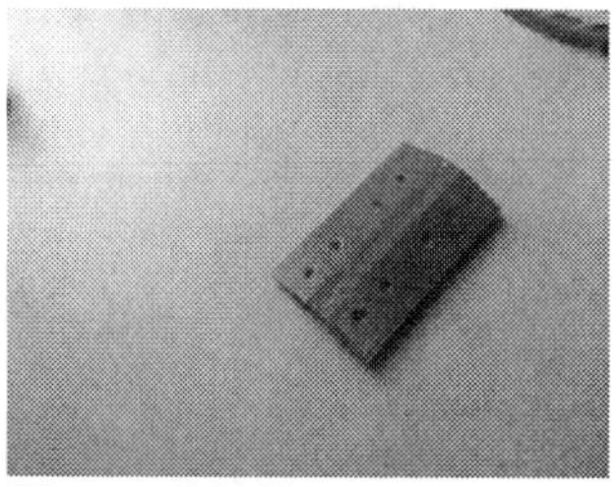

Figure 24. Connector prototype

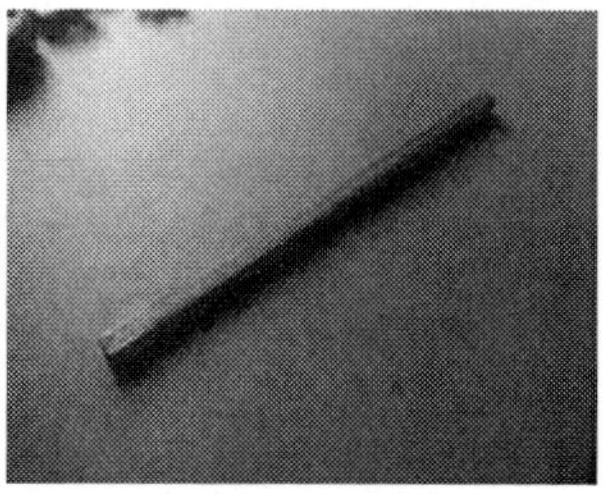

Figure 25. Shaft prototype

The 3D printer shows the materials used for fabricate the prototype. The materials used is a polylactic acid which can be fermented from crops such as maize. The 3D printer has no problem printing complex shape especially the sandwich layer. Moreover it is a fast and simple process when it comes to operation. Table 4 shows the time required to fabricate the parts and weight of the prototype respectively.

Table 4. Rapid prototyping details process

Part	Time (m)	Weight (g)
Disc	29	5
Shaft	100	17
Connector	19	3

5. Conclusion And Recommendation

The study has successfully shown that the application of CAD, CAE and CAM was able to design, analyze and fabricate the prototype of sandwich disc butterfly valve. The design of the valve structure has been modulated using the application of CAD. The design based on influences and considerations that will improved the performance of the structure. The application of CAE helps to analyse the stress concentration and the stress configuration region. A comparison analysis was done using the finite element to justify the performance of the structure. The second application of CAE is the computational fluid flow analysis. The upstream pressure and the downstream pressure was analysed to calculate the cavitation index and determine the performance throughout each opening position of the valve. The CAM process was done using rapid prototyping to produce a prototype and analysed the structure in form of prototype. The structure was fabricated based on the model designed initially through the application of CAD. The structure was downscale to fit the area required based on the specification of the 3D printer. The polyhedral mesh is a core volume mesh model that dictates the main aspects of the entire mesh to be constructed. Polyhedral cells created typically have an average of 14 cell faces and provide a balanced solution. A large advantage that the polyhedral meshing model has compared to tetrahedral meshes is that they are relatively easy and efficient to generate, and contain approximately five times fewer cells than a tetrahedral mesh, thus alleviating computational burdens. In order to properly apply computational fluid analysis, the pipe length downstream from the installed butterfly valve must be long enough that the flow has become fully developed so as not to prematurely force the flow to a zero gradient condition.

6. References

[1] Akiri, 2012 Disk Failure of Butterfly Valve. Petronas Valve and Actuator.
[2] American Water Work Association 2015 Rubber-Seated Butterfly Valve . ANSI/AWWA C504-15 11.
[3] Association A W 2012 Butterfly Valves: Torque,Head Loss and Cavitation Analysis. West Quincy Avenue: Knovel.
[4] Batchu S 2015 Solid Metal Versus Sandwich Panels *Composite Materials Application* 15
[5] Brochure. 2013 Hobbs Valves (Caerphilly: Trecenydd Business Park)
[6] Control J 2008 Valve and Actuator Manual , *Valve Basics and Sizing Information Section Engineering Data Book Section Vb2: Butterfly Valves*, (Milwaukee: Johnsion Inc) 6.
[7] Dickenson T 1999 *Valve,Piping & Pipeline Handbook 3rd Edition* (Amsterdam: Elsevier Advance Technology) 6
[8] Francis T 1992 *Introduction of CAM: Organizational change and innovation* 240
[9] Kuzmin D 2007 *Introduction to Computational Fluid Dynamics* (Dortmund:Institute of Applied Mathematics)
[10] Nilsen V B 1991 *Naval Engineer Journal* 47
[11] OSHA E A 1997 *Chemical Safety Alert* 46
[12] Plantema F J 1966 *Sandwich Construction: The Bending and Buckling of Sandwich Beams, Plates, and Shells,* (New York: Jon Wiley and Sons) 8
[13] Subhash Verma V K 2006 Water Supply Engineering (Mumbai: Vikas) 32

The RSM approach to develop a new correlation for density of metal-oxide aqueous nanofluids

E Montazer*, E Salami, H Yarmand, S N Kazi*, A Badarudin

Department of Mechanical Engineering, Faculty of Engineering, University of Malaya, Kuala Lumpur 50603, Malaysia

E-mail: *elham.montazer@gmail.com and salimnewaz@um.edu.my

Abstract
Density is one of the basic thermo-physical properties of a fluid that should be analysed before conducting performance analysis of the fluid. In the present study, the influence of design parameters on the density of nanofluids was investigated using response surface methodology (RSM). The quadratic model generated by RSM is used to predict the performance parameters, i.e. temperature and mass concentration with reasonably good accuracy. Improved empirical correlations were proposed based on the experimental data for density prediction of the metal oxide nanofluids. Experimentally measured densities of two different nanofluids at the nanoparticle mass fraction of up to 0.2% and the temperature range of 20°C-40°C were examined. The enhancement of densities compared to the density of base fluid at 20°C and 40°C, are about 0.217% for 0.2% fraction of ZnO nanoparticles. In addition, the densities of SiO_2 nanofluids are improved 0.117% compared to the density of distilled water. Finally the RSM outputs were compared with the results obtained from the experimental data. It was observed that the optimal RSM model is accurate and the absolute maximum deviations measured values from the predicted densities of ZnO and SiO_2 nanofluids are 0.008% and 0.014% respectively.

1. Introduction

Nanofluids were invented by Choi [1], after understanding mixing nanoparticles (1-100 nm) with a base fluid improves heat transfer. Materials generally used as nanoparticles include chemically stable metals (e.g. gold, copper), metal dioxides (e.g., alumina, silica, zirconia, titania), dioxide ceramics (e.g. Al2O3, CuO), metal carbides (e.g. SiC), metal nitrides (e.g. SiN), carbon in various forms (e.g., diamond, graphite, carbon nanochannels, fullerene) and functionalized nanoparticles, Cheng [2].
In the previous decade, scientists had tested different kinds of nanomaterials for making nanofluids. Al_2O_3, SiO_2, ZnO, CuO, zirconia and silver are more common metal and metalloid oxides for preparing nanofluids compared to other metal oxide nanofluids [3-6].

Although nanofluids have high potential to improve heat transfer compared to water, more pumping power is required in order to achieve a desired flow rate. This is due to the fact that nanoparticles, which are used to make nanofluids are weighty. It is important to understand certain fluid properties that are essential for thermal performance. Such properties are primarily density, viscosity, and thermal conductivity.

Amongst the different thermo-physical properties of nanofluids, viscosity and thermal conductivity were more interested for the researchers [7-10]. A few number of researchers worked on other properties such as specific heat and density. Viscosity and thermal conductivity are not enough to evaluate heat transfer coefficient theoretically. Specific heat and density are also essential for heat transfer information. Since these properties have significant influence on nanofluid flow and heat transfer characteristics, so they should be determined accurately [11]. Therefore, some density measurements have been presented for different nanofluids at various conditions [12], while suggesting the reported theoretical correlation for the estimation of nanofluids density is based on simple model [13]. Nanoparticle concentration increment displays the role of the interaction on the fluid density since the effect of particle size and size distribution on density is almost negligible [11]. Attempt to obtain models based on accurate experimental data in a proper manner would be a powerful technique for understanding and predicting a process. In these cases, novel approaches such as a response surface methodology (RSM) and an artificial neural network (ANN) are efficient in predicting the density of nanofluids based on the available data [14].

Recently, Sha and Edwards [15] comprehensively investigated the applications of ANNs in material science. Hassan et al. [16] predicted the density, porosity, and hardness of an aluminium-copper based composite material using ANNs. Singh et al. [17] applied ANNs to predict the effective thermal conductivity of moist porous materials. A review paper [18] showed that the implementation of ANN technique has greatly promoted research in materials science and technology. Karimi and Yousefi [14] applied the hybrid model, including back-propagation network (BPN) and genetic algorithm (GA) to estimate the nanofluids density. Sadollah et al. [19] investigated the effect of different process factors on the stability of nanofluid by using optimization methods. They had compared the results predicted by utilizing RSM and ANNs. Thus the purpose of this paper is to minimize the total mean squared error between net output and desired output through optimization. The presented model determines the effective density of the nanofluids as a function of temperature, nanoparticle mass fraction, nanoparticle and the base fluid densities.

2. Methodology:

2.1. Materials

Silicon dioxide (SiO_2) nanopowder with assay 99.5% and particulate size ranges from 10 to 20 nm were purchased from Sigma-Aldrich Co., Selangor, Malaysia. SiO_2 nanofluid has been prepared by dispersing nanoparticles in base fluid. Distilled water was used as a base fluid for the preparation of SiO_2 nanofluids. An ultrasonic disruptor has used to produce nanofluids from the specific nanoparticles and the selected base fluids. No surfactant additives were added during the preparation process. In the present case after 30 minutes of intensive sonication, the stable suspensions were obtained according to the desired mass concentrations (wt.%) of SiO_2.

Zinc oxide (ZnO) spherical shaped nanopowder with assay 97% and particulate size was less than 50 nm were purchased from Sigma-Aldrich Co., Selangor, Malaysia. The ZnO nanoparticles sedimented within several minutes as zinc oxide is insoluble in water and also, the particles remained within the clusters without being dispersed. Enhanced stability of the ZnO nanofluids is observed with sodium hexametaphosphate (*SHMP*). *SHMP* with purity 96% was purchased from Sigma-Aldrich Co., Selangor, Malaysia. After 60 minutes of intensive sonication, the stable suspensions were obtained according to the desired mass concentrations (wt%) of ZnO. The most stable suspension as per our sedimentation studies contained surfactant and nanoparticles in wt% ratio 1:10 (*SHMP: ZnO*).

2.2. Experimental works

The density of a fluid is defined as a ratio of the mass to the volume, which it occupies (equation 1).

$$\rho = m/v \tag{1}$$

The density of the liquid is higher than the density of gas for the same quantity of the fluid, because gases are more elastic than liquid and therefore they tend to occupy the maximum space (volume).

The densities of distilled water and nanofluids were measured experimentally by KEM Density/Specific Gravity Meter DA-645. The accuracy of density measurement and temperature are 0.00005 g/cm^3 and 0.3°C respectively. The measurements were recorded 3 times for each temperature. The average results show in figure 1 and figure 2 which are display the density measurements of distilled water, SiO_2 and ZnO nanofluids for different weight concentrations at the temperature range of 20°C to 40°C.

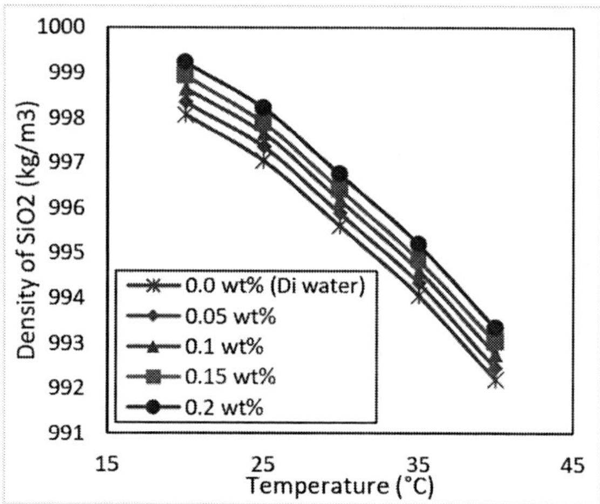

Figure 1. Density of *SiO2* nanofluids as a function of temperature at different weight concentrations.

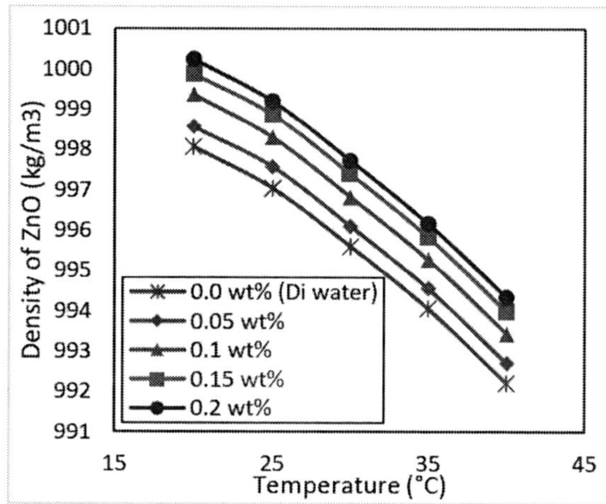

Figure 2. Density of *ZnO* nanofluids as a function of temperature at different weight concentrations.

Nanofluids exhibit higher density compared to the conventional fluid such as water. The density of nanofluids depends on the concentration of the nanoparticles. Density of a fluid decreases with the increasing of its temperature and also increases with the increasing of its pressure. It is an important property of the fluid. Based on the density, the flow of the fluid can be classified as either incompressible or compressible. For incompressible flow, the density of the fluid is constant; whereas for compressible flow the density changes with respect to the temperature and pressure of the fluid.

2.3. RSM optimization Process

RSM is an empirical statistical modeling technique employed for multiple regression analysis using quantitative data obtained from properly designed experiments to solve multivariate equations simultaneously [20, 21]. RSM is used to determine the optimum density, for the nanofluids. A central composite full of three independent variables was used. The independent variables selected for the optimization were temperature, mass concentration and density ratio of the base fluid to the nanoparticles. Regression analysis was performed on the data obtained from the experiments. Coding of the variables was done according to the following equation (equation 2):

$$x_i = \frac{X_i - X_{CP}}{\Delta X_i} \qquad i = 1,2, \ldots, k \tag{2}$$

Where x_i, is the dimensionless value of an independent variable; X_i, refers to the real value of an independent variable; X_{CP}, is the real value of an independent variable at the center point; and ΔX_i, is the step change of the real value of the variable i corresponding to a variation of a unit for the dimensionless value of the variable i. The experiments were carried out in duplicate, which was necessary to estimate the variability of the measurements. Replicates at the center of the domain in three blocks permit the checking of the absence of bias between several sets of experiments. The relationship of the independent variables and the response was calculated by the second-order polynomial equation (3).

$$Y_{Pre} = \alpha_0 + \sum_{i=1}^{k} \alpha_i X_i + \sum_{i=1}^{k-1} \sum_{j=i+1}^{k} \alpha_{ij} X_i X_j + \sum_{i=1}^{k} \alpha_{ii} X_{ii}^2 \quad (3)$$

Where, Y_{Pre} is the predicted response; α_0 a constant; α_i the linear coefficient; α_{ii} the squared coefficient; and α_{ij} the cross-product coefficient, k is the number of factors.

The mean squared error (MSE) was used to quantify the difference between predicted and actual values. The MSE can be expressed as follows:

$$MSE = \frac{1}{k} \sum_{i=1}^{k} (Y_{Pre} - Y_{Exp})^2 \quad (4)$$

The response surface method is a sequential process and its procedure can be summarized as shown in Figure 3.

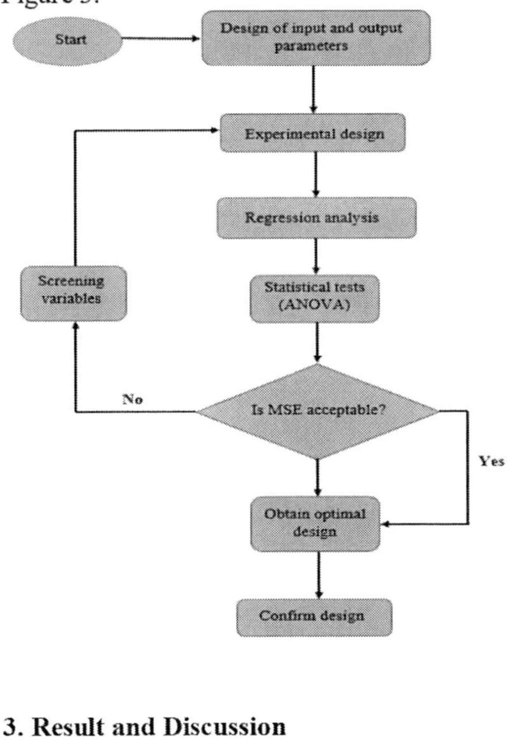

Figure 3. Flowchart of RSM modeling approach for optimal design

3. Result and Discussion

The response surface methodology is employed using Minitab Version 17.3.1 software for the optimization of a problem with 3 optimization variables. A quadratic response surface is the approximated density for SiO_2 and ZnO based on the Least Square method. Weight fraction ($\emptyset$), Temperature (T), base fluid density (ρ_f) and density ratio of base fluid to nanoparticle (ρ_f/ρ_{np}) are input variables and nanofluid density (ρ_{nf}) is the output variable.

The quadratic equation for finding the best-fitted curves to experimental data points for SiO_2 and ZnO are represented by equations 5 and 6, respectively.

$$\rho_{Pre(Sio2)} = 1000.02 + 5.182\ \emptyset + 0.0055\ T + 2.2\ \emptyset^2 - 0.005\ T^2 - 0.0008\ \emptyset T \quad (5)$$

$$\rho_{Pre(ZnO)} = 999.640 + 21.112\ \emptyset - 0.001\ T - 39.60\ \emptyset^2 - 0.005\ T^2 - 0.011\emptyset T \quad (6)$$

Where, T is a temperature and $\emptyset$ is the nanoparticle weight fraction.

Thus the trained of the net model has used to predict the nanofluid density and the comparison of the predicted and the experimental values are presented here (figure 4 & figure 5). As it can be seen that good agreements were achieved between the predicted and the experimental values of the densities with the mean squared error of MSE = 0.251% for SiO_2 and MSE = 1.24% for ZnO nanofluids and the high correlation coefficients of R = 0.9992 for SiO_2 and R = 0.9986 for ZnO.

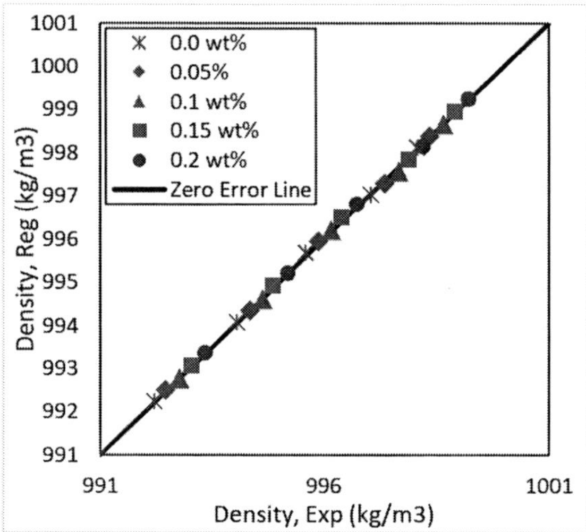

Figure 4. Validity of the modeling of the RSM optimization method to predicate the density of SiO_2 nanofluid according to Equation (5).

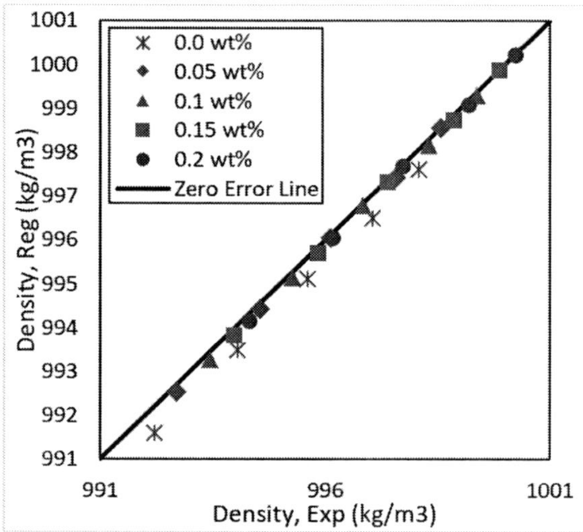

Figure 5. Validity of modeling of the RSM optimization method to predicate of the density of ZnO nanofluid according to Equation (6).

4. Conclusions

In this study the response surface methodology has successfully been applied for estimating the densities of SiO_2 and ZnO nanofluids. The proposed model could evaluate the densities as a function of temperature, nanoparticle mass fraction, base fluid densities and the density ratios of the base fluid to the nanoparticles. Based on the experimental data, the trained net has shown good agreement between the predicted and the experimental values of the densities with the mean squared error, (MSE) and the high correlation coefficients, (R) for SiO_2 nanofliuds. Similarly, acceptable R and MSE were achieved for ZnO nanofluids. Furthermore, experimental data demonstrate the density increases with the increase of mass concentrations and decreases with the increase of temperatures following the conventional trend of nanofluids.

Acknowledgment

The authors gratefully acknowledge High Impact Research Grant UM.C/HIR/MOHE/ENG/45 and UMRG Grant RP012D-13AET, University of Malaya, Malaysia, for support to conduct this research work.

IOP Conf. Series: Materials Science and Engineering **210** (2017) 012071 doi:10.1088/1757-899X/210/1/012071

References

[1] Chol S 1995 *ASME-Publications-Fed* **231** 99-106

[2] Cheng L 2009 *Recent Patents on Engineering* **3** 1-7

[3] Yarmand H, Gharehkhani S, Ahmadi G, Shirazi S F S, Baradaran S, Montazer E, Zubir M N M, Alehashem M S, Kazi S and Dahari M 2015 *Energy Conversion and Management* **100** 419-28

[4] Khoshvaght-Aliabadi M 2014 *Energy conversion and management* **88** 96-105

[5] Yarmand H, Ahmadi G, Gharehkhani S, Kazi S N, Safaei M R, Alehashem M S and Mahat A B 2014 *Entropy* **16** 6116-32

[6] Peyghambarzadeh S, Hashemabadi S, Chabi A and Salimi M 2014 *Energy Conversion and Management* **86** 28-38

[7] Ren Y, Xie H and Cai A 2005 *Journal of Physics D: Applied Physics* **38** 3958

[8] Mintsa H A, Roy G, Nguyen C T and Doucet D 2009 *International Journal of Thermal Sciences* **48** 363-71

[9] Chandrasekar M, Suresh S and Bose A C 2010 *Experimental Thermal and Fluid Science* **34** 210-6

[10] Papari M M, Yousefi F, Moghadasi J, Karimi H and Campo A 2011 *International Journal of Thermal Sciences* **50** 44-52

[11] Maiga S E B, Palm S J, Nguyen C T, Roy G and Galanis N 2005 *International Journal of Heat and Fluid Flow* **26** 530-46

[12] Pastoriza-Gallego M, Casanova C, Legido J a and Piñeiro M 2011 *Fluid Phase Equilibria* **300** 188-96

[13] Vajjha R, Das D and Mahagaonkar B 2009 *Petroleum Science and Technology* **27** 612-24

[14] Karimi H and Yousefi F 2012 Application of artificial neural network–genetic algorithm (ANN–GA) to correlation of density in nanofluids *Fluid Phase Equilibria* **336** 79-83

[15] Sha W and Edwards K 2007 *Materials & design* **28** 1747-52

[16] Hassan A M, Alrashdan A, Hayajneh M T and Mayyas A T 2009 *Journal of materials processing technology* **209** 894-9

[17] Singh R, Bhoopal R and Kumar S 2011 *Building and Environment* **46** 2603-8

[18] Sumpter B G and Noid D W 1996 *Annual Review of Materials Science* **26** 223-77

[19] Sadollah A, Ghadimi A, Metselaar I H and Bahreininejad A 2013 *Science and Engineering of Composite Materials* **20** 319-30

[20] Montazer E, Mirzaei M, Salami E, Ward T, Romli F and Kazi S 2016 *IOP Conference Series: Materials Science and Engineering*: IOP Publishing) p 012023

[21] Rao K J, Kim C-H and Rhee S-K 2000 *Process Biochemistry* **35** 639-47

IOP Conf. Series: Materials Science and Engineering **210** (2017) 012072 doi:10.1088/1757-899X/210/1/012072

Analytical study on different blade-shape design of HAWT for wasted kinetic energy recovery system (WKERS)

J B Goh*[1,2], Z Jamaludin[2], F A Jafar[2], M Mat Ali[2], M N Ali Mokhtar[2], and C H Tan[3]

[1] Centre of Graduates Studies, Universiti Teknikal Malaysia Melaka, Hang Tuah Jaya, 76100 Durian Tunggal, Melaka, Malaysia
[2] Faculty of Manufacturing Engineering, Universiti Teknikal Malaysia Melaka, Hang Tuah Jaya, 76100 Durian Tunggal, Melaka, Malaysia
[3] Sales Department, Technojis Enterprise Sdn. Bhd., Setiawalk, Persiaran Wawasan, 47160, Puchong, Selangor, Malaysia

E-mail: *Jaybee-Goh@hotmail.com and M051510046@student.utem.edu.my

Abstract. Wasted kinetic energy recovery system (WKERS) is a wind renewable gadget installed above a cooling tower outlet to harvest the discharged wind for electrical regeneration purpose. The previous WKERS is operated by a horizontal axis wind turbine (HAWT) with delta blade design but the performance is still not at the optimum level. Perhaps, a better blade-shape design should be determined to obtain the optimal performance, as it is believed that the blade-shape design plays a critical role in HAWT. Hence, to determine a better blade-shape design for a new generation of WKERS, elliptical blade, swept blade and NREL Phase IV blade are selected for this benchmarking process. NREL Phase IV blade is a modern HAWT's blade design by National Renewable Energy Laboratory (NREL) research lab. During the process of benchmarking, Computational Fluid Dynamics (CFD) analysis was ran by using SolidWorks design software, where all the designs are simulated with linear flow simulation. The wind speed in the simulation is set at 10.0 m/s, which is compatible with the average wind speed produced by a standard size cooling tower. The result is obtained by flow trajectories of air motion, surface plot and cut plot of the applied blade-shape. Besides, the aspect ratio of each blade is calculated and included as one of the reference in the comparison. Hence, the final selection of the best blade-shape design will bring to the new generation of WKERS.

1. Introduction

The world is full of energy. Most of the energy are extracted from the fossil fuels (eg: coal, natural gas, petroleum etc.), which are also the primary source for energy in the world [1]. Overharvesting of fossil fuels brings negative impacts on human health and environment by the emissions of greenhouse gas [2]. For developing sustainability in long term, the utilization of renewable energy such as solar, wind, rain, tides and waves are encouraged. To minimize the negative impacts on energy supply chain in Malaysia, renewable energy is considered as the fifth energy under the Fifth Fuel Policy which was launched in 2000 [3].

Renewable energy industry is still new and unfamiliar in Malaysia; the professionality of the local experts is still a distant to what the country needs. The common renewable energy available in Malaysia are solar power, hydropower, and wind power [4,5]. However, Malaysia is not a great geographical location for expansion of wind power since the wind speed is unstable. National Wind

Watch defines that the cut-in speed of a standard modern HAWT is about 3 m/s to 4 m/s (7-9 mph), but the average annual wind speed in Malaysia is analyzed to be less than 2.0 m/s [6–8]. On the other hand, a standard industrial cooling tower can produce a constant wind speed up to 11.0 m/s. The blade design is the most important part to be investigated to harvest the best lift effect from the cooling tower outlet. Therefore, three different blade-shape designs are analyzed in this study.

2. New HAWT's Blade-shape Design

The three blade-shape designs are elliptical blade, swept blade, and NREL Phase VI blade. The ideas of both elliptical and swept blade-shape are from the aircraft wing, whereas the NREL Phase IV blade is from a standard modern HAWT. To complete this research work, three blade-shapes are designed and analyzed then further compared to select the best design for future research work.

2.1. The Consideration of Three Blades HAWT

The number of blade is one of the important issues that need to be considered before developing a wind turbine [9,10]. A turbine with only one blade will be the cheapest costing for development, but the presence of unbalancing is one of the problem. Anyway, the one-bladed turbine is able to rotate at high tip-speed ratio. The even number of blade will affect the stability of the turbine and has a lower inertia moment. Thus theoretically, the lesser the number of blade, the higher the aerodynamical efficiency configuration of a wind turbine [9,11], but is not advisable in reality due to its stability issue.

Rationally, wind turbine with two blades is better to overcome the stability problem, but the wind turbine itself will face a wobbling phenomenon. In other words, it is similar to gyroscopic precession. Since the wind turbine is facing the wind all the time, the blades direction need to change perpendicularly to the wind direction. So, according to yawing theory, when the blades are in line with the tower and the axis of rotation (vertical position), the resistance to the yawing force is very low. On the other hand, when the blades are not in line with the tower (horizontal position), the blades spin at longer distance from the axis of rotation so it experiences much resistance to yawing motion. Therefore, the yawing motion is experiencing starts and stops twice for every revolution, and this causes the turbine to stress due to blade chattering [11]. Figure 1 shows the graph of power coefficient Cp variation against tip-speed ratio for the computational result of two, three and four blades of Savonius wind turbine.

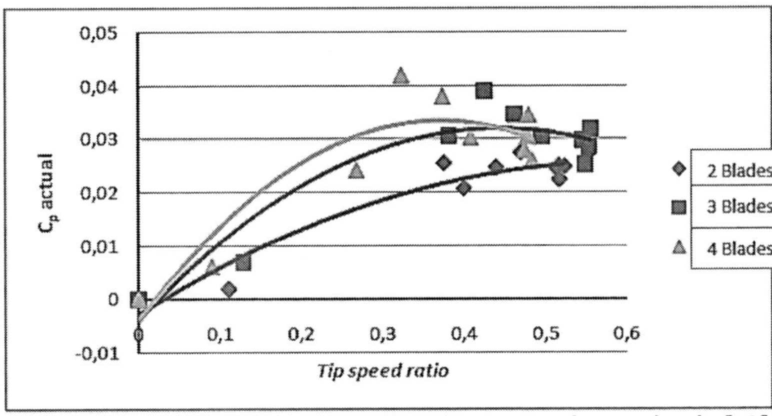

Figure 1. Power coefficient C_p variation versus tip speed ratio [10].

The wind turbine with four blades perform well at lower tip-speed ratio, but the three blades design has the best performance at high tip-speed ratio. In other words, the four-bladed design is the best choice for wind speed of 6 m/s; but when coming to the wind speed range of 6 m/s to 8 m/s, the three blades turbine will have the advantages [10]. Besides, the three-bladed wind turbine will only have a tiny vibration or chatter. This is because the resistance to the yawing motion will be balanced by two

IOP Conf. Series: Materials Science and Engineering **210** (2017) 012072 doi:10.1088/1757-899X/210/1/012072

blades when one blade is at the horizontal position (i.e. not in line with the axis of rotation and the tower) [11]. Therefore, a total number of three blades is proven to become a criterion wind turbine for the current industries based on the aerodynamic concept [9–12]. So, a wind turbine with three blades shows the perfect combination of high rotational speed and minimum stress.

2.2. Blade-shape Design

In this study, two aircrafts' wing design (elliptical & swept) and one modern HAWT's blade design (NREL Phase VI) are selected for the benchmarking analysis. Elliptical wing is the best for wind gliding purpose especially applied on Supermarine Spitfire fighter aircraft, which using aerofoil NACA 2209.4. Meanwhile, the swept wing on Boeing 747-8 commercial airplane using aerofoil BACxxx-il is the largest airbus until now. Even though Boeing 747-8 is huge and heavy, but its take off time and length is shorter compared to others. As mentioned, for the third blade-shape, NREL Phase VI is the exact design of a modern HAWT which is referring to aerofoil S809. For a better view, the 3D wireframe and solid modelling of three blade-shapes design are drafted in Figure 2, Figure 3 and Figure 4.

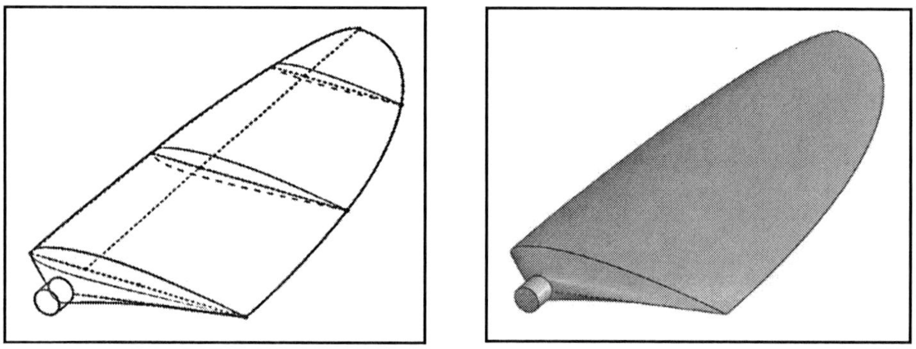

(a) (b)

Figure 2. 3D Wireframe (a) and Solid (b) Modelling of Elliptical Blade-shape Design.

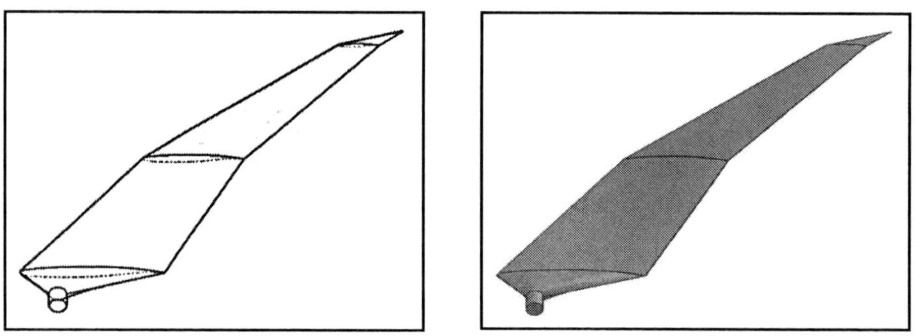

(a) (b)

Figure 3. 3D Wireframe (a) and Solid (b) Modelling of Swept Blade-shape Design.

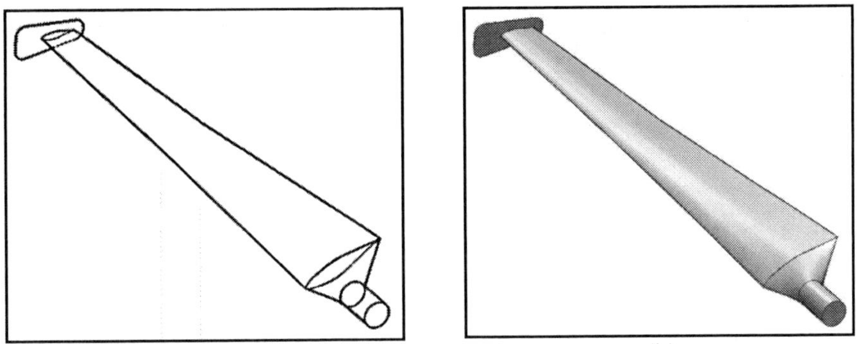

(a) (b)

Figure 4. 3D Wireframe (a) and Solid (b) Modelling of NREL Phase VI Blade-shape Design.

There are some standard formulae needed for calculation of benchmarking in the analytical process. For example, to calculate the aspect ratio, AR as in Equation (2), the value of wingspan, b and chord, C are needed in Equation (1);

$$S = \frac{\pi \times b \times C}{4} \tag{1}$$

where S is the wing area of the blade, MAC illustrates the mean aerodynamic chord shown in Equation (4).

$$AR = \frac{b^2}{S} = \frac{b}{MAC} \tag{2}$$

For some reasons, the blade-shape is tapered to its tip, so the Equation (3) is needed for calculating taper ratio, λ for the blade. Hence, in order to find the length of MAC as in Equation (4) for swept blade and NREL Phase VI blade, the taper ratio is calculated;

$$\lambda = \frac{C_{Tip}}{C_{Root}} \tag{3}$$

$$MAC = C_{Root} \times \frac{2}{3} \times \left[\frac{(1+\lambda+\lambda^2)}{(1+\lambda)}\right] \tag{4}$$

whereas the taper ratio is defined as the chord at the tip, C_{Tip} to the chord at the root, C_{Root}. After all, the calculated aspect ratio is included as one of the criteria for the overall analysis process, as well as the twisted idea for the blade design since it is proven to optimize the blade performance [12–15].

3. Linear Flow Simulation

Computational Fluid Dynamics (CFD) analysis was ran by using SolidWorks software, all the designs are simulated with linear flow simulation. Linear flow simulation ran through every single blade-shape designs with gas/air velocity of 10 m/s as the fix variable. This 10 m/s wind speed is obtained through a calculation of the average wind speed measured from the outlet of the cooling tower. The cooling tower outlet is divided into several concentric parts to obtain the average wind speed from each band of measuring point as shown in Figure 5.

To analyze the elliptical blade by 3D SolidWorks linear flow simulation, all the desired information such as air velocity, air direction and air density were inserted into the project simulation. In this study, force is the main parameter for measuring the pressures applied on the surface of the blade. When each simulation had successfully run, a few features such as flow trajectories of air

IOP Conf. Series: Materials Science and Engineering **210** (2017) 012072 doi:10.1088/1757-899X/210/1/012072

motion, surface plots and cut plot were selected for the purpose of showing a clear view of flow and pressure analysis on each part of the blade.

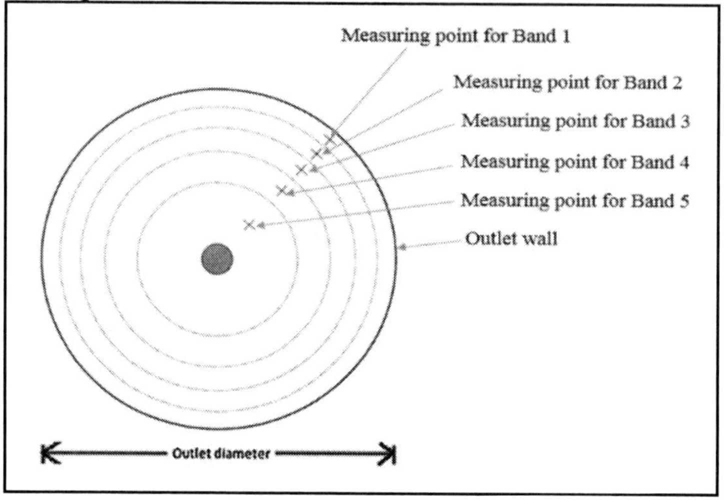

Figure 5. The Methodology of Measuring Discharged Air Velocity for each Band at the Cooling Tower Outlet.

3.1. *Elliptical Blade Flow Discussion*

In order to demonstrate the study of linear flow simulation on the elliptical blade, the flow trajectories, surface plot and cut plot and static pressure were used to visualize the wind flow from leading edge to trailing edge through the surface of the elliptical blade. Referring to Figure 6 and Figure 7, the leading edge of the elliptical blade experienced a higher pressure (≥ 101.345 kPa) and lower pressure (101.295-101.310 kPa) on the blade body. However, there is a slight higher pressure compared to the blade body along the trailing edge (101.325-101.330 kPa), which is showing that drag forces had exerted on the blade before the wind pass through the blade surface. Anyway, from the cut plot in Figure 8, the pressure is uniform along the blade surface, so the elliptical blade is very good for gliding purpose. Consequently, this is the reason that elliptical blade is usually being used on aircraft to perform aerobatic maneuver show.

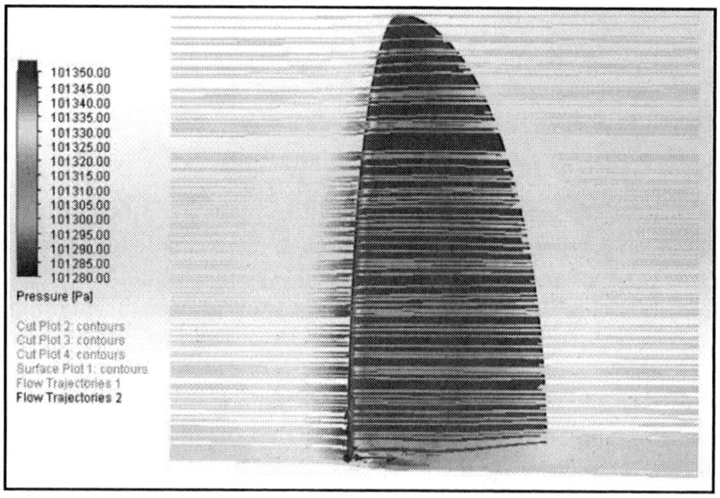

Figure 6. Flow Trajectories of Elliptical Blade.

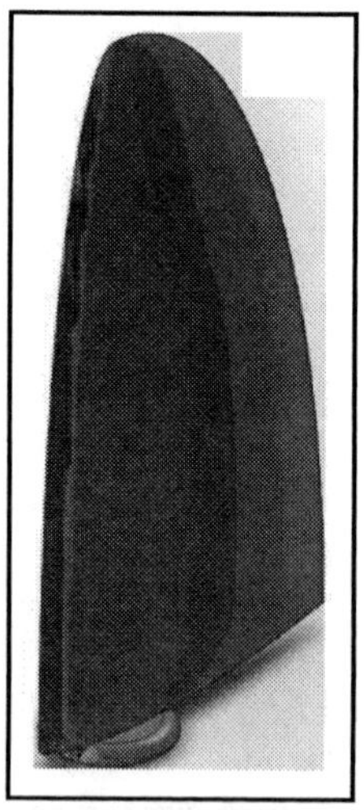

Figure 7. Surface Plot of Elliptical Blade.

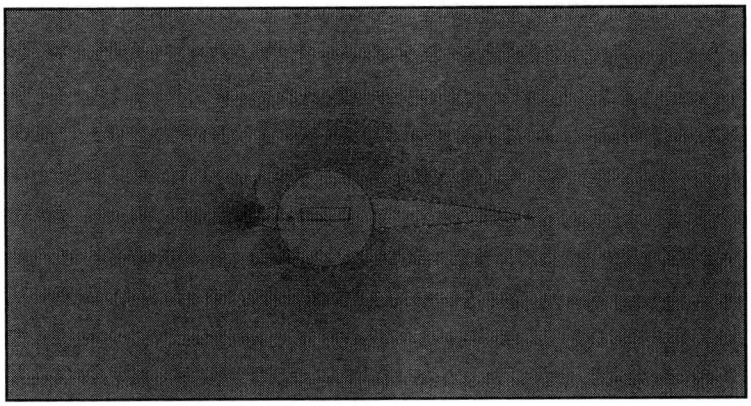

Figure 8. Cut Plot of Elliptical Blade.

According to the Bernoulli's Principal, the visualization of the flow trajectories, surface plot and cut plot are clear to be understood. However, a graphical method will help to interpret more information on the mechanical motion of the elliptical blade. Pressure analysis of linear flow simulation along the leading edge to trailing edge on the elliptical blade is plotted as in Figure 9, with the position of the particular edges as reference.

The line chart indicates the pressure between the leading edge and trailing edge. Based on the fluctuating of the line, Edge<1> has a higher pressure than Edge<2> as the air divides at the leading edge and flow neatly at the trailing edge. Besides, the line chart shows the pressure from the root chord to tip chord of elliptical blade is decreasing due to the different thickness. So, the thicker the aerofoil, the higher the pressure as the division of air occurs at the leading edge of the blade. After all, the line variation on Edge<2> is fairly consistent and shows the air flows smoothly along the trailing edge on elliptical blade.

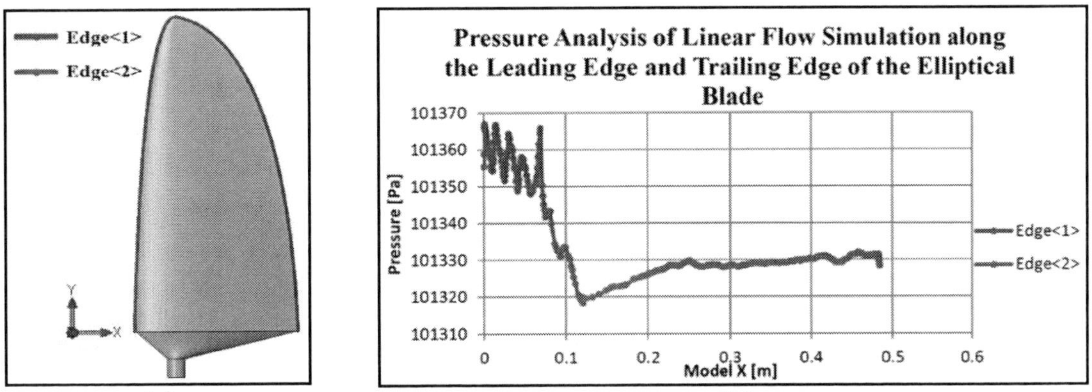

(a) (b)

Figure 9. Pressure Analysis of Elliptical Blade (b) with the Edges Reference (a).

3.2. Swept Blade Flow Discussion

The process of swept blade linear flow simulation is similar to elliptical blade. The result of flow trajectories, surface plot and cut plot for swept blade have been achieved as shown in Figure 10, Figure 11 and Figure 12. Referring to both Figure 10 and Figure 11, most leading edge and part of the trailing edge along the swept blade experience a high pressure ($\geq$ 101.328 kPa) compared to other parts. However, a large region of blade body experiences a low pressure ($\leq$ 101.32 kPa). According to

IOP Conf. Series: Materials Science and Engineering **210** (2017) 012072 doi:10.1088/1757-899X/210/1/012072

Figure 12, the performance of swept blade is ineffective while undergoing the air flow. This shows that swept blade is not advisable for the use of wind turbine due to the low wind speed source.

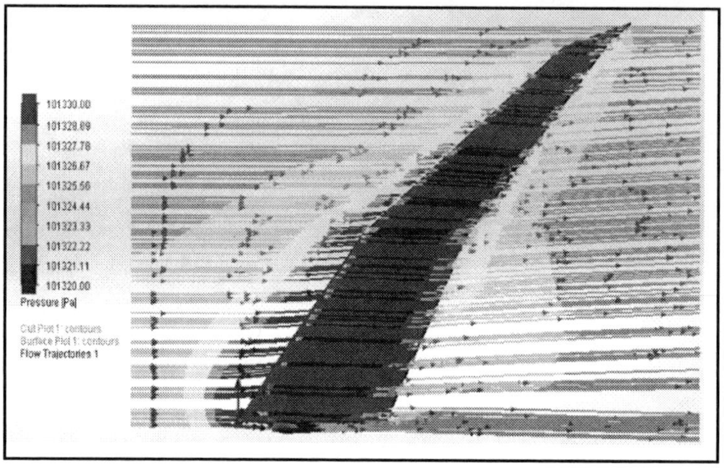

Figure 10. Flow Trajectories of Swept Blade.

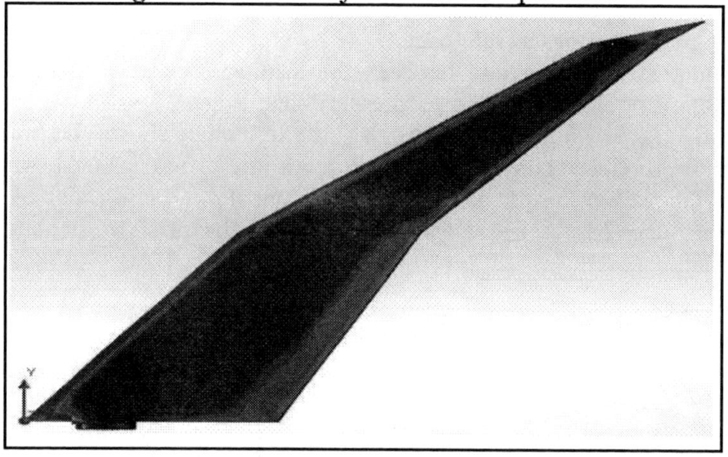

Figure 11. Surface Plot of Swept Blade.

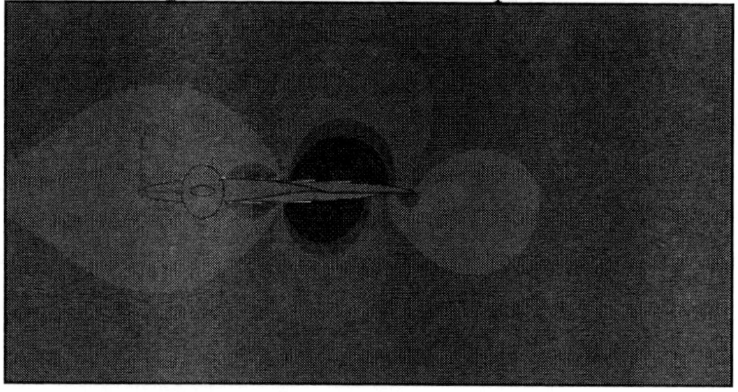

Figure 12. Cut Plot of Swept Blade.

In order to get a better view and understand the linear flow simulation of swept blade, Edge<5>, Edge<6>, Edge<7>, Edge<8> and Edge<9> play a critical role in the simulation process. Edge<5> presented in Figure 13 shows a high pressure occurring at the root chord of leading edge. However, the pressure is decreasing towards the tip chord along the leading edge. Edge<9> illustrates the pressure from the root chord slightly increasing to tip chord along the trailing edge. Lastly, the wingtip of the trailing edge showing pressure drop again which is presented by Edge<7>. This blade shows lower lifting force and greater dragging force compared to elliptical blade. Therefore, the swept blade can perform greatly on aircraft but may not be a suitable blade-shape design for wind turbine, since it needs a higher leading speed to provide lifting force.

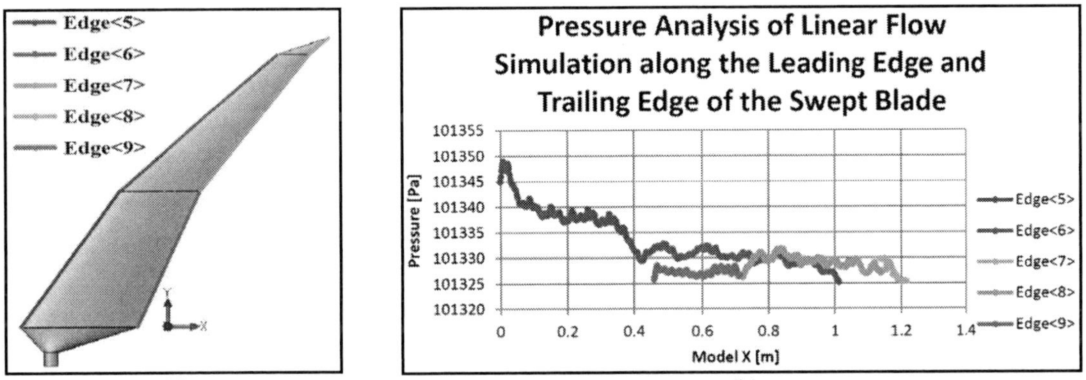

Figure 13. Pressure Analysis of Elliptical Blade (b) with the Edges Reference (a).

3.3. NREL Blade Flow Discussion

The same processes as both elliptical and swept blade are done to NREL Phase VI blade when gaining the linear flow simulation result. Start from the leading edge of NREL blade, the air spreads the flow to both suction and pressure surface at the moment it experiencs high pressure around 101.35 kPa then escapes with extremely low pressure which almost as low as 101.305 kPa through trailing edge as illustrated in Figure 14 and Figure 15. Based on the cut plot in Figure 16, the pressure at the leading edge is high but the pressure on the blade body and trailing edge is low and stable. Again, when referring to the trailing edge in Figure 14, the flow trajectories can be observed from root chord to tip chord which is effective smooth while passing through the blade surface.

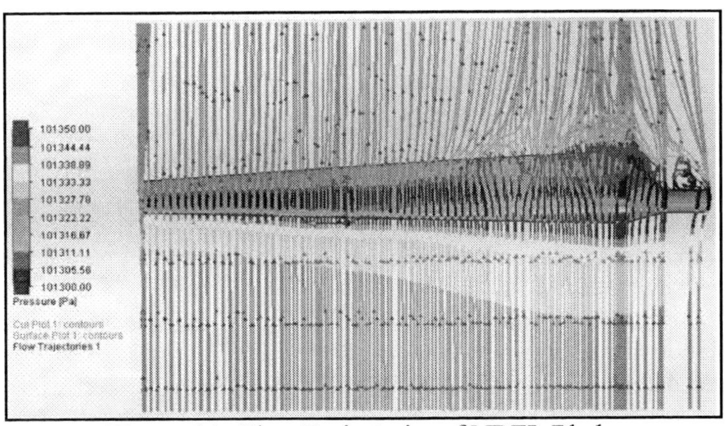

Figure 14. Flow Trajectories of NREL Blade.

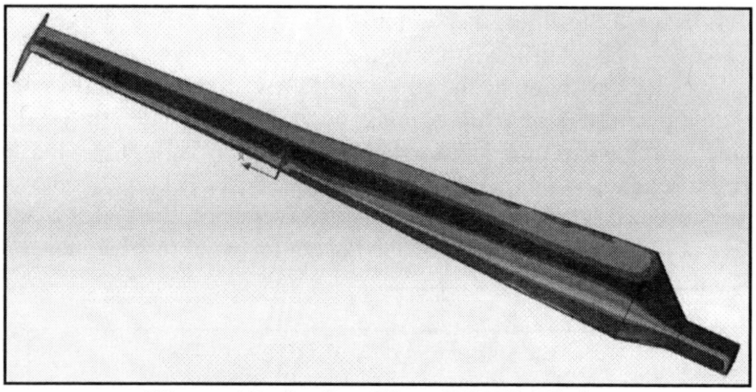

Figure 15. Surface Plot of NREL Blade.

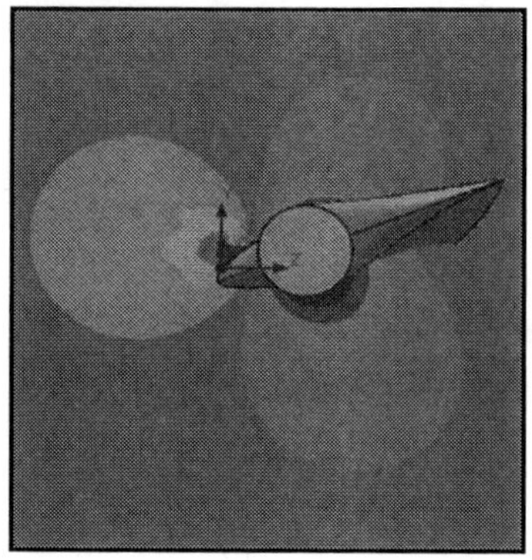

Figure 16. Cut Plot of NREL Blade.

In order to translate the animation of flow simulation, the graph of pressure analysis of linear flow simulation along the leading edge to trailing edge on NREL Phase Vi blade is produced as in Figure 17, together with the references of specific edges position. The line graph illustrated the Edge<1> with unusually high pressure experience along the leading edge, in the meantime, excessively low pressure along the trailing edge which presented in Edge<2>. Besides, the Edge<2> also show the pressure slightly decreasing near the root chord of trailing edge. Hence, there are some turbulence flow near the root chord as shown in Figure 14.

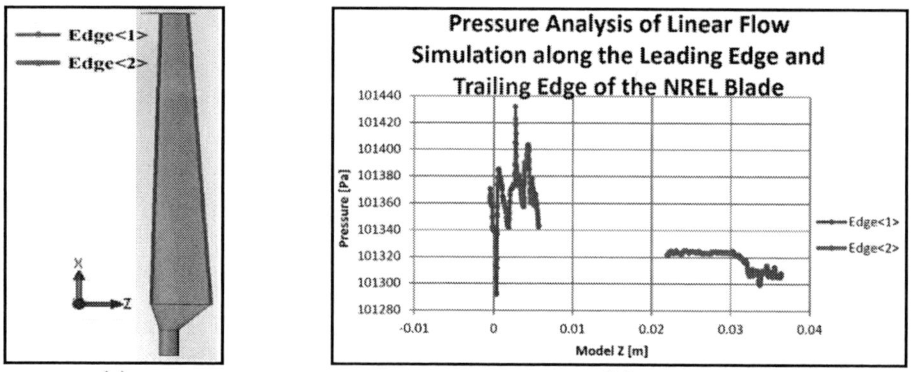

(a) (b)

Figure 17. Pressure Analysis of NREL Blade (b) with the Edges Reference (a).

4. Conclusion

This research work highlights the ability to maximize the horizontal axis wind turbine (HAWT) to improve the performance of wasted kinetic recovery system (WKERS) by benchmarking different blade-shape design. By analyzing the flow trajectories, pressure of surface plot & cut plot and pressure analysis graph along leading edge and trailing edge, NREL Phase VI blade is selected as the best blade design in this linear flow simulation as the lift is easily formed since it experiences the lowest pressure among three blades.

Moreover, NREL blade is the only blade which uses a wingtip and twisted blade in its design. Regarding to the supported documents by National Renewable Energy Laboratory, it is proven that the wingtip and twisted blade help the wind turbine itself to increase the performance. Therefore, it is believed that the winglet will reduce or block the vortices form at the wingtip in order to reduce the undesirable drag force. Therefore, this study conclude that the NREL Phase VI blade with a high aspect ratio, high lift and low drag will proceed to new revolution of WKERS.

In addition, with the selection of best blade design for WKERS, more energy can be harvested from the wind and more electricity can be generated. As a conclusion, WKERS with NREL Phase VI blade design is practical and can be implemented. Further enhancements and modifications will be carried out to improve the efficiency of the whale-inspired wind turbine:

- Conduct a rotational flow simulation for a better view of the wind flow before and after the WKERS.
- Organize another flow simulation by using another professional software to compare with the result obtained in this 3D SolidWorks design software.
- Apply the idea of whale-inspired design on the leading-edge of the blade to increase the performance of WKERS.
- Develop a real-scale prototype to validate the performance of WKERS.

Acknowledgement
The authors appreciate this product and its idea of WKERS is protected by legal copyright laws by the Intellectual Property Corporation of Malaysia under application number: LY2016000528. Thanks to the Center of Graduate Studies of Universiti Teknikal Malaysia Melaka (UTeM) for the support and approval of this study topic. Special appreciation is also credited to:

- Cabaran Inovasi Inklusif 2016 Zon Tengah/Selatan
- Malaysia Technology Expo (MTE 2016)
- The International Conference & Exposition on Inventions by Institutions of Higher Learning (Pecipta'15)
- UTeM Research & Innovation Expo (UTeMEX2015)

for affirmation of this project potential through the awards given.

References

[1] U.S. Energy Information Administration 2015 *Monthly Energy Review - December 2015* (Washington: U.S. Energy Information Administration)
[2] Chong W T, Hew W P, Yip S Y, Fazlizan a., Poh S C, Tan C J and Ong H C 2014 *Energy Convers. Manag.* **87** 145–55
[3] Chong C, Ni W, Ma L, Liu P and Li Z 2015 *energies* **8** 2828–66
[4] Azman A Y, Rahman A A, Bakar N A, Hanaffi F and Khamis A 2011 *2011 IEEE 1st Conf. Clean Energy Technol. CET 2011* 170–6
[5] Rahim N A, Saidur R and Wadi Abbas Al-Fatlawi A 2014 *3rd IET Int. Conf. Clean Energy Technol. 2014* 65
[6] Basil G 2013 Possibility of electricity from wind energy in Malaysia : Some rough *New Straits Times* 7
[7] Sopian K, Othman M Y H and Wirsat A 1995 *Renew. Energy* **6** 1005–16
[8] Najib A M, Mustafa M A, Jamaludin Z, Zulfattah A M and Hashim A R 2015 **699** 498–503
[9] Muluk S 2014 *Natl. High Sch. J. Sci.* 1–7
[10] Wenehenubun F, Saputra A and Sutanto H 2015 *Energy Procedia* **68** 297–304
[11] Megraw S 2012 WHY DON â€™ T WIND TURBINES HAVE MORE THAN 3 BLADES? 1–6
[12] Schubel P J and Crossley R J 2012 *Energies* **5** 3425–49
[13] Johansen J and Sørensen N 2006 *Riso-R-1543(EN)* **1543**
[14] Burton T, Jenkins N, Sharpe D and Bossanyi E 2011 *Wind Turbine Blade Aerodynamics* (West Sussex, England: John Wiley & Sons, Ltd)
[15] Elfarra M A 2011 Horizontal Axis Wind Turbine Rotor Blade: Winglet and Twist Aerodynamic Design and Optimization Using CFD (Natural and Applied Sciences of Middle East Technicak University)

IOP Conf. Series: Materials Science and Engineering **210** (2017) 012073 doi:10.1088/1757-899X/210/1/012073

Nano-mechanical properties and structural of a 3D-printed biodegradable biomimetic micro air vehicle wing

E Salami[1*], E Montazer[1], T A Ward[2], P B Ganesan[1]

[1] Department of Mechanical Engineering, Faculty of Engineering, University of Malaya, Kuala Lumpur 50603, Malaysia

[2] School of Engineering and Physical Science, Heriot-Watt University, Putrajaya, Malaysia

E-mail: *erfansalami@hotmail.com

Abstract

The biomimetic micro air vehicles (BMAV) are unmanned, micro-scaled aircraft that are bio-inspired from flying organisms to achieve the lift and thrust by flapping their wings. The main objectives of this study are to design a BMAV wing (inspired from the dragonfly) and analyse its nano-mechanical properties. In order to gain insights into the flight mechanics of dragonfly, reverse engineering methods were used to establish three-dimensional geometrical models of the dragonfly wings, so we can make a comparative analysis. Then mechanical test of the real dragonfly wings was performed to provide experimental parameter values for mechanical models in terms of nano-hardness and elastic modulus. The mechanical properties of wings were measured by nanoindentre. Finally, a simplified model was designed and the dragonfly-like wing frame structure was bio-mimicked and fabricated using a 3D printer. Then mechanical test of the BMAV wings was performed to analyse and compare the wings under a variety of simplified load regimes that are concentrated force, uniform line-load and a torque. This work opened up the possibility towards developing an engineering basis for the biomimetic design of BMAV wings.

1. Introduction

Being very manoeuvrable and having high load bearing capacities make insect flight more interesting subject for aerodynamic researchers [1-5]. Their superb flight agility can be largely attributed to their complex wing structure. The wings of a flying insect generally contribute only 2% of the total body mass of an insect [6]. These lightweight wings must be able to flap millions of times throughout an insect's lifespan; while enduring collisions, torsional loads, and many other forms of deformation [7]. Biomimicry of these attributes in an aircraft is highly desired. This has motivated research into a new class of micro-sized unmanned aircraft called Biomimetic Micro Air Vehicles (BMAV). Due to their small size and weight, BMAV would be ideal for flying indoors or in enclosed spaces. Fitted with micro cameras or sensors, they could be used to examine potentially hazardous areas (e.g. toxin spills, high voltage power lines, criminal activities, hostile forces, etc.) without endangering a human pilot [8].

Natural biomaterials are not only constructional materials but also functional materials. Research on the relationship between the structure function and performance of natural biomaterials is the basis for the development of biomimetic structures and materials. The structures, functions and materials of insect wings have optimised properties which have evolved over millions of years. The wings of dragonflies are mainly composed of veins and membranes.

The measurements of the mechanical properties of the wing of insects were mainly carried out using traditional tensile tests [9-11]. However, the tensile test cannot be appropriate to small scale samples of complex compartment. The new technique of nanoindentation is an excellent tool for the study of the mechanical behaviour of materials with nanoscale sizes [12-15]. Nanoindentation technique has been established the near surface mechanical properties of materials and it was used to measure the characteristics of the surface of the wing of dragonfly. Few studies were focus on the whole wing membranes only using nanoindentation [16-19]. Before improvements in nanoindentation technique, dragonfly wing research was primarily carried out through tensile tests [20]. At present, nanoindentation has become a useful tool for characterising the mechanical properties of thin elastic and viscoelastic materials [21, 22]. Thus the aim of this study will carefully concern the comparison of nanomechanical properties of veins and critical points of the dragonfly wings and bio-mimicked ones by nanoindentation.

2. Methodology

According to our previous study [20] the simplified dragonfly wing was modelled by using CAD software. PLA (Polylactic acid) is one of the most commonly used 3D printing materials. This biodegradable material has the virtue of being odorless and does not warp significantly like the other 3D printing materials. Due to its low melting point, using PLA as a 3D printing material does not require a heated bed. The 3D printer was utilized to "fine printing method" for this research since the sample size was approximately 8 cm with a thickness of 0.5 mm. Both forewing and hindwing were fabricated using PLA. Figure 1 shows the 3D model of a simplified dragonfly wing structure and figure 1 presents the fabricated dragonfly wings model. MarkerBot Replicator 2X experimental 3D printer that was used in this study was purchased from Markerbot (Brooklyn, USA). Grey PLA printing filament was purchased from MyDuino and used as received. CAD designs were drawn and visualized in Solidworks Software and transferred to 3D printable format using the Makerbot Desktop Software supplied with the Markerbot Replicator 2X.

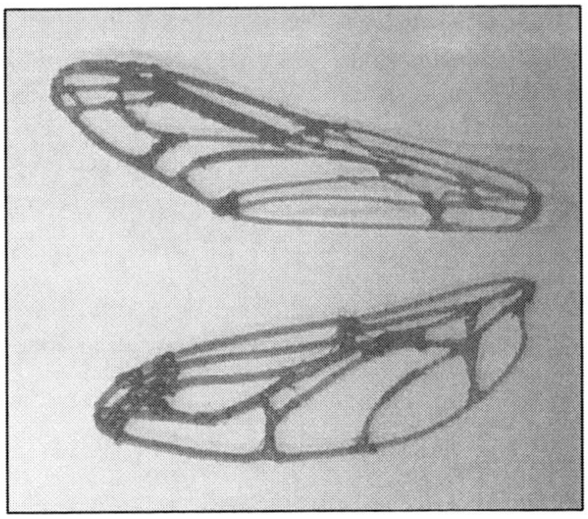

Figure 1. Fabricated dragonfly forewing and hindwing

Complete wing design (PLA based structure) and actual dragonfly wings were tested using Hysitron TI 750 UBI nanoindenter machine. Five points on a simplified BMAV wing and actual dragonfly wing were examined based on the crucial part in a wing structure as mentioned by Sun and Bhushan [1]. Figure 2 and Figure 3 display the location of the critical veins and points which are chosen for experimental tests.

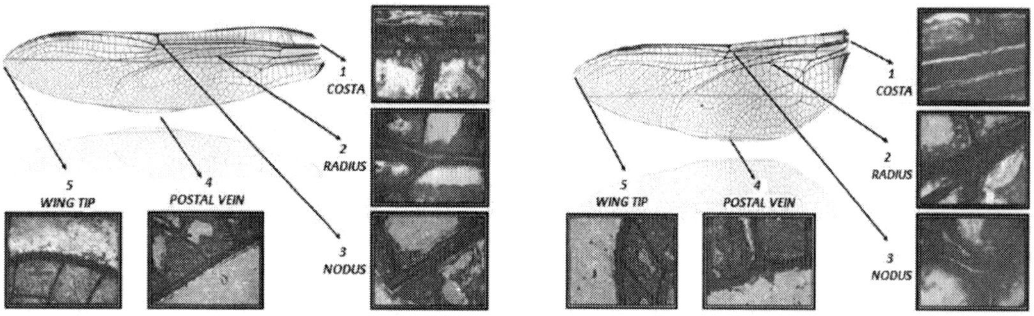

(a) (b)

Figure 2. Location of the veins and the critical point of the dragonfly wings; (a) forewing, (b) hindwing

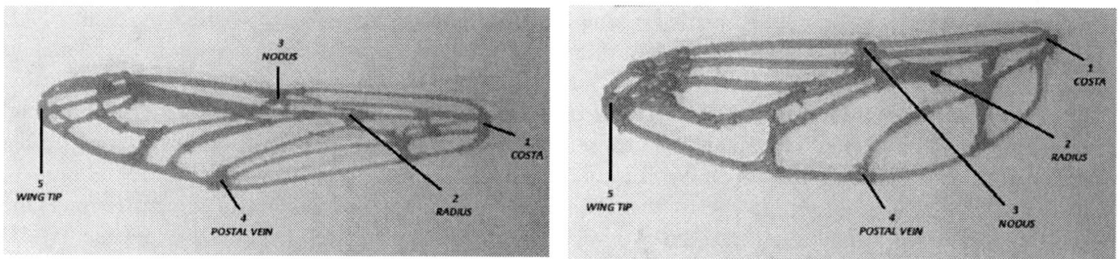

(a) (b)

Figure 3. Location of the veins and the critical point of the BMAV wings; (a) forewing, (b) hindwing

The two mechanical properties measured most frequently using load and depth sensing indentation techniques are elastic modulus, E, and hardness H. In a commonly used method, data are obtained from complete cycle of loading and unloading.

Average values of hardness and modulus of actual dragonfly wing and simplified BMAV wing were acquired using a fixed displacement method using Berkovich indenter instead of Vickers pyramid. The Berkovich indenter is a good choice for standard testing because it produces plasticity at very low loads, and minimizes the influence of friction. According to the theory on nanoindentation test [12, 15, 23] the reduced modulus is expressed as:

$$E_r = \frac{\sqrt{\pi}}{2\beta}\frac{S}{\sqrt{A}} \tag{1}$$

Where S is the contact stiffness that is defined by $S=(dP/dh)$ unload and β is coefficient of Berkovich tip $\beta = 1.034$. The load and displacement during indentation process are analysed to obtain the contact area. The contact area, A is related to the contact depth h_c as:

$$A = \sum_{n=0}^{8}\left(C_n h_c^{1/2^{n-1}}\right) \tag{2}$$

Where C_n are constant depends on the indenter and given by machine calibration. In equation (2) h_c is determined by:

$$h_c = h_{max} - \varepsilon\frac{P_{max}}{S} \tag{3}$$

For the Berkovich tip, $\varepsilon = 0.75$ is a constant. The relationship between indentation load, P, and penetration depth, h, is as shown in figure 4. Maximum load P_{max} and maximum penetration depth h_{max} were recorded automatically by the indenter. From equations (1) and (2) we can obtain the reduced modulus E_r. The effects of a non-rigid indenter on the load displacement behaviour can be consider by defining an effective modulus E_r, as follows:

$$\frac{1}{E_r} = \frac{1-v^2}{E} + \frac{1-v_i^2}{E_i} \tag{4}$$

Where E is the Young's modulus and v is the Poisson's ratio of the specimen. E_i and v_i are the values from the indenter. Using diamond tip on the indenter the value for E_i is 1141 GPa and v_i as

0.07 respectively. Poisson's ratio of the specimen is taken as 0.3 by default. The hardness of the material is defined by:

$$H = \frac{P_{max}}{A} \quad (5)$$

From equations (1) to (5), the modulus and the hardness of the samples can be obtained.

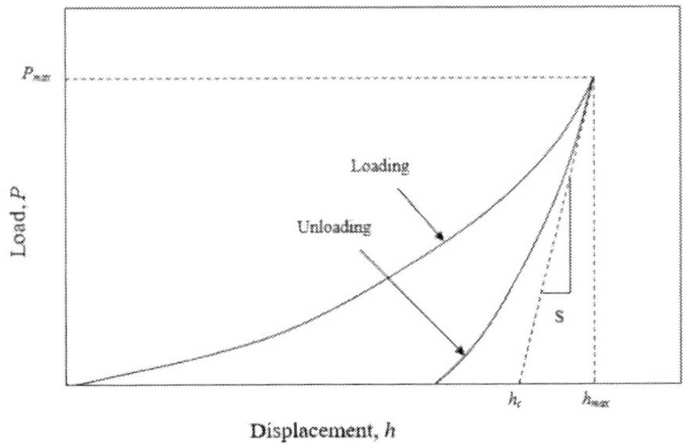

Figure 4. Schematic graph of the load via displacement of the nanoindentation test

3. Result and discussions

The force–displacement curve is shown in figure 4. The nanoindentation tests were done in two different settings that were called load control and displacement control. Therefore, in load control tests the maximum load for each test is fixed to 9999.25 μ N and in displacement control approach, the maximum depth is fixed to 499.93 nm.

Figure 5 shows the modulus and hardness of dragonfly forewing and hindwing and simplified PLA forewing and hindwing. Different points on the wing bear different loads on actual dragonfly wings, therefore their mechanical properties were different because of adaptation. Relatively speaking, the mechanical properties of a real dragonfly wing vary with the position. Group of researchers [24] examined the mechanical properties of the costa, radius and postal veins of a dragonfly wing and concluded that variation in modulus were in between 0.5 GPa to 3.5 Gpa. The modulus and hardness we retrieved from different points were comparable to previously published article [7, 25, 26].

The hardness of actual dragonfly wings varies in all the points. A study conducted by Smith et al. [11] discovered that dragonfly wing veins have an air-filled hollow section. Hardness might be influenced by the diameter of the hollow section in a particular point on the vein. The simplified BMAV wing structure is drawn by fine lines of melted PLA in a course of outer layer flowed by stacking layer by layer until complete. The surface profile is jaggy, this leads to a lower hardness than actual dragonfly veins that grow throughout its lifespan and strengthen key points on the wing profile.

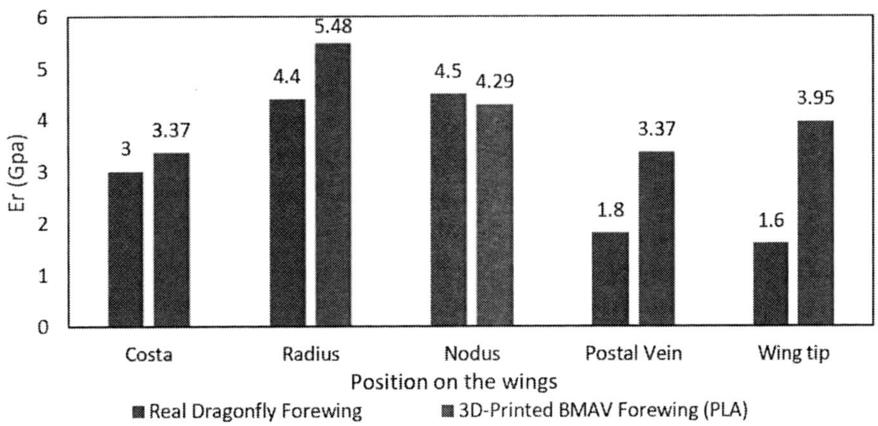

(a)

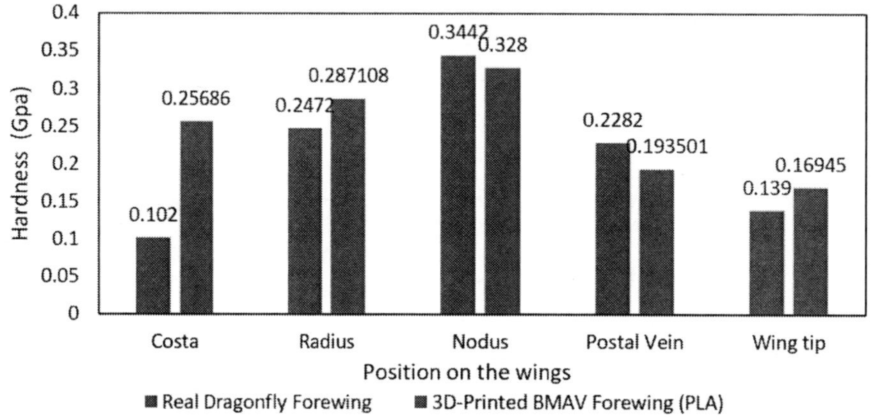

(b)

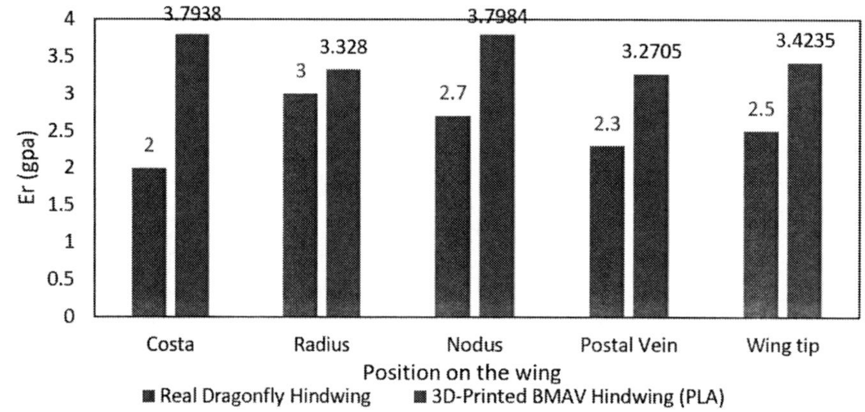

(c)

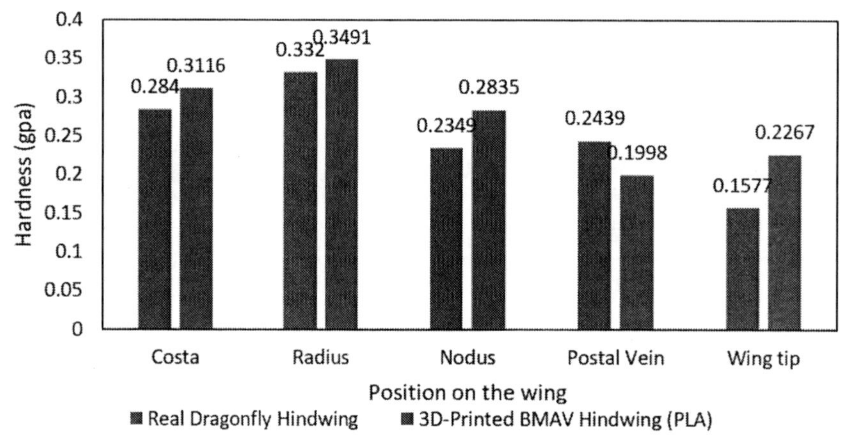

(d)

Figure 5. The comprasion of nanoindentation results of real dragonfly wings and BMAV wings based on critical points; (a) elastic modulus of forewings, (b) hardness of forewings, (c) elastic modulus of hindwings and (d) hardness of forewings.

Generally, the nanoindentation results for simplified BMAV wings are higher than real dragonfly's wings. As figure 5 (a) and (c) show, Radius and Nodus in both forewing and hindwing of real dragonfly displayed the highest modulus compared to other points. The Wing tip and the Costa points in the real dragonfly's forewing has the lowest modulus and hardness, respectively. According to figure 5 (a) and (b), same points for the lowest amount of Er and H were achieved for the BMAV forewing.

The maximum of Er and H appears at the Nodus for forewing, they are equal to 4.5 Gpa and 0.3442 Gpa, respectively. This shows that the various mechanical properties correspond to the flight mechanism. The Nodus is a location that serves to bear torsion. It plays an essential role in stability, greatly reducing the danger of buckling and structure failure [7]. The arc structure at the leading edge has the effect of preventing torsional deformation from becoming large, and the Nodus prevents the bending moment at the centre of the wing from becoming large [27].

4. Conclusion

Bionics is the imitation of the principles of biosystems and the creation corresponding techniques or artificial systems possessing the characteristics of biosystems. The main idea of bionics is to learn from natural systems and apply those lessons to technical innovation. Accordingly, in the current study, the nanoindentation method has successfully been used for testing the mechanical properties of real dragonfly wings and biodegradable BMAV wings. The mechanical test was performed to provide experimental parameter values for mechanical models in terms of nano-hardness and elastic modulus. The mechanical properties vary with the position on the wings. Based on the experimental data, in most of the points the nanoindentation results for simplified BMAV wings are higher than real dragonfly's wings. These results indicate that the 3-D printed PLA wings are comparable in hardness and elastic modulus to the real dragonfly wings. Therefore, the results can help to proceed further simulation and analysis in Bionics and gain more breakthrough in the future.

References

[1] Sun J and Bhushan B 2012 *Comptes Rendus Mécanique* **340** 3-17
[2] Jongerius S and Lentink D 2010 *Experimental Mechanics* **50** 1323-34
[3] Mingallon M and Ramaswamy S 2011 *ASME 2011 International Mechanical Engineering Congress and Exposition*: American Society of Mechanical Engineers) pp 723-30
[4] Rajabi H, Moghadami M and Darvizeh A 2011 *Journal of Bionic Engineering* **8** 165-73
[5] Wootton R, Evans K, Herbert R and Smith C 2000 *Journal of experimental Biology* **203** 2921-31
[6] Combes S A 2010 *Encyclopedia of aerospace engineering*
[7] Kesel A B, Philippi U and Nachtigall W 1998 *Computers in biology and medicine* **28** 423-37
[8] Lewitowicz J, Kowalczyk G, Sibilski K and Zurek J 2008 *26th International Congress of the Aeronautical Sciences (ICAS), Anchorage, AK,* pp 1-24
[9] Combes S and Daniel T 2003 *Journal of experimental biology* **206** 2979-87
[10] Newman D and Wootton R 1986 *Journal of Experimental Biology* **125** 361-72
[11] Smith C, Herbert R, Wootton R and Evans K 2000 *Journal of Experimental Biology* **203** 2933-43
[12] Oliver W C and Pharr G M 1992 *Journal of materials research* **7** 1564-83
[13] Nix W D 1997 *Materials Science and Engineering: A* **234** 37-44
[14] Hay J C, Bolshakov A and Pharr G 1999 *Journal of materials Research* **14** 2296-305
[15] Saha R and Nix W D 2002 *Acta Materialia* **50** 23-38
[16] Xiao K, Bai K, Wang W and Song F 2007 *Acta Mechanica Sinica* **23** 281-5
[17] Hooper I R, Vukusic P and Wootton R 2006 *Optics express* **14** 4891-7
[18] Wang X-S, Li Y and Shi Y-F 2008 *Composites Science and Technology* **68** 186-92
[19] Wan Y-l, Cong Q, Wang X-j and Yan Z 2008 *Journal of Bionic Engineering* **5** 40-5
[20] Salami E, Ganesan P, Ward T, Viyapuri R and Romli F 2016 *IOP Conference Series: Materials Science and Engineering*: IOP Publishing) p 012014
[21] Ahearne M, Yang Y, Then K Y and Liu K-K 2007 *Annals of Biomedical Engineering* **35** 1608-16
[22] Tong J, Zhao Y, Sun J and Chen D 2007 *Journal of Materials Science* **42** 2894-8
[23] Joslin D and Oliver W 1990 *Journal of Materials Research* **5** 123-6
[24] Sun J, Pan C, Tong J and Zhang J 2010 *IET nanobiotechnology* **4** 10-8
[25] Song F, Xiao K, Bai K and Bai Y 2007 *Materials Science and Engineering: A* **457** 254-60
[26] Chang C H, Ting K E and Chen K T 2009 *Advanced Materials Research* 1325-8

IOP Conf. Series: Materials Science and Engineering **210** (2017) 012073 doi:10.1088/1757-899X/210/1/012073

[27] Machida K and Shimanuki J 2005 Structure analysis of the wing of a dragonfly. In: *Microtechnologies for the New Millennium 2005*: International Society for Optics and Photonics) 671-6

Comparative study on microstructure, crystallite size and lattice strain of as-deposited and thermal treatment silver silicon nitride coating on Ti6Al4V alloy

Umi Zalilah*[1], R Mahmoodian[1,2]

[1]Department of Mechanical Engineering, Faculty of Engineering, University of Malaya, 50603 Kuala Lumpur, Malaysia
[2]Center of Advanced Manufacturing and Material Processing, Department of Engineering, University of Malaya, 50603 Kuala Lumpur, Malaysia.

E-mail: *zalilahzaidi_92@yahoo.com

Abstract. Silver silicon nitride coating were deposited on Ti6Al4V alloy using physical vapor deposition magnetron sputtering technique. Field Emission Spectroscopy (FESEM), Electron Dispersive Spectroscopy (EDS) and X-ray diffraction (XRD) were used to characterize as-deposited and after heat treatment of AgSiN coatings in order to understand the morphology, compositions and structure. Meanwhile, in determining the crystallite size and lattice strain, the simplified Williamson-Hall plot method was utilized. The heat treated coated sample shown to reveal granular surface structure, bigger crystallite size and lattice strain as compared to the as-deposited coated sample.

1. Introduction

Titanium and its alloys had been used in wide range of applications such as machining [1], medical and aerospace. The choice of this alloy is due to its excellent properties such as high strength to weight ratio, corrosion resistance [2] and good biocompatibility [3]. Nevertheless, taking into consideration releasing of Al and V elements as well as implant associated bacterial infections, investigation and modifying titanium alloy surface are still needed and ongoing [4]. The investigations including during fabrication process like coating the surface as well as the post- treatment process like heating.

In coating, structure morphology, particle size and lattice strain play a significant role while x-ray diffraction peaks has become a very powerful tool to extract the information and to study the coating properties. There are several analysis methods that could be used to analyze the crystallite size and lattice strain of the coating. Some of the analysis technique are Scherer's equation, Warren-Averbach analysis (W-A), Williamson-Hall plot analysis (W-H), Rietveld refinement [5, 6] and Fourier analysis technique. Scherer's equation is a well-known equation use to estimate the crystallite size by using the full width at half maximum (FWHM) of diffraction peaks. However, the Scherer's formula give the lower bound of the crystallite size since the equation neglect the broadening from other important factors such as inhomogeneous strain and instrumental effect. Meanwhile, Warren-Averbach analysis need at least two reflections along the same crystallographic direction [7]. Warren-Averbach followed by Fourier technique, which need to convolute size and strain broadening make these method quite difficult to carry out [8]. At the same time, Williamson-Hall plot analysis which consider to be an average method but still hold an unavoidable position for crystallite size determination [9].

In this study, the titanium alloy (Ti6Al4V) substrates were coated with silver silicon nitride coating using a reactive magnetron sputtering physical vapor deposition (PVD) [10, 11]. In order to investigate and compare the crystallite size and lattice strain for the as-deposited and heat treatment sample, Williamson-hall plot analysis method was employed.

2. Materials and Method

The material selected for this study was Ti6Al4V and they were cut into sizes of 15 x 15 x 2 mm using wire cut electrical discharge machining (EDM) machine. The samples then were grounded with silica papers starting from 800, 1000, 1500, 2000 and 2500 grit by rotation 90^0 at every change of silica paper grit. Next step, samples were polished using polishing cloth with the help of diamond liquid of 6 µm. In order to remove grease, contaminants and dirt, samples were ultrasonically cleaned in ethanol for 15 mins and rinse with distilled water. Samples were dried using blow dryer before loaded into the deposition chamber.

Deposition process was done using SG Control Engineering Pte. Ltd machine which utilized the physical vapor deposition magnetron sputtering technique (PVDMS). The chamber was vacuum to base pressure of $2x10^{-5}$ Torr and working pressure was maintained around $6.5x10^{-3}$ Torr after flow in of argon gas. Silver and silicon target both with 99.99% purity were connected to DC and RF power supply respectively. Oxide layer and dust on target were removed by pre-sputtering in pure argon gas for 15 minutes with the shutter shut the targets. Silver interlayer was deposit for 15 mins meanwhile the second layer was co-sputtering both silver and silicon target for 20 mins. Other deposition parameters can be referred to Table 1.

Table 1. PVD deposition parameter variable.

Parameter variables	Values
DC power (Watt)	20
RF power (Watt)	100
Bias voltage (V)	150
Argon flow rate (sccm)	20
Nitrogen flow rate (sccm)	8
Temperature (^{0}C)	100
Deposition time (mins)	20

The specimens were taken out of the chamber after thin film deposition completed. In order to investigate the effect of heat treatment on coatings, the as-deposited coatings were put inside an Argon purged furnace. With continuous monitoring, the samples were heated up to 400°C in pure argon gas environment with heating rate of 1°C/min. The samples were kept in the 400°C conditions for 60 minutes. Samples then were let cooled down naturally before they were taken out for characterization. The schematic diagram of temperature as function of time of thermal treatment was shown as figure 1.

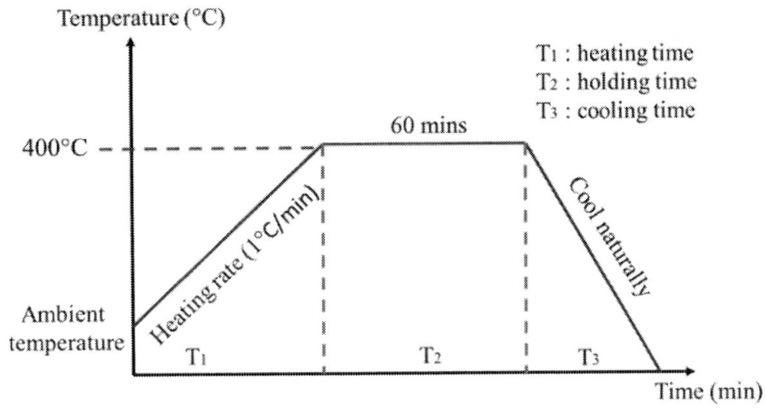

Figure 1. Schematic diagram of temperature as function of time of thermal treatment.

3. Results and discussion

Figure 2 (c)(d) illustrated the digital photograph and FESEM image of as –deposited coating and after thermal treatment at 400⁰C, respectively. In digital photograph, we can see that no obvious changes were observed on the coating surface. In order to get detailed image, we further investigated samples' microstructures using field emission electron microscope (FESEM) attached with electron dispersive spectroscopy (EDS). The detailed image presented in Figure 2 (b)(d) showed all coatings had granular surface structure background with some white particles were spotted that labelled as a, b and c on the surface. Using point and spot analysis from EDS, white particles were confirmed to come from silver element. Meanwhile, the backgrounds of coatings were mixture of silver, silicon and nitrogen elements with atomic percentage of 33.1, 1.9 and 1.9 atomic%, respectively. It looks like there were no significant difference between the compositions of as-deposited and after heat treatment coating. At the same time, other elements were also noticed like carbon, insignificant gases and moisture which believed to come from the deposition chamber and environment thus will not further discussed.

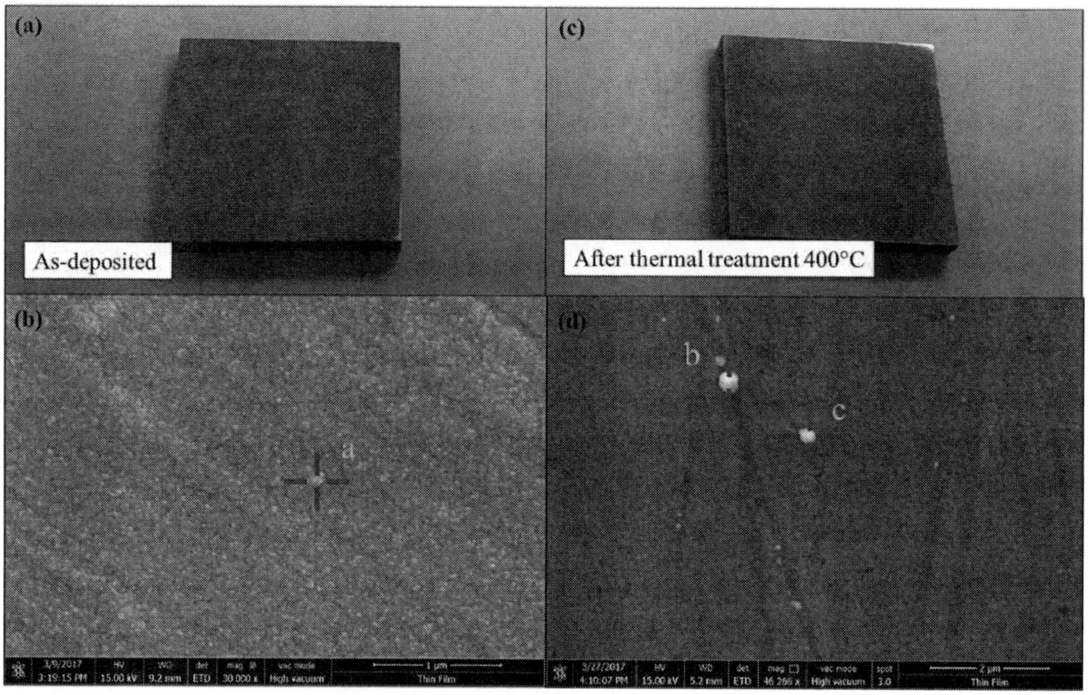

Figure 2. Digital photograph and surface FESEM image of (a-b) as-deposited coating and (c-d) after 400^0C of heat treatment.

Cross-sectional FESEM image and EDS analysis for both as-deposited and after thermal treatment were presented in Figure 3 and Figure 4 accordingly. It was observed that the coating structures were changed from columnar and loose structure bilayer coating for as-deposited sample to dense, compact and homogeneous monolayer of coating after heat treatment sample. It was believed that heat treatment causing the interlayer to diffuse into the subsequent layer or into the substrate therefore forming the monolayer. This can be proved from the mapping EDS analysis in Figure 4 which showed the dispersion of each element in the coating layer. By comparing both coatings, it is assured that after thermal treatment the coating became more compact, dense and displaying small granular size.

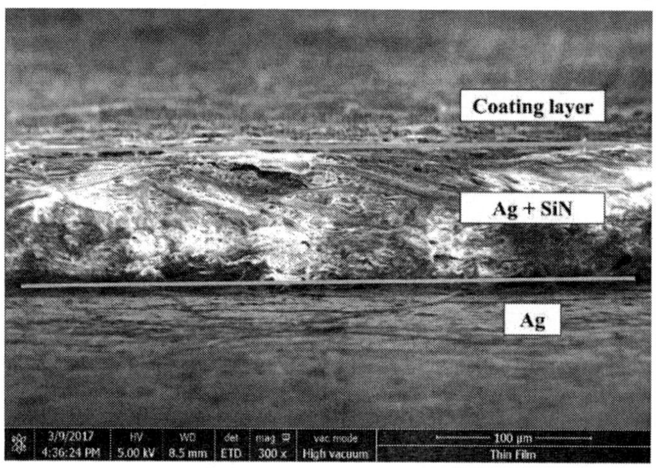

Figure 3. FESEM image cross section of as-deposited silver silicon nitride coating at -150V.

IOP Conf. Series: Materials Science and Engineering **210** (2017) 012074 doi:10.1088/1757-899X/210/1/012074

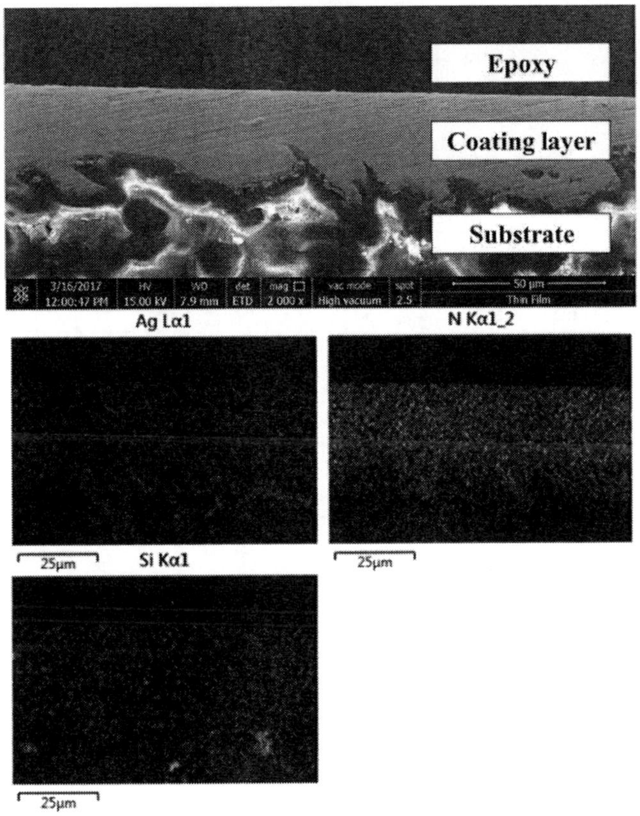

Figure 4. FESEM image and EDS analysis cross section of silver silicon nitride coating after heat treatment.

The X-ray diffraction profile analysis were conducted for all samples and all peaks with their respective phases and h, k, l (hkl) values were presented in Figure 5. For both samples of as-deposited coating (Figure 5.a) and heat treatment (Figure 5.b), phases detected in the diffractogram were identical. Silver phase with ICSD No. 98-004-4387 was found at peak positioned 2θ = 38.1°, 44.19°, 64.48° and 77.46° with crystal orientation of (111), (002) and (113) respectively. The interaction between substrate which is titanium alloy (Ti6Al4V) with the coating was noted with the present of intermetallic phase of $AgTi_3$ (ICSD No. 98-005-8370) at positioned 2θ = 38.1°. In addition, silver nitride, AgN peak was identified at 2θ = 53.17° (ICSD No. 98-018-5555). Silicon nitride, Si_3N_4 was noted at 2θ = 40.35° with ICSD No. 98-003-5660. The silicon nitride amorphous phase contributed to the relatively low intensity peak in the diffraction pattern in Figure 5 (a) and (b). Whereas, there was one peak that belong to the combination of all three elements namely silver Bis (silicate) nitrate with chemical formula Ag_9 (SiO_4) NO_3 or $Ag_9NO_{11}Si_2$ (ICSD No. 98-006-2402). Clearly, after heat treatment was applied to the samples, all peaks in diffractogram become well defined while the intensity was doubled (sharper). Applying heat treatment usually may cause the materials to reorganize and re-oriented themselves, which then contributed to the well-defined structure of peaks.

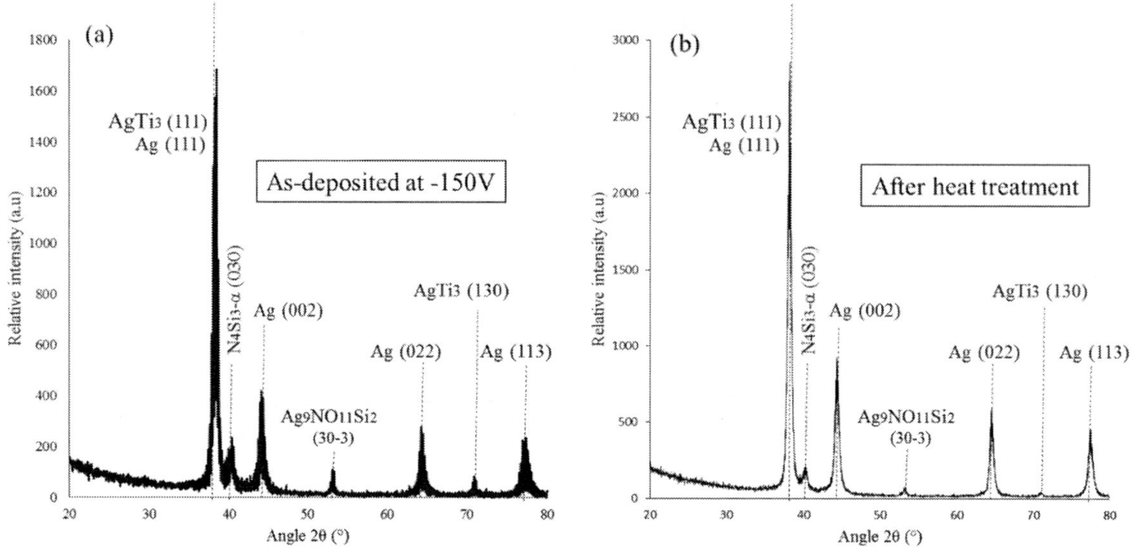

Figure 5. XRD diffractogram of silver silicon nitride coating on Ti6Al4V alloy (a) as-deposited and (b) after heat treatment.

From the XRD peaks, the value of crystallite size and lattice strain can be obtained. There were many methods can be used to calculate these values such as Williamson and Hall method (W-H), Warren Averbach analysis (W-A), Rietveld refinement and pseudoVoigt function. In this study, we employed the W-H method which is a simplified method and most commonly used. The Scherer's equation, a well-known equation in determining crystallite size can be expressed as,

$$\tau = \frac{k\,\lambda}{\beta\ \cos\theta} \tag{1}$$

Where τ is the mean of grain size, *k* is a dimensionless shape factor (typical value of 0.9), λ is the x-ray wavelength (Cu Kα cathode wavelength λ=15.405 nm), **β** is the broadening of full width at half maximum (FWHM) of the peak and θ is the Bragg's angle in degree unit. The broadening peaks can be attributed to crystallite size as in equation (2),

$$\beta crystallite = \frac{k\,\lambda}{L\cos\theta} \tag{2}$$

And broadening due to strain as in equation (3),

$$\beta\varepsilon = C\tan\theta \tag{3}$$

Therefore, the sum of broadening due to crystallite and lattice strain in equation (4) is gained by subtracting the broadening of the instrumental effect and broadening of diffraction peak.

$$\beta total = \beta crystallites + \beta\varepsilon \tag{4}$$

Substituting the equation of $\boldsymbol{\beta}_{crystallites}$ and $\boldsymbol{\beta}_{strain}$ into $\boldsymbol{\beta}_{total}$, we will obtain equation as follows.

$$\beta total = \frac{k\,\lambda}{L\cos\theta} + C\sin\theta \tag{5}$$

Rearrange the equation by dividing both sides with cos θ, the finalize equation in (6) which known as Williamson-hall plot equation in the form of Y= m X + C is shown as follows,

$$\beta\ total\cos\theta = \frac{k\,\lambda}{L} + C\sin\theta \tag{6}$$

Using this equation, we plot cos θ against sin θ for as-deposited coating samples presented in Figure 6 accompanied by tabulated FWHM data in Table 2 meanwhile for heat treatment coating samples presented in Figure 7 with FWHM data were tabulated in Table 3.

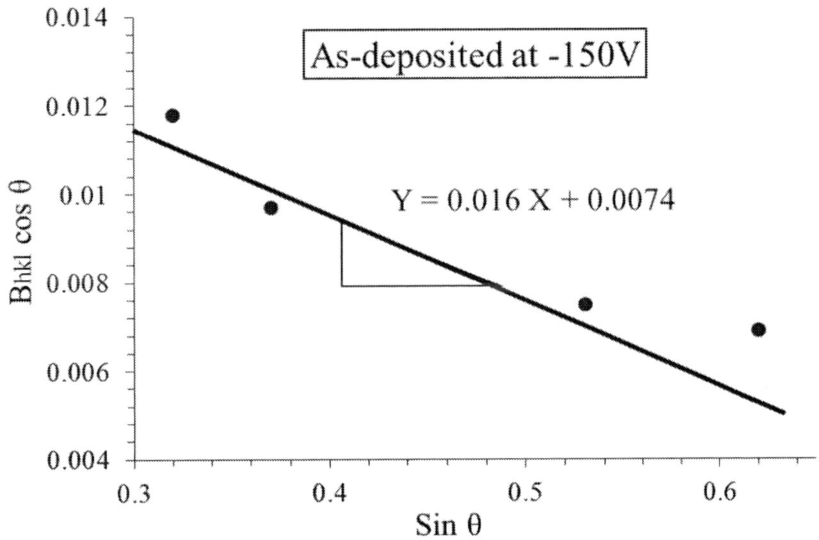

0

Figure 6. Plot of B_{hkl} cos θ versus sin θ for as-deposited silver silicon nitride coating on Ti6Al4V alloy.

Table 2. Full width at half maximum (FWHM) of as-deposited silver silicon nitride coating, Cu Kα-1, λ = 1.54060Å.

Peak No.	2θ (°)	*h k l*	FWHM (°)	FWHM (rad)
1	38.21	1 1 1	0.7164	0.01250
2	44.18	0 0 2	0.6140	0.01071
3	64.41	0 2 2	0.5117	0.00893
4	77.28	1 1 3	0.5117	0.00893

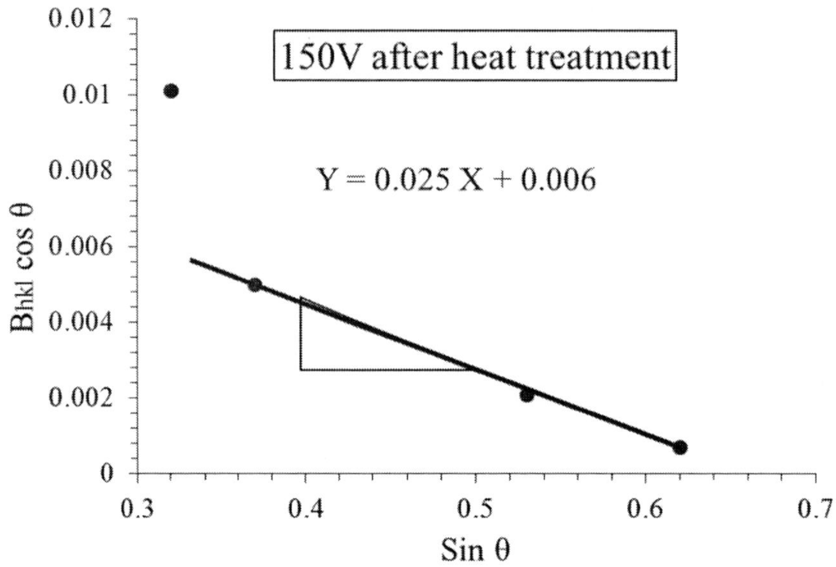

Figure 7. Plot of B_{hkl} cos θ versus sin θ after heat treatment sample of silver silicon nitride (AgSiN) coating on Ti6Al4V alloy.

Table 3. Full width at half maximum (FWHM) after heat treatment of silver silicon nitride coating, Annealed at 400°C, Cu Kα-1, λ = 1.54060Å.

Peak No.	2θ (°)	*h k l*	FWHM (°)	FWHM (rad)
1	38.18	1 1 1	0.6140	0.01071
2	44.37	0 0 2	0.5117	0.00548
3	64.55	0 2 2	0.4605	0.00252
4	77.48	1 1 3	0.3582	0.00090

From the linear fit graph obtained in Figure 6 and Figure 7, the estimation value of crystallite size and microstrain can be obtained by compared with the Williamson-Hall plot equation as in equation 6. Y-intercept will give the estimation of crystallite size value while the slope of the graph represents the lattice strain of the coating. Positive slope is an indication of tensile stress meanwhile negative slope indicate the presence of compressive stress within the material. After comparing with W-H equation, estimated crystallite size and microstrain obtained for as-deposited coating were 187nm and 0.016. The crystallite size estimation and lattice strain of coating after heat treatment were 165nm and 0.031. By comparison, the crystallite size of coating after heat treatment was smaller and the lattice strain was higher than as-deposited coating. The W-H plot crystallite size values showed good agreement with values obtain using Scherer's equation.

4. Conclusions

We prepared silver silicon nitride coating on Ti6Al4V alloy using physical vapor deposition magnetron sputtering technique. Comparison in terms of structure between as-deposited coating and thermal treatment coating showed the crystallite size of coating after heat treatment was bigger and the lattice strain was higher than as-deposited coating.

Acknowledgment

The author would like to acknowledge the research support from research PPP grant University of Malaya (PG185-2016A).

References

[1] Liu W, Li A, Wu H et.al 2016 *Ceramics Int* . **45** 17693–697
[2] Chen S, A D Usta ,M Eriten, et.al 2017 *Surf.Coat.Technol.* **315** 220-231
[3] Saleem S, Ahmad R, Ayub R et.al 2017 *App.Surf.Sci.* **394** 586-597
[4] Wang R, H Xiaojing, Gao Yet.al 2017 *Mater.Sci.Eng.,C*, 2017 **75** 7-15
[5] Mahmoodian R, et al 2015 *Plos One.* **10**
[6] Mahmoodian R, M A Hassan, and M Hamdi 2015 *Mechanical properties and performance of engineering ceramics and composites x, (*Daytona Beach, FL, USA John Wiley & Sons, Inc.) ch5:41-8
[7] Ghosh J, S K Chattopadhayay, Meikap A K, et al 2008 *J.Alloys Compd.* **453** 131-137
[8] Warren B E and B L Averbach *1950 J.Appl.Phys* **21** 595-599
[9] Venkateswarlu K, A Chandra Bose and N Rameshbabu 2010 *Physica B: J.Phys:Condens Matter* **405** 4256-4261
[10] Hassan M A, A Bushroa and R Mahmoodian 2015 *Surf.Coat.Technol* **277** 216–221
[11] Hassan M A , et al. 2015 *Adv.Mater.Res* **1105** 69-73

Development of an end-of-life vehicle recovery model using system dynamics and future research needs

N Mohamad-Ali[1], R A R Ghazilla[1*], S H Abdul-Rashid[1], N Sakundarini[2], A Ahmad-Yazid[1] and L Stephenie[1]

[1]Centre for Product Design and Manufacturing, Department of Mechanical Engineering, Faculty of Engineering, University of Malaya, 50603 Kuala Lumpur, Malaysia
[2]Department of Mechanical, Materials & Manufacturing Engineering, The University of Nottingham Malaysia Campus, Jalan Broga, 43500 Semenyih, Selangor, Malaysia

E-mail:* r_ariffin@um.edu.my and nurhasyimahali@um.edu.my

Abstract. The implementation of end-of-life vehicle (ELV) recovery policy in Malaysia has led vehicle manufacturers to look at different ways to improve design and development of vehicles. Nowadays, it is crucial to incorporate end-of-life (EOL) design strategies into the vehicle design in order to enhance the effectiveness of the ELV recovery network. Although recent studies have shown that product design has a significant effect on the product recovery rate, there is a lack of studies on how EOL design strategies affects the effectiveness of ELV recovery, particularly when there are dynamic changes in the behaviour of the product recovery network. Thus, in this study, we developed a preliminary model based on the system dynamics approach in order to predict the effectiveness of ELV recovery in response to dynamic changes of various factors (including EOL design strategies) in the business environment. We developed this model based on preliminary data that we had gathered from unstructured interviews with the key stakeholders of ELV management in Malaysia. We believe that our model will greatly benefit product designers in incorporating the appropriate EOL design strategies in order to boost ELV recovery effectiveness in Malaysia.

1. Introduction

It is known that the disposal of end-of-life vehicles (ELVs) creates hazardous wastes and toxic emissions. The dramatic increase in the number of vehicle ownership throughout the world has led to a significant increase in global fuel demands, material requirements, and carbon emissions [1]. The results of a survey on 27 European countries in 2012 revealed that ELVs contribute to 6,230,000 tonnes of wastes, of which the highest amount of ELV wastes was generated in France (1,229,096 tonnes), followed by United Kingdom (1,129,392 tonnes) and Italy (874,887 tonnes) [2]. In 2012, 84.1 million vehicles were manufactured in order to fulfil consumer demands, with China, United States of America and Japan being the countries with the highest statistics of vehicle production and usage [3]. Malaysia was ranked as the third country in Asia in terms of the highest vehicle production and sales in 2013. According to Amelia et al. [4], most of the vehicle sales in Malaysia was dominated by local manufacturers (i.e. PROTON and PERODUA), and both of these manufacturers use new parts in vehicle design and development. However, owing to the detrimental impact of vehicle production, usage and wastes on the environment, it is more crucial than ever for manufacturers to be proactive in tackling sustainability issues in order to improve the quality of life of the people while simultaneously reduce environmental impact. Hence, the sustainable manufacturing concept was introduced to assist manufacturers to achieve this purpose [5]. Sustainable manufacturing is implemented throughout the product life cycle, beginning from the acquisition of raw materials until the end of life (retirement) of the product, taking into consideration the levels, processes, and systems involved in manufacturing [6]. In general, the product life cycle consists of four stages: (1) pre-manufacturing, (2) manufacturing, (3) use, and (4) post-use.

Product recovery elements are an essential part of the cradle-to-cradle philosophy, in which wastes recovered at the product end of life are used as the raw materials to manufacture new products. According to Jawahir et al. [6], the term 'recover' refers to the activity of collecting EOL products for subsequent post-use activities. According to these researchers, the main goal of redesign, remanufacturing, reuse, recover, recycle and reduce (i.e. the so-called 6R concept) is to reduce wastes, particularly at the product end of life. The product design stage appears to be a critical stage to incorporate EOL design strategies into the product which will facilitate product recovery after the product has reached its end of life [7]. Nowadays, manufacturers should implement the product recovery concept in order to boost their environmental image, which in turn, gives them a competitive edge, considering that consumers are becoming more environmentally conscious. By implementing the product recovery concept, new products can be launched into the market while conserving natural resources, maximizing the consumption of materials and minimizing wastes [6]. It is evident here that the product design plays a crucial role in incorporating product recovery strategies into the product.

There are many challenges in incorporating product recovery strategies (or rather, EOL design strategies) in vehicle design in Malaysia. The concept of ELV recovery is relatively new in this country due to several barriers which hinder its effective implementation. At present, it is a norm for manufacturers in Malaysia to use new parts in manufacturing new products [4]. In addition, there is a lack of knowledge in eco-design [8] as well as ELV management. There are no reliable statistics on the number of ELVs that have been recovered in Malaysia to date [9]. Furthermore, there is a lack of a proper ELV management system [10]. It is found in a recent study [11] that most local manufacturers do not consider design for environment and design for disassembly as part of their strategy to reduce the environmental impact of their products. All of these issues need to be addressed and hence, there is a critical need to understand how EOL design strategies can contribute towards a positive environmental impact of products, specifically ELVs.

In this study, we propose a preliminary model for ELV recovery in Malaysia based on the system dynamics approach in order to gain a better understanding on the impact of EOL design strategies (particularly design for reuse, design for remanufacturing, and design for recycling) on ELV recovery effectiveness. We believe that our model will greatly facilitate product designers in the automotive industry to predict the ELV recovery effectiveness when there are dynamic changes in the behaviour of the ELV recovery system.

2. Literature review

At present, there is a large number of studies published in the literature related to the application of system dynamics (SD) in analysing EOL design strategies, and several key works in this field are summarized in table 1.

Table 1. Summary of key works related to the application of SD in analysing EOL design strategies.

Field of study	EOL design strategy	Key findings of the study	Limitations of the study	Reference
Policy	Refurbishing	• In general, closed-loop supply chain policies require a well-calibrated return rate for various different types of products. • A sustainable return policy increases sales significantly. • Product design as well as social awareness and acceptance of refurbished products are critical for aftermarket activities.	The study is only focused on cost benefit scenario and analysis.	[12]
	Recycling and remanufacturing	• Increasing the number of remanufacturers is greatly beneficial during the initial stages of industrial development. • There is a need for recycling subsidy for insufficient cores.	• The case study is only carried out for auto parts in China. • The authors assumed that 100% of the recycling subsidy fee goes to the recyclers.	[13]
	Recycling	• Auto recyclers and operators who are able to adapt their premises and workforce are more likely to remain longer in the business.	• The study is focused only on three areas: (1) premises, (2) workforce, and (3) sourcing of ELVs.	[14]
Design processes	–	• The authors are able to correctly model the workflow in and between the different phases of processes.	• The study is focused on the engineering design process.	[15]
Sustainable manufacturing	–		• The components of the framework are limited to manufacturing, environmental, financial, and social domains.	[16]

Legislation	Recycling and disposal	• Legislation has a significant effect on the sum of waste disposal.	• The case study is carried out on waste electrical and electronic equipment (WEEE) and refrigerators. • Delays in legislation are not considered in the activities of the closed-loop supply chain.	[17]
	Reduce, reuse and disposal	• Levying tax on used car exports will facilitate control of used car exports and promote economic opportunities for manufacturers, remanufacturers, recyclers, consumers and the Government of Japan.	• The study is limited only to the automotive industry in Japan. • This study does not address the joint effect of refurbished and recycled products towards sustainability.	[18]
Materials	Recycling	• The recovery rate is influenced by the material composition of the vehicle. • It is expected that the amount of shredder residue per vehicle will continue to rise over the years.	• The study is limited to aluminium and plastic materials. • The study is only focused on economic impact.	[19]
	Recycling	• Product design affects product 'mineralogy', recycling of cars, and the recycling rate.	This study is limited to wrought and cast aluminium steel, copper, as well as remainder materials including organic fraction.	[20]

Design for product recovery has a long history, as evidenced from the number of key works related to EOL design strategies shown in table 1. In general, the design strategies can be classified as policy, design processes, legislation and materials. However, based on our literature survey, none of these studies are focused on improving the ELV recovery effectiveness. In addition, there are only a few studies focused on simultaneous implementation of multiple EOL design strategies (e.g. reuse, recycling and remanufacturing) [13, 18–19, 21–24]. Even though [25] considered all of these strategies, their study is centred on searching for an optimal solution and they did not use dynamic and stochastic approaches such as SD. Based on the results of our literature survey, we perceived that it is crucial to develop a model for ELV recovery in Malaysia. We envisioned that our model will assist product designers in the automotive industry in Malaysia to predict ELV recovery effectiveness, taking into consideration the dynamic changes of interrelated factors involved in ELV recovery. This model will indeed be useful since it can be used by product designers to assess how the EOL design strategies (particularly design for reuse, design for remanufacturing, and design for recycling) affects ELV recovery effectiveness during the early stages of product design and development. We used the SD approach to develop our model.

2.1. Causal loop diagrams

The SD concept was introduced by Jay Wright Forrester. This technique involves creating a causal loop diagram, which is a pictorial representation of the cause and effect of the system. This diagram is developed based on the development of stock and flow. In this study, we developed the causal loop diagram by identifying the variables of the ELV recovery system from literature survey. Figure 1 shows an example of a causal loop diagram, which shows how an automotive recycling programme affects the workforce of the automotive recycling industry [14]. The cause and effect of this causal loop diagram is described briefly as follows:

The starting point of this causal loop diagram is 'Business turnover', which increases (↑) the 'Workforce size' of auto recyclers, which in turn, increases (↑) the 'Labour cost'. The increase in 'Labour cost' decreases (↓) 'Business profits' while at the same time increases (↑) 'Optimized operations'. The increase in 'Optimized operations' increases (↑) the 'Workforce skill level', which in turn, increases (↑) the 'Workforce efficiency'. The increase in 'Workforce efficiency' increases (↑) the 'Workforce size'.

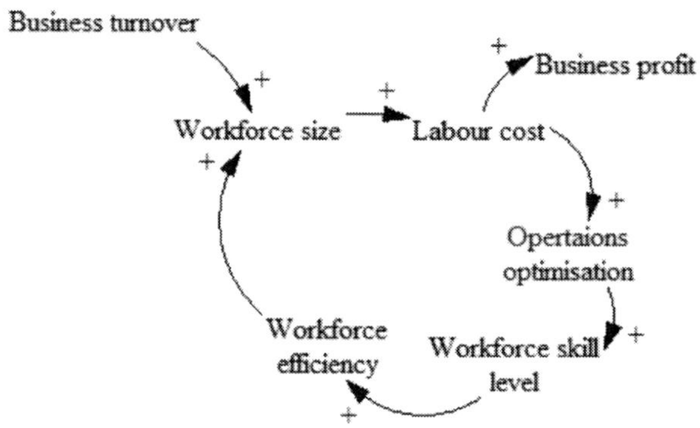

Figure 1. Example of a causal loop diagram for workforce dynamics [14]

In this study, we created causal loop diagrams for ELVs and reusable parts, as shown in figure 2, respectively. We extracted the variables to construct the causal loop diagrams after reviewing the works of [26] and [27].

There are two primary loops in figure 2(a), which are described as follows:

1. 'Dismantling activities' will decrease (↓) 'ELVs', which will increase (↑) the 'Operation cost'. The increase in 'Operation cost' (↑) will reduce (↓) the 'Number of ELV parts removed', which will reduce (↓) 'Recycling activities'. The decrease in 'Recycling activities' will increase (↑) 'ELVs'.
2. The increase in 'Operation cost' will reduce (↓) 'ELV management', which will increase (↑) 'Car imports'. The increase in 'Car imports' will increase (↑) 'Disposal cost', which in turn reduces (↓) the 'Number of participating firms' and consequently decreases (↓) 'ELV management'. The decrease in 'ELV management' will increase (↑) 'ELVs'.

The causal loop diagram shown in figure 2(b) is described as follows: An increase in the 'Value of reusable parts' will increase (↑) the 'Profits gained from reusable parts', which leads to an increase (↑) in the 'Collection of parts'. The increase in the 'Collection of parts' will decrease (↓) the 'Demand for products from original equipment manufacturers', which consequently increases (↑) the 'Demand for reusable parts' and 'Value of reusable parts'. The increase in the 'Value of reusable parts' will also promote (↑) the 'Number of remanufacturers', which will eventually increase (↑) the 'Demand for reusable parts'.

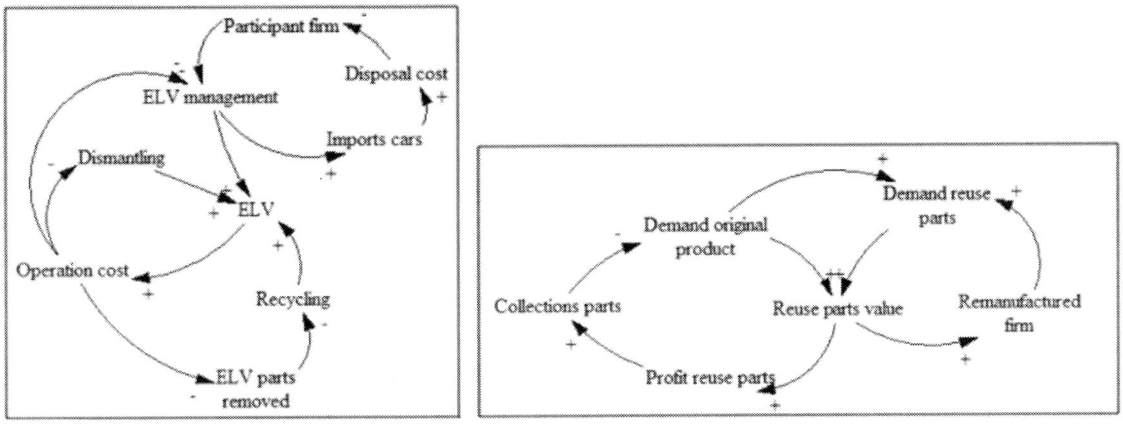

a) b)

Figure 2. Causal loop diagram for (a) ELVs and (b) reusable parts

The relationship between figure 2(a) and figure 2(b) can be established by connecting the effect of 'Demand for reusable parts' on 'Recycling activities'. We expected that there will be an increase in recycling activities when there is an increase in the demand for reusable parts. This in turn, will boost effectiveness of the ELV recovery system in Malaysia. However, we perceived that it is not possible to achieve this goal without incorporating design for environment strategies during the early stages of product design and development. Selecting the appropriate EOL design strategies is a challenging task since product designers need to anticipate the long-term environmental impact of the product. Moreover, this task is influenced by a combination of factors which include policies stipulated by the local government and international authorities. Thus, careful planning is needed in order to select the appropriate EOL design strategies to be incorporated into the product.

3. Proposed model

There are various factors that need to be considered when selecting a suitable EOL design strategy. In this case, we need to consider the types of EOL design strategies available for ELVs, as well as identify the key performance indicators, and select endogenous and exogenous variables. This task is complicated by the fact that these factors are interrelated, and changing one factor may have significant effect on other factors within the product recovery network as well as ELV recovery effectiveness. For example, if the product designer selects design for reuse as the EOL design strategy, the product will have a specific recovery chain once it has reached its end of life. This recovery chain will affect other chains in the product recovery network such as economic growth for automotive spare part dealers as well as competition between remanufactured parts and those manufactured by original equipment manufacturers. Even though the ELV recovery scenario appears simple, it is actually complex in nature because of a variety of interrelated factors, and each of these factors may change because of market trends, which will have impact the ELV recovery effectiveness as a whole.

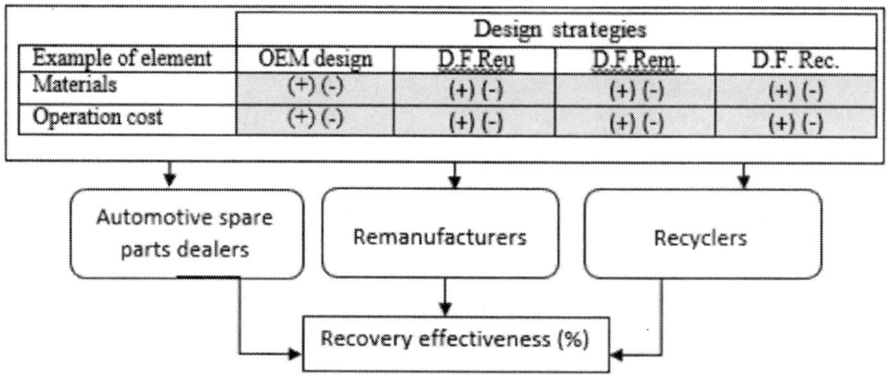

Figure 3. Elements of design strategies

Figure 3 shows an example of the elements of design strategies. In this example, there are four design strategies, namely, design by original equipment manufacturers (OEM design), design for reuse (D.F.Reu), design for remanufacturing (D.F.Rem.) and design for recycling (D.F.Rec.). Each of these design strategies consist of two elements: (1) materials, and (2) operation cost. These elements are set up by the product designers during the early stages of product design and development to suit each design strategy. These elements have their own cause and effect towards the design strategies, as indicated by the shaded cells. This cause and effect will also affect other chains in the recovery network such as automotive spare part dealers, remanufacturers and recyclers. For instance, figure 4 shows the plot of a basic stock and flow analysis, which indicates how monthly recovery improvement affects the recovery effectiveness. We assumed here that the recovery activities improve at a rate of 6% per month. After one year, the recovery effectiveness is 72% instead of 0% at the start of the year. However, it shall be noted that this stock and flow analysis is carried out based on our understanding on the current ELV recovery scenario in Malaysia using the SD approach. For this reason, we implemented SD to link the EOL design strategies and ELV recovery factors in order to evaluate and predict the ELV recovery effectiveness. We believe that this technique is suitable to evaluate the impact of EOL design strategies on ELV recovery effectiveness based on the interrelationship between factors, as shown in figure 5. However, we still need to assess the reliability and capability of this model by taking into account the key performance ELV recovery indicators, controlled and uncontrolled factors, as well as design strategy elements (i.e. designer's policy).

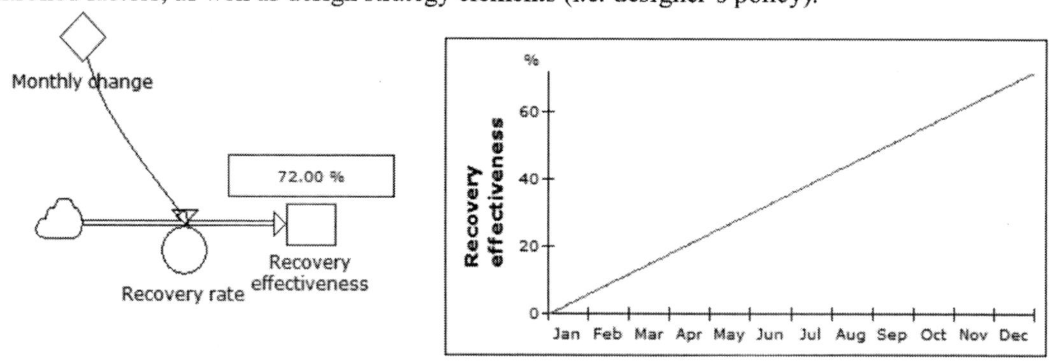

Figure 4. Basic stock and flow analysis

Based on our preliminary analysis, the recovery effectiveness plunges to –948% if the number of parts disposed per month is 84%, as shown in figure 5. Hence, we anticipate that incorporating EOL design strategies during the early stages of product design and development will enhance recovery effectiveness, as shown in figure 4. However, we need to conduct a critical assessment of the factors involved to ensure that our model is representative of the actual ELV recovery scenario in Malaysia. It shall be highlighted that we developed this initial model based on literature survey as well as preliminary data that we gathered from unstructured interviews with the key stakeholders involved in ELV management in Malaysia. Our model is still incomplete, which is why a number of elements in the model are indicated by a question mark or hashtag, as shown in figure 5. At present, we have not yet obtained information regarding the data for reuse, data for recycling, compatibility data, collection of parts, collection compatibility rate, materials used, level of contamination from fasteners, availability of economical reprocessing, and market value of materials. We need to further refine our model in order to determine the best EOL strategies that should be implemented by product designers in order to enhance the effectiveness of ELV recovery in Malaysia, taking into account other factors such as stakeholders, economic factors and infrastructures.

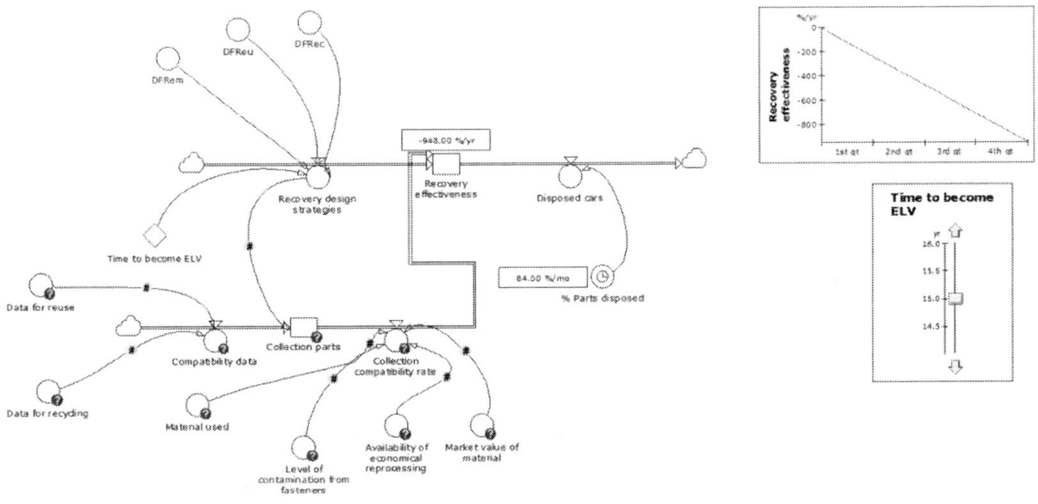

Figure 5. Our preliminary model for ELV recovery in Malaysia

4. Conclusion

At present, ELV management is still relatively new in Malaysia due to several barriers which hinder its effective implementation including lack of knowledge on eco-design and other EOL design strategies such as design for reuse, design for remanufacturing, and design for recycling. Hence, it is imperative to gain a better understanding on how EOL design strategies contribute towards a positive environmental impact for ELVs. In order to fulfil this need, we propose a preliminary model for ELV recovery based on the system dynamics approach in order to understand the impact of different EOL design strategies (specifically design for reuse, design for remanufacturing, and design for recycling) on the effectiveness of ELV recovery in Malaysia.

We also present a brief review of studies pertaining to the application of system dynamics in analysing EOL design strategies. Selecting a suitable EOL design strategy to enhance ELV recovery effectiveness is a challenging task since the ELV recovery system involves a large number of factors with complex interrelationships that are subject to changes in market trends. Therefore, predicting the recovery effectiveness in such uncertainties is crucial in order to improve the current ELV management system in Malaysia. Our model offers a potential solution whereby product designers can analyse changing trends in the automotive industry from a holistic point of view. This tool will be beneficial for product designers to enhance the recovery effectiveness of ELVs in Malaysia.

Acknowledgments
The authors graciously acknowledge the financial support provided by University of Malaya and Ministry of Education Malaysia under the Fundamental Research Grant Scheme (Project no.: FP0592016).

References

[1] Mcauley J W 2003 *Environmental Science & Technology.* **23** 5414-5416
[2] European Comission (2015, December 11) End-of-life vehicle statistics *Overview of Statistical Information.*
[3] NAP 2014 National Automotive Policy Malaysia.
[4] Amelia, Wahab, Haron C C, Muhamad N and Azhari C 2009 *Journal of Cleaner Production.* **17** 1572–1579
[5] Rusinko C A 2007 *IEEE Transactions on Engineering Management* **3** 445-454
[6] Jawahir I, O W Dillon, J Rouch, K Joshi, K J Venkatachalam and Jaafar I H 2006 *Int Research/Expert Conf.* 1-10
[7] Ishii, K, Eubanks, C F and Marco P D 1994 *Materials & Design.* **15** 225-233
[8] Taha Z, Sakundarini N, Ghazila R A and Gonzales J 2010 *Journal of Applied Sciences Research* **6** 2143-2150
[9] Azmi M, Saman M Z and Sharif S 2013 *Proc. of the 11th Global Conf. on Sustainable Manufacturing - Innovative Solutions* (Berlin: Universitätsverlag der TU Berlin) 187-193
[10] MAARA 2015 *Automotive Recycling Transformation Through 4Rs* Malaysia automotive recyclers association
[11] Ghazilla R A, Sakundarini N, Taha Z, Abdul-Rashid S H and Yusoff S 2015 *Journal of Cleaner Production* **108** 331-342
[12] Bhattacharjee S, and Cruz J 2015 **77** 67–86
[13] Wang Y, Chang, X, Chen Z, Zhong Y, and Fan T 2014 *Cleaner Production* **74** 161-171
[14] Halabi E E and Doolan M 2012 *International Society for the Systems Sciences (*California: Creative Commons Attribution) 1-16
[15] Kasperek D, Lindinger M, Maisenbacher S and Maurer M 2014 *Risk and Change management in complex systems (*Paris: DSM 14 Proc) 234-242
[16] Kibira, D., Jain, S., & McLean, C. R. 2009 *Proc of the 27th International Conf* (New Mexico: The System Dynamics Society) 94
[17] Georgiadis P and Besiou M 2008 *Cleaner Production* **16** 1665–1678
[18] Kumar S and Yamaoka T 2007 *Manufacturing Technology Management* **18** 115-138
[19] Bandivadekar A P, Kumar V, Gunter K L and Sutherland J W 2004 *Journal of Manufacturing Systems 23* 22-29
[20] Schaik A V, Reuter M, Boin U and Dalmijn W 2002 *Minerals Engineering.* **15** 1001–1016
[21] Dong X, Li C, Li J, Huang W, Wang J and Liao R 2012 *Journal of Cleaner Production* **20** 72-81
[22] Georgiadis P and Besiou M 2010 *Int J Adv Manuf Technol.* **47** 475–493
[23] Abdulrahman M D A, Subramanian N, Liu C and Shu C 2015 *Journal of Cleaner Production.* **105** 311-323
[24] Rose C M, Beiter K A, Ishii K and Masui K 1998 *ASME Design for Manufacturing Symposium* (Atlanta: ASME) 1-9
[25] Remery M, Mascle C and Agard B 2012 *Journal of Engineering Design* **23** 419-441
[26] Cassells S, Holland J and Meister A 2005 *Journal of Environmental Policy & Planning* **7** 107-124
[27] Linton J 1999 *Journal of Electronics Manufacturing* **9** 29-40

IOP Conf. Series: Materials Science and Engineering **210** (2017) 012076 doi:10.1088/1757-899X/210/1/012076

Subjective responses of mental workload during real time driving: A pilot field study

N I A Rahman[1*], S Z M Dawal[1] and N Yusoff[1]
Department of Mechanical Engineering, Faculty of Engineering, University of Malaya, Kuala Lumpur, Malaysia.

Email: *izzahrahman@um.edu.my

Abstract. This study evaluated drivers' mental workload in real time driving to identify the driving situation's complexity influences in an attempt to further design on a complete experimental study. Three driving settings were prepared: Session A (simple situation); Session B (moderately complex situation); Session C (very complex situation). To determine the mental workload, the NASA-Task Load Index (TLX) was administered to four drivers after each experimental driving session. The results showed that the Own Performance (OP) was the highest for session A (highway), while Physical Demand (PD) recorded the highest mean workload score across the session B (rural road) and C (city road). Based on the overall results of the study, it can be concluded that the highway is less demanding compared to rural and city road. It can be highlighted in this study that in the rural and city road driving situation, the timing must be set correctly to assure the relevant traffic density. Thus, the sensitivity of the timing must be considered in the future experiment. A larger number of experience drivers must be used in evaluating the driving situations to provide results that can be used to draw more realistic experiments and conclusions.

1. Introduction

Driving can be pictured as the relationship between human and machine system environment (driver–vehicle) from the user's point of view. It symbolized an enclosed section that ends with an environmental modification (e.g., "free driving" turns over into "following") [1]. There are several elements of the environment affect the complexity of a driving situation. For example road design (motorways vs. rural roads vs. city roads), road layout (straight vs. with curves, even vs. inclined, junction vs. no junction) and traffic flow (high density vs. low density) [2]. Fastenmeier and Gstalter [1] discussed on the taxonomy of the situation complexity which then categorized a very complex situation as an urban road, with curves or junctions, and with a high traffic density. It is important and crucial to consider these situation characterizing elements in performing the driving task.

Mental workload is one of the critical elements involved in a driving task. Some definition of mental workload was simplified as the relationship between the mental demand of a task and the worker being [3, 4]. Demand can be defined as the effort to be made in order to achieve a certain goal. A human that was given a goal to achieve will allocate an amount of effort in terms of mental, physical and emotional effort. In a driving task, the destination here can be described as the goal of a demand. Ergonomists are concerned about the drivers' mental workload since an overload of information may lead to distraction while the driving task is performed [5, 6].

Generally, drivers of any vehicle have to go through various different types of conditions and environments while driving. A driving situation contributes as the main factor of the mental workload of drivers [6]. There are situations when a driver is overload and underload in performing the driving task. For example, when a driver experiencing an overload situation such as he/she has to carry out several tasks concurrently while also having to pay attention to other vehicles in a high density traffic. This kind of situation can lead to stress and deteriorated performance [7]. On the other hand, when mental workload is much lower than the proper level, people will become bored and tend to make mistakes [8]. In an underload driving situation, monotonous driving can cause drowsiness for example long distance driving in a low density highway driving. Sleepiness when driving possibly will occur in situations such as stuck in traffic jams where sometimes vehicles on the road does not move at all for hours. Normal mental workload occur when a driver has enough capacity to perform the main driving activity, awake, focus with less distractions and confusion along the driving task.

Consequently, it is significant to evaluate mental workload in real driving conditions to identify more on the driving situations influences. This critical finding is important not only for young drivers, but also different population such as elderly drivers. Furthermore, it is valuable to have early identification of mental workload level in this pilot study to further design on the complete experimental study with more highlighted parameter and critical settings. In this study, it is hypothesized that city driving contributes low workload and considered as less complex, followed by rural road driving and city driving situation.

2. Procedure

A. Participants

The participants consisted of four healthy male volunteers. None of them have health problems that could potentially interfere with their ability to perform the driving task. The requirement of becoming a participant is holding a driver's license for at a year and having driven at least 10,000 km.

B. Experimental setup

The goal of the experimental study was to measure mental workload in real time and in real operational environments. The traffic intensity was not controllable and depends on the type of road setting. The route are identical to all participants (total driving distance is about 30 km), which took on the average about 45 minutes to complete. The classification of the driving environments is based on the classification pattern in terms of different complexities (low, medium, high), as proposed by Paxion [2] and partly from Fastenmeier [1]. Details of the environment setting are as in table 1.

Table 1. The routes and environment setting

	Session A: Simple Situation	**Session B :Moderately Complex Situation**	**Session C :Very complex Situation**
Road design	Highway	Rural Road	City Road
Speed limits	110km/h	90km/h	60 to 90km/h
Road layout	Straight	Few curves	Many curves
Traffic flow	Low density	Moderate density	High density

C. Task description

The protocol was approved by the Medical Ethics Committee, University Malaya Medical Centre. Upon arriving at the setup station, the participants were informed on details of the experiment and signed an informed consent and filled out a demographic questionnaire. Participants were then given the opportunity to familiarize themselves with the vehicle and environment by driving around on a two-lane practice road without any other traffic. Participants were told that their first priority was to drive safely. Subsequently, the participants were instructed to drive according to the route that has been set. A research associate was seated in the rear of the vehicle and was responsible for providing driving directions, ensuring safe vehicle operation. In each end of the A, B and C session, participants received the instruction to pull-over and park the car. This break lasted around five minutes throughout which the participants were asked to subjectively rate the NASA- Task Load Index (TLX) scales based on that particular session.

D. Subjective measures – NASA Task Load Index (TLX)

The NASA-TLX [9] is used to evaluate the operators' subjective workload. The participants will be asked to self-report his mental workload after finishing every primary task of each route session. The participant takes approximately 5 to 10 minutes to complete the subjective assessment. The NASA-TLX method includes six sub-scales (table 2). The mean (raw TLX) and weighted mean (weighted workload) of these six sub-scales were calculated using paired comparisons.

Table 2. NASA-TLX scales [9].

Title	Descriptions
Mental Demand	How much mental and perceptual activity was required (e.g., thinking, deciding, calculating, remembering, looking, searching, etc.)? Was the task easy or demanding, simple or complex, exacting, or forgiving?
Physical Demand	How much physical activity was required (e.g., turning, controlling, activating, etc.)? Was the task easy or demanding, slow or brisk, slack or strenuous, restful or laborious?
Temporal Demand	How much time pressure did you feel due to the rate or pace at which the tasks or task elements occurred? Was the pace slow and leisurely or rapid and frantic?
Performance	How successful do you think you were in accomplishing the goals of the task set by the experimenter (or yourself)? How satisfied were you with your performance in accomplishing these goals?
Effort	How hard did you have to work (mentally and physically) to accomplish your level of performance?
Frustration Level	How insecure, discouraged, irritated, stressed, and annoyed versus secure, gratified, content, relaxed, and complacent did you feel during the task?

E. Statistical Analysis

The descriptive statistics and difference between subjective responses of the three conditions during the driving task for all sub-scales and overall workload of the NASA-TLX were analyzed using the IBM Statistical Package for Social Science (SPSS) for Windows version 23.0 (IBM SPSS Statistics for Windows Version 23.0, Armonk, NY: IBM Corp).

IOP Conf. Series: Materials Science and Engineering **210** (2017) 012076 doi:10.1088/1757-899X/210/1/012076

3. Results

A. *Subjective workload attributed to situation complexity*

The average age and driving experience of the participants were 38.5 ± 14.5 (SD) and 17.5 ± 11.7 (SD) years, respectively. The results of the subjective ratings of workload measured by NASA-TLX across the A, B and C sessions are summarized in table 1. The NASA-TLX scores ranged from 0, representing no demand, to 100, representing maximum demand. The mean subjective workload ratings for each subscale for Session A, B and C is summarized in tables 3 and the graphical compositions are shown in figure 1 to 3. For session A (highway), Own Performance (OP) was the highest score followed by Temporal Demand (TD). Overall, the Physical Demand (PD) recorded the highest mean workload score across the session B (rural road) and C (city road). The y-axis of the figure 1 to 3 shows the weighted ratings as the dependent measure while the x-axis represent the six workload subscales as the independent measures. Meanwhile, the width of the subscale bar reflects the weight (or the relative importance) assigned to it, with the value of the weight on top of the bar; the height of the bar represents its rating. The adjusted workload rating is, therefore, the area of the subscale bar obtained as the product of the weight and the rating. On the other hand, the overall weighted rating for each driving session was acquired by dividing the sum of the adjusted workload score across the six subscales by 15. These are summarized shown graphically in figure 4. Session B had the highest overall workload score, followed by Session C and Session A respectively.

Table 3. NASA-TLX data summary for Session A, B and C.

Session	Subscale	Weight	Raw Rating	Ajusted Rating (Weight x Raw)
A	MD	2	8.75	17.50
	PD	3	7.50	22.50
	TD	4	11.25	45.00
	OP	4	23.75	95.00
	EF	2	10.00	20.00
	FR	0	8.75	0.00
B	MD	1	27.50	27.50
	PD	4	43.75	175.00
	TD	3	33.75	101.25
	OP	5	13.75	68.75
	EF	3	32.50	97.50
	FR	1	17.50	17.50
C	MD	1	10.00	10.00
	PD	4	35.00	140.00
	TD	3	28.75	86.25
	OP	5	11.25	56.25
	EF	3	12.50	37.50
	FR	1	10.00	10.00

MD: Mental Demand, PD: Physical Demand, TD: Temporal Demand, OP: Own Performance, EF: Effort, FR: Frustration.

IOP Conf. Series: Materials Science and Engineering **210** (2017) 012076 doi:10.1088/1757-899X/210/1/012076

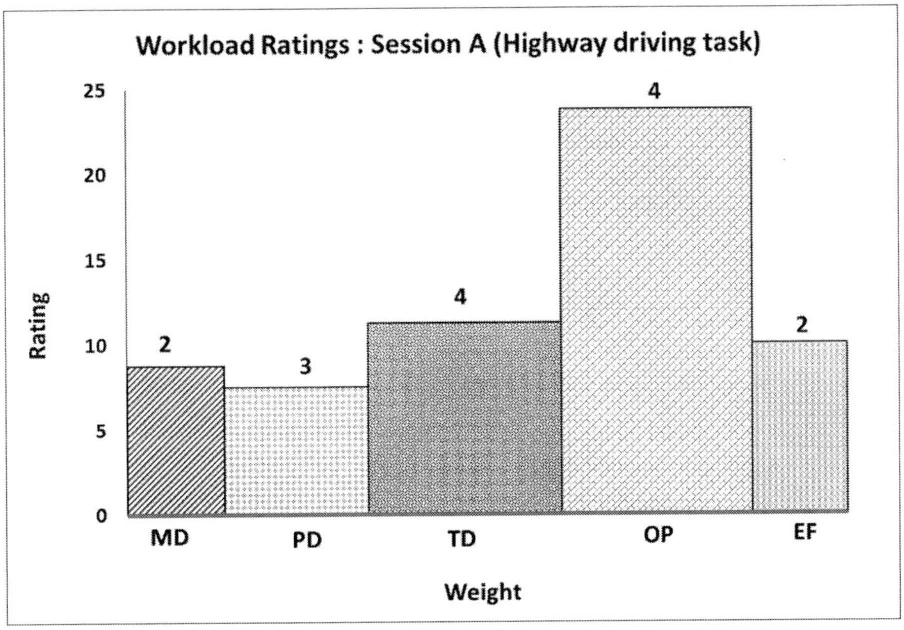

Figure 1. Plot of NASA-TLX workload ratings for Session A.

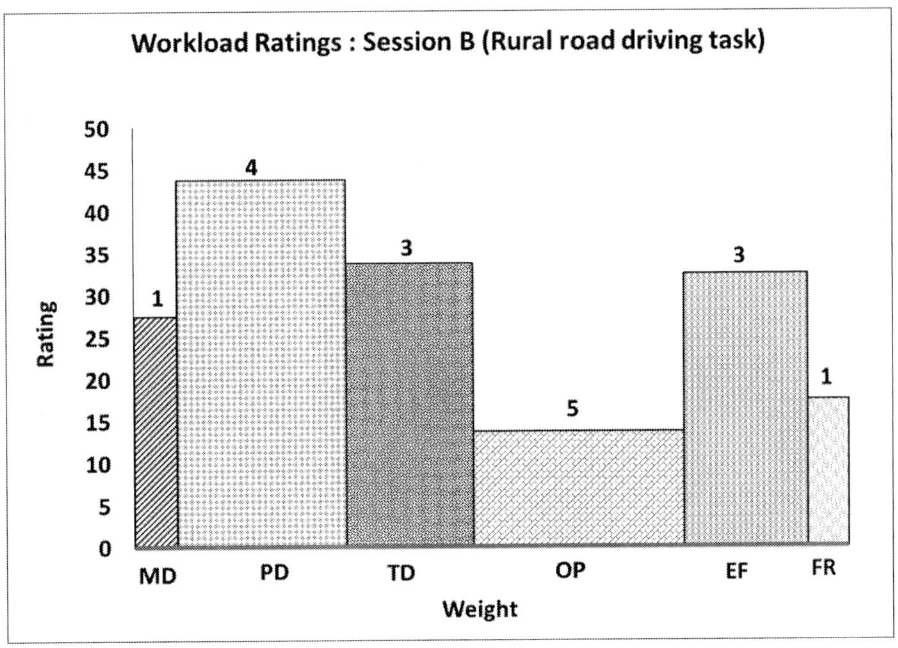

Figure 2. Plot of NASA-TLX workload ratings for Session B.

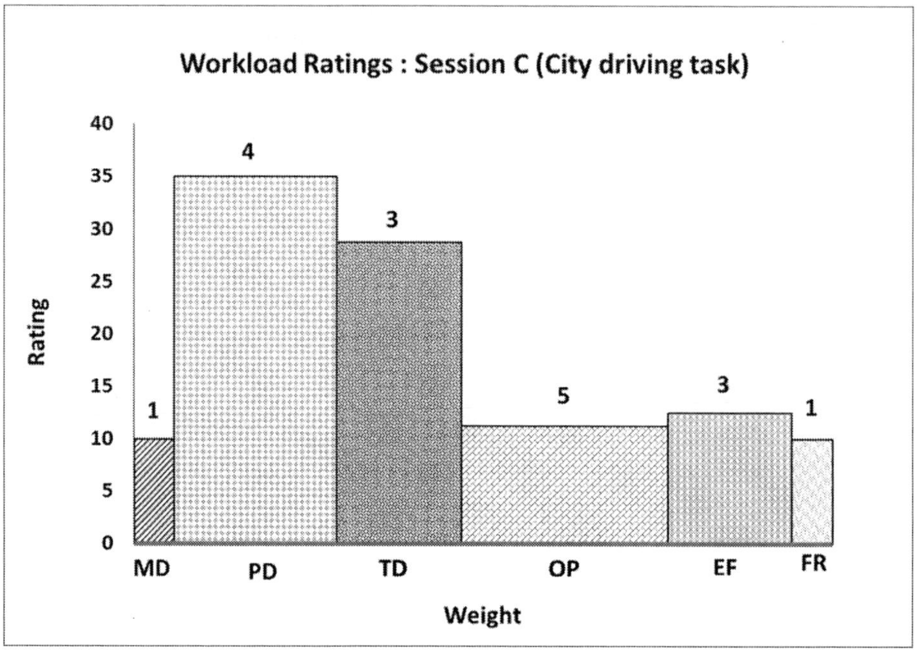

Figure 3. Plot of NASA TLX workload ratings for Session C.

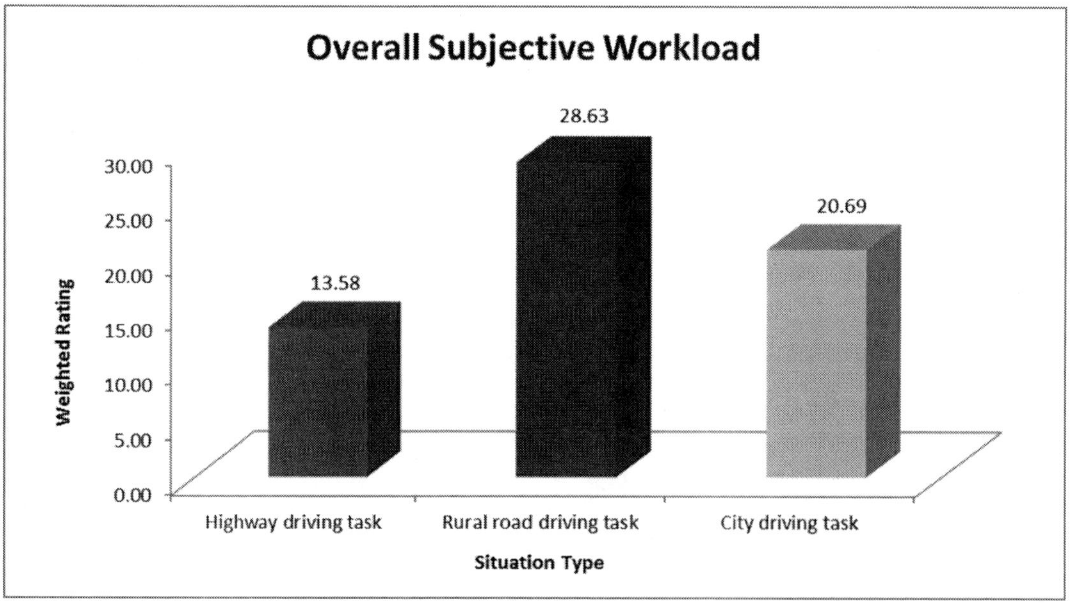

Figure 4. Plot of overall workload ratings for Session A, B and C.

4. Discussion

The results of the subjective workload scores from the NASA-TLX showed that Own Performance (OP) induced the highest workload in Session A. However, the score is still much lower from the highest score of the other two sessions. This partially confirms the research hypothesis that the highway is less demanding compared to the rural and city driving. The result highlighted that there is no frustration at all during the session while the participant rated low workload on the mental demand.

Mental demand is expected to be low due to the straight road layout and high speed limit. However, Thompson [10] found in their study that percentage of drivers seriously injured or killed by crashes increased with the increase of speed limits in South Australia. It is reflected here that when the situation is low demanding, the performance may impaired due to underload condition.

Meanwhile, for rural (Session B) and city (Session C) driving situation, it is surprisingly recorded that physical demand scored the highest workload. The score translated to users feeling that they had to exert a substantial amount of physical movement to accomplish the goals of the task. The reason that can be attributed to this outcome is that the rural area may have car, pedestrian, cyclist crossing and appearing during the driving sessions plus houses on both sides of the road. The city driving situation was extended from those conditions with more shops and houses, traffic lights and parked busses and cars along the road. Overall, there was significant different in the mental workload of the three types of situations which Session B scored the highest overall subjective workload. When compared to Session B, the mental workload was reduced by 52.57% for Session A, and 27.73 % for Session C. This finding is consistent with the emphasized findings by Verwey [6] stating that a driving situation constitutes the major determinant of the mental workload of drivers. Similarly, Paxion [11] reported that an increase of subjective workload was globally observed when the situation complexity increased, with a tendency to higher self-assessment of the workload in very complex situation than in the simple one. It is anticipated in this study that city driving will score the highest workload, however, the result is contradict which it is probably due to lack complexity of the traffic density of the city during the experimental session. City road traffic density usually is at most congested condition in the weekday morning where people are travelling to work. It can be highlighted here that the experiment timing is very important to see clearly driving situation complexity differences. Furthermore, the lack number of participants in this study probably led to the result achieved.

5. Conclusion

This study described the effect of driving situations on subjective workload demands. The goal of the research was to obtain an early identification of mental workload level with respect to the situation complexity to further design on a complete experimental study. Three driving situations were formed and compared. The six subscales of the NASA-TLX subjective workload assessment tool was used to obtain the workload ratings for characterizing each driving situation setting. The results showed Own Performance (OP) was the highest for Session A (highway), Physical Demand (PD) recorded the highest mean workload score across the Session B (rural road) and C (city road). Based on the overall results of the study, it can be concluded that the highway is less demanding compared to rural and city road. It is highlighted that the timing must be set correctly in the rural and city road driving situation. Thus, future experiment will have to consider the sensitivity in the experimental timing. More number of experience drivers must be used in evaluating the driving situations in order to provide results that can be used to draw more realistic experiments and conclusions.

Acknowledgement

This work is financially supported by University of Malaya under the Postgraduate Research Grant (PPP) grant no.: PG1162016A.

References

[1] Fastenmeier W and Gstalter H 2007 *Saf. Sci.* **9** 952-79
[2] Paxion J, Galy E and Berthelon C 2014 *Frontiers in Psychology* **5** 1344
[3] de Waard D 1996 *The Measurement of Drivers' Mental Workload* (Groningen: Groningen University)
[4] Leung A W 2006 *Neurophysiological Correlates of Performance and Fatigue in Study of Mental Workload* (Hong Kong: The Hong Kong Polytechnic University)
[5] Mayser C, Piechulla W, Weiss K E and König W 2003 Driver workload monitoring *50th-Anni. Conf. of the GfA and XVII Ann. ISOES Conf.*
[6] Verwey W B 2000 *Ergo* **2** 187-209
[7] Gáspár P , Szalay Z and Aradi S 2014 *Eng. Psychol. and Cog. Ergo.: 7th Int. Conf. EPCE 2007 Proc.* 316-23
[9] Hart S G and Staveland L E 1988 *Adv. Psychol.* **52** 139-83
[10] Thompson J P, Baldock M R J, Mathias J L and Wundersitz L N 2013 *Accid. Anal. Prev.* **50** 768-75
[11] Paxion J, Galy E and Berthelon C 2015 *App. Ergo.* **51** 343-49

The effect of elevated die temperature on deformation of deep drawn round metal cup

M A M Basril[*1], Y. M. Hafsyam[1], M. Azuddin[1], I.A. Choudhury[1]
[*1]Department of Mechanical Engineering, Faculty of Engineering, University of Malaya, 50603 Kuala Lumpur, Malaysia

E-mail: *muhammadbasril@siswa.um.edu.my

Abstract. One of the major considerations in the current deep drawing practice is the product quality. In this research, the effect of heating temperature on the drawability of the round metal cup has been investigated. Firstly, round metal cups of aluminium and mild steel were drawn from the blank diameters of 60 mm, 65 mm and 70 mm. The experiment conducted at room temperature first, then at 50°C and 100°C. The elongation of the major and minor strains along the cup profile after the process is measured and analysed. On the other hand, the defects from the experiment output and ABAQUS/CAE simulation are compared. The result from experiment shows that the highest major elongation is 11.64 mm and it is happened to a deep drawn aluminium round cup with LDR of 1.69 at temperature of 100°C. On the other hand, for deep drawn mild steel round cup, shows highest major elongation of 12.44 mm for a cup with LDR of 1.56 at 100°C. Both of these statements indicates that the higher temperature could improve the formability of the deep drawn parts besides reducing the probability of the defect to be happened.

1. Introduction

Deep drawing is a manufacturing process in which sheet metal is progressively formed into a three-dimensional shape through the mechanical action of a die forming the metal around a punch. The deep drawn metal stamping process work hardens the metal, resulting in a seamless finished part that is stronger than the base material. Deep drawing can produce precise and intricate parts, including both symmetrical and asymmetrical parts.

Deep drawing operation for complex part geometries can only be realised by expensive multi-step production processes. In order to overcome this problem, deep drawing can be conducted at elevated temperatures so that the formability of the drawn cup in term of limiting drawing ratio (LDR) and strain elongation can improved significantly. Referring to Lean et al. [1], the application of temperature increment towards punch and die can produce several outcomes such as increasing limiting drawing ratio, increment on depth of drawn cup and reduce the fracture effect. Therefore, to study the effect of heating temperature on round deep drawing by utilizing aluminium and mild steel blank specimen is important in order to prove the indictment that has been made. On top of that, an experimental-numerical method also introduced to provide a vivid comparison between experimental outcomes and simulation. Not only that, a numerical analysis also useful for determining the safe condition also can be obtained which can act as the reference for the experimental works as indicated in previous research [2].

Next, the limiting drawing ratio (LDR) is ratio of the maximum diameter of the workpiece that is drawn without tearing also can resembled the formability of the drawn cup [3]. Referring to Kim et al. [4], there were several factors that can influence the formability of a material such as characteristics of sheet metal die, blank holding force, thickness of blank, forming temperature and heating technique, the bending and unbending effect, frictional force and geometrical parameters of the setup. This is because, as known before, the deep drawing process will involve almost all the parts of the cup. According to Zeng et al. [5], there will be huge difference for stress and strain distribution in regions caused by the punch and die in deep drawing process that utilize round blank. For the blank that situated around punch region, it much easier to expose the risk of tearing as the biaxial tensile stress that exists around the punch region, especially the punch corner area. Not only that, over the die region, the will be radial tensile stress while compressive stress existed during the deep drawing process.

The warm forming procedure is planned to overcome this low formability problem by utilizing elevated temperature which is under the recrystallization temperature. According to the works of Jayahari et al. [6], which conduct a study on austenitic stainless steel (ASS)-304 of under warm deep drawing with a different blank diameter. Based on the findings, the limiting drawing ratio (LDR) at room temperature was 2.1 and increase to 2.5 as the process conducted at 150°C. In this experiment, different blank diameters were tested with different temperatures up until 300°C and it show positive results in term of limiting drawing ratio. Not only that, Moon et al. [7] finds that the variation of material flow stress with temperature is substantial even with slight difference of tools temperature change. Poor decision on working parameters could leads to several kinds of failure such as wrinkling.

Wrinkling effect was formed when the radial tensile stress and tangential compressive stress occur simultaneously. It was causes by worst case of BHP pressure and it is preventable if the die set geometrical dimension and the forming forces were planned and tested beforehand. Not only that, Padmanabhan et al. [8], has state that in order to confine wrinkling on the drawn cup, the least blank holding pressure (BHP) is needed for the deep drawing operation. As the BHP pressure is too high, it will causes fracture at the top edge of the cup while if BHP pressure is low, it will causes wrinkling effect on the drawn cup. Hence, proper working parameters have to be decided first so that the relevant outcomes for the analysis purpose can be obtained.

2. Methodology

2.1 Set up of Experiment

The set up for the experimental procedures for warm deep drawing has been shown in figure 1. A single acting hydraulic cylinder is attached to a die set and it is controlled by hydraulic pump as shown in figure 2. On top of that, a temperature controller which shown in figure 3, is used to control the die set temperature while the die set is heated up by heater. A set of type K thermocouple is used to sense temperature and the reading is shown on the controller. The punching force generated from single acting hydraulic cylinder and being control by hydraulic pump. It has 101 kN capacity of force with 105 mm stroke which will press sheet metal downward into die cavity when the process takes place.

IOP Conf. Series: Materials Science and Engineering **210** (2017) 012077 doi:10.1088/1757-899X/210/1/012077

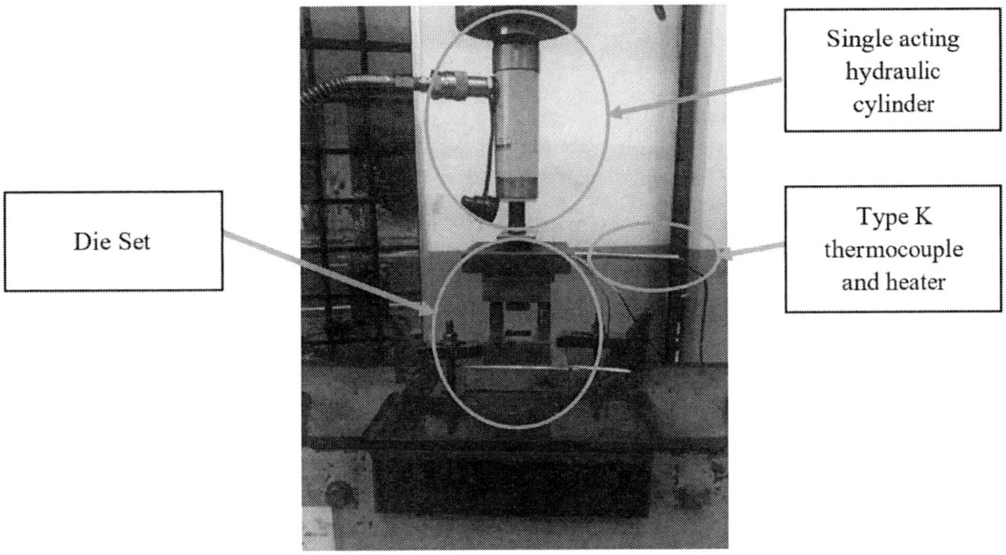

Figure 1. Experimental setup

Figure 2. Hydraulic pump

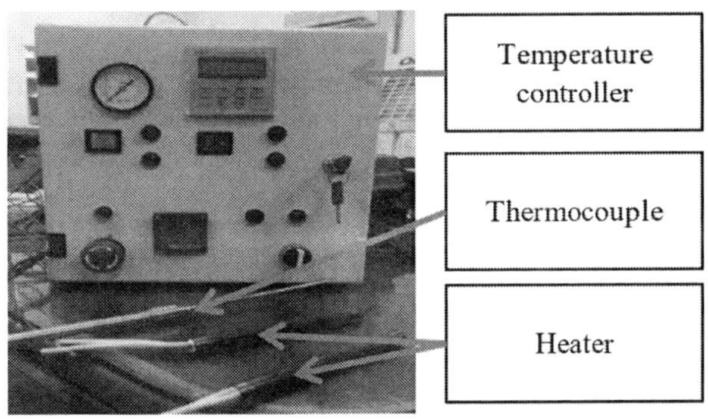

Figure 3. Temperature controller, heater and type K thermocouple

2.2 Procedure

The experimental works for the round deep drawing were conduct as follows:

1. Blank will be prepared into 3 different diameter sizes which are 60 mm, 65 mm and 70 mm. The blank also provided with circular grid with 10 mm diameter for analysis purposes as shown in figure 4.
2. After the preparation, the aluminium blank will be placed on the die set and being clamped using blank holder.
3. The die set will attached to the hydraulic cylinder with pump.
4. The experiment for room temperature condition is started as pump is activated and generates punching force onto the blank until it is fully deform.
5. The drawn cup being extracted from the die set. The changes in shape of engraved circular grid are being measured and tabulated. After that, the next blank diameter size undergoes the same process.
6. The experiment is repeated from step 1 to 6 but implementing the warm temperature and heating technique. The warm temperature decided to be at 50°C and 100°C. On top of that, the experiment will be conducted in 3 different heating technique:

a. Heating both upper and lower dies
b. Heating upper dies
c. Heating lower dies

7. After that, all the procedure mention above is once again repeated but, by using mild steel blank. All the results data has been tabulated were discussed and analysed.

(a) (b)

Figure 4. Circular Grid engraved on (a) aluminium and (b) mild steel blank specimen surface

3. Results and discussion

3.1 Condition of deep drawn round cup

Before proceed with the described procedure, a series of test runs has been conducted in order to decide the best heating condition From observation, most of the cup drawn from the experiment was in sound condition as the warm temperature was applied on lower side of the die set. These outcomes were the same as the indictment of Ghaffari Tari, Worswick, and Winkler [9] that the deep drawing process can be enhance by introducing the non-isothermal condition. This condition is achieved, as there is temperature difference between punch nose and flange regions. Therefore, based on the test runs outcomes, the lower die set been decided to be heated for this study. As per observation, it can be clearly seen that the earing effect exist at all sound cup that has been drawn from bot room and warm temperature condition. This can be referred in figure 5 shown below.

(a) (b)

Figure 5. Example of the test runs outcomes of (a) aluminium and (b) mild steel round deep drawing

Next, the drawability condition of the drawn round aluminium cup is tabulated as shown in table 1 while for drawn round mild steel cup can be refer from table 2. Besides that, the limiting drawing ratio (LDR) of the successful drawn cup has been calculated and shown in the same table.

Table 1. Drawability condition for drawn round aluminium cup

No.	Blank size (mm)	LDR	Temperature (°C)	Heating Technique	Conditions
1	60	1.56	Room Temperature	-	Successful
2	65	1.69	Room Temperature	-	Successful
3	70	1.82	Room Temperature	-	Successful
4	60	1.56	50°C	Lower	Successful
5	65	1.69	50°C	Lower	Successful
6	70	1.82	50°C	Lower	Failed
7	60	1.56	100°C	Lower	Successful
8	65	1.69	100°C	Lower	Successful
9	70	1.82	100°C	Lower	Failed

IOP Conf. Series: Materials Science and Engineering **210** (2017) 012077 doi:10.1088/1757-899X/210/1/012077

Table 2. Drawability condition for drawn round mild steel cup

No.	Blank size (mm)	LDR	Temperature (°C)	Heating Technique	Conditions
1	60	1.56	Room Temperature	-	Successful
2	65	1.69	Room Temperature	-	Successful
3	70	1.82	Room Temperature	-	Successful
4	60	1.56	50°C	Lower	Successful
5	65	1.69	50°C	Lower	Successful
6	70	1.82	50°C	Lower	Successful
7	60	1.56	100°C	Lower	Successful
8	65	1.69	100°C	Lower	Successful
9	70	1.82	100°C	Lower	Successful

3.2 Major and Minor elongation of drawn round cup (Aluminium)

The outcomes data after elongation of the circular grid is measured and recorded. For each drawn cups, there must be at least 10 clearly visible circles from the circular pattern, which, experience major and minor elongation. Then, it is calculated and compared to the original diameter of the circlular grid pattern, which is 10 mm. The increment and decrement of the circle is recorded as shown in table 3 and 4.

Table 3. Elongation of drawn round aluminium cup with 60 mm blank diameter (LDR=1.56)

Number of Element	Temperature (°C)					
	Room Temperature		50		100	
	Major (mm)	Minor (mm)	Major (mm)	Minor (mm)	Major (mm)	Minor (mm)
1	10.49	9.42	10.75	8.92	10.24	9.67
2	10.25	8.43	10.05	9.10	10.88	8.73
3	10.49	9.90	10.49	9.20	11.98	9.35
4	10.72	9.13	10.36	9.73	11.59	8.92
5	10.41	8.93	11.16	9.26	11.19	8.45
6	10.40	8.70	10.16	9.11	10.76	8.77
7	10.71	9.48	11.02	8.27	10.44	9.21
8	10.43	8.86	10.81	8.86	10.23	8.42
9	10.53	8.93	11.19	8.87	11.09	8.36
10	10.49	9.11	10.89	8.95	11.97	9.41
Average	**10.49**	**8.99**	**10.69**	**9.03**	**11.04**	**8.93**

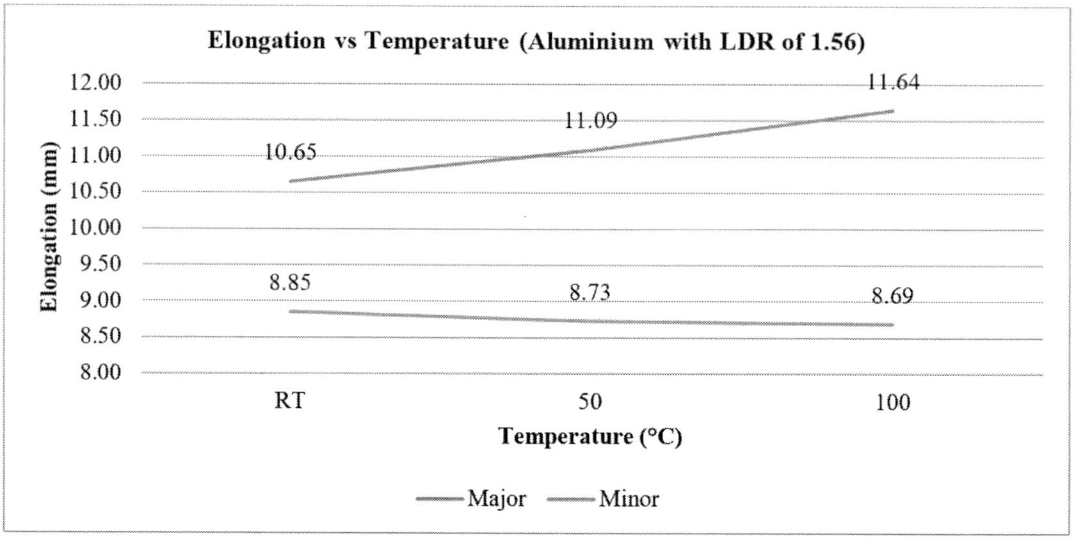

Figure 6. Elongation for deep drawn round aluminium cup with LDR of 1.56

Table 3 indicates the average major elongation of deep drawn round aluminium cup with LDR 1.56 at room temperature was 10.49 mm and minor elongation was 8.99 mm. The major elongation was increase to 10.69 mm, as warm temperature of 50°C was applied to lower die set. This was also applicable to the minor elongation that increases to 9.13mm. As the warm temperature increase up until 100°C, the major elongation increase to 11.04 mm while minor elongation decrease to 8.93 mm. This also can be observed from figure 6 shown above. It shows that, at warm temperature of 50°C, the major elongation experience 1.9% of increment compare to room temperature condition, while 5.2% for warm temperature of 100°C. On the other hand, the minor elongation for cup drawn at 50°C suffers 0.4% decrement and continues to decrease to 0.6% as the cup being drawn at 100°C.

Table 4. Elongation of drawn round aluminium cup with 65 mm blank diameter (LDR=1.69)

Number of Element	Temperature (°C)					
	Room Temperature		50		100	
	Major (mm)	Minor (mm)	Major (mm)	Minor (mm)	Major (mm)	Minor (mm)
1	11.71	8.58	11.98	8.32	12.24	8.02
2	10.06	9.57	10.45	9.05	10.84	8.53
3	10.77	8.27	11.48	8.94	12.18	8.24
4	10.69	8.23	11.57	8.58	12.45	9.1
5	10.39	9.29	10.56	9.17	10.72	8.57
6	10.64	9.37	11.22	8.18	11.79	8.82
7	10.52	9.13	11.41	9.3	12.29	8.34
8	10.68	8.41	10.88	8.22	11.08	9.27
9	10.53	8.7	11.07	8.17	11.6	8.73
10	10.49	8.91	10.83	9.33	11.17	9.32
Average	**10.65**	**8.85**	**11.09**	**8.73**	**11.64**	**8.69**

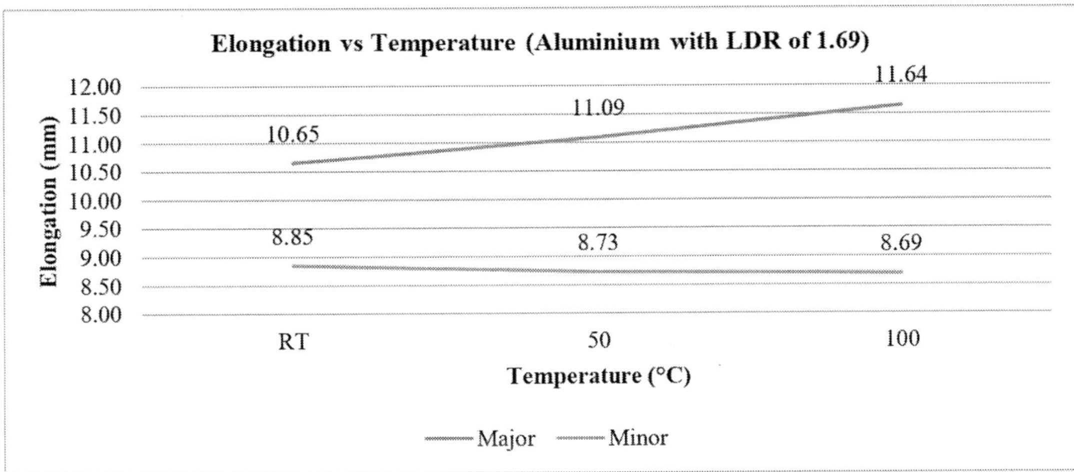

Figure 7. Elongation for deep drawn round aluminium cup with LDR of 1.69

Table 4 show the average major elongation of deep drawn round aluminium cup with LDR 1.69 at room temperature was 10.65 mm and minor elongation was 8.85 mm. The major elongation is increase to 11.09 mm as warm temperature of 50°C was applied to lower die set. On the other hand, the minor elongation showed increase to 8.73 mm. As the warm temperature increase up until 100°C, the major elongation increase to 11.64 mm while minor elongation decreasing to 8.69 mm. This also can be observed from figure 7 shown above. It shows that, at warm temperature of 50°C, the major elongation experience 4.1% of increment compares to room temperature condition, while 5.0% for warm temperature of 100°C. On the other hand, the minor elongation for cup drawn at 50°C suffers 0.4% decrement and continues to decrease to 0.5% as the cup being drawn at 100°C.

3.3 Major and Minor elongation of drawn round cup (Mild Steel)

After all of the aluminium blank has been successfully being drawn, the experiment move on by proceed to warm deep drawing process of mild steel blanks. The outcomes data after elongation of the circular grid is measured and recorded. For each drawn cups, there must be at least 10 clearly visible circles from the circular pattern, which, experience major and minor elongation, which then will be calculated and compared to the original diameter of the circular grid pattern, which is 10 mm. The increment and decrement of the circle is recorded as shown in table 5, 6, and 7.

Table 5. Elongation of drawn round mild steel cup with 60 mm blank diameter (LDR=1.56)

Number of Element	Temperature (°C)					
	Room Temperature		50		100	
	Major (mm)	Minor (mm)	Major (mm)	Minor (mm)	Major (mm)	Minor (mm)
1	9.51	8.48	10.98	8.89	12.44	8.29
2	9.67	6.87	10.45	8.42	11.43	9.33
3	10.13	9.31	11.43	8.60	12.67	7.88
4	11.43	8.34	11.25	8.73	11.07	8.13
5	10.58	9.59	10.47	9.06	10.36	8.52
6	10.71	8.15	10.71	8.33	10.71	8.39
7	10.08	9.03	10.32	8.43	10.45	8.07
8	10.23	9.54	10.53	7.65	10.82	8.45
9	11.02	8.12	10.83	8.71	10.64	8.30
10	10.32	8.93	11.54	9.15	11.56	9.12
Average	**9.51**	**8.48**	**10.98**	**8.89**	**12.44**	**8.29**

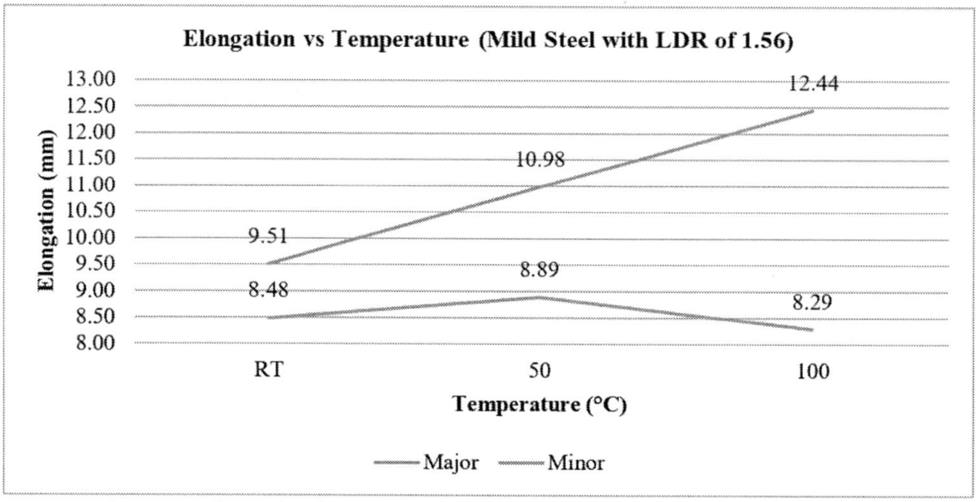

Figure 8. Elongation for deep drawn round mild steel cup with LDR of 1.56

Table 5 shows the average major elongation of deep drawn round mild steel cup with LDR 1.56 at room temperature was 9.51 mm and minor elongation was 8.48 mm. The major elongation is increase to 10.98 mm as warm temperature of 50°C was applied to lower die set. On the other hand, the minor elongation showed increase to 8.89 mm. As the warm temperature increase up until 100°C, the major elongation increase to 12.44 mm while minor elongation decreasing to 8.29 mm. This also can be observed from figure 8 shown above. It shows that, at warm temperature of 50°C, the major elongation experience 15.46% of increment compares to room temperature condition, while 13.3% for warm temperature of 100°C. On the other hand, the minor elongation for cup drawn at 50°C increase 4.8% compare to room temperature condition. After that, the minor elongation continues to suffer decrement about 6.75% as the cup being drawn at 100°C.

Table 6. Elongation of drawn round mild steel cup with 65 mm blank diameter (LDR=1.69)

Number of Element	Temperature (°C)					
	Room Temperature		50		100	
	Major (mm)	Minor (mm)	Major (mm)	Minor (mm)	Major (mm)	Minor (mm)
1	10.91	9.13	11.21	9.10	11.49	8.84
2	11.1	8.39	11.23	8.49	11.31	8.51
3	11.07	9.17	12.82	8.75	10.57	8.32
4	11.38	8.75	10.42	8.81	11.33	8.87
5	11.09	9.06	11.29	8.70	11.48	8.53
6	10.7	9.45	10.92	9.23	11.13	8.75
7	10.47	9.35	10.71	8.80	10.91	8.25
8	10.84	8.39	11.58	8.47	11.32	8.54
9	11.27	9.51	11.18	9.09	12.09	8.67
10	11.44	8.85	11.39	8.88	11.45	8.91
Average	**11.03**	**9.01**	**11.27**	**8.83**	**11.31**	**8.62**

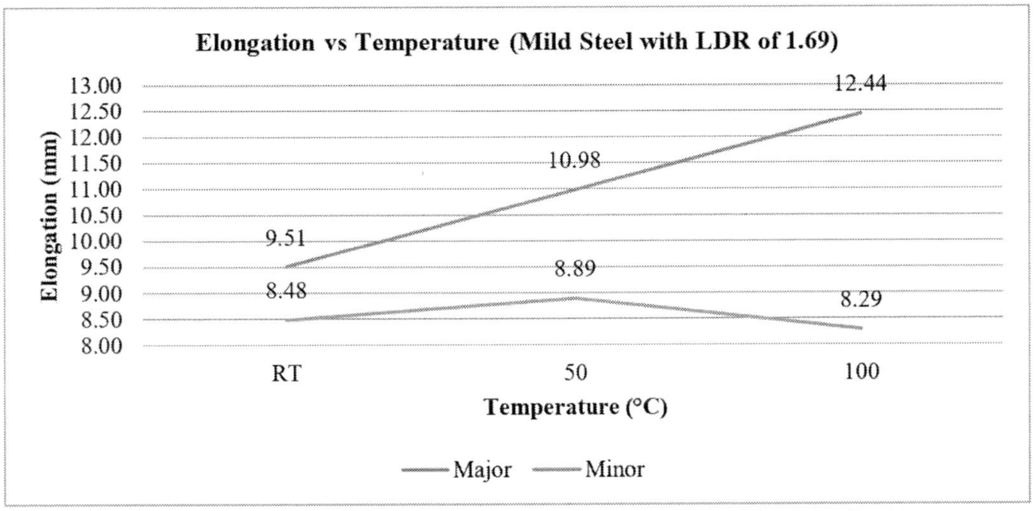

Figure 9. Elongation for deep drawn round mild steel cup with LDR of 1.69

Table 6 shows the average major elongation of deep drawn round mild steel cup with LDR 1.69 at room temperature was 11.03 mm and minor elongation was 9.01 mm. The major elongation is increase to 11.27 mm as warm temperature of 50°C was applied to lower die set. On the other hand, the minor elongation showed decrease to 8.83 mm. As the warm temperature increase up until 100°C, the major elongation increase to 11.31 mm while minor elongation decreasing to 8.62 mm. This also can be observed from figure 9 shown above. It shows that, at warm temperature of 50°C, the major elongation experience 2.18% of increment compares to room temperature condition, while 0.35% decrement for warm temperature of 100°C. On the other hand, the minor elongation for cup drawn at 50°C decrease 1.99% compare to room temperature condition. After that, the minor elongation continues to suffer decrement about 2.38% as the cup being drawn at 100°C.

Table 7. Elongation of drawn round mild steel cup with 70 mm blank diameter (LDR=1.82)

Number of Element	Temperature (°C)					
	Room Temperature		50		100	
	Major (mm)	Minor (mm)	Major (mm)	Minor (mm)	Major (mm)	Minor (mm)
1	13.25	7.84	12.77	7.91	11.89	7.98
2	11.93	9.06	12.25	8.98	12.56	8.89
3	10.58	9.1	11.42	9.21	12.07	9.21
4	10.86	9.24	11.21	8.67	11.53	8.1
5	12.61	7.89	12.16	7.57	12.51	7.25
6	11.01	9.15	11.28	8.38	11.55	7.61
7	10.82	8.23	10.96	8.42	12.1	8.61
8	12.23	8.88	11.82	8.65	11.41	8.46
9	11.98	7.98	11.73	8.40	11.79	8.82
10	12.43	9.12	12.64	9.10	12.84	8.47
Average	**11.77**	**8.65**	**11.82**	**8.53**	**12.03**	**8.34**

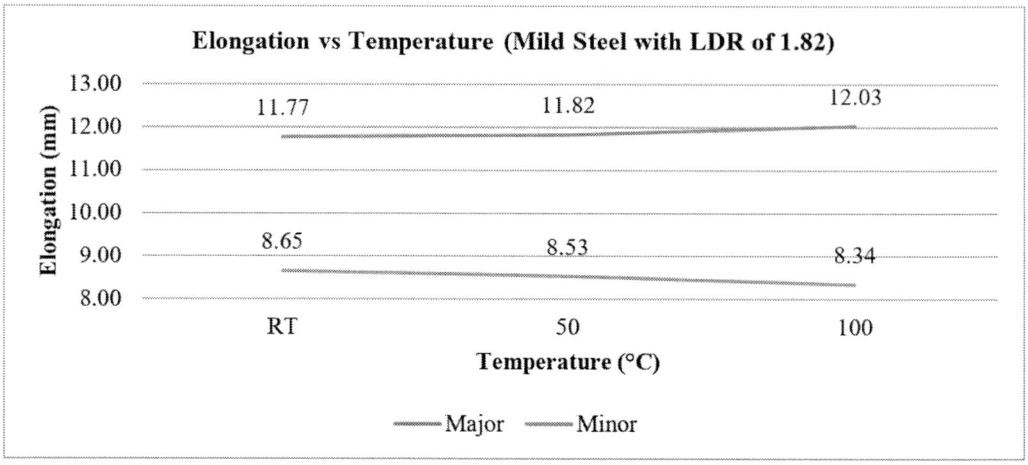

Figure 10. Elongation for deep drawn round mild steel cup with LDR of 1.82

Table 11 shows the average major elongation of deep drawn round mild steel cup with LDR 1.82 at room temperature was 11.77 mm and minor elongation was 8.65 mm. The major elongation is increase to 11.82 mm as warm temperature of 50°C was applied to lower die set. On the other hand, the minor elongation showed decrease to 8.53 mm. As the warm temperature increase up until 100°C, the major elongation increase to 12.03 mm while minor elongation decreasing to 8.34 mm. This also can be observed from figure 10 shown above. It shows that, at warm temperature of 50°C, the major elongation experience 0.42% of increment compares to room temperature condition and 1.78% increment for warm temperature of 100°C. On the other hand, the minor elongation for cup drawn at 50°C decrease 1.39% compare to room temperature condition. After that, the minor elongation continues to suffer decrement about 2.23% as the cup being drawn at 100°C.

3.4 Discussion on major and minor elongation of drawn round cup (Aluminium)

Based on observation, the drawn cup using blank with 60 mm diameter where the major elongation is increase with temperature while the minor elongation is decrease with temperature and the same condition happen as blank with 65 mm diameter.

However, we can notice that bigger increment happen for drawn cup with LDR of 1.69 at warm temperature. Not only have that, from the results also we notice that the major elongation shows more percentage increment compared to minor elongation. Therefore, figure 11 shown below will shows a comparison between major elongation and limiting drawing ratio for all temperature condition has been made. This is due to achieve a better understanding regarding the phenomena as correlation between the parameters can be analysed in details.

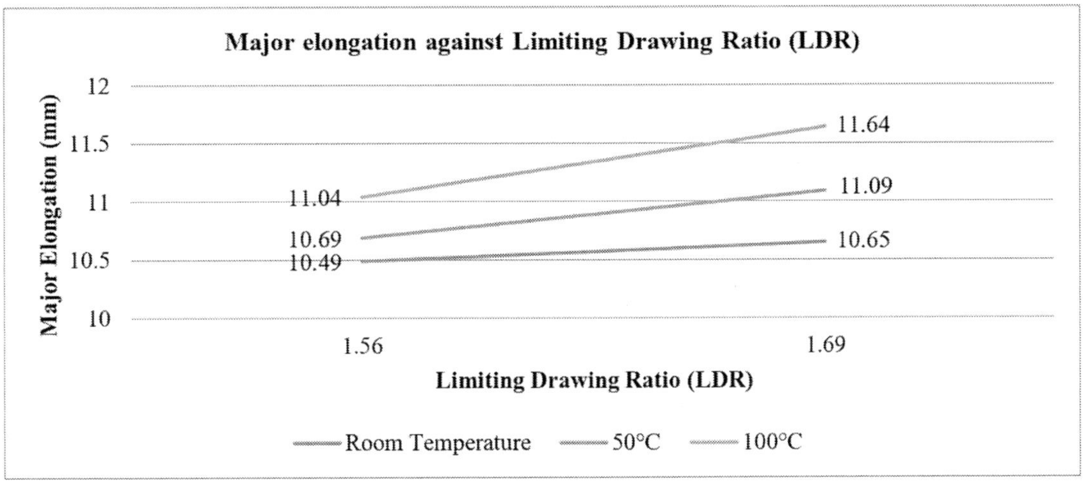

Figure 11. Major elongation of drawn aluminium cup against LDR in different temperature condition

In figure 11, the major elongation against LDR in different temperature condition was show. From this graph, we could notice that major elongation for drawn cup is increasing as the temperature increase. On top of that, as the LDR increase, the difference between major elongations among different temperature condition is higher. This is most likely to happen due to the warm temperature that applied during the process was almost similar the recrystallization temperature of the aluminium. Thus, it helps to improve the material flowability during the process. Moreover, the material does not repel to deformation, as much is it in room temperature condition.

In addition to that, compare to these three temperature condition, the warm temperature of 100°C show the highest increment percentage in term of major elongation. This indicates that the temperature difference can improve the drawability in term of elongation. In addition to that, the possibility to produce sound cup with higher LDR is increased with this approach which is proves the outcomes from Palumbo et al. works [10].

3.5 Discussion on major and minor elongation of drawn round cup (Mild Steel)

As we can see from the outcomes, the drawn mild steel cup manages to be drawn at higher LDR compare to aluminium. From the graph, the drawn cup with LDR of 1.56 and 1.69 shows the major elongation is spiking up with a steep line gradient as the temperature increase. Different from the others, the drawn cup with LDR of 1.89 shows a low major elongation increment, which appear as a mild slope on figure 10.

Therefore, as the major elongation appears to have more percentage increment compared to minor elongation, a comparison between major elongation and limiting drawing ratio for all temperature condition has been made. This is because; the correlation between the parameters is much clearer and can be seen in figure 12.

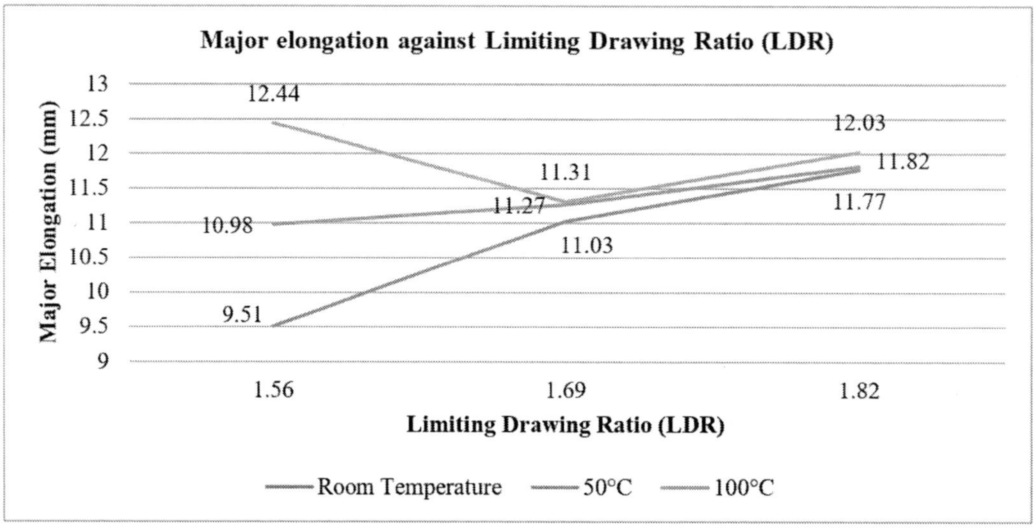

Figure 12. Major elongation of drawn mild steel cup against LDR in different temperature condition

Figure 12 shows the major elongation against LDR in different temperature condition for deep drawing process of mild steel cup. From this graph, we could notice that major elongation for drawn cup with LDR of 1.56 in increasing as the temperature increase except for round mild steel cup that drawn at 100°C. On the other hand, for drawn cup with LDR of 1.69, all of them were increase as the temperature being increase. However, when we compare the major elongation for the cup drawn at 100°C, we can notice that the major elongation at LDR 1.82 is much lower compare to LDR 1.56.

According to Lean et al. [1], the deep drawing process exhibit the principal plane of a plate sheet is positive while the other side is negative. Furthermore, the elevated temperature will affect the sign of principal plane. Changing sign of plate without changing the punching direction will shows a difference in collecting major strain data. Thus, this explains why the major elongation for drawn cup with LDR of 1.69 is lower when compares to the other cup with different LDR. Not only that, another reason that can contribute to this phenomena is the warm temperature applied to the process is not near enough to the recrystallization temperature of the material. Hence, the reduction of strength of the material were still not achieved and the materials still hard to flow during the process.

For deep drawing process that conducted at 100°C, we also can notice that the major elongation at lower LDR that is 1.56 is higher when compare to the others and shows a contradiction with the previous discussion. This is most likely to happen due to long heating time that being experience by each cup with different LDR. During the test run, the average time for the heating process was around 30 minutes. Based on this, 30 minutes has been set up as heating time throughout the experiment. The smaller the LDR with smaller blank diameter tends to absorb heat much more faster compare to the bigger one. Hence, the cup with smaller blank can achieve recrystallization stage while the other cannot and results better material flow and elongation.

3.6 Comparison between experimental and ABAQUS/CAE outcomes

ABAQUS/CAE Finite Element Analysis is a software for finite element analysis which already being utilized to analyse the part of the car, aviation, and mechanical items. In this research, the simulation of ABAQUS/CAE only conducted in 2D form. In this simulation, the comparison among different blank size is conducted. The comparison between simulation and experimental result also being carried out and shown in table 8.

Table 8. Comparison between experimental outcomes and ABAQUS/CAE simulation

Blank Diameter (mm)	**Experimental**	**ABAQUS Simulation**
60		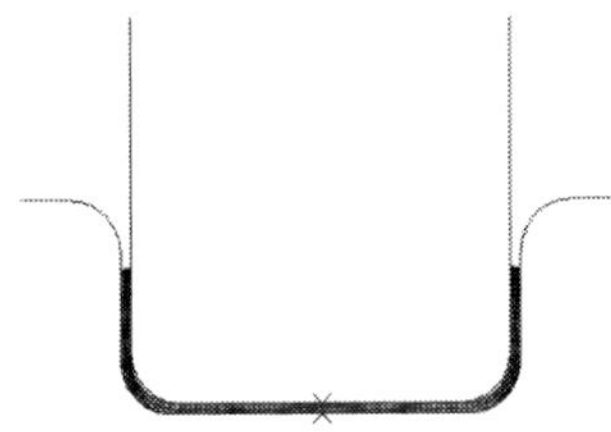
65		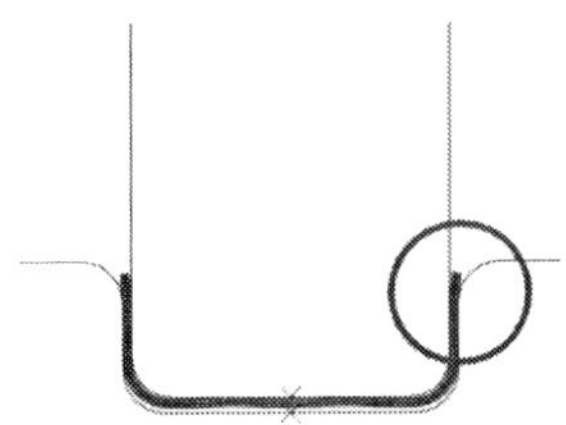
70		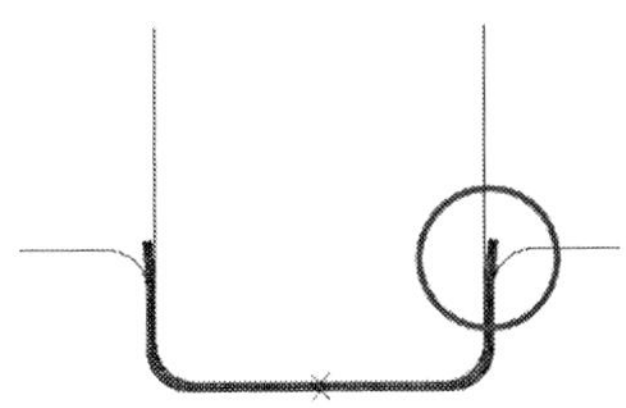

The comparison between experimental and ABAQUS/CAE simulation were carried out in order to predict and verify the process outcomes that has been yield. Based on observation from experimental conduct, there was no wrinkling effect present onto the cup flange. This indicates that a sound cup has been produced through the process and this also has been verified by the simulation outcomes from ABAQUS/CAE as there were also no wrinkling occurs. As the LDR being increase, we can observe that there is a mild wrinkling effect present at top of the cups. This trait also noticed on the outcomes from the simulation of ABAQUS/CAE as the drawn cup also being slightly wrinkled. The reason behind this was the insufficient die height that causes the cup not being fully deformed. On top of that, the blank holding force that exerted onto the blank could be insufficient and cause the radial tensile stress and tangential compressive stress occur simultaneously.

4. Conclusion

From the experimental work conduct and finite element analysis simulation, we can deduce that:

1. The deep drawing process in elevated temperature will give a good formability for aluminium and mild steel. The increment in temperature will increase the major elongation and the LDR.
2. The physical, mechanical and thermal properties of material can affect warm deep drawing as the parameter need to be set-up properly according to the specification.
3. The working temperature increment could decrease the probability of defect such as fracture and wrinkling effect to be happen. Therefore, a sound cup with better aesthetic value could be produce with high temperature value.
4. There is result from experimental works could be predicted and compared through finite element analysis simulation method by using ABAQUS/CAE software.

Acknowledgement

The acknowledgements should have been given to the Ministry of Higher Education and the University of Malaya under the FRGS grant (FP040-2013B) for the financial support.

References

[1] Y W Lean and M Azuddin 2016 *IOP Conference Series Material Science Engineering* **114** 12009
[2] G Palumbo and L Tricarico 2007 *J. Mater. Process. Technol.* **184** 115–123
[3] D Banabic 2010 *Sheet Metal Forming Processes: Constitutive Modelling and Numerical Simulation*
[4] H S Kim, M Koç, and J Ni 2008 *J. Mater. Process. Technol.* **197** 393–407
[5] X Zeng and S Mahdavian 1998 *J. Mater. Process. Technol.* **84** 38–46
[6] L Jayahari, P V Sasidhar, P P Reddy, B BaluNaik, A K Gupta, and S K Singh 2014 *J. King Saud Univ. - Eng. Sci.* **26** 21–31
[7] Y Moon, Y Kang, J Park, and S Gong 2001 *Int. J. Mach. Tools Manuf.* **41** 1283–1294.
[8] R Padmanabhan, M C Oliveira, J L Alves, and L F Menezes 2007 *Finite Elem. Anal. Des.* **43** 1062–1067
[9] D Ghaffari Tari, M J Worswick, and S Winkler 2013 *J. Mater. Process. Technol.* **213** 1337–1347
[10] G Palumbo, D Sorgente, L Tricarico, S H Zhang, and W T Zheng 2006 *J. Achiev. Mater. Manuf. Eng.* **14** 111–118

An experimental investigation of piezoelectric P(VDF-TrFE) thick film on flexible substrate as energy harvester

Chow Khoon Keat[1,2*], Kok Swee Leong[2], Lau Kok Tee[3]

[1] Department of Electrical Engineering, Politeknik Ungku Omar, Jalan Raja Musa Mahadi, 31400, Ipoh, Perak, Malaysia
[2] Faculty of Electronics and Computer Engineering, Univerisiti Teknikal Malaysia Melaka, Hang Tuah Jaya, 76100 Durian Tunggal, Melaka, Malaysia.
[3]Faculty of Manufacturing Engineering, Universiti Teknikal Melaysia Melaka, Hang Tuah Jaya, 76100 Durian Tunggal, Melaka, Malaysia.

E-mail: *markchowmy@yahoo.co.uk

Abstract. This paper proposes an experimental inves tigation of energy harvester using poly(vinylidene fluoride-trifluoroethylene) or P(VDF-TrFE) thick-film on flexible substrate by using print screen and rod method. Polyester film being used as the substrate where a sandwiched layer of electrode-piezopolymer-electrode thick film is deposited on. The thick- film is then annealed at 100°C and polarized at 100 V for the film with a thickness of about 18μm, being inspected under EDX, FESEM and XRD. The fabricated energy harvester piezoelectric is able to generate a maximum output power of 4.36 μW at an externa l electrical load of 1 kΩ with a maximum peak-to-peak of about 3.0V when an impact free-fall force of 0.2N was applied on the thick-film.

1. Introduction

In 21st century, energy harvesting applications have become a great interest in wearable technology and flexible electronics. Most of these applications need low power consumption in the range of mW or μW level of power source and currently there are using conventional batteries as their power source. Therefore, many researchers have been conducted for energy harvesting as self-power source of wearable technology and flexible electronics application using piezoelectric material such as piezoceramics and piezopolymers. The piezoceramic materials such as barium titanate ($BaTiO_3$), lead zirconatetitanate (PZT), sodium potassium niobate (KNN) and piezopolymers materials are not suitable for wearable technology because they need to be cured at higher temperature and on rigid substrate. Therefore piezopolymer such as polyvinylidenefluoride or P(VDF) and poly(vinylidene fluoride-trifluoroethelene) or P(VDF-TrFE) are discovered for the purposes of wearable technology. [1-3].

The piezopolymer materials have many advantages as compared to piezoceramic materials such as PZT and its become popular after the discovery of piezoelectricity in polyvinylidene by Kawai in the early 1964 [4]. One major advantages of polymer piezoelectric materials is that it is easily fabricated at lower temperatures compared to ceramic piezoelectric such as KNN and $BaTiO_3$. This property make them attractive for piezoelectric applications on flexible substrate, although there are few known fact that piezopolymers have much lower piezoelectric coefficients compared to piezoceramic. The other advantages are PVDF and PVDF-trifluoroethylene (PVDF-TrFE) piezopolymers with lower piezoelectricity compared to piezoceramics material such as PZT but it can conform to complex structural surfaces due to its flexibility nature [5].

The main disadvantage of the PVDF piezopolymer materials is mechanical stretching, which is not suitable for conventional micro-fabrication processes [6]. As a solution, P(VDF-TrFE) materials is chosen where it consists of crystalline structure and therefore the piezoelectric properties depend on the molecular proportion x of vinyledene fluoride in P(VDFxTrFE1$-x$). The presence of TrFE element in the copolymer of the PVDF-TrFE film introduces significant features to the PVDF homo-polymer. One of the advantages of PVDF-TrFE is that it increases the tendency to crystallize in the polar β- phase without the requirement of mechanical stretching to transform the non polar α-phase to the polar β-phase as in the case of PVDF, when $0.6 < x < 0.85$ [7].Many studies has been carried out on PVDF- TrFE copolymers, and the copolymer at composition near 75/25 mol.% exhibits the highest ferroelectric responses for energy harvesting application [8].

This paper reports the experimental investigation of energy harvester by using poly(vinylidene fluoride-trifluoroethylene) or P(VDF-TrFE) as thick-film on flexible substrate, the result of screen-printing techniques with rod method for fabrication of flexible thick film by using Field-emission Scanning Electron Microscope and the voltage response to free-fall force on P(VDF- TrFE) thick film was studied for micro energy harvester application.

2. Experiment

2.1 Materials

In this work, Poly(vinylidene fluoride-trifluoroethylene) or P(VDF-TrFE) is prepared with a molecular weight of 350,000 gmol^{-1}[75:25mol%] manufactured by Kureha, Japan.

2.2 Fabrication

P(VDF-TrFE) (75:25mol %) powder were dissolved in *N,N*-dimethylformamide (DMF, Sigma-Aldrich Co.), weight in percentage of 15wt% [9-11]. In order to ensure good adhesion in between active layer or P(VDF-TrFE) was sandwiched between top electrode and bottom electrode on the flexible substrate (Melinex with thickness of 75 µm)by using screen printing and rod method. The powder was dissolved in DMF and mechanically stirred at 100 °C for 1 hour. The solution was then immersed in an ultrasonic bath for 20min to ensure that the solution was fully dissolved.

Prior to the P(VDF-TrFE) deposition, first Ag electrode layer were screen-printed using screen stencil onto the blank Melinex substrate. The second Ag electrode layer was applied after the infra-red light treatment. The thick film was dried under infra-red light for 15 min with a temperature of 60°C in order to remove the residual solvent. Then, the fabricated thick film were annealed in an oven for 4h at 80°C to increase crystallinity [12]. The thick film was subsequently annealed in an ambient environment oven at 100°C for 1 hour to obtain a smooth and crystalline surface with reduced porosity.

The P(VDF-TrFE) thick film was electrical polled using high voltage polarizer at 100 V for 20 min at 100°C (near to the Curie temperature of P(VDF-TrFE). This polarization is to align the domain dipole according to the polarity of the high voltage. The dimension of the fabricated energy harvested and electrode layers is shown in table 1 with step of process parameters are summarised in table 2.The fabricated device of the sandwiched P(VDF-TrFE) thick film is illustrated in figure 1 which is being used for the experimental testing in next step.

Table 1. Dimension of fabricated sensor device

Design	Parameter, L x W (mm)
P(VDF-TRFE) layer	2.5 x 2.0
Top Electrode	3.4 x 1.0
Bottom Electrode	3.4 x 0.8

IOP Conf. Series: Materials Science and Engineering **210** (2017) 012078 doi:10.1088/1757-899X/210/1/012078

Table 2. Summary of paste properties and printed layer process parameters

Name	Materials	Printing methods	Curing conditions
Top and bottom electrode layer	Silver Conductor Paste	Print screen	Infa Red 80°C,30min
P(VDF-TrFE) layer	Poly(vinyli-dene fluoride trifluoroethylene	Rod Method	Infa Red 80°C, 30 min, at plate annealing 100°C
Melinex layer	Polyester film	-	-

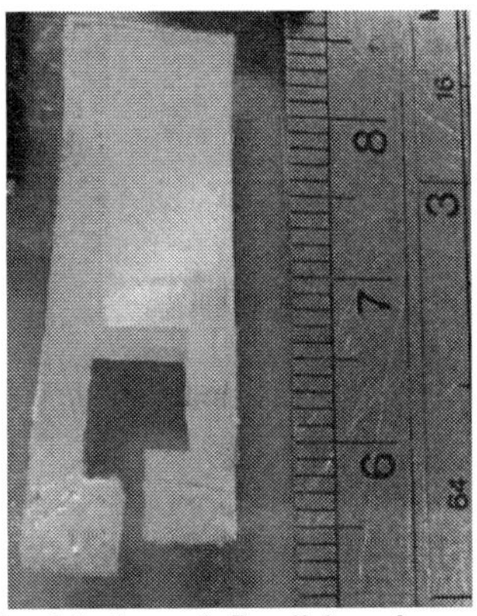

Figure 1. Photograph of a fabricated flexible piezoelectric P(VDF-TrFE) energy harvester

2.3 Experimental setup

A free-fall drop impact test was carried out on the fabricated energy harvester by dropping a known mass of plasticine with varying weights from 0.03 N to 1N (refrains from falling out of the target by using a plastic cylinder as shown in figure 1) at a fixed height of 20cm. Figure shows instantaneous of electrical voltage output voltage from the P(VDF-TrFE) thick film when the plasticine is released to be free-fall onto the P(VDF-TrFE) thick film.

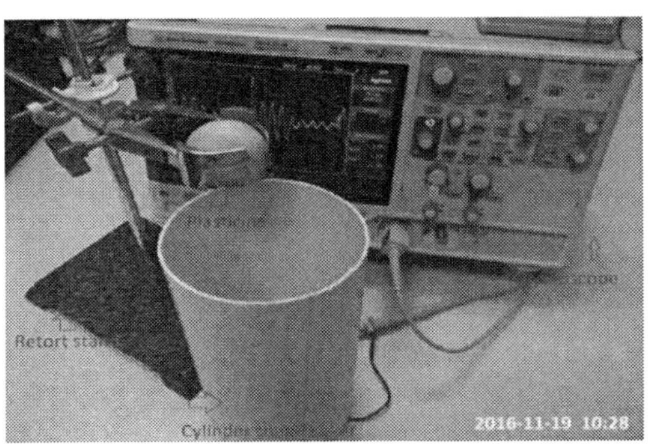

Figure 2. Experiment setup

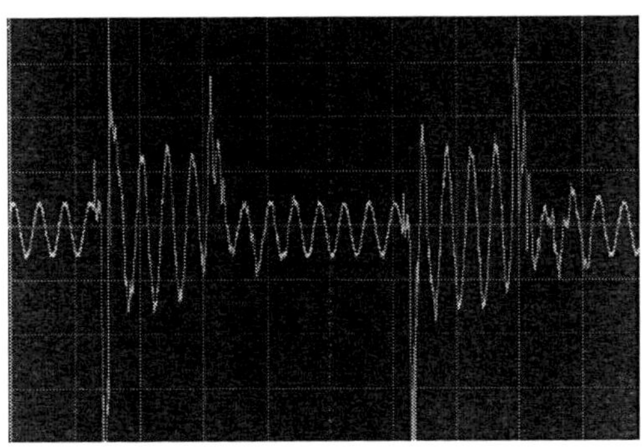

Figure 3. Output AC voltage peak when impact force

3. Results and discussion

3.1 Microstructure inspection and characterization

The surface microstructures of the deposited layers of P(VDF-TrFE) energy harvester were characterized by Ultra High Resolution Field Emission Scanning Electron Microscopy (FE-SEM, accelerating voltage = 3 kV, Model: Merlin, Carl Zeis AG). Figure 4 (a) and (b) shows FESEM top surface of dried thick film which had layers of the Ag/P(VDF-TrFE)/Ag/Melinex composite structure, while (b) displays the cross-section of P(VDF-TrFE) thick film with thickness about 60μm.

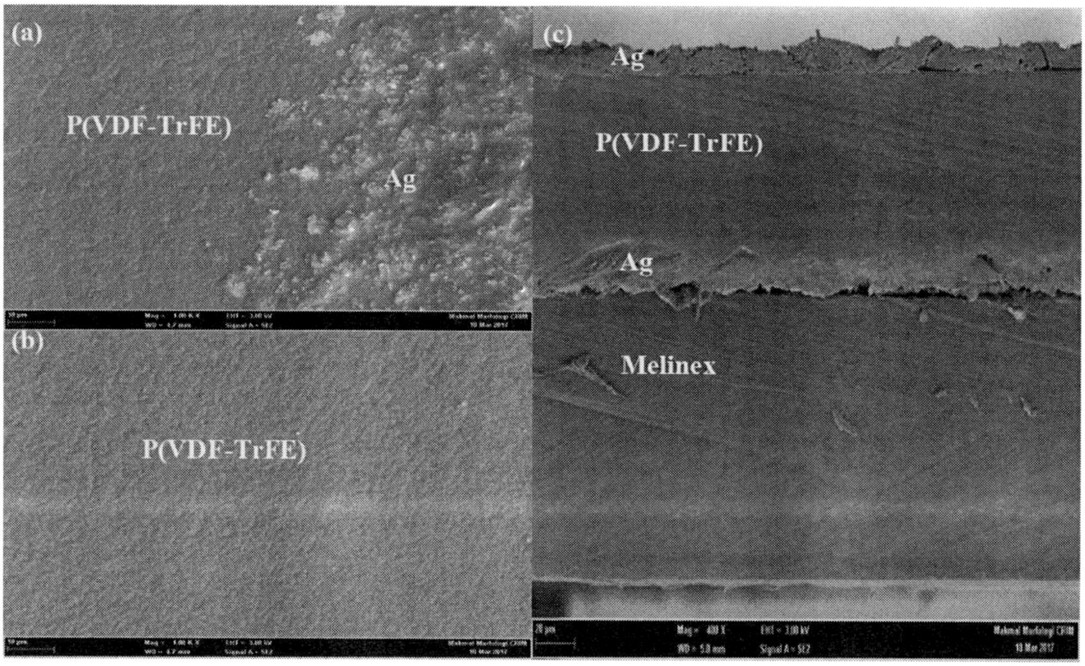

Figure 4. (a)FESEM of the surface layers P(VDF-TrFE)/Ag, (b)FESEM of the surface of a P(VDF-TrFE) thick film dried at 80°C and annealed at 100°C for 15 mins and (c) Cross-section micrograph of electrical harvester structure

3.2 Electrical performance

To verify piezopolymers responses, an impact force test using free-fall plasticine were conducted and the electrical voltage outputs were measured. The peak-to-peak AC voltage versus the free-fall impact force shows a saturation at about 3V when a impact force of 0.2N is applied (see figure 5) A highest output voltage of 3.06V was generated at 0.2N load. When the force exceeding 0.7 N, the average output voltage settled down to 2.70V.

Figure 6 shows the electrical output power of piezopolymers P(VDF-TrFE) thick film when the output terminal of the film is connected to external resistive load varied from 700Ω to 5 kΩ. This was after the impact force was fixed at 0.2N. A maximum output power of 4.36 μW is being measured across an optimum external load of 1kΩ.

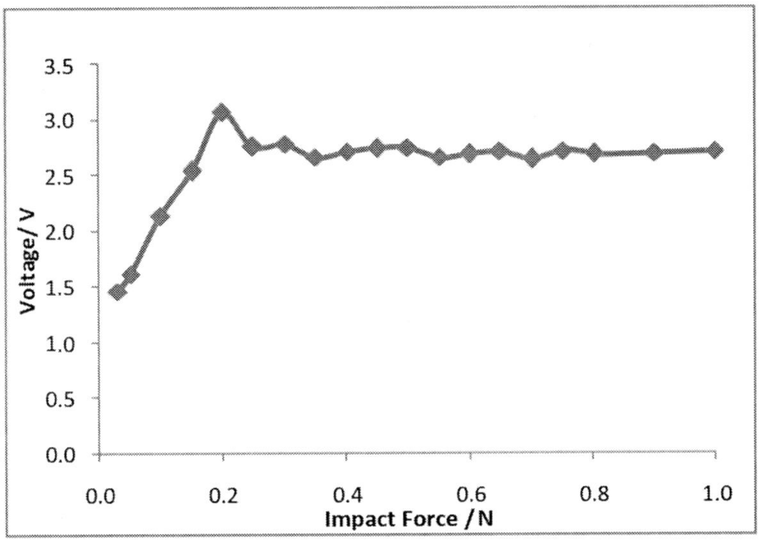

Figure 5. Voltage output vs impact force.

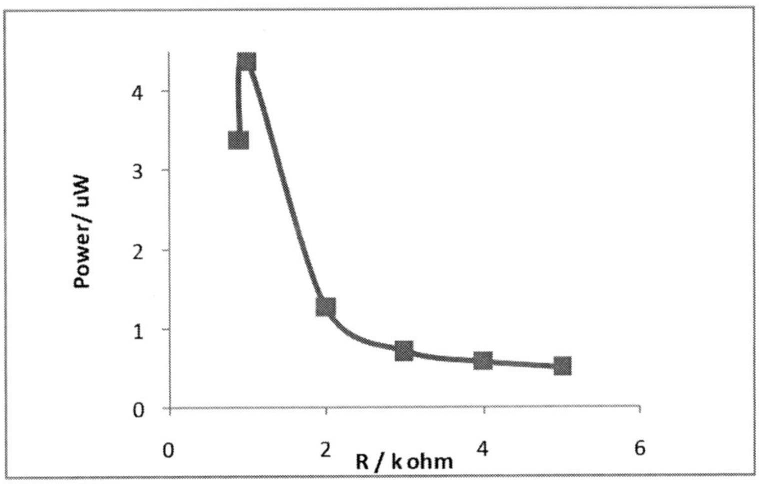

Figure 6. Power vs Resistive load.

4. Conclusion

In this paper, the piezopolymers P(VDF-TrFE) thick film were successfully developed onto flexible substrate as an energy harvester which it provides a promising solution to harvest a consistent electric voltage and power output when exerted a range of free-fall impact force. The output voltage achieved a maximum of about 3V with an impact force of 0.2N which derived a power of 4.36 μW when connected with 1kΩ external resistive load.

Acknowledgements

This work was supported by Ministry of Higher Education of Malaysia for research grant of FRGS/2/2014/SG02/FKEKK/02/F00244. The authors also acknowledge Morphology Laboratory, Centre for Research and Innovation Management, Universiti Kebangsaan Malaysia for the FESEM characterization service. The authors also appreciate the supports given by Advanced Sensors & Embedded Control (ASECs) research group.

References

[1] Shanshank Priya and Dainel J Inman 2009 *Energy Harvesting Technologies* (Springer US)

[2] Sukwisute P, Muensit N, Soontaranon S and Rugmai S, 2013 *Applied Physics Letters*, **6** 063905

[3] Kim H S, Kim J H and Kim J 2011 *International journal of precision engineering and manufacturing* **6** 1129-1141

[4] Kawai H 1969 *Japanese Journal of Applied Physics* **7** 975

[5] Wang Z L 2004 *Journal of Physics: Condensed Matter* **25** R829

[6] Li C, Wu P M, Lee S, Gorton A, Schulz M J and Ahn C H 2008 *Journal of Microelectromechanical Systems* **2** 334-341

[7] Ohigashi H 1985 *Japanese Journal of Applied Physics* **2** 23

[8] Furukawa T 1989 *Phase Transitions: A Multinational Journal* **3** 143-211

[9] Lau K T, Suan S M., Zaimi M, Razak J A, Asyadi M A and Mohamad N 2016 *Journal of Advanced Manufacturing Technology (JAMT)* **2**

[10] Pi Z, Zhu L, Zhang J, Wu D, Zhang DW, Zhang Z B and Zhang S L 2013 *ASIC (ASICON), 2013 IEEE 10th International Conference* 1-4

[11] Lau K T, Kok S L, Razak J A, Munawar R F 2016 *Malaysian Journal of Mircroscopy* **10** 138-144

[12] Mahdi R I, Gan W C and Majid W H 2014 *Sensors,* **10** 19115-19127

IOP Conf. Series: Materials Science and Engineering **210** (2017) 012079 doi:10.1088/1757-899X/210/1/012079

The ergonomics body posture on repetitive and heavy lifting activities of workers in aerospace manufacturing warehouse

S. R. Kamat[1], N. E. N. Md Zula[1], N. S. Rayme*[1], S. Shamsuddin[1] and K. Husain[2]
[1]Faculty of Manufacturing Engineering, University of Technical Malaysia Melaka, Hang Tuah Jaya 76100 Durian Tunggal, Melaka, Malaysia
[2]Kolej Unversiti Islam Antarabangsa Selangor (KUIS) Bandar Seri Putra, 43000 Kajang, Selangor, Malaysia

E-mail: *syafiqah.rayme@yahoo.com

Abstract. Warehouse is an important entity in manufacturing organizations. It usually involves working activities that relate ergonomics risk factors including repetitive and heavy lifting activities. Aerospace manufacturing workers are prone of having musculoskeletal disorder (MSD) problems because of the manual handling activities. From the questionnaires is states that the workers may have experience discomforts experience during manual handling work. Thus, the objectives of this study are; to investigate the body posture and analyze the level of discomfort for body posture of the workers while performing the repetitive and heavy lifting activities that cause MSD problems and to suggest proper body posture and alternatives to reduce the MSD related problems. Methodology of this study involves interviews, questionnaires distribution, anthropometry measurements, RULA (Right Upper Limb Assessment) assessment sheet and CATIA V5 RULA analysis, NIOSH lifting index (LI) and recommended weight limit (RWL). Ten workers are selected for pilot study and as for anthropometry measurement all workers in the warehouse department were involved. From the first pilot study, the RULA assessment score in CATIA V5 shows the highest score which is 7 for all postures and results after improvement of working posture is very low hence, detecting weight of the material handling is not in recommendation. To reduce the risk of MSD through the improvisation of working posture, the weight limit is also calculated in order to have a RWL for each worker. Therefore, proposing a guideline for the aerospace workers involved with repetitive movement and excessive lifting will help in reducing the risk of getting MSD.

1. Introduction

Ergonomic comes from Greek word which means "work law", it also describes as "the effort to fit the system to the human" which means that to fit the unique human limitation and abilities by selecting and designing the informed decision, tasks, environment, tools and equipment. The dimensions that define ergonomics discipline include philosophy, theory, technology or environment, management, design, practice and education [1]. The main focus of this study is to improve the ergonomics body posture of workers in the aerospace manufacturing company.

The warehouse of the aerospace manufacturing company had been chosen because the working process design in the building involves repetitive task and heavy lifting activities. The risk of ergonomics can be seen affecting the workers in the building as they had complained about them experiencing fatigue and pains at certain areas of the body especially at the lower back of the body. The complaints had been retrieved through the distributed questionnaires. All those symptoms and effects will greatly contribute to repetitive stress injuries (RSIs) and musculoskeletal disorder (MSDs). Both are known as parts of ergonomic injuries. Ergonomics injuries are the bad effects that caused by the existence of the ergonomics risk factors such as awkward postures, sustained postures, contact pressure, forceful exertion, forceful strain and exposure to vibration, heat or cold. When the risk factors combined, and exerted on the worker through a continuous period, the risk factors will lead to injury, pain and disability. As example, in a manufacturing company, if an injury occurs, the dangerous single event will place a stress on body tissues. Although the tissues are capable to recover its condition, the repetitions of the hard manufacturing activity that cause the injury will slower the healing process.

1.1. Problem Statement

Based on the literature review and previous studies, the repetitive and heavy lifting activities can contribute to ergonomics injury. Besides, there are problems detected through the interview session with a manufacturing engineer and data collections of the questionnaires distributed to the workers. The activities in the warehouse involved manual handling task that needs the workers to lift and move heavy objects repetitively. The objects can be referred to tools and aircraft panels which are the products of the company. Although there is equipment such as forklift had been provided by the company to lift big panels, there are still many panels that are cannot be lifted by the equipment. It is due to their sizes and the space provided in the warehouse. All the intentions stated required the worker to lift the panels manually. Besides, the worker had to risk their body to muscle injuries while lifting the panels in order to follow the existing design of work process. According to the questionnaires data collections, majority of the workers had already been experiencing the back pain, shoulder pain and several other pains that related to muscle fatigue when there are too many panels that needs to be lifted. Anyhow, the lack of training about the importance of right manual handling techniques, the least awareness about the serious injuries like MSDs and RSIs can be seen as the main reason why the workers were maintaining their bad manual handling technique although they had already been experiencing the symptoms of the injuries.

1.2. Objectives

Based on the problems arise related to the repetitive and heavy manual lifting activities in the warehouse, the objectives of the study are;

i) To investigate about the body posture of the workers while doing repetitive and heavy lifting activities in the warehouse.
ii) To analyse the discomfort body posture of the workers while undertaken the repetitive and heavy lifting activities that cause the musculoskeletal disorder problems.
iii) To suggest the proper body posture and ways to reduce the musculoskeletal disorder problems.

1.3. Scope and Limitations of Study

This study only focus on the activities that involved the most repetitive and heavy lifting activities that was seen as the critical activities among all the activities involved in the warehouse. The data about the body posture of the affected worker had been retrieved. The musculoskeletal disorder problems of the workers had been retrieved by the distribution of questionnaires to the workers and RULA analysis methods. The working body posture of the workers had been reclaimed by taking the photos and videos of the workers while they are working. The working postures of the workers while performing the repetitive and heavy lifting activities had been observed and determined to be analysed. The ergonomics risk factors that had been studied include; awkward body posture, repetition movement, heavy lifting, contact stress and poor design work process. Besides that, other important things have been considered in this study are the requirement of the working activities, ergonomics obligation, ergonomics problems and ergonomic principles.

In advance, other aspects such as NIOSH lifting equation, anthropometric measurement, muscle fatigue, musculoskeletal disorder (MSDs) and repetitive stress injuries (RSIs) had been enclosed in this study. The software that had been used to analyse the working body posture is CatiaV5. The working body posture analysis had been done to prove the effectiveness of the body posture improvement. However, the result of the study is only based on the simulation. There are no fabrication activities and real implementation occurred in order to test the suggested improvement. This study did not cover the labour productivity issue.

2. Literature reviews

Like in any other manufacturing industry, an aerospace manufacturing industry is also exposed to manufacturing activities that can risk the workers to the ergonomic risk factors [2]. It is important for a manufacturing industry to implement ergonomics technique in the working area because with the existence of proper ergonomics application; cost can be reduced, productivity level can be improved, quality of working can be increased, employee engagement can be improved and better safety culture can be created.

Ergonomics risk factor conferred by the United Stated Occupational Safety and Health Administration includes work environment, body posture, forceful exertion, prolonged static load, repetition, lack of rest and improper aid [3]. Thus, it is necessary for the employer to make sure that the ergonomics risk factor in the worker workstation is at the possible lowest term. In addition to musculoskeletal disorder leading to medical expenses, the ergonomics risk factor can leads to overtime wage, cost of training and hiring for replacement, loss of productivity, management cost and other miscellaneous charges.

In order to offer the worker the finest man-machine boundary and the required adjustability with the aim to prevent discomfort and workplace injuries, a good ergonomic design principles must be practised [4]. An article named "Ergonomic: The backbone of a more productive warehouse" mentioned that workstation design can give big influence on the ergonomic health of the workers and the ergonomics workstation is all about body posture. Body posture while working can be categorised to three types which are standing, sitting and both. The ability of the workstation's features such as length and height to be adjusted so that its features are compatible to workers anthropometry measurement is a very good ergonomics workstation.

Rapid Upper Limb Assessment (RULA) is an assessment that used to analyse the disclosure of workers to ergonomic risk related to upper extremity musculoskeletal disorder [5]. The things that are mattered by the RULA ergonomic assessment tool are the postural load requirement and the biomechanical of job demand especially on the trunk, upper extremities and neck. The evaluation of the required body posture, repetition and force can be made by only single page worksheet. Both scores for section A which is for the arm and wrist areas and for section B which is for the neck and truck are must be inserted in order to obtain the evaluation score.

The manual handling operations regulation that was altered in 2002 is applied to the extensive assortment of manual handling activities include pulling, pushing, lifting, carrying or lowering. The type of loads may include living thing for instance animal or person, or non-living thing such as trolley or box. It was discussed that the improper manual handling activity is one of the utmost typical causes of the injury happened at workplace [6].

The musculoskeletal disorder can be the effect of the improper manual handling. The risk factors of the MSDs development include; manual material handling, manual labour, awkward postures and existing injury. Repetitive movement is a simple harmonic motion through back and forth movement placed at the equilibrium or the centre point where the optimum displacement on one side is equal to the optimum displacement on the other side as well [7]. The force used for the motion is always focus on the equilibrium location and is directly proportional to the distance from it, and the time interval of each complete vibration is equal to one another. In manufacturing industries, there are many repetition movements that have to be carried out in order to achieve the requirement of the manufacturing activities. There are manufacturing activities like fitting, saw cutting, manual threading, painting, and machining that requires the worker to work in repetitive movement for the task completion. In warehouse, the manual material handling activity, wrapping activity and lifting activity can be seen as the activities that require the worker to work in repetition motion. However, rapid movement of the repetitive motion activities can risk the worker to the ergonomics injury.

Lifting is an activity in order to bring or move something upward generally from the ground or other support elements to a higher location. In warehouse, it is the main activity that is involved. Lifting in warehouse means to lift the inventory or equipment within the warehouse for receiving, storing, wrapping and shipping purpose. However, if the object to be lifted is heavy and not compatible with the worker, the situation can lead to injury. The biggest risk factors regarding to lifting activity are the cumulative trauma and overexertion [8]. It is advisable for the workers to use smart lifting techniques and only work in their own abilities so that they will be less exposed to muscle pulls, wrist injuries, back sprains, elbow injuries, spinal injuries and any other related injury. There are many causes that can contribute to all those injuries. Some of the causes are incompatible weightiness of loads, awkward body postures, high frequency of materials to be lifted and long durations needed for the lifting activities, inadequate handhold and environmental factors. The important aspects of design are the human fit, function and form in the demanding global marketplace. The best in class human modelling which is the human activity analysis permitted detailed investigation into human centred designation issue regarding to workplace's environment before it is existed. The way a human being interacts with objects in working environment, the effects of lifting activities; pulling, carrying task, lowering and pushing were the focus points of the human activities analysis. The equation used to measure the effects of all those activities is the NIOSH 1981/1991 and Snook and Ciriello equation. CATIA V5 promotes the solution towards ergonomic body posture in two main software tools [9]. The first software tool which is the human measurement editor is used to create detailed digital human model to extended analysis. It is capable of creating human manikin by using complex series of advance anthropomorphic tools and by adaptation of smaller number of crucial variables. The software human builder in CATIA which is the human builder consist from complex sets of tools that are devoted to create the human digital model which called manikin, the manikin manipulation and the interaction analysis with the product. In order to generate the view form, function and comfort, the manikin can be used.

Other than the mentioned tools, CATIA had also provides the analysis of workstation assembly. There are two other software modules has been created in order to fulfil the purpose which are the human activity analysis and the human posture analysis. The interaction between the human and the objects in the working area and the effects of the sinking, lifting, pressing and pulling can be analysed by the human activity analysis software. The software is able to evaluate all the aspects of the human effectiveness through the analysis of the static position to the complex activities by the execution of tasks. The software enables the user to maximize the safety, the comfort and the efficiency by the usage of wide spectrum of trolls to ergonomic analysis which involve complex evaluate the interaction between the manikin and the objects in the virtual environment.

The RULA (Rapid Upper Limb Assessment), push and pull analysis, loading analysis by load lifting according to NIOSH 1991, NIOSH 1981 and Snook & Ciriello, carry and biomechanical analysis. The workstation can be created in accordance with health and safety standards by this tool in order to maximize the safety and comfort. Next software tool provided by CATIA is the human posture analysis which analyse the way human posture influences the productivity by the execution of task. The tool can analyse preferred angles, the local and total postures and convenience. The user can do the quantitative and quality analysis of all the aspects of the manikin postures through the module. In addition, the model allows the user to test postures, generate the evaluation, optimized and iterated the data.

IOP Conf. Series: Materials Science and Engineering **210** (2017) 012079 doi:10.1088/1757-899X/210/1/012079

3. Methodology

The data collection started by the observation, interview sessions with manufacturing engineer and the distribution of questionnaires to the workers in warehouse department. These are to contemplate with the problem statement of the study. The next phase is the collection of anthropometry measurement of the workers, Nordic questionnaires, photos and videos of the workers while doing their tasks in the warehouse. The flowchart of the study is as in figure 1.

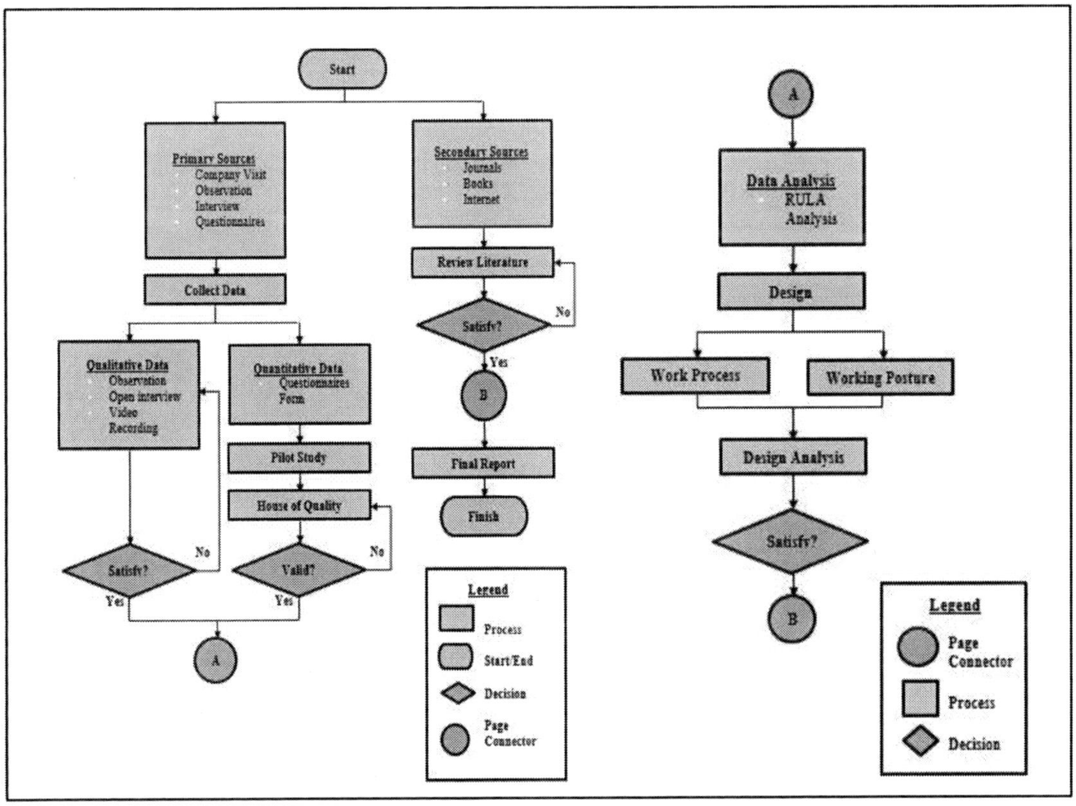

Figure 1. Overall flowchart of the study

Several observations had been done in the warehouse including the procedure of working tasks, layout and environmental of warehouse, type of body postures needed to conduct tasks and tools equipped to the workers. Another method that had been used in order to gather the qualitative data is by distributing questionnaires. The questionnaires were distributed to the workers in the company's warehouse. As for the pilot study, ten respondents has been chose for the questionnaires distribution. The respondents were divided into two different categories which are for workers under 30 years old and workers 30 years old and above. For the second phase of the study, more detailed questionnaires had been distributed to twenty workers involving in manual handling activities in the warehouse. The selected type of the questionnaire that will be distributed for the second phase of the project is the Nordic type of questionnaire. The Nordic questionnaire is one of the most known survey tool to detect musculoskeletal disorder [10]. Initially, the standardized Nordic questionnaires were created by an organized Nordic researcher's team. The main objective of the Nordic questionnaires creation is to screen the musculoskeletal disorders and for epidemiological studies.

4. Results and Discussion

The first part of data collection was the Nordic questionnaire data, which divided into two parts, Part A for respondents' basic data and Part B for respondent's cognitive data.

4.1 Analysis of Body Posture of the Worker While Handling the Tasks

figure 2 shows the first worker is lifting 30 kilogrammes spoiler core up to 190meters height. The critical working posture involve repetitive heavy lifting activities and awkward working body posture. The frequency of lifting is 2 lift per minute. The middle worker in figure 2 shows a worker lifting a box of vacuum bag sealant weighed 8 kilogrammes from the lower rack. The critical working posture involve repetitive heavy lifting activities and awkward lifting body posture. The frequency of lifting is 3 lift per minute. The worker on the right in figure 2 shows worker is lifting a 7 kilogrammes box from the top storey of a rack. The critical working posture involve heavy lifting activities and awkward lifting body posture. The frequency of lifting is 3 lift per minute.

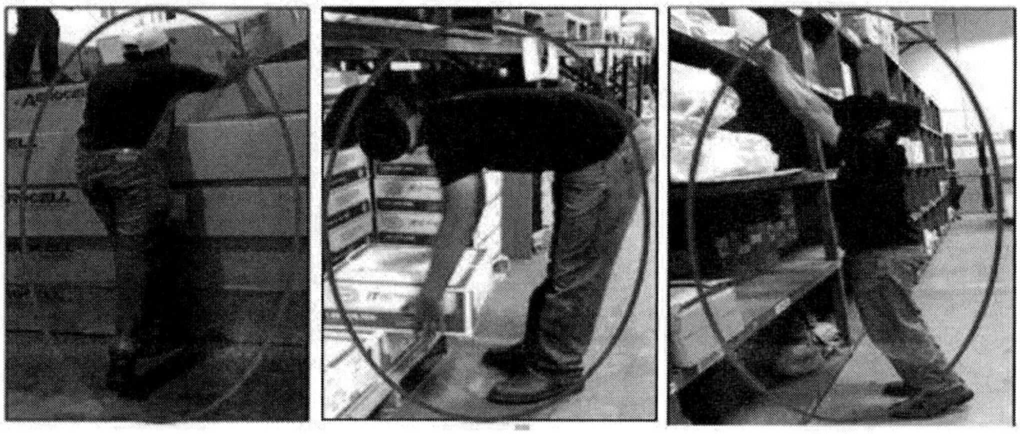

Figure 2. From left to right (Posture A, B and C respectively)

4.2 CATIA V5 RULA Analysis and Body Postures Improvement

Besides analysing the body postures manually using the RULA assessment worksheet, the body postures had also been analysed using the CATIA V5 RULA analysis application software. The alignment of the body postures, the type of working mode involved and the weight of the load involved had been set up and inserted in the software to achieve more accurate RULA score. Then, the body postures had been improved as the suggestion to reduce the musculoskeletal disorder problem.

4.2.1 Working Posture A before Improvement

Figure 3 shows the working Posture A before improvement and can be considered as awkward body posture as the needs of the worker to raise his arm, hunching his back and involving excess use of the back muscle. Improper lifting or overstressing weak back muscles is the cause of an acute ligament sprain or muscle strain [8]. Figure 3 shows the high RULA score retrieved from the working posture A for right side of body.

IOP Conf. Series: Materials Science and Engineering **210** (2017) 012079 doi:10.1088/1757-899X/210/1/012079

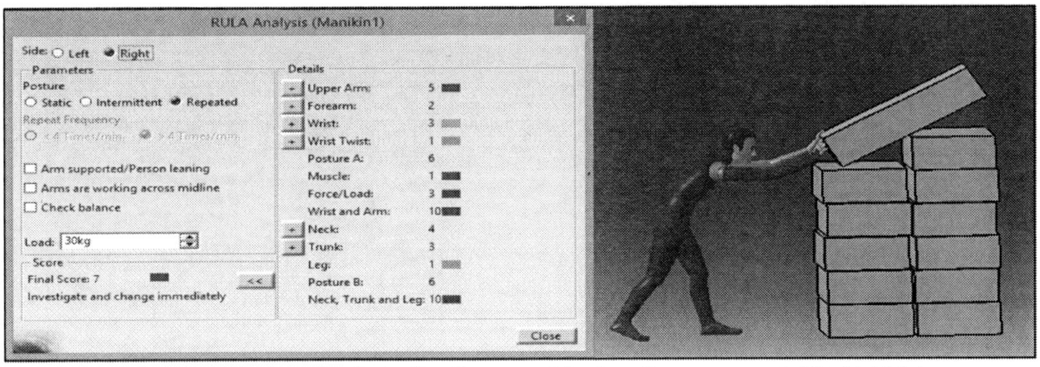

Figure 3. *RULA Analysis for Posture A before Improvement (Right Side)*

4.2.2 Working Posture A after Improvement

The working posture A involve the repetitive heavy lifting of 30 kilogrammes load. The load that has to be lifted manually should not exceed 50 pounds which is 22.67 kilogrammes [8]. Otherwise, the manual heavy lifting task can lead to ergonomic related injuries such as back sprains. That is why even the working posture had been improved, the RULA score of the improved working posture is still high for both sides of the body which is 7 as shown in figure 4. The most affected body parts for the lifting activities are the muscle, wrist and arms.

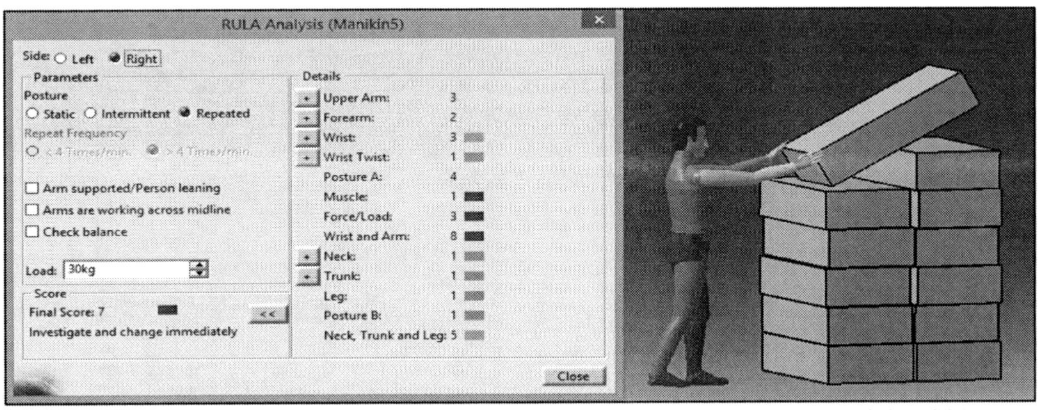

Figure 4. RULA Analysis for Posture A after Improvement (Right Side)

4.2.3 Working Posture B before Improvement

The working posture B as shown in the figure 5 involved awkward working body postures. The trunk part is bent too low which allow excessing pressure on the lower back of the body. The improper working posture can leads to herniation. Herniation is the condition occurred when a vertebral disk position misallocated or breaks within doing the inappropriate lifting activities [11].

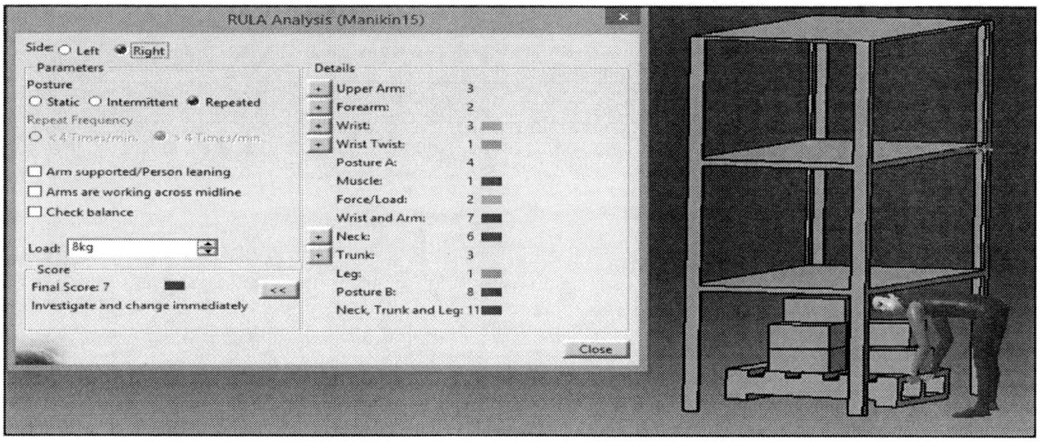

Figure 5. RULA Analysis for Posture B before Improvement (Right Side)

4.2.4 Working Posture O after Improvement

The RULA score after the working body posture are still high as shown in the figure 6. This is due to the repetitive working mode required to accomplish the lifting task. One of the factors that lead to the lifting related injuries that have to be avoided is the repetitive working mode which requires the worker to exerting force to their muscle repeatedly [8].

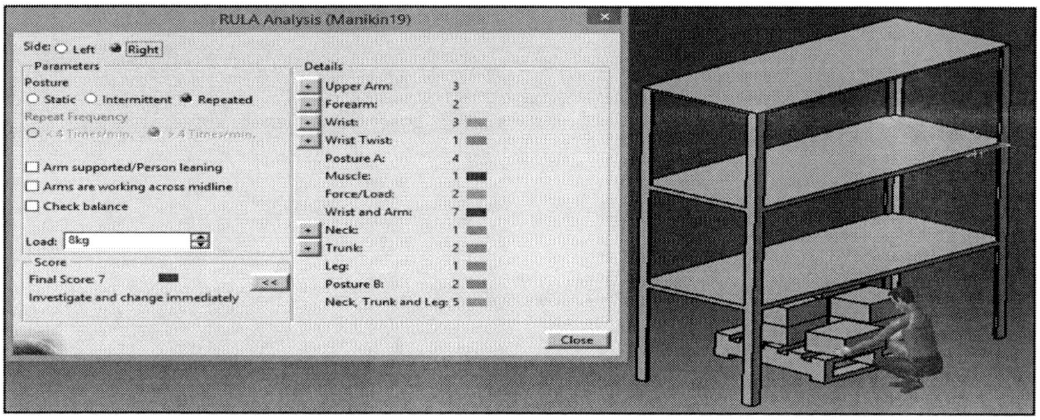

Figure 6. RULA Analysis for Posture B after Improvement (Right Side)

4.2.5 Working Posture C before Improvement

The working body postures shown in the figure 7 involved the overuse of wrist and arm of the worker. The increment pressure inside the nonflexible anatomical structure in the wrist can leads to the entrapment of neuropathy effect such as carpal tunnel syndrome [12]. Tennis elbow is a painful injury caused of the overuse of the elbow [13].

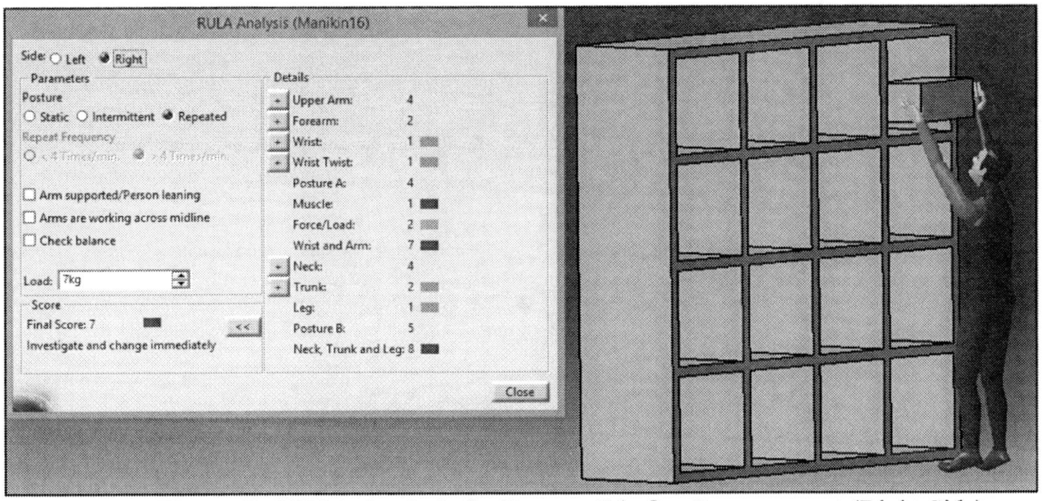

Figure 7. RULA Analysis for Posture C before Improvement (Right Side)

4.2.6 Working Posture C after Improvement

The RULA scores for the improved working posture C for right side as shown in the figure 8 are still high because of the workstation design is too high which requires the worker to lift up their arm to the maximum height. The employer should implement provide the suitable workstation design to minimize the ergonomic risk factor [14]. In Figure 8, the score is still not improved even though the colour of muscle had changed to green. This is cause by the repetitive movement and suggestions of additional aid device for the workers in order to reduce the score.

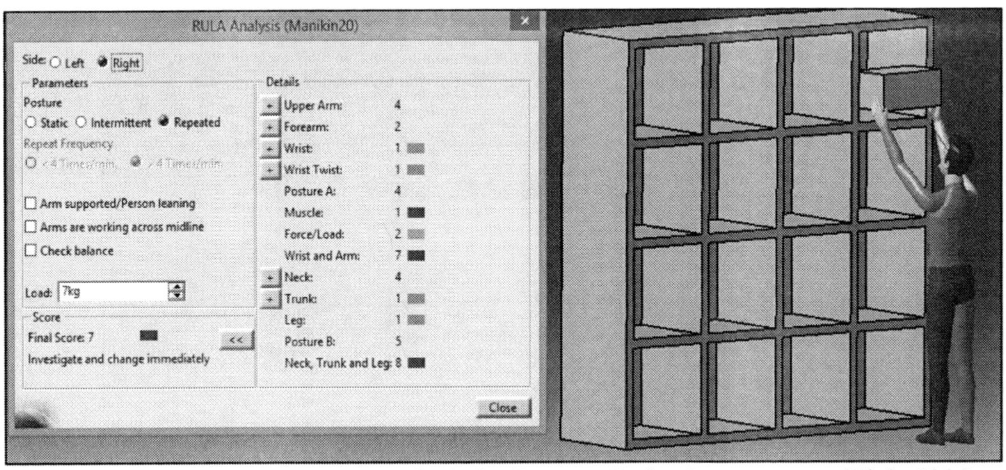

Figure 8. RULA Analysis for Posture C after Improvement (Right Side)

4.3 Suggestion Aid Portable Gantry Crane

The lifting equipment that had been suggested to replace the manual lifting task is the portable gantry crane. It is a small version of gantry crane that can be moved to replace the manual lifting task inside the warehouse. There are racks located inside of the warehouse that can be the constraints for the overhead crane to reach object in the rack. Therefore, this portable gantry crane can be used to lift objects placed in the rack or at the narrower space which cannot be reached by the forklift. Figure 9 shows the suggested design of the portable gantry crane.

IOP Conf. Series: Materials Science and Engineering **210** (2017) 012079 doi:10.1088/1757-899X/210/1/012079

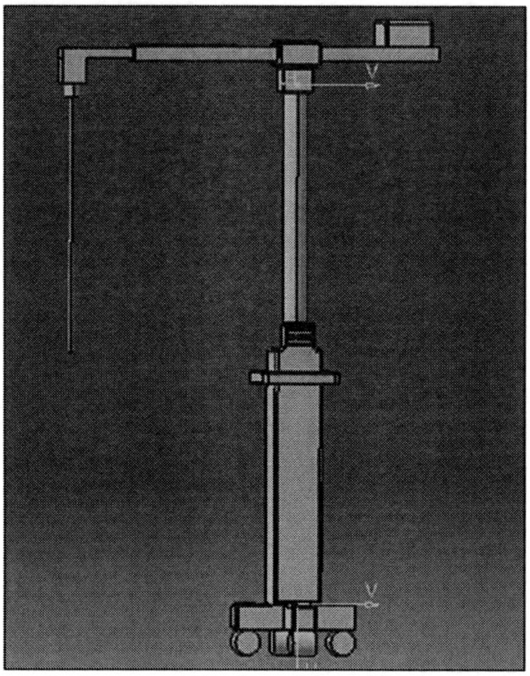

Figure 9. Suggested Design of Portable Gantry Crane.

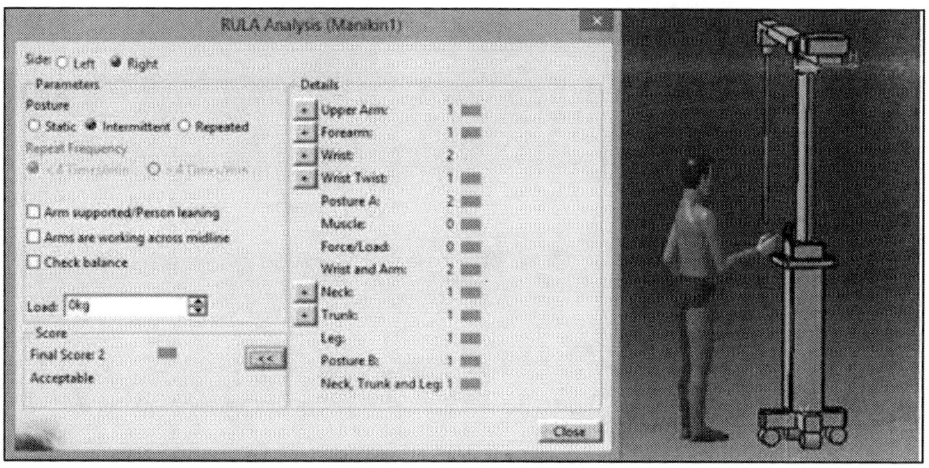

Figure 10. RULA Analysis for Worker's Body Posture While Using Portable Gantry Crane (Right Side)

Figure 10 shows the RULA score which is 2 for right side of the body that can be obtained if the manual lifting task inside the warehouse is replaced by the usage of the portable gantry crane. The RULA score 2 is the desired RULA score as it indicates that the posture is acceptable if it is not maintained or repeated for long periods of time [9]. It is favourable if the manual lifting task can be replaced by the usage of automation system to avoid from musculoskeletal disorder problem [6]. The portable gantry crane is an automated lifting equipment. The workers just have to control its axis movement by clicking the switches on the crane.

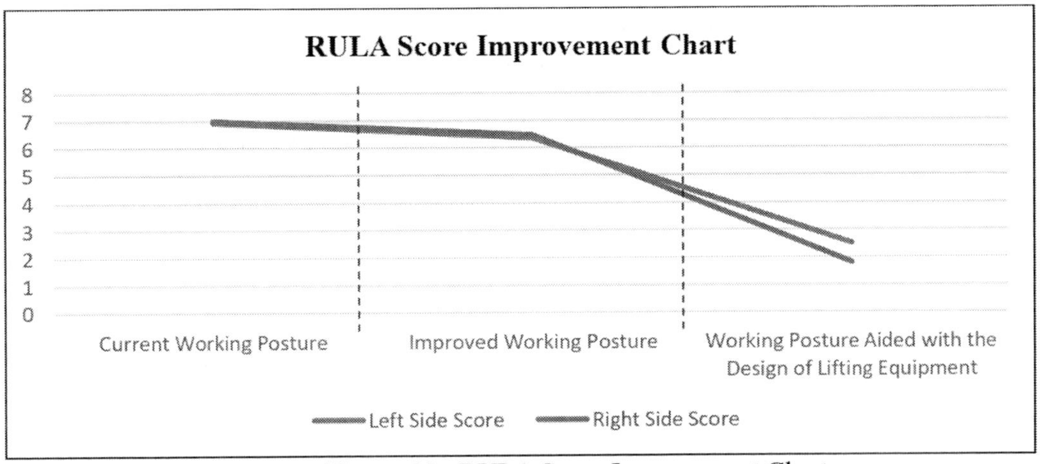

Figure 11. RULA Score Improvement Chart

Figure 11 shows the RULA score pattern of the current working posture, improvised working posture and the working posture aided with the design of lifting equipment. The decreasing pattern of the RULA score along the improvement effort is good as the smaller the RULA score, the safer the working posture is toward the ergonomic risk factor [9]. The decreasing value of RULA score from current working posture to the improved working posture is lesser than the decreasing value of RULA score from the improved working posture to working poster aided with the design of lifting equipment. This is due to involvement of heavy lifting task although the working posture is improved. There are many causes that can contribute to ergonomic injuries including muscle pulls, wrist injuries, back sprains, elbow injuries, spinal injuries and other related injury [13]. Some of the causes are incompatible weightiness of loads, awkward body postures, high frequency of materials to be lifted and long durations needed for the lifting activities, inadequate handhold and environmental factors.

5. Conclusion

The main idea of this study is to investigate and improve the ergonomics working postures of the workers working in the warehouse which focused on the task related to manual lifting activities. The current working body postures was analysed using CATIA V5 software application to identify whether the working body postures to identify the possible areas of improvement that can be made to reduce the ergonomic risk factor and to reduce the risk of musculoskeletal disorder among the workers. From the RULA analysis by the CATIA V5, even though the posture of the workers is improved by changing the posture, the score is still 7 and its convey a risk of getting ergonomic injuries because of the load of the task is still heavy and repetitions of work still occurs. However, when a design of helping device is added into the working environment, the score of RULA is decreases from 7 to 2. Thus, proved that the suggested working posture and helping device are required in order to reduce the risk of getting ergonomic injuries.

References

[1] Salvendy G 2012 *Handbook of human factors and ergonomics* (New Jersey: John Wiley & Sons, Inc.)
[2] Douglas F 2011 *Aerospace Manufacture and Maintenance* Encyclopedia of Occupational Health and Safety International Labor Organization Geneva Retrieved from http://www.ilo.org
[3] Wang M L & Lin H F 2004 *Industrial Engineering & System Management* (Hsinchu: Chung Hua University)
[4] Harry C 2002 *Constant Force Technology* (Trapp Road: St. Paul)

[5] Middlesworth M 2007 *A Step-by-Step Guide: Rapid Upper Limb Assessment (RULA)* Ergonomics Plus Retrieved from www. Ergo-Plus.Com.
[6] Tony W and Hunter L 2012 *Health and Safety Executive (HSE)* **3**
[7] Colombini D 2002 Risk *Assessment and Management of Repetitive Movements and Exertions of Upper Limbs* (Kidlington: Elsevier Science Ltd) 43- 45
[8] Thomas R, Colombini D, Occhipinti E & Alvarez E 2013 *Manual Lifting: A Guide to the Study of Simple and Complex Lifting Tasks* (London: Taylor and Francis Group) 24-30
[9] Stefan V, Katarina S, Albert M 2008 *Design of Manual Assembly Workstations in CATIA* (Trnava: Department of Machining and Assembly, Institute of Production Technologies Faculty of Materials Science and Technology)
[10] Miyashita K, Morioka I, Seyed M, Yamamoto H, Miyai N, & Kaewboonchoo O 1998 *J Occup Health* **40** 218–222
[11] Szepietowski M and Cronin R 2013 Health and Safety Executive for Northern Ireland (HSENI) and the Health and Safety Authority (HSA) ISBN: 978-1-84496-172-6
[12] Ibrahim I, W S Khan, N Goddard & P Smitham 2012 *The Open Orthopaedics Journal* **6** 69-79.
[13] Joshua S 2013 American Academy of Orthopaedic Surgeons
[14] Robert R 1996 *Preventing Repetitive Stress Injuries.* (US: Occupational Safety and Health Administration (OSHA))

Influence of injection temperatures and fiberglass compositions on mechanical properties of polypropylene

Tony Tiew Chun Keey* and M. Azuddin

Department of Mechanical Engineering Faculty of Engineering, University of Malaya, Kuala Lumpur 50603, Malaysia

E-mail: *louissac@siswa.um.edu.my and azuddin@um.edu.my

Abstract. Injection molding process appears to be one of the most suitable mass and cost efficiency manufacturing processes for polymeric parts nowadays due to its high efficiency of large scale production. When down-scaling the products and components, the limits of conventional injection molding process are reached. These constraints had initiated the development of conventional injection molding process into a new era of micro injection molding technology. In this study, fiberglass reinforced polypropylenes (PP) with various glass fiber percentage materials were used. The study start with fabrication of micro tensile specimens at three different injection temperature, 260°C, 270°C and 280°C for different percentage by weight of fiberglass reinforced PP. Then evaluate the effects of various injection temperatures on the tensile properties of micro tensile specimens. Different percentage by weight of fiberglass reinforced PP were tested as well and it was found that 20% fiberglass reinforced PP possessed the greatest percentage increase of tensile strength with increasing temperatures.

1. Introduction

Injection molding (IM) process is one of the conventional manufacturing techniques used widely nowadays due to its high efficiency of large scale production of polymeric parts. In this process, raw plastic materials are melted and formed in a cavity known as mold in order to produce three dimensional parts. Large quantities of small to medium size objects can be produced from low melting point materials through this process. Nowadays, many products that we use are manufactured by this injection molding process. Raw materials normally melted polymer is injected into a mold through injection nozzle and the melted polymer solidifies within a cavity and retains the shape of the mold. After the part has cooled, the mold opens up and the part is ejected.

The development of micro injection molding technology started in the middle of 80s and has become one of the most important technologies to the population nowadays. When down-scaling the products and components, the limits of conventional injection molding process are reached and these constraints had initiated the development of conventional injection molding into a new era of micro injection molding technology. Tosello and Hansen (2010) [1] said that a new micro injection molding machine with a screw extruder and a plunger injection unit has been developed over the last ten years. The main difference between the micro injection molding machine and the conventional macro machine with the conventional reciprocating screw injection system is by separating melt plastification and injection. A small injection plunger of a few millimeters in diameter is used for melt injection to control metering accuracy. This micro injection molding machine consists of the splitting of the four functions of the reciprocating screw which are plasticisation, metering, locking and injecting into different components.

Chang et al. (2007) [2] presented a concept of external module of micro injection molding system in 2007. This external module equipped with a hot runner plunger module can apply about 30-100 tonnes reciprocating screw hydraulic or fully electric injection molding machines. This external module was designed similar to an injection mold to fit the geometry of the general injection molding machine. This external module could be applied to most reciprocating injection molding machines where a precise metering for micro injection molding can be achieved.

Ultimate tensile strength (UTS) is defined as a stress measured as force per unit area. In another words, it is the maximum stress a material can withstand before breaking when it is being stretched. UTS can be determined by performing tensile test to the samples. Tensile testing, also known as tensile test is most commonly used mechanical test to determine the mechanical characteristics of a material. In tensile testing, a sample is subjected to tension until breaking (Czichos et al. 2006) [3]. In 1997, Ilzhofer et al. [4] developed a micro tensile testing machine in order to investigate the mechanical properties of micro specimens. In 2010, Zrida et al. [5] conducted dynamic high-speed tensile tests on a polypropylene (PP) with a servo-hydraulic tensile machine with an original grip system. The purpose of this study was to investigate the effects of strain rate on mechanical behaviour of polypropylene. Experimental results showed that Young's modulus increased with strain rate and the surface becomes rougher.

Arabi et al. [6] also conducted an experimental study on investigating effects of specimen shape on residual mechanical properties of polypropylene (PP) fiber self-compacting concrete (SCC) exposed to elevated temperatures from 200°C to 600°C. The experimental results revealed that PP fiber can enhance residual strength of concrete subjected to thermal shock induced by air cooling from high temperatures up to 600°C to room temperature. Al-Maadeedet al. 2014 [7] also conducted a study on polypropylene/fiberglass with concentration on processing, characterization and modelling

It can be summarized that injection molding process plays a vital role in mass production of micro parts nowadays. However, to identify the suitability of micro tensile specimen for micro sizes are challenging issues. The mechanical behaviour of materials is described by their deformation and fracture characteristics under applied stresses. Tensile test is the most commonly used mechanical testing method to investigate the mechanical characteristics of micro tensile specimens. In my study, tensile test is performed in order to determine the tensile behaviour of the micro tensile specimens of fiberglass reinforced polypropylene with various glass fiber percentages.

2. Experimental Design

This section explains the material, mold, machine and testing method used in this work. This was done to evaluate performance of micro tensile specimens by conducting tensile test. Experimental details such as equipment/instruments, materials and setting of parameters are further explained in the following sections.

2.1. Fiberglass Reinforced Polypropylene (PP)

Glass fiber or more commonly known as fiberglass are made from numerous fine fibers of glass and normally lightweight. Fiberglass reinforced PP is the modified plastics with good mechanical properties, good resistance towards heat and chemical. Fiberglass reinforced PP is normally used for structure and function parts as well as industry of centrifuge and axial fans for its strong resistance against heat and chemical. The Fiberglass reinforced PP used in this study was supply by Ferro Corporation, USA (http://www.ferro.com/). Percentage glass fiber use in the experiment was 10%, 20%, 25% and 30%. The properties of Fiberglass Reinforced Polypropylene (PP) are shown in table 1

Table 1. Properties of Fiberglass Reinforced Polypropylene (PP)

Properties		10% Fiberglass-Reinforced Polypropylene	20% Fiberglass-Reinforced Polypropylene	25% Fiberglass-Reinforced Polypropylene	30% Fiberglass-Reinforced Polypropylene
Physical Properties	Density, g/cc	0.930 - 1.45	0.970 - 1.46	1.00 - 1.51	1.09 - 1.56
	Melt Flow, g/10 min	1.00 - 15.0	0.500 - 34.0	0.400 - 25.0	1.00 - 25.0
Mechanical Properties	Hardness, Rockwell R	76.0 - 110	53.0 - 105	57.0 - 107	58.0 - 72.0
	Tensile Strength, Ultimate, MPa	20.0 - 55.8	22.0 - 115	6.89 - 130	30.0 - 150
	Elongation at Break, %	1.50 - 34.0	1.00 - 100	0.900 - 30.0	1.00 - 10.0
	Modulus of Elasticity, GPa	1.70 - 5.17	1.03 - 6.55	2.25 - 15.0	3.59 - 12.4
	Izod Impact, Notched, J/cm	0.267 - 3.74	0.135 - 4.50	0.320 - 3.26	0.427 - 7.52
Thermal Properties	Thermal Conductivity, W/m-K	0.159 - 0.187	0.288 - 1.21	0.270 - 0.331	0.346 - 0.360
Processing Properties	Melt Temperature, °C	180 - 260	145 - 302	145 - 302	190 - 290
	Mold Temperature, °C	20.0 - 80.0	18.0 - 100	4.00 - 100	15.0 - 100

2.2. Mold for Microtensile Specimen

The mold cavity used will be fixed and the same dimensions as the test specimen by following the standard of the microtensile specimen, D1708-02a (Standard Test Method for Tensile Properties of Plastics By Use of Microtensile Specimens) as shown in figure 1. Figure 2 shows the mold used to produce microtensile specimen.

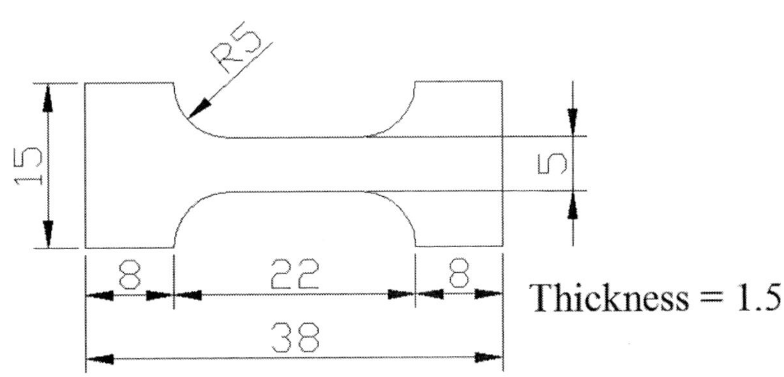

Figure 1. the dimensions microtensile specimen in mm

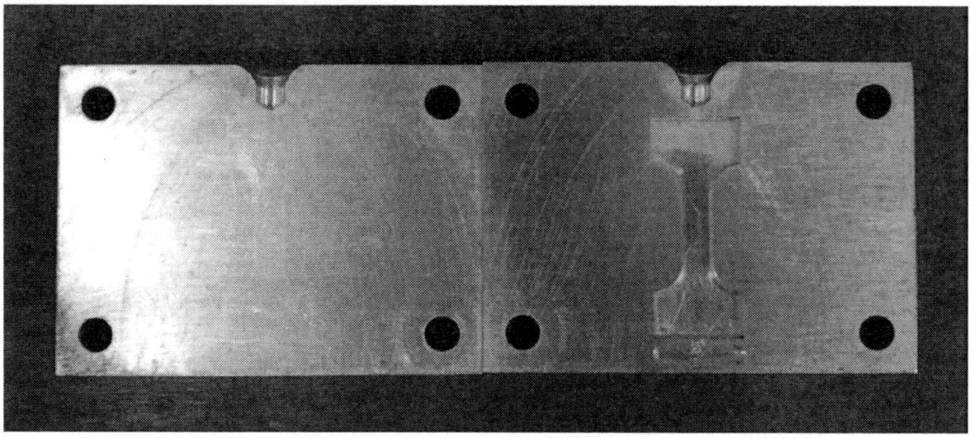

Figure 2. Mold for micro tensile specimen (74mm x 32mm x 53mm)

2.3. Injection Molding Apparatus

The test specimens produced in this study is produced by a vertical injection molding machine as shown in figure 3 (Azuddin et. al, 2015) [8]. The machine use pneumatic system to generate plunger injection motion. The machine also include temperature controller to control various melt temperature. Figure 3 shows the vertical injection molding machine.

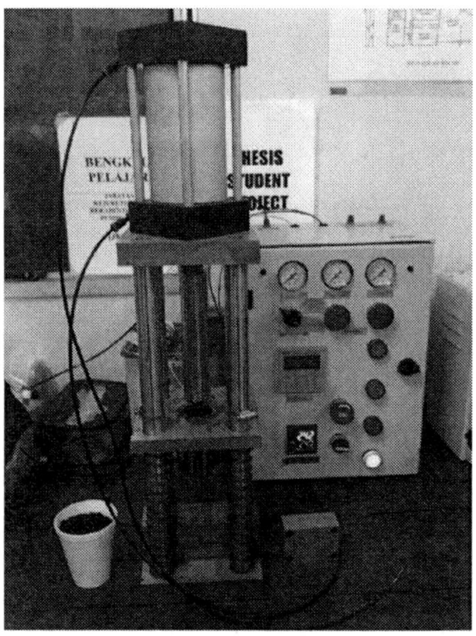

Figure 3. Vertical injection molding machine

2.4. Procedure to Prepare Micro Tensile Specimens

1) Start with injection molding process. Heat up the hopper of vertical injection molding machine.
2) Put materials into heated hopper and allow for melting.
3) Inject pressure to push the plunger down in order to inject molten material into micro tensile cavity.
4) Produce five samples or more of micro tensile specimens for three different injection temperatures.
5) Repeat steps 1 - 4 with different percentage by weight of glass reinforced PP.

2.5. Evaluation of Micro tensile specimens

Tensile test is conducted to evaluate the performance of micro tensile specimens. Tensile strength is the output parameter of experiments to evaluate the performance of micro tensile specimens. The INSTRON 3369 universal testing machine was used to perform tensile test and measure the tensile strength. Five micro tensile samples named A, B, C, D and E was tested.

3. Result and Discussion

3.1. Tensile strength of 10% fiberglass reinforced PP

Figure 4 shows a 10% fiberglass reinforced PP strength at various injection temperatures. At injection temperature of 260°C, highest tensile stress value recorded is 55.55MPa for sample B and lowest value is 51.83MPa. Tensile strength at injection temperature of 270°C shows unstable tensile behaviour where it fluctuates in the range from 52MPa to 55MPa. At 280°C, tensile strength values show less fluctuation of tensile behaviour compared to 260°C and 270°C.

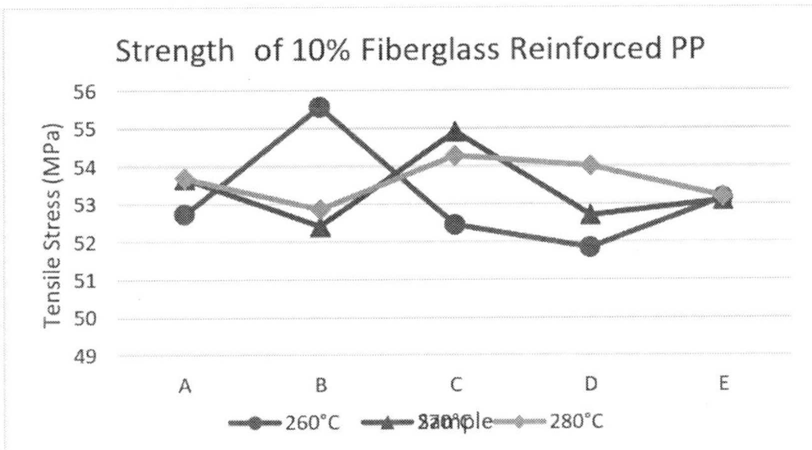

Figure 4. Graph of tensile strength for 10% fiberglass reinforced PP

3.2. Tensile strength of 20% fiberglass reinforced PP

A tensile performance of 20% fiberglass reinforced PP is shown in figure 5. 20% fiberglass reinforced PP shows uniform tensile behaviour at three different injection temperatures. At injection temperature of 260°C, the tensile strength falls in the range from 57MPa to 65MPa. When injection temperature rises to 270°C, the range gets higher where tensile strength falls between 72MPa and 76MPa. There is a drastic increase in tensile strength. For injection temperature at 280°C, the range falls the same as 270°C.

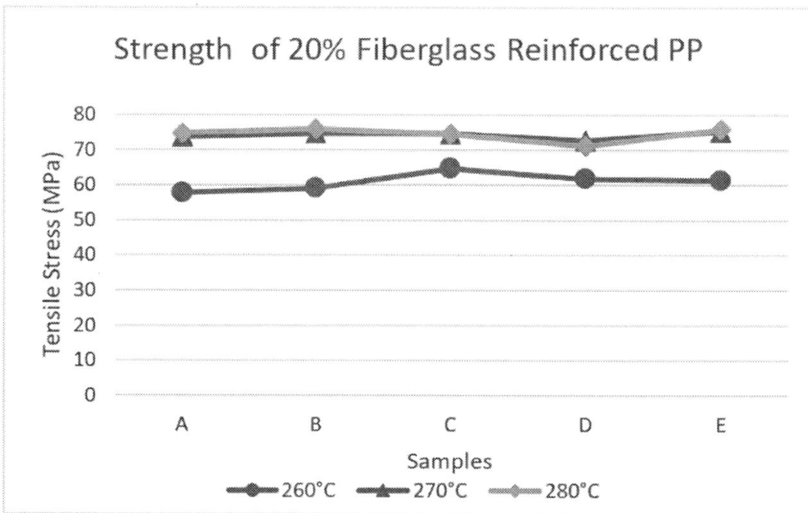

Figure 5. Graph of tensile strength for 20% fiberglass reinforced PP

3.3. Tensile strength of 25% fiberglass reinforced PP

Figure 6 shows a 25 % fiberglass reinforced PP strength at various injection temperatures. For 25% fiberglass reinforced PP, sample A shows the lowest tensile strength 56.16MPa at injection temperature 260°C. At 270°C, tensile strength falls in the range from 59MPa to 62MPa. A less fluctuation of behaviour at 270°C. When injection temperature rises to 280°C, tensile strength increases with the highest value recorded 63.83MPa for sample E

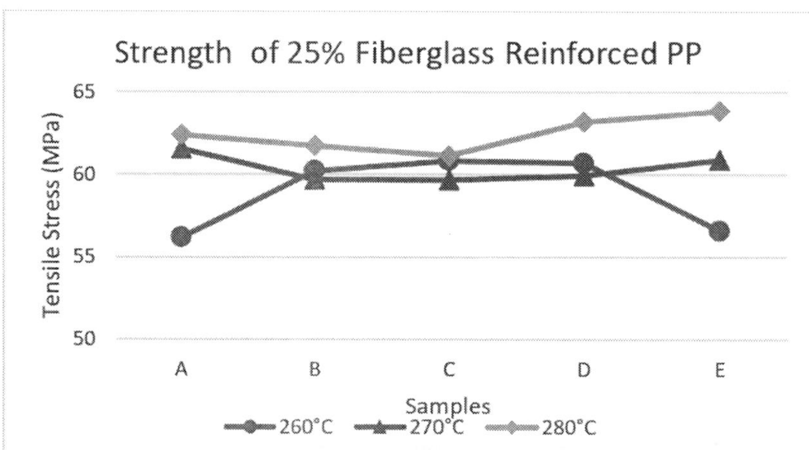

Figure 6. Graph of tensile strength for 25% fiberglass reinforced PP

3.4. Tensile strength of 30% fiberglass reinforced PP

A tensile performance of 30% fiberglass reinforced PP is shown in figure 7. At 260°C, 30% fiberglass reinforced PP shows decrease in tensile strength from sample A to C and increases slightly for sample D before decreasing to 45.88MPa. When injection temperature increases to 270°C, tensile strength shows fluctuation and falls in the range from 49MPa to 52MPa. At 280°C, it shows uniform behaviour at the range from 51MPa to 52MPa.

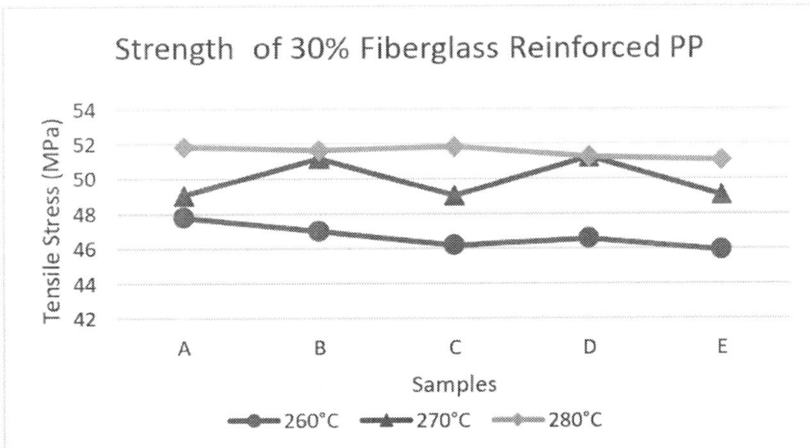

Figure 7. Graph of tensile strength for 30% fiberglass reinforced PP

3.5. Comparison between all Fiberglass reinforced PP

Figure 10 show the comparison of tensile strength of various percentage Fiberglass reinforced PP with melt temperature. Average tensile strength for 10% fiberglass reinforced PP is 53.13MPa at injection temperature of 260°C. When at 270°C, tensile strength increases to 53.35MPa. For injection temperature of 280°C, the tensile strength recorded is 53.60MPa. Overall, 10% fiberglass reinforced PP shows uniform increase in tensile strength with temperatures. Average tensile strength of 20% fiberglass reinforced PP at 260°C is 60.95MPa. When injection temperature rises to 270°C, there is a drastic increase in tensile strength. Tensile strength recorded at 270°C is 74.20MPa, an increase of 22.15% in tensile strength. At injection temperatures of 280°C, tensile strength increases slightly to 74.45MPa. 25% fiberglass reinforced PP shows an average tensile strength of 58.88MPa at 260°C. Then, tensile strength increases to 60.33MPa when injection temperature is raised to 270°C. An increase of 2.46% in tensile strength is recorded. At injection temperature of 280°C, tensile strength shows 62.44MPa which is an increase of 3.50% in tensile strength. Average tensile strength of 30% fiberglass reinforced PP at 260°C is 46.66571MPa. When at 270°C, tensile strength increases to 49.90MPa. An increase of 6.95% in tensile strength is recorded. For injection temperature at 280°C, tensile strength increases 3.23% to 51.52MPa.

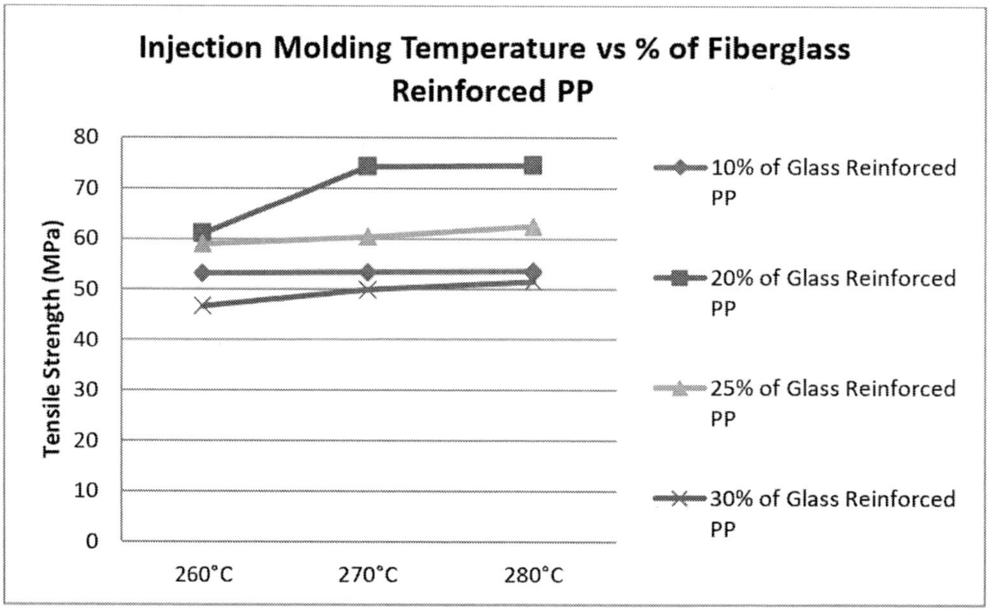

Figure 8. Average of all Fiberglass Reinforced PP

4. Conclusion

In this study, fiberglass reinforced PP with various fiberglass percentage are evaluated in terms of tensile strength. The following conclusions are made based on the analysis of the tensile tests:

1) Tensile strength increased with increasing temperatures.
2) Tensile tests for different percentage by weight of fiberglass reinforced PP showed increase in tensile strength.
3) Tensile strength of 10% fiberglass reinforced PP increased from 53.13MPa to 53.60MPa, an increase of 0.87% when injection temperature increased to 280°C.
4) 20% fiberglass reinforced PP showed greatest increase of 22.15% in tensile strength from 60.95MPa to 74.45MPa when injection temperature raised from 260°C to 280°C.
5) An increase of 6.05% in tensile strength was found in 25% fiberglass reinforced PP when temperature increased to 280°C. A uniform increase in tensile strength from 58.88669MPa (260°C) to 60.33 (270°C) and lastly 62.44MPa at 280°C.
6) 30% fiberglass reinforced PP showed uniform increase in tensile strength from 46.66MPa at 260°C to 49.90MPa at 270°C and then 51.52MPa at 280°C. It was found that a total of 10.41% increase in tensile strength.

Acknowledgments
The authors acknowledge the financial support from the Ministry of Higher Education, Malaysia, under Research University grant scheme project number RG027/09AET.

References

[1] Tosello, G Hansen, H N & Guerrier P 2010 *International Conference on Multi-Material Micro Manufacture* 283-286
[2] Chang P C, Hwang S J, Lee H H, & Huang D Y 2007 *Journal of Materials Processing Technology* **10** 163-172
[3] Czichos H, Saito T, and Smith L 2006 Measurement methods for mechanical properties *Materials measurement methods* (London: Springer) chapter 1 283-387
[4] Ilzhofer A, Schneider H, & Tsakmakis Ch 1997 *Microsystem Technologies* **5** 46-50
[5] Zrida M, Laurent H, Grolleau V, Rio G, Khlif M, Guines D, Masmoudi N, & Bradai C 2010 *Polymer Testing* **8** 685-692
[6] Arabi N S Al Qadi & Sleiman M Al-Zaidyeen 2014 *Journal of King Saud University – Engineering Sciences* **26** 33-39
[7] M A Al-Maadeed, Yasser M Shabana, P Noorunnisa Khanam 2014 *Materials and Design* **58** 374-380
[8] Azuddin M, Choudhury I A and Taha Z 2015 *Journal of the Brazilian Society of Mechanical Sciences and Engineering* **37** 79-86.

Institute of Physics
Dirac House, Temple Back
Bristol BS1 6BE UK

ISSN: 1757-8981
ISBN 978-1-5108-4294-6